AF706510

Mechanism of the Heavens

MARY SOMERVILLE
PIERRE SIMON MARQUIS DE LAPLACE

CAMBRIDGE
UNIVERSITY PRESS

CAMBRIDGE UNIVERSITY PRESS

Cambridge, New York, Melbourne, Madrid, Cape Town, Singapore,
São Paolo, Delhi, Dubai, Tokyo

Published in the United States of America by Cambridge University Press, New York

www.cambridge.org
Information on this title: www.cambridge.org/9781108001571

© in this compilation Cambridge University Press 2009

This edition first published 1831
This digitally printed version 2009

ISBN 978-1-108-00157-1 Paperback

This book reproduces the text of the original edition. The content and language reflect the beliefs, practices and terminology of their time, and have not been updated.

Cambridge University Press wishes to make clear that the book, unless originally published by Cambridge, is not being republished by, in association or collaboration with, or with the endorsement or approval of, the original publisher or its successors in title.

CAMBRIDGE LIBRARY COLLECTION

Books of enduring scholarly value

Mathematical Sciences

From its pre-historic roots in simple counting to the algorithms powering modern desktop computers, from the genius of Archimedes to the genius of Einstein, advances in mathematical understanding and numerical techniques have been directly responsible for creating the modern world as we know it. This series will provide a library of the most influential publications and writers on mathematics in its broadest sense. As such, it will show not only the deep roots from which modern science and technology have grown, but also the astonishing breadth of application of mathematical techniques in the humanities and social sciences, and in everyday life.

Mechanism of the Heavens

Published in 1831, this work forms part of a collection of introductory volumes published by the Society of the Diffusion of Useful Knowledge. It is a greatly adapted and simplified translation of Laplace's multi-volume Mécanique Céleste and was written to give popular audiences an introduction to astronomical methods and theories. Its author was remarkable for her mathematical accomplishments at a time when formal training in the subject for women was virtually unknown. Laplace famously told her that 'There have been only three women who have understood me. These are yourself, Mrs Somerville, Caroline Herschel and a Mrs Greig of whom I know nothing.' Mary Somerville was in fact Mrs Greig.

Cambridge University Press has long been a pioneer in the reissuing of out-of-print titles from its own backlist, producing digital reprints of books that are still sought after by scholars and students but could not be reprinted economically using traditional technology. The Cambridge Library Collection extends this activity to a wider range of books which are still of importance to researchers and professionals, either for the source material they contain, or as landmarks in the history of their academic discipline.

Drawing from the world-renowned collections in the Cambridge University Library, and guided by the advice of experts in each subject area, Cambridge University Press is using state-of-the-art scanning machines in its own Printing House to capture the content of each book selected for inclusion. The files are processed to give a consistently clear, crisp image, and the books finished to the high quality standard for which the Press is recognised around the world. The latest print-on-demand technology ensures that the books will remain available indefinitely, and that orders for single or multiple copies can quickly be supplied.

The Cambridge Library Collection will bring back to life books of enduring scholarly value (including out-of-copyright works originally issued by other publishers) across a wide range of disciplines in the humanities and social sciences and in science and technology.

MECHANISM

OF

THE HEAVENS.

BY

MRS. SOMERVILLE.

LONDON:

JOHN MURRAY, ALBEMARLE-STREET.

MDCCCXXXI

Academiæ Cantabrigiensis
Liber.

TO

HENRY, LORD BROUGHAM AND VAUX,

LORD HIGH CHANCELLOR OF GREAT BRITAIN,

This Work, undertaken at His Lordship's request, is inscribed as a testimony of the Author's esteem and regard.

Although it has unavoidably exceeded the limits of the Publications of the Society for the Diffusion of Useful Knowledge, for which it was originally intended, his Lordship still thinks it may tend to promote the views of the Society in its present form. To concur with that Society in the diffusion of useful knowledge, would be the highest ambition of the Author,

MARY SOMERVILLE.

Royal Hospital, Chelsea,
21*st July*, 1831.

PRELIMINARY DISSERTATION.

In order to convey some idea of the object of this work, it may be useful to offer a few preliminary observations on the nature of the subject which it is intended to investigate, and of the means that have already been adopted with so much success to bring within the reach of our faculties, those truths which might seem to be placed so far beyond them.

All the knowledge we possess of external objects is founded upon experience, which furnishes a knowledge of facts, and the comparison of these facts establishes relations, from which, induction, the intuitive belief that like causes will produce like effects, leads us to general laws. Thus, experience teaches that bodies fall at the surface of the earth with an accelerated velocity, and proportional to their masses. Newton proved, by comparison, that the force which occasions the fall of bodies at the earth's surface, is identical with that which retains the moon in her orbit; and induction led him to conclude that as the moon is kept in her orbit by the attraction of the earth, so the planets might be retained in their orbits by the attraction of the sun. By such steps he was led to the discovery of one of those powers with which the Creator has ordained that matter should reciprocally act upon matter.

Physical astronomy is the science which compares and identifies the laws of motion observed on earth with the motions that take place in the heavens, and which traces, by an uninterrupted chain of deduction from the great principle that governs the universe, the revolutions and rotations of the planets, and the oscillations of the fluids at their surfaces, and which estimates the changes the system has hitherto undergone or may hereafter experience, changes which require millions of years for their accomplishment.

The combined efforts of astronomers, from the earliest dawn of civilization, have been requisite to establish the mechanical theory of astronomy: the courses of the planets have been observed for ages with a degree of perseverance that is astonishing, if we consider the imperfection, and even the want of instruments. The real motions of the earth have been separated from the apparent motions of the planets; the laws of the planetary revolutions have been discovered; and the discovery of these laws has led to the knowledge of the gravitation of matter. On the other hand, descending from the principle of gravitation, every motion in the system of the world has been so completely explained, that no astronomical phenomenon can now be transmitted to posterity of which the laws have not been determined.

Science, regarded as the pursuit of truth, which can only be attained by patient and unprejudiced investigation, wherein nothing is too great to be attempted, nothing so minute as to be justly disregarded, must ever afford occupation of consummate interest and of elevated meditation. The contemplation of the works of creation elevates the mind to the admiration of whatever is great and noble, accomplishing the object of all study, which in the elegant language of Sir James Mackintosh is to inspire the love of truth, of wisdom, of beauty, especially of goodness, the highest beauty, and of that supreme and eternal mind, which contains all truth and wisdom, all beauty and goodness. By the love or delightful contemplation and pursuit of these transcendent aims for their own sake only, the mind of man is raised from low and perishable objects, and prepared for those high destinies which are appointed for all those who are capable of them.

The heavens afford the most sublime subject of study which can be derived from science: the magnitude and splendour of the objects, the inconceivable rapidity with which they move, and the enormous distances between them, impress the mind with some notion of the energy that maintains them in their motions with a durability to which we can see no limits. Equally conspicuous is the goodness of the great First Cause in having endowed man with faculties by which he can not only appreciate the magnificence of his works, but trace, with precision, the operation of his laws, use the globe he inhabits as a base wherewith to measure the magnitude and distance of the sun and planets, and make the diameter of the earth's orbit the first step of a scale by which he may ascend to the starry firmament. Such pursuits, while they ennoble the mind, at the same time inculcate humility, by showing that there is a barrier, which no energy, mental or physical, can ever enable us to pass: that however profoundly we may penetrate the depths of space, there still remain innumerable systems, compared with which those which seem so mighty to us must dwindle into insignificance, or even become invisible; and that not only man, but the globe he inhabits, nay the whole system of which it forms so small a part, might be annihilated, and its extinction be unperceived in the immensity of creation.

A complete acquaintance with Physical Astronomy can only be attained by those who are well versed in the higher branches of mathematical and mechanical science: such alone can appreciate the extreme beauty of the results, and of the means by which these results are obtained. Nevertheless a sufficient skill in analysis to follow the general outline, to see the mutual dependence of the different parts of the system, and to comprehend by what means some of the most extraordinary conclusions have been arrived at, is within the reach of many who shrink from the task, appalled by difficulties, which perhaps are not more formidable than those incident to the study of the elements of every branch of knowledge, and possibly overrating them by not making a sufficient distinction between the degree of mathematical acquirement necessary for making discoveries, and that which is requisite for understanding what others have done. That the study of mathematics and their application to astronomy are full of interest will be allowed by all who have devoted their time and attention to these pursuits, and they only can estimate the delight of arriving at truth, whether it be in the discovery of a world, or of a new property of numbers.

It has been proved by Newton that a particle of matter placed without the surface of a hollow sphere is attracted by it in the same manner as if its mass, or the whole matter it contains, were collected in its centre. The same is therefore true of a solid sphere which may be supposed to consist of an infinite number of concentric hollow spheres. This however is not the case with a spheroid, but the celestial bodies are so nearly spherical, and at such remote distances from each other, that they attract and are attracted as if each were a dense point situate in its centre of gravity, a circumstance which greatly facilitates the investigation of their motions.

The attraction of the earth on bodies at its surface in that latitude, the square of whose sine is $\frac{1}{3}$, is the same as if it were a sphere; and experience shows that bodies there fall through 16.0697 feet in a second. The mean distance of the moon from the earth is about sixty times the mean radius of the earth. When the number 16.0697 is diminished in the ratio of 1 to 3600, which is the square of the moon's distance from the earth, it is found to be exactly the space the moon would fall through in the first second of her descent to the earth, were she not prevented by her centrifugal force, arising from the velocity with which she moves in her orbit. So that the moon is retained in her orbit by a force having the same origin and regulated by the same law with that which causes a stone to fall at the earth's surface. The earth may therefore be regarded as the centre of a force which extends to the moon; but as experience shows that the action and reaction of matter are equal and contrary, the moon must attract the earth with an equal and contrary force.

Newton proved that a body projected in space will move in a conic section, if it be attracted by a force directed towards a fixed point, and having an intensity inversely as the square of the distance; but that any deviation from that law will cause it to move in a curve of a different nature. Kepler ascertained by direct observation that the planets describe ellipses round the sun, and later observations show that comets also move in conic sections: it consequently follows that the sun attracts all the planets and comets inversely as the square of their distances from his centre; the sun therefore is the centre of a force extending indefinitely in space, and including all the bodies of the system in its action.

Kepler also deduced from observation, that the squares of the periodic times of the planets, or the times of their revolutions round the sun, are proportional to the cubes of their mean distances from his centre: whence it follows, that the intensity of gravitation of all the bodies towards the sun is the same at equal distances; consequently gravitation is proportional to the masses, for if the planets and comets be supposed to be at equal distances from the sun and left to the effects of gravity, they would arrive at his surface at the same time. The satellites also gravitate to their primaries according to the same law that their primaries do to the sun. Hence, by the law of action and reaction, each body is itself the centre of an attractive force extending indefinitely in space, whence proceed all the mutual disturbances that render the celestial motions so complicated, and their investigation so difficult.

The gravitation of matter directed to a centre, and attracting directly as the mass, and inversely as the square of the distance, does not belong to it when taken in mass; particle acts on particle according to the same law when at sensible distances from each other. If the sun acted on the centre of the earth without attracting each of its particles, the tides would be very much greater than they now are, and in other respects they also would be very different. The gravitation of the earth to the sun results from the gravitation of all its particles, which in their turn attract the sun in the ratio of their respective masses. There is a reciprocal action likewise between the earth and every particle at its surface; were this not the case, and were any portion of the earth, however small, to attract another portion and not be itself attracted, the centre of gravity of the earth would be moved in space, which is impossible.

The form of the planets results from the reciprocal attraction of their component particles. A detached fluid mass, if at rest, would assume the form of a sphere, from the reciprocal attraction of its particles; but if the mass revolves about an axis, it becomes flattened at the poles, and bulges at the equator, in consequence of the centrifugal force arising from the velocity of rotation. For, the centrifugal force diminishes the gravity of the particles at the equator, and equilibrium can only exist when these two forces are balanced by an increase of gravity; therefore, as the attractive force is the same on all particles at equal distances from the centre of a sphere, the equatorial particles would recede from the centre till their increase in number balanced the centrifugal force by their attraction, consequently the sphere would become an oblate spheroid; and a fluid partially or entirely covering a solid, as the ocean and atmosphere cover the earth, must assume that form in order to remain in equilibrio. The surface of the sea is therefore spheroidal, and the surface of the earth only deviates from that figure where it rises above or sinks below the level of the sea; but the deviation is so small that it is unimportant when compared with the magnitude of the earth. Such is the form of the earth and planets, but the compression or flattening at their poles is so small, that even Jupiter, whose rotation is the most rapid, differs but little from a sphere. Although the planets attract each other as if they were spheres on account of their immense distances, yet the satellites are near enough to be sensibly affected in their motions by the forms of their primaries. The moon for example is so near the earth, that the reciprocal attraction between each of her particles and each of the particles in the prominent mass at the terrestrial equator, occasions considerable disturbances in the motions of both bodies. For, the action of the moon on the matter at the earth's equator produces a nutation in the axis of rotation, and the reaction of that matter on the moon is the cause of a corresponding nutation in the lunar orbit.

If a sphere at rest in space receives an impulse passing through its centre of gravity, all its parts will move with an equal velocity in a straight line; but if the impulse does not pass through the centre of gravity, its particles having unequal velocities, will give it a rotatory motion at the same time that it is translated in space. These motions are independent of one another, so that a contrary impulse passing through its centre of gravity will impede its progression, without interfering with its rotation. As the sun rotates about an axis, it seems probable if an impulse in a contrary direction has not been given to his centre of gravity, that he moves in space accompanied by all those bodies which compose the solar system, a circumstance that would in no way interfere with their relative motions; for, in consequence of our experience that force is proportional to velocity, the reciprocal attractions of a system remain the same, whether its centre of gravity be at rest, or moving uniformly in space. It is computed that had the earth received its motion from a single impulse, such impulse must have passed through a point about twenty-five miles from its centre.

Since the motions of the rotation and translation of the planets are independent of each other, though probably communicated by the same impulse, they form separate subjects of investigation.

A planet moves in its elliptical orbit with a velocity varying every instant, in consequence of two forces, one tending to the centre of the sun, and the other in the direction of a tangent to its orbit, arising from the primitive impulse given at the time when it was launched into space: should the force in the tangent cease, the planet would fall to the sun by its gravity; were the sun not to attract it, the planet would fly off in the tangent. Thus, when a planet is in its aphelion or at the point where the orbit is farthest from the sun, his action overcomes its velocity, and brings it towards him with such an accelerated motion, that it at last overcomes the sun's attraction, and shoots past him; then, gradually decreasing in velocity, it arrives at the aphelion where the sun's attraction again prevails. In this motion the radii vectores, or imaginary lines joining the centres of the sun and planets, pass over equal areas in equal times.

If the planets were attracted by the sun only, this would ever be their course; and because his action is proportional to his mass, which is immensely larger than that of all the planets put together, the elliptical is the nearest approximation to their true motions, which are extremely complicated, in consequence of their mutual attraction, so that they do not move in any known or symmetrical curve, but in paths now approaching to, and now receding from the elliptical form, and their radii vectores do not describe areas exactly proportional to the time. Thus the areas become a test of the existence of disturbing forces.

To determine the motion of each body when disturbed by all the rest is beyond the power of analysis; it is therefore necessary to estimate the disturbing action of one planet at a time, whence arises the celebrated problem of the three bodies, which originally was that of the moon, the earth, and the sun, namely,—the masses being given of three bodies projected from three given points, with velocities given both in quantity and direction; and supposing the bodies to gravitate to one another with forces that are directly as their masses, and inversely as the squares of the distances, to find the lines described by these bodies, and their position at any given instant.

By this problem the motions of translation of all the celestial bodies are determined. It is one of extreme difficulty, and would be of infinitely greater difficulty, if the disturbing action were not very small, when compared with the central force. As the disturbing influence of each body may be found separately, it is assumed that the action of the whole system in disturbing any one planet is equal to the sum of all the particular disturbances it experiences, on the general mechanical principle, that the sum of any number of small oscillations is nearly equal to their simultaneous and joint effect.

On account of the reciprocal action of matter, the stability of the system depends on the intensity of the primitive momentum of the planets, and the ratio of their masses to that of the sun: for the nature of the conic sections in which the celestial bodies move, depends on the velocity with which they were first propelled in space; had that velocity been such as to make the planets move in orbits of unstable equilibrium, their mutual attractions might have changed them into parabolas or even hyperbolas; so that the earth and planets might ages ago have been sweeping through the abyss of space: but as the orbits differ very little from circles, the momentum of the planets when projected, must have been exactly sufficient to ensure the permanency and stability of the system. Besides the mass of the sun is immensely greater than those of the planets; and as their inequalities bear the same ratio to their elliptical motions as their masses do to that of the sun, their mutual disturbances only increase or diminish the eccentricities of their orbits by very minute quantities; consequently the magnitude of the sun's mass is the principal cause of the stability of the system. There is not in the physical world a more splendid example of the adaptation of means to the accomplishment of the end, than is exhibited in the nice adjustment of these forces.

The orbits of the planets have a very small inclination to the plane of the ecliptic in which the earth moves; and on that account, astronomers refer their motions to it at a given epoch as a known and fixed position. The paths of the planets, when their mutual disturbances are omitted, are ellipses nearly approaching to circles, whose planes, slightly inclined to the ecliptic, cut it in straight lines passing through the centre of the sun; the points where the orbit intersects the plane of the ecliptic are its nodes.

The orbits of the recently discovered planets deviate more from the ecliptic: that of Pallas has an inclination of 35° to it: on that account it will be more difficult to determine their motions. These little planets have no sensible effect in disturbing the rest, though their own motions are rendered very irregular by the proximity of Jupiter and Saturn.

The planets are subject to disturbances of two distinct kinds, both resulting from the constant operation of their reciprocal attraction, one kind depending upon their positions with regard to each other, begins from zero, increases to a maximum, decreases and becomes zero again, when the planets return to the same relative positions. In consequence of these, the troubled planet is sometimes drawn away from the sun, sometimes brought nearer to him; at one time it is drawn above the plane of its orbit, at another time below it, according to the position of the disturbing body. All such changes, being accomplished in short periods, some in a few months, others in years, or in hundreds of years, are denominated Periodic Inequalities.

The inequalities of the other kind, though occasioned likewise by the disturbing energy of the planets, are entirely independent of their relative positions; they depend on the relative positions of the orbits alone, whose forms and places in space are altered by very minute quantities in immense periods of time, and are therefore called Secular Inequalities.

In consequence of disturbances of this kind, the apsides, or extremities of the major axes of all the orbits, have a direct, but variable motion in space, excepting those of Venus, which are retrograde; and the lines of the nodes move with a variable velocity in the contrary direction. The motions of both are extremely slow; it requires more than 109770 years for the major axis of the earth's orbit to accomplish a sidereal revolution, and 20935 years to complete its tropical motion. The major axis of Jupiter's orbit requires no less than 197561 years to perform its revolution from the disturbing action of Saturn alone. The periods in which the nodes revolve are also very great. Beside these, the inclination and eccentricity of every orbit are in a state of perpetual, but slow change. At the present time, the inclinations of all the orbits are decreasing; but so slowly, that the inclination of Jupiter's orbit is only six minutes less now than it was in the age of Ptolemy. The terrestrial eccentricity is decreasing at the rate of 3914 miles in a century; and if it were to decrease equably, it would be 36300 years before the earth's orbit became a circle. But in the midst of all these vicissitudes, the major axes and mean motions of the planets remain permanently independent of secular changes; they are so connected by Kepler's law of the squares of the periodic times being proportional to the cubes of the mean distances of the planets from the sun, that one cannot vary without affecting the other.

With the exception of these two elements, it appears, that all the bodies are in motion, and every orbit is in a state of perpetual change. Minute as these changes are, they might be supposed liable to accumulate in the course of ages sufficiently to derange the whole order of nature, to alter the relative positions of the planets, to put an end to the vicissitudes of the seasons, and to bring about collisions, which would involve our whole system, now so harmonious, in chaotic confusion. The consequences being so dreadful, it is natural to inquire, what proof exists that creation will be preserved from such a catastrophe? for nothing can be known from observation, since the existence of the human race has occupied but a point in duration, while these vicissitudes embrace myriads of ages. The proof is simple and convincing. All the variations of the solar system, as well secular as periodic, are expressed analytically by the sines and cosines of circular arcs, which increase with the time; and as a sine or cosine never can exceed the radius, but must oscillate between zero and unity, however much the time may increase, it follows, that when the variations have by slow changes accumulated in however long a time to a maximum, they decrease by the same slow degrees, till they arrive at their smallest value, and then begin a new course, thus for ever oscillating about a mean value. This, however, would not be the case if the planets moved in a resisting medium, for then both the eccentricity and the major axes of the orbits would vary with the time, so that the stability of the system would be ultimately destroyed. But if the planets do move in an ethereal medium, it must be of extreme rarity, since its resistance has hitherto been quite insensible.

Three circumstances have generally been supposed necessary to prove the stability of the system: the small eccentricities of the planetary orbits, their small inclinations, and the revolution of all the bodies, as well planets as satellites, in the same direction. These, however, are not necessary conditions: the periodicity of the terms in which the inequalities are expressed is sufficient to assure us, that though we do not know the extent of the limits, nor the period of that grand cycle which probably embraces millions of years, yet they never will exceed what is requisite for the stability and harmony of the whole, for the preservation of which every circumstance is so beautifully and wonderfully adapted.

The plane of the ecliptic itself, though assumed to be fixed at a given epoch for the convenience of astronomical computation, is subject to a minute secular variation of 52″.109, occasioned by the reciprocal action of the planets; but as this is also periodical, the terrestrial equator, which is inclined to it at an angle of about 23° 28′, will never coincide with the plane of the ecliptic; so there never can be perpetual spring. The rotation of the earth is uniform; therefore day and night, summer and winter, will continue their vicissitudes while the system endures, or is untroubled by foreign causes.

> Yonder starry sphere
> Of planets, and of fix'd, in all her wheels
> Resembles nearest, mazes intricate,
> Eccentric, intervolved, yet regular
> Then most, when most irregular they seem.

The stability of our system was established by La Grange, 'a discovery,' says Professor Playfair, 'that must render the name for ever memorable in science, and revered by those who delight in the contemplation of whatever is excellent and sublime. After Newton's discovery of the elliptical orbits of the planets, La Grange's discovery of their periodical inequalities is without doubt the noblest truth in physical astronomy; and, in respect of the doctrine of final causes, it may be regarded as the greatest of all.'

Notwithstanding the permanency of our system, the secular variations in the planetary orbits would have been extremely embarrassing to astronomers, when it became necessary to compare observations separated by long periods. This difficulty is obviated by La Place, who has shown that whatever changes time may induce either in the orbits themselves, or in the plane of the ecliptic, there exists an invariable plane passing through the centre of gravity of the sun, about which the whole system oscillates within narrow limits, and which is determined by this property; that if every body in the system be projected on it, and if the mass of each be multiplied by the area described in a given time by its projection on this plane, the sum of all these products will be a maximum. This plane of greatest inertia, by no means peculiar to the solar system, but existing in every system of bodies submitted to their mutual attractions only, always remains parallel to itself, and maintains a fixed position, whence the oscillations of the system may be estimated through unlimited time. It is situate nearly half way between the orbits of Jupiter and Saturn, and is inclined to the ecliptic at an angle of about 1° 35′ 31″.

All the periodic and secular inequalities deduced from the law of gravitation are so perfectly confirmed by observations, that analysis has become one of the most certain means of discovering the planetary irregularities, either when they are too small, or too long in their periods, to be detected by other methods. Jupiter and Saturn, however, exhibit inequalities which for a long time seemed discordant with that law. All observations, from those of the Chinese and Arabs down to the present day, prove that for ages the mean motions of Jupiter and Saturn have been affected by great inequalities of very long periods, forming what appeared an anomaly in the theory of the planets. It was long known by observation, that five times the mean motion of Saturn is nearly equal to twice that of Jupiter; a relation which the sagacity of La Place perceived to be the cause of a periodic inequality in the mean motion of each of these planets, which completes its period in nearly 929 Julian years, the one being retarded, while the other is accelerated. These inequalities are strictly periodical, since they depend on the configuration of the two planets; and the theory is perfectly confirmed by observation, which shows that in the course of twenty centuries, Jupiter's mean motion has been accelerated by 3° 23′, and Saturn's retarded by 5°.13′.

It might be imagined that the reciprocal action of such planets as have satellites would be different from the influence of those that have none; but the distances of the satellites from their primaries are incomparably less than the distances of the planets from the sun, and from one another, so that the system of a planet and its satellites moves nearly as if all those bodies were united in their common centre of gravity; the action of the sun however disturbs in some degree the motion of the satellites about their primary.

The changes that take place in the planetary system are exhibited on a small scale by Jupiter and his satellites; and as the period requisite for the development of the inequalities of these little moons only extends to a few centuries, it may be regarded as an epitome of that grand cycle which will not be accomplished by the planets in myriads of centuries. The revolutions of the satellites about Jupiter are precisely similar to those of the planets about the sun; it is true they are disturbed by the sun, but his distance is so great, that their motions are nearly the same as if they were not under his influence. The satellites like the planets, were probably projected in elliptical orbits, but the compression of Jupiter's spheroid is very great in consequence of his rapid rotation; and as the masses of the satellites are nearly 100000 times less than that of Jupiter, the immense quantity of prominent matter at his equator must soon have given the circular form observed in the orbits of the first and second satellites, which its superior attraction will always maintain. The third and fourth satellites being further removed from its influence, move in orbits with a very small eccentricity. The same cause occasions the orbits of the satellites to remain nearly in the plane of Jupiter's equator, on account of which they are always seen nearly in the same line; and the powerful action of that quantity of prominent matter is the reason why the motion of the nodes of these little bodies is so much more rapid than those of the planet. The nodes of the fourth satellite accomplish a revolution in 520 years, while those of Jupiter's orbit require no less than 50673 years, a proof of the reciprocal attraction between each particle of Jupiter's equator and of the satellites. Although the two first satellites sensibly move in circles, they acquire a small ellipticity from the disturbances they experience.

The orbits of the satellites do not retain a permanent inclination, either to the plane of Jupiter's equator, or to that of his orbit, but to certain planes passing between the two, and through their intersection ; these have a greater inclination to his equator the further the satellite is removed, a circumstance entirely owing to the influence of Jupiter's compression.

A singular law obtains among the mean motions and mean longitudes of the three first satellites. It appears from observation, that the mean motion of the first satellite, plus twice that of the third, is equal to three times that of the second, and that the mean longitude of the first satellite, minus three times that of the second, plus twice that of the third, is always equal to two right angles. It is proved by theory, that if these relations had only been approximate when the satellites were first launched into space, their mutual attractions would have established and maintained them. They extend to the synodic motions of the satellites, consequently they affect their eclipses, and have a very great influence on their whole theory. The satellites move so nearly in the plane of Jupiter's equator, which has a very small inclination to his orbit, that they are frequently eclipsed by the planet. The instant of the beginning or end of an eclipse of a satellite marks the same instant of absolute time to all the inhabitants of the earth ; therefore the time of these eclipses observed by a traveller, when compared with the time of the eclipse computed for Greenwich or any other fixed meridian, gives the difference of the meridians in time, and consequently the longitude of the place of observation. It has required all the refinements of modern instruments to render the eclipses of these remote moons available to the mariner ; now however, that system of bodies invisible to the naked eye, known to man by the aid of science alone, enables him to traverse the ocean, spreading the light of knowledge and the blessings of civilization over the most remote regions, and to return loaded with the productions of another hemisphere. Nor is this all : the eclipses of Jupiter's satellites have been the means of a discovery, which, though not so immediately applicable to the wants of man, unfolds a property of light, that medium, without whose cheering influence all the beauties of the creation would have been to us a blank. It is observed, that those eclipses of the first satellite which happen when Jupiter is near conjunction, are later by 16′ 26″ than those which take place when the planet is in opposition. But as Jupiter is nearer to us when in opposition by the whole breadth of the earth's orbit than when in conjunction, this circumstance was attributed to the time employed by the rays of light in crossing the earth's orbit, a distance of 192 millions of miles; whence it is estimated, that light travels at the rate of 192000 miles in one second. Such is its velocity, that the earth, moving at the rate of nineteen miles in a second, would take two months to pass through a distance which a ray of light would dart over in eight minutes. The subsequent discovery of the aberration of light confirmed this astonishing result.

Objects appear to be situate in the direction of the rays that proceed from them. Were light propagated instantaneously, every object, whether at rest or in motion, would appear in the direction of these rays; but as light takes some time to travel, when Jupiter is in conjunction, we see him by means of rays that left him 16′ 26″ before; but during that time we have changed our position, in consequence of the motion of the earth in its orbit; we therefore refer Jupiter to a place in which he is not. His true position is in the diagonal of the parallelogram, whose sides are in the ratio of the velocity of light to the velocity of the earth in its orbit, which is as 192000 to 19. In consequence of aberration, none of the heavenly bodies are in the place in which they seem to be. In fact, if the earth were at rest, rays from a star would pass along the axis of a telescope directed to it; but if the earth were to begin to move in its orbit with its usual velocity, these rays would strike against the side of the tube; it would therefore be necessary to incline the telescope a little, in order to see the star. The angle contained between the axis of the telescope and a line drawn to the true place of the star, is its aberration, which varies in quantity and direction in different parts of the earth's orbit; but as it never exceeds twenty seconds, it is insensible in ordinary cases.

The velocity of light deduced from the observed aberration of the fixed stars, perfectly corresponds with that given by the eclipses of the first satellite. The same result obtained from sources so different, leaves not a doubt of its truth. Many such beautiful coincidences, derived from apparently the most unpromising and dissimilar circumstances, occur in physical astronomy, and prove dependences which we might otherwise be unable to trace. The identity of the velocity of light at the distance of Jupiter and on the earth's surface shows that its velocity is uniform; and if light consists in the vibrations of an elastic fluid or ether filling space, which hypothesis accords best with observed phenomena, the uniformity of its velocity shows that the density of the fluid throughout the whole extent of the solar system, must be proportional to its elasticity. Among the fortunate conjectures which have been confirmed by subsequent experience, that of Bacon is not the least remarkable. 'It produces in me,' says the restorer of true philosophy, 'a doubt, whether the face of the serene and starry heavens be seen at the instant it really exists, or not till some time later; and whether there be not, with respect to the heavenly bodies, a true time and an apparent time, no less than a true place and an apparent place, as astronomers say, on account of parallax. For it seems incredible that the species or rays of the celestial bodies can pass through the immense interval between them and us in an instant; or that they do not even require some considerable portion of time.'

As great discoveries generally lead to a variety of conclusions, the aberration of light affords a direct proof of the motion of the earth in its orbit; and its rotation is proved by the theory of falling bodies, since the centrifugal force it induces retards the oscillations of the pendulum in going from the pole to the equator. Thus a high degree of scientific knowledge has been requisite to dispel the errors of the senses.

The little that is known of the theories of the satellites of Saturn and Uranus is in all respects similar to that of Jupiter. The great compression of Saturn occasions its satellites to move nearly in the plane of its equator. Of the situation of the equator of Uranus we know nothing, nor of its compression. The orbits of its satellites are nearly perpendicular to the plane of the ecliptic.

Our constant companion the moon next claims attention. Several circumstances concur to render her motions the most interesting, and at the same time the most difficult to investigate of all the bodies of our system. In the solar system planet troubles planet, but in the lunar theory the sun is the great disturbing cause; his vast distance being compensated by his enormous magnitude, so that the motions of the moon are more irregular than those of the planets; and on account of the great ellipticity of her orbit and the size of the sun, the approximations to her motions are tedious and difficult, beyond what those unaccustomed to such investigations could imagine. Neither the eccentricity of the lunar orbit, nor its inclination to the plane of the ecliptic, have experienced any changes from secular inequalities; but the mean motion, the nodes, and the perigee, are subject to very remarkable variations.

From an eclipse observed at Babylon by the Chaldeans, on the 19th of March, seven hundred and twenty-one years before the Christian era, the place of the moon is known from that of the sun at the instant of opposition; whence her mean longitude may be found; but the comparison of this mean longitude with another mean longitude, computed back for the instant of the eclipse from modern observations, shows that the moon performs her revolution round the earth more rapidly and in a shorter time now, than she did formerly; and that the acceleration in her mean motion has been increasing from age to age as the square of the time; all the ancient and intermediate eclipses confirm this result. As the mean motions of the planets have no secular inequalities, this seemed to be an unaccountable anomaly, and it was at one time attributed to the resistance of an ethereal medium pervading space; at another to the successive transmission of the gravitating force: but as La Place proved that neither of these causes, even if they exist, have any influence on the motions of the lunar perigee or nodes, they could not affect the mean motion, a variation in the latter from such a cause being inseparably connected with variations in the two former of these elements. That great mathematician, however, in studying the theory of Jupiter's satellites, perceived that the secular variations in the elements of Jupiter's orbit, from the action of the planets, occasion corresponding changes in the motions of the satellites: this led him to suspect that the acceleration in the mean motion of the moon might be connected with the secular variation in the eccentricity of the terrestrial orbit; and analysis has proved that he assigned the true cause.

If the eccentricity of the earth's orbit were invariable, the moon would be exposed to a variable disturbance from the action of the sun, in consequence of the earth's annual revolution; but it would be periodic, since it would be the same as often as the sun, the earth, and the moon returned to the same relative positions: on account however of the slow and incessant diminution in the eccentricity of the terrestrial orbit, the revolution of our planet is performed at different distances from the sun every year. The position of the moon with regard to the sun, undergoes a corresponding change; so that the mean action of the sun on the moon varies from one century to another, and occasions the secular increase in the moon's velocity called the acceleration, a name which is very appropriate in the present age, and which will continue to be so for a vast number of ages to come; because, as long as the earth's eccentricity diminishes, the moon's mean motion will be accelerated; but when the eccentricity has passed its minimum and begins to increase, the mean motion will be retarded from age to age. At present the secular acceleration is about 10″, but its effect on the moon's place increases as the square of the time. It is remarkable that the action of the planets thus reflected by the sun to the moon, is much more sensible than their direct action, either on the earth or moon. The secular diminution in the eccentricity, which has not altered the equation of the centre of the sun by eight minutes since the earliest recorded eclipses, has produced a variation of 1° 48′ in the moon's longitude, and of 7° 12′ in her mean anomaly.

The action of the sun occasions a rapid but variable motion in the nodes and perigee of the lunar orbit; the former, though they recede during the greater part of the moon's revolution, and advance during the smaller, perform their sidereal revolutions in $6793^{days}.4212$, and the latter, though its motion is sometimes retrograde and sometimes direct, in $3232^{days}.5807$, or a little more than nine years: but such is the difference between the disturbing energy of the sun and that of all the planets put together, that it requires no less than 109770 years for the greater axis of the terrestrial orbit to do the same. It is evident that the same secular variation which changes the sun's distance from the earth, and occasions the acceleration in the moon's mean motion, must affect the motion of the nodes and perigee; and it consequently appears, from theory as well as observation, that both these elements are subject to a secular inequality, arising from the variation in the eccentricity of the earth's orbit, which connects them with the acceleration; so that both are retarded when the mean motion is anticipated. The secular variations in these three elements are in the ratio of the numbers 3, 0.735, and 1; whence the three motions of the moon, with regard to the sun, to her perigee, and to her nodes, are continually accelerated, and their secular equations are as the numbers 1, 4, and 0.265, or according to the most recent investigations as 1, 4, 6776 and 0.391. A comparison of ancient eclipses observed by the Arabs, Greeks, and Chaldeans, imperfect as they are, with modern observations, perfectly confirms these results of analysis.

Future ages will develop these great inequalities, which at some most distant period will amount to many circumferences. They are indeed periodic; but who shall tell their period? Millions of years must elapse before that great cycle is accomplished; but 'such changes, though rare in time, are frequent in eternity.'

The moon is so near, that the excess of matter at the earth's equator occasions periodic variations in her longitude and latitude; and, as the cause must be proportional to the effect, a comparison of these inequalities, computed from theory, with the same given by observation, shows that the compression of the terrestrial spheroid, or the ratio of the difference between the polar and equatorial diameter to the diameter of the equator is $\frac{1}{305 \cdot 05}$. It is proved analytically, that if a fluid mass of homogeneous matter, whose particles attract each other inversely as the square of the distance, were to revolve about an axis, as the earth, it would assume the form of a spheroid, whose compression is $\frac{1}{230}$. Whence it appears, that the earth is not homogeneous, but decreases in density from its centre to its circumference. Thus the moon's eclipses show the earth to be round, and her inequalities not only determine the form, but the internal structure of our planet; results of analysis which could not have been anticipated. Similar inequalities in Jupiter's satellites prove that his mass is not homogeneous, and that his compression is $\frac{1}{13 \cdot 8}$.

The motions of the moon have now become of more importance to the navigator and geographer than those of any other body, from the precision with which the longitude is determined by the occultations of stars and lunar distances. The lunar theory is brought to such perfection, that the times of these phenomena, observed under any meridian, when compared with that computed for Greenwich in the Nautical Almanack, gives the longitude of the observer within a few miles. The accuracy of that work is obviously of extreme importance to a maritime nation; we have reason to hope that the new Ephemeris, now in preparation, will be by far the most perfect work of the kind that ever has been published.

From the lunar theory, the mean distance of the sun from the earth, and thence the whole dimensions of the solar system are known; for the forces which retain the earth and moon in their orbits, are respectively proportional to the radii vectores of the earth and moon, each being divided by the square of its periodic time; and as the lunar theory gives the ratio of the forces, the ratio of the distance of the sun and moon from the earth is obtained: whence it appears that the sun's distance from the earth is nearly 396 times greater than that of the moon.

The method however of finding the absolute distances of the celestial bodies in miles, is in fact the same with that employed in measuring distances of terrestrial objects. From the extremities of a known base the angles which the visual rays from the object form with it, are measured; their sum subtracted from two right-angles gives the angle opposite the base; therefore by trigonometry, all the angles and sides of the triangle may be computed; consequently the distance of the object is found. The angle under which the base of the triangle is seen from the object, is the parallax of that object; it evidently increases and decreases with the distance; therefore the base must be very great indeed, to be visible at all from the celestial bodies. But the globe itself whose dimensions are ascertained by actual admeasurement, furnishes a standard of measures, with which we compare the distances, masses, densities, and volumes of the sun and planets.

The courses of the great rivers, which are in general navigable to a considerable extent, prove that the curvature of the land differs but little from that of the ocean; and as the heights of the mountains and continents are, at any rate, quite inconsiderable when compared with the magnitude of the earth, its figure is understood to be determined by a surface at every point perpendicular to the direction of gravity, or of the plumb-line, and is the same which the sea would have if it were continued all round the earth beneath the continents. Such is the figure that has been measured in the following manner:—

A terrestrial meridian is a line passing through both poles, all the points of which have contemporaneously the same noon. Were the lengths and curvatures of different meridians known, the figure of the earth might be determined; but the length of one degree is sufficient to give the figure of the earth, if it be measured on different meridians, and in a variety of latitudes; for if the earth were a sphere, all degrees would be of the same length, but if not, the lengths of the degrees will be greatest where the curvature is least; a comparison of the length of the degrees in different parts of the earth's surface will therefore determine its size and form.

An arc of the meridian may be measured by observing the latitude of its extreme points, and then measuring the distance between them in feet or fathoms; the distance thus determined on the surface of the earth, divided by the degrees and parts of a degree contained in the difference of the latitudes, will give the exact length of one degree, the difference of the latitudes being the angle contained between the verticals at the extremities of the arc. This would be easily accomplished were the distance unobstructed, and on a level with the sea; but on account of the innumerable obstacles on the surface of the earth, it is necessary to connect the extreme points of the arc by a series of triangles, the sides and angles of which are either measured or computed, so that the length of the arc is ascertained with much laborious computation. In consequence of the inequalities of the surface, each triangle is in a different plane; they must therefore be reduced by computation to what they would have been, had they been measured on the surface of the sea; and as the earth is spherical, they require a correction to reduce them from plane to spherical triangles.

Arcs of the meridian have been measured in a variety of latitudes, both north and south, as well as arcs perpendicular to the meridian. From these measurements it appears that the length of the degrees increase from the equator to the poles, nearly as the square of the sine of the latitude; consequently, the convexity of the earth diminishes from the equator to the poles. Many discrepancies occur, but the figure that most nearly follows this law is an ellipsoid of revolution, whose equatorial radius is 3962.6 miles, and the polar radius 3949.7; the difference, or 12.9 miles, divided by the equatorial radius, is $\frac{1}{308 \cdot 7}$, or $\frac{1}{309}$ nearly; this fraction is called the compression of the earth, because, according as it is greater or less, the terrestrial ellipsoid is more or less flattened at the poles; it does not differ much from that given by the lunar inequalities. If we assume the earth to be a sphere, the length of a degree of the meridian is $69\frac{1}{22}$ British miles; therefore 360 degrees, or the whole circumference of the globe is 24856, and the diameter, which is something less than a third of the circumference, is 7916 or 8000 miles nearly. Eratosthenes, who died 194 years before the Christian era, was the first to give an approximate value of the earth's circumference, by the mensuration of an arc between Alexandria and Syene.

But there is another method of finding the figure of the earth, totally independent of either of the preceding. If the earth were a homogeneous sphere without rotation, its attraction on bodies at its surface would be everywhere the same; if it be elliptical, the force of gravity theoretically ought to increase, from the equator to the pole, as the square of the sine of the latitude; but for a spheroid in rotation, by the laws of mechanics the centrifugal force varies as the square of the sine of the latitude from the equator where it is greatest, to the pole where it vanishes; and as it tends to make bodies fly off the surface, it diminishes the effects of gravity by a small quantity. Hence by gravitation, which is the difference of these two forces, the fall of bodies ought to be accelerated in going from the equator to the poles, proportionably to the square of the sine of the latitude; and the weight of the same body ought to increase in that ratio. This is directly proved by the oscillations of the pendulum; for if the fall of bodies be accelerated, the oscillations will be more rapid; and that they may always be performed in the same time, the length of the pendulum must be altered. Now, by numerous and very careful experiments, it is proved that a pendulum, which makes 86400 oscillations in a mean day at the equator, will do the same at every point of the earth's surface, if its length be increased in going to the pole, as the square of the sine of the latitude. From the mean of these it appears that the compression of the terrestrial spheroid is about $\frac{1}{342}$, which does not differ much from that given by the lunar inequalities, and from the arcs of the meridian. The near coincidence of these three values, deduced by methods so entirely independent of each other, shows that the mutual tendencies of the centres of the celestial bodies to one another, and the attraction of the earth for bodies at its surface, result from the reciprocal attraction of all their particles. Another proof may be added; the nutation of the earth's axis, and the precession of the equinoxes, are occasioned by the action of the sun and moon on the protuberant matter at the earth's equator; and although these inequalities do not give the absolute value of the terrestrial compression, they show that the fraction expressing it is comprised between the limits $\frac{1}{279}$ and $\frac{1}{578}$.

It might be expected that the same compression should result from each, if the different methods of observation could be made without error. This, however, is not the case; for such discrepancies are found both in the degrees of the meridian and in the length of the pendulum, as show that the figure of the earth is very complicated; but they are so small when compared with the general results, that they may be disregarded. The compression deduced from the mean of the whole, appears to be about $\frac{1}{320}$; that given by the lunar theory has the advantage of being independent of the irregularities at the earth's surface, and of local attractions. The form and size of the earth being determined, it furnishes a standard of measure with which the dimensions of the solar system may be compared.

The parallax of a celestial body is the angle under which the radius of the earth would be seen if viewed from the centre of that body; it affords the means of ascertaining the distances of the sun, moon, and planets. Suppose that, when the moon is in the horizon at the instant of rising or setting, lines were drawn from her centre to the spectator and to the centre of the earth, these would form a right-angled triangle with the terrestrial radius, which is of a known length; and as the parallax or angle at the moon can be measured, all the angles and one side are given; whence the distance of the moon from the centre of the earth may be computed. The parallax of an object may be found, if two observers under the same meridian, but at a very great distance from one another, observe its zenith distances on the same day at the time of its passage over the meridian. By such contemporaneous observations at the Cape of Good Hope and at Berlin, the mean horizontal parallax of the moon was found to be 3454″.2; whence the mean distance of the moon is about sixty times the mean terrestrial radius, or 240000 miles nearly. Since the parallax is equal to the radius of the earth divided by the distance of the moon; under the same parallel of latitude it varies with the distance of the moon from the earth, and proves the ellipticity of the lunar orbit; and when the moon is at her mean distance, it varies with the terrestrial radii, thus showing that the earth is not a sphere.

Although the method described is sufficiently accurate for finding the parallax of an object so near as the moon, it will not answer for the sun which is so remote, that the smallest error in observation would lead to a false result; but by the transits of Venus that difficulty is obviated. When that planet is in her nodes, or within $1\frac{1}{4}°$ of them, that is, in, or nearly in the plane of the ecliptic, she is occasionally seen to pass over the sun like a black spot. If we could imagine that the sun and Venus had no parallax, the line described by the planet on his disc, and the duration of the transit, would be the same to all the inhabitants of the earth; but as the sun is not so remote but that the semidiameter of the earth has a sensible magnitude when viewed from his centre, the line described by the planet in its passage over his disc appears to be nearer to his centre or farther from it, according to the position of the observer; so that the duration of the transit varies with the different points of the earth's surface at which it is observed. This difference of time, being entirely the effect of parallax, furnishes the means of computing it from the known motions of the earth and Venus, by the same method as for the eclipses of the sun. In fact the ratio of the distances of Venus and the sun from the earth at the time of the transit, are known from the theory of their elliptical motion; consequently, the ratio of the parallaxes of these two bodies, being inversely as their distances, is given; and as the transit gives the difference of the parallaxes, that of the sun is obtained. In 1769, the parallax of the sun was determined by observations of a transit of Venus made at Wardhus in Lapland, and at Otaheite in the South Sea, the latter observation being the object of Cook's first voyage. The transit lasted about six hours at Otaheite, and the difference in the duration at these two stations was eight minutes; whence the sun's parallax was found to be $8''.72$: but by other considerations it has subsequently been reduced to $8''.575$; from which the mean distance of the sun appears to be about 95996000, or ninety-six millions of miles nearly. This is confirmed by an inequality in the motion of the moon, which depends on the parallax of the sun, and which when compared with observation gives $8''.6$ for the sun's parallax.

The parallax of Venus is determined by her transits, that of Mars by direct observation. The distances of these two planets from the earth are therefore known in terrestrial radii; consequently their mean distances from the sun may be computed;

and as the ratios of the distances of the planets from the sun are known by Kepler's law, their absolute distances in miles are easily found.

Far as the earth seems to be from the sun, it is near to him when compared with Uranus; that planet is no less than 1843 millions of miles from the luminary that warms and enlivens the world; to it, situate on the verge of the system, the sun must appear not much larger than Venus does to us. The earth cannot even be visible as a telescopic object to a body so remote; yet man, the inhabitant of the earth, soars beyond the vast dimensions of the system to which his planet belongs, and assumes the diameter of its orbit as the base of a triangle, whose apex extends to the stars.

Sublime as the idea is, this assumption proves ineffectual, for the apparent places of the fixed stars are not sensibly changed by the earth's annual revolution; and with the aid derived from the refinements of modern astronomy and the most perfect instruments, it is still a matter of doubt whether a sensible parallax has been detected, even in the nearest of these remote suns. If a fixed star had the parallax of one second, its distance from the sun would be 20500000 millions of miles. At such a distance not only the terrestrial orbit shrinks to a point, but, where the whole solar system, when seen in the focus of the most powerful telescope, might be covered by the thickness of a spider's thread. Light, flying at the rate of 200000 miles in a second, would take three years and seven days to travel over that space; one of the nearest stars may therefore have been kindled or extinguished more than three years before we could have been aware of so mighty an event. But this distance must be small when compared with that of the most remote of the bodies which are visible in the heavens. The fixed stars are undoubtedly luminous like the sun; it is therefore probable that they are not nearer to one another than the sun is to the nearest of them. In the milky way and the other starry nebulæ, some of the stars that seem to us to be close to others, may be far behind them in the boundless depth of space; nay, may rationally be supposed to be situate many thousand times further off: light would therefore require thousands of years to come to the earth from those myriads of suns, of which our own is but 'the dim and remote companion.'

The masses of such planets as have no satellites are known by comparing the inequalities they produce in the motions of the earth and of each other, determined theoretically, with the same inequalities given by observation, for the disturbing cause must necessarily be proportional to the effect it produces. But as the quantities of matter in any two primary planets are directly as the cubes of the mean distances at which their satellites revolve, and inversely as the squares of their periodic times, the mass of the sun and of any planets which have satellites, may be compared with the mass of the earth. In this manner it is computed that the mass of the sun is 354936 times greater than that of the earth; whence the great perturbations of the moon and the rapid motion of the perigee and nodes of her orbit. Even Jupiter, the largest of the planets, is 1070.5 times less than the sun. The mass of the moon is determined from four different sources,—from her action on the terrestrial equator, which occasions the nutation in the axis of rotation; from her horizontal parallax, from an inequality she produces in the sun's longitude, and from her action on the tides. The three first quantities, computed from theory, and compared with their observed values, give her mass respectively equal to the $\frac{1}{71}$, $\frac{1}{74\cdot2}$, and $\frac{1}{69\cdot2}$ part of that of the earth, which do not differ very much from each other; but, from her action in raising the tides, which furnishes the fourth method, her mass appears to be about the seventy-fifth part of that of the earth, a value that cannot differ much from the truth.

The apparent diameters of the sun, moon, and planets are determined by measurement; therefore their real diameters may be compared with that of the earth; for the real diameter of a planet is to the real diameter of the earth, or 8000 miles, as the apparent diameter of the planet to the apparent diameter of the earth as seen from the planet, that is, to twice the parallax of the planet. The mean apparent diameter of the sun is 1920″, and with the solar parallax 8″.65, it will be found that the diameter of the sun is about 888000 miles; therefore, the centre of the sun were to coincide with the centre of the earth, his volume would not only include the orbit of the moon, but would extend nearly as far again, for the moon's mean distance from the earth is about sixty times the earth's mean radius or 240000 miles; so that twice the distance of the moon is 480000 miles, which differs but little from the solar radius; his equatorial radius is probably not much less than the major axis of the lunar orbit.

The diameter of the moon is only 2160 miles; and Jupiter's diameter of 88000 miles is incomparably less than that of the sun. The diameter of Pallas does not much exceed 71 miles, so that an inhabitant of that planet, in one of our steam-carriages, might go round his world in five or six hours.

The oblate form of the celestial bodies indicates rotatory motion, and this has been confirmed, in most cases, by tracing spots on their surfaces, whence their poles and times of rotation have been determined. The rotation of Mercury is unknown, on account of his proximity to the sun; and that of the new planets has not yet been ascertained. The sun revolves in twenty-five days ten hours, about an axis that is directed towards a point half way between the pole star and Lyra, the plane of rotation being inclined a little more than 70° to that on which the earth revolves. From the rotation of the sun, there is every reason to believe that he has a progressive motion in space, although the direction to which he tends is as yet unknown: but in consequence of the reaction of the planets, he describes a small irregular orbit about the centre of inertia of the system, never deviating from his position by more than twice his own diameter, or about seven times the distance of the moon from the earth.

The sun and all his attendants rotate from west to east on axes that remain nearly parallel to themselves in every point of their orbit, and with angular velocities that are sensibly uniform. Although the uniformity in the direction of their rotation is a circumstance hitherto unaccounted for in the economy of Nature, yet from the design and adaptation of every other part to the perfection of the whole, a coincidence so remarkable cannot be accidental; and as the revolutions of the planets and satellites are also from west to east, it is evident that both must have arisen from the primitive causes which have determined the planetary motions.

The larger planets rotate in shorter periods than the smaller planets and the earth; their compression is consequently greater, and the action of the sun and of their satellites occasions a nutation in their axes, and a precession of their equinoxes, similar to that which obtains in the terrestrial spheroid from the attraction of the sun and moon on the prominent matter at the equator. In comparing the periods of the revolutions of Jupiter and Saturn with the times of their rotation, it appears that a year of Jupiter contains nearly ten thousand of his days, and that of Saturn about thirty thousand Saturnian days.

The appearance of Saturn is unparalleled in the system of the world; he is surrounded by a ring even brighter than himself, which always remains in the plane of his equator, and viewed with a very good telescope, it is found to consist of two concentric rings, divided by a dark band. By the laws of mechanics, it is impossible that this body can retain its position by the adhesion of its particles alone; it must necessarily revolve with a velocity that will generate a centrifugal force sufficient to balance the attraction of Saturn. Observation confirms the truth of these principles, showing that the rings rotate about the planet in 10½ hours, which is considerably less than the time a satellite would take to revolve about Saturn at the same distance. Their plane is inclined to the ecliptic at an angle of 31°; and in consequence of this obliquity of position they always appear elliptical to us, but with an eccentricity so variable as even to be occasionally like a straight line drawn across the planet. At present the apparent axes of the rings are as 1000 to 160; and on the 29th of September, 1832, the plane of the rings will pass through the centre of the earth when they will be visible only with superior instruments, and will appear like a fine line across the disc of Saturn. On the 1st of December in the same year, the plane of the rings will pass through the centre of the sun.

It is a singular result of the theory, that the rings could not maintain their stability of rotation if they were everywhere of uniform thickness; for the smallest disturbance would destroy the equilibrium, which would become more and more deranged, till at last they would be precipitated on the surface of the planet. The rings of Saturn must therefore be irregular solids of unequal breadth in the different parts of the circumference, so that their centres of gravity do not coincide with the centres of their figures.

Professor Struve has also discovered that the centre of the ring is not concentric with the centre of Saturn; the interval between the outer edge of the globe of the planet and the outer edge of the ring on one side, is 11″.073, and on the other side the interval is 11″.288; consequently there is an eccentricity of the globe in the ring of 0″.215.

If the rings obeyed different forces, they would not remain in the same plane, but the powerful attraction of Saturn always maintains them and his satellites in the plane of his equator. The rings, by their mutual action, and that of the sun and satellites, must oscillate about the centre of Saturn, and produce phenomena of light and shadow, whose periods extend to many years.

The periods of the rotation of the moon and the other satellites are equal to the times of their revolutions, consequently these bodies always turn the same face to their primaries; however, as the mean motion of the moon is subject to a secular inequality which will ultimately amount to many circumferences, if the rotation of the moon were perfectly uniform, and not affected by the same inequalities, it would cease exactly to counterbalance the motion of revolution; and the moon, in the course of ages, would successively and gradually discover every point of her surface to the earth. But theory proves that this never can happen; for the rotation of the moon, though it does not partake of the periodic inequalities of her revolution, is affected by the same secular variations, so that her motions of rotation and revolution round the earth will always balance each other, and remain equal. This circumstance arises from the form of the lunar spheroid, which has three principal axes of different lengths at right angles to each other. The moon is flattened at the poles from her centrifugal force, therefore her polar axis is least; the other two are in the plane of her equator, but that directed towards the earth is the greatest. The attraction of the earth, as if it had drawn out that part of the moon's equator, constantly brings the greatest axis, and consequently the same hemisphere towards us, which makes her rotation participate in the secular variations in her mean motion of revolution. Even if the angular velocities of rotation and revolution had not been nicely balanced in the beginning of the moon's motion, the attraction of the earth would have recalled the greatest axis to the direction of the line joining the centres of the earth and moon; so that it would vibrate on each side of that line in the same manner as a pendulum oscillates on each side of the vertical from the influence of gravitation.

No such libration is perceptible; and as the smallest disturbance would make it evident, it is clear that if the moon has ever been touched by a comet, the mass of the latter must have been extremely small; for if it had been only the hundred-thousandth part of that of the earth, it would have rendered the libration sensible. A similar libration exists in the motions of Jupiter's satellites; but although the comet of 1767 and 1779 passed through the midst of them, their libration still remains insensible. It is true, the moon is liable to librations depending on the position of the spectator; at her rising, part of the western edge of her disc is visible, which is invisible at her setting, and the contrary takes place with regard to her eastern edge. There are also librations arising from the relative positions of the earth and moon in their respective orbits, but as they are only optical appearances, one hemisphere will be eternally concealed from the earth. For the same reason, the earth, which must be so splendid an object to one lunar hemisphere, will be for ever veiled from the other. On account of these circumstances, the remoter hemisphere of the moon has its day a fortnight long, and a night of the same duration not even enlightened by a moon, while the favoured side is illuminated by the reflection of the earth during its long night. A moon exhibiting a surface thirteen times larger than ours, with all the varieties of clouds, land, and water coming successively into view, would be a splendid object to a lunar traveller in a journey to his antipodes.

The great height of the lunar mountains probably has a considerable influence on the phenomena of her motion, the more so as her compression is small, and her mass considerable.

In the curve passing through the poles, and that diameter of the moon which always points to the earth, nature has furnished a permanent meridian, to which the different spots on her surface have been referred, and their positions determined with as much accuracy as those of many of the most remarkable places on the surface of our globe.

The rotation of the earth which determines the length of the day may be regarded as one of the most important elements in the system of the world. It serves as a measure of time, and forms the standard of comparison for the revolutions of the celestial bodies, which by their proportional increase or decrease would soon disclose any changes it might sustain. Theory and observation concur in proving, that among the innumerable vicissitudes that prevail throughout creation, the period of the earth's diurnal rotation is immutable. A fluid, as Mr. Babbage observes, in falling from a higher to a lower level, carries with it the velocity due to its revolution with the earth at a greater distance from its centre. It will therefore accelerate, although to an almost infinitesimal extent, the earth's daily rotation. The sum of all these increments of velocity, arising from the descent of all the rivers on the earth's surface, would in time become perceptible, did not nature, by the process of evaporation, raise the waters back to their sources; and thus again by removing matter to a greater distance from the centre, destroy the velocity generated by its previous approach; so that the descent of the rivers does not affect the earth's rotation. Enormous masses projected by volcanoes from the equator to the poles, and the contrary, would indeed affect it, but there is no evidence of such convulsions. The disturbing action of the moon and planets, which has so powerful an effect on the revolution of the earth, in no way influences its rotation: the constant friction of the trade winds on the mountains and continents between the tropics does not impede its velocity, which theory even proves to be the same, as if the sea together with the earth formed one solid mass. But although these circumstances be inefficient, a variation in the mean temperature would certainly occasion a corresponding change in the velocity of rotation: for in the science of dynamics, it is a principle in a system of bodies, or of particles revolving about a fixed centre, that the momentum, or sum of the products of the mass of each into its angular velocity and distance from the centre is a constant quantity, if the system be not deranged by an external cause. Now since the number of particles in the system is the same whatever its temperature may be, when their distances from the centre are diminished, their angular velocity must be increased in order that the preceding quantity may still remain constant. It follows then, that as the primitive momentum of rotation with which the earth was projected into space must necessarily remain the same, the smallest decrease in heat, by contracting the terrestrial spheroid, would accelerate its rotation, and consequently diminish the length of the day. Notwithstanding the constant accession of heat from the sun's rays, geologists have been induced to believe from the nature of fossil remains, that the mean temperature of the globe is decreasing.

The high temperature of mines, hot springs, and above all, the internal fires that have produced, and do still occasion such devastation on our planet, indicate an augmentation of heat towards its centre; the increase of density in the strata corresponding to the depth and the form of the spheroid, being what theory assigns to a fluid mass in rotation, concur to induce the idea that the temperature of the earth was originally so high as to reduce all the substances of which it is composed to a state of fusion, and that in the course of ages it has cooled down to its present state; that it is still becoming colder, and that it will continue to do so, till the whole mass arrives at the temperature of the medium in which it is placed, or rather at a state of equilibrium between this temperature, the cooling power of its own radiation, and the heating effect of the sun's rays. But even if this cause be sufficient to produce the observed effects, it must be extremely slow in its operation; for in consequence of the rotation of the earth being a measure of the periods of the celestial motions, it has been proved, that if the length of the day had decreased by the three hundredth part of a second since the observations of Hipparchus two thousand years ago, it would have diminished the secular equation of the moon by 4″.4. It is therefore beyond a doubt, that the mean temperature of the earth cannot have sensibly varied during that time; if then the appearances exhibited by the strata are really owing to a decrease of internal temperature, it either shows the immense periods requisite to produce geological changes to which two thousand years are as nothing, or that the mean temperature of the earth had arrived at a state of equilibrium before these observations. However strong the indications of the primitive fluidity of the earth, as there is no direct proof, it can only be regarded as a very probable hypothesis; but one of the most profound philosophers and elegant writers of modern times has found, in the secular variation of the eccentricity of the terrestrial orbit, an evident cause of decreasing temperature. That accomplished author, in pointing out the mutual dependences of phenomena, says—'It is evident that the mean temperature of the whole surface of the globe, in so far as it is maintained by the action of the sun at a higher degree than it would have were the sun extinguished, must depend on the mean quantity of the sun's rays which it receives, or, which comes to the same thing, on the total quantity received in a given invariable time: and the length of the year being unchangeable in all the fluctuations of the planetary system, it follows, that the total amount of solar radiation will determine, *cæteris paribus*, the general climate of the earth. Now it is not difficult to show, that this amount is inversely proportional to the minor axis of the ellipse described by the earth about the sun, regarded as slowly variable; and that, therefore, the major axis remaining, as we know it to be, constant, and the orbit being actually in a state of approach to a circle, and consequently the minor axis being on the increase, the mean annual amount of solar radiation received by the whole earth must be actually on the decrease. We have, therefore, an evident real cause to account for the phenomenon.' The limits of the variation in the eccentricity of the earth's orbit are unknown; but if its ellipticity has ever been as great as that of the orbit of Mercury or Pallas, the mean temperature of the earth must have been sensibly higher than it is at present; whether it was great enough to render our northern climates fit for the production of tropical plants, and for the residence of the elephant, and the other inhabitants of the torrid zone, it is impossible to say.

The relative quantity of heat received by the earth at different moments during a single revolution, varies with the position of the perigee of its orbit, which accomplishes a tropical revolution in 20935 years. In the year 1250 of our era, and 29653 years before it, the perigee coincided with the summer solstice; at both these periods the earth was nearer the sun during the summer, and farther from him in the winter than in any other position of the apsides: the extremes of temperature must therefore have been greater than at present; but as the terrestrial orbit was probably more elliptical at the distant epoch, the heat of the summers must have been very great, though possibly compensated by the rigour of the winters; at all events, none of these changes affect the length of the day.

It appears from the marine shells found on the tops of the highest mountains, and in almost every part of the globe, that immense continents have been elevated above the ocean, which must have engulphed others. Such a catastrophe would be occasioned by a variation in the position of the axis of rotation on the surface of the earth; for the seas tending to the new equator would leave some portions of the globe, and overwhelm others.

But theory proves that neither nutation, precession, nor any of the disturbing forces that affect the system, have the smallest influence on the axis of rotation, which maintains a permanent position on the surface, if the earth be not disturbed in its rotation by some foreign cause, as the collision of a comet which may have happened in the immensity of time. Then indeed, the equilibrium could only have been restored by the rushing of the seas to the new equator, which they would continue to do, till the surface was every where perpendicular to the direction of gravity. But it is probable that such an accumulation of the waters would not be sufficient to restore equilibrium if the derangement had been great; for the mean density of the sea is only about a fifth part of the mean density of the earth, and the mean depth even of the Pacific ocean is not more than four miles, whereas the equatorial radius of the earth exceeds the polar radius by twenty-five or thirty miles; consequently the influence of the sea on the direction of gravity is very small; and as it appears that a great change in the position of the axes is incompatible with the law of equilibrium, the geological phenomena must be ascribed to an internal cause. Thus amidst the mighty revolutions which have swept innumerable races of organized beings from the earth, which have elevated plains, and buried mountains in the ocean, the rotation of the earth, and the position of the axis on its surface, have undergone but slight variations.

It is beyond a doubt that the strata increase in density from the surface of the earth to its centre, which is even proved by the lunar inequalities; and it is manifest from the mensuration of arcs of the meridian and the lengths of the seconds pendulum that the strata are elliptical and concentric. This certainly would have happened if the earth had originally been fluid, for the denser parts must have subsided towards the centre, as it approached a state of equilibrium; but the enormous pressure of the superincumbent mass is a sufficient cause for these phenomena. Professor Leslie observes, that air compressed into the fiftieth part of its volume has its elasticity fifty times augmented; if it continue to contract at that rate, it would, from its own incumbent weight, acquire the density of water at the depth of thirty-four miles. But water itself would have its density doubled at the depth of ninety-three miles, and would even attain the density of quicksilver at a depth of 362 miles. In descending therefore towards the centre through 4000 miles, the condensation of ordinary materials would surpass the utmost powers of conception. But a density so extreme is not borne out by astronomical observation. It might seem therefore to follow, that our planet must have a widely cavernous structure, and that we tread on a crust or shell, whose thickness bears a very small proportion to the diameter of its sphere. Possibly too this great condensation at the central regions may be counterbalanced by the increased elasticity due to a very elevated temperature. Dr. Young says that steel would be compressed into one-fourth, and stone into one-eighth of its bulk at the earth's centre. However we are yet ignorant of the laws of compression of solid bodies beyond a certain limit; but, from the experiments of Mr. Perkins, they appear to be capable of a greater degree of compression than has generally been imagined.

It appears then, that the axis of rotation is invariable on the surface of the earth, and observation shows, that were it not for the action of the sun and moon on the matter at the equator, it would remain parallel to itself in every point of its orbit.

The attraction of an exterior body not only draws a spheroid towards it; but, as the force varies inversely as the square of the distance, it gives it a motion about its centre of gravity, unless when the attracting body is situated in the prolongation of one of the axes of the spheroid.

The plane of the equator is inclined to the plane of the ecliptic at an angle of about 23° 28′, and the inclination of the lunar orbit on the same is nearly 5°; consequently, from the oblate figure of the earth, the sun and moon acting obliquely and unequally on the different parts of the terrestrial spheroid, urge the plane of the equator from its direction, and force it to move from east to west, so that the equinoctial points have a slow retrograde motion on the plane of the ecliptic of about 50″.412 annually. The direct tendency of this action would be to make the planes of the equator and ecliptic coincide; but in consequence of the rotation of the earth, the inclination of the two planes remains constant, as a top in spinning preserves the same inclination to the plane of the horizon. Were the earth spherical this effect would not be produced, and the equinoxes would always correspond to the same points of the ecliptic, at least as far as this kind of action is concerned. But another and totally different cause operates on this motion, which has already been mentioned. The action of the planets on one another and on the sun, occasions a very slow variation in the position of the plane of the ecliptic, which affects its inclination on the plane of the equator, and gives the equinoctial points a slow but direct motion on the ecliptic of 0″.312 annually, which is entirely independent of the figure of the earth, and would be the same if it were a sphere. Thus the sun and moon, by moving the plane of the equator, cause the equinoctial points to retrograde on the ecliptic; and the planets, by moving the plane of the ecliptic, give them a direct motion, but much less than the former; consequently the difference of the two is the mean precession, which is proved, both by theory and observation, to be about 50″.1 annually. As the longitudes of all the fixed stars are increased by this quantity, the effects of precession are soon detected; it was accordingly discovered by Hipparchus, in the year 128 before Christ, from a comparison of his own observations with those of Timocharis, 155 years before. In the time of Hipparchus the entrance of the sun into the constellation Aries was the beginning of spring, but since then the equinoctial points have receded 30°; so that the constellations called the signs of the zodiac are now at a considerable distance from those divisions of the ecliptic which bear their names. Moving at the rate of 50″.1 annually, the equinoctial points will accomplish a revolution in 25868 years; but as the precession varies in different centuries, the extent of this period will be slightly modified. Since the motion of the sun is direct, and that of the equinoctial points retrograde, he takes a shorter time to return to the equator than to arrive at the same stars; so that the tropical year of 365.242264 days must be increased by the time he takes to move through an arc of 50″.1, in order to have the length of the sidereal year. By simple proportion it is the 0.014119th part of a day, so that the sidereal year is 365.256383.

The mean annual precession is subject to a secular variation; for although the change in the plane of the ecliptic which is the orbit of the sun, be independent of the form of the earth, yet by bringing the sun, moon and earth into different relative positions from age to age, it alters the direct action of the two first on the prominent matter at the equator; on this account the motion of the equinox is greater by 0″.455 now than it was in the time of Hipparchus; consequently the actual length of the tropical year is about 4″.154 shorter than it was at that time. The utmost change that it can experience from this cause amounts to 43″.

Such is the secular motion of the equinoxes, but it is sometimes increased and sometimes diminished by periodic variations, whose periods depend on the relative positions of the sun and moon with regard to the earth, and occasioned by the direct action of these bodies on the equator. Dr. Bradley discovered that by this action the moon causes the pole of the equator to describe a small ellipse in the heavens, the diameters of which are 16″ and 20″. The period of this inequality is nineteen years, the time employed by the nodes of the lunar orbit to accomplish a revolution. The sun causes a small variation in the description of this ellipse; it runs through its period in half a year. This nutation in the earth's axis affects both the precession and obliquity with small periodic variations; but in consequence of the secular variation in the position of the terrestrial orbit, which is chiefly owing to the disturbing energy of Jupiter on the earth, the obliquity of the ecliptic is annually diminished by 0″.52109. With regard to the fixed stars, this variation in the course of ages may amount to ten or eleven degrees; but the obliquity of the ecliptic to the equator can never vary more than two or three degrees, since the equator will follow in some measure the motion of the ecliptic.

It is evident that the places of all the celestial bodies are affected by precession and nutation, and therefore all observations of them must be corrected for these inequalities.

The densities of bodies are proportional to their masses divided by their volumes; hence if the sun and planets be assumed to be spheres, their volumes will be as the cubes of their diameters. Now the apparent diameters of the sun and earth at their mean distance, are 1922″ and 17″.08, and the mass of the earth is the $\frac{1}{354936}$th part of that of the sun taken as the unit; it follows therefore, that the earth is nearly four times as dense as the sun; but the sun is so large that his attractive force would cause bodies to fall through about 450 feet in a second; consequently if he were even habitable by human beings, they would be unable to move, since their weight would be thirty times as great as it is here. A moderate sized man would weigh about two tons at the surface of the sun. On the contrary, at the surface of the four new planets we should be so light, that it would be impossible to stand from the excess of our muscular force, for a man would only weigh a few pounds. All the planets and satellites appear to be of less density than the earth. The motions of Jupiter's satellites show that his density increases towards his centre; and the singular irregularities in the form of Saturn, and the great compression of Mars, prove the internal structure of these two planets to be very far from uniform.

Astronomy has been of immediate and essential use in affording invariable standards for measuring duration, distance, magnitude, and velocity. The sidereal day, measured by the time elapsed between two consecutive transits of any star at the same meridian, and the sidereal year, are immutable units with which to compare all great periods of time; the oscillations of the isochronous pendulum measure its smaller portions. By these invariable standards alone we can judge of the slow changes that other elements of the system may have undergone in the lapse of ages.

The returns of the sun to the same meridian, and to the same equinox or solstice, have been universally adopted as the measure of our civil days and years. The solar or astronomical day is the time that elapses between two consecutive noons or midnights; it is consequently longer than the sidereal day, on account of the proper motion of the sun during a revolution of the celestial sphere; but as the sun moves with greater rapidity at the winter than at the summer solstice, the astronomical day is more nearly equal to the sidereal day in summer than in winter. The obliquity of the ecliptic also affects its duration, for in the equinoxes the arc of the equator is less than the corresponding arc of the ecliptic, and in the solstices it is greater. The astronomical day is therefore diminished in the first case, and increased in the second. If the sun moved uniformly in the equator at the rate of $59' \, 8''.3$ every day, the solar days would be all equal; the time therefore, which is reckoned by the arrival of an imaginary sun at the meridian, or of one which is supposed to move in the equator, is denominated mean solar time, such as is given by clocks and watches in common life: when it is reckoned by the arrival of the real sun at the meridian, it is apparent time, such as is given by dials. The difference between the time shown by a clock and a dial is the equation of time given in the *Nautical Almanac*, and sometimes amounts to as much as sixteen minutes. The apparent and mean time coincide four times in the year.

Astronomers begin the day at noon, but in common reckoning the day begins at midnight. In England it is divided into twenty-four hours, which are counted by twelve and twelve; but in France, astronomers adopting decimal division, divide the day into ten hours, the hour into one hundred minutes, and the minute into a hundred seconds, because of the facility in computation, and in conformity with their system of weights and measures. This subdivision is not used in common life, nor has it been adopted in any other country, though their scientific writers still employ that division of time. The mean length of the day, though accurately determined, is not sufficient for the purposes either of astronomy or civil life. The length of the year is pointed out by nature as a measure of long periods; but the incommensurability that exists between the lengths of the day, and the revolutions of the sun, renders it difficult to adjust the estimation of both in whole numbers. If the revolution of the sun were accomplished in 365 days, all the years would be of precisely the same number of days, and would begin and end with the sun at the same point of the ecliptic; but as the sun's revolution includes the fraction of a day, a civil year and a revolution of the sun have not the same duration. Since the fraction is nearly the fourth of a day, four years are nearly equal to four revolutions of the sun, so that the addition of a supernumerary day every fourth year nearly compensates the difference; but in process of time further correction will be necessary, because the fraction is less than the fourth of a day. The period of seven days, by far the most permanent division of time, and the most ancient monument of astronomical knowledge, was used by the Brahmins in India with the same denominations employed by us, and was alike found in the Calendars of the Jews, Egyptians, Arabs, and Assyrians; it has survived the fall of empires, and has existed among all successive generations, a proof of their common origin.

The new moon immediately following the winter solstice in the 707th year of Rome was made the 1st of January of the first year of Cæsar; the 25th of December in his 45th year, is considered as the date of Christ's nativity; and Cæsar's 46th year is assumed to be the first of our era. The preceding year is called the first year before Christ by chronologists, but by astronomers it is called the year 0. The astronomical year begins on the 31st of December at noon ; and the date of an observation expresses the days and hours which actually elapsed since that time.

Some remarkable astronomical eras are determined by the position of the major axis of the solar ellipse. Moving at the rate of 61″.906 annually, it accomplishes a tropical revolution in 20935 years. It coincided with the line of the equinoxes 4000 or 4089 years before the Christian era, much about the time chronologists assign for the creation of man. In 6485 the major axis will again coincide with the line of the equinoxes, but then the solar perigee will coincide with the equinox of spring; whereas at the creation of man it coincided with the autumnal equinox. In the year 1250 the major axis was perpendicular to the line of the equinoxes, and then the solar perigee coincided with the solstice of winter, and the apogee with the solstice of summer. On that account La Place proposed the year 1250 as a universal epoch, and that the vernal equinox of that year should be the first day of the first year.

The variations in the positions of the solar ellipse occasion corresponding changes in the length of the seasons. In its present position spring is shorter than summer, and autumn longer than winter; and while the solar perigee continues as it now is, between the solstice of winter and the equinox of spring, the period including spring and summer will be longer than that including autumn and winter: in this century the difference is about seven days. These intervals will be equal towards the year 6485, when the perigee comes to the equinox of spring. Were the earth's orbit circular, the seasons would be equal; their differences arise from the eccentricity of the earth's orbit, small as it is; but the changes are so gradual as to be imperceptible in the short space of human life.

No circumstance in the whole science of astronomy excites a deeper interest than its application to chronology. 'Whole nations,' says La Place, 'have been swept from the earth, with their language, arts and sciences, leaving but confused masses of ruin to mark the place where mighty cities stood; their history, with the exception of a few doubtful traditions, has perished; but the perfection of their astronomical observations marks their high antiquity, fixes the periods of their existence, and proves that even at that early period they must have made considerable progress in science.'

The ancient state of the heavens may now be computed with great accuracy; and by comparing the results of computation with ancient observations, the exact period at which they were made may be verified if true, or if false, their error may be detected. If the date be accurate, and the observation good, it will verify the accuracy of modern tables, and show to how many centuries they may be extended, without the fear of error. A few examples will show the importance of this subject.

At the solstices the sun is at his greatest distance from the equator, consequently his declination at these times is equal to the obliquity of the ecliptic, which in former times was determined from the meridian length of the shadow of the style of a dial on the day of the solstice. The lengths of the meridian shadow at the summer and winter solstice are recorded to have been observed at the city of Layang, in China, 1100 years before the Christian era. From these, the distances of the sun from the zenith of the city of Layang are known. Half the sum of these zenith distances determines the latitude, and half their difference gives the obliquity of the ecliptic at the period of the observation; and as the law of the variation in the obliquity is known, both the time and place of the observations have been verified by computation from modern tables. Thus the Chinese had made some advances in the science of astronomy at that early period; the whole chronology of the Chinese is founded on the observations of eclipses, which prove the existence of that empire for more than 4700 years. The epoch of the lunar tables of the Indians, supposed by Bailly to be 3000 before the Christian era, was proved by La Place from the acceleration of the moon, not to be more ancient than the time of Ptolemy. The great inequality of Jupiter and Saturn whose cycle embraces 929 years, is peculiarly fitted for marking the civilization of a people. The Indians had determined the mean motions of these two planets in that part of their periods when the apparent mean motion of Saturn was at the slowest, and that of Jupiter the most rapid. The periods in which that happened were 3102 years before the Christian era, and the year 1491 after it.

The returns of comets to their perihelia may possibly mark the present state of astronomy to future ages.

The places of the fixed stars are affected by the precession of the equinoxes; and as the law of that variation is known, their positions at any time may be computed. Now Eudoxus, a contemporary of Plato, mentions a star situate in the pole of the equator, and from computation it appears that κ Draconis was not very far from that place about 3000 years ago; but as Eudoxus lived only about 2150 years ago, he must have described an anterior state of the heavens, supposed to be the same that was determined by Chiron, about the time of the siege of Troy. Every circumstance concurs in showing that astronomy was cultivated in the highest ages of antiquity.

A knowledge of astronomy leads to the interpretation of hieroglyphical characters, since astronomical signs are often found on the ancient Egyptian monuments, which were probably employed by the priests to record dates. On the ceiling of the portico of a temple among the ruins of Tentyris, there is a long row of figures of men and animals, following each other in the same direction; among these are the twelve signs of the zodiac, placed according to the motion of the sun: it is probable that the first figure in the procession represents the beginning of the year. Now the first is the Lion as if coming out of the temple; and as it is well known that the agricultural year of the Egyptians commenced at the solstice of summer, the epoch of the inundations of the Nile, if the preceding hypothesis be true, the solstice at the time the temple was built must have happened in the constellation of the lion; but as the solstice now happens 21°.6 north of the constellation of the Twins, it is easy to compute that the zodiac of Tentyris must have been made 4000 years ago.

The author had occasion to witness an instance of this most interesting application of astronomy, in ascertaining the date of a papyrus sent from Egypt by Mr. Salt, in the hieroglyphical researches of the late Dr. Thomas Young, whose profound and varied acquirements do honour not only to his country, but to the age in which he lived. The manuscript was found in a mummy case; it proved to be a horoscope of the age of Ptolemy, and its antiquity was determined from the configuration of the heavens at the time of its construction.

The form of the earth furnishes a standard of weights and measures for the ordinary purposes of life, as well as for the determination of the masses and distances of the heavenly bodies. The length of the pendulum vibrating seconds in the latitude of London forms the standard of the British measure of extension. Its length oscillating in vacuo at the temperature of 62° of Fahrenheit, and reduced to the level of the sea, was determined by Captain Kater, in parts of the imperial standard yard, to be 39.1387 inches. The weight of a cubic inch of water at the temperature of 62° Fahrenheit, barometer 30, was also determined in parts of the imperial troy pound, whence a standard both of weight and capacity is deduced. The French have adopted the metre for their unit of linear measure, which is the ten millionth part of that quadrant of the meridian passing through Formentera and Greenwich, the middle of which is nearly in the forty-fifth degree of latitude. Should the national standards of the two countries be lost in the vicissitudes of human affairs, both may be recovered, since they are derived from natural standards presumed to be invariable. The length of the pendulum would be found again with more facility than the metre; but as no measure is mathematically exact, an error in the original standard may at length become sensible in measuring a great extent, whereas the error that must necessarily arise in measuring the quadrant of the meridian is rendered totally insensible by subdivision in taking its ten millionth part. The French have adopted the decimal division not only in time, but in their degrees, weights, and measures, which affords very great facility in computation. It has not been adopted by any other people; though nothing is more desirable than that all nations should concur in using the same division and standards, not only on account of the convenience, but as affording a more definite idea of quantity. It is singular that the decimal division of the day, of degrees, weights and measures, was employed in China 4000 years ago;

and that at the time Ibn Junis made his observations at Cairo, about the year 1000, the Arabians were in the habit of employing the vibrations of the pendulum in their astronomical observations.

One of the most immediate and striking effects of a gravitating force external to the earth is the alternate rise and fall of the surface of the sea twice in the course of a lunar day, or 24^{h} 50^{m} 48^{s} of mean solar time. As it depends on the action of the sun and moon, it is classed among astronomical problems, of which it is by far the most difficult and the least satisfactory. The form of the surface of the ocean in equilibrio, when revolving with the earth round its axis, is an ellipsoid flattened at the poles; but the action of the sun and moon, especially of the moon, disturbs the equilibrium of the ocean.

If the moon attracted the centre of gravity of the earth and all its particles with equal and parallel forces, the whole system of the earth and the waters that cover it, would yield to these forces with a common motion, and the equilibrium of the seas would remain undisturbed. The difference of the forces, and the inequality of their directions, alone trouble the equilibrium.

It is proved by daily experience, as well as by strict mechanical reasoning, that if a number of waves or oscillations be excited in a fluid by different forces, each pursues its course, and has its effect independently of the rest. Now in the tides there are three distinct kinds of oscillations, depending on different causes, producing their effects independently of each other, which may therefore be estimated separately.

The oscillations of the first kind which are very small, are independent of the rotation of the earth; and as they depend on the motion of the disturbing body in its orbit, they are of long periods. The second kind of oscillations depends on the rotation of the earth, therefore their period is nearly a day: and the oscillations of the third kind depend on an angle equal to twice the angular rotation of the earth; and consequently happen twice in twenty-four hours. The first afford no particular interest, and are extremely small; but the difference of two consecutive tides depends on the second. At the time of the solstices, this difference which, according to Newton's theory, ought to be very great, is hardly sensible on our shores. La Place has shown that this discrepancy arises from the depth of the sea, and that if the depth were uniform, there would be no difference in the consecutive tides, were it not for local circumstances: it follows therefore, that as this difference is extremely small, the sea, considered in a large extent, must be nearly of uniform depth, that is to say, there is a certain mean depth from which the deviation is not great. The mean depth of the Pacific ocean is supposed to be about four miles, that of the Atlantic only three. From the formulæ which determine the difference of the consecutive tides it is also proved that the precession of the equinoxes, and the nutation in the earth's axis, are the same as if the sea formed one solid mass with the earth.

The third kind of oscillations are the semidiurnal tides, so remarkable on our coasts; they are occasioned by the combined action of the sun and moon, but as the effect of each is independent of the other, they may be considered separately.

The particles of water under the moon are more attracted than the centre of gravity of the earth, in the inverse ratio of the square of the distances; hence they have a tendency to leave the earth, but are retained by their gravitation, which this tendency diminishes. On the contrary, the moon attracts the centre of the earth more powerfully than she attracts the particles of water in the hemisphere opposite to her; so that the earth has a tendency to leave the waters but is retained by gravitation, which this tendency again diminishes. Thus the waters immediately under the moon are drawn from the earth at the same time that the earth is drawn from those which are diametrically opposite to her; in both instances producing an elevation of the ocean above the surface of equilibrium of nearly the same height; for the diminution of the gravitation of the particles in each position is almost the same, on account of the distance of the moon being great in comparison of the radius of the earth. Were the earth entirely covered by the sea, the water thus attracted by the moon would assume the form of an oblong spheroid, whose greater axis would point towards the moon, since the columns of water under the moon and in the direction diametrically opposite to her are rendered lighter, in consequence of the diminution of their gravitation; and in order to preserve the equilibrium, the axes 90° distant would be shortened. The elevation, on account of the smaller space to which it is confined, is twice as great as the depression, because the contents of the spheroid always remain the same. The effects of the sun's attraction are in all respects similar to those of the moon's, though greatly less in degree, on account of his distance; he therefore only modifies the form of this spheroid a little. If the waters were capable of instantly assuming the form of equilibrium, that is, the form of the spheroid, its summit would always point to the moon, notwithstanding the earth's rotation; but on account of their resistance, the rapid motion produced in them by rotation prevents them from assuming at every instant the form which the equilibrium of the forces acting on them requires. Hence, on account of the inertia of the waters, if the tides be considered relatively to the whole earth and open sea, there is a meridian about 30° eastward of the moon, where it is always high water both in the hemisphere where the moon is, and in that which is opposite. On the west side of this circle the tide is flowing, on the east it is ebbing, and on the meridian at 90° distant, it is everywhere low water. It is evident that these tides must happen twice in a day, since in that time the rotation of the earth brings the same point twice under the meridian of the moon, once under the superior and once under the inferior meridian.

In the semidiurnal tides there are two phenomena particularly to be distinguished, one that happens twice in a month, and the other twice in a year.

The first phenomenon is, that the tides are much increased in the syzigies, or at the time of new and full moon. In both cases the sun and moon are in the same meridian, for when the moon is new they are in conjunction, and when she is full they are in opposition. In each of these positions their action is combined to produce the highest or spring tides under that meridian, and the lowest in those points that are 90° distant. It is observed that the higher the sea rises in the full tide, the lower it is in the ebb. The neap tides take place when the moon is in quadrature, they neither rise so high nor sink so low as the spring tides. The spring tides are much increased when the moon is in perigee. It is evident that the spring tides must happen twice a month, since in that time the moon is once new and once full.

The second phenomenon in the tides is the augmentation which occurs at the time of the equinoxes when the sun's declination is zero, which happens twice every year. The greatest tides take place when a new or full moon happens near the equinoxes while the moon is in perigee. The inclination of the moon's orbit on the ecliptic is 5° 9′; hence in the equinoxes the action of the moon would be increased if her node were to coincide with her perigee. The equinoctial gales often raise these tides to a great height. Beside these remarkable variations, there are others arising from the declination of the moon, which has a great influence on the ebb and flow of the waters.

Both the height and time of high water are thus perpetually changing; therefore, in solving the problem, it is required to determine the heights to which they rise, the times at which they happen, and the daily variations.

The periodic motions of the waters of the ocean on the hypothesis of an ellipsoid of revolution entirely covered by the sea, are very far from according with observation; this arises from the very great irregularities in the surface of the earth, which is but partially covered by the sea, the variety in the depths of the ocean, the manner in which it is spread out on the earth, the position and inclination of the shores, the currents, the resistance the waters meet with, all of them causes which it is impossible to estimate, but which modify the oscillations of the great mass of the ocean. However, amidst all these irregularities, the ebb and flow of the sea maintain a ratio to the forces producing them sufficient to indicate their nature, and to verify the law of the attraction of the sun and moon on the sea. La Place observes, that the investigation of such relations between cause and effect is no less useful in natural philosophy than the direct solution of problems, either to prove the existence of the causes, or trace the laws of their effects. Like the theory of probabilities, it is a happy supplement to the ignorance and weakness of the human mind. Thus the problem of the tides does not admit of a general solution; it is certainly necessary to analyse the general phenomena which ought to result from the attraction of the sun and moon, but these must be corrected in each particular case by those local observations which are modified by the extent and depth of the sea, and the peculiar circumstances of the port.

Since the disturbing action of the sun and moon can only become sensible in a very great extent of water, it is evident that the Pacific ocean is one of the principal sources of our tides; but in consequence of the rotation of the earth, and the inertia of the ocean, high water does not happen till some time after the moon's southing. The tide raised in that world of waters is transmitted to the Atlantic, and from that sea it moves in a northerly direction along the coasts of Africa and Europe, arriving later and later at each place. This great wave however is modified by the tide raised in the Atlantic, which sometimes combines with that from the Pacific in raising the sea, and sometimes is in opposition to it, so that the tides only rise in proportion to their difference. This great combined wave, reflected by the shores of the Atlantic, extending nearly from pole to pole, still coming northward, pours through the Irish and British channels into the North sea, so that the tides in our ports are modified by those of another hemisphere. Thus the theory of the tides in each port, both as to their height and the times at which they take place, is really a matter of experiment, and can only be perfectly determined by the mean of a very great number of observations including several revolutions of the moon's nodes.

The height to which the tides rise is much greater in narrow channels than in the open sea, on account of the obstructions they meet with. In high latitudes where the ocean is less directly under the influence of the luminaries, the rise and fall of the sea is inconsiderable, so that, in all probability, there is no tide at the poles, or only a small annual and monthly one. The ebb and flow of the sea are perceptible in rivers to a very great distance from their estuaries. In the straits of Pauxis, in the river of the Amazons, more than five hundred miles from the sea, the tides are evident. It requires so many days for the tide to ascend this mighty stream, that the returning tides meet a succession of those which are coming up; so that every possible variety occurs in some part or other of its shores, both as to magnitude and time. It requires a very wide expanse of water to accumulate the impulse of the sun and moon, so as to render their influence sensible; on that account the tides in the Mediterranean and Black Sea are scarcely perceptible.

These perpetual commotions in the waters of the ocean are occasioned by forces that bear a very small proportion to terrestrial gravitation: the sun's action in raising the ocean is only the $\frac{1}{38448000}$ of gravitation at the earth's surface, and the action of the moon is little more than twice as much, these forces being in the ratio of 1 to 2.35333. From this ratio the mass of the moon is found to be only $\frac{1}{75}$th part of that of the earth. The initial state of the ocean has no influence on the tides; for whatever its primitive conditions may have been, they must soon have vanished by the friction and mobility of the fluid. One of the most remarkable circumstances in the theory of the tides is the assurance that in consequence of the density of the sea being only one-fifth of the mean density of the earth, the stability of the equilibrium of the ocean never can be subverted by any physical cause whatever. A general inundation arising from the mere instability of the ocean is therefore impossible.

The atmosphere when in equilibrio is an ellipsoid flattened at the poles from its rotation with the earth: in that state its strata are of uniform density at equal heights above the level of the sea, and it is sensibly of finite extent, whether it consists of particles infinitely divisible or not. On the latter hypothesis it must really be finite; and even if the particles of matter be infinitely divisible, it is known by experience to be of extreme tenuity at very small heights. The barometer rises in proportion to the superincumbent pressure. Now at the temperature of melting ice, the density of mercury is to that of air as 10320 to 1; and as the mean height of the barometer is 29.528 inches, the height of the atmosphere by simple proportion is 30407 feet, at the mean temperature of 62°, or 34153 feet, which is extremely small, when compared with the radius of the earth. The action of the sun and moon disturbs the equilibrium of the atmosphere, producing oscillations similar to those in the ocean, which occasion periodic variations in the heights of the barometer. These, however, are so extremely small, that their existence in latitudes so far removed from the equator is doubtful; a series of observations within the tropics can alone decide this delicate point. La Place seems to think that the flux and reflux distinguishable at Paris may be occasioned by the rise and fall of the ocean, which forms a variable base to so great a portion of the atmosphere.

The attraction of the sun and moon has no sensible effect on the trade winds; the heat of the sun occasions these aërial currents, by rarefying the air at the equator, which causes the cooler and more dense part of the atmosphere to rush along the surface of the earth to the equator, while that which is heated is carried along the higher strata to the poles, forming two currents in the direction of the meridian. But the rotatory velocity of the air corresponding to its geographical situation decreases towards the poles; in approaching the equator it must therefore revolve more slowly than the corresponding parts of the earth, and the bodies on the surface of the earth must strike against it with the excess of their velocity, and by its reaction they will meet with a resistance contrary to their motion of rotation; so that the wind will appear, to a person supposing himself to be at rest, to blow in a contrary direction to the earth's rotation, or from east to west, which is the direction of the trade winds. The atmosphere scatters the sun's rays, and gives all the beautiful tints and cheerfulness of day. It transmits the blue light in greatest abundance; the higher we ascend, the sky assumes a deeper hue, but in the expanse of space the sun and stars must appear like brilliant specks in profound blackness.

The sun and most of the planets appear to be surrounded with atmospheres of considerable density. The attraction of the earth has probably deprived the moon of hers, for the refraction of the air at the surface of the earth is at least a thousand times as great as at the moon. The lunar atmosphere, therefore, must be of a greater degree of rarity than can be produced by our best air-pumps; consequently no terrestrial animal could exist in it.

Many philosophers of the highest authority concur in the belief that light consists in the undulations of a highly elastic ethereal medium pervading space, which, communicated to the optic nerves, produce the phenomena of vision. The experiments of our illustrious countryman, Dr. Thomas Young, and those of the celebrated Fresnel, show that this theory accords better with all the observed phenomena than that of the emission of particles from the luminous body. As sound is propagated by the undulations of the air, its theory is in a great many respects similar to that of light. The grave or low tones are produced by very slow vibrations, which increase in frequency progressively as the note becomes more acute. When the vibrations of a musical chord, for example, are less than sixteen in a second, it will not communicate a continued sound to the ear; the vibrations or pulses increase in number with the acuteness of the note, till at last all sense of pitch is lost. The whole extent of human hearing, from the lowest notes of the organ to the highest known cry of insects, as of the cricket, includes about nine octaves.

The undulations of light are much more rapid than those of sound, but they are analogous in this respect, that as the frequency of the pulsations in sound increases from the low tones to the higher, so those of light augment in frequency, from the red rays of the solar spectrum to the extreme violet. By the experiments of Sir William Herschel, it appears that the heat communicated by the spectrum increases from the violet to the red rays; but that the maximum of the hot invisible rays is beyond the extreme red. Heat in all probability consists, like light and sound, in the undulations of an elastic medium. All the principal phenomena of heat may actually be illustrated by a comparison with those of sound. The excitation of heat and sound are not only similar, but often identical, as in friction and percussion; they are both communicated by contact and by radiation; and Dr. Young observes, that the effect of radiant heat in raising the temperature of a body upon which it falls, resembles the sympathetic agitation of a string, when the sound of another string, which is in unison with it, is transmitted to it through the air. Light, heat, sound, and the waves of fluids are all subject to the same laws of reflection, and, indeed, their undulating theories are perfectly similar. If, therefore, we may judge from analogy, the undulations of the heat producing rays must be less frequent tnan those of the extreme red of the solar spectrum; but if the analogy were perfect, the interference of two hot rays ought to produce cold, since darkness results from the interference of two undulations of light, silence ensues from the interference of two undulations of sound; and still water, or no tide, is the consequence of the interference of two tides.

The propagation of sound requires a much denser medium than that of either light or heat; its intensity diminishes as the rarity of the air increases; so that, at a very small height above the surface of the earth, the noise of the tempest ceases, and the thunder is heard no more in those boundless regions where the heavenly bodies accomplish their periods in eternal and sublime silence.

What the body of the sun may be, it is impossible to conjecture; but he seems to be surrounded by an ocean of flame, through which his dark nucleus appears like black spots, often of enormous size. The solar rays, which probably arise from the chemical processes that continually take place at his surface, are transmitted through space in all directions; but, notwithstanding the sun's magnitude, and the inconceivable heat that must exist where such combustion is going on, as the intensity both of his light and heat diminishes with the square of the distance, his kindly influence can hardly be felt at the boundaries of our system. Much depends on the manner in which the rays fall, as we readily perceive from the different climates on our globe. In winter the earth is nearer the sun by $\frac{1}{30}$th than in summer, but the rays strike the northern hemisphere more obliquely in winter than in the other half of the year. In Uranus the sun must be seen like a small but brilliant star, not above the hundred and fiftieth part so bright as he appears to us; that is however 2000 times brighter than our moon to us, so that he really is a sun to Uranus, and probably imparts some degree of warmth. But if we consider that water would not remain fluid in any part of Mars, even at his equator, and that in the temperate zones of the same planet even alcohol and quicksilver would freeze, we may form some idea of the cold that must reign in Uranus, unless indeed the ether has a temperature. The climate of Venus more nearly resembles that of the earth, though, excepting perhaps at her poles, much too hot for animal and vegetable life as they exist here; but in Mercury the mean heat, arising only from the intensity of the sun's rays, must be above that of boiling quicksilver, and water would boil even at his poles. Thus the planets, though kindred with the earth in motion and structure, are totally unfit for the habitation of such a being as man.

The direct light of the sun has been estimated to be equal to that of 5563 wax candles of a moderate size, supposed to be placed at the distance of one foot from the object: that of the moon is probably only equal to the light of one candle at the distance of twelve feet; consequently the light of the sun is more than three hundred thousand times greater than that of the moon; for which reason the light of the moon imparts no heat, even when brought to a focus by a mirror.

In adverting to the peculiarities in the form and nature of the earth and planets, it is impossible to pass in silence the magnetism of the earth, the director of the mariner's compass, and his guide through the ocean. This property probably arises from metallic iron in the interior of the earth, or from the circulation of currents of electricity round it: its influence extends over every part of its surface, but its accumulation and deficiency determine the two poles of this great magnet, which are by no means the same as the poles of the earth's rotation. In consequence of their attraction and repulsion, a needle freely suspended, whether it be magnetic or not, only remains in equilibrio when in the magnetic meridian, that is, in the plane which passes through the north and south magnetic poles. There are places where the magnetic meridian coincides with the terrestrial meridian; in these a magnetic needle freely suspended, points to the true north, but if it be carried successively to different places on the earth's surface, its direction will deviate sometimes to the east and sometimes to the west of north. Lines drawn on the globe through all the places where the needle points due north and south, are called lines of no variation, and are extremely complicated. The direction of the needle is not even constant in the same place, but changes in a few years, according to a law not yet determined. In 1657, the line of no variation passed through

London. In the year 1819, Captain Parry, in his voyage to discover the north-west passage round America, sailed directly over the magnetic pole; and in 1824, Captain Lyon, when on an expedition for the same purpose, found that the variation of the compass was 37° 30′ west, and that the magnetic pole was then situate in 63° 26′ 51″ north latitude, and in 80° 51′ 25″ west longitude. It appears however from later researches that the law of terrestrial magnetism is of considerable complication, and the existence of more than one magnetic pole in either hemisphere has been rendered highly probable. The needle is also subject to diurnal variations; in our latitudes it moves slowly westward from about three in the morning till two, and returns to its former position in the evening.

A needle suspended so as only to be moveable in the vertical plane, dips or becomes more and more inclined to the horizon the nearer it is brought to the magnetic pole. Captain Lyon found that the dip in the latitude and longitude mentioned was 86° 32′. What properties the planets may have in this respect, it is impossible to know, but it is probable that the moon has become highly magnetic, in consequence of her proximity to the earth, and because her greatest diameter always points towards it.

The passage of comets has never sensibly disturbed the stability of the solar system; their nucleus is rare, and their transit so rapid, that the time has not been long enough to admit of a sufficient accumulation of impetus to produce a perceptible effect. The comet of 1770 passed within 80000 miles of the earth without even affecting our tides, and swept through the midst of Jupiter's satellites without deranging the motions of those little moons. Had the mass of that comet been equal to the mass of the earth, its disturbing action would have shortened the year by the ninth of a day; but, as Delambre's computations from the Greenwich observations of the sun, show that the length of the year has not been sensibly affected by the approach of the comet, La Place proved that its mass could not be so much as the 5000th part of that of the earth. The paths of comets have every possible inclination to the plane of the ecliptic, and unlike the planets, their motion is frequently retrograde. Comets are only visible when near their perihelia. Then their velocity is such that its square is twice as great as that of a body moving in a circle at the same distance; they consequently remain a very short time within the planetary orbits; and as all the conic sections of the same focal distance sensibly coincide through a small arc on each side of the extremity of their axis, it is difficult to ascertain in which of these curves the comets move, from observations made, as they necessarily must be, at their perihelia: but probably they all move in extremely eccentric ellipses, although, in most cases, the parabolic curve coincides most nearly with their observed motions. Even if the orbit be determined with all the accuracy that the case admits of, it may be difficult, or even impossible, to recognise a comet on its return, because its orbit would be very much changed if it passed near any of the large planets of this or of any other system, in consequence of their disturbing energy, which would be very great on bodies of so rare a nature. Halley and Clairaut predicted that, in consequence of the attraction of Jupiter and Saturn, the return of the comet of 1759 would be retarded 618 days, which was verified by the event as nearly as could be expected.

The nebulous appearance of comets is perhaps occasioned by the vapours which the solar heat raises at their surfaces in their passage at the perihelia, and which are again condensed as they recede from the sun. The comet of 1680 when in its perihelion was only at the distance of one-sixth of the sun's diameter, or about 148000 miles from its surface; it consequently would be exposed to a heat 27500 times greater than that received by the earth. As the sun's heat is supposed to be in proportion to the intensity of his light, it is probable that a degree of heat so very intense would be sufficient to convert into vapour every terrestrial substance with which we are acquainted.

In those positions of comets where only half of their enlightened hemisphere ought to be seen, they exhibit no phases even when viewed with high magnifying powers. Some slight indications however were once observed by Hevelius and La Hire in 1682; and in 1811 Sir William Herschel discovered a small luminous point, which he concluded to be the disc of the comet. In general their masses are so minute, that they have no sensible diameters, the nucleus being principally formed of denser strata of the nebulous matter, but so rare that stars have been seen through them. The transit of a comet over the sun's disc would afford the best information on this point. It was computed that such an event was to take place in the year 1827; unfortunately the sun was hid by clouds in this country, but it was observed at Viviers and at Marseilles at the time the comet must have been on it, but no spot was seen. The tails are often of very great length, and are generally situate in the planes of their orbits; they follow them in their descent towards the sun, but precede them in their return, with a small degree of curvature; but their extent and form must vary in appearance, according to the position of their orbits with regard to the ecliptic. The tail of the comet of 1680 appeared, at Paris, to extend over sixty-two degrees. The matter of which the tail is composed must be extremely buoyant to precede a body moving with such velocity; indeed the rapidity of its ascent cannot be accounted for. The nebulous part of comets diminishes every time they return to their perihelia; after frequent returns they ought to lose it altogether, and present the appearance of a fixed nucleus; this ought to happen sooner in comets of short periods. La Place supposes that the comet of 1682 must be approaching rapidly to that state. Should the substances be altogether or even to a great degree evaporated, the comet will disappear for ever. Possibly comets may have vanished from our view sooner than they otherwise would have done from this cause. Of about six hundred comets that have been seen at different times, three are now perfectly ascertained to form part of our system; that is to say, they return to the sun at intervals of 76, $6\frac{1}{3}$, and $3\frac{1}{4}$ years nearly.

A hundred and forty comets have appeared within the earth's orbit during the last century that have not again been seen; if a thousand years be allowed as the average period of each, it may be computed by the theory of probabilities, that the whole number that range within the earth's orbit must be 1400; but Uranus being twenty times more distant, there may be no less than 11,200,000 comets that come within the known extent of our system. In such a multitude of wandering bodies

It is just possible that one of them may come in collision with the earth; but even if it should, the mischief would be local, and the equilibrium soon restored. It is however more probable that the earth would only be deflected a little from its course by the near approach of the comet, without being touched. Great as the number of comets appears to be, it is absolutely nothing when compared to the number of the fixed stars. About two thousand only are visible to the naked eye, but when we view the heavens with a telescope, their number seems to be limited only by the imperfection of the instrument. In one quarter of an hour Sir William Herschel estimated that 116000 stars passed through the field of his telescope, which subtended an angle of 15′. This however was stated as a specimen of extraordinary crowding; but at an average the whole expanse of the heavens must exhibit about a hundred millions of fixed stars that come within the reach of telescopic vision.

Many of the stars have a very small progressive motion, especially μ Cassiopeia and 61 Cygni, both small stars; and, as the sun is decidedly a star, it is an additional reason for supposing the solar system to be in motion. The distance of the fixed stars is too great to admit of their exhibiting a sensible disc; but in all probability they are spherical, and must certainly be so, if gravitation pervades all space. With a fine telescope they appear like a point of light; their twinkling arises from sudden changes in the refractive power of the air, which would not be sensible if they had discs like the planets. Thus we can learn nothing of the relative distances of the stars from us and from one another, by their apparent diameters; but their annual parallax being insensible, shows that we must be one hundred millions of millions of miles from the nearest; many of them however must be vastly more remote, for of two stars that appear close together, one may be far beyond the other in the depth of space. The light of Sirius, according to the observations of Mr. Herschel, is 324 times greater than that of a star of the sixth magnitude; if we suppose the two to be really of the same size, their distances from us must be in the ratio of 57.3 to 1, because light diminishes as the square of the distance of the luminous body increases.

Of the absolute magnitude of the stars, nothing is known, only that many of them must be much larger than the sun, from the quantity of light emitted by them. Dr. Wollaston determined the approximate ratio that the light of a wax candle bears to that of the sun, moon, and stars, by comparing their respective images reflected from small glass globes filled with mercury, whence a comparison was established between the quantities of light emitted by the celestial bodies themselves. By this method he found that the light of the sun is about twenty millions of millions of times greater than that of Sirius, the brightest, and supposed to be the nearest of the fixed stars. If Sirius had a parallax of half a second, its distance from the earth would be 525481 times the distance of the sun from the earth; and therefore Sirius, placed where the sun is, would appear to us to be 3.7 times as large as the sun, and would give 13.8 times more light; but many of the fixed stars must be immensely greater than Sirius. Sometimes stars have all at once appeared, shone with a brilliant light, and then vanished. In 1572 a star was discovered in Cassiopeia, which rapidly increased in brightness till it even surpassed that of Jupiter; it then gradually diminished in splendour, and after exhibiting all the variety of tints that indicates the changes of combustion, vanished sixteen months after its discovery, without altering its position. It is impossible to imagine any thing more tremendous than a conflagration that could be visible at such a distance. Some stars are periodic, possibly from the intervention of opaque bodies revolving about them, or from extensive spots on their surfaces. Many thousands of stars that seem to be only brilliant points, when carefully examined are found to be in reality systems of two or more suns revolving about a common centre. These double and multiple stars are extremely remote, requiring the most powerful telescopes to show them separately.

The first catalogue of double stars in which their places and relative positions are determined, was accomplished by the talents and industry of Sir William Herschel, to whom astronomy is indebted for so many brilliant discoveries, and with whom originated the idea of their combination in binary and multiple systems, an idea which his own observations had completely established, but which has since received additional confirmation from those of his son and Sir James South, the former of whom, as well as Professor Struve of Dorpat, have added many thousands to their numbers. The motions of revolution round a common centre of many have been clearly established, and their periods determined with considerable accuracy. Some have already since their first discovery accomplished nearly a whole revolution, and one, if the latest observations can be depended on, is actually considerably advanced in its second period. These interesting systems thus present a species of sidereal chronometer, by which the chronology of the heavens will be marked out to future ages by epochs of their own, liable to no fluctuations from planetary disturbances such as obtain in our system.

Possibly among the multitudes of small stars, whether double or insulated, some may be found near enough to exhibit distinct parallactic motions, or perhaps something approaching to planetary motion, which may prove that solar attraction is not confined to our system, or may lead to the discovery of the proper motion of the sun. The double stars are of various hues, but most frequently exhibit the contrasted colours. The large star is generally yellow, orange, or red; and the small star blue, purple, or green. Sometimes a white star is combined with a blue or purple, and more rarely a red and white are united. In many cases, these appearances are due to the influences of contrast on our judgment of colours. For example, in observing a double star where the large one is of a full ruby red, or almost blood colour, and the small one a fine green, the latter lost its colour when the former was hid by the cross wires of the telescope. But there are a vast number of instances where the colours are too strongly marked to be merely imaginary. Mr. Herschel observes in one of his papers in the *Philosophical Transactions*, as a very remarkable fact, that although red single stars are common enough, no example of an insulated blue, green, or purple one has as yet been produced.

In some parts of the heavens, the stars are so near together as to form clusters, which to the unassisted eye appear like thin white clouds: such is the milky way, which has its brightness from the diffused light of myriads of stars. Many of these clouds, however, are never resolved into separate stars, even by the highest magnifying powers. This nebulous matter exists in vast abundance in space. No fewer than 2500 nebulæ were observed by Sir William Herschel, whose places have been computed from his observations, reduced to a common epoch, and arranged into a catalogue in order of right ascension by his sister Miss Caroline Herschel, a lady so justly celebrated for astronomical knowledge and discovery. The nature and use of this matter scattered over the heavens in such a variety of forms is involved in the greatest obscurity. That it is a self-luminous, phosphorescent material substance, in a highly dilated or gaseous state, but gradually subsiding by the mutual gravitation of its particles into stars and sidereal systems, is the hypothesis which seems to be most generally received; but the only way that any real knowledge on this mysterious subject can be obtained, is by the determination of the form, place, and present state of each individual nebula, and a comparison of these with future observations will show generations to come the changes that may now be going on in these rudiments of future systems. With this view, Mr. Herschel is now engaged in the difficult and laborious investigation, which is understood to be nearly approaching its completion, and the results of which we may therefore hope ere long to see made public. The most conspicuous of these appearances are found in Orion, and in the girdle of Andromeda. It is probable that light must be millions of years travelling to the earth from some of the nebulæ.

So numerous are the objects which meet our view in the heavens, that we cannot imagine a part of space where some light would not strike the eye: but as the fixed stars would not be visible at such distances, if they did not shine by their own light, it is reasonable to infer that they are suns; and if so, they are in all probability attended by systems of opaque bodies, revolving about them as the planets do about ours. But although there be no proof that planets not seen by us revolve about these remote suns, certain it is, that there are many invisible bodies wandering in space, which, occasionally coming within the sphere of the earth's attraction, are ignited by the velocity with which they pass through the atmosphere, and are precipitated with great violence on the earth. The obliquity of the descent of meteorites, the peculiar matter of which they are composed, and the explosion with which their fall is invariably accompanied, show that they are foreign to our planet. Luminous spots altogether independent of the phases have occasionally appeared on the dark part of the moon, which have been ascribed to the light arising from the eruption of volcanoes; whence it has been supposed that meteorites have been projected from the moon by the impetus of volcanic eruption; it has even been computed, that if a stone were projected from the moon in a vertical line, and with an initial velocity of 10992 feet in a second, which is more than four times the velocity of a ball when first discharged from a cannon, instead of falling back to the moon by the attraction of gravity, it would come within the sphere of the earth's attraction, and revolve about it like a satellite. These bodies, impelled either by the direction of the primitive impulse, or by the disturbing action of the sun, might ultimately penetrate the earth's atmosphere, and arrive at its surface. But from whatever source meteoric stones may come, it seems highly probable, that they have a common origin, from the uniformity, we may almost say identity, of their chemical composition.

The known quantity of matter bears a very small proportion to the immensity of space. Large as the bodies are, the distances that separate them are immeasurably greater; but as design is manifest in every part of creation, it is probable that if the various systems in the universe had been nearer to one another, their mutual disturbances would have been inconsistent with the harmony and stability of the whole. It is clear that space is not pervaded by atmospheric air, since its resistance would long ere this have destroyed the velocity of the planets; neither can we affirm it to be void, when it is traversed in all directions by light, heat, gravitation, and possibly by influences of which we can form no idea; but whether it be replete with an ethereal medium, time alone will show.

Though totally ignorant of the laws which obtain in the more distant regions of creation, we are assured, that one alone regulates the motions of our own system; and as general laws form the ultimate object of philosophical research, we cannot conclude these remarks without considering the nature of that extraordinary power, whose effects we have been endeavouring to trace through some of their mazes. It was at one time imagined, that the acceleration in the moon's mean motion was occasioned by the successive transmission of the gravitating force; but it has been proved, that, in order to produce this effect, its velocity must be about fifty millions of times greater than that of light, which flies at the rate of 200000 miles in a second: its action even at the distance of the sun may therefore be regarded as instantaneous; yet so remote are the nearest of the fixed stars, that it may be doubted whether the sun has any sensible influence on them.

The analytical expression for the gravitating force is a straight line; the curves in which the celestial bodies move by the force of gravitation are only lines of the second order; the attraction of spheroids according to any other law would be much more complicated; and as it is easy to prove that matter might have been moved according to an infinite variety of laws, it may be concluded, that gravitation must have been selected by Divine wisdom out of an infinity of other laws, as being the most simple, and that which gives the greatest stability to the celestial motions.

It is a singular result of the simplicity of the laws of nature, which admit only of the observation and comparison of ratios, that the gravitation and theory of the motions of the celestial bodies are independent of their absolute magnitudes and distances; consequently if all the bodies in the solar system, their mutual distances, and their velocities, were to diminish proportionally, they would describe curves in all respects similar to those in which they now move; and the system might be successively reduced to the smallest sensible dimensions, and still exhibit the same appearances. Experience shows that a very different law of attraction prevails when the particles of matter are placed within inappreciable distances from each other, as in chemical and capillary attractions, and the attraction of cohesion; whether it be a modification of gravity, or that some new and unknown power comes into action, does not appear; but as a change in the law of the force takes place at one end of the scale, it is possible that gravitation may not remain the same at the immense distance of the fixed stars. Perhaps the day may come when even gravitation, no longer regarded as an ultimate principle, may be resolved into a yet more general cause, embracing every law that regulates the material world.

The action of the gravitating force is not impeded by the intervention even of the densest substances. If the attraction of the sun for the centre of the earth, and for the hemisphere diametrically opposite to him, was diminished by a difficulty in penetrating the interposed matter, the tides would be more obviously affected. Its attraction is the same also, whatever the substances of the celestial bodies may be, for if the action of the sun on the earth differed by a millionth part from his action on the moon, the difference would occasion a variation in the sun's parallax amounting to several seconds, which is proved to be impossible by the agreement of theory with observation. Thus all matter is pervious to gravitation, and is equally attracted by it.

As far as human knowledge goes, the intensity of gravitation has never varied within the limits of the solar system; nor does even analogy lead us to expect that it should; on the contrary, there is every reason to be assured, that the great laws of the universe are immutable like their Author. Not only the sun and planets, but the minutest particles in all the varieties of their attractions and repulsions, nay even the imponderable matter of the electric, galvanic, and magnetic fluids are obedient to permanent laws, though we may not be able in every case to resolve their phenomena into general principles. Nor can we suppose the structure of the globe alone to be exempt from the universal fiat, though ages may pass before the changes it has undergone, or that are now in progress, can be referred to existing causes with the same certainty with which the motions of the planets and all their secular variations are referable to the law of gravitation. The traces of extreme antiquity perpetually occurring to the geologist, give that information as to the origin of things which we in vain look for in the other parts of the universe. They date the beginning of time; since there is every reason to believe, that the formation of the earth was contemporaneous with that of the rest of the planets; but they show that creation is the work of Him with whom 'a thousand years are as one day, and one day as a thousand years.'

PHYSICAL ASTRONOMY.

THE infinite varieties of motion in the heavens, and on the earth, obey a few laws, so universal in their application, that they regulate the curve traced by an atom which seems to be the sport of the winds, with as much certainty as the orbits of the planets. These laws, on which the order of nature depends, remained unknown till the sixteenth century, when Galileo, by investigating the circumstances of falling bodies, laid the foundation of the science of mechanics, which Newton, by the discovery of gravitation, afterwards extended from the earth to the farthest limits of our system.

This original property of matter, by means of which we ascertain the past and anticipate the future, is the link which connects our planet with remote worlds, and enables us to determine distances, and estimate magnitudes, that might seem to be placed beyond the reach of human faculties. To discern and deduce from ordinary and apparently trivial occurrences the universal laws of nature, as Galileo and Newton have done, is a mark of the highest intellectual power.

Simple as the law of gravitation is, its application to the motions of the bodies of the solar system is a problem of great difficulty, but so important and interesting, that the solution of it has engaged the attention and exercised the talents of the most distinguished mathematicians; among whom La Place holds a distinguished place by the brilliancy of his discoveries, as well as from having been the first to trace the influence of this property of matter from the elliptical motions of the planets, to its most remote effects on their mutual perturbations. Such was the object contemplated by him in his splendid work on the Mechanism of the Heavens; a work which may be considered as a great problem of dynamics, wherein it is required to deduce all the phenomena of the solar system from the abstract laws of motion, and to confirm the truth of those laws, by comparing theory with observation.

Tables of the motions of the planets, by which their places may be determined at any instant for thousands of years, are computed from the analytical formulæ of La Place. In a research so profound and complicated, the most abstruse analysis is required, the higher branches of mathematical science are employed from the first, and approximations are made to the most intricate series. Easier methods, and more convergent series, may probably be discovered in process of time, which will supersede those now in use; but the work of La Place, regarded as embodying the results of not only his own researches, but those of so many of his illustrious predecessors and contemporaries, must ever remain, as he himself expressed it to the writer of these pages, a monument to the genius of the age in which it appeared.

Although physical astronomy is now the most perfect of sciences, a wide range is still left for the industry of future astronomers. The whole system of comets is a subject involved in mystery; they obey, indeed, the general law of gravitation, but many generations must be swept from the earth before their paths can be traced through the regions of space, or the periods of their return can be determined. A new and extensive field of investigation has lately been opened in the discovery of thousands of double stars, or, to speak more strictly, of systems of double stars, since many of them revolve round centres in various and long periods. Who can venture to predict when their theories shall be known, or what laws may be revealed by the knowledge of their motions?—but, perhaps, *Veniet tempus, in quo ista quæ nunc latent, in lucem dies extrahat et longioris ævi diligentia: ad inquisitionem tantorum ætas una non sufficit. Veniet tempus, quo posteri nostri tam aperta nos nescisse mirentur.*

It must, however, be acknowledged that many circumstances seem to be placed beyond our reach. The planets are so remote, that observation discloses but little of their structure; and although their similarity to the earth, in the appearance of their surfaces, and in their annual and diurnal revolutions producing the vicissitudes of seasons, and of day and night, may lead us to fancy that they are peopled with inhabitants like ourselves; yet, were it even permitted to form an analogy from the single instance of the earth, the only one known to us, certain it is that the physical nature of the inhabitants of the planets, if such there be, must differ essentially from ours, to enable them to endure every gradation of temperature, from the intensity of heat in Mercury, to the extreme cold that probably reigns in Uranus. Of the use of Comets in the economy of nature it is impossible to form an idea; still less of the Nebulæ, or cloudy appearances that are scattered through the immensity of space; but instead of being surprised that much is unknown, we have reason to be astonished that the successful daring of man has developed so much.

In the following pages it is not intended to limit the account of the *Mécanique Céleste* to a detail of results, but rather to endeavour to explain the methods by which these results are deduced from one general equation of the motion of matter. To accomplish this, without having recourse to the higher branches of mathematics, is impossible; many subjects, indeed, admit of geometrical demonstration; but as the object of this work is rather to give the spirit of La Place's method than to pursue a regular system of demonstration, it would be a deviation from the unity of his plan to adopt it in the present case.

Diagrams are not employed in La Place's works, being unnecessary to those versed in analysis; some, however, will be occasionally introduced for the convenience of the reader.

BOOK I.

CHAPTER I.

DEFINITIONS, AXIOMS, &c.

1. The activity of matter seems to be a law of the universe, as we know of no particle that is at rest. Were a body absolutely at rest, we could not prove it to be so, because there are no fixed points to which it could be referred; consequently, if only one particle of matter were in existence, it would be impossible to ascertain whether it were at rest or in motion. Thus, being totally ignorant of absolute motion, relative motion alone forms the subject of investigation: a body is, therefore, said to be in motion, when it changes its position with regard to other bodies which are assumed to be at rest.

2. The cause of motion is unknown, force being only a name given to a certain set of phenomena preceding the motion of a body, known by the experience of its effects alone. Even after experience, we cannot prove that the same consequents will invariably follow certain antecedents; we only believe that they will, and experience tends to confirm this belief.

3. No idea of force can be formed independent of matter; all the forces of which we have any experience are exerted by matter; as gravity, muscular force, electricity, chemical attractions and repulsions, &c. &c., in all which cases, one portion of matter acts upon another.

4. When bodies in a state of motion or rest are not acted upon by matter under any of these circumstances, we know by experience that they will remain in that state: hence a body will continue to move uniformly in the direction of the force which caused its motion, unless in some of the cases enumerated, in which we have ascertained by experience that a change of motion will take place, then a force is said to act.

5. Force is proportional to the differential of the velocity, divided

by the differential of the time, or analytically $F = \frac{dv}{dt}$, which is all we know about it.

6. The direction of a force is the straight line in which it causes a body to move. This is known by experience only.

7. In dynamics, force is proportional to the indefinitely small space caused to be moved over in a given indefinitely small time.

8. Velocity is the space moved over in a given time, how small soever the parts may be into which the interval is divided.

9. The velocity of a body moving uniformly, is the straight line or space over which it moves in a given interval of time; hence if the velocity v be the space moved over in one second or unit of time, vt is the space moved over in t seconds or units of time; or representing the space by s, $s = vt$.

10. Thus it is proved that the space described with a uniform motion is proportional to the product of the time and the velocity.

11. Conversely, v, the space moved over in one second of time, is equal to s, the space moved over in t seconds of time, multiplied by $\frac{1''}{t}$, or $v = s\frac{1''}{t} = \frac{s}{t}$.

12. Hence the velocity varies directly as the space, and inversely as the time; and because $t = \frac{s}{v}$,

13. The time varies directly as the space, and inversely as the velocity.

14. Forces are proportional to the velocities they generate in equal times.

The intensity of forces can only be known by comparing their effects under precisely similar circumstances. Thus two forces are equal, which in a given time will generate equal velocities in bodies of the same magnitude; and one force is said to be double of another which, in a given time, will generate double the velocity in one body that it will do in another body of the same magnitude.

15. The intensity of a force may therefore be expressed by the ratios of numbers, or both its intensity and direction by the ratios of lines, since the direction of a force is the straight line in which it causes the body to move.

16. In general, a line expressing the intensity of a force is taken in the direction of the force, beginning from the point of application.

17. Since motion is the change of rectilinear distance between two points, it appears that force, velocity, and motion are expressed by the ratios of spaces; we are acquainted with the ratios of quantities only.

Uniform Motion.

18. A body is said to move uniformly, when, in equal successive intervals of time, how short soever, it moves over equal intervals of space.

19. Hence in uniform motion the space is proportional to the time.

20. The only uniform motion that comes under our observation is the rotation of the earth upon its axis; all other motions in nature are accelerated or retarded. The rotation of the earth forms the only standard of time to which all recurring periods are referred. To be certain of the uniformity of its rotation is, therefore, of the greatest importance. The descent of materials from a higher to a lower level at its surface, or a change of internal temperature, would alter the length of the radius, and consequently the time of rotation: such causes of disturbance do take place; but it will be shown that their effects are so minute as to be insensible, and that the earth's rotation has suffered no sensible change from the earliest times recorded.

21. The equality of successive intervals of time may be measured by the recurrence of an event under circumstances as precisely similar as possible: for example, from the oscillations of a pendulum. When dissimilarity of circumstances takes place, we rectify our conclusions respecting the presumed equality of the intervals, by introducing an equation, which is a quantity to be added or taken away, in order to obtain the equality.

Composition and Resolution of Forces.

fig. 1.

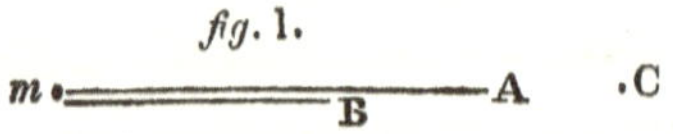

22. Let m be a particle of matter which is free to move in every direction; if two forces, represented both in intensity and direction by the lines mA and mB, be applied to it, and urge it towards C, the particle will move by the combined action of these two forces, and it will require a force equal

to their sum, applied in a contrary direction, to keep it at rest. It is then said to be in a state of equilibrium.

fig. 2.

23. If the forces mA, mB, be applied to a particle m in contrary directions, and if mB be greater than mA, the particle m will be put in motion by the difference of these forces, and a force equal to their difference acting in a contrary direction will be required to keep the particle at rest.

24. When the forces mA, mB are equal, and in contrary directions, the particle will remain at rest.

fig. 3.

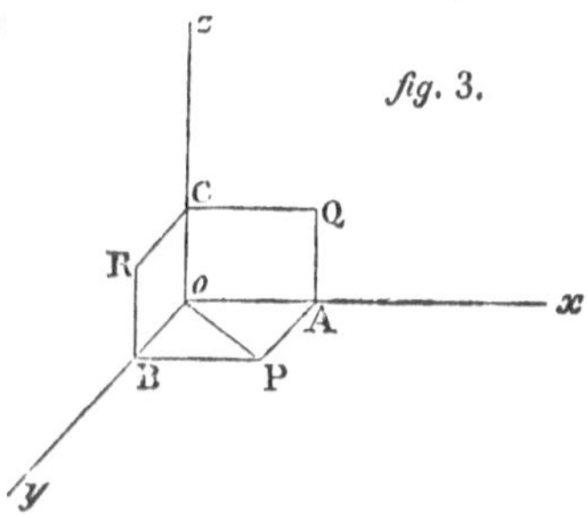

25. It is usual to determine the position of points, lines, surfaces, and the motions of bodies in space, by means of three plane surfaces, oP, oQ, oR, fig. 3, intersecting at given angles. The intersecting or co-ordinate planes are generally assumed to be perpendicular to each other, so that xoy, xoz, yoz, are right angles. The position of ox, oy, oz, the axes of the co-ordinates, and their origin o, are arbitrary; that is, they may be placed where we please, and are therefore always assumed to be known. Hence the position of a point m in space is determined, if its distance from each co-ordinate plane be given; for by taking oA, oB, oC, fig. 4, respectively equal to the given distances, and drawing three planes through A, B, and C, parallel to the co-ordinate planes, they will intersect in m.

fig. 4.

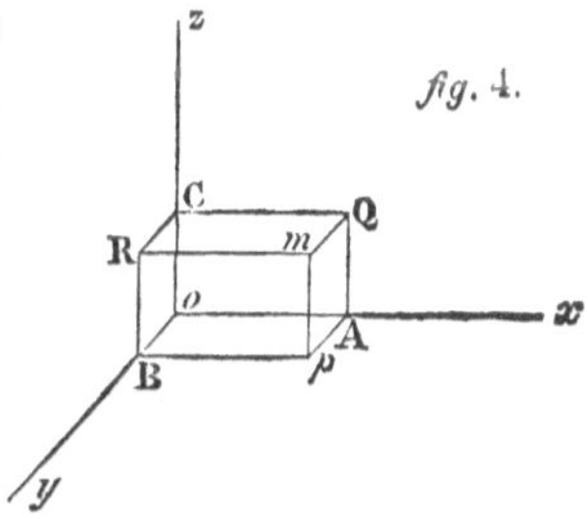

26. If a force applied to a particle of matter at m, (fig. 5,) make it approach to the plane oQ uniformly by the space mA, in a given time t; and if another force applied to m cause it to approach the plane oR uniformly by the space mB, in the same time t, the particle will move in the diagonal

fig. 5.

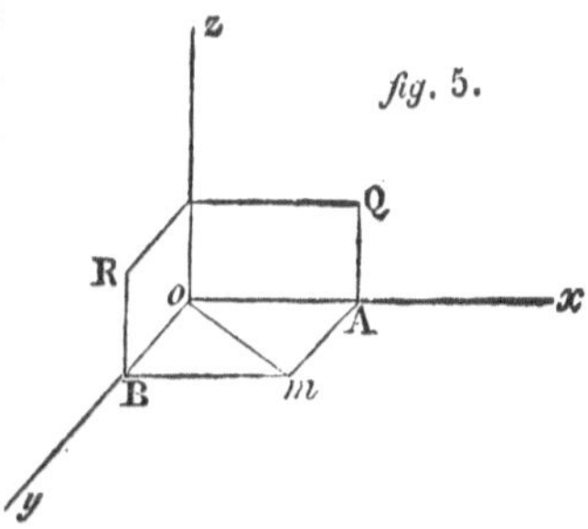

mo, by the simultaneous action of these two forces. For, since the forces are proportional to the spaces, if a be the space described in one second, at will be the space described in t seconds; hence if at be equal to the space mA, and bt equal to the space mB, we have $t = \frac{m\text{A}}{a} = \frac{m\text{B}}{b}$; whence $m\text{A} = \frac{a}{b}\, m\text{B}$ which is the equation to a straight line *mo*, passing through *o*, the origin of the co-ordinates. If the co-ordinates be rectangular, $\frac{a}{b}$ is the tangent of the angle *mo*A, for $m\text{B} = o\text{A}$, and $o\text{A}m$ is a right angle; hence $o\text{A} : \text{A}m :: 1 : \tan \text{A}om$; whence $m\text{A} = o\text{A} \times \tan \text{A}om = m\text{B} \,.\, \tan \text{A}om$. As this relation is the same for every point of the straight line *mo*, it is called its equation. Now since forces are proportional to the velocities they generate in equal times, *m*A, *m*B are proportional to the forces, and may be taken to represent them. The forces *m*A, *m*B are called component or partial forces, and *mo* is called the resulting force. The resulting force being that which, taken in a contrary direction, will keep the component forces in equilibrio.

27. Thus the resulting force is represented in magnitude and direction by the diagonal of a parallelogram, whose sides are *m*A, *m*B the partial ones.

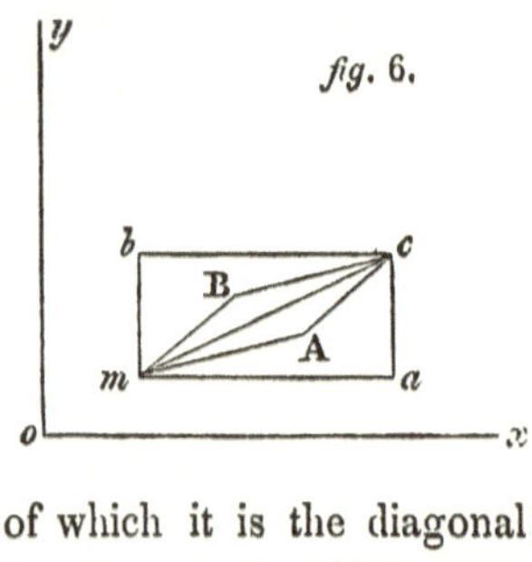

fig. 6.

28. Since the diagonal *cm*, fig. 6, is the resultant of the two forces *m*A, *m*B, whatever may be the angle they make with each other, so, conversely these two forces may be used in place of the single force *mc*. But *mc* may be resolved into any two forces whatever which form the sides of a parallelogram of which it is the diagonal; it may, therefore, be resolved into two forces *ma*, *mb*, which are at right angles to each other. Hence it is always possible to resolve a force *mc* into two others which are parallel to two rectangular axes *ox*, *oy*, situate in the same plane with the force; by drawing through *m* the lines *ma*, *mb*, respectively, parallel to *ox*, *oy*, and completing the parallelogram *macb*.

29. If from any point C, fig. 7, of the direction of a resulting force *m*C, perpendiculars CD, CE, be drawn on the directions of the

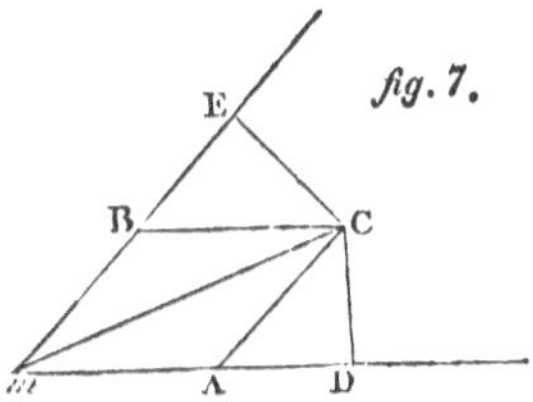

fig. 7.

component forces mA, mB, these perpendiculars are reciprocally as the component forces. That is, CD is to CE as CA to CB, or as their equals mB to mA.

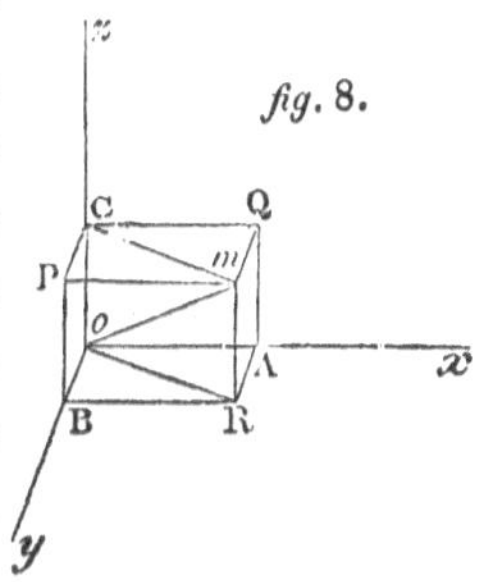

fig. 8.

30. Let BQ, fig. 8, be a figure formed by parallel planes seen in perspective, of which mo is the diagonal. If mo represent any force both in direction and intensity, acting on a material point m, it is evident from what has been said, that this force may be resolved into two other forces, mC, mR, because mo is the diagonal of the parallelogram $mCoR$. Again mC is the diagonal of the parallelogram $mQCP$, therefore it may be resolved into the two forces mQ, mP; and thus the force mo may be resolved into three forces, mP, mQ, and mR; and as this is independent of the angles of the figure, the force mo may be resolved into three forces at right angles to each other. It appears then, that any force mo may be resolved into three other forces parallel to three rectangular axes given in position: and conversely, three forces mP, mQ, mR, acting on a material point m, the resulting force mo may be obtained by constructing the figure BQ with sides proportional to these forces, and drawing the diagonal mo.

fig. 9.

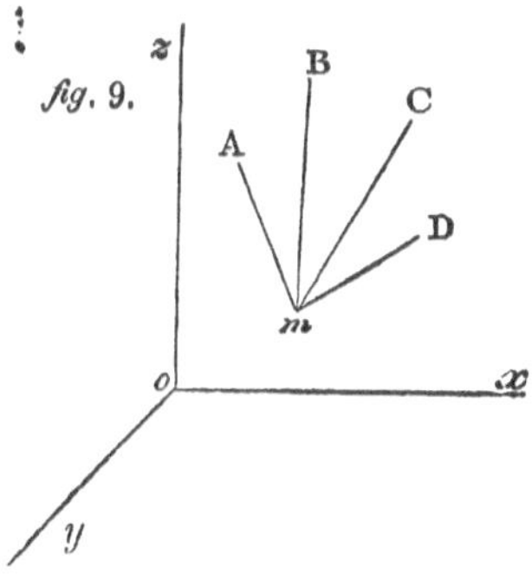

31. Therefore, if the directions and intensities with which any number of forces urge a material point be given, they may be reduced to one single force whose direction and intensity is known. For example, if there were four forces, mA, mB, mC, mD, fig. 9, acting on m, if the resulting force of mA and mB be found, and then that of mC and mD; these four forces would be reduced to two, and by finding the resulting force of these two, the four forces would be reduced to one.

32. Again, this single resulting force may be resolved into three

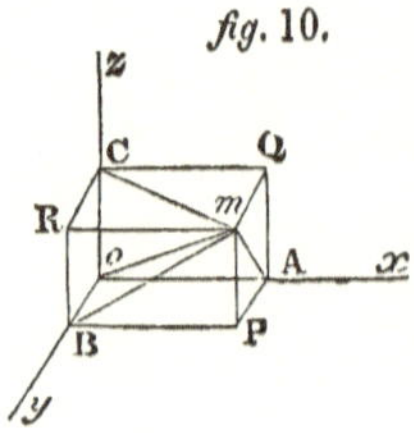

fig. 10.

forces parallel to three rectangular axes ox, oy, oz, fig. 10, which would represent the action of the forces mA, mB, &c., estimated in the direction of the axes; or, which is the same thing, each of the forces mA, mB, &c. acting on m, may be resolved into three other forces parallel to the axes.

33. It is evident that when the partial forces act in the same direction, their sum is the force in that axis; and when some act in one direction, and others in an opposite direction, it is their difference that is to be estimated.

34. Thus any number of forces of any kind are capable of being resolved into other forces, in the direction of two or of three rectangular axes, according as the forces act in the same or in different planes.

35. If a particle of matter remain in a state of equilibrium, though acted upon by any number of forces, and free to move in every direction, the resulting force must be zero.

36. If the material point be in equilibrio on a curved surface, or on a curved line, the resulting force must be perpendicular to the line or surface, otherwise the particle would slide. The line or surface resists the resulting force with an equal and contrary pressure.

37. Let $o\text{A}=\text{X}$, $o\text{B}=\text{Y}$, $o\text{C}=\text{Z}$, fig. 10, be three rectangular component forces, of which $om=\text{F}$ is their resulting force. Then, if mA, mB, mC be joined, $om=\text{F}$ will be the hypothenuse common to three rectangular triangles, $o\text{A}m$, $o\text{B}m$, and $o\text{C}m$. Let the angles $mo\text{A}=a$, $mo\text{B}=b$, $mo\text{C}=c$; then

$$\text{X}=\text{F}\cos a,\quad \text{Y}=\text{F}\cos b,\quad \text{Z}=\text{F}\cos c. \qquad (1).$$

Thus the partial forces are proportional to the cosines of the angles which their directions make with their resultant. But BQ being a rectangular parallelopiped

$$\text{F}^2=\text{X}^2+\text{Y}^2+\text{Z}^2. \qquad (2).$$

Hence

$$\frac{\text{X}^2+\text{Y}^2+\text{Z}^2}{\text{F}^2}=\cos^2 a+\cos^2 b+\cos^2 c=1.$$

When the component forces are known, equation (2) will give a value of the resulting force, and equations (1) will determine its direction by the angles a, b, and c; but if the resulting force be given, its resolution into the three component forces X, Y, Z, making with it the angles a, b, c, will be given by (1). If one of the component forces as Z be zero, then

$$c = 90°, \ \mathrm{F} = \sqrt{\mathrm{X}^2 + \mathrm{Y}^2}, \ \mathrm{X} = \mathrm{F} \cos a, \ \mathrm{Y} = \mathrm{F} \cos b.$$

38. Velocity and force being each represented by the same space, whatever has been explained with regard to the resolution and composition of the one applies equally to the other.

The general Principles of Equilibrium.

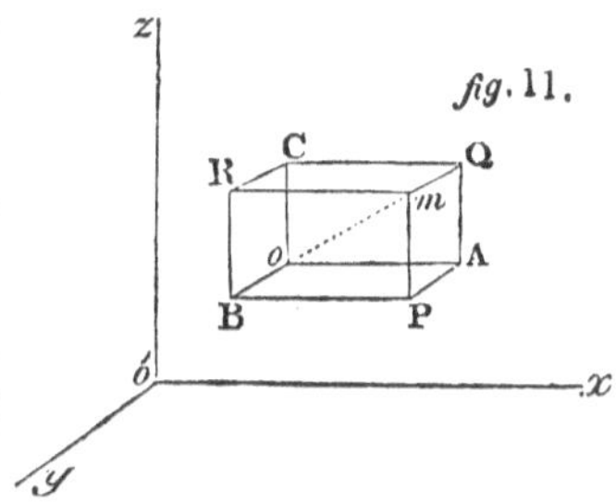

fig. 11.

39. The general principles of equilibrium may be expressed analytically, by supposing o to be the origin of a force F, acting on a particle of matter at m, fig. 11, in the direction om. If o' be the origin of the co-ordinates; a, b, c, the co-ordinates of o, and x, y, z those of m; the diagonal om, which may be represented by r, will be

$$r = \sqrt{(x-a)^2 + (y-b)^2 + (z-c)^2}$$

But F, the whole force in om, is to its component force in

$$o\mathrm{A} :: r : a - x,$$

hence the component force parallel to the axis ox is

$$\mathrm{F}\,\frac{(x-a)}{r}.$$

In the same manner it may be shown, that

$$\mathrm{F}\,\frac{(y-b)}{r}; \quad \mathrm{F}\,\frac{(z-c)}{r}$$

are the component forces parallel to oy and oz. Now the equation of the diagonal gives

$$\frac{\delta r}{\delta x} = \frac{(x-a)}{r} \quad \frac{\delta r}{\delta y} = \frac{(y-b)}{r}; \quad \frac{\delta r}{\delta z} = \frac{(z-c)}{r};$$

hence the component forces of F are

$$\mathrm{F}\left(\frac{\delta r}{\delta x}\right); \ \mathrm{F}\left(\frac{\delta r}{\delta y}\right); \ \mathrm{F}\left(\frac{\delta r}{\delta z}\right).$$

Again, if F′ be another force acting on the particle at m in another direction r', its component forces parallel to the co-ordinates will be,

$$\mathrm{F}'\left(\frac{\delta r'}{\delta x}\right); \ \mathrm{F}'\left(\frac{\delta r'}{\delta y}\right); \ \mathrm{F}'\left(\frac{\delta r'}{\delta z}\right).$$

And any number of forces acting on the particle m may be resolved in the same manner, whatever their directions may be. If Σ be employed to denote the sum of any number of finite quantities, represented by the same general symbol

$$\Sigma.\text{F}.\left(\frac{\delta r}{\delta x}\right)=\text{F}\left(\frac{\delta r}{\delta x}\right)+\text{F}'\left(\frac{\delta r'}{\delta x}\right)+\text{F}''\left(\frac{\delta r''}{\delta x}\right)+\ \&\text{c}.$$

is the sum of the partial forces urging the particle parallel to the axis ox. Likewise $\Sigma.\text{F}.\left(\frac{\delta r}{\delta y}\right)$; $\Sigma.\text{F}\left(\frac{\delta r}{\delta z}\right)$; are the sums of the partial forces that urge the particle parallel to the axis oy and oz. Now if $\text{F}_{\prime}$ be the resulting force of all the forces F, F′, F″, &c. that act on the particle m, and if u be the straight line drawn from the origin of the resulting force to m, by what precedes

$$\text{F}_{\prime}\left(\frac{\delta u}{\delta x}\right);\ \text{F}_{\prime}\left(\frac{\delta u}{\delta y}\right);\ \text{F}_{\prime}\left(\frac{\delta u}{\delta z}\right),$$

are the expressions of the resulting force $\text{F}_{\prime}$, resolved in directions parallel to the three co-ordinates; hence

$$\text{F}_{\prime}\left(\frac{\delta u}{\delta x}\right)=\Sigma.\text{F}\left(\frac{\delta r}{\delta x}\right);\ \text{F}_{\prime}\left(\frac{\delta u}{\delta y}\right)=\Sigma.\text{F}\left(\frac{\delta r}{\delta y}\right);\text{F}_{\prime}\left(\frac{\delta u}{\delta z}\right)=\Sigma.\text{F}\left(\frac{\delta r}{\delta z}\right),$$

or if the sums of the component forces parallel to the axis x, y, z, be represented by X, Y, Z, we shall have

$$\text{F}_{\prime}\left(\frac{\delta u}{\delta x}\right)=\text{X};\ \text{F}_{\prime}\left(\frac{\delta u}{\delta y}\right)=\text{Y};\ \text{F}_{\prime}\left(\frac{\delta u}{\delta z}\right)=\text{Z}.$$

If the first of these be multiplied by δx, the second by δy, and the third by δz, their sum will be

$$\text{F}_{\prime}\delta u=\text{X}\delta x+\text{Y}\delta y+\text{Z}\delta z.$$

40. If the intensity of the force can be expressed in terms of the distance of its point of application from its origin, X, Y, and Z may be eliminated from this equation, and the resulting force will then be given in functions of the distance only. All the forces in nature are functions of the distance, gravity for example, which varies inversely as the square of the distance of its origin from the point of its application. Were that not the case, the preceding equation could be of no use.

41. When the particle is in equilibrio, the resulting force is zero; consequently

$$\text{X}\delta x+\text{Y}\delta y+\text{Z}\delta z=0 \qquad (3),$$

which is the general equation of the equilibrium of a free particle.

42. Thus, when a particle of matter urged by any forces whatever remains in equilibrio, the sum of the products of each force by the element of its direction is zero. As the equation is true, whatever be the values of δx, δy, δz, it is equivalent to the three partial equations in the direction of the axes of the co-ordinates, that is to

$$X = 0, \; Y = 0, \; Z = 0,$$

for it is evident that if the resulting force be zero, its component forces must also be zero.

On Pressure.

43. A pressure is a force opposed by another force, so that no motion takes place.

44. Equal and proportionate pressures are such as are produced by forces which would generate equal and proportionate motions in equal times.

45. Two contrary pressures will balance each other, when the motions which the forces would separately produce in contrary directions are equal; and one pressure will counterbalance two others, when it would produce a motion equal and contrary to the resultant of the motions which would be produced by the other forces.

46. It results from the comparison of motions, that if a body remain at rest, by means of three pressures, they must have the same ratio to one another, as the sides of a triangle parallel to the directions.

On the Normal.

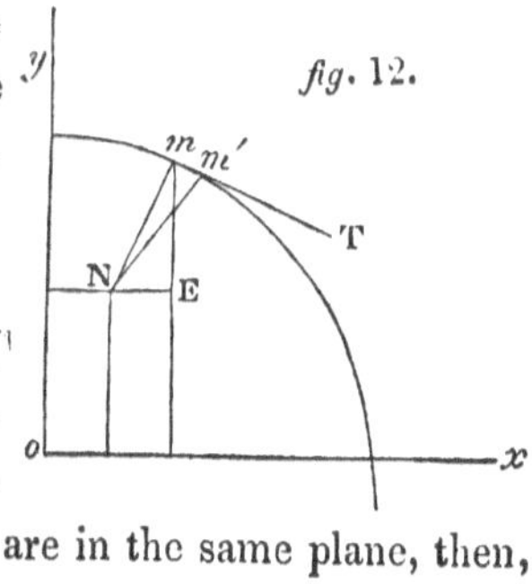

fig. 12.

47. The normal to a curve, or surface in any point m, fig. 12, is the straight line mN perpendicular to the tangent mT. If mm' be a plane curve

$$mN = \sqrt{(x-a)^2+(y-b)^2}$$

x and y being the co-ordinates of m, a and b those of N. If the point m be on a surface, or curve of double curvature, in which no two of its elements are in the same plane, then,

$$mN = \sqrt{(x-a)^2 + (y-b)^2 + (z-c)^2}$$

x, y, z being the co-ordinates of m, and a, b, c those of N. The centre of curvature N, which is the intersection of two consecutive normals mN, m'N, never varies in the circle and sphere, because the curvature is every where the same; but in all other curves and surfaces the position of N changes with every point in the curve or surface, and a, b, c, are only constant from one point to another. By this property, the equation of the radius of curvature is formed from the equation of the curve, or surface. If r be the radius of curvature, it is evident, that though it may vary from one point to another, it is constant for any one point m where $\delta r = 0$.

Equilibrium of a Particle on a curved Surface.

48. The equation (3) is sufficient for the equilibrium of a particle of matter, if it be free to move in any direction; but if it be constrained to remain on a curved surface, the resulting force of all the forces acting upon it must be perpendicular to the surface, otherwise it would slide along it; but as by experience it is found that re-action is equal and contrary to action, the perpendicular force will be resisted by the re-action of the surface, so that the re-action is equal, and contrary to the force destroyed; hence if $R_{\prime}$ be the resistance of the surface, the equation of equilibrium will be

$$X\delta x + Y\delta y + Z\delta z = -R_{\prime}\delta r.$$

δx, δy, δz are arbitrary; these variations may therefore be assumed to take place in the direction of the curved surface on which the particle moves: then by the property of the normal, $\delta r = 0$; which reduces the preceding equation to

$$X\delta x + Y\delta y + Z\delta z = 0.$$

But this equation is no longer equivalent to three equations, but to two only, since one of the elements δx, δy, δz, must be eliminated by the equation of the surface.

49. The same result may be obtained in another way. For if $u = 0$ be the equation of the surface, then $\delta u = 0$; but as the equation of the normal is derived from that of the surface, the equation $\delta r = 0$ is connected with the preceding, so that $\delta r = N\delta u$. But

$$r = \sqrt{(x-a)^2 + (y-b)^2 + (z-c)^2}$$

whence

$$\frac{\delta r}{\delta x} = \frac{x-a}{r}; \ \frac{\delta r}{\delta y} = \frac{y-b}{r}; \ \frac{\delta r}{\delta z} = \frac{z-c}{r};$$

consequently,

$$\left\{\left(\frac{\delta r}{\delta x}\right)^2 + \left(\frac{\delta r}{\delta y}\right)^2 + \left(\frac{\delta r}{\delta z}\right)^2\right\} = 1,$$

on account of which, the equation

$$\delta r = \mathrm{N}\delta u \text{ gives } \mathrm{N}^2 \left\{\left(\frac{\delta u}{\delta r}\right)^2 + \left(\frac{\delta u}{\delta y}\right)^2 + \left(\frac{\delta u}{\delta z}\right)^2\right\} = 1,$$

or

$$\mathrm{N}. = \frac{1}{\sqrt{\left(\frac{\delta u}{\delta x}\right)^2 + \left(\frac{\delta u}{\delta y}\right)^2 + \left(\frac{\delta u}{\delta z}\right)^2}},$$

for u is a function of x, y, z; hence,

$$\mathrm{R}_{,}\delta r = \frac{\mathrm{R}_{,}\delta u}{\sqrt{\left(\frac{\delta u}{\delta x}\right)^2 + \left(\frac{\delta u}{\delta y}\right)^2 + \left(\frac{\delta u}{\delta z}\right)^2}}; \text{ and if}$$

$$\lambda = \frac{\mathrm{R}_{,}}{\sqrt{\left(\frac{\delta u}{\delta x}\right)^2 + \left(\frac{\delta u}{\delta y}\right)^2 + \left(\frac{\delta u}{\delta z}\right)^2}},$$

then $\mathrm{R}_{,}\delta r$ becomes $\lambda\delta u$, and the equation of the equilibrium of a particle m, on a curved line or surface, is

$$\mathrm{X}\delta x + \mathrm{Y}\delta y + \mathrm{Z}\delta z + \lambda\delta u = 0 \qquad (4),$$

where δu is a function of the elements δx, δy, δz: and as this equation exists whatever these elements may be, each of them may be made zero, which will divide it into three equations; but they will be reduced to two by the elimination of λ. And these two, with the equation of the surface $u = 0$, will suffice to determine x, y, z, the co-ordinates of m in its position of equilibrium. These found, N and consequently λ become known. And since $\mathrm{R}_{,}$ is the resistance

$$\mathrm{R}_{,} = \lambda\sqrt{\left(\frac{\delta u}{\delta x}\right)^2 + \left(\frac{\delta u}{\delta y}\right)^2 + \left(\frac{\delta u}{\delta z}\right)^2}$$

is the pressure, which is equal and contrary to the resistance, and is therefore determined.

50. Thus if a particle of matter, either free or obliged to remain on a curved line or surface, be urged by any number of forces, it will continue in equilibrio, if the sum of the products of each force by the element of its direction be zero.

Virtual Velocities.

51. This principle, discovered by John Bernouilli, and called the principle of virtual velocities, is perfectly general, and may be expressed thus :—

If a particle of matter be arbitrarily moved from its position through an indefinitely small space, so that it always remains on the curve or surface, which it ought to follow, if not entirely free, the sum of the forces which urge it, each multiplied by the element of its direction, will be zero in the case of equilibrium.

On this general law of equilibrium, the whole theory of statics depends.

fig. 13.

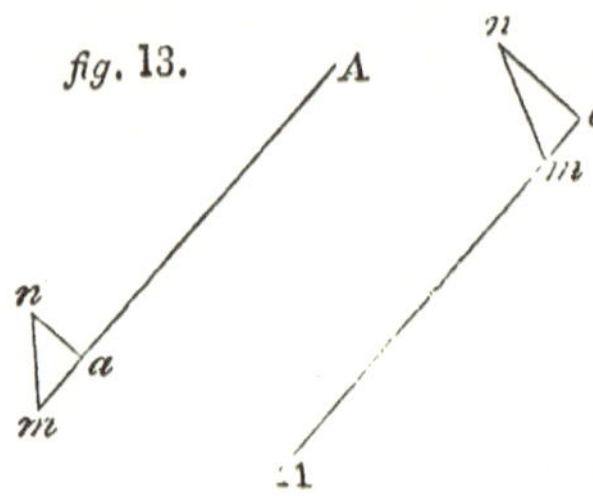

52. An idea of what virtual velocity is, may be formed by supposing that a particle of matter *m* is urged in the direction *m*A by a force applied to *m*. If *m* be arbitrarily moved to any place *n* indefinitely near to *m*, then *mn* will be the virtual velocity of *m*.

53. Let *na* be drawn at right angles to *m*A, then *ma* is the virtual velocity of *m* resolved in the direction of the force *m*A: it is also the projection of *mn* on *m*A; for

$$mn : ma :: 1 : \cos nma \text{ and } ma = mn \cos nma.$$

54. Again, imagine a polygon ABCDM of any number of sides, either in the same plane or not, and suppose the sides MA, AB, &c., to represent, both in magnitude and direction, any forces applied to a particle at M. Let these forces be resolved in the direction of the axis *o x*, so that *ma*, *ab*, *bc*, &c. may be the projections of the sides of the polygon, or the cosines of the angles made by the sides of the polygon with *o x* to the several radii MA, AB, &c., then will the segments *ma*, *ab*, *bc*, &c. of the axis represent the resolved portions of the forces estimated in that single direction, and calling α, β, γ, &c. the angles above mentioned,

fig. 14.

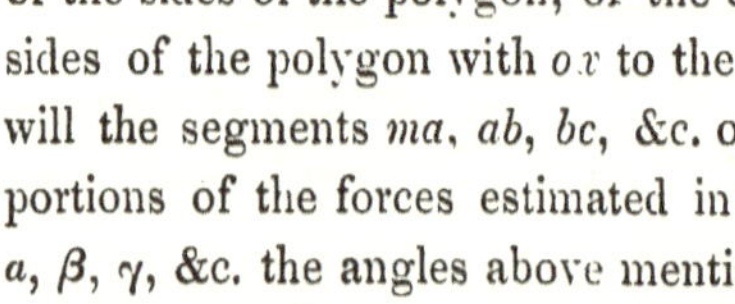

$$ma = \text{MA} \cos \alpha;\ ab = \text{AB} \cos \beta;\ \text{and } bc = \text{BC} \cos \gamma,$$

&c. and the sum of these partial forces will be

$$MA\cos\alpha + AB\cos\beta + BC\cos\gamma + \&c. = 0$$

by the general property of polygons, as will also be evident if we consider that *dm*, *ma*, *ab* lying towards *o* are to be taken positively, and *bc, cd* lying towards *x* negatively; and the latter making up the same whole *bd* as the former, their sums must be zero. Thus it is evident, that if any number of forces urge a particle of matter, the sum of these forces when estimated in any given direction, must be zero when the particle is in equilibrio; and *vice versâ*, when this condition holds, the equilibrium will take place. Hence, we see that a point will rest, if urged by forces represented by the sides of a polygon, taken in order.

In this case also, the sum of the virtual velocities is zero; for, if M be removed from its place through an infinitely small space in any direction, since the position of *ox* is arbitrary, it may represent that direction, and *ma*, *ab*, *bc*, *cd*, *dm*, will therefore represent the virtual velocities of M in directions of the several forces, whose sum, as above shown, is zero.

55. The principle of virtual velocities is the same, whether we consider a material particle, a body, or a system of bodies.

Variations.

56. The symbol δ is appropriated to the calculus of variations, whose general object is to subject to analytical investigation the changes which quantities undergo when the relations which connect them are altered, and when the functions which are the objects of discussion undergo a change of form, and pass into other functions by the gradual variation of some of their elements, which had previously been regarded as constant. In this point of view, variations are only differentials on another hypothesis of constancy and variability, and are therefore subject to all the laws of the differential calculus.

57. The variation of a function may be illustrated by problems of maxima and minima, of which there are two kinds, one not subject to the law of variations, and another that is. In the former case, the quantity whose maximum or minimum is required

fig. 15.

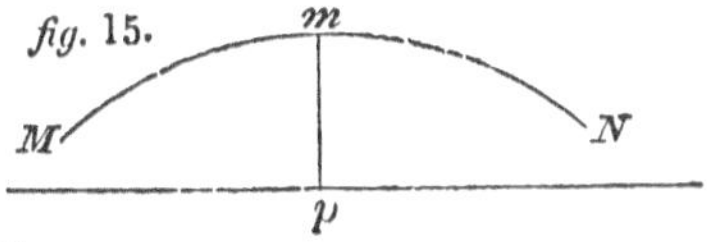

depends by known relations on some arbitrary independent variable; —for example, in a *given* curve MN, fig. 15, it is required to determine the point in which the ordinate $p\ m$ is the greatest possible. In this case, the curve, or function expressing the curve, remains the same; but in the other case, the form of the function whose maximum or minimum is required, is variable; for,

fig. 16.

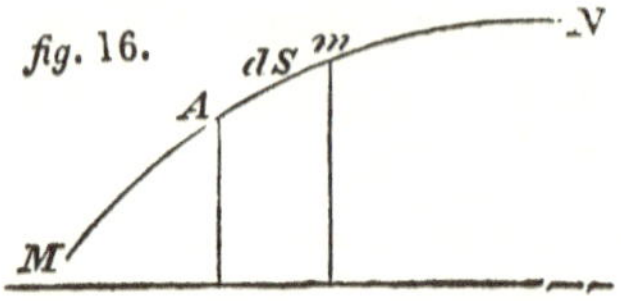

let M, N, fig. 16, be any two given points in space, and suppose it were required, among the infinite number of curves that can be drawn between these two points, to determine that whose length is a minimum. If ds be the element of the curve, $\int ds$ is the curve itself; now as the required curve must be a minimum, the variation of $\int ds$ when made equal to zero, will give that curve, for when quantities are at their maxima or minima, their increments are zero. Thus the form of the function $\int ds$ varies so as to fulfil the conditions of the problem, that is to say, in place of retaining its general form, it takes the form of that particular curve, subject to the conditions required.

58. It is evident from the nature of variations, that the variation of a quantity is independent of its differential, so that we may take the differential of a variation as $d.\delta y$, or the variation of a differential as $\delta.dy$, and that $d.\delta y = \delta.dy$.

59. From what has been said, it appears that virtual velocities are real variations; for if a body be moving on a curve, the virtual velocity may be assumed either to be on the curve or not on the curve; it is consequently independent of the law by which the co-ordinates of the curve vary, unless when we choose to subject it to that law.

CHAPTER II.

VARIABLE MOTION.

60. When the velocity of a moving body changes, the cause of that change is called an accelerating or retarding force; and when the increase or diminution of the velocity is uniform, its cause is called a continued, or uniformly accelerating or retarding force, the increments of space which would be described in a given time with the initial velocities being always equally increased or diminished.

Gravitation is a uniformly accelerating force, for at the earth's surface a stone falls $16\frac{1}{11}$ feet nearly, during the first second of its motion, $48\frac{3}{11}$ during the second, $80\frac{5}{11}$ during the third, &c., falling every second $32\frac{2}{11}$ feet more than during the preceding second.

61. The action of a continued force is uninterrupted, so that the velocity is either gradually increased or diminished; but to facilitate mathematical investigation it is assumed to act by repeated impulses, separated by indefinitely small intervals of time, so that a particle of matter moving by the action of a continued force is assumed to describe indefinitely small but unequal spaces with a uniform motion, in indefinitely small and equal intervals of time.

62. In this hypothesis, whatever has been demonstrated regarding uniform motion is equally applicable to motion uniformly varied; and X, Y, Z, which have hitherto represented the components of an impulsive force, may now represent the components of a force acting uniformly.

Central Force.

63. If the direction of the force be always the same, the motion will be in a straight line; but where the direction of a continued force is perpetually varying it will cause the particle to describe a curved line.

Demonstration.—Suppose a particle impelled in the direction mA, fig. 17, and at the same time attracted by a continued force whose origin is in o, the force being supposed to act impulsively at equal successive infinitely small times. By the first impulse alone, in any given time the particle would move equably to A: but in the same time the action of the continued, or as it must now be considered the impulsive force alone, would cause it to move uniformly through

ma; hence at the end of that time the particle would be found in B, having described the diagonal mB. Were the particle now left to itself, it would move uniformly to C in the next equal interval of time; but the action of the second impulse of the attractive force would bring it equably to b in the same time. Thus at the end of the second interval it would be found in D, having described the diagonal BD, and so on. In this manner the particle would describe the polygon mBDE; but if the intervals between the successive impulses of the attractive force be indefinitely small, the diagonals mB, BD, DE, &c., will also be indefinitely small, and will coincide with the curve passing through the points m, B, D, E, &c.

fig. 17.

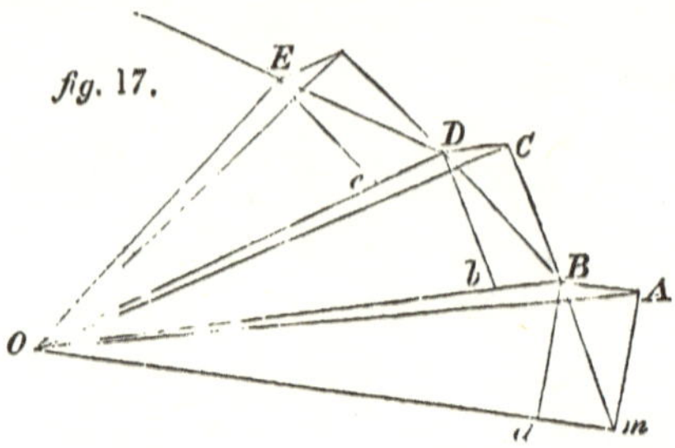

64. In this hypothesis, no error can arise from assuming that the particle describes the sides of a polygon with a uniform motion; for the polygon, when the number of its sides is indefinitely multiplied, coincides entirely with the curve.

65. The lines mA, BC, &c., fig. 17, are tangents to the curve in the points, m, B, &c.; it therefore follows that when a particle is moving in a curved line in consequence of any continued force, if the force should cease to act at any instant, the particle would move on in the tangent with an equable motion, and with a velocity equal to what it had acquired when the force ceased to act.

fig. 18.

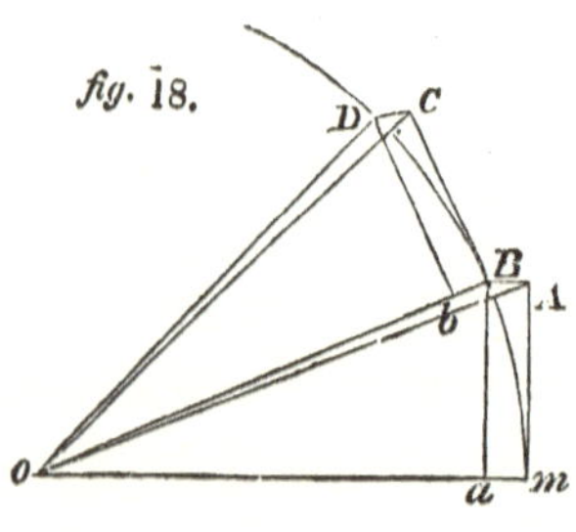

66. The spaces ma, Bb, CD, fig. 18, &c., are the sagittæ of the indefinitely small arcs mB, BD, DE, &c. Hence the effect of the central force is measured by ma, the sagitta of the arc mB described in an indefinitely small given time, or by $\frac{(\text{arc } m\text{B})^2}{2 \cdot om} = ma$, om being the radius of the circle coinciding with the curve in m.

67. We shall consider the element or differential of time to be a constant quantity; the element of space to be the indefinitely small

space moved over in an element of time, and the element of velocity to be the velocity that a particle would acquire, if acted on by a constant force during an element of time. Thus, if t, s and v be the time, space, and velocity, the elements of these quantities are dt, ds, and dv; and as each element is supposed to express an arbitrary unit of its kind, these heterogeneous quantities become capable of comparison. As a decrement only differs from an increment by its sign, any expressions regarding increasing quantities will apply to those that decrease by changing the signs of the differentials; and thus the theory of retarded motion is included in that of accelerated motion.

68. In uniformly accelerated motion, the force at any instant is directly proportional to the second element of the space, and inversely as the square of the element of the time.

Demonstration.—Because in uniformly accelerated motion, the velocity is only assumed to be constant for an indefinitely small time, $v = \frac{ds}{dt}$, and as the element of the time is constant, the differential of the velocity is $dv = \frac{d^2s}{dt}$; but since a constant force, acting for an indefinitely small time, produces an indefinitely small velocity, $\mathrm{F}dt = dv$; hence $\mathrm{F} = \frac{d^2s}{dt^2}$.

General Equations of the Motions of a Particle of Matter.

69. The general equation of the motion of a particle of matter, when acted on by any forces whatever, may be reduced to depend on the law of equilibrium.

Demonstration.—Let m be a particle of matter perfectly free to obey any forces X, Y, Z, urging it in the direction of three rectangular co-ordinates x, y, z. Then regarding velocity as an effect of force, and as its measure, by the laws of motion these forces will produce in the instant dt, the velocities $\mathrm{X}dt$, $\mathrm{Y}dt$, $\mathrm{Z}dt$, proportional to the intensities of these forces, and in their directions. Hence when m is free, by article 68,

$$d \,.\, \frac{dx}{dt} = \mathrm{X}dt\,; \quad d \,.\, \frac{dy}{dt} = \mathrm{Y}dt\,; \quad d \,.\, \frac{dz}{dt} = \mathrm{Z}dt\,; \qquad (5)$$

for the forces X, Y, Z, being perpendicular to each other, each one is independent of the action of the other two, and may be regarded as if it acted alone. If the first of these equations be multiplied by δx, the second by δy, and the third by δz, their sum will be

$$0 = \left(X - \frac{d^2x}{dt^2}\right)\delta x + \left(Y - \frac{d^2y}{dt^2}\right)\delta y + \left(Z - \frac{d^2z}{dt^2}\right)\delta z, \qquad (6)$$

and since $X - \frac{d^2x}{dt^2}$; $Y - \frac{d^2y}{dt^2}$; $Z - \frac{d^2z}{dt^2}$; are separately zero, δx, δy, δz, are absolutely arbitrary and independent; and *vice versâ*, *if* they *are* so, this one equation will be equivalent to the three separate ones.

This is the general equation of the motion of a particle of matter, when free to move in every direction.

2nd case.—But if the particle m be not free, it must either be constrained to move on a curve, or on a surface, or be subject to a resistance, or otherwise subject to some condition. But matter is not moved otherwise than by force; therefore, whatever constrains it, or subjects it to conditions, is a force. If a curve, or surface, or a string constrains it, the force is called reaction: if a fluid medium, the force is called resistance: if a condition however abstract, (as for example that it move in a tautochrone,) still this condition, by obliging it to move out of its free course, or with an unnatural velocity, must ultimately resolve itself into *force;* only that in this case it is an implicit and not an explicit function of the co-ordinates. This new force may therefore be considered first, as involved in X, Y, Z; or secondly, as added to them when it is resolved into X′, Y′, Z′.

In the first case, if it be regarded as included in X, Y, Z, these really contain an indeterminate function: but the equations

$$Xdt = \frac{d^2x}{dt}; \quad Ydt = \frac{d^2y}{dt}; \quad Zdt = \frac{d^2z}{dt},$$

still subsist; and therefore also equation (6).

Now however, there are not enough of equations to determine x, y, z, in functions of t, because of the unknown forms of X′, Y′, Z′; but if the equation $u = 0$, which expresses the condition of restraint, with all its consequences $du = 0$, $\delta u = 0$, &c., be superadded to these, there will then be enough to determine the problem. Thus the equations are

$$u = 0; \quad X - \frac{d^2x}{dt^2} = 0; \quad Y - \frac{d^2y}{dt^2} = 0; \quad Z - \frac{d^2z}{dt^2} = 0.$$

u is a function of x, y, z, X, Y, Z, and t. Therefore the equation $u = 0$ establishes the existence of a relation

$$\delta u = p\delta x + q\delta y + r\delta z = 0$$

between the variations δx, δy, δz, which can no longer be regarded as arbitrary; but the equation (6) subsists whether they be so or not, and may therefore be used simultaneously with $\delta u = 0$ to eliminate one; after which the other two being *really* arbitrary, their co-efficients *must* be separately zero.

In the second case; if we do not regard the forces arising from the conditions of constraint as involved in X, Y, Z, let $\delta u = 0$ be that condition, and let X′, Y′, Z′, be the unknown forces brought into action by that condition, by which the action of X, Y, Z, is modified; then will the whole forces acting on m be X+X′, Y+Y′, Z+Z′, and under the influence of these the particle will move as a *free particle;* and therefore δx, δy, δz, being any variations

$$0 = \left(X + X' - \frac{d^2x}{dt^2}\right)\delta x + \left(Y + Y' - \frac{d^2y}{dt^2}\right)\delta y + \left(Z + Z' - \frac{d^2z}{dt^2}\right)\delta z$$

or,

$$0 = \left(X - \frac{d^2x}{dt^2}\right)\delta x + \left(Y - \frac{d^2y}{dt^2}\right)\delta y + \left(Z - \frac{d^2z}{dt^2}\right)\delta z \qquad (7)$$
$$+ X'\delta x + Y'\delta y + Z'\delta z;$$

and this equation is independent of any particular relation between δx, δy, δz, and holds good whether they subsist or not. But the condition $\delta u = 0$ establishes a relation of the form $p\delta x + q\delta y + r\delta z = 0$,

where $$p = \left(\frac{du}{dx}\right),\quad q = \left(\frac{du}{dy}\right),\quad r = \left(\frac{du}{dz}\right);$$

and since this is true, it is so when multiplied by any arbitrary quantity λ; therefore,

$$\lambda\,(p\delta x + q\delta y + r\delta z) = 0, \text{ or } \lambda\delta u = 0;$$

because $$\delta u = p\delta x + q\delta y + r\delta z = 0.$$

If this be added to equation (7), it becomes

$$0 = \left(X - \frac{d^2x}{dt^2}\right)\delta x + \left(Y - \frac{d^2y}{dt^2}\right)\delta y + \left(Z - \frac{d^2z}{dt^2}\right)\delta z$$
$$+ X'\delta x + Y'\delta y + Z'\delta z - \lambda\delta u,$$

which is true whatever δx, δy, δz, or λ may be.

Now since X′, Y′, Z′, are forces acting in the direction x, y, z, (though unknown) they may be compounded into one resultant $R_{\prime}$, which must have one direction, whose element may be represented

by δs. And since the single force $R_{,}$ is resolved into X', Y', Z', we must have $$X'\delta x + Y'\delta y + Z'\delta z = R_{,}\delta s;$$
so that the preceding equation becomes
$$0 = \left(X - \frac{d^2x}{dt^2}\right)\delta x + \left(Y - \frac{d^2y}{dt^2}\right)\delta y + \left(Z - \frac{d^2z}{dt^2}\right)\delta z$$
$$+ R_{,}\delta s - \lambda\delta u \qquad (8)$$
and this is true whatever λ may be.

But λ being thus left arbitrary, we are at liberty to determine it by any convenient condition. Let this condition be $R_{,}\delta s - \lambda\delta u = 0$, or $\lambda = R_{,}\,.\,\frac{\delta s}{\delta u}$, which reduces equation (8) to equation (6). So when X, Y, Z, are the only acting forces explicitly given, this equation still suffices to resolve the problem, provided it be taken in conjunction with the equation $\delta u = 0$, or, which is the same thing,
$$p\delta x + q\delta y + r\delta z = 0,$$
which establishes a relation between δx, δy, δz.

Now let the condition $\lambda = s.\frac{\delta s}{\delta u}$ be considered which determines λ.

Since $R_{,}$ is the resultant of the forces X', Y', Z', its magnitude must be represented by $\sqrt{X'^2 + Y'^2 + Z'^2}$ by article 37, and since $R_{,}\delta s = \lambda\delta u$, or
$$X'\delta x + Y'\delta y + Z'\delta z = \lambda.\frac{du}{dx}\delta x + \lambda\,.\,\frac{du}{dy}\delta y + \lambda\,.\,\frac{du}{dz}\delta z,$$
therefore, in order that dx, dy, dz, may remain arbitrary, we must have $$X' = \lambda\frac{du}{dx};\quad Y' = \lambda\frac{du}{dy};\quad Z' = \lambda\frac{du}{dz};$$
and consequently
$$R_{,} = \sqrt{X'^2 + Y'^2 + Z'^2} = \lambda.\sqrt{\left(\frac{du}{dx}\right)^2 + \left(\frac{du}{dy}\right)^2 + \left(\frac{du}{dz}\right)^2} \qquad (9)$$
and $$\lambda = \frac{R_{,}}{\sqrt{\left(\frac{du}{dx}\right)^2 + \left(\frac{du}{dy}\right)^2 + \left(\frac{du}{dz}\right)^2}}$$
and if to abridge $\dfrac{1}{\sqrt{\left(\frac{du}{dx}\right)^2 + \left(\frac{du}{dy}\right)^2 + \left(\frac{du}{dz}\right)^2}} = K$; then if α, β, γ, be the angles that the normal to the curve or surface makes with the co-ordinates, $K\frac{du}{dx} = \cos\alpha$, $K\frac{du}{dy} = \cos\beta$, $K\frac{du}{dz} = \cos\gamma$,
and $$X' = R_{,}.\cos\alpha,\quad Y_{,} = R_{,}.\cos\beta,\quad Z_{,} = R_{,}.\cos\gamma.$$

Thus if u be given in terms of x, y, z; the four quantities λ, X', Y', and Z', will be determined. If the condition of constraint expressed by $u = 0$ be pressure against a surface, R, is the re-action.

Thus the general equation of a particle of matter moving on a curved surface, or subject to any given condition of constraint, is proved to be

$$0 = \left(X - \frac{d^2x}{dt^2}\right)\delta x + \left(Y - \frac{d^2y}{dt^2}\right)\delta y + \left(Z - \frac{d^2z}{dt^2}\right)\delta z + \lambda\delta u \quad (10).$$

70. The whole theory of the motion of a particle of matter is contained in equations (6) and (10); but the finite values of these equations can only be found when the variations of the forces are expressed at least implicitly in functions of the distance of the moving particle from their origin.

71. When the particle is free, if the forces X, Y, Z, be eliminated from

$$X - \frac{d^2x}{dt^2} = 0; \quad X - \frac{d^2y}{dt^2} = 0; \quad Z - \frac{d^2z}{dt^2} = 0$$

by functions of the distance, these equations, which then may be integrated at least by approximation, will only contain space and time; and by the elimination of the latter, two equations will remain, both functions of the co-ordinates which will determine the curve in which the particle moves.

72. Because the force which urges a particle of matter in motion, is given in functions of the indefinitely small increments of the co-ordinates, the path or trajectory of the particle depends on the nature of the force. Hence if the force be given, the curve in which the particle moves may be found; and if the curve be given, the law of the force may be determined.

73. Since one constant quantity may vanish from an equation at each differentiation, so one must be added at each integration; hence the integral of the three equations of the motion of a particle being of the second order, will contain six arbitrary constant quantities, which are the data of the problem, and are determined in each case either by observation, or by some known circumstances peculiar to each problem.

74. In most cases finite values of the general equation of the motion of a particle cannot be obtained, unless the law according to which the force varies with the distance be known; but by assuming from experience, that the intensity of the forces in nature

varies according to some law of the distance and leaving them otherwise indeterminate, it is possible to deduce certain properties of a moving particle, so general that they would exist whatever the forces might in other respects be. Though the variations differ materially, and must be carefully distinguished from the differentials dx, dy, dz, which are the spaces moved over by the particle parallel to the co-ordinates in the instant dt; yet being arbitrary, we may assume them to be equal to these, or to any other quantities consistent with the nature of the problem under consideration. Therefore let δx, δy, δz, be assumed equal to dx, dy, dz, in the general equation of motion (6), which becomes in consequence

$$\mathrm{X}dx + \mathrm{Y}dy + \mathrm{Z}dz = \frac{dxd^2x + dyd^2y + dzd^2z}{dt^2}.$$

75. The integral of this equation can only be obtained when the first member is a complete differential, which it will be if all the forces acting on the particle, in whatever directions, be functions of its distance from their origin.

Demonstration.—If F be a force acting on the particle, and s the distance of the particle from its origin, $\mathrm{F}\frac{x}{s}$ is the resolved portion parallel to the axis x; and if F′, F″, &c., be the other forces acting on the particle, then $\mathrm{X} = \Sigma.\mathrm{F}\frac{x}{s}$ will be the sum of all these forces resolved in a direction parallel to the axis x. In the same manner, $\mathrm{Y} = \Sigma.\mathrm{F}\frac{y}{s}$; $\mathrm{Z} = \Sigma.\mathrm{F}\frac{z}{s}$ are the sums of the forces resolved in a direction parallel to the axes y and z, so that $\mathrm{X}dx + \mathrm{Y}dy + \mathrm{Z}dz = \Sigma.\mathrm{F}\frac{xdx + ydy + zdz}{s} = \Sigma.\mathrm{F}\frac{sds}{s} = \Sigma\,.\,\mathrm{F}ds$, which is a complete differential when F, F′, &c., are functions of s.

76. In this case, the integral of the first member of the equation is $\int(\mathrm{X}dx + \mathrm{Y}dy + \mathrm{Z}dz)$, or $f(x, y, z,)$ a function of x, y, z; and by integration the second is $\frac{1}{2}\frac{dx^2+dy^2+dz^2}{dt^2}$ which is evidently the half of the square of the velocity; for if any curve MN, fig. 19, be represeented by s, its first differential ds or Am is $\sqrt{\mathrm{AD}^2 + \mathrm{D}m^2} = \sqrt{dx^2 + dy^2}$; hence, $ds^2 = dx^2 + dy^2$ when the curve

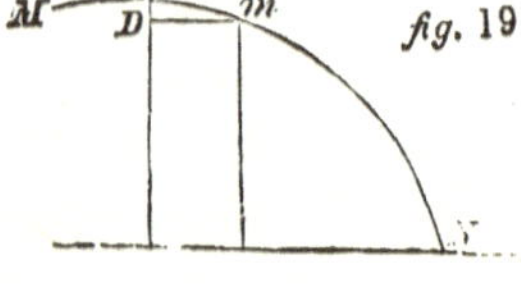

fig. 19

is in one plane, but when in space it is $ds^2 = dx^2 + dy^2 + dz^2$: and as $\frac{ds}{dt}$, [the element of the space divided by the element of the time is the velocity: therefore,

$$\frac{1}{2}\,\frac{dx^2 + dy^2 + dz^2}{dt^2} = \frac{1}{2}v^2;$$

consequently, $$2f(x, y, z) + c = v^2,$$
c being an arbitrary constant quantity introduced by integration.

77. This equation will give the velocity of the particle in any point of its path, provided its velocity in any other point be known: for if A be its velocity in that point of its trajectory whose co-ordinates are a, b, c, then

$$A^2 = c + 2f(a, b, c),$$

and $$v^2 - A^2 = 2f(x, y, z) - 2f(a, b, c);$$

whence v will be found when A is given, and the co-ordinates a, b, c, x, y, z, are known.

It is evident, from the equation being independent of any particular curve, that if the particle begins to move from any given point with a given velocity, it will arrive at another given point with the same velocity, whatever the curve may be that it has described.

78. When the particle is not acted on by any forces, then X, Y, and Z are zero, and the equation becomes $v^2 = c$. The velocity in this case, being occasioned by a primitive impulse, will be constant; and the particle, in moving from one given point to another, will always take the shortest path that can be traced between these points, which is a particular case of a more general law, called the principle of Least Action.

Principle of Least Action.

fig. 20.

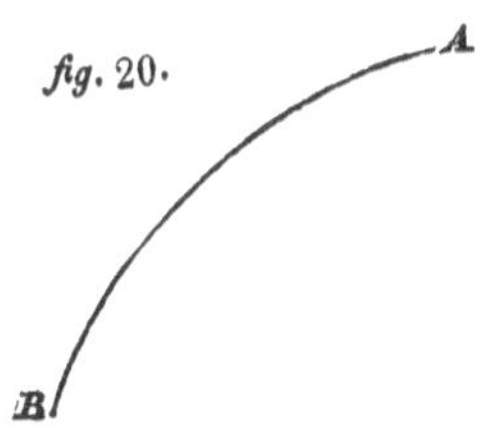

79. Suppose a particle beginning to move from a given point A, fig. 20, to arrive at another given point B, and that its velocity at the point A is given in magnitude but not in direction. Suppose also that it is urged by accelerating forces X, Y, Z, such, that the finite value of $Xdx + Ydy + Zdz$ can be obtained. We may then determine v the velocity of the particle in terms of x, y, z, without knowing the curve described by the

particle in moving from A to B. If ds be the element of the curve, the finite value of vds between A and B will depend on the nature of the path or curve in which the body moves. The principle of Least Action consists in this, that if the particle be free to move in every direction between these two points, except in so far as it obeys the action of the forces X, Y, Z, it will in virtue of this action, choose the path in which the integral $\int vds$ is a minimum; and if it be constrained to move on a given surface, it will still move in the curve in which $\int vds$ is a minimum among all those that can be traced on the surface between the given points.

To demonstrate this principle, it is required to prove the variation of $\int vds$ to be zero, when A and B, the extreme points of the curve are fixed.

By the method of variations $\delta\int vds = \int \delta . vds$: for $\int$ the mark of integration being relative to the differentials, is independent of the variations.

Now $\qquad \delta . vds = \delta v . ds + v\delta ds$, but $v = \dfrac{ds}{dt}$ or $ds = vdt$;

hence $\qquad \delta v . ds = v\delta v dt = dt \tfrac{1}{2} \delta . v^2$,

and therefore $\qquad \delta . vds = dt . \tfrac{1}{2}\delta . v^2 + v . \delta . ds.$

The values of the two last terms of this equation must be found separately. To find $dt . \frac{1}{2}\delta . v^2$. It has been shown that

$$v^2 = c + 2\int (Xdx + Ydy + Zdz),$$

its differential is $\quad vdv = (Xdx + Ydy + Zdz)$,

and changing the differentials into variations,

$$\tfrac{1}{2}\delta . v^2 = X\delta x + Y\delta y + Z\delta z.$$

If $\frac{1}{2}\delta . v^2$ be substituted in the general equation of the motion of a particle on its surface, it becomes

$$\tfrac{1}{2}\delta . v^2 = \frac{d^2x}{dt^2}\delta x + \frac{d^2y}{dt^2}\delta y + \frac{d^2z}{dt^2}\delta z + \lambda\delta u = 0.$$

But $\lambda\delta u$ does not enter into this equation when the particle is free; and when it must move on the surface whose equation is $u = 0$, δu is also zero; hence in every case the term $\lambda\delta u$ vanishes; therefore

$$dt . \tfrac{1}{2}\delta . v^2 = \frac{d^2x}{dt^2}\delta x + \frac{d^2y}{dt^2}\delta y + \frac{d^2z}{dt^2}\delta z$$

is the value of the first term required.

A value of the second term $v . \delta . ds$ must now be found. Since

$$ds^2 = dx^2 + dy^2 + dz^2,$$

its variation is $ds.\delta ds = dx.\delta dx + dy.\delta dy + dz.\delta dz$, but $ds = vdt$, hence

$$v.\delta ds = \frac{dx}{dt}\delta dx + \frac{dy}{dt}\delta dy + \frac{dz}{dt}\delta dz,$$

which is the value of the second term; and if the two be added, their sum is

$$\delta.vds = d\left\{\frac{dx}{dt}\delta x + \frac{dy}{dt}\delta y + \frac{dz}{dt}\delta z\right\},$$

as may easily be seen by taking the differential of the last member of this equation. Its integral is

$$\delta\int vds = \frac{dx}{dt}\delta x + \frac{dy}{dt}\delta y + \frac{dz}{dt}\delta z.$$

If the given points A and B be moveable in space, the last member of this equation will determine their motion; but if they be fixed points, the last member which is the variation of the co-ordinates of these points is zero: hence also $\delta\int vds = 0$, which indicates either a maximum or minimum, but it is evident from the nature of the problem that it can only be a minimum. If the particle be not urged by accelerating forces, the velocity is constant, and the integral is vs. Then the curve s described by the particle between the points A and B is a minimum; and since the velocity is uniform, the particle will describe that curve in a shorter time than it would have done any other curve that could be drawn between these two points.

80. The principle of least action was first discovered by Euler: it has been very elegantly applied to the reflection and refraction of light. If a ray of light IS, fig. 21, falls on any surface CD, it will be turned back or reflected in the direction Sr, so that ISA $=$ rSA. But if the medium whose surface is CD be diaphanous, as glass or water, it will be broken or refracted at S, and will enter the denser medium in the direction SR, so that the sine of the angle of incidence ISA will be to the sine of the angle of refraction RSB, in a constant ratio for any one medium. Ptolemy discovered that light, when reflected from any surface, passed from one given point to another by the shortest path, and in the shortest time possible, its velocity being uniform.

fig. 21.

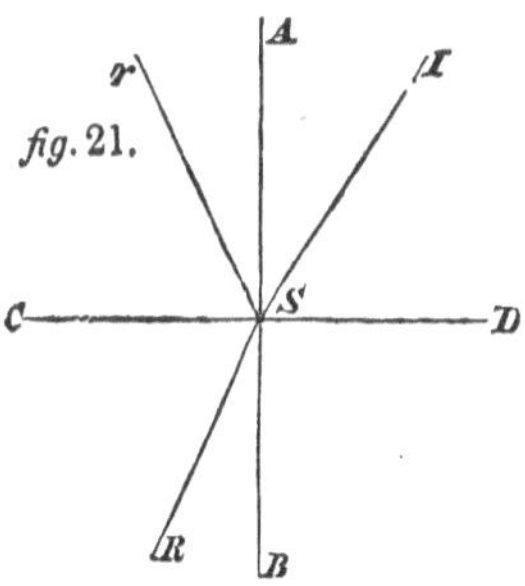

Fermat extended the same principle to the refraction of light; and supposing the velocity of a ray of light to be less in the denser medium, he found that the ratio of the sine of the angle of incidence to that of the angle of refraction, is constant and greater than unity. Newton however proved by the attraction of the denser medium on the ray of light, that in the corpuscular hypothesis its velocity is greater in that medium than in the rarer, which induced Maupertuis to apply the theory of maxima and minima to this problem. If IS, a ray of light moving in a rare medium, fall obliquely on CD the surface of a medium that is more dense, it moves uniformly from I to S; but at the point S both its direction and velocity are changed, so that at the instant of its passage from one to the other, it describes an indefinitely small curve, which may be omitted without sensible error: hence the whole trajectory of the light is ISR; but IS and SR are described with different velocities; and if these velocities be v and v', then the variation of $\text{IS} \times v + \text{SR} \times v'$ must be zero, in order that the trajectory may be a minimum: hence the general expression $\delta\int v ds = 0$ becomes in this case $\delta.(\text{IS} \times v + \text{SR} \times v') = 0$, when applied to the refraction of light; from whence it is easily found, by the ordinary analysis of maxima and minima, that $v \sin \text{ISA} = v' \sin \text{RBS}$. As the ratio of these sines depends on the ratio of the velocities, it is constant for the transition out of any one medium into another, but varies with the media, on account of the velocity of light being different in different media. If the denser medium be a crystallized diaphanous substance, the velocity of light in it will depend on the direction of the luminous ray; it is constant for any one ray, but variable from one ray to another. Double refraction, as in Iceland spar and in crystallized bodies, arises from the different velocities of the rays; in these substances two images are seen instead of one. Huygens first gave a distinct account of this phenomenon, which has since been investigated by others.

Motion of a Particle on a curved Surface.

81. The motion of a particle, when constrained to move on a curve or surface, is easily determined from equation (7); for if the

variations be changed into differentials, and if X′, Y′, Z′ be eliminated by their values in the end of article 69, that equation becomes

$$\frac{dx \,.\, d^2x + dy \,.\, d^2y + dz \,.\, d^2z}{dt^2} = Xdx + Ydy + Zdz$$

$$+ \, R_{,} \{dx \,.\, \cos\alpha + dy \,.\, \cos\beta + dz \,.\, \cos\gamma\},$$

$R_{,}$ being the reaction in the normal, and α, β, γ the angles made by the normal with the co-ordinates. But the equation of the surface being $u = 0$,

$$du = \frac{du}{dx} \,.\, dx + \frac{du}{dy} \,.\, dy + \frac{du}{dz} \,.\, dz = 0;$$

consequently, by article 69,

$$\lambda du = dx \,.\, \cos\alpha + dy \,.\, \cos\beta + dz \,.\, \cos\gamma = 0;$$

so that the pressure vanishes from the preceding equation; and when the forces are functions of the distance, the integral is

$$2f(x, y, z) + c = v^2,$$

and $$A^2 - v^2 = 2f(x, y, z) - 2f(a, b, c),$$

as before. Hence, if the particle be urged by accelerating forces, the velocity is independent of the curve or surface on which the particle moves; and if it be not urged by accelerating forces, the velocity is constant. Thus the principle of Least Action not only holds with regard to the curves which a particle describes in space, but also for those it traces when constrained to move on a surface.

82. It is easy to see that the velocity must be constant, because a particle moving on a curve or surface only loses an indefinitely small part of its velocity of the second order in passing from one indefinitely small plane of a surface or side of a curve to the consecutive; for if the particle be moving on *ab* with

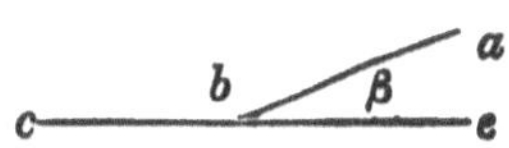

the velocity v; then if the angle $abe = \beta$, the velocity in *bc* will be $v \cos\beta$; but $\cos\beta = 1 - \frac{1}{2}\beta^2 - \text{\&c.}$; therefore the velocity on *bc* differs from the velocity on *ab* by the indefinitely small quantity $\frac{1}{2}v \,.\, \beta^2$. In order to determine the pressure of the particle on the surface, the analytical expression of the radius of curvature must be found.

Radius of Curvature.

83. The circle AmB, fig. 22, which coincides with a curve or curved surface through an indefinitely small space on each side of m the point of contact, is called the curve of equal curvature, or the osculating circle of the curve MN, and om is the radius of curvature.

fig. 22.

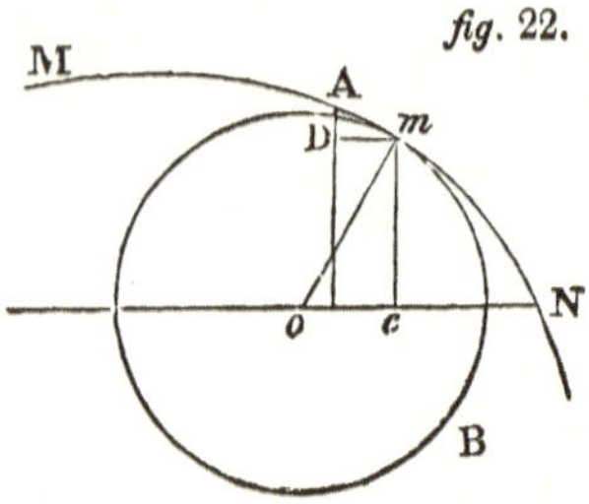

In a plane curve the radius of curvature r, is expressed by

$$r = \frac{ds^2}{\sqrt{(d^2x)^2 + (d^2y)^2}}$$

and in a curve of double curvature it is

$$r = \frac{ds^2}{\sqrt{(d^2x)^2 + (d^2y)^2 + (d^2z)^2}},$$

ds being the constant element of the curve.

Let the angle com be represented by θ, then if Am be the indefinitely small but constant element of the curve MN, the triangles com and ADm are similar; hence $mA : mD :: om : mc$, or $ds : dx :: 1 : \sin\theta$, and $\sin\theta = \frac{dx}{ds}$. In the same manner $\cos\theta = \frac{dy}{ds}$. But $d \,.\, \cos\theta = - d\theta \sin\theta$, and $d\theta = - \frac{d.\cos\theta}{\sin\theta}$; also $d \,.\, \sin\theta = d\theta\cos\theta$, and $d\theta = \frac{d\,.\,\sin\theta}{\cos\theta}$; but these evidently become

$$d\theta = + \frac{ds}{dy} \,.\, d\frac{dx}{ds} \text{ and } d\theta = - \frac{ds}{dx} \,.\, d\frac{dy}{ds}; \text{ or}$$

$$d\theta = + \frac{d^2x}{dy}, \text{ and } d\theta = - \frac{d^2y}{dx}.$$

Now if om the radius of curvature be represented by r, then moA being the indefinitely small increment $d\theta$ of the angle com, we have $r : ds :: 1 : d\theta$; for the sine of the infinitely small angle is to be considered as coinciding with the arc: hence $d\theta = \frac{ds}{r}$, whence $r = - \frac{ds\,.\,dy}{d^2x} = \frac{dsdx}{d^2y}$. But $dx^2 + dy^2 = ds^2$, and as ds is constant $dx\,.\,d^2x + dyd^2y = 0$. Whence $\frac{d^2x}{d^2y} = - \frac{dy}{dx}$, or $\left(\frac{d^2x}{d^2y}\right)^2 = \frac{dy^2}{dx^2}$, and adding one to each side of the last equation, it becomes

$$\frac{dx^2+dy^2}{dx^2}=\frac{ds^2}{dx^2}=\frac{(d^2x)^2+(d^2y)^2}{(d^2y)^2}.$$

Whence

$$\frac{dx}{d^2y}=\frac{ds}{\sqrt{(d^2x)^2+(d^2y)^2}}.$$

But it has been shown that $r=\frac{dxds}{d^2y}$; hence in a plane curve the radius of curvature is

$$r=\frac{ds^2}{\sqrt{(d^2x)^2+(d^2y)^2}}.$$

We may imagine MN to be the projection of a curve of double curvature on the plane xoy, then

fig. 23.

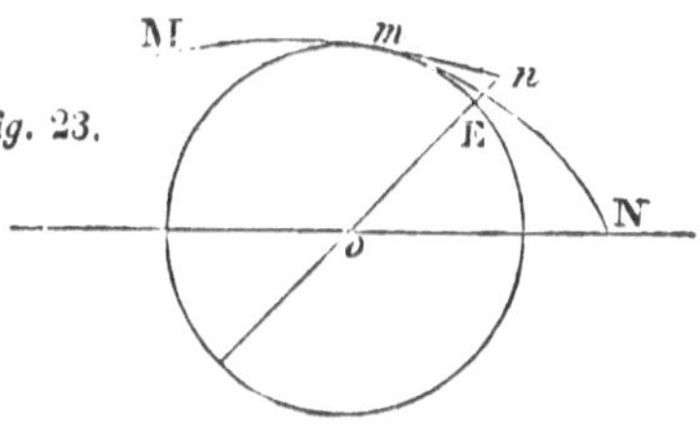

$$r=\frac{ds^2}{\sqrt{(d^2x)^2+(d^2y)^2}}$$ will be

the projection of the radius of curvature on $x\,o\,y$, and it is evident that a similar expression will be found for the projection of the radius of curvature on each of the other co-ordinate planes. In fact $\frac{1}{2}\sqrt{(d^2x)^2+(d^2y)^2}$ is the sagitta of curvature nE; for

$$(nm)^2=2r.n\text{E},$$

or

$$r=\frac{(nm)^2}{2n\text{E}}=\frac{ds^2}{2n\text{E}}=\frac{ds^2}{\sqrt{(d^2x)^2+(d^2y)^2}}$$

for the arc being indefinitely small, the tangent may be considered as coinciding with it. Thus the three projections of the sagitta of curvature of the surface, or curve of double curvature, are

$$\tfrac{1}{2}\sqrt{(d^2x)^2+(d^2y)^2};\ \tfrac{1}{2}\sqrt{(d^2x)^2+(d^2z)^2};\ \tfrac{1}{2}\sqrt{(d^2y)^2+(d^2z)^2};$$

hence the sum of their squares is

$$\tfrac{1}{2}\sqrt{(d^2x)^2+(d^2y)^2+(d^2z)^2};$$

and the radius of curvature of a surface, or curve of double curvature, is

$$r=\frac{ds^2}{\sqrt{(d^2x)^2+(d^2y)^2+(d^2z)^2}}.$$

Pressure of a Particle moving on a curved Surface.

84. If the particle be moving on a curved surface, it exerts a pressure which the surface opposes with an equal and contrary pressure.

Demonstration.—For if F be the resulting force of the partial accelerating forces X, Y, Z, acting on the particle at m, it may be resolved into two forces, one in the direction of the tangent mT, and the other in the normal mN, fig. 12. The forces in the tangent have their full effect, and produce a change in the velocity of the particle, but those in the normal are destroyed by the resistance of the surface. If the particle were in equilibrio, the whole pressure would be that in the normal; but when the particle is in motion, the velocity in the tangent produces another pressure on the surface, in consequence of the continual effort the particle makes to fly off in the tangent. Hence when the particle is in motion, its whole pressure on the surface is the difference of these two pressures, which are both in the direction of the normal, but one tends to the interior of the surface and the other from it. The velocity in the tangent is variable in consequence of the accelerating forces X, Y, Z, and becomes uniform if we suppose them to cease.

Centrifugal Force.

85. When the particle is not urged by accelerating forces, its motion is owing to a primitive impulse, and is therefore uniform. In this case X, Y, Z, are zero, the pressure then arising from the velocity only, tends to the exterior of the surface.

And as v the velocity is constant, if ds be the element of the curve described in the time dt, then

$$ds = vdt, \text{ whence } dt = \frac{ds}{v},$$

therefore ds is constant; and when this value of dt is substituted in equation (7), in consequence of the values of X′, Y′, Z′, in the end of article 69, it gives

$$v^2 \,.\, \frac{d^2x}{ds^2} = \mathrm{R}_{\prime} \cos\alpha$$

$$v^2 \,.\, \frac{d^2y}{ds^2} = \mathrm{R}_{\prime} \cos\beta$$

$$v^2 \,.\, \frac{d^2z}{ds^2} = \mathrm{R}_{\prime} \cos\gamma$$

for by article 81 the particle may be considered as free, whence

$$\mathrm{R}_{\prime} = \frac{v^2 \sqrt{(d^2x)^2+(d^2y)^2+(d^2z)^2}}{ds^2};$$

and as the osculating radius is

$$r = \frac{ds^2}{\sqrt{(d^2x)^2+(d^2y)^2+(d^2z)^2}},$$

so

$$\mathrm{R}_{\prime} = \frac{v^2}{r}.$$

The first member of this equation was shown to be the pressure of the particle on the surface, which thus appears to be equal to the square of the velocity, divided by the radius of curvature.

86. It is evident that when the particle moves on a surface of unequal curvature, the pressure must vary with the radius of curvature.

87. When the surface is a sphere, the particle will describe that great circle which passes through the primitive direction of its motion. In this case the circle A*m*B is itself the path of the particle; and in every part of its motion, its pressure on the sphere is equal to the square of the velocity divided by the radius of the circle in which it moves; hence its pressure is constant.

88. Imagine the particle attached to the extremity of a thread assumed to be without mass, whereof the other extremity is fixed to the centre of the surface; it is clear that the pressure which the particle exerts against the circumference is equal to the tension of the thread, provided the particle be restrained in its motion by the thread alone. The effort made by the particle to stretch the thread, in order to get away from the centre, is the centrifugal force.

Hence the centrifugal force of a particle revolving about a centre, is equal to the square of its velocity divided by the radius.

89. The plane of the osculating circle, or the plane that passes through two consecutive and indefinitely small sides of the curve described by the particle, is perpendicular to the surface on which the particle moves. And the curve described by the particle is the shortest line that can be drawn between any two points of the surface, consequently this singular law in the motion of a particle on a surface depends on the principle of least action. With regard to the Earth, this curve drawn from point to point on its surface is called a perpendicular to the meridian; such are the lines which have been measured both in France and England, in order to ascertain the true figure of the globe.

90. It appears that when there are no constant or accelerating forces, the pressure of a particle on any point of a curved surface is equal to the square of the velocity divided by the radius of curvature at that point. If to this the pressure due to the accelerating forces be added, the whole pressure of the particle on the surface will be obtained, when the velocity is variable.

91. If the particle moves on a surface, the pressure due to the centrifugal force will be equal to what it would exert against the curve it describes resolved in the direction of the normal to the surface in that point; that is, it will be equal to the square of the velocity divided by the radius of the osculating circle, and multiplied by the sine of the angle that the plane of that circle makes with the tangent plane to the surface.

fig. 24.

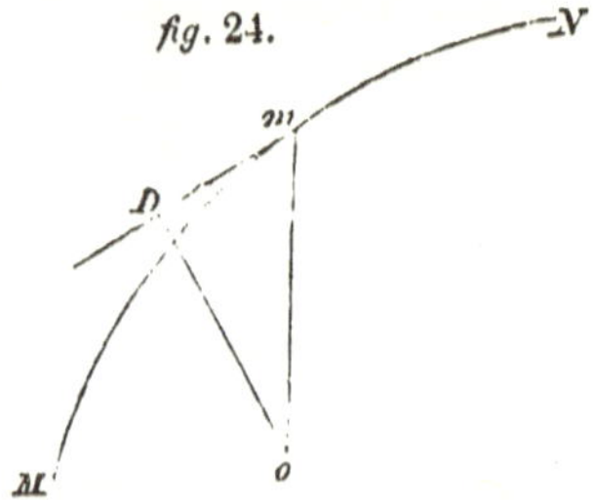

Let MN, fig. 24, be the path of a particle on the surface; *mo* the radius of the osculating circle at *m*, and *m*D a tangent to the surface at *m*; then *om* being radius, *o*D is the sine of the inclination of the plane of the osculating circle on the plane that is tangent to the surface at *m*, the centrifugal force is equal to

$$\frac{v^2 \times o\mathrm{D}}{om}.$$

If to this, the part of the pressure which is due to the accelerating forces be added, the sum will be the whole pressure on the surface.

92. It appears that the centrifugal force is that part of the pressure which depends on velocity alone; and when there are no accelerating forces it is the pressure itself.

93. It is very easy to show that in a circle, the centrifugal force is equal and contrary to the central force.

fig. 25.

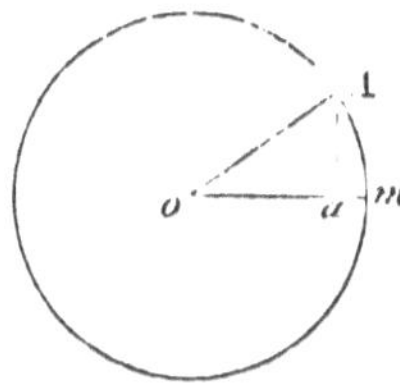

Demonstration.—By article 63 a central force F combined with an impulse, causes a particle to describe an indefinitely small arc mA, fig. 25, in the time dt. As the sine may be taken for the tangent, the space described from the impulse alone is

$$a\text{A} = vdt;$$

but

$$(a\text{A})^2 = 2r \,.\, am,$$

so

$$am = \frac{v^2\, dt^2}{2r},$$

r being radius. But as the central force causes the particle to move through the space

$$am = \tfrac{1}{2}\text{F} \,.\, dt^2,$$

in the same time,

$$\frac{v^2}{r} = \text{F}.$$

94. If v and v' be the velocities of two bodies, moving in circles whose radii are r and r', their velocities are as the circumferences divided by the times of their revolutions; that is, directly as the space, and inversely as the time, since circular motion is uniform. But the radii are as their circumferences, hence

$$v^2 : v'^2 :: \frac{r^2}{t^2} : \frac{r'^2}{t'^2},$$

t and t' being the times of revolution. If c and c' be the centrifugal forces of the two bodies, then

$$c : c' :: \frac{v^2}{r} : \frac{v'^2}{r'},$$

or, substituting for v^2 and v'^2, we have

$$c : c' :: \frac{r}{t^2} : \frac{r'}{t'^2}.$$

Thus the centrifugal forces are as the radii divided by the squares of the times of revolution.

95. With regard to the Earth the times of rotation are everywhere the same; hence the centrifugal forces, in different latitudes, are as the radii of these parallels. These elegant theorems discovered by Huygens, led Newton to the general theory of motion in curves, and to the law of universal gravitation.

Motion of Projectiles.

96. From the general equation of motion is also derived the motion of projectiles.

Gravitation affords a perpetual example of a continued force; its influence on matter is the same whether at rest or in motion; it penetrates its most intimate recesses, and were it not for the resistance of the air, it would cause all bodies to fall with the same velocity: it is exerted at the greatest heights to which man has been able to ascend, and in the most profound depths to which he has penetrated. Its direction is perpendicular to the horizon, and therefore varies for every point on the earth's surface; but in the motion of projectiles it may be assumed to act in parallel straight lines; for, any curves that projectiles could describe on the earth may be esteemed as nothing in comparison of its circumference.

The mean radius of the earth is about 4000 miles, and MM. Biot and Gay Lussac ascended in a balloon to the height of about four miles, which is the greatest elevation that has been attained, but even that is only the 1000th part of the radius.

The power of gravitation at or near the earth's surface may, without sensible error, be considered as a uniform force; for the decrease of gravitation, inversely as the square of the distance, is hardly perceptible at any height within our reach.

97. *Demonstration.*—If a particle be projected in a straight line MT, fig. 26, forming any angle whatever with the horizon, it will constantly deviate from the direction MT by the action of the gravitating force, and will describe a curve MN, which is concave towards the horizon, and to which MT is tangent at M. On this particle there

fig. 26.

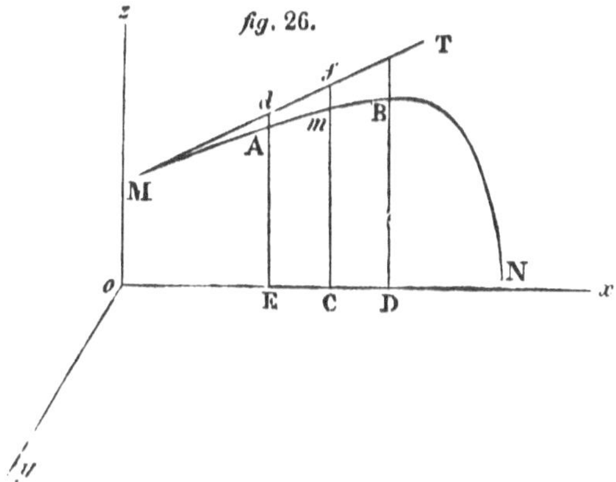

are two forces acting at every instant of its motion: the resistance of the air, which is always in a direction contrary to the motion of the particle; and the force of gravitation, which urges it with an accelerated motion, according to the perpendiculars E*d*, C*f*, &c. The resistance of the air may be resolved into three partial forces, in the direction of the three axes *ox*, *oy*, *oz*, but gravitation acts on the particle in the direction of *oz* alone. If A represents the resistance of the air, its component force in the axis *ox* is evidently $-\text{A}\,\frac{dx}{ds}$; for if A*m* or *ds* be the space proportional to the resistance, then

$$\text{A}m : \text{E}c :: \text{A} : \text{A}\,\frac{\text{E}c}{\text{A}m} = \text{A}\,\frac{dx}{ds};$$

but as this force acts in a direction contrary to the motion of the particle, it must be taken with a negative sign. The resistance in the axes *oy* and *oz* are $-\text{A}\,\frac{dy}{ds}, -\text{A}\,\frac{dz}{ds}$; hence if g be the force of gravitation, the forces acting on the particle are

$$\text{X} = -\text{A}\,\frac{dx}{ds};\quad \text{Y} = -\text{A}\,\frac{dy}{ds};\quad \text{Z} = g - \text{A}\,\frac{dz}{ds}.$$

As the particle is free, each of the virtual velocities is zero; hence we have

$$\frac{d^2x}{dt^2} = -\text{A}\,\frac{dx}{ds};\quad \frac{d^2y}{dt^2} = -\text{A}\,\frac{dy}{ds};\quad \frac{d^2z}{dt^2} = g - \text{A}\,\frac{dz}{ds};$$

for the determination of the motion of the projectile. If A be eliminated between the two first, it appears that

$$\frac{d^2x}{dt}\cdot\frac{dy}{dt} = \frac{d^2y}{dt}\cdot\frac{dx}{dt}, \quad \text{or } d\log\frac{dx}{dt} = d\log\frac{dy}{dt};$$

and integrating, $\log\frac{dx}{dt} = \log C + \log\frac{dy}{dt}$. Whence $\frac{dx}{dt} = C\frac{dy}{dt}$, or $dx = Cdy$, and if we integrate a second time,

$$x = Cy + D,$$

in which C and D are the constant quantities introduced by double integration. As this is the equation to a straight line, it follows that the projection of the curve in which the body moves on the plane xoy is a straight line, consequently the curve MN is in the plane zox, that is at right angles to xoy; thus MN is a plane curve, and the motion of the projectile is in a plane at right angles to the horizon. Since the projection of MN on xoy is the straight line ED, therefore $y = 0$, and the equation $\frac{d^2y}{dt^2} = -A\frac{dy}{dt}$ is of no use in the solution of the problem, there being no motion in the direction oy. Theoretical reasons, confirmed to a certain extent by experience, show that the resistance of the air supposed of uniform density is proportional to the square of the velocity;

hence, $$A = hv^2 = h\frac{ds^2}{dt^2},$$

h being a quantity that varies with the density, and is constant when it is uniform; thus the general equations become

$$(a)\quad \frac{d^2x}{dt^2} = -h\,.\,\frac{ds}{dt}\,.\,\frac{dx}{dt};\qquad \frac{d^2z}{dt^2} = g - h\,.\,\frac{ds}{dt}\,.\,\frac{dz}{dt};$$

the integral of the first is

$$\frac{dx}{dt} = C\,.\,c^{-hs},$$

C being an arbitrary constant quantity, and c the number whose hyperbolic logarithm is unity.

In order to integrate the second, let $dz = udx$, u being a function of z; then the differential according to t gives

$$\frac{d^2z}{dt^2} = \frac{du}{dt}\,.\,\frac{dx}{dt} + u\,.\,\frac{d^2x}{dt^2}.$$

If this be put in the second of equations (a), it becomes, in consequence of the first,

$$\frac{du}{dt}\,.\,\frac{dx}{dt} = -g;$$

or, eliminating dt by means of the preceding integral, and making

$$-\frac{g}{2C^2} = a,$$

it becomes

$$\frac{du}{dx} = 2ac^{2\lambda}.$$

The integral of this equation will give u in functions of x, and when substituted in

$$dz = udx,$$

it will furnish a new equation of the first order between z, x, and t, which will be the differential equation of the trajectory.

If the resistance of the medium be zero, $h = 0$, and the preceding equation gives

$$u = 2ax + b,$$

and substituting $\frac{dz}{dx}$ for u, and integrating again

$$z = ax^2 + bx + b'$$

b and b' being arbitrary constant quantities. This is the equation to a parabola whose axis is vertical, which is the curve a projectile would describe in vacuo. When

$$h = 0, \; d^2z = gdt^2 ;$$

and as the second differential of the preceding integral gives

$$d^2z = 2adx^2 ; \; dt = dx\sqrt{\frac{2a}{g}},$$

therefore

$$t = x\sqrt{\frac{2a}{g}} + a'.$$

If the particle begins to move from the origin of the co-ordinates, the time as well as x, y, z, are estimated from that point; hence b' and a' are zero, and the two equations of motion become

$$z = ax^2 + bx ; \text{ and } t = x\sqrt{\frac{2a}{g}} ;$$

whence

$$z = g\frac{t^2}{2} + tb\sqrt{\frac{g}{2a}}.$$

These three equations contain the whole theory of projectiles in vacuo; the second equation shows that the horizontal motion is uniform, being proportional to the time; the third expresses that the motion in the perpendicular is uniformly accelerated, being as the square of the time.

Theory of Falling Bodies.

99. If the particle begins to move from a state of rest, $b=0$, and the equations of motion are

$$\frac{dz}{dt} = gt, \text{ and } z = \tfrac{1}{2}gt^2.$$

The first shows that the velocity increases as the time; the second shows that the space increases as the square of the time, and that the particle moving uniformly with the velocity it has acquired in the time t, would describe the space $2z$, that is, double the space it has moved through. Since gt expresses the velocity v, the last of the preceding equations gives

$$2gz = g^2t^2 = v^2,$$

where z is the height through which the particle must have descended from rest, in order to acquire the velocity v. In fact, were the particle projected perpendicularly upwards, the parabola would then coincide with the vertical: thus the laws of parabolic motion include those of falling bodies; for the force of gravitation overcomes the force of projection, so that the initial velocity is at length destroyed, and the body then begins to fall from the highest point of its ascent by the force of gravitation, as from a state of rest. By experience it is found to acquire a velocity of nearly 32.19 feet in the first second of its descent at London, and in two seconds it acquires a velocity of 64.38, having fallen through 16.095 feet in the first second, and in the next $32.19 + 16.095 = 48.285$ feet, &c. The spaces described are as the odd numbers 1, 3, 5, 7, &c.

These laws, on which the whole theory of motion depends, were discovered by Galileo.

Comparison of the Centrifugal Force with Gravity.

100. The centrifugal force may now be compared with gravity, for if v be the velocity of a particle moving in the circumference of a circle of which r is the radius, its centrifugal force is $f = \frac{v^2}{r}$. Let h be the space or height through which a body must fall in order to acquire a velocity equal to v; then by what was shown in article 99, $v^2 = 2hg$, for the accelerating force in the present case is gravity; hence $f = \frac{2 \cdot h \cdot g}{r}$. If we suppose $h = \frac{1}{2} r$, the centrifugal force becomes equal to gravity.

101. Thus, if a heavy body be attached to one extremity of a thread, and if it be made to revolve in a horizontal plane round the other extremity of the thread fixed to a point in the plane; if the velocity of revolution be equal to what the body would acquire by falling through a space equal to half the length of the thread, the body will stretch the thread with the same force as if it hung vertically.

102. Suppose the body to employ the time T to describe the circumference whose radius is r; then π being the ratio of the circumference to the diameter, $v = \frac{2\pi r}{\mathrm{T}}$, whence

$$f = \frac{4\pi^2 r}{\mathrm{T}^2}.$$

Thus the centrifugal force is directly proportional to the radius, and in the inverse ratio of the square of the time employed to describe the circumference. Therefore, with regard to the earth, the centrifugal force increases from the poles to the equator, and gradually diminishes the force of gravity. The equatorial radius, computed from the mensuration of degrees of the meridian, is 20920600 feet, $\mathrm{T} = 365^{d}.2564$, and as it appears, by experiments with the pendulum, that bodies fall at the equator 16.0436 feet in a second, the preceding formulæ give the ratio of the centrifugal force to gravity at the equator equal to $\frac{1}{289}$. Therefore if the rotation of the earth were 17 times more rapid, the centrifugal force would be equal to gravity, and at the equator bodies would be in equilibrio from the action of these two forces.

Simple Pendulum.

103. A particle of matter suspended at the extremity of a thread, supposed to be without weight, and fixed at its other extremity, forms the simple pendulum.

fig. 27.

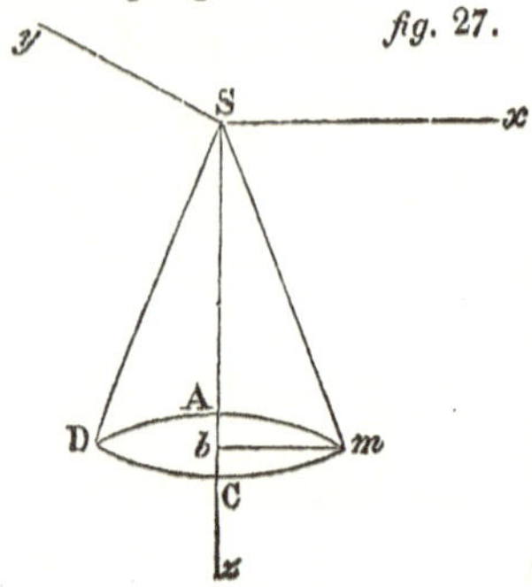

104. Let m, fig. 27, be the particle of matter, Sm the thread, and S the point of suspension. If an impulse be given to the particle, it will move in a curve mADC, as if it were on the surface of the sphere of which S is the centre; and the greatest deviation from the vertical Sz would be measured by the sine of the angle CSm. This motion arises from the combined action of gravitation and the impulse.

105. The impulse may be such as to make the particle describe a curve of double curvature; or if it be given in the plane xSz, the particle will describe the arc of a circle DCm, fig. 28; but it is evident that the extent of the arc will be in proportion to the intensity of the impulse, and it may be so great as to cause the particle to describe an indefinite number of circumferences. But if the impulse be small, or if the particle be drawn from the vertical to a point B and then left to itself, it will be urged in the vertical by gravitation, which will cause it to describe the arc mC with an accelerated velocity; when at C it will have acquired so much velocity that it will overcome the force of gravitation, and having passed that point, it will proceed to D; but in this half of the arc its motion will be as much retarded by gravitation as it was accelerated in the other half; so that on arriving at D it will have lost all its velocity, and it will descend through DC with an accelerated motion which will carry it to B again. In this manner it would continue to move for ever, were it not for the resistance of the air. This kind of motion is called oscillation.

fig. 28.

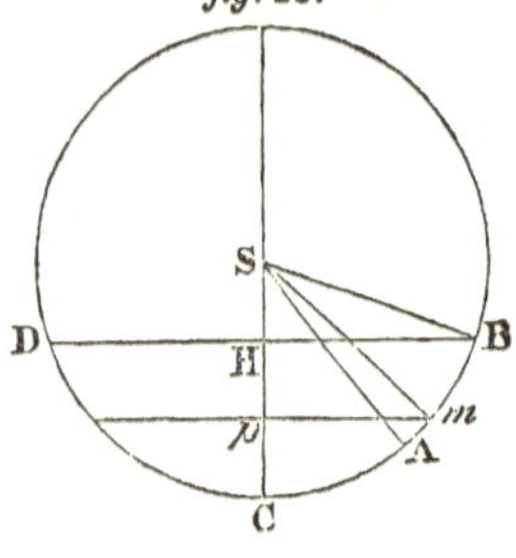

The time of an oscillation is the time the particle employs to move through the arc BCD.

106. *Demonstration.*—Whatever may be the nature of the curve, it has already been shown in article 99, that at any point m, $v^2 = 2gz$, g being the force of gravitation, and $z = \text{H}p$, the height through which the particle must have descended in order to acquire the velocity v. If the particle has been impelled instead of falling from rest, and if I be the velocity generated by the impulse, the equation becomes $v^2 = \text{I} + 2gz$. The velocity at m is directly as the element of the space, and inversely as the element of the time; hence

$$v^2 = \frac{(\text{A}m)^2}{dt^2} = \frac{ds^2}{dt^2} = \text{I} + 2g \cdot z;$$

whence
$$dt = \frac{-\,ds}{\sqrt{\text{I} + 2g \cdot z}}.$$

The sign is made negative, because z diminishes as t augments. If the equation of the trajectory or curve mCD be given, the value of $ds = \text{A}m$ may be obtained from it in terms of $z = \text{H}p$, and then the finite value of the preceding equation will give the time of an oscillation in that curve.

107. The case of greatest importance is that in which the trajectory is a circle of which Sm is the radius; then if an impulse be given to the pendulum at the point B perpendicular to SB, and in the plane xoz, it will oscillate in that plane. Let h be the height through which the particle must fall in order to acquire the velocity given by the impulse, the initial velocity I will then be $2gh$; and if BSC $= \alpha$ be the greatest amplitude, or greatest deviation of the pendulum from the vertical, it will be a constant quantity. Let the variable angle mSC $= \theta$, and if the radius be r, then S$p = r\cos\theta$; SH $= r\cos\alpha$; Hp = Sp − SH $= r(\cos\theta - \cos\alpha)$, and the elementary arc $m\text{A} = rd\theta$; hence the expression for the time becomes

$$dt = \frac{-\,rd\theta}{\sqrt{2g(h + r\cos\theta - r\cos\alpha)}}.$$

This expression will take a more convenient form, if $x = \text{C}p = (1 - \cos\theta)$ be the versed sine of mSC, and $\beta = (1 - \cos\alpha)$ the versed sine of BSC; then $d\theta = \dfrac{dx}{\sqrt{2x - x^2}}$, and

$$dt = \frac{-rdx}{\sqrt{2x - x^2} \cdot \sqrt{2g(h + r\beta - rx)}}$$

$$v = \sqrt{2g(h + r\beta - rx)}.$$

Since the versed sine never can surpass 2, if $h + r\beta > 2r$, the velocity will never be zero, and the pendulum will describe an indefinite number of circumferences; but if $h + r\beta < 2r$, the velocity v will be zero at that point of the trajectory where $x = \frac{h + r\beta}{r}$, and the pendulum will oscillate on each side of the vertical.

If the origin of motion be at the commencement of an oscillation, $h = 0$, and

$$dt = -\tfrac{1}{2}\sqrt{\frac{r}{g}} \cdot \frac{dx}{\sqrt{\beta x - x^2}\sqrt{1 - \frac{x}{2}}}.$$

Now

$$\left(1 - \frac{x}{2}\right)^{-\frac{1}{2}} = 1 + \frac{1}{2} \cdot \frac{x}{2} + \frac{1.3}{2.4} \cdot \frac{x^2}{4} + \frac{1.3.5}{2.4.6} \cdot \frac{x^2}{8} + \&c.$$

therefore,

$$dt = -\tfrac{1}{2}\sqrt{\frac{r}{g}} \cdot \frac{dx}{\sqrt{\beta x - x^2}}\left\{1 + \frac{1}{2} \cdot \frac{x}{2} + \frac{1.3}{2.4} \cdot \frac{x^2}{4} + \&c.\right\}$$

By La Croix' *Integral Calculus*,

$$\int \frac{-dx}{\sqrt{\beta x - x^2}} = \text{arc}\left(\cos = \frac{2x - \beta}{\beta}\right) + \text{constant.}$$

But the integral must be taken between the limits $x = \beta$ and $x = 0$, that is, from the greatest amplitude to the point C. Hence

$$\int \frac{-dx}{\sqrt{\beta x - x^2}} = \pi;$$

π being the ratio of the circumference to the diameter. From the same author it will be found that

$$\int \frac{-xdx}{\sqrt{\beta x - x^2}} = \tfrac{1}{2}\beta\pi; \int \frac{-x^2dx}{\sqrt{\beta x - x^2}} = \tfrac{1}{2} \cdot \tfrac{3}{4}\beta^2\pi, \&c. \&c.$$

between the same limits. Hence, if $\frac{1}{2}$T be the time of half an oscillation,

$$T = \pi\sqrt{\frac{r}{g}}\left\{1 + \left(\frac{1}{2}\right)^2\frac{\beta}{2} + \left(\frac{1.3}{2.4}\right)^2\frac{\beta^2}{4} + \left(\frac{1.3.5}{2.4.6}\right)^2\frac{\beta^3}{8} + \&c.\right\}$$

This series gives the time whatever may be the extent of the oscillations; but if they be very small, $\frac{\beta}{2}$ may be omitted in most cases; then

$$T = \pi \sqrt{\frac{r}{g}}. \qquad (11)$$

As this equation does not contain the arcs, the time is independent of their amplitude, and only depends on the length of the thread and the intensity of gravitation; and as the intensity of gravitation is invariable for any one place on the earth, the time is constant at that place. It follows, that the small oscillations of a pendulum are performed in equal times, whatever their comparative extent may be.

The series in which the time of an oscillation is given however, shows that it is not altogether independent of the amplitude of the arc. In very delicate observations the two first terms are retained; so that

$$T = \pi \sqrt{\frac{r}{g}} \left\{1 + \left(\frac{1}{2}\right)^2 \frac{\beta}{2}\right\}, \text{ or } T = \pi \sqrt{\frac{r}{g}} \left\{1 + \left(\frac{1}{2}\right)^2 \frac{\alpha^2}{4}\right\} \qquad (12)$$

for as β is the versed sine of the arc α, when the arc is very small, $\beta = \frac{\alpha^2}{2}$ nearly. The term $\pi \sqrt{\frac{r}{g}} \left(\frac{1}{2}\right)^2 \frac{\alpha^2}{4}$, which is very small, is the correction due to the magnitude of the arc described, and is the equation alluded to in article 9, which must be applied to make the times equal. This correction varies with the arc when the pendulum oscillates in air, therefore the resistance of the medium has an influence on the duration of the oscillation.

108. The intensity of gravitation at any place on the earth may be determined from the time and the corresponding length of the pendulum. If the earth were a sphere, and at rest, the intensity of gravitation would be the same in every point of its surface; because every point in its surface would then be equally distant from its centre. But as the earth is flattened at the poles, the intensity of gravitation increases from the equator to the poles; therefore the pendulum that would oscillate in a second at the equator, must be lengthened in moving towards the poles.

If h be the space a body would describe by its gravitation during the time T, then $2h = g\text{T}^2$, and because $\text{T}^2 = \pi^2 \cdot \frac{r}{g}$; therefore

$$h = \tfrac{1}{2}\pi^2 \cdot r. \qquad (13)$$

If r be the length of a pendulum beating seconds in any latitude, this expression will give h, the height described by a heavy body during the first second of its fall.

The length of the seconds pendulum at London is 39.1387 inches; consequently in that latitude gravitation causes a heavy body to fall through 16.0951 feet during the first second of its descent.

Huygens had the merit of discovering that the rectilinear motion of heavy bodies might be determined by the oscillations of the pendulum. It is found by experiments first made by Sir Isaac Newton, that the length of a pendulum vibrating in a given time is the same, whatever the substance may be of which it is composed; hence gravitation acts equally on all bodies, producing the same velocity in the same time, when there is no resistance from the air.

Isochronous Curve.

109. The oscillations of a pendulum in circular arcs being isochronous only when the arc is very small, it is now proposed to investigate the nature of the curve in which a particle must move, so as to oscillate in equal times, whatever the amplitude of the arcs may be.

The forces acting on the pendulum at any point of the curve are the force of gravitation resolved in the direction of the arc, and the resistance of the air which retards the motion. The first is $-g\frac{\text{A}p}{\text{A}m}$, or $-g \cdot \frac{dz}{ds}$, the arc Am being indefinitely small; and the second, which is proportional to the square of the velocity, is expressed by $-n\left(\frac{ds}{dt}\right)^2$, in which n is any number, for the velocity is directly as the element of the space, and inversely as the element of the time. Thus $-g.\frac{dz}{ds} - n\frac{ds^2}{dt^2}$ is the whole force acting on the pendulum, hence the equation $F = \frac{d^2s}{dt^2}$ article 68, becomes $-g\frac{dz}{ds} - n\frac{ds^2}{dt^2} = \frac{d^2s}{dt^2}$.

fig. 29.

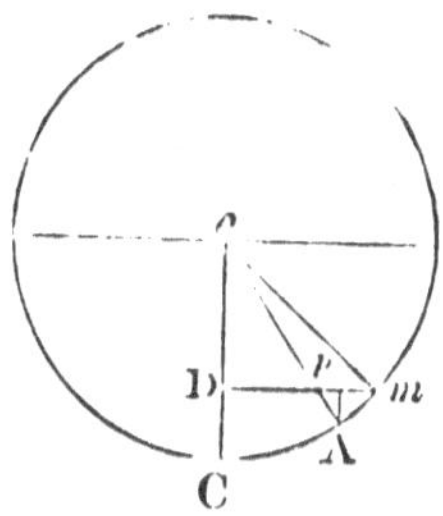

The integral of which will give the isochronous curve in air; but the most interesting results are obtained when the particle is assumed to move in vacuo; then $n = 0$, and the equation becomes $\frac{d^2s}{dt^2} = -g\frac{dz}{ds}$,

which, multiplied by $2ds$ and integrated, gives $\frac{ds^2}{dt^2} = c - 2gz$, c being an arbitrary constant quantity.

Let $z = h$ at m, fig. 29, where the motion begins, the velocity being zero at that point, then will $c = 2gh$, and therefore

$$\frac{ds^2}{dt^2} = 2g\,(h - z);$$

whence
$$dt = -\frac{ds}{\sqrt{2g(h-z)}}:$$

the sign is negative, because the arc diminishes as the time increases. When the radical is developed,

$$dt = -\frac{ds}{\sqrt{2gh}}\left\{1 + \tfrac{1}{2}\frac{z}{h} + \frac{1\,.\,3}{2\,.\,4}\cdot\frac{z^2}{h^2} + \&c.\right\}$$

Whatever the nature of the required curve may be, s is a function of z; and supposing this function developed according to the powers of z, its differential will have the form,

$$\frac{ds}{dz} = az^{i} + bz^{ii} + \&c.$$

Substituting this value of ds in the preceding equation, it becomes

$$dt = -\frac{a}{\sqrt{2g}}\cdot\frac{z^{i}}{h^{\frac{1}{2}}}\left\{1 + \frac{1}{2}\cdot\frac{z}{h} + \frac{1\,.\,3}{2\,.\,4}\cdot\frac{z^2}{h^2} + \&c.\right\}dz$$
$$-\frac{b}{\sqrt{2g}}\cdot\frac{z^{ii}}{h^{\frac{1}{2}}}\left\{1 + \frac{1}{2}\cdot\frac{z}{h} + \frac{1\,.\,3}{2\,.\,4}\cdot\frac{z^2}{h^2} + \&c.\right\}dz.$$

The integral of this equation, taken from $z = h$ to $z = 0$, will give the time employed by the particle in descending to C, the lowest point of the curve. But according to the conditions of the problem, the time must be independent of h, the height whence the particle has descended; consequently to fulfil that condition, all the terms of the value of dt must be zero, except the first; therefore b must be zero, and $i + 1 = \frac{1}{2}$, or $i = -\frac{1}{2}$; thus $ds = az^{-\frac{1}{2}}dz$; the integral of which is $s = 2az^{\frac{1}{2}}$, the equation to a cycloid DzE, fig. 30, with a horizontal base, the only curve in vacuo having the property required. Hence the oscillations of a pendulum moving in a cycloid are rigorously isochronous in vacuo. If $r = 2\text{BC}$, by the properties of the cycloid $r = 2a^2$, and if the preceding value of ds be put in

$$dt = -\frac{ds}{\sqrt{2g(h-z)}}$$

its integral is $$t = \tfrac{1}{2}\sqrt{\frac{r}{g}}\,.\,\text{arc}\left(\cos = \frac{2z-h}{h}\right).$$

It is unnecessary to add a constant quantity if $z = h$ when $t = 0$. If $\frac{1}{2}$T be the time that the particle takes to descend to the lowest point in the curve where $z = 0$, then

$$\text{T} = \sqrt{\frac{r}{g}}\,.\,\text{arc}\,(\cos = -1) = \pi.\sqrt{\frac{r}{g}}.$$

Thus the time of descent through the cycloidal arc is equal to a semi-oscillation of the pendulum whose length is r, and whose oscillations are very small, because at the lowest point of the curve the cycloidal arc ds coincides with the indefinitely small arc of the osculating circle whose vertical diameter is $2r$.

110. The cycloid in question is formed by supposing a circle ABC, fig. 30, to roll along a straight line ED. The curve EAD traced by a point A in its circumference is a cycloid. In the same manner the cycloidal arcs SD, SE, may be traced by a point in a circle, rolling on the other side of DE. These arcs are such, that if we imagine a thread fixed at S to be applied to SD, and then unrolled so that it may always be tangent to SD, its extremity D will trace the cycloid DzE; and the tangent zS is equal to the corresponding arc DS. It is evident also, that the line DE is equal to the circumference of the circle ABC. The curve SD is called the involute, and the curve Dz the evolute. In applying this principle to the construction of clocks, it is so difficult to make the cycloidal arcs SE, SD, round which the thread of the pendulum winds at each vibration, that the motion in small circular arcs

fig. 30.

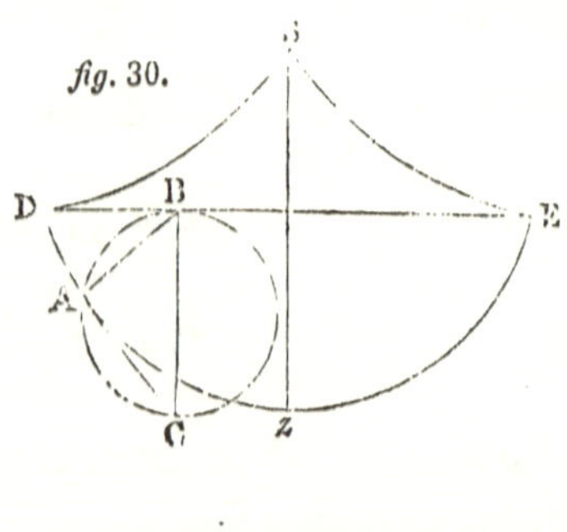

is preferred. The properties of the isochronous curve were discovered by Huygens, who first applied the pendulum to clocks.

111. The time of the very small oscillation of a circular pendulum is expressed by $T=\pi\sqrt{\frac{r}{g}}$ r being the length of the pendulum, and consequently the radius of the circle AmB, fig. 31. Also $t=\sqrt{\frac{2z}{g}}$ is the time employed by a heavy body to fall by the force of gravitation through a height equal to z.

fig. 31.

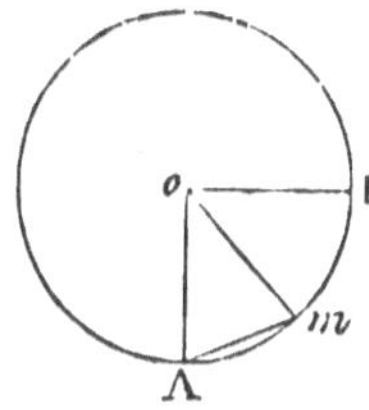

Now the time employed by a heavy body to fall through a space equal to twice the length of the pendulum will be $t=\sqrt{\frac{4r}{g}}$

hence $$\tfrac{1}{2}T : t :: \tfrac{1}{2}\pi\sqrt{\frac{r}{g}} : \sqrt{\frac{4r}{g}},$$

or $$\frac{T}{2} : t :: \frac{\pi}{2} : 2$$

that is, the time employed to move through the arc Am, which is half an oscillation, is to the time of falling through twice the length of the pendulum, as a fourth of the circumference of the circle AmB to its diameter. But the times of falling through all chords drawn to the lowest point A, fig. 32, of a circle are equal: for the accelerating force F in any chord AB, is to that of gravitation as AC : AB, or as AB to AD, since the triangles are similar. But the forces being as the spaces, the times are equal: for as

fig. 32.

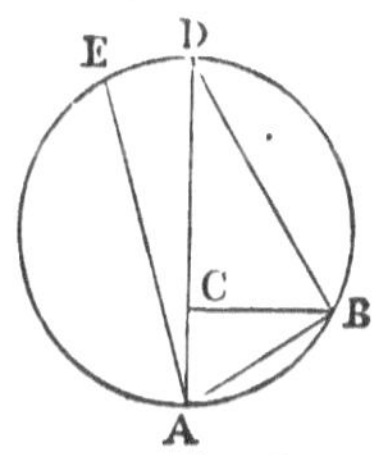

$$F : g : : AB : AD \text{ and } T : t :: \frac{AB}{F} : \frac{AD}{g},$$

it follows that $T=t$.

112. Hence the time of falling through the chord AB, is the same with that of falling through the diameter; and thus the time of falling through the arc AB is to the time of falling through the chord AB as $\frac{\pi}{2}$: 2, that is, as one-fourth of the circumference to the dia-

meter, or as 1.57079 to 2. Thus the straight line AB, though the shortest that can be drawn between the points B and A, is not the line of quickest descent.

Curve of quickest Descent.

113. In order to find the curve in which a heavy body will descend from one given point to another in the shortest time possible, let $CP = z$, $PM = y$, and $CM = s$, fig. 33. The velocity of a body moving in the curve at M will be $\sqrt{2gz}$, g being the force of gravitation. Therefore

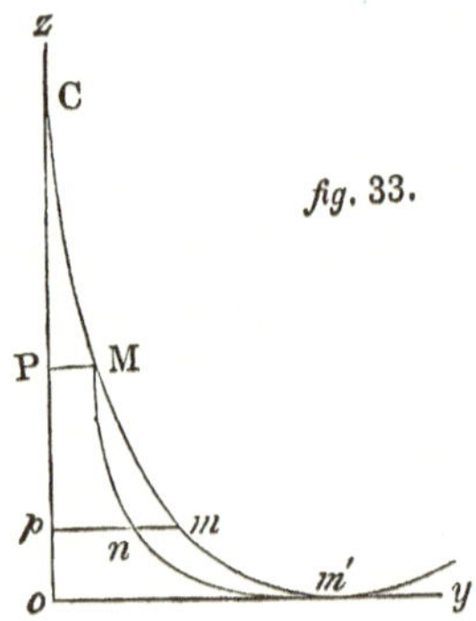

fig. 33.

$$\sqrt{2gz} = \frac{ds}{dt} \text{ or } dt = \frac{ds}{\sqrt{2gz}}$$

the time employed in moving from M to m. Now let

$Cp = z + dz = z'$, $pm = y + dy = y'$, and $Cm = ds + s = s'$.

Then the time of moving through mm' is $\frac{ds'}{\sqrt{2gz'}}$. Therefore the time of moving from M to m' is $\frac{ds}{\sqrt{2gz}} + \frac{ds'}{\sqrt{2gz'}}$, which by hypothesis must be a minimum, or, by the method of variations,

$$\delta \frac{ds}{\sqrt{z}} + \delta \frac{ds'}{\sqrt{z'}} = 0.$$

The values of z and z' are the same for any curves that can be drawn between the points M and m': hence $\delta dz = 0$ $\delta dz' = 0$. Besides, whatever the curves may be, the ordinate om' is the same for all; hence $dy + dy'$ is constant, therefore $\delta(dy + dy') = 0$: whence $\delta dy = -\delta dy'$; and $\delta \frac{ds}{\sqrt{z}} + \delta \frac{ds'}{\sqrt{z'}} = 0$, from these considerations, becomes $\frac{dy}{ds\sqrt{z}} - \frac{dy'}{ds\sqrt{z'}} = 0$. Now it is evident, that the second term of this equation is only the first term in which each variable quantity is augmented by its increment, so that

$$\frac{dy}{ds\sqrt{z}} - \frac{dy'}{ds\sqrt{z'}} = d.\ \frac{dy}{ds\sqrt{z}} = 0,$$

whence
$$\frac{dy}{ds\sqrt{z}} = A.$$

But $\frac{dy}{ds}$ is the sine of the angle that the tangent to the curve makes with the line of the abscissæ, and at the point where the tangent is horizontal this angle is a right angle, so that $\frac{dy}{ds} = 1$: hence if a be the value of z at that point, $A = \frac{1}{\sqrt{a}}$, and $\frac{dy}{ds} = \sqrt{\frac{z}{a}}$, but, $ds^2 = dy^2 + dz^2$, therefore
$$\frac{dy}{dz} = \sqrt{\frac{z}{a-z}},$$
the equation to the cycloid, which is the curve of quickest descent.

CHAPTER III.

ON THE EQUILIBRIUM OF A SYSTEM OF BODIES.

Definitions and Axioms.

114. Any number of bodies which can in any way mutually affect each other's motion or rest, is a system of bodies.

115. Momentum is the product of the mass and the velocity of a body.

116. Force is proportional to velocity, and momentum is proportional to the product of the velocity and the mass; hence the only difference between the equilibrium of a particle and that of a solid body is, that a particle is balanced by equal and contrary forces, whereas a body is balanced by equal and contrary momenta.

117. For the same reason, the motion of a solid body differs from the motion of a particle by the mass alone, and thus the equation of the equilibrium or motion of a particle will determine the equilibrium or motion of a solid body, if they be multiplied by its mass.

118. A moving force is proportional to the quantity of momentum generated by it.

Reaction equal and contrary to Action.

119. The law of reaction being equal and contrary to action, is a general induction from observations made on the motions of bodies when placed within certain distances of one another; the law is, that the sum of the momenta generated and estimated in a given direction is zero. It is found by experiment, that if two spheres A and B of the same dimensions and of homogeneous matter, as of gold, be suspended by two threads so as to touch one another when at rest, then if they be drawn aside from the perpendicular to equal heights and let fall at the same instant, they will strike one another centrically, and will destroy each other's motion, so as to remain at rest in the perpendicular. The experiment being repeated with spheres of homogeneous matter, but of different dimensions, if the velocities be inversely as the quantities of matter, the bodies

after impinging will remain at rest. It is evident, that in this case, the smaller sphere must descend through a greater space than the larger, in order to acquire the necessary velocity. If the spheres move in the same or in opposite directions, with different momenta, and one strike the other, the body that impinges will lose exactly the quantity of momentum that the other acquires. Thus, in all cases, it is known by experience that reaction is equal and contrary to action, or that equal momenta in opposite directions destroy one another. Daily experience shows that one body cannot acquire motion by the action of another, without depriving the latter body of the same quantity of motion. Iron attracts the magnet with the same force that it is attracted by it; the same thing is seen in electrical attractions and repulsions, and also in animal forces; for whatever may be the moving principle of man and animals, it is found they receive by the reaction of matter, a force equal and contrary to that which they communicate, and in this respect they are subject to the same laws as inanimate beings.

Mass proportional to Weight.

120. In order to show that the mass of bodies is proportional to their weight, a mode of defining their mass without weighing them must be employed; the experiments that have been described afford the means of doing so, for having arrived at the preceding results, with spheres formed of matter of the same kind, it is found that one of the bodies may be replaced by matter of another kind, but of different dimensions from that replaced. That which produces the same effects as the mass replaced, is considered as containing the same mass or quantity of matter. Thus the mass is defined independent of weight, and as in any one point of the earth's surface every particle of matter tends to move with the same velocity by the action of gravitation, the sum of their tendencies constitutes the weight of a body; hence the mass of a body is proportional to its weight, at one and the same place.

Density.

121. Suppose two masses of different kinds of matter, A, of hammered gold, and B of cast copper. If A in motion will destroy the

motion of a third mass of matter C, and twice B is required to produce the same effect, then the density of A is said to be double the density of B.

Mass proportional to the Volume into the Density.

122. The masses of bodies are proportional to their volumes multiplied by their densities; for if the quantity of matter in a given cubical magnitude of a given kind of matter, as water, be arbitrarily assumed as the unit, the quantity of matter in another body of the same magnitude of the density ρ, will be represented by ρ; and if the magnitude of the second body to that of the first be as m to 1, the quantity of matter in the second body will be represented by $m \times \rho$.

Specific Gravity.

123. The densities of bodies of equal volumes are in the ratio of their weights, since the weights are proportional to their masses; therefore, by assuming for the unit of density the maximum density of distilled water at a constant temperature, the density of a body will be the ratio of its weight to that of a like volume of water reduced to this maximum.

This ratio is the specific gravity of a body.

Equilibrium of two Bodies.

124. If two heavy bodies be attached to the extremities of an inflexible line without mass, which may turn freely on one of its points; when in equilibrio, their masses are reciprocally as their distances from the point of motion.

Demonstration.—For, let two heavy bodies, m and m', fig. 34, be attached to the extremities of an inflexible line, free to turn round one of its points n, and suppose the line to be bent in n, but so little, that $m'nm$ only differs from two right angles by an indefinitely small angle amn, which may be represented by ω. If g be the force of gravitation, gm, gm' will be the gravitation of the two bodies. But the gravitation gm acting in the direction na may be resolved into two forces, one in the

fig. 34.

direction mn, which is destroyed by the fixed point n, and another acting on m' in the direction $m'm$. Let $mn = f$, $m'n = f'$; then $m'm = f + f'$ very nearly. Hence the whole force gm is to the part acting on m' :: $na : mm'$, and the action of m on m', is $\dfrac{gm\,(f+f')}{na}$; but $m'n : na :: 1 : \omega$, for the arc is so small that it may be taken for its sine. Hence $na = \omega . f'$, and the action of m on m' is $\dfrac{gm.(f+f')}{\omega f'}$.

In the same manner it may be shown that the action of m' on m is $\dfrac{gm'\,(f+f')}{\omega f}$; but when the bodies are in equilibrio, these forces must be equal: therefore $\dfrac{gm\,(f+f')}{\omega f'} = \dfrac{gm'\,(f+f')}{\omega f}$, whence $gm.f = gm'.f'$, or $gm : gm' :: f' : f$, which is the law of equilibrium in the lever, and shows the reciprocal action of parallel forces.

Equilibrium of a System of Bodies.

125. The equilibrium of a system of bodies may be found, when the system is acted on by any forces whatever, and when the bodies also mutually act on, or attract each other.

Demonstration.—Let m, m', m'', &c., be a system of bodies attracted by a force whose origin is in S, fig. 35; and suppose each body to act on all the other bodies, and also to be itself subject to the action of each,—the action of all these forces on the bodies m, m', m'', &c., are as the masses of these bodies and the intensities of the forces conjointly.

Let the action of the forces on one body, as m, be first considered; and, for simplicity, suppose the number of bodies to be only three—m, m', and m''. It is evident that m is attracted by the force at S, and also urged by the reciprocal action of the bodies m' and m''.

fig. 35.

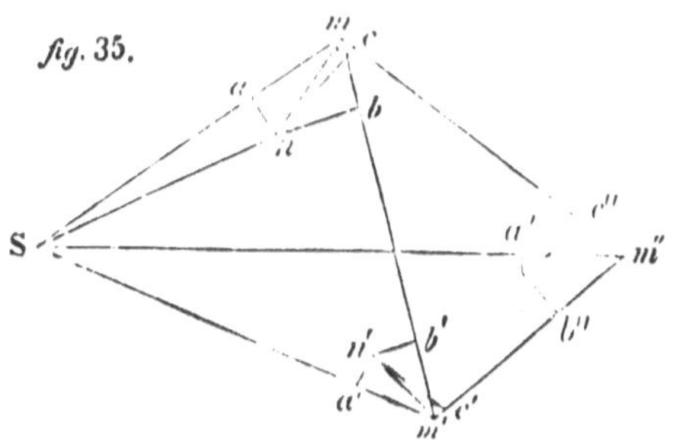

Suppose m' and m'' to remain fixed, and that m is arbitrarily moved to n: then mn is the virtual velocity of m; and if the perpendiculars na, nb, nc be drawn, the lines ma, mb, mc, are the virtual velocities of m resolved in the direction of the forces which act on m. Hence, by the principle of virtual velocities, if the action of the force at S on m be multiplied by ma, the mutual action of m and m' by mb, and the mutual action of m and m'' by mc, the sum of these products must be zero when the point m is in equilibrio; or, m being the mass, if the action of S on m be F.m, and the reciprocal actions of m on m' and m'' be p, p', then

$$m\text{F} \times ma + p \times mb + p' \times mc = 0.$$

Now, if m and m'' remain fixed, and that m' is moved to n', then

$$m'\text{F}' \times m'a' + p \times m'b' + p'' \times m'c' = 0.$$

And a similar equation may be found for each body in the system. Hence the sum of all these equations must be zero when the system is in equilibrio. If, then, the distances Sm, Sm', Sm'', be represented by s, s', s'', and the distances mm', mm'', $m'm''$, by f, f', f'', we shall have

$$\Sigma . m\text{F}\delta s + \Sigma . p\delta f + \Sigma . p\delta f' \pm, \text{\&c.} = 0,$$

Σ being the sum of finite quantities; for it is evident that

$$\delta f = mb + m'b', \ \delta f' = mc + m''c'', \text{ and so on.}$$

If the bodies move on surfaces, it is only necessary to add the terms Rδr, R$'\delta r'$, &c., in which R and R$'$ are the pressures or resistances of the surfaces, and δr $\delta r'$ the elements of their directions or the variations of the normals. Hence in equilibrio

$$\Sigma . m\text{F}\delta s + \Sigma . p\delta f + \text{\&c.} + \text{R}\delta r + \text{R}'\delta r', \text{\&c.} = 0.$$

Now, the variation of the normal is zero; consequently the pressures vanish from this equation: and if the bodies be united at fixed distances from each other, the lines mm', $m'm''$, &c., or f, f', &c., are constant:—consequently $\delta f = 0$, $\delta f' = 0$, &c.

The distance f of two points m and m' in space is

$$f = \sqrt{(x' - x)^2 + (y' - y)^2 + (z' - z)^2},$$

x, y, z, being the co-ordinates of m, and x', y', z', those of m'; so that the variations may be expressed in terms of these quantities: and if they be taken such that $\delta f = 0$, $\delta f' = 0$, &c., the mutual action of the bodies will also vanish from the equation, which is reduced to

$$\Sigma . m\text{F} . \delta s = 0. \qquad (14).$$

126. Thus in every case the sum of the products of the forces into the elementary variations of their directions is zero when the system is in equilibrio, provided the conditions of the connexion of the

system be observed in their variations or virtual velocities, which are the only indications of the mutual dependence of the different parts of the system on each other.

127. The converse of this law is also true—that when the principle of virtual velocities exists, the system is held in equilibrio by the forces at S alone.

Demonstration.—For if it be not, each of the bodies would acquire a velocity v, v', &c., in consequence of the forces $m\text{F}$, $m'\text{F}'$, &c. If δn, $\delta n'$, &c., be the elements of their direction, then

$$\Sigma . m\text{F}\delta s - \Sigma . mv\delta n = 0.$$

The virtual velocities δn, $\delta n'$, &c., being arbitrary, may be assumed equal to vdt, $v'dt$, &c., the elements of the space moved over by the bodies; or to v, v', &c., if the element of the time be unity. Hence

$$\Sigma . m\text{F}\delta s - \Sigma . mv^2 = 0.$$

It has been shown that in all cases $\Sigma . m\text{F}\delta s = 0$, if the virtual velocities be subject to the conditions of the system. Hence, also, $\Sigma . mv^2 = 0$; but as all squares are positive, the sum of these squares can only be zero if $v = 0$, $v' = 0$, &c. Therefore the system must remain at rest, in consequence of the forces $\text{F}m$, &c., alone.

Rotatory Pressure.

128. Rotation is the motion of a body, or system of bodies, about a line or point. Thus the earth revolves about its axis, and billiard-ball about its centre.

129. A rotatory pressure or moment is a force that causes a system of bodies, or a solid body, to rotate about any point or line. It is expressed by the intensity of the motive force or momentum, multiplied by the distance of its direction from the point or line about which the system or solid body rotates.

On the Lever.

130. The lever first gave the idea of rotatory pressure or moments, for it revolves about the point of support or fulcrum.

When the lever mm', fig. 36, is in equilibrio, in consequence of forces applied to two heavy bodies at its extremities, the rotatory pressure of these forces, with regard to N, the point of support, must be equal and contrary.

Demonstration.—Let ma, $m'a'$, fig. 36, which are proportional to the velocities, represent the forces acting on m and m' during the indefinitely small time in which the bodies m and m' describe the indefinitely small spaces ma, $m'a'$. The distance of the direction of the forces ma, $m'a'$ from the fixed point N, are Nm, Nm'; and the momentum of m into Nm, must be equal to the momentum of m' into Nm'; that is, the product of ma by Nm and the mass m, must be equal to the product of $m'a'$ by Nm' and the mass m' when the lever is in equilibrio;

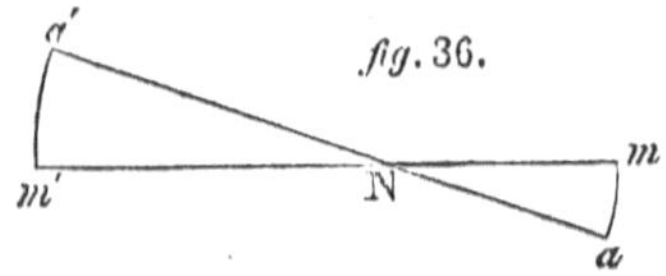

fig. 36.

or, $ma \times \text{N}m \times m = m'a' \times \text{N}m' \times m'$. But

$ma \times \text{N}m$ is twice the triangle Nma, and

$m'a' \times \text{N}m'$ is twice the triangle N$m'a'$;

hence twice the triangle Nma into the mass m, is equal to twice the triangle N$m'a'$ into the mass m', and these are the rotatory pressures which cause the lever to rotate about the fulcrum; thus, in equilibrio, the rotatory pressures are equal and contrary, and the moments are inversely as the distances from the point of support.

Projection of Lines and Surfaces.

131. Surfaces and areas may be projected on the co-ordinate planes by letting fall perpendiculars from every point of them on these planes. For let oMN, fig. 37, be a surface meeting the plane xoy in o, the origin of the co-ordinates, but rising above it towards MN. If perpendiculars be drawn from every point of the area oMN on the plane xoy, they will trace the area omn, which is the projection of oMN.

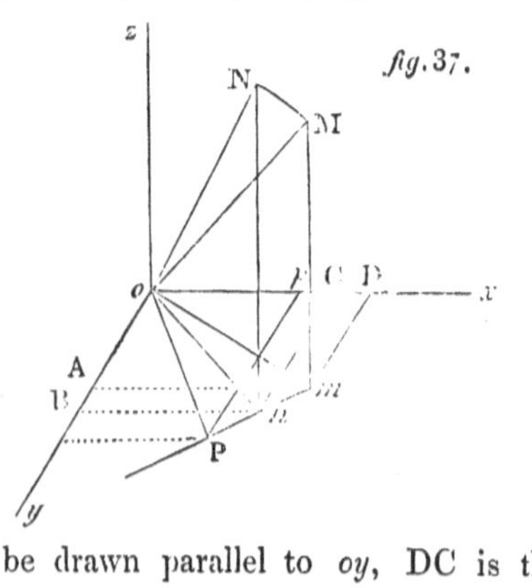

fig. 37.

Since, by hypothesis, xoy is a right angle, if the lines mD, nC, be drawn parallel to oy, DC is the projection of mn on the axis ox. In the same manner AB is the projection of the same line on oy.

Equilibrium of a System of Bodies invariably united.

132. A system of bodies invariably united will be in equilibrio upon a point, if the sum of the moments of rotation of all the forces that act upon it vanish, when estimated parallel to three rectangular co-ordinates.

Demonstration.—Suppose a system of bodies invariably united, moving about a fixed point o in consequence of an impulse and a force of attraction; o being the origin of the attractive force and of the co-ordinates.

Let one body be considered at a time, and suppose it to describe the indefinitely small arc MN, fig. 37, in an indefinitely small time, and let mn be the projection of this arc on the plane xoy. If m be the mass of the body, then $m \times mn$ is its momentum, estimated in the plane xoy; and if oP be perpendicular to mn, it is evident that $m \times mn \times o\text{P}$ is its rotatory pressure. But $mn \times o\text{P}$ is twice the triangle mon; hence the rotatory pressure is equal to the mass m into twice the triangle mon that the body could describe in an element of time. But when m is at rest, the rotatory pressure must be zero; hence in equilibrio, $m \times mn \times o\text{P} = 0$.

Let omn, fig. 38, be the projected area, and complete the parallelogram oDEB; then if oD, oA, the co-ordinates of m, be represented by x and y, it is evident that y increases, while x diminishes; hence

$$\text{CD} = -\,dx, \text{ and AB} = dy.$$

Join OE, then $$no\text{E} = \tfrac{1}{2}n\text{D},$$

because the triangle and parallelogram are on the same base and between the same parallels; also $mo\text{E} = \tfrac{1}{2}\text{AE}$: hence the triangle

$$mon = \tfrac{1}{2}\{n\text{D} + \text{AE}.\}$$

fig. 38.

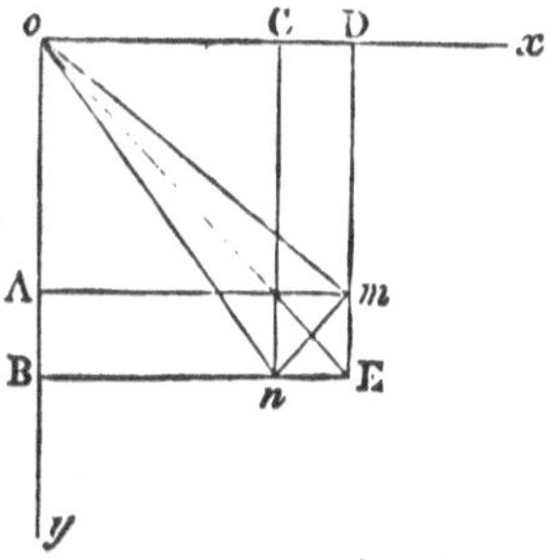

Now $n\text{D} = -\,dx\,(y + dy)$

and $\text{AE} = xdy$,

therefore $mon = \tfrac{1}{2}(xdy - ydx) - \tfrac{1}{2}dxdy$; but when the arc mn is indefinitely small, $\tfrac{1}{2}dxdy = \tfrac{1}{2}n\text{E}.m\text{E}$ may be omitted in comparison of the first powers of these quantities, hence the triangle

$$mon = \tfrac{1}{2}(xdy - ydx),$$

therefore $m(xdy - ydx) = 0$ is the rotatory pressure in the plane xoy

when m is in equilibrio. A similar equation must exist for each co-ordinate plane when m is in a state of equilibrium with regard to each axis, therefore also

$$m\,(xdz - zdx) = 0,\ m\,(ydz - zdy) = 0.$$

The same may be proved for every body in the system, consequently when the whole is in equilibrio on the point o

$$\Sigma m(xdy - ydx) = 0 \quad \Sigma m\,(xdz - zdx) = 0$$

$$\Sigma m\,(ydz - zdy) = 0. \qquad (15).$$

133. This property may be expressed by means of virtual velocities, namely, that a system of bodies will be at rest, if the sum of the products of their momenta by the elements of their directions be zero, or by article 125

$$\Sigma m \mathrm{F} \delta s = 0.$$

Since the mutual distances of the parts of the system are invariable, if the whole system be supposed to be turned by an indefinitely small angle about the axis oz, all the co-ordinates z', z'', &c., will be invariable. If $\delta\varpi$ be any arbitrary variation, and if

$$\delta x = y\delta\varpi \qquad \delta y = -\,x\delta\varpi$$

$$\delta x' = y'\delta\varpi \qquad \delta y' = -\,x'\delta\varpi;$$ then f being the mutual distance of the bodies m and m' whose co-ordinates are x, y, z; x', y', z', there will arise

$$\delta f = \delta\sqrt{(x' - x)^2 + (y' - y)^2 + (z' - z)^2} =$$

$$\frac{x' - x}{f}(\delta x' - \delta x) + \frac{y' - y}{f}(\delta y' - \delta y) =$$

$$\tfrac{1}{2}\{(x' - x)(y' - y)\delta\varpi - (y' - y)(x' - x)\delta\varpi\} = 0.$$

So that the values assumed for δx, δy, $\delta x'$, $\delta y'$ are not incompatible with the invariability of the system. It is therefore a permissible assumption.

Now if s be the direction of the force acting on m, its variation is

$$\delta s = \frac{\delta s}{\delta x}\,\delta x + \frac{\delta s}{\delta y}\,\delta y,$$

since z is constant; and substituting the preceding values of δx, δy, the result is

$$\delta s = \frac{\delta s}{\delta x}\cdot y\delta\varpi - \frac{\delta s}{\delta y}\cdot x\delta\varpi = \delta\varpi\left\{\frac{\delta s}{\delta x}\cdot y - \frac{\delta s}{\delta y}\,x\right\}$$

or, multiplying by the momentum $\mathrm{F}m$,

$$\mathrm{F}m\delta s = \mathrm{F}m\left\{y\frac{\delta s}{\delta x} - x\frac{\delta s}{\delta y}\right\}\delta\varpi.$$

In the same manner with regard to the body m'

$$F'm'\delta s' = F'm'\left\{y'\frac{\delta s'}{\delta x'} - x'\frac{\delta s'}{\delta y'}\right\}\delta\varpi,$$

and so on; and thus the equation $\Sigma mF\delta s = 0$ becomes

$$\Sigma mF\left\{y\frac{\delta s}{\delta x} - x\frac{\delta s}{\delta y}\right\} = 0.$$

It follows, from the same reasoning, that

$$\Sigma mF\left\{z\frac{\delta s}{\delta x} - x\frac{\delta s}{\delta z}\right\} = 0,$$

$$\Sigma mF\left\{z\frac{\delta s}{\delta y} - y\frac{\delta s}{\delta z}\right\} = 0.$$

In fact, if X, Y, Z be the components of the force F in the direction of the three axes, it is evident that

$$X = F\frac{\delta s}{\delta x}; \quad Y = F\frac{\delta s}{\delta y}; \quad Z = F\frac{\delta s}{\delta z};$$

and these equations become

$$\left.\begin{array}{l}\Sigma my.X - \Sigma mx.Y = 0\\ \Sigma mz.X - \Sigma mx.Z = 0\\ \Sigma mz.Y - \Sigma my.Z = 0\end{array}\right\} \quad (16).$$

But $\Sigma mFy\frac{\delta s}{\delta x}$ expresses the sum of the moments of the forces parallel to the axis of x to turn the system round that of z, and $\Sigma mFx\frac{\delta s}{\delta y}$ that of the forces parallel to the axis of y to do the same, but estimated in the contrary direction;—and it is evident that the forces parallel to z have no effect to turn the system round z. Therefore the equation $\Sigma mF\left(y\frac{\delta s}{\delta x} - x\frac{\delta s}{\delta y}\right) = 0$, expresses that the sum of the moments of rotation of the whole system relative to the axis of z must vanish, that the equilibrium of the system may subsist. And the same being true for the other rectangular axes (whose positions are arbitrary), there results this general theorem, viz., that in order that a system of bodies may be in equilibro upon a point, the sum of the moments of rotation of all the forces that act on it must vanish when estimated parallel to any three rectangular co-ordinates.

134. These equations are sufficient to ensure the equilibrium of the system when o is a fixed point; but if o, the point about which it rotates, be not fixed, the system, as well as the origin o, may be car-

ried forward in space by a motion of translation at the same time that the system rotates about o, like the earth, which revolves about the sun at same time that it turns on its axis. In this case it is not only necessary for the equilibrium of the system that its rotatory pressure should be zero, but also that the forces which cause the translation when resolved in the direction of the axis ox, oy, oz, should be zero for each axis separately.

On the Centre of Gravity.

135. If the bodies m, m', m'', &c., be only acted on by gravity, its effect would be the same on all of them, and its direction may be considered the same also; hence

$$F = F' = F'' = \&c.,$$

and also the directions

$$\frac{\delta s}{\delta x} = \frac{\delta s}{\delta x'} = \&c. \quad \frac{\delta s}{\delta y} = \frac{\delta s}{\delta y'} = \&c. \quad \frac{\delta s}{\delta z} = \frac{\delta s}{\delta z'} = \&c.,$$

are the same in this case for all the bodies, so that the equations of rotatory pressure become

$$F\left\{\frac{\delta s}{\delta x}.\Sigma my - \frac{\delta s}{\delta y}\Sigma mx\right\} = 0$$

$$F\left\{\frac{\delta s}{\delta z}.\Sigma my - \frac{\delta s}{\delta y}\Sigma mz\right\} = 0$$

$$F\left\{\frac{\delta s}{\delta x}.\Sigma mz - \frac{\delta s}{\delta z}.\Sigma mx\right\} = 0$$

or, if X, Y, Z, be considered as the components of gravity in the three co-ordinate axes by article 133

$$\left.\begin{array}{l} X.\Sigma my - Y.\Sigma mx = 0 \\ Z.\Sigma my - Y.\Sigma mz = 0 \\ X.\Sigma mz - Z.\Sigma mx = 0 \end{array}\right. \qquad (17).$$

It is evident that these equations will be zero, whatever the direction of gravity may be, if

$$\Sigma mx = 0, \quad \Sigma my = 0, \quad \Sigma mz = 0. \qquad (18).$$

Now since $F\frac{\delta s}{\delta x}$, $F\frac{\delta s}{\delta y}$, $F\frac{\delta s}{\delta z}$, are the components of the force of gravity in the three co-ordinates ox, oy, oz,

$$F.\frac{\delta s}{\delta x}.\Sigma m; \; F.\frac{\delta s}{\delta y}.\Sigma m; \; F.\frac{\delta s}{\delta z}.\Sigma m;$$

are the forces which translate the system parallel to these axes. But if o be a fixed point, its reaction would destroy these forces.

By article 49, $\left(\frac{\delta s}{\delta x}\right)^2 + \left(\frac{\delta s}{\delta y}\right)^2 + \left(\frac{\delta s}{\delta z}\right)^2 = 1$

is the diagonal of a parallelopiped, of which

$$\frac{\delta s}{\delta x}, \frac{\delta s}{\delta y}, \frac{\delta s}{\delta z},$$

are the sides; therefore these three compose one resulting force equal to $F.\Sigma m$. This resulting force is the weight of the system which is thus resisted or supported by the reaction of the fixed point o.

136. The point o round which the system is in equilibrio, is the centre of gravity of the system, and if that point be supported, the whole will be in equilibrio.

On the Position and Properties of the Centre of Gravity.

137. It appears from the equations (18), that if any plane passes through the centre of gravity of a system of bodies, the sum of the products of the mass of each body by its distance from that plane is zero. For, since the axes of the co-ordinates are arbitrary, any one of them, as $x\,o\,x'$, fig. 39, may be assumed to be the section of the plane in question, the centre of gravity of the system of bodies m, m', &c., being in o. If the perpendiculars ma, $m'b$, &c., be drawn from each body on the plane $x\,o\,x'$, the product of the mass m by the distance ma plus the product of m' by $m'b$ plus, &c., must be zero; or, representing the distances by z, z', z'', &c., then

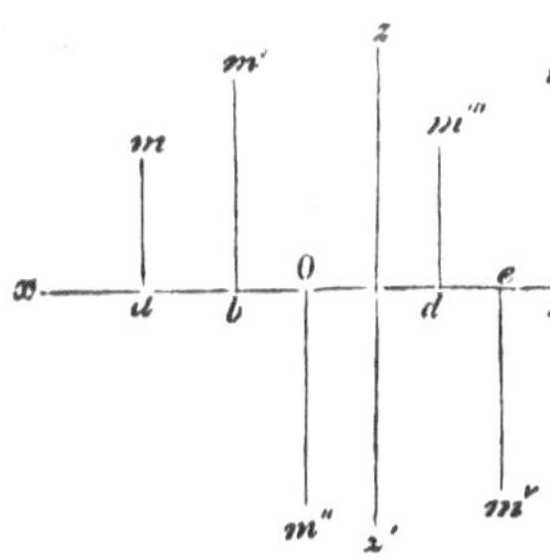

fig. 39.

$$mz + m'z - m''\,z'' + m'''\,z''' + \&c. = 0\,;$$

or, according to the usual notation,

$$\Sigma.mz = 0.$$

And the same property exists for the other two co-ordinate planes Since the position of the co-ordinate planes is arbitrary, the property obtains for every set of co-ordinate planes having their origin in *o*. It is clear that if the distances *ma*, *m'b*, &c., be positive on one side of the plane, those on the other side must be negative, otherwise the sum of the products could not be zero.

138. When the centre of gravity is not in the origin of the co-ordinates, it may be found if the distances of the bodies m, m', m'', &c., from the origin and from each other be known.

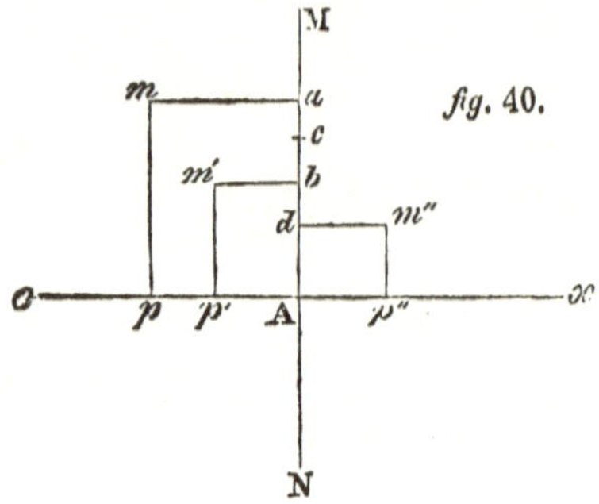

fig. 40.

Demonstration.—For let *o*, fig. 40, be the origin, and *c* the centre of gravity of the system m, m', &c. Let MN be the section of a plane passing through *c*; then by the property of the centre of gravity just explained,

$$m.ma + m'.m'b - m''.m''d + \&c. = 0;$$

but $$ma = oA - op;\ m'b = oA - op', \&c. \&c.,$$

hence $$m(oA - op) + m'(oA - op') + \&c. = 0;$$

or if A*o* be represented by $\bar{x}$, and $op\ op'\ op''$, &c., by $x\ x'\ x''$, &c., then will

$$m(\bar{x} - x) + m'(\bar{x} - x') - m''(\bar{x} - x'') + \&c. = 0.$$

Whence

$$\bar{x}(m + m' - m'' + \&c.) = mx + m'x' - m''x'' + \&c.,$$

$$\text{and, } \bar{x} = \frac{mx + m'x' + \&c.}{m + m' - m'' + \&c.} = \frac{\Sigma.mx}{\Sigma m}. \qquad (19).$$

Thus, if the masses of the bodies and their respective distances from the origin of the co-ordinates be known, this equation will give the distance of the centre of gravity from the plane *yoz*. In the same manner its distances from the other two co-ordinate planes are found to be

$$\bar{y} = \frac{\Sigma.my}{\Sigma m} \qquad \bar{z} = \frac{\Sigma.mz}{\Sigma m}. \qquad (20).$$

139. Thus, because the centre of gravity is determined by its three co-ordinates $\bar{x}$, $\bar{y}$, $\bar{z}$, it is a single point.

140. But these three equations give

$$\bar{x}^2+\bar{y}^2+\bar{z}^2 = \frac{(\Sigma mx)^2 + (\Sigma my)^2 + (\Sigma mz)^2}{(\Sigma m)^2}, \text{ or}$$

$$\bar{x}^2+\bar{y}^2+\bar{z}^2 = \frac{\Sigma m(x^2+y^2+z^2)}{\Sigma m} - \frac{\Sigma mm'\{(x'-x)^2+(y'-y)^2+(z'-z)^2\}}{(\Sigma m)^2}$$

The last term of the second member is the sum of all the products similar to those under Σ when all the bodies of the system are taken in pairs.

141. It is easy to show that the two preceding values of $\bar{x}^2+\bar{y}^2+\bar{z}^2$ are identical, or that

$$\frac{(\Sigma mx)^2}{(\Sigma m)^2} = \frac{\Sigma mx^2}{\Sigma m} - \frac{\Sigma mm'\ (x'-x)^2}{(\Sigma m)^2}$$

or $$(\Sigma mx)^2 = \Sigma m . \Sigma mx^2 - \Sigma mm'\ (x' - x)^2.$$

Were there are only two planets, then

$$\Sigma m = m + m', \ \Sigma mx = mx + m'x', \ \Sigma mm' = mm';$$

consequently

$$(\Sigma mx)^2 = (mx + m'x')^2 = m^2x^2 + m'^2x'^2 + 2mm'xx'.$$

With regard to the second member

$$\Sigma m . \Sigma mx^2 = (m+m')\ (mx^2+m'x'^2) = m^2x^2+m'^2x'^2+mm'x^2+mm'x'^2,$$

and $$\Sigma mm'\ (x'-x)^2 = mm'x'^2 + mm'x^2 - 2mm'xx';$$

consequently

$$\Sigma m . \Sigma mx^2 - \Sigma mm'\ (x'-x)^2 = m^2x^2 + m'^2x'^2 + 2mm'xx' = (\Sigma mx)^2.$$

This will be the case whatever the number of planets may be; and as the equations in question are symmetrical with regard to x, y, and z, their second members are identical.

Thus the distance of the centre of gravity from a given point may be found by means of the distances of the different points of the system from this point, and of their mutual distances.

142. By estimating the distance of the centre of gravity from any three fixed points, its position in space will be determined.

Equilibrium of a Solid Body.

143. If the bodies m, m', m'', &c., be indefinitely small, infinite in number, and permanently united together, they will form a solid mass, whose equilibrium may be determined by the preceding equations.

For if x, y, z, be the co-ordinates of any one of its indefinitely small particles dm, and X, Y, Z, the forces urging it in the direction of these axes, the equations of its equilibrium will be

$$\int \mathrm{X}dm = 0 \quad \int \mathrm{Y}dm = 0 \quad \int \mathrm{Z}dm = 0$$

$$\int (\mathrm{X}y - \mathrm{Y}x)\, dm = 0\,; \int (\mathrm{X}z - \mathrm{Z}x)\, dm = 0\,; \int (\mathrm{Z}y - \mathrm{Y}z)\, dm = 0.$$

The three first are the equations of translation, which are destroyed when the centre of gravity is a fixed point; and the last three are the sums of the rotatory pressures.

CHAPTER IV.

MOTION OF A SYSTEM OF BODIES.

144. It is known by observation, that the relative motions of a system of bodies, are entirely independent of any motion common to the whole; hence it is impossible to judge from appearances alone, of the absolute motions of a system of bodies of which we form a part; the knowledge of the true system of the world was retarded, from the difficulty of comprehending the relative motions of projectiles on the earth, which has the double motion of rotation and revolution. But all the motions of the solar system, determined according to this law, are verified by observation.

By article 117, the equation of the motion of a body only differs from that of a particle, by the mass; hence, if only one body be considered, of which m is the mass, the motion of its centre of gravity will be determined from equation (6), which in this case becomes

$$m\left\{X - \frac{d^2x}{dt^2}\right\}\delta x + m\left\{Y - \frac{d^2y}{dt^2}\right\}\delta y + m\left\{Z - \frac{d^2z}{dt^2}\right\}\delta z = 0.$$

A similar equation may be found for each body in the system, and one condition to be fulfilled is, that the sum of all such equations must be zero;—hence the general equation of a system of bodies is

$$0 = \Sigma m\left(X - \frac{d^2x}{dt^2}\right)\delta x + \Sigma m\left(Y - \frac{d^2y}{dt^2}\right)\delta y + \Sigma m\left(Z - \frac{d^2z}{dt^2}\right)\delta z, \quad (21.)$$

in which $\Sigma mX,\ \Sigma mY,\ \Sigma mZ,$

are the sums of the products of each mass by its corresponding component force, for

$$\Sigma mX = mX + m'X' + m''X'' + \&c.;$$

and so for the other two.

Also $$\Sigma m\frac{d^2x}{dt^2},\ \Sigma m\frac{d^2y}{dt^2},\ \Sigma m\frac{d^2z}{dt^2},$$

are the sums of the products of each mass, by the second increments of the space respectively described by them, in an element of time in the direction of each axis, since

$$\Sigma m\frac{d^2x}{dt^2} = m\frac{d^2x}{dt^2} + m'\frac{d^2x'}{dt^2} + \&c.$$

the expressions $\Sigma m \dfrac{d^2y}{dt^2}, \Sigma m \dfrac{d^2z}{dt^2}$

have a similar signification.

From this equation all the motions of the solar system are directly obtained.

145. If the forces be invariably supposed to have the same intensity at equal distances from the points to which they are directed, and to vary in some ratio of that distance, all the principles of motion that have been derived from the general equation (6), may be obtained from this, provided the sum of the masses be employed instead of the particle.

146. For example, if the equation, in article 74, be multiplied by Σm, its finite value is found to be

$$\Sigma m V^2 = C + 2\Sigma \int m\,(Xdx + Ydy + Zdz).$$

This is the Living Force or Impetus of a system, which is the sum of the masses into the square of their respective velocities, and is analogous to the equation

$$V^2 = C + 2v,$$

relating to a particle.

147. When the motion of the system changes by insensible degrees, and is subject to the action of accelerating forces, the sum of the indefinitely small increments of the impetus is the same, whatever be the path of the bodies, provided that the points of departure and arrival be the same.

148. When there is a primitive impulse without accelerating forces, the impetus is constant.

149. Impetus is the true measure of labour; for if a weight be raised ten feet, it will require four times the labour to raise an equal weight forty feet. If both these weights be allowed to descend freely by their gravitation, at the end of their fall their velocities will be as 1 to 2; that is, as the square roots of their heights. But the effects produced will be as their masses into the heights from whence they fell, or as their masses into 1 and 4; but these are the squares of the velocities, hence the impetus is the mass into the square of the velocity. Thus the impetus is the true measure of the labour employed to raise the weights, and of the effects of their descent, and is entirely independent of time.

150. The principle of least action for a particle was shown, in article 80, to be expressed by $\delta \int v ds = 0$,

when the extreme points of its path are fixed; hence, for a system of bodies, it is

$$\Sigma\delta\int mvds = 0, \qquad \text{or } \Sigma\delta\int mv^2dt = 0.$$

Thus the sum of the living forces of a system of bodies is a minimum, during the time that it takes to pass from one position to another.

If the bodies be not urged by accelerating forces, the impetus of the system during a given time, is proportional to that time, therefore the system moves from one given position to another, in the shortest time possible: which is the principle of least action in a system of bodies.

On the Motion of the Centre of Gravity of a System of Bodies.

151. In a system of bodies the common centre of gravity of the whole either remains at rest or moves uniformly in a straight line, as if all the bodies of the system were united in that point, and the concentrated forces of the system applied to it.

Demonstration.—These properties are derived from the general equation (21) by considering that, if the centre of gravity of the system be moved, each body will have a corresponding and equal motion independent of any motions the bodies may have among themselves: hence each of the virtual velocities δx, δy, δz, will be increased by the virtual velocity of the centre of gravity resolved in the direction of the axes; so that they become

$$\delta x + \delta\bar{x},\ \delta y + \delta\bar{y},\ \delta z + \delta\bar{z}:$$

thus the equation of the motion of a system of bodies is increased by the term,

$$\Sigma.m\left\{X - \frac{d^2x}{dt^2}\right\}\delta\bar{x} + \Sigma.m\left\{Y - \frac{d^2y}{dt^2}\right\}\delta\bar{y}$$
$$+ \Sigma.m\left\{Z - \frac{d^2z}{dt^2}\right\}\delta\bar{z}$$

arising from the consideration of the centre of gravity. If the system be free and unconnected with bodies foreign to it, the virtual velocity of the centre of gravity, is independent of the connexion of the bodies of the system with each other; therefore $\delta\bar{x}$, $\delta\bar{y}$, $\delta\bar{z}$ may each be zero, whatever the virtual velocity of the bodies themselves may be; hence

$$\Sigma.m\left\{X - \frac{d^2x}{dt^2}\right\} = 0$$

$$\Sigma . m \left\{ Y - \frac{d^2y}{dt^2} \right\} = 0, \qquad \Sigma . m \left\{ Z - \frac{d^2z}{dt^2} \right\} = 0.$$

But it has been shewn that the co-ordinates of the centre of gravity are,

$$\bar{x} = \frac{\Sigma . mx}{\Sigma . m}; \quad \bar{y} = \frac{\Sigma . my}{\Sigma . m}; \quad \bar{z} = \frac{\Sigma . mz}{\Sigma . m}.$$

Consequently,

$$d^2\bar{x} = \frac{\Sigma . md^2x}{\Sigma . m}; \quad d^2\bar{y} = \frac{\Sigma . md^2y}{\Sigma . m}; \quad d^2\bar{z} = \frac{\Sigma . md^2z}{\Sigma . m}.$$

Now $\Sigma . md^2x = dt^2 . \Sigma . mX$; $\Sigma . md^2y = dt^2 . \Sigma mY$;

$$\Sigma . md^2z = dt^2 \Sigma . mZ;$$

hence

$$\frac{d^2\bar{x}}{dt^2} = \frac{\Sigma . mX}{\Sigma . m}; \quad \frac{d^2\bar{y}}{dt^2} = \frac{\Sigma . mY}{\Sigma . m}; \quad \frac{d^2\bar{z}}{dt^2} = \frac{\Sigma . mZ}{\Sigma . m}. \qquad (22).$$

These three equations determine the motion of the centre of gravity.

152. Thus the centre of gravity moves as if all the bodies of the system were united in that point, and as if all the forces which act on the system were applied to it.

153. If the mutual attraction of the bodies of the system be the only accelerating force acting on these bodies, the three quantities ΣmX, ΣmY, ΣmZ are zero.

Demonstration.—This evidently arises from the law of reaction being equal and contrary to action; for if F be the action that an element of the mass m exercises on an element of the mass m', whatever may be nature of this action, $m'F$ will be the accelerating force with which m is urged by the action of m'; then if f be the mutual distance of m and m', by this action only

$$X = \frac{m'F(x' - x)}{f}; \quad Y = \frac{m'F(y' - y)}{f}; \quad Z = \frac{m'F(z' - z)}{f}. \qquad (23).$$

For the same reasons, the action of m' on m will give

$$X' = \frac{mF(x - x')}{f}; \quad Y' = \frac{mF(y - y')}{f}; \quad Z' = \frac{mF(z - z')}{f};$$

hence

$$mX + m'X' = 0; \quad mY + m'Y' = 0; \quad mZ + m'Z' = 0;$$

and as all the bodies of the system, taken two and two, give the same results, therefore generally

$$\Sigma . mX = 0; \quad \Sigma . mY = 0; \quad \Sigma . mZ = 0.$$

154. Consequently

$$\frac{d^2\bar{x}}{dt^2} = 0; \quad \frac{d^2\bar{y}}{dt^2} = 0; \quad \frac{d^2\bar{z}}{dt^2} = 0;$$

and integrating,

$$\bar{x} = a\,t + b; \quad \bar{y} = a't + b'; \quad \bar{z} = a''t + b'';$$

in which a, a', a''; b, b', b'', are the arbitrary constant quantities introduced by the double integration. These are equations to straight lines; for, suppose the centre of gravity to begin to move at A, fig. 41, in the direction ox, the distance oA is invariable, and is represented by b; and as at increases with the time t, it represents the straight line AB.

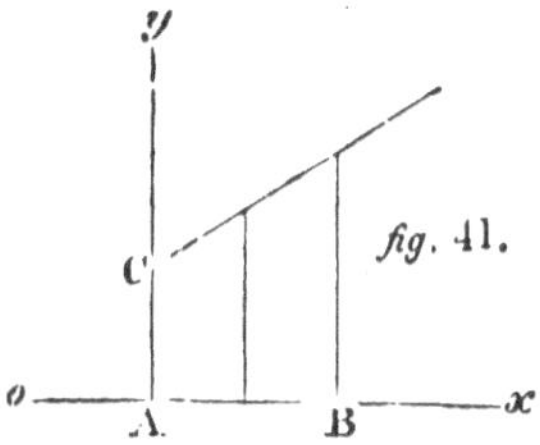

fig. 41.

155. Thus the motion of the centre of gravity in the direction of each axis is a straight line, and by the composition of motions it describes a straight line in space; and as the space it moves over increases with the time, its velocity is uniform; for the velocity, being directly as the element of the space, and inversely as the element of the time, is

$$\sqrt{\left(\frac{d\bar{x}}{dt}\right)^2 + \left(\frac{d\bar{y}}{dt}\right)^2 + \left(\frac{d\bar{z}}{dt}\right)^2};$$

or

$$\sqrt{a^2 + a'^2 + a''^2}.$$

Thus the velocity is constant, and therefore the motion uniform.

156. These equations are true, even if some of the bodies, by their mutual action, lose a finite quantity of motion in an instant.

157. Thus, it is possible that the whole solar system may be moving in space; a circumstance which can only be ascertained by a comparison of its position with regard to the fixed stars at very distant periods. In consequence of the proportionality of force to velocity, the bodies of the solar system would maintain their relative motions, whether the system were in motion or at rest.

On the Constancy of Areas.

158. If a body propelled by an impulse describe a curve A M B, fig. 42, in consequence of a force of attraction in the point o, that force may be resolved into two, one in the direction of the normal AN,

and the other in the direction of the element of the curve or tangent: the first is balanced by the centrifugal force, the second augments or diminishes the velocity of the body; but the velocity is always such that the areas AoM, MoB, described by the radius vector Ao, are proportional to the time; that is, if the body moves from A to M in the same time that it would move from M to B, the area AoM will be equal to the area MoB.

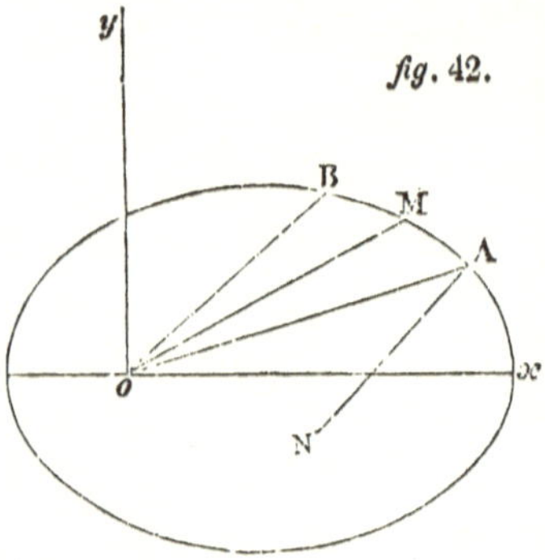

fig. 42.

If a system of bodies revolve about any point in consequence of an impulse and a force of attraction directed to that point, the sums of their masses respectively multiplied by the areas described by their radii vectores, when projected on the three co-ordinate planes, are proportional to the time. *Demonstration.*—For if we only consider the areas that are projected on the plane xoy, fig. 43, the forces in the direction oz, which are perpendicular to that plane, must be zero; hence

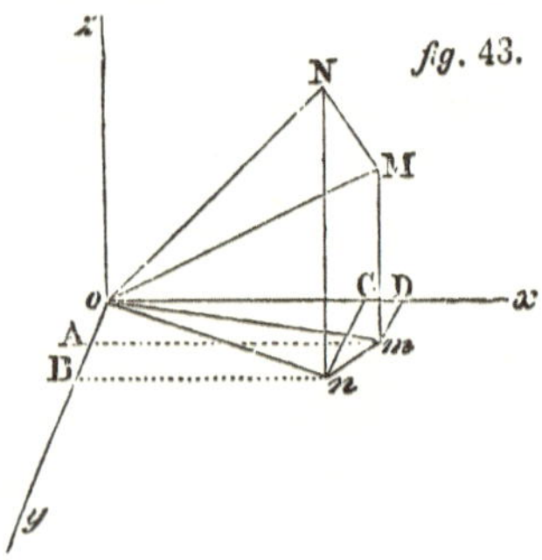

fig. 43.

$$Z = 0, \ Z' = 0, \ \&c.;$$

and the general equation of the motion of a system of bodies becomes

$$\Sigma m \left\{ X - \frac{d^2x}{dt^2} \right\} \delta x + \Sigma . m \left\{ Y - \frac{d^2y}{dt^2} \right\} \delta y = 0.$$

If the same assumptions be made here as in article **133**, namely,

$$\delta x = y \delta \varpi \qquad \delta y = - x \delta \varpi$$
$$\delta x' = y' \delta \varpi \qquad \delta y' = - x' \delta \varpi, \ \&c. \ \&c.,$$

and if these be substituted in the preceding equation, it becomes, with regard to the plane xoy,

$$\Sigma . m \left(\frac{x d^2y - y d^2x}{dt^2} \right) = \Sigma m . (xY - yX).$$

In the same manner

$$\Sigma m.\left(\frac{zd^2x - xd^2z}{dt^2}\right) = \Sigma m.(zX - xZ) \qquad (24).$$

$$\Sigma m.\left(\frac{yd^2z - zd^2y}{dt^2}\right) = \Sigma m.(yZ - zY)$$

are obtained for the motions of the system with regard to the planes xoz, yoz. These three equations, together with

$$\Sigma m.\frac{d^2x}{dt^2} = \Sigma mX,\ \Sigma m.\frac{d^2y}{dt^2} = \Sigma mY,\ \Sigma m.\frac{d^2z}{dt^2} = \Sigma mZ, \quad (25).$$

are the general equations of the motions of a system of bodies which does not contain a fixed point.

159. When the bodies are independent of foreign forces, and only subject to their reciprocal attraction and to the force at o, the sum of the terms

$$m\{Xy - Yx\} + m'\{X'y' - Y'x'\},$$

arising from the mutual action of any two bodies in the system, m, m', is zero, by reason of the equality and opposition of action and reaction; and this is true for every such pair as m and m'', m' and m'', &c. If f be the distance of m from o, F the force which urges the body m towards that origin, then

$$X = -F\frac{x}{f}, \quad Y = -F\frac{y}{f}, \quad Z = -F\frac{z}{f}$$

are its component forces; and when substituted in the preceding equations, F vanishes; the same may be shown with regard to m', m'', &c. Hence the equations of areas are reduced to

$$\Sigma m\left\{\frac{yd^2x - xd^2y}{dt^2}\right\} = 0,$$

$$\Sigma m\left\{\frac{zd^2x - xd^2z}{dt^2}\right\} = 0,$$

$$\Sigma m\left\{\frac{yd^2z - zd^2y}{dt^2}\right\} = 0,$$

and their integrals are

$$\Sigma m\{xdy - ydx\} = cdt.$$
$$\Sigma m\{zdx - xdz\} = c'dt. \qquad (26).$$
$$\Sigma m\{ydz - zdy\} = c''dt.$$

As the first members of these equations are the sum of the masses of all the bodies of the system, respectively multiplied by the projections of double the areas they describe on the co-ordinate planes, this sum is proportional to the time.

If the centre of gravity be the origin of the co-ordinates, the preceding equations may be expressed thus,

$$cdt = \frac{\Sigma mm' \{ (x' - x)\ (dy' - dy) - (y' - y)\ (dx' - dx) \}}{\Sigma m}$$

$$c'dt = \frac{\Sigma mm' \{ (z' - z)\ (dx' - dx) - (x' - x)\ (dz' - dz) \}}{\Sigma m}$$

$$c''dt = \frac{\Sigma mm' \{ (y' - y)\ (dz' - dz) - (z' - z)\ (dy' - dy) \}}{\Sigma m}$$

So that the principle of areas is reduced to depend on the co-ordinates of the mutual distances of the bodies of the system.

160. These results may be expressed by a diagram. Let m, m', m'', fig. 44, &c., be a system of bodies revolving about o, the origin of the co-ordinates, in consequence of a central force and a primitive impulse.—Suppose that each of the radii vectores, om, om', om'', &c., describes the indefinitely small areas, MoN, $M'oN'$, &c.,

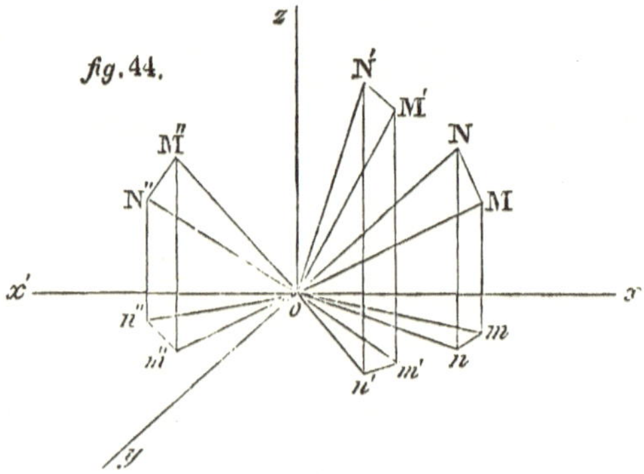

fig. 44.

in an indefinitely small time, represented by dt; and let mon, $m'on'$, &c., be the projections of these areas on the plane xoy. Then the equation

$$\Sigma m \{ xdy - ydx \} = cdt,$$

shows that the sum of the products of twice the area mon by the mass m, twice the area $m'on'$ by the mass m', twice $m''on''$ by the mass m'', &c., is proportional to the element of the time in which they are described: whence it follows that the sum of the projections of the areas, each multiplied by the corresponding mass, is proportional to the finite time in which they are described. The other two equations express similar results for the areas projected on the planes xoz, yoz.

161. The constancy of areas is evidently true for any plane whatever, since the position of the co-ordinate planes is arbitrary. The

three equations of areas give the space described by the bodies on each co-ordinate plane in value of the time: hence, if the time be known or assumed, the corresponding places of the bodies will be found on the three planes, and from thence their true positions in space may be determined, since that of the co-ordinate planes is supposed to be known. It was shown, in article 132, that

$$\Sigma m \{ xdy - ydx \},$$
$$\Sigma m \{ zdx - xdz \},$$
$$\Sigma m \{ zdy - ydz \},$$

are the pressures of the system, tending to make it turn round each of the axes of the co-ordinates: hence the principle of areas consists in this—that the sum of the rotatory pressures which cause a system of bodies to revolve about a given point, is zero when the system is in equilibrio, and proportional to the time when the system is in motion.

162. Let us endeavour to ascertain whether any planes exist on which the sums of the areas are zero when the system is in motion. To solve this problem it is necessary to determine one set of co-ordinates in values of another.

fig. 45.

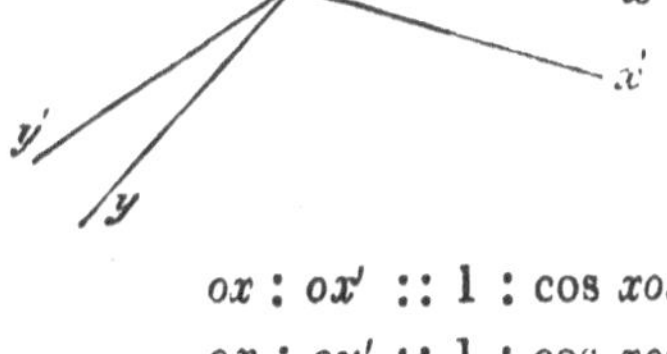

163. If ox, oy, oz, fig. 45, be the co-ordinates of a point m, it is required to determine the position of m by means of ox', oy', oz', three new rectangular co-ordinates, having the same origin as the former.

We shall find a value of ox or x first. Now,

$$ox : ox' :: 1 : \cos xox' \quad \text{or} \quad x' = x \cos xox'.$$
$$ox : oy' :: 1 : \cos xoy' \quad \text{or} \quad y' = x \cos xoy'.$$
$$ox : oz' :: 1 : \cos xoz' \quad \text{or} \quad z' = x \cos xoz'.$$

If the sum of these quantities be taken, after multiplying the first by $\cos xox'$, the second by $\cos xoy'$, and the third by $\cos xoz'$, we shall have $x' \cos xox' + y' \cos xoy' + z' \cos xoz' =$

$$x \{ \cos^2 xox' + \cos^2 xoy' + \cos^2 xoz' \} = x.$$

Let $o\gamma$, fig. 46, be the intersection of the old plane xoy with the new $x'oy'$; and let θ be the inclination of these two planes; also let γox, $\gamma ox'$ be represented by ψ and ϕ. Values of the

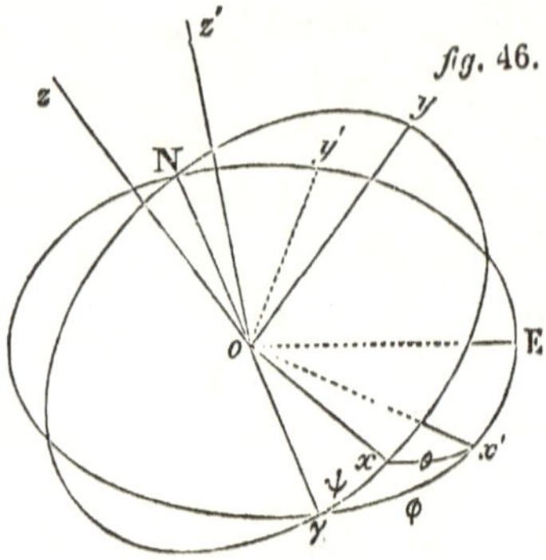

fig. 46.

cosines of xox', xoy', xoz', must be found in terms of θ, ψ, and ϕ. In the right-angled triangle $\gamma xx'$, the sides γx, $\gamma x'$, are ψ and ϕ, and the angle opposite the side xx' is θ:— hence, by spherical trigonometry,

$$\cos xox' = \cos\theta \sin\phi \sin\psi + \cos\psi \cos\phi.$$

This equation exists, whatever the values of ϕ and ψ may be; hence, if $\phi + 90°$ be put for ϕ, the line ox' will take the place of oy', the angle xox' will become xoy', and the preceding equation will give

$$\cos xoy' = \cos\theta \sin\psi \cos\phi - \cos\psi \sin\phi.$$

Cos xoz' is found from the triangle whose three sides are the arcs intercepted by the angles $\gamma oz'$, γox, and xoz'. The angle opposite to the last side is $90° - \theta$, $\gamma oz' = 90°$ $\gamma ox = \psi$,
then the general equation becomes

$$\cos xoz' = \sin\theta \sin\psi.$$

If these expressions for the cosines be substituted in the value of x, it becomes $x = x'\{\cos\theta \sin\phi \sin\psi + \cos\psi \cos\phi\} +$

$$y'\{\cos\theta \cos\phi \sin\psi - \cos\psi \sin\phi\} + z' \sin\theta \sin\psi.$$

In the same manner, the values of y and z are found to be

$$y = x'\{\cos\theta \cos\psi \sin\phi - \sin\psi \cos\phi\} +$$

$$y'\{\cos\theta \cos\psi \cos\phi + \sin\psi \sin\phi\} + z'\{\sin\theta \cos\psi\}$$

$$z = -x'\{\sin\theta \sin\phi\} - y'\{\sin\theta \cos\phi\} + z' \cos\theta.$$

By substituting these values of x, y, z, in any equation, it will be transformed from the planes xoy, xoz, yoz, to the new planes $x'oy'$, $x'oz'$, $y'oz'$.

164. We have now the means of ascertaining whether, among the infinite number of co-ordinate planes whose origin is in o, the centre of gravity of a system of bodies, there be any on which the sums of the areas are zero. This may be known by substituting the preceding values of x, y, z, and their differentials in the equations of areas: for the angles θ, ψ, and ϕ being arbitrary, such values may be assumed for two of them as will make the sums of the projected areas on two of the co-ordinate planes zero; and if there be any

incongruity in this assumption, it will appear in the determination of the third angle, which in that case would involve some absurdity in the areas on the third plane. That, however, is by no means the case, for the sum of the areas on the third plane is then found to be a maximum. If the substitution be made, and the angles ψ and θ so assumed that

$$\sin\theta\sin\psi = \frac{c''}{\sqrt{c^2 + c'^2 + c''^2}}, \quad \sin\theta\cos\psi = \frac{-c'}{\sqrt{c^2 + c'^2 + c''^2}},$$

it follows that
$$\cos\theta = \frac{c}{\sqrt{c^2 + c'^2 + c''^2}},$$

whence
$$\Sigma m \frac{x'dy' - y'dx'}{dt} = \sqrt{c^2 + c'^2 + c''^2}, \qquad (27).$$

$$\Sigma m \frac{x'dz' - z'dx'}{dt} = 0, \qquad \Sigma m \frac{y'dz' - z'dy'}{dt} = 0.$$

Thus, in every system of revolving bodies, there does exist a plane, on which the sum of the projected areas is a maximum; and on every plane at right angles to it, they are zero. One plane alone possesses that property.

165. If the attractive force at *o* were to cease, the bodies would move by the primitive impulse alone, and the principle of areas would be also true in this case; it even exists independently of any abrupt changes of motion or velocity, among the bodies; and also when the centre of gravity has a rectilinear motion in space. Indeed it follows as a matter of course, that all the properties which have been proved to exist in the motions of a system of bodies, whose centre of gravity is at rest, must equally exist, if that point has a uniform and rectilinear motion in space, since experience shows that the relative motions of a system of bodies, is independent of any motion common to them all.

Demonstration.—However, that will readily appear, if $\bar{x}, \bar{y}, \bar{z}$, be assumed, as the co-ordinates of *o*, the moveable centre of gravity estimated from a fixed point P, fig. 47, and if *o*A, AB, B*m*, or x', y', z', be the co-ordinates of *m*, one of the bodies of the system with regard to the moveable point *o*. Then the co-ordinates of *m* relatively

fig. 47.

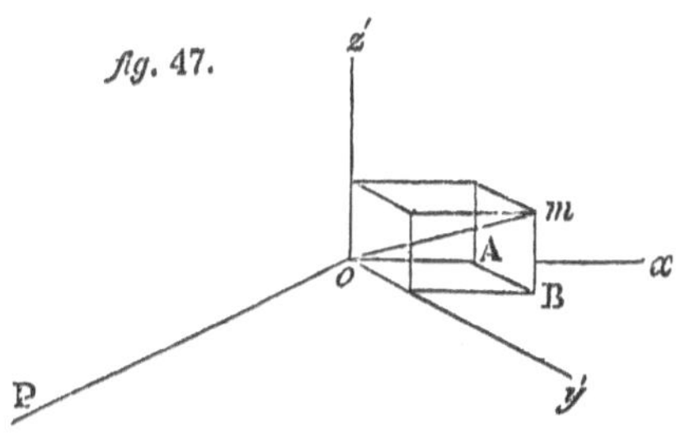

to P will be $\bar{x} + x'$, $\bar{y} + y'$, $\bar{z} + z'$. If these be put instead of x, y, z, in the different equations relative to the motions of a system, by attending to the properties of the centre of gravity, $\bar{x}$, $\bar{y}$, $\bar{z}$, vanish from these equations, which then become independent of them. If $\bar{x} + x'$, $\bar{y} + y'$, $\bar{z} + z'$ be put for x, y, z, in equations (25), they become

$$\Sigma.m \{ d^2\bar{x} + d^2x' \} - \Sigma m\, \text{X}dt^2 = 0.$$
$$\Sigma.m \{ d^2\bar{y} + d^2y' \} - \Sigma m\, \text{Y}dt^2 = 0.$$
$$\Sigma.m \{ d^2\bar{z} + d^2z' \} - \Sigma.m\, \text{Z}dt^2 = 0.$$

But when the centre of gravity has a rectilinear and uniform motion in space, it has been shown, that

$$\frac{d^2\bar{x}}{dt^2} = 0; \quad \frac{d^2\bar{y}}{dt} = 0; \quad \frac{d^2\bar{z}}{dt^2} = 0;$$

which reduces the preceding equations to their original form, namely,

$$\Sigma.m\, \frac{d^2x'}{dt^2} = \Sigma m\text{X}, \ \Sigma m.\, \frac{d^2y'}{dt^2} = \Sigma m\text{Y}, \ \Sigma m.\, \frac{d^2z'}{dt^2} = \Sigma m\text{Z}.$$

If the same substitution be made in

$$\Sigma m \left\{\frac{xd^2y - yd^2x}{dt^2}\right\} = \Sigma m\, (x\text{Y} - y\text{X})$$

it becomes

$$\frac{\bar{x}\Sigma.md^2y' - \bar{y}\Sigma md^2x'}{dt^2} + \Sigma m.\, \frac{x'd^2y' - y'd^2x'}{dt^2}$$
$$= \Sigma m.\, (\text{Y}x' - \text{X}y') + \bar{x}.\Sigma m\text{Y} - \bar{y}.\Sigma m\text{X}.$$

But in consequence of the preceding equations it is reduced to

$$\Sigma.m.\left\{\frac{x'd^2y' - y'd^2x'}{dt^2}\right\} = \Sigma.m.(x'\text{Y} - y'\text{X}.$$

In the same manner it may be shown that

$$\Sigma.m.\left\{\frac{z'd^2x' - x'd^2z'}{dt^2}\right\} = \Sigma.m.\, (z'\text{X} - x'\text{Z}),$$

$$\Sigma.m.\left\{\frac{y'd^2z' - z'd^2y'}{dt^2}\right\} = \Sigma.m.\, (y'\text{Z} - z'\text{Y}).$$

Thus the equations that determine the motions of a system of bodies are the same, whether the centre of gravity be at rest, or moving uniformly in a straight line; consequently the principles of Impetus, of Least Action, and of the Conservation of Areas, exist in either case.

166. Let the effect produced by the motion of the centre of gravity

on the position of the plane for which the areas are a maximum, be now determined.

If $\bar{x} + x$, $\bar{y} + y$, $\bar{z} + z$, be put for x, y, z in equations (26), they will retain the same form, namely,

$$\Sigma m\,(x'dy' - y'dx') = c\,dt,$$
$$\Sigma m\,(z'dx' - x'dz') = c'dt,$$
$$\Sigma m\,(y'dz' - z'dy') = c''dt;$$

for, in consequence of the rectilinear motion of the origin,

$$\bar{x}d\bar{y} - \bar{y}d\bar{x} = 0,\ \bar{z}d\bar{x} - \bar{x}d\bar{z} = 0,\ \bar{y}d\bar{z} - \bar{z}d\bar{y} = 0.$$

And as the position of the plane in question is determined by the constant quantities c, c', and c'', it will always remain parallel to itself during the motion of the system; on that account it is called the Invariable Plane.

167. Thus, when there are no foreign forces acting on the system, the centre of gravity either remains at rest, or moves uniformly in a straight line; and if that point be assumed as the origin of the co-ordinates, the principles of the conservation of areas and living forces will exist with regard to it; and the invariable plane, always passing through that point, will remain parallel to itself, and will be carried along with the centre of gravity in the general motion of the system.

On the Motion of a System of Bodies in all possible Mathematical relations between Force and Velocity.

168. In nature, force is proportional to velocity; but as a matter of speculation, La Place has investigated the motions of a system of bodies in every possible relation between these two quantities. It is rather singular that such an hypothesis should involve no contradiction; on the contrary, principles similar to the preservation of impetus, the constancy of areas, the motion of the centre of gravity, and the least action, actually exist.

CHAPTER V

THE MOTION OF A SOLID BODY OF ANY FORM WHATEVER.

169. If a solid body receives an impulse in a direction passing through its centre of gravity, all its parts will move with an equal velocity; but if the direction of the impulse passes on one side of that centre, the different parts of the body will have unequal velocities, and from this inequality results a motion of rotation in the body round its centre of gravity, at the same time that the centre is moved forward, or translated with the same velocity it would have taken, had the impulse passed through it. Thus the double motions of rotation and translation are produced by one impulse.

170. If a body rotates about its centre of gravity, or about an axis, and is at the same time carried forward in space; and if an equal and contrary impulse be given to the centre of gravity, so as to stop its progressive motion, the rotation will go on as before it received the impulse.

171. If a body revolves about a fixed axis, each of its particles will describe a circle, whose plane is perpendicular to that axis, and its radius is the distance of the particle from the axis. It is evident, that every point of the solid will describe an arc of the same number of degrees in the same time; hence, if the velocity of each particle be divided by its radius or distance from the axis, the quotient will be the same for every particle of the body. This is called the angular velocity of the solid.

172. The axis of rotation may change at every instant, the angular velocity is therefore the same for every particle of the solid for any one instant, but it may vary from one instant to another.

173. The general equations of the motion of a solid body are the same with those of a system of bodies, provided we assume the bodies m, m', m'', &c. to be a system of particles, infinite in number, and united into a solid mass by their mutual attraction.

Let x, y, z, be the co-ordinates of dm, a particle of a solid body urged by the forces X, Y, Z, parallel to the axes of the co-ordinates; then if S the sign of ordinary integrals be put for Σ, and dm for m, the general equations of the motion of a system of bodies in article 158 become

$$S \, . \, \frac{d^2x}{dt^2} dm = S \, . \, Xdm,$$

$$S \, . \, \frac{d^2y}{dt^2} dm = S \, . \, Ydm, \qquad (28)$$

$$S \, . \, \frac{d^2z}{dt^2} dm = S \, . \, Zdm,$$

$$S \left(\frac{xd^2y - yd^2x}{dt^2} \right) dm = S \, . \, (xY - yX) dm,$$

$$S \left(\frac{zd^2x - xd^2z}{dt^2} \right) dm = S \, . \, (zX - xZ) dm, \qquad (29)$$

$$S \left(\frac{yd^2z - zd^2y}{dt^2} \right) dm = S \, . \, (yZ - zY) dm,$$

which are the general equations of the motion of a solid, of which m is the mass.

Determination of the Equations of the Motion of the Centre of Gravity of a Solid in Space.

174. Let $\bar{x}+x'$, $\bar{y}+y'$, $\bar{z}+z'$, be put for x, y, z, in equations (28) then

$$S \, . \, dm \left\{ \frac{d^2\bar{x} + d^2x'}{dt^2} \right\} = S \, . \, Xdm$$

$$S \, . \, dm \left\{ \frac{d^2\bar{y} + d^2y'}{dt^2} \right\} = S \, . \, Ydm \qquad (30)$$

$$S \, . \, dm \left\{ \frac{d^2\bar{z} + d^2z'}{dt^2} \right\} = S \, . \, Zdm$$

in which $\bar{x}$, $\bar{y}$, $\bar{z}$, are the co-ordinates of o the moveable centre of gravity of the solid referred to P a fixed point, and x' y' z' are the co-ordinates of dm referred to o, fig. 47. Now the co-ordinates of the centre of gravity being the same for all the particles of the solid,

$$S \, . \, dm \frac{d^2\bar{x}}{dt^2} = m \, \frac{d^2\bar{x}}{dt^2}$$

$$S \, . \, dm \frac{d^2\bar{y}}{dt^2} = m \, \frac{d^2\bar{y}}{dt^2}$$

$$S \, . \, dm \frac{d^2\bar{z}}{dt^2} = m \, \frac{d^2\bar{z}}{dt^2}.$$

And, with regard to the centre of gravity,

$$S . x'dm = 0$$

$$S . y'dm = 0$$

$$S . z'dm = 0$$

which denote the sum of the particles of the body into their respective distances from the origin; therefore their differentials are

$$S . dm \frac{d^2x'}{dt^2} = 0$$

$$S . dm \frac{d^2y'}{dt^2} = 0$$

$$S . dm \frac{d^2z'}{dt^2} = 0.$$

This reduces the equations (30) to

$$m \frac{d^2\bar{x}}{dt^2} = S . Xdm$$

$$m \frac{d^2\bar{y}}{dt^2} = S . Ydm \qquad (31)$$

$$m \frac{d^2\bar{z}}{dt^2} = S . Zdm.$$

These three equations determine the motion of the centre of gravity of the body in space, and are similar to those which give the motion of the centre of gravity of a system of bodies.

The solid therefore moves in space as if its mass were united in its centre of gravity, and all the forces that urge the body applied to that point.

175. If the same substitution be made in the first of equations (29), and if it be observed that as $\bar{x}$, $\bar{y}$, $\bar{z}$, are the same for all the particles

$$S\,(\bar{x}d^2\bar{y} - \bar{y}d^2\bar{x})\,dm = m\,(\bar{x}d^2\bar{y} - \bar{y}d^2\bar{x})$$

$$S\,(\bar{x}Y - \bar{y}X)\,dm = \bar{x}.S.Ydm - \bar{y}.S.Xdm;$$

also $S\,(x'd^2\bar{y} - y'd^2\bar{x} + \bar{x}d^2y' - \bar{y}d^2x')\,dm =$

$$d^2\bar{y}.S.x'dm - d^2\bar{x}.S.y'dm + \bar{x}.S.d^2y'.dm - \bar{y}.S.d^2x'.dm = 0,$$

because x', y', z', are referred to the centre of gravity as the origin of the co-ordinates; consequently the co-ordinates $\bar{x}$, $\bar{y}$, $\bar{z}$, and their differentials vanish from the equation, which therefore retains its original form. Similar results will be obtained for the areas on the other two co-ordinate planes, and thus equations (29) retain the same forms, whether the centre of gravity be in motion or at rest, proving the motions of rotation and translation to be independent of one another.

Rotation of a Solid.

176. If to abridge

$$\begin{aligned} S\,(yZ - zY)\,dm &= M, \\ S\,(zX - xZ)\,dm &= M', \\ S\,(xY - yX)\,dm &= M''. \end{aligned}$$

The integrals of equations (29), with regard to the time, will be

$$\begin{aligned} S\left(\frac{ydz - zdy}{dt}\right) dm &= \int M dt, \\ S\left(\frac{zdx - xdz}{dt}\right) dm &= \int M' dt, \qquad (32) \\ S\left(\frac{xdy - ydx}{dt}\right) dm &= \int M'' dt. \end{aligned}$$

These equations, which express the properties of areas, determine the rotation of the solid;—equations (31) give the motion of its centre of gravity in space. S expresses the sum of the particles of the body, and $\int$ relates to the time alone.

177. Impetus is the mass into the square of the velocity, but the velocity of rotation depends on the distance from the axis, the angle being the same; hence the impetus of a revolving body is the sum of the products of each particle, multiplied by the square of its distance from the axis of rotation. Suppose oA, oB, oC, fig. 10, to be the co-ordinates of a particle dm, situate in m, and let them be represented by x, y, z; then because $mA = Ro$, $mB = Qo$, $mC = Po$, the squares of the distances of dm from the three axes ox, oy, oz, are respectively

$$(mA)^2 = y^2 + z^2, \quad (mB)^2 = x^2 + z^2, \quad (mC)^2 = x^2 + y^2.$$

Hence if A′, B′, C′, be the impetus or moments of inertia of a solid with regard to the axes ox, oy, oz, then

$$\begin{aligned} A' &= S\,.\,dm\,(y^2 + z^2) \\ B' &= S\,.\,dm\,(x^2 + z^2) \qquad (33) \\ C' &= S\,.\,dm\,(x^2 + y^2). \end{aligned}$$

178. If an impulse be given to a sphere of uniform density, in a

direction which does not pass through its centre of gravity, it will revolve about an axis perpendicular to the plane passing through the centre of the sphere and the direction of the force; and it will continue to rotate about the same axis even if new forces act on the sphere, provided they act equally on all its particles; and the areas which each of its particles describes will be constant.

179. If the solid be not a sphere, it may change its axis of rotation at every instant; it is therefore of importance, to ascertain if any axes exist in the solid, about which it would rotate permanently.

180. If a body rotates permanently about an axis, the rotatory pressures arising from the centrifugal forces of the solid are equal and contrary in each point of the axis, so that their sum is zero, and the areas described by every particle in the solid are proportional to the time; but if foreign forces disturb the rotation, the rotatory pressures on the axis of rotation are unequal, which causes a perpetual change of axis, and a variation in the areas described by the particles of the body, so that they are no longer proportional to the time. Thus the inconstancy of areas becomes a test of disturbing forces. In this disturbed rotation the body may be considered to have a permanent rotation during an instant only.

181. When three axes of a solid body are permanent axes of rotation, the rotatory pressures on them are zero; this is expressed by the equations $S.xydm = 0$; $S.xzdm = 0$; $S.yzdm = 0$; which characterize such axes. To show this, it is necessary to prove that when these equations hold, the rotation of the body round any one axis causes no twisting effort to displace that axis; for example, that the centrifugal forces developed by rotation round z, produce no rotatory pressure round y and x; and so for the other, and *vice versâ*.

Demonstration.—Let $r = \sqrt{x^2 + y^2}$ be the distance of a particle dm from z the axis of rotation, and let ω be the angular velocity of the particle. By article 171 $\omega = \dfrac{v}{r}$, therefore $\omega^2.r = \dfrac{v^2}{r}$ is the centrifugal force arising from rotation round z, and acting in the direction r. When resolved in the direction x, and multiplied by dm, it gives

$$\omega^2 rdm.\frac{x}{r} = \omega^2 xdm,$$

which, regarded as a force tending to turn the system round y, gives rotatory pressure $= w^2xzdm$, because it acts at the distance z from the axis y. Therefore when $S.xzdm = 0$, the whole effect is zero. Similarly, when $S.yzdm = 0$, the whole effect of the revolving system to turn round x vanishes. Therefore, in order that z should be permanent axis of rotation,

$$S.xzdm = 0, \quad Syzdm = 0.$$

In like manner, in order that y should be so,

$$S.xydm = 0, \quad Szydm = 0$$

must exist; and in order that x should be so

$$Syxdm = 0, \quad S.zxdm = 0$$

must exist, all of which are in fact only three different equations, namely, $S.xydm = 0$, $Sxzdm = 0$, $S.yzdm = 0$; (34)
and if these hold at once, x, y, z, will all be permanent axes of rotation.

Thus the impetus is as the square of the distance from the axis of rotation, and the rotatory pressures are simply as the distance from the same axis.

182. In order to ascertain whether a solid possesses any permanent axes of rotation, let the origin be a fixed point, and let x', y', z', be the co-ordinates of a particle dm, fixed in the solid, but revolving with it about its centre of gravity. The whole theory of rotation is derived from the equations (32) containing the principle of areas. These are the areas projected on the fixed co-ordinate planes xoy, xoz, yoz, fig. 48; but if ox', oy', oz' be the new axes that revolve with the solid, and if the values of x, y, z, given in article 163, be substituted, they will be the same sums, when projected on the new co-ordinate planes $x'oy'$, $x'oz'$, $y'oz'$. The angles θ, ψ, and ϕ, introduced by this change are arbitrary, so that the position of the new axes ox', oy', oz', in the solid, remains indeterminate; and these three angles may be made to fulfil any conditions of the problem.

183. The equations of rotation will take the most simple form if we suppose $x'\,y'\,z'$ to be the principal axes of rotation, which they will become if the values of θ, ψ, and ϕ can be so assumed as to make the rotatory pressures $S.x'z'dm$, $Sx'y'dm$, $Sy'z'dm$, zero at once, then the equations (32) of the areas, when transformed to functions of x', y', z', and deprived of these terms, will determine the rotation of the body about its principal, or permanent axes of rotation, x', y', z'.

184. If the body has no principal axes of rotation, it will be impossible to obtain such values of θ, ϕ and ψ, as will make the rotatory pressures zero; it must therefore be demonstrated whether or not it be possible to determine the angles in question, so as to fulfil the requisite condition.

185. To determine the existence and position of the principal axes of the body, or the angles θ, ϕ, and ψ, so that

$$\text{S}.x'y'dm = 0; \quad \text{S}x'z'dm = 0; \quad \text{S}y'z'dm = 0.$$

Let values of x', y', z', in functions of x, y, z, determined from the equations in article 163 be substituted in the preceding expressions, then if to abridge,

$$\text{S}.x^2dm = l^2 \quad \text{S}y^2dm = n^2 \quad \text{S}z^2dm = s^2$$
$$\text{S}.xydm = f \quad \text{S}xzdm = g \quad \text{S}yzdm = h,$$

there will result

$$\cos\phi.\text{S}.x'z'dm - \sin\phi.\text{S}.y'z'dm =$$
$$(l^2 - n^2)\sin\theta\sin\psi\cos\psi + f\sin\theta(\cos^2\psi - \sin^2\psi)$$
$$+ \cos\theta(g\cos\psi - h\sin\psi); \qquad (35)$$
$$\sin\phi\,\text{S}.x'z'dm + \cos\phi\,\text{S}.y'z'dm =$$
$$\sin\theta\cos\theta\{l^2\sin^2\psi + n^2\cos^2\psi - s^2 + 2f\sin\psi\cos\psi\}$$
$$+ (\cos^2\theta - \sin^2\theta).(g\sin\psi + h\cos\psi).$$

If the second members of these be made equal to zero, there will be

$$\tan\theta = \frac{h\sin\psi - g\cos\psi}{(l^2 - n^2)\sin\psi\cos\psi + f(\cos^2\psi - \sin^2\psi)}, \text{ and}$$

$$\tfrac{1}{2}\tan 2\theta = \frac{g\sin\psi + h\cos\psi}{s^2 - l^2\sin^2\psi - n^2\cos^2\psi - 2f\sin\psi\cos\psi},$$

but $$\tfrac{1}{2}\tan 2\theta = \frac{\tan\theta}{1 - \tan^2\theta},$$

by the arithmetic of sines; hence, equating these two values of $\frac{1}{2}\tan 2\theta$, and substituting for $\tan\theta$ its value in ψ; then if to abridge, $u = \tan\psi$, after some reduction it will be found that

$$0 = (gu + h)(hu - g)^2 +$$
$$\{(l^2 - n^2)u + f(1 - u^2)\}.\{(hs^2 - hl^2 + fg)u + gn^2 - gs^2 - hf\};$$

where u is of the third degree. This equation having at least one real root, it is always possible to render the first members of the two equations (35) zero at the same time, and consequently

$$(\text{S}.x'z'dm)^2 + (\text{S}.y'z'dm)^2 = 0.$$

But that can only be the case when $Sx'z'dm = 0$, $Sy'z'dm = 0$. The value of $u = \tan\psi$, gives ψ, consequently $\tan\theta$ and θ become known.

It yet remains to determine the condition $S.x'y'dm = 0$, and the angle ϕ. If substitution be made in $S.x'y'dm = 0$, for x' and y' from article 163, it will take the form $H\sin 2\phi + L\cos 2\phi$, H and L being functions of the known quantities θ and ψ; as it must be zero, it gives

$$\tan 2\phi = -\frac{L}{H};$$

and thus the three axes ox', oy', oz', determined by the preceding values of θ, ψ, and ϕ, satisfy the equations

$$Sx'z'dm = 0, \quad Sy'z'dm = 0, \quad Sx'y'dm = 0.$$

186. The equation of the third degree in u seems to give three systems of principal axes, one for each value of u; but u is the tangent of the angle formed by the axis x with the line of intersection of the plane xy with that of $x'y'$; and as any one of the three axes, x', y', z', may be changed into any other of them, since the preceding equations will still be satisfied, therefore the equation in u will determine the tangent of the angle formed by the axis x with the line of intersection of xy and $x'y'$, with that of xy and $x'z'$, or with that of xy and $y'z'$. Consequently the three roots of the equation in u are real, and belong to the same system of axes.

187. Whence every body has at least one system of principal and rectangular axes, round any one of which if the body rotates, the opposite centrifugal forces balance each other. This theorem was first proposed by Segner in the year 1755, and was demonstrated by Albert Euler in 1760.

188. The position of the principal axes ox', oy', oz', in the interior of the solid, is now completely fixed; and if there be no disturbing forces, the body will rotate permanently about any one of them, as oz', fig. 48; but if the rotation be disturbed by foreign forces, the solid will only rotate for an instant about oz', and in the next element of time it will rotate about oz'', and so on, perpetually changing. Six equations are therefore required to

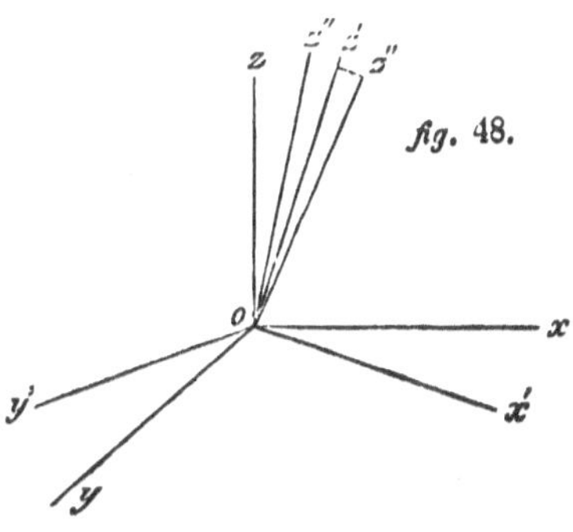

fig. 48.

fix the position of the instantaneous axis oz''; three will determine its place with regard to the principal axes ox', oy', oz', and three more are necessary to determine the position of the principal axes themselves in space, that is, with regard to the fixed co-ordinates ox, oy, oz. The permanency of rotation is not the same for all the three axes, as will now be shown.

189. The principal axes possess this property—that the moment of inertia of the solid is a maximum for one of these, and a minimum for another. Let x', y', z', be the co-ordinates of dm, relative to the three principal axes, and let x, y, z, be the co-ordinates of the same element referred to any axes whatever having the same origin. Now if

$$C' = \text{S}\,(x^2 + y^2)\,dm$$

be the moment of inertia relatively to one of these new axes, as z, then substituting for x and y their values from article 163, and making

$$A = S\,(y'^2 + z'^2).dm;\quad B = S\,(x'^2 + z'^2)dm;\quad C = S\,(x'^2 + y'^2)dm;$$

the value of C' will become

$$C' = A\sin^2\theta\,\sin^2\phi + B\sin^2\theta\cos^2\phi + C\cos^2\theta,$$

in which $\sin^2\theta\sin^2\phi$, $\sin^2\theta\cos^2\phi$, $\cos^2\theta$,

are the squares of the cosines of the angles made by ox', oy', oz', with oz; and A, B, C, are the moments of inertia of the solid with regard to the axes x', y', and z', respectively. The quantity C' is less than the greatest of the three quantities A, B, C, and exceeds the least of them; the greatest and the least moments of inertia belong therefore, to the principal axes. In fact, C' must be less than the greatest of the three quantities A, B, C, because their joint coefficients are always equal to unity; and for a similar reason it is always greater than the least.

190. When $A = B = C$, then all the axes of the solid are principal axes, and it will rotate permanently about any one of them. The sphere of uniform density is a solid of this kind, but there are many others.

191. When two of the moments of inertia are equal, as $A = B$, then

$$C' = A\sin^2\theta + C\cos^2\theta;$$

and all the moments of inertia in the same plane with these are equal: hence all the axes situate in that plane are principal axes. The ellipsoid of revolution of uniform density is of this kind; all the axes in the plane of its equator being principal axes.

192. An ellipsoid of revolution is formed by the rotation of an ellipse ABCD about its minor axis BD. Then AC is its equator.

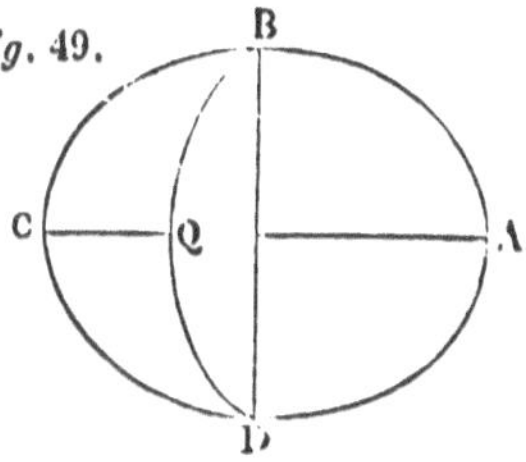

fig. 49.

When the moments of inertia are unequal, the rotation round the axes which have their moment of inertia a maximum or minimum is stable, that is, round the least or greatest axis; but the rotation is unstable round the third, and may be destroyed by the slightest cause. If stable rotation be slightly deranged, the body will never deviate far from its equilibrium; whereas in unstable rotation, if it be disturbed, it will deviate more and more, and will never return to its former state.

193. This theorem is chiefly of importance with regard to the rotation of the earth. If xoy (*fig.* 46) be the plane of the ecliptic, and z its pole; $x'oy'$ the plane of the equator, and z' its pole: then oz' is the axis of the earth's rotation, $zoz' = \theta$ is the obliquity of the ecliptic, γN the line of the equinoxes, and γ the first point of Aries: hence $xo\gamma = \psi$ is the longitude of ox, and $x'o\gamma = \phi$ is the longitude of the principal revolving axis ox', or the measure of the earth's rotation: oz' is therefore one of the permanent axes of rotation.

The earth is flattened at the poles, therefore oz' is the least of the permanent axes of rotation, and the moment of inertia with regard to it, is a maximum. Were there no disturbing forces, the earth would rotate permanently about it; but the sun and moon, acting unequally on the different particles, disturb its rotation. These disturbing forces do not sensibly alter the velocity of rotation, in which neither theory nor observation have detected any appreciable variation; nor do they sensibly displace the poles of rotation on the surface of the earth; that is to say, the axis of rotation, and the plane of the equator which is perpendicular to it, always meet the surface in the same points; but these forces alter the direction of the polar axis in space, and produce the phenomena of precession and nutation; for the earth rotates about oz'', fig. 50, while oz'' revolves about its mean place oz', and at the same time oz' describes a cone about oz; so that the motion of the axis of rotation

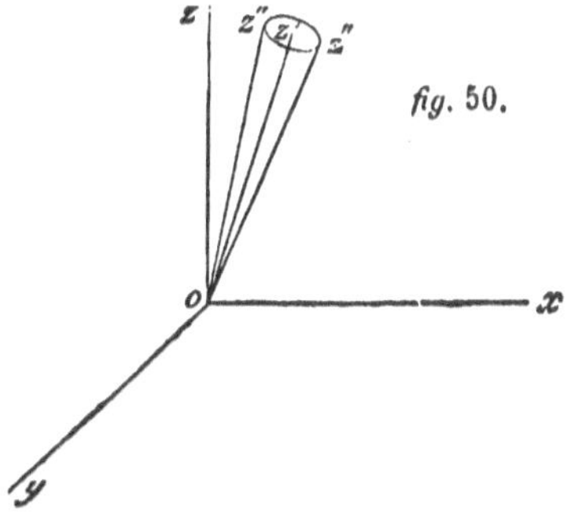

fig. 50.

is very complicated. That axis of rotation, of which all the points remain at rest during the time dt, is called an instantaneous axis of rotation, for the solid revolves about it during that short interval, as it would do about a fixed axis.

The equations (32) must now be so transformed as to give all the circumstances of rotatory motion.

194. The equations in article 163, for changing the co-ordinates, will become

$$\begin{aligned} x &= ax' + by' + cz' \\ y &= a'x' + b'y' + c'z' \\ z &= a''x' + b''y' + c''z'. \end{aligned} \qquad (36)$$

If to abridge

$$\begin{aligned} a &= \cos\theta \sin\psi \sin\phi + \cos\psi \cos\phi \\ b &= \cos\theta \sin\psi \cos\phi - \cos\psi \sin\phi \\ c &= \sin\theta \sin\psi \\ a' &= \cos\theta \cos\psi \sin\phi - \sin\psi \cos\phi \\ b' &= \cos\theta \cos\psi \cos\phi + \sin\psi \sin\phi \\ c' &= \sin\theta \cos\psi \\ a'' &= -\sin\theta \sin\phi \\ b'' &= -\sin\theta \cos\phi \\ c'' &= \cos\theta, \end{aligned}$$

where a, b, c are the cosines of the angles made by x with x', y', z'; a', b', c' are the cosines of the angles made by y with x', y', z'; and a'', b'', c'' are the cosines of the angles made by z with the same axes

Whatever the co-ordinates of dm may be, since they have the same origin, $x^2 + y^2 + z^2 = x'^2 + y'^2 + z'^2$.

By means of these it may be found that

$$\begin{aligned} a^2 + a'^2 + a''^2 &= 1 & ab + a'b' + a''b'' &= 0 \\ b^2 + b'^2 + b''^2 &= 1 & ac + a'c' + a''c'' &= 0 \\ c^2 + c'^2 + c''^2 &= 1 & bc + b'c' + b''c'' &= 0. \end{aligned}$$

In the same manner, to obtain x', y', z', in functions of x, y, z,

$$\begin{aligned} x' &= ax + a'y + a''z \\ y' &= bx + b'y + b''z \\ z' &= cx + c'y + c''z, \end{aligned} \qquad (37)$$

whence the equations of condition,

$$\begin{aligned} a^2 + b^2 + c^2 &= 1 & aa' + bb' + cc' &= 0 \\ a'^2 + b'^2 + c'^2 &= 1 & aa'' + bb'' + cc'' &= 0 \\ a''^2 + b''^2 + c''^2 &= 1 & a'a'' + b'b'' + c'c'' &= 0, \end{aligned}$$

six of the quantities a, b, c, a', b', c', a'', b'', c'', are determined by the preceding equations, and three remain arbitrary.

If values of x', y', z', found from equations (36) be compared with their values in equations (37), there will result

$$\begin{array}{lll} a = b'c'' - b''c' & a' = b''c - bc'' & a'' = bc' - b'c \\ b = a''c' - a'c'' & b' = ac'' - a''c & b'' = a'c - ac' \quad (38) \\ c = a'b'' - a''b' & c' = a''b - ab'' & c'' = ab' - a'b. \end{array}$$

195. The axes x', y', z' retain the same position in the interior of the body during its rotation, and are therefore independent of the time; but the angles $a, b, c, a', b', c', a'', b'', c''$, vary with the time; hence, if values of y, z, $\frac{dy}{dt}$ $\frac{dz}{dt}$, from equations (36,) be substituted in the first of equations (32), it will become

$$\begin{aligned} \text{S}\Big\{&\left(\frac{a'da''-a''da'}{dt}\right)x'^2+\left(\frac{b'db''-b''db'}{dt}\right)y'^2+\left(\frac{c'dc''-c''dc'}{dt}\right)z'^2 \\ &+\left(\frac{a'db''-b''da'+b'da''-a''db'}{dt}\right)x'y' \\ &+\left(\frac{a'dc''-c''da'+c'da''-a''dc'}{dt}\right)x'z' \\ &+\left(\frac{b'dc''-c''db'+c'db''-b''dc'}{dt}\right)y'z'\Big\}\,dm = \int M.dt. \end{aligned}$$

If a', a'', b', &c. be eliminated from this equation by their values in (38), and if to abridge

$$\begin{aligned} cdb + c'db' + c''db'' &= -bdc - b'dc' - b''dc'' = pdt \\ adc + a'dc' + a''dc'' &= -cda - c'da' - c''da'' = qdt \qquad (39) \\ bda + b'da' + b''da'' &= -adb - a'db' - a''db'' = rdt \end{aligned}$$

$A = \text{S}\,(y'^2 + z'^2)\,dm$; $B = \text{S}\,(x'^2 + z'^2)\,dm$; $C = \text{S}\,(x'^2 + y'^2)\,dm$.

And if $\quad \text{S}.x'y'\,dm = 0 \quad \text{S}.x'zdm = 0 \quad \text{S}.y'z'dm = 0$,

it will be found that

$$aAp + bBq + cCr = \int Mdt;$$

by the same process it may be found that

$$a'Ap + b'Bq + c'Cr = \int M'dt,$$
$$a''Ap + b''Bq + c''Cr = \int M''dt.$$

196. If the differentials of these three equations be taken, making all the quantities vary except A, B, and C, then the sum of the first differential multiplied by a, plus the second multiplied by a', plus the third multiplied by a'', will be

$$A\frac{dp}{dt} + (C - B).qr = aM + a'M' + a''M'',$$

in consequence of the preceding relations between $a\ a'\ a''$, $b\ b'\ b''$, $c\ c'\ c''$, and their differentials. By a similar process the coefficients $b\ b'\ b''$, &c., may be made to vanish, and then if

$$aM + a'M' + a''M'' = N$$
$$bM + b'M' + b''M'' = N'$$
$$cM + c'M' + c''M'' = N''$$

the equations in question are transformed to

$$A\frac{dp}{dt} + (C - B).qr = N$$

$$B\frac{dq}{dt} + (A - C).rp = N' \qquad (40)$$

$$C\frac{dr}{dt} + (B - A).pq = N''$$

And if a, a', a'', b, b', &c., and their differentials, be replaced by their functions in ϕ, ψ, and θ, given in article 194, the equations (39) become

$$pdt = \sin\phi\ \sin\theta.d\psi - \cos\phi.d\theta$$
$$qdt = \cos\phi\ \sin\theta.d\psi + \sin\phi.d\theta \qquad (41)$$
$$rdt = d\phi - \cos\theta.d\psi.$$

197. These six equations contain the whole theory of the rotation of the planets and their satellites, and as they have been determined in the hypothesis of the rotatory pressures being zero, they will give their rotation nearly about their principal axes.

198. The quantities p, q, r determine oz'', the position of the real and instantaneous axis of rotation, with regard to its principal axis oz'; when a body has no motion but that of rotation, all the points in a permanent axis of rotation remain at rest; but in an instantaneous axis of rotation the axis can only be regarded as at rest from one instant to another.

If the equations (36) for changing the co-ordinates, be resumed, then with regard to the axis of rotation,

$$dx = 0,\ dy = 0,\ dz = 0,$$

since all its points are at rest; therefore the indefinitely small spaces moved over by that axis in the direction of these co-ordinates being zero, the equations in question become,

$$x'da + y'db + z'dc = 0,$$
$$x'da' + y'db' + z'dc' = 0,$$
$$x'da'' + y'db'' + z'dc'' = 0,$$

which will determine x', y', z', and consequently oz'' the axis in question.

For if the first of these equations be multiplied by c, the second by c', and the third by c'', their sum is

$$py' - qz' = 0. \qquad (42)$$

Again, if the first be multiplied by b, the second by b', and the third by b'', their sum is

$$rx' - p'z = 0. \qquad (43)$$

Lastly, if the first equation be multiplied by a, the second by a', and the third by a'', their sum is

$$qz' - ry' = 0.$$

The last of these is contained in the two first, which are the equations to a straight line oz'', which forms, with the principal axes x', y', z', angles whose cosines are

$$\frac{p}{\sqrt{p^2+q^2+r^2}}; \quad \frac{q}{\sqrt{p^2+q^2+r^2}}; \quad \frac{r}{\sqrt{p^2+q^2+r^2}}; \qquad (44)$$

for the two last give

$$x'^2 = z'^2\,\frac{p^2}{r^2};\ y'^2 = z'^2\,\frac{q^2}{r^2};$$

whence $$x'^2 + y'^2 + z'^2 = z'^2\left\{\frac{q^2+r^2+p^2}{r^2}\right\};$$

and therefore

$$\frac{z'}{\sqrt{x'^2+y'^2+z'^2}} = \frac{r}{\sqrt{p^2+q^2+r^2}}.$$

But $$oz'' = \sqrt{x'^2+y'^2+z'^2},$$

and $$oz'' : oc :: 1 : \cos z''oc;$$

then if x', y', z', be the co-ordinates of the point z'',

$$\cos z''oc = \frac{z'}{\sqrt{x'^2+y'^2+z'^2}} = \frac{r}{\sqrt{p^2+q^2+r^2}}.$$

In the same manner

$$\cos z''ox' = \frac{p}{\sqrt{p^2+q^2+r^2}}$$

and $$\cos z''oy' = \frac{q}{\sqrt{p^2+q^2+r^2}}.$$

Consequently oz'' is the instantaneous axis of rotation.

fig. 51.

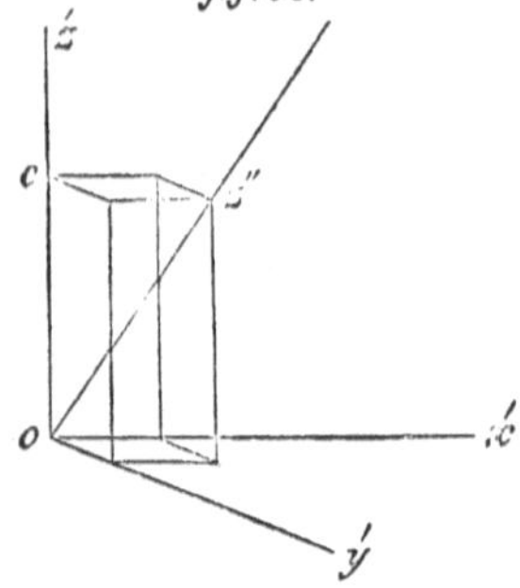

201. The angular velocity of rotation is also given by these quantities. If the object be to determine it for a point in the axis, as for example where $oc = 1$, then

$$x' = 0,\ y' = 0,$$

and the values of dx, dy, dz give, when divided by dt,

$$\frac{d\psi}{dt}\sin\theta,\ \frac{d\theta}{dt}\cos\theta,\ -\frac{d\theta}{dt}\sin\theta,$$

for the components of the velocity of a particle; hence the resulting velocity is

$$\frac{\sqrt{d\theta^2 + d\psi^2 \sin^2\theta}}{dt} = \sqrt{q^2 + r^2},$$

which is the sum of the squares of the two last of equations (41).

199. But in order to obtain the angular velocity of the body, this quantity must be divided by the distance of the particle at c' from the axis oz''; but this distance is evidently equal to the sine of $z''oc$, the angle between oz' and oz'', the principal and instantaneous axes of rotation; but

$$\frac{r}{\sqrt{p^2 + q^2 + r^2}}$$

is the cosine of this angle; hence

$$\sqrt{1 - \frac{r^2}{p^2 + q^2 + r^2}},\ \text{or}\ \frac{\sqrt{q^2 + p^2}}{\sqrt{p^2 + q^2 + r^2}},\ \text{is the sine;}$$

and therefore

$$\sqrt{p^2 + q^2 + r^2}$$

is the angular velocity of rotation. Thus, whatever may be the rotation of a body about a point that is fixed, or one considered to be fixed, the motion can only be rotation about an axis that is fixed during an instant, but may vary from one instant to another.

200. The position of the instantaneous axis with regard to the three principal axes, and the angular velocity of rotation, depend on p, q, r, whose determination is very important in these researches; and as they express quantities independent of the situation of the fixed plane xoy, they are themselves independent of it.

201. Equations (40) determine the rotation of a solid troubled by the action of foreign forces, as for example, that of the earth when

disturbed by the sun and moon. But the same equations will also determine the rotation of a solid, when not disturbed in its rotation.

Rotation of a Solid not subject to the action of Disturbing Forces, and at liberty to revolve freely about a Fixed Point, being its Centre of Gravity, or not.

202. Values of p, q, r in terms of the time must be obtained, in order to ascertain all the circumstances of rotation at every instant.

If we suppose that there are no disturbing forces, the areas are constant: hence the equations (40) become

$$\begin{aligned} A.dp + (C - B).q.r.dt &= 0; \\ B.dq + (A - C).r.p.dt &= 0; \\ C.dr + (B - A).p.q.dt &= 0. \end{aligned} \qquad (45)$$

If the first be multiplied by p, the second by q, and the third by r, their sum is

$$Apdp + Bqdq + Crdr = 0,$$

and its integral is

$$Ap^2 + Bq^2 + Cr^2 = k^2, \qquad (46)$$

k^2 being a constant quantity. Again, if the three equations be multiplied respectively by Ap, Bq, Cr, and integrated, they give

$$A^2p^2 + B^2q^2 + C^2r^2 = h^2, \qquad (47)$$

a constant quantity. This equation contains the principle of the preservation of impetus or living force which is constant in conformity with article 148. From these two integrals are obtained:

$$p^2 = \frac{h^2 - Bk + (B - C) . Cr^2}{A(A-B)} \qquad (48)$$

$$q^2 = \frac{h^2 - Ak + (A - C) . Cr^2)}{B(B-A)}.$$

By the substitution of these values of p and q, the last of equations (45) when resolved according to dt, gives

$$dt = \frac{Cdr . \sqrt{AB}}{\sqrt{\{(h^2 - Bk + (B - C).Cr^2).(-h^2 + Ak + (C - A)).Cr^2\}}} \qquad (49)$$

This equation will give by quadratures the value of t in r, and reciprocally the value of r in t; and thus by the substitution of this value of r in equations (48) the three quantities p, q and r become known in functions of the time.

This equation can only be integrated when any two of the moments of inertia are equal, either when

$$A = B, \quad B = C, \quad \text{or } A = C;$$

in each of these cases the solid is a spheroid of revolution.

203. Thus p, q, r, being known functions of the time, the angular velocity of the solid, and its rotation with regard to the principal axes, are known at every instant.

204. This however is not sufficient. To become acquainted with all the circumstances of rotation, it is requisite to know the position of the principal axes themselves with regard to quiescent space, that is, their position relatively to the fixed axes x, y, z. But for that purpose the angles ϕ, ψ, and θ, must be determined in functions of the time, or, which is the same thing, in functions of p, q, r, which may now be regarded as known quantities.

If the first of equations (45) be multiplied by a, the second by b, and the third by c, their sum when integrated, in consequence of the relations between the angles in article 194, is

$$aAp + bBq + cCr = l, \text{ by a similar process}$$

$$a'Ap + b'Bq + c'Cr = l', \qquad (50)$$

$$a''Ap + b''Bq + c''Cr = l'',$$

l, l', l'', being arbitrary constant quantities. These equations coincide with those in article 195, and contain the principle of areas. They are not however three distinct integrals, for the sum of their squares is

$$A^2p^2 + B^2q^2 + C^2r^2 = l^2 + l'^2 + l''^2,$$

in consequence of the equations in article 194. But this is the same with (47); hence $\quad l^2 + l'^2 + l''^2 = h^2$

being an equation of condition, equations (50) will only give values of two of the angles ϕ, ψ, and θ.

The constant quantities l, l', l'', correspond with c, c', c'', in article 164, therefore $\quad \frac{1}{2}t\sqrt{l^2 + l'^2 + l''^2}$

is the sum of the areas described in the time t by the projection of each particle of the body on the plane on which that sum is a maximum. If xoy be that plane, l and l' are zero: therefore, in every solid body in rotation about an axis, there exists a plane, on which the sum of the areas described by the projections of the particles of the solid during a finite time is a maximum. It is called the Invariable Plane, because it remains parallel to itself during the motion of the body: it is also named the plane of the Greatest Rotatory Pressure.

Since $$l = 0, \quad l' = 0, \quad l'' = h,$$
if the first of equations (50) be multiplied by a, the second by a', and the third by a'', in consequence of the equations in article 194, their sum is
$$a'' = \frac{Ap}{h};$$
in the same manner it will be found that
$$b'' = \frac{Bq}{h}, \quad c'' = \frac{Cr}{h};$$
or, substituting the values of a'', b'', c'', from article 194,
$$\sin\theta' \sin\phi' = -\frac{Ap}{h}, \sin\theta' \cos\phi' = -\frac{Bq}{h}, \cos\theta' = \frac{Cr}{h}. \quad (51)$$
The accented angles θ', ϕ', ψ', relate to the invariable plane, and θ, ϕ, ψ, to the fixed plane.

Because p, q, r, are known functions of the time, ϕ' and θ' are determined, and if $d\theta$ be eliminated between the two first of equation (41), the result will be
$$\sin^2\theta'.d\psi' = \sin\theta' . \sin\phi'.pdt + \sin\theta' . \cos\phi'.qdt.$$
But in consequence of equations (51), and because
$$Ap^2 + Bq^2 = k - Cr^2,$$
$$d\psi' = \frac{Cr^2 - k}{h^2 - C^2r^2} . hdt;$$
and as r is given in functions of the time by equation (49), ψ' is determined.

Thus, p, q, r, ψ', θ', and ϕ', are given in terms of the time: so that the position of the three principal axes with regard to the fixed axes, ox, oy, oz; and the angular velocity of the body, are known at every instant.

205. As there are six integrations, there must be six arbitrary constant quantities for the complete solution of the problem. Besides h and k, two more will be introduced by the integration of dt and $d\psi'$. Hence two are still required, because by the assumption of xoy for the invariable plane, l and l' become zero.

Now the three angles, ψ', ϕ', θ', are given in terms of p, q, r, and these last are known in terms of the time; hence ψ', ϕ', θ', (fig. 49,) are known with regard to the invariable plane xoy: and by trigonometry it will be easy to determine values of ψ, ϕ, θ, with regard to any fixed plane whatever, which will introduce two new arbitrary quantities, making in all six, which are requisite for the complete solution of the problem.

206. These two new arbitrary quantities are the inclination of the invariable plane on the fixed plane in question, and the angular distance of the line of intersection of these two planes from a line arbitrarily assumed on the fixed plane; and as the initial position of the fixed plane is supposed to be given, the two arbitrary quantities are known.

If the position of the three principal axes with regard to the invariable plane be known at the origin of the motion, ϕ', θ', will be given, and therefore p, q, r, will be known at that time; and then equation (46) will give the value of k.

The constant quantity arising from the integration of dt depends on the arbitrary origin or instant whence the time is estimated, and that introduced by the integration of $d\psi'$ depends on the origin of the angle ψ', which may be assumed at pleasure on the invariable plane.

207. The determination of the sixth constant quantity h is very interesting, as it affords the means of ascertaining the point in which the sun and planets may be supposed to have received a primitive impulse, capable of communicating to them at once their rectilinear and rotatory motions.

The sum of the areas described round the centre of gravity of the solid by the radius of each particle projected on a fixed plane, and respectively multiplied by the particles, is proportional to the moment of the primitive force projected on the same plane; but this moment is a maximum relatively to the plane which passes through the point of primitive impulse and centre of gravity, hence it is the invariable plane.

208. Let G, fig. 52, be the centre of gravity of a body of which ABC is a section, and suppose that it has received an impulse in the plane ABC at the distance GF, from its centre of gravity; it will move forward in space at the same time that it will rotate about an axis perpendicular to the plane ABC. Let v be the velocity generated in the centre of gravity by the primitive impulse; then if m be the mass of the body, $m.v.$GF will be the moment of this

fig. 52.

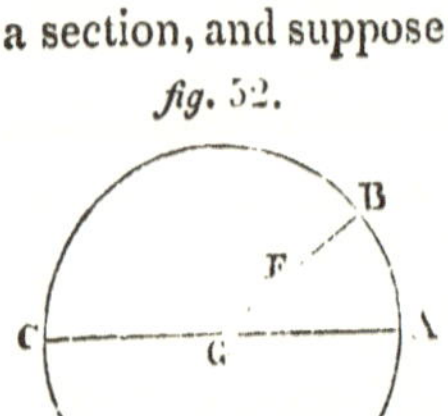

impulse, and multiplying it by $\frac{1}{2}t$, the product will be equal to the sum of the areas described during the time t; but this sum was shown to be $\frac{1}{2}t\sqrt{l^2+l'^2+l''^2}$;

hence $$\sqrt{l^2+l'^2+l''^2} = m.v.\text{GF} = h;$$

which determines the sixth arbitrary constant quantity h. Were the angular velocity of rotation, the mass of the body and the velocity of its centre of gravity known, the distance GF, the point of primitive impulse, might be determined.

209. It is not probable that the primitive impulse of the planets and other bodies of the system passed exactly through their centres of gravity; most of them are observed to have a rotatory motion, though in others it has not been ascertained, on account of their immense distances, and the smallness of their volumes. As the sun rotates about an axis, he must have received a primitive impulse not passing through his centre of gravity, and therefore it would cause him to move forward in space accompanied by the planetary system, unless an impulse in the contrary direction had destroyed that motion, which is by no means likely. Thus the sun's rotation leads us to presume that the solar system may be in motion.

210. Suppose a planet of uniform density, whose radius is R, to be a sphere revolving round the sun in S, at the distance SG or $\bar{r}$, with an angular velocity represented by u, then the velocity of the centre of gravity will be $v = u\bar{r}$.

Imagine the planet to be put in motion by a primitive impulse, passing through the point F, fig. 53, then the sphere will rotate about an axis perpendicular to the invariable plane, with an angular velocity equal to r, for the components q and p at right angles to that plane will be zero; hence, the equation

fig. 53.

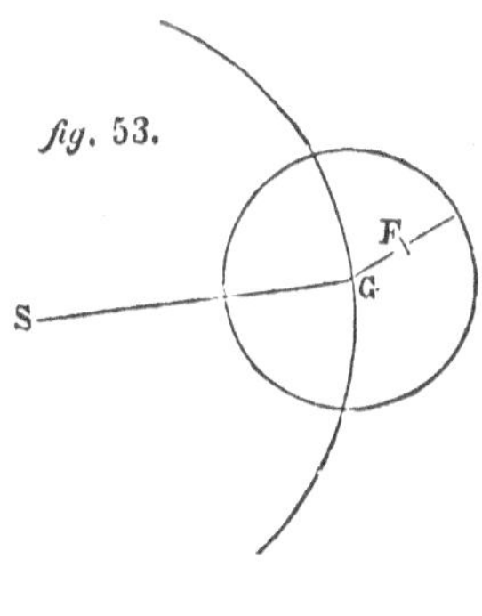

$$\sqrt{l^2+l'^2+l''^2} = m.v.\text{GF}.$$

becomes $l'' = m\,u\,\bar{r}.\text{FG}$; and $l'' = r\text{C}$.

The centre of gyration is that point of a body in rotation, into which, if all the particles were condensed, it would retain the same degree of rotatory power. It is found that the square of the radius of gyration in a sphere, is equal to $\frac{2}{5}$ of the square of its semi-diameter;

hence the rotatory inertia C becomes $\frac{2}{5} m \text{R}^2$,

thus $l'' = r \times \frac{2}{5} m \text{R}^2$, and $\text{GF} = \frac{2}{5} \cdot \frac{\text{R}^2}{r} \cdot \frac{r}{u}$.

211. Hence, if the ratio of the mean radius of a planet to its mean distance from the sun, and the ratio of its angular velocity of rotation to its angular velocity in its orbit, could be ascertained, the point in which the primitive impulse was given, producing its motion in space, might be determined.

212. Were the earth a sphere of uniform density, the ratio $\frac{\text{R}}{r}$ would be 0.000042665 ; and the ratio of its rotatory velocity to that in its orbit is known by observation to be 366.25638, whence GF $= \frac{\text{R}}{160}$; and as the mean radius of the earth is about 4000 miles, the primitive impulse must have been given at the distance of 25 miles from the centre. However, as the density of the earth is not uniform, but decreases from the centre to the surface, the distance of the primitive impulse from its centre of gravity must have been something less.

213. The rotation of the earth has established a relation between time and the arcs of a circle. Every point in the surface of the earth passes through 360° in 24 hours; and as the rotation is uniform, the arcs described are proportional to the time, so that one of these quantities may represent the other. Thus, if α be an arc of any number of degrees, and t the time employed to describe it, $360° : \alpha :: 24 : t$: hence $\alpha = \frac{360}{24} t$; or, if the constant co-efficient $\frac{360}{24}$ be represented by n, $\alpha = nt$, and $\sin \alpha = \sin nt$, $\cos \alpha = \cos nt$.

In the same manner the periodic time of the moon being 27.3 days nearly, an arc of the moon's orbit would be $\frac{360}{27.3} t$, and may also be expressed by nt. Thus, n may have all values, so that nt is a general expression for any arc that increases uniformly with the time.

214. The motions of the planets are determined by equations of these forms,

$$\frac{d^2u}{dt^2} + n^2u = R$$

$$\frac{d^2u}{dt^2} + n^2u = 0,$$

which are only transformations of the general equation of the motions of a system of bodies. The integrals of both give a value of u in terms of the sines and cosines of circular arcs increasing with the time; the first by approximation, but the integral of the second will be obtained by making $u = c^x$, c being the number whose Napierian logarithm is unity.

Whence $$d^2u = c^x(d^2x + dx^2),$$

and the equation in question becomes

$$d^2x + dx^2 + n^2dt^2 = 0.$$

Let $$dx = ydt, \text{ then } d^2x = dydt,$$

since the element of the time is constant, which changes the equation to $$dy + dt\,(n^2 + y^2) = 0.$$

If $y = m$ a constant quantity, $dm = dy = 0$,

hence $$n^2 + m^2 = 0;$$

whence $$m = \mp n\sqrt{-1},$$

but $$dx = ydt = \mp ndt\sqrt{-1},$$

the integral of which is

$$x = \mp nt\sqrt{-1}.$$

As x has two values, $u = c^x$ gives

$$u = bc^{nt\sqrt{-1}}, \quad \text{and } u = b'c^{-nt\sqrt{-1}};$$

and because either of these satisfies the conditions of the problem, their sum $$u = bc^{nt\sqrt{-1}} + b'c^{-nt\sqrt{-1}},$$

also satisfies the conditions and is the general solution, b and b' being arbitrary constant quantities.

But $$c^{nt\sqrt{-1}} = \cos nt + \sqrt{-1}\sin nt,$$

$$c^{-nt\sqrt{-1}} = \cos nt - \sqrt{-1}\sin nt.$$

Hence $$u = (b + b')\cos nt + (b - b')\sqrt{-1}\sin nt.$$

Let $$b + b' = M\sin\epsilon; \quad (b - b')\sqrt{-1} = M\cos\epsilon;$$

and then $$u = M\{\sin\epsilon\cos nt + \cos\epsilon\sin nt)$$

or $$u = M\sin(nt + \epsilon),$$

*

which is the integral required, because M and ϵ are two arbitrary constant quantities.

215. Since a sine or cosine never can exceed the radius, $\sin.(nt + \epsilon)$ never can exceed unity, however much the time may increase; therefore u is a periodic quantity whose value oscillates between fixed limits which it never can surpass. But that would not be the case were n an imaginary quantity; for let

$$n = \alpha \pm \beta\sqrt{-1};$$

then the two values of x become

$$x = \beta t + \alpha t\sqrt{-1} \quad x = \beta t - \alpha t\sqrt{-1},$$

consequently,

$$c^{\beta t + \alpha t\sqrt{-1}} = c^{\beta t}.c^{\alpha t\sqrt{-1}} = c^{\beta t}\{\cos \alpha t + \sqrt{-1}\sin \alpha t\}$$

$$c^{\beta t - \alpha t\sqrt{-1}} = c^{\beta t}.c^{-\alpha t\sqrt{-1}} = c^{\beta t}\{\cos \alpha t - \sqrt{-1}\sin \alpha t\}$$

whence $u = c^{\beta t}\{(b + b')\cos \alpha t + (b - b')\sqrt{-1}\sin \alpha t\}$

or substituting for $b + b'$; $(b - b')\sqrt{-1}$;

$$u = c^{\beta t}.M.\sin(\alpha t + \epsilon);$$

But $$c^{\beta t} = 1 + \beta t + \tfrac{1}{2}\beta^2 t^2 + \frac{1}{2.3}\beta^3 t^3 + \&c.$$

therefore $c^{\beta t}$ increases indefinitely with the time, and u is no longer a periodic function, but would increase to infinity.

Were the roots of n^2 equal, then $x = \beta t$, and

$$u = C.c^{\beta t}, \text{ C being constant.}$$

Thus it appears that if the roots of n^2 be imaginary or equal, the function u would increase without limit.

These circumstances are of the highest importance, because the stability of the solar system depends upon them.

Rotation of a Solid which turns nearly round one of its principal Axes, as the Earth and the Planets, but not subject to the action of accelerating Forces.

216. Since the axis of rotation oz'' is very near oz', fig. 50, the angle $z'oz''$ is so small, that its cosine $\frac{r}{\sqrt{p^2 + q^2 + r^2}}$ differs but little from unity; hence p and q are so minute that their product may be omitted, which reduces equations (45) to

$$Cdr = 0,$$
$$Adp + (C - B)\,qrdt = 0,$$
$$Bdq + (A - C)\,prdt = 0;$$

the first shows the angular velocity to be uniform, and the two last give

$$\frac{d^2q}{dt^2} + \frac{(A-C)}{B}r\cdot\frac{dp}{dt} = 0; \quad \frac{dp}{dt} = \frac{(B-C)}{A}qr;$$

hence if the constant quantity

$$\frac{(A-C)(B-C)}{AB}r^2 = n^2,$$

the result will be $$\frac{d^2q}{dt^2} + n^2q = 0;$$

and by article 214, $$q = M'\cos(nt+g).$$

In the same manner $$p = M\sin(nt+g);$$

whence $$M' = M\,.\sqrt{\frac{A(A-C)}{B(B-C)}}.$$

217. If oz'' the real axis of rotation coincides with oz', the principal axis in the beginning of the motion, then q and p are zero; hence also, $M = 0$, and $M' = 0$. It follows therefore, that in this case q and p will always be zero, and the axis oz'' will always coincide with oz'; whence, if the body begins to turn round one of its principal axes, it will continue to rotate uniformly about that axis for ever. On account of this remarkable property these are called the natural axes of rotation; it belongs to them exclusively, for if the position of the real axis of rotation oz'' be invariable on the surface of the body, the angular velocity will be constant;

hence $$dp = 0, \quad dq = 0, \quad dr = 0,$$

and $(C-B)\,qrdt = 0$, $(A-C)\,rpdt = 0$, $(B-A)\,pqdt = 0$.

218. If A, B, C, be unequal, these equations will only be zero in every case when two of the quantities p, q, r, are zero; but then, the real axis coincides with one of the principal axes.

If two of the moments of inertia be equal, as $A = B$, the three equations are reduced to $rp = 0$, $qr = 0$; both of which will be satisfied, that is, they will both be zero for every value of q and p, if $r = 0$. The axis of rotation is, therefore, in a plane at right angles to the third principal axis; but as the body is then a solid of revolution, every axis in that plane is a principal axis.

219. When $A = B = C$, the three preceding equations are zero, whatever may be the values of p, q, r, then all the axes of the body will be principal axes. Thus the principal axes alone have the property of permanent rotation, though they do not possess that property in the same degree.

*

220. Suppose the real axis of rotation oz'', fig. 50, to deviate by an indefinitely small quantity from oz', the third principal axis, the coefficients M and M' will then be indefinitely small, since $q = M' \times \cos(nt + g)$, and $p = M \sin(nt + g)$ are indefinitely small. Now if n be a real quantity, $\sin(nt + g)$, $\cos(nt + g)$, will never exceed very narrow limits, therefore q and p will remain indefinitely small; so that the real axis oz'' will make indefinitely small oscillations about the third principal axis. But if n be imaginary, by article 215,

$$\sin(nt + g), \quad \cos(nt + g),$$

will be changed into quantities which increase with the time, and the real axis of rotation will deviate more and more from the third principal axis, so that the motion will have no stability. The value of n will decide that important point.

Since $$n = r\sqrt{\frac{(A - C)(B - C)}{AB}},$$

it will be a real quantity when C the moment of inertia with regard to oz', is either the greatest or the least of the three moments of inertia A, B, C, for then the product $(A - C)(B - C)$ will be positive; but if C have a value that is between those of A and B, that product will be negative, and n imaginary. Hence the rotation will be stable about the greatest and least of the principal axes, but unstable about the third.

221. Having determined the rotation of the solid, it only remains to ascertain the position of the principal axis with regard to quiescent space, that is, with regard to the fixed axes ox, oy, oz. That evidently depends on the angles ϕ, ψ, and θ.

If the third principal axis oz', fig. 50, be assumed to be nearly at right angles to the plane xoy, the angle zoz', or θ, will be so very small that its square may be omitted, and its cosine assumed equal to unity; then the equations (41)

give $d\phi - d\psi = rdt$; or if $r = \alpha$, be a constant quantity,

the integral is, $$\psi = \phi - \alpha t + \epsilon.$$

If $\sin\theta\cos\phi = s$, $\sin\theta\sin\phi = u$, the two first of equations (41), after the elimination of $d\psi$, give

$$\frac{ds}{dt} + \alpha u = -p, \quad \frac{du}{dt} - \alpha s = q.$$

The integrals of these two quantities are obtained by the method in article 214, and are

$$s = \mathcal{C} \cos(\alpha t + \lambda) - \frac{BM'}{C\alpha} \cos(nt + g),$$

$$u = \mathcal{C} \sin(\alpha t + \lambda) - \frac{AM}{C\alpha} \sin(nt + g),$$

$\mathcal{C}$ and λ being new arbitrary quantities introduced by integration. The problem is completely solved, since s and u give θ and ϕ in values of the time, and ψ is given in values of ϕ and the time.

Compound Pendulums.

222. Hitherto the rotation of a solid about its centre of gravity has been considered either when the body is free, or when the centre of gravity is fixed; but imagine a solid OP, fig. 54, to revolve about a fixed axis in o which does not pass through its centre of gravity. If the body be drawn aside from the vertical oz, and then left to itself, it will oscillate about that axis by the action of gravitation alone. This solid body of any form whatever is the compound pendulum, and its motion is perfectly similar to that of the simple pendulum already described, depending on the property of areas.

fig. 54.

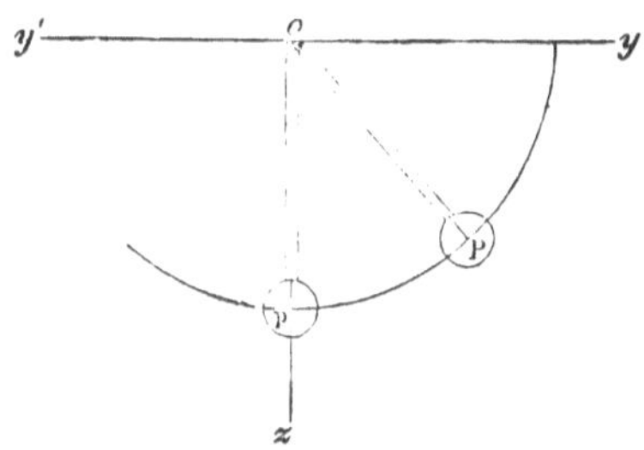

The motion being in the plane zoy, the sums of the areas in the other two planes are zero; so that the motion of the pendulum is derived from the equation $S\left(\frac{yd^2z - zd^2y}{dt^2}\right) dm = S(yZ - zY)\, dm$.

In order to adapt that equation to the motion of the pendulum, let $oy = y$, $oP = z$, $Ao = z'$, $Ay = y'$, hence $PA = -y'$, fig. 55; and let the angle PoA be represented by θ. P is the centre of gravity of the pendulum, which is supposed to rotate about the axis ox, passing through o at right angles to the plane zoy, and therefore it cannot be represented in the diagram.

fig. 55.

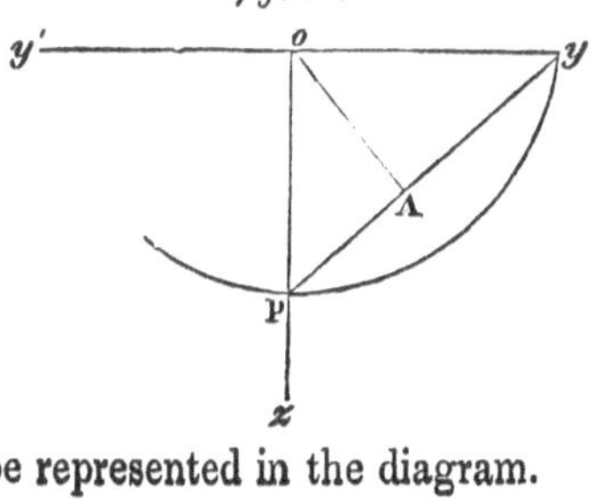

Now
$$-y' = z \sin \theta$$
$$z' = z \cos \theta$$
$$z' = y \sin \theta$$
$$y' = y \cos \theta$$

If the first of these four equations be multiplied by $\sin \theta$, and the second by $\cos \theta$, their sum is

$$z = z' \cos \theta - y' \sin \theta;$$

in the same way $\qquad y = z' \sin \theta + y' \cos \theta.$

If these values be substituted in the equation of areas it becomes

$$A \frac{d^2\theta}{dt^2} = - \text{S}\,(yZ - zY)\, dm,$$

for $\qquad A = \text{S}\,(y'^2 + z'^2)\, dm.$

If the pendulum moves by the force of gravitation alone in the direction oz, $\qquad Y = 0 \qquad Z = g.$

Hence
$$A \frac{d^2\theta}{dt^2} = - \text{S}gydm.$$

If the value of y be substituted in this it becomes,

$$A \frac{d^2\theta}{dt^2} = - g \sin \theta.\ \text{S}\, z'dm - g \cos \theta.\ \text{S}.\, y'dm.$$

Because z' passes through the centre of gravity of the pendulum, the rotatory pressure $\text{S}.y'dm$ is zero; hence

$$A \frac{d^2\theta}{dt^2} = - g \sin \theta.\ \text{S}.z'dm.$$

If L be the distance of the centre of gravity of the pendulum from the axis of rotation ox, the rotatory pressure $\text{S}.\ z'dm$ becomes Lm, in which m is the whole mass of the pendulum; hence

$$A \frac{d^2\theta}{dt^2} = - Lmg \sin \theta,$$

or
$$\frac{d\theta^2}{dt^2} = \frac{2Lmg}{A}.\ \cos \theta + \text{C},$$

C being an arbitrary constant quantity.

223. If a simple pendulum be considered, of which all the atoms are united in a point at the distance of l from the axis of rotation ox, its rotatory inertia will be $A = ml^2$, m being the mass of the body,

and $l^2 = x^2 + y^2$. In this case $l = L$. Substituting this value for A, we find

$$\frac{d\theta^2}{dt^2} = \frac{2g}{l}\cos\theta + C.$$

224. Thus it appears, that if the angular velocities of the compound and simple pendulums be equal when their centres of gravity are in the vertical, their oscillations will be exactly the same, provided also that the length of the simple pendulum be equal to the rotatory inertia of the solid body with regard to the axis of motion, divided by the product of the mass by the distance of its centre of gravity from the axis, or $l = \frac{A}{mL}$.

Thus such a relation is established between the lengths of the two pendulums, that the length of a simple pendulum may be found, whose oscillations are performed in the same time with those of a compound pendulum.

In this manner the length of the simple pendulum beating seconds has been determined from observations on the oscillations of the compound pendulum.

CHAPTER VI.

ON THE EQUILIBRIUM OF FLUIDS.

Definitions, &c.

225. A FLUID is a mass of particles which yield to the slightest pressure, and transmit that pressure in every direction.

226. Mobility of the particles constitutes the difference between fluids and solids.

227. There are, indeed, fluids in nature whose particles adhere more or less to each other, called viscous fluids; but those only whose particles do not adhere in any degree, but possess perfect mobility, are the subject of this investigation.

228. Strictly speaking, all fluids are compressible, for even liquids under very great pressure change their volume; but as the compression is insensible in ordinary circumstances, fluids of perfect mobility are divided into compressible or elastic fluids, and incompressible.

229. The elastic and compressible fluids are atmospheric air, the gases, and steam. When compressed, these fluids change both form and volume, and regain their primitive state as soon as the pressure is removed. Some of the gases are found to differ from atmospheric air in losing their elastic form, and becoming liquid when compressed to a certain degree, as lately proved by Mr. Faraday, and steam is reduced to water when its temperature is diminished; but atmospheric air, and others of the gases, always retain their gaseous form, whatever the degree of pressure may be.

230. It is impossible to ascertain the forms of the particles of fluids, but as all of them, considered in mass, afford the same phenomena, it can have no influence on the laws of their motions.

Equilibrium of Fluids.

231. When a fluid mass is in equilibrio, each particle must itself be held in equilibrio by the forces acting upon it, together with the pressures of the surrounding particles.

232. It is evident, that whatever the accelerating forces or pressures may be, they can all be resolved into component forces parallel to three rectangular co-ordinates, ox, oy, oz.

Equation of Equilibrium.

233. Imagine a system of fluid particles, forming a rectangular parallelopipedon A B C D, fig. 56, and suppose its sides parallel to the co-ordinate axes. Suppose also, that it is pressed on all sides by the surrounding fluid, at the same time that it is urged by accelerating forces.

fig. 56.

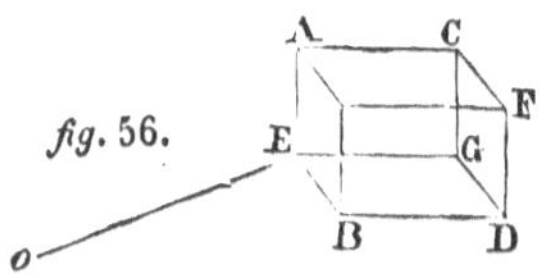

234. It is evident, that the pressure on the face A B, must be in a contrary direction to the pressure on the face C D; hence the mass will be urged by the difference of these pressures: but this difference may be considered as a single force acting either on the face A B or C D; consequently the difference of the pressures multiplied by the very small area A B will be the whole pressure, urging the mass parallel to the side EG. In the same manner, the pressures urging the mass in a direction parallel to E B and E A, are the area E C into the difference of the pressures on the faces E C and BF; and the area ED into the difference of the pressures on E D and A F.

235. Because the mass is indefinitely small, if x, y, z, be the co-ordinates of E, the edges E G, E B, E A, may be represented by dx, dy, dz. Then p being the pressure on a unit of surface, $pdydz$ will be the pressure on the face A B, in the direction E G. At G, x becomes $x + dx$, y and z remaining the same; hence as p is considered a function of x, y, z, it becomes

$$p' = p + \left(\frac{dp}{dx}\right) dx \text{ at the point G;}$$

hence
$$p - p' = -\left(\frac{dp}{dx}\right)dx,$$

and
$$pdydz - p'dydz = -\left(\frac{dp}{dx}\right) dx \,.\, dydz.$$

Now $pdydz$ is the pressure on AB, and $p'dydz$ is the pressure on CD;

hence
$$-\left(\frac{dp}{dx}\right) dx\,.\,dydz = (p - p')\,dydz$$

is the difference of the pressures on the faces A B and C D. In the same manner it may be proved that

$$-\left(\frac{dp}{dy}\right) dy.dxdz, \quad \text{and} -\left(\frac{dp}{dz}\right) dz.dydx$$

are the differences of the pressures on the faces B F, A G, and on E D, A F.

236. But if X, Y, Z, be the accelerating forces in the direction of the axes, when multiplied by the volume $dx\ dy\ dz$, and by ρ its density, they become the momenta

$$\rho.\text{X}\,dx\ dy\ dz,$$
$$\rho.\text{Y}\,dx\ dy\ dz,$$
$$\rho.\text{Z}\,dx\ dy\ dz.$$

But these momenta must balance the pressures in the same directions when the fluid mass is in equilibrio; hence, by the principle of virtual velocities

$$\left\{\rho\text{X} - \frac{dp}{dx}\right\}\delta x + \left\{\rho\text{Y} - \frac{dp}{dy}\right\}\delta y + \left\{\rho\text{Z} - \frac{dp}{dz}\right\}\delta z = 0, \text{ or}$$

$$\frac{dp}{dx}\delta x + \frac{dp}{dy}\delta y + \frac{dp}{dz}\delta z = \rho\,\{\text{X}\delta x + \text{Y}\delta y + \delta\text{Z}z\}.$$

As the variations are arbitrary, they may be made equal to the differentials, and then

$$dp = \rho\,\{\text{X}dx + \text{Y}dy + \text{Z}dz\} \qquad (52)$$

is the general equation of the equilibrium of fluids, whether elastic or incompressible. It shows, that the indefinitely small increment of the pressure is equal to the density of the fluid mass multiplied by the sum of the products of each force by the element of its direction.

237. This equation will not give the equilibrium of a fluid under all circumstances, for it is evident that in many cases equilibrium is impossible; but when the accelerating forces are attractive forces directed to fixed centres, it furnishes another equation, which shows the relation that must exist among the component forces, in order that equilibrium may be possible at all. It is called an equation of condition, because it expresses the general condition requisite for the existence of equilibrium.

Equations of Condition.

238. Assuming the forces X, Y, Z, to be functions of the distance,

by article 75. The second member of the preceding equation is an exact differential; and as p is a function of x, y, z, it gives the partial equations

$$\frac{dp}{dx} = \rho X\,;\ \frac{dp}{dy} = \rho Y\,;\ \frac{dp}{dz} = \rho Z\,;$$

but the differential of the first, according to y, is

$$\frac{d^2p}{dxdy} = \frac{d.\rho X}{dy}$$

and the differential of the second, according to x, is

$$\frac{d^2p}{dydx} = \frac{d.\rho Y}{dx};$$

hence

$$\frac{d\,.\,\rho X}{dy} = \frac{d\,.\,\rho Y}{dx}.$$

By a similar process, it will be found that

$$\frac{d\,.\,\rho Y}{dz} = \frac{d\,.\,\rho Z}{dy};\ \frac{d\,.\rho X}{dz} = \frac{d\,.\,\rho Z}{dx}.$$

These three equations of condition are necessary, in order that the equation (52) may be an exact differential, and consequently integrable. If the differentials of these three equations be taken, the sum of the first multiplied by Z, of the second multiplied by X, and of the third multiplied by $-$ Y, will be

$$0 = X.\frac{dY}{dz} - Y.\frac{dX}{dz} + Z.\frac{dX}{dy} - X.\frac{dZ}{dy} + Y.\frac{dZ}{dx} - Z.\frac{dY}{dx}$$

an equation expressing the relation that must exist among the forces X, Y, Z, in order that equilibrium may be possible.

Equilibrium will always be possible when these conditions are fulfilled; but the exterior figure of the mass must also be determined.

Equilibrium of homogeneous Fluids.

239. If the fluid be free at its surface, the pressure must be zero in every point of the surface when the mass is in equilibrio; so that $p = 0$, and

$$\rho\{Xdx + Ydy + Zdz\} = 0,$$

whence

$$\int(Xdx + Ydy + Zdz) = \text{constant},$$

supposing it an exact differential, the density being constant.

The resulting force on each particle must be directed to the interior of the fluid mass, and must be perpendicular to the surface; for were it not, it might be resolved into two others, one perpendicular, and one horizontal; and in consequence of the latter, the particle would slide along the surface.

If $u=0$ be the equation of the surface, by article 69 the equation of equilibrium at the surface will be

$$Xdx + Ydy + Zdz = \lambda du,$$

λ being a function of x, y, z; and by the same article, the resultant of the forces X, Y, Z, must be perpendicular to those parts of the surface where the fluid is free, and the first member must be an exact differential.

Equilibrium of heterogeneous Fluids.

240. When the fluid mass is heterogeneous, and when the forces are attractive, and their intensities functions of the distances of the points of application from their origin, then the density depends on the pressure; and all the strata or layers of a fluid mass in which the pressure is the same, have the same density throughout their whole extent.

Demonstration. Let the function

$$Xdx + Ydy + Zdz$$

be an exact difference, which by article 75 will always be the case when the forces X, Y, Z, are attractive, and their intensities functions of the mutual distances of the particles. Assume

$$\phi = \int(Xdx + Ydy + Zdz), \tag{53}$$

ϕ being a function of x, y, z; then equation (52) becomes

$$dp = \rho . d\phi. \tag{54}$$

The first member of this equation is an exact differential, and in order that the second member may also be an exact differential, the density ρ must be a function of ϕ. The pressure p will then be a function of ϕ also; and the equation of the free surface of the fluid will be $\phi =$ constant quantity, as in the case of homogeneity. Thus the pressure and the density are the same for all the points of the same layer. The law of the variation of the density in passing from one layer to another depends on the function in ϕ which expresses it. And when that function is given, the pressure will be obtained by integrating the equation $dp = \rho d\phi$.

241. It appears from the preceding investigation, that a homogeneous liquid will remain in equilibrio, if all its particles act on each other, and are attracted towards any number of fixed centres; but in that case, the resulting force must be perpendicular to the surface of the liquid, and must tend to its interior. If there be but one force or attraction directed to a fixed point, the mass would become a sphere, having that point in its centre, whatever the law of the force might be.

242. When the centre of the attractive force is at an infinite distance, its direction becomes parallel throughout the whole extent of the fluid mass; and the surface, when in equilibrio, is a plane perpendicular to the direction of the force. The surface of a small extent of stagnant water may be estimated plane, but when it is of great extent, its surface exhibits the curvature of the earth.

243. A fluid mass that is not homogeneous but free at its surface will be in equilibrio, if the density be uniform throughout each indefinitely small layer or stratum of the mass, and if the resultant of all the accelerating forces acting on the surface be perpendicular to it, and tending towards the interior. If the upper strata of the fluid be most dense, the equilibrium will be unstable; if the heaviest is undermost, it will be stable.

244. If a fixed solid of any form be covered by fluid as the earth is by the atmosphere, it is requisite for the equilibrium of the fluid that the intensity of the attractive forces should depend on their distances from fixed centres, and that the resulting force of all the forces which act at the exterior surface should be perpendicular to it, and directed towards the interior.

245. If the surface of an elastic fluid be free, the pressure cannot be zero till the density be zero; hence an elastic fluid cannot be in equilibrio unless it be either shut up in a close vessel, or, like the atmosphere, it extend in space till its density becomes insensible.

Equilibrium of Fluids in Rotation.

246. Hitherto the fluid mass has been considered to be at rest; but suppose it to have a uniform motion of rotation about a fixed axis, as for example the axis oz. Let ω be the velocity of rotation common to all the particles of the fluid, and r the distance of a par-

ticle dm from the axis of rotation, the co-ordinates of dm being x, y, z. Then ωr will be the velocity of dm, and its centrifugal force resulting from rotation, will be $\omega^2 r$, which must therefore be added to the accelerating forces which urge the particle; hence equation (53) will become

$$d\phi = Xdx + Ydy + Zdz + \omega^2 rdr.$$

And the differential equation of the strata, and of the free surface of the fluid, will be

$$Xdx + Ydy + Zdz + \omega^2 . rdr = 0. \tag{55}$$

The centrifugal force, therefore, does not prevent the function ϕ from being an exact differential, consequently equilibrium will be possible, provided the condition of article 238 be fulfilled.

247. The regularity of gravitation at the surface of the earth; the increase of density towards its centre; and, above all, the correspondence of the form of the earth and planets with that of a fluid mass in rotation, have led to the supposition that these bodies may have been originally fluid, and that their parts, in consolidating, have retained nearly the form they would have acquired from their mutual attractions, together with the centrifugal force induced by rotation when fluid. In this case, the laws expressed by the preceding equations must have regulated their formation.

CHAPTER VII.

MOTION OF FLUIDS.

General Equation of the Motion of Fluids.

248. THE mass of a fluid particle being $\rho\, dx\, dy\, dz$, its momentum in the axis x arising from the accelerating forces is, by article 144,

$$\left\{X - \frac{d^2x}{dt^2}\right\}\rho\, dx\, dy\, dz.$$

But the pressure resolved in the same direction is

$$\left(\frac{dp}{dx}\right) dx\ dy\ dz.$$

Consequently the equation of the motion of a fluid mass in the axis ox, when free, is

$$\left\{X - \frac{d^2x}{dt^2}\right\}\rho - \frac{dp}{dx} = 0. \qquad (56)$$

In the same manner its motions in the axes y and z are

$$\left.\begin{aligned}\left\{Y - \frac{d^2y}{dt^2}\right\}\rho - \frac{dp}{dy} &= 0,\\ \left\{Z - \frac{d^2z}{dt^2}\right\}\rho - \frac{dp}{dz} &= 0.\end{aligned}\right\} \qquad (56)$$

And by the principle of virtual velocities the general equation of fluids in motion is

$$\{X\delta x + Y\delta y + Z\delta z\} - \frac{\delta p}{\rho} = \frac{d^2x}{dt^2}\,\delta x + \frac{d^2y}{dt^2}\,\delta y + \frac{d^2z}{dt^2}\,\delta z. \qquad (57)$$

This equation is not rigorously true, because it is formed in the hypothesis of the pressures being equal on all sides of a particle in motion, which Poisson has proved not to be the case; but, as far as concerns the following analysis, the effect of the inequality of pressure is insensible.

249. The preceding equation, however, does not express all the circumstances of the motion of a fluid. Another equation is requisite.

A solid always preserves the same form whatever its motion may be, which is by no means the case with fluids; for a mass

ABCD, fig. 57, formed of particles possessing perfect mobility, changes its form by the action of the forces, so that it always continues to fit into the intervals of the surrounding molecules without leaving any void. In this consists the continuity of fluids, a property which furnishes the other equation necessary for the determination of their motions.

Equation of Continuity.

250. Suppose at any given time the form of a very small fluid mass to be that of a rectangular parallelopiped A B C D, fig. 57. The action of the forces will change it into an oblique angled figure N E F K, during the indefinitely small time that it moves from its first to its second position. Now N E F G may differ from A B C D both in form and density, but not in mass; for if the density depends on the pressure, the same forces that change the form may also produce a change in the pressure, and, consequently, in the density; but it is evident that the mass must always remain the same, for the number of molecules in A B C D can neither be increased nor diminished by the action of the forces; hence the volume of A B C D into its primitive density must still be equal to volume of N E F G into the new density; hence, if

fig. 57.

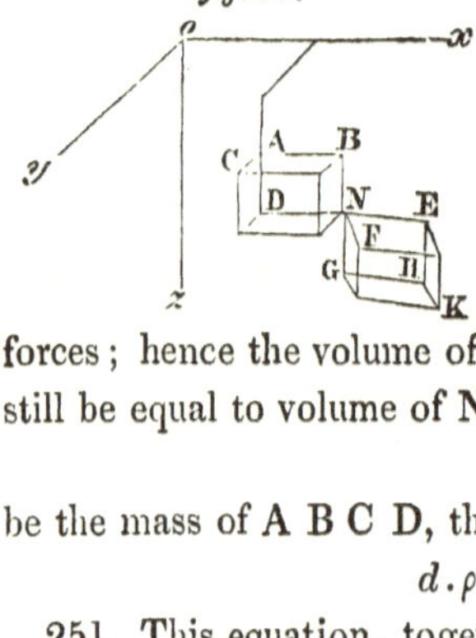

$$\rho \, dx \, dy \, dz,$$

be the mass of A B C D, the equation of continuity will be

$$d.\rho \, dx \, dy \, dz = 0. \quad (58)$$

251. This equation, together with equations (56), will determine the four unknown quantities x, y, z, and p, in functions of the time, and consequently the motion of the fluid.

Developement of the Equation of Continuity.

252. The sides of the small parallelopiped, after the time dt, become

$$dx + d.dx, \quad dy + d.dy, \quad dz + d.dz;$$

or, observing that the variation of dx only arises from the increase of x, the co-ordinates y and z remaining the same, and that the variations of dy, dz, arise only from the similar increments of y and z;

hence the edges of the new mass are

$$NE = dx\left(1 + \frac{d^2x}{dx}\right)$$

$$NG = dy\left(1 + \frac{d^2y}{dy}\right)$$

$$NF = dz\left(1 + \frac{d^2z}{dz}\right)$$

fig. 58.

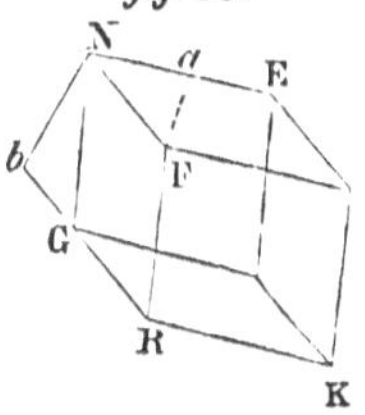

If the angles GNF and FNE, fig. 58, be represented by θ and ψ, the volume of the parallelopiped NK will be $NE.NG \sin\theta.NF \sin\psi$;

for $$Fa = NF.\ \sin\psi$$
$$Nb = NG.\ \sin\theta,$$

Fa, Nb being at right angles to NE and RG; but as θ and ψ were right angles in the primitive volume, they could only vary by indefinitely small arcs in the time dt; hence in the new volume

$$\theta = 90^\circ \pm d\theta,\ \psi = 90^\circ \pm d\psi,$$

consequently,

$$\sin\theta = \sin(90^\circ \pm d\theta) = \cos d\theta = 1 - \tfrac{1}{2}d\theta^2 + \&c.$$
$$\sin\psi = \sin(90^\circ \pm d\psi) = \cos d\psi = 1 - \tfrac{1}{2}d\psi^2 + \&c.$$

and omitting $d\theta^2,\ d\psi^2,\ \sin\theta = \sin\psi = 1,$

and the volume becomes NE . NG . NF; substituting for the three edges their preceding values, and omitting indefinitely small quantities of the fifth order, the volume after the time dt is

$$dxdydz\left\{1 + \frac{d^2x}{dx} + \frac{d^2y}{dy} + \frac{d^2z}{dz}\right\}.$$

The density varies both with the time and space; hence ρ, the primitive density, is a function of t, x, y and z, and after the time dt, it is

$$\rho + \frac{d\rho}{dt}dt + \frac{d\rho}{dx}dx + \frac{d\rho}{dy}dy + \frac{d\rho}{dz}dz;$$

consequently, the mass, being the product of the volume and density, is, after the time dt, equal to

$$dm = \rho.dx\,dy\,dz\left(1 + \frac{d\rho}{dt}dt + \frac{d\rho}{dx}dx + \frac{d\rho}{dy}dy + \frac{d\rho}{dz}dz\right.$$
$$\left. + \rho\frac{d^2x}{dx} + \rho\frac{d^2y}{dy} + \rho\frac{d^2z}{dz}\right);$$

And the equation

$$d.\rho.\,(dx\,dy\,dz)=0$$

becomes
$$\frac{d\rho}{dt}+\frac{d.\rho\frac{dx}{dt}}{dx}+\frac{d.\rho\frac{dy}{dt}}{dy}+\frac{d.\rho\frac{dz}{dt}}{dz}=0 \qquad (59)$$

as will readily appear by developing this quantity, which is the general equation of continuity.

253. The equations (56) and (59) determine the motions both of incompressible and elastic fluids.

254. When the fluid is incompressible, both the volume and density remain the same during the whole motion; therefore the increments of these quantities are zero; hence, with regard to the volume

$$\frac{d^2x}{dx}+\frac{d^2y}{dy}+\frac{d^2z}{dz}=0; \qquad (60)$$

and with regard to the density,

$$\frac{d\rho}{dt}+\frac{d\rho}{dx}dx+\frac{d\rho}{dy}dy+\frac{d\rho}{dz}dz=0. \qquad (61)$$

255. By means of these two equations and the three equations (56), the five unknown quantities p, ρ, x, y and z, may be determined in functions of t, which remains arbitrary; and therefore all the circumstances of the motion of the fluid mass may be known for any assumed time.

256. If the fluid be both incompressible and homogeneous, the density is constant, therefore $d\rho=0$, and as the last equation becomes identical, the motion of the fluid is obtained from the other four.

Second form of the Equation of the Motions of Fluids.

257. It is occasionally more convenient to regard x, y, z, the co-ordinates of the fluid particle dm, as known quantities, and

$$\frac{dx}{dt},\ \frac{dy}{dt},\ \frac{dz}{dt},$$

its velocities in the direction of the co-ordinates, as unknown. In order to transform the equations (56) and (59) to suit this case, let

$$s=\frac{dx}{dt},\ u=\frac{dy}{dt},\ v=\frac{dz}{dt};$$

these quantities being functions of x, y, z, and t. The differentials of these equations when x, y, z, and t, vary all at once; and when

sdt, udt, vdt, are put for dx, dy, dz, become

$$ds = \frac{ds}{dt}dt + \frac{ds}{dx}.sdt + \frac{ds}{dy}.udt + \frac{ds}{dz}.vdt,$$

$$du = \frac{du}{dt}dt + \frac{du}{dx}.sdt + \frac{du}{dy}.udt + \frac{ds}{dz}.vdt, \tag{62}$$

$$dv = \frac{dv}{dt}dt + \frac{dv}{dx}.sdt + \frac{dv}{dy}.udt + \frac{dv}{dz}.vdt,$$

And as $\quad ds = \dfrac{d^2x}{dt}, \; du = \dfrac{d^2y}{dt}, \; dv = \dfrac{d^2z}{dt},$

the equations (56) become, by the substitution of the preceding quantities,

$$\frac{dp}{dx} = \rho\left\{X - \frac{ds}{dt} - \frac{ds}{dx}.s - \frac{ds}{dy}u - \frac{ds}{dz}.v\right\}$$

$$\frac{dp}{dy} = \rho\left\{Y - \frac{du}{dt} - \frac{du}{dx}.s - \frac{du}{dy}.u - \frac{du}{dz}.v\right\} \tag{63}$$

$$\frac{dp}{dz} = \rho\left\{Z - \frac{dv}{dt} - \frac{dv}{dx}.s - \frac{dv}{dy}.u - \frac{dv}{dz}.v\right\}$$

and by the same substitution, the equation (59) of continuity becomes

$$\frac{d\rho}{dt} + \frac{d.\rho s}{dx} + \frac{d.\rho u}{dy} + \frac{d.\rho v}{dz} = 0, \tag{64}$$

which, for incompressible and homogeneous fluids, is

$$\frac{ds}{dx} + \frac{du}{dy} + \frac{dv}{dz} = 0. \tag{65}$$

The equations (63) and (64) will determine s, u, and v, in functions of x, y, x, t, and then the equations

$$dx = sdt \quad dy = udt \quad dz = vdt$$

will give x, y, z, in functions of the time. The whole circumstances of the fluid mass will therefore be known.

Integration of the Equations of the Motions of Fluids.

258. The great difficulty in the theory of the motion of fluids, consists in the integration of the partial equations (63) and (64), which has not yet been surmounted, even in the most simple problems. It may, however, be effected in a very extensive case, in which

$$sdx + udy + vdx$$

is a complete differential of a function ϕ, of the three variable quantities x, y, z; so that

$$sdx + udy + vdz = d\phi.$$

259. If in the equation (57) the variations which are arbitrary, be made equal to the differentials of the same quantities; and if, as in nature, the accelerating forces X, Y, Z be functions of the distance, then

$$Xdx + Ydy + Zdz$$

will be a complete differential, and may be expressed by dV, so that the equation in question becomes

$$\frac{dp}{\rho} = dV - dx.\frac{d^2x}{dt^2} - dy.\frac{d^2y}{dt^2} - dz.\frac{d^2z}{dt^2} \qquad (66)$$

But the function ϕ gives the velocities of the fluid mass in the directions of the axes, *viz.*

$$s = \frac{d\phi}{dx}, u = \frac{d\phi}{dy}, v = \frac{d\phi}{dz}.$$

By the substitution of these values in equation (62), ds, du, dv, and consequently

$$\frac{d^2x}{dt}, \frac{d^2y}{dt}, \frac{d^2z}{dt},$$

will be obtained in functions of ϕ, by which the preceding equation becomes

$$\frac{dp}{\rho} = dV - \frac{ds}{dt}.dx - \frac{du}{dt}.dy - \frac{dv}{dt}.dz - \tfrac{1}{2}d\left(\frac{d\phi^2}{dx^2} + \frac{d\phi^2}{dy^2} + \frac{d\phi^2}{dz^2}\right)$$

Now
$$\frac{ds}{dt}.dx + \frac{du}{dt}dy + \frac{dv}{dt}dz = d.\frac{d\phi}{dt};$$

consequently,

$$\int\frac{dp}{\rho} = V - \frac{d\phi}{dt} - \tfrac{1}{2}\left(\frac{d\phi^2}{dx^2} + \frac{d\phi^2}{dy^2} + \frac{d\phi^2}{dz^2}\right) \qquad (67)$$

The constant quantity introduced by integration is included in the function ϕ. By the same substitution, the equation of continuity becomes

$$\frac{d\rho}{dt} + \frac{d.\rho\frac{d\phi}{dx}}{dx} + \frac{d.\rho\frac{d\phi}{dy}}{dy} + \frac{d.\rho\frac{d\phi}{dz}}{dz} = 0. \qquad (68)$$

The two last equations determine the motion of the fluid mass in the case under consideration.

260. It is impossible to know all the cases in which the function $sdx + udy + vdz$ is an exact differential, but it may be proved that

if it be so at any one instant, it will be an exact differential during the whole motion of a fluid. *Demonstration.*—Suppose that at any one instant it is a complete differential, it will then be integrable, and may be expressed by $d\phi$; in the following instant it will become

$$d\phi + \frac{ds}{dt}.dx + \frac{du}{dt}.dy + \frac{dv}{dt}\,dz$$

It will still be an exact differential, if

$$\frac{ds}{dt}\,dx + \frac{du}{dt}\,dy + \frac{dv}{dt}\,dz \text{ be one.}$$

Now the latter quantity being equal to $d.\frac{d\phi}{dt}$, equation (67) gives

$$\frac{ds}{dt}dx + \frac{du}{dt}dy + \frac{ds}{dt}dz = d\text{V} - \frac{dp}{\rho} - \tfrac{1}{2}d\left(\frac{d\phi^2}{dx^2} + \frac{d\phi^2}{dy^2} + \frac{d\phi^2}{dz^2}\right).$$

And if the density ρ be a function of p the pressure, the second member of this equation will be an exact differential, consequently the first member will be one also, and thus the function

$$sdx + udy + vdz$$

is a complete differential in the second instant, if it be one in the first; it will therefore be a complete differential during the whole motion of the fluid.

Theory of small Undulations of Fluids.

261. If the oscillations of a fluid be very small, the squares and products of the velocities s, u, v, may be neglected: then the preceding equation becomes

$$d\text{V} - \frac{dp}{\rho} = \frac{ds}{dt}\,dx + \frac{du}{dt}\,dy + \frac{dv}{dt}\,dz.$$

If ρ be a function of p, the first member will be a complete differential, therefore the second member, and consequently

$$sdx + udy + vdz$$

is one also, so that the equation is capable of integration; and as its last member is equal to $d.\frac{d\phi}{dt}$, the integral is

$$\text{V} - \int\frac{dp}{\rho} = \frac{d\phi}{dt}. \qquad (69)$$

This equation, together with equation (68) of continuity, contain the whole theory of the small undulations of fluids.

262. An idea may be formed of these undulations by the effect

of a stone dropped into still water; a series of small concentric circular waves will appear, extending from the point where the stone fell. If another stone be let fall very near the point where the first fell, a second series of concentric circular waves will be produced; but when the two series of undulations meet, they will cross, each series continuing its course independently of the other, the circles cutting each other in opposite points. An infinite number of such undulations may exist without disturbing the progress of one another. In sound, which is occasioned by undulations in the air, a similar effect is produced: in a chorus, the melody of one voice may be distinguished from the general harmony. Coexisting vibrations may also be excited in solid bodies, each undulation having its perfect effect, independently of the others. If the directions of the undulations coincide, their joint motions will be the sum or the difference of the separate motions, according as similar or dissimilar parts of the undulations are coincident. In undulations of equal frequency, when two series exactly coincide in point of time, the united velocity of the particular motions will be the greatest or least;—and if the undulations are of equal strength, they will totally destroy each other, when the time of the greatest direct motion of one undulation coincides with that of the greatest retrograde motion of the other.

The general principle of Interferences was first shown by Dr. Young to be applicable to all vibratory motions, which he illustrated beautifully by the remarkable phenomena of two rays of light producing darkness, and the concurrence of two musical sounds producing silence. The first may be seen by looking at the flame of a candle through two extremely narrow parallel slits in a card; and the latter is rendered evident by what are termed beats in music.

The same principle serves to explain why neither flood nor ebb tides take place at Batsham in Tonquin on the day following the moon's passage across the equator; the flood tide arrives by one channel at the same instant that the ebb arrives by another, so that the interfering waves destroy each other.

Co-existing vibrations show the comprehensive nature and elegance of analytical formulæ. The general equation of small undulations is the sum of an infinite number of equations, each of which gives a single series of undulations, like the surface of water in a shower, which at once contains an infinite number of undulations, and yet exhibits each independently of the rest.

Rotation of a homogeneous Fluid.

263. If a fluid mass rotates uniformly about an axis, its component velocity in the axis of rotation is zero; the velocities in the other two axes are angular velocities—independent of the time, the motion being uniform: indeed, the motion is the same with that of a solid body revolving about a fixed axis. If the mass revolves about the axis z, and if ω be the angular velocity at the distance of unity from that axis, the component velocities will be

$$s = -\omega y, \quad u = \omega x, \quad v = 0;$$

and from equations (63) it will be easily found that

$$\frac{dp}{\rho} = dV + \omega^2 (xdx + ydy);$$

and if ρ be constant, the integral is

$$\frac{p}{\rho} = V + \frac{\omega^2}{2}(x^2 + y^2).$$

The equation (65) of continuity will be satisfied, since

$$\frac{ds}{dx} = 0, \qquad \frac{du}{dy} = 0, \qquad \frac{dv}{dz} = 0.$$

264. This motion of a fluid mass is therefore possible, although it is a case in which the function

$$sdx + udy + vdz$$

is not an exact differential; for by the substitution of the preceding values of the velocities, it becomes

$$sdx + udy + vdz = \omega (xdy - ydx),$$

an expression that cannot be integrated. Therefore, in the theory of the tides caused by the disturbing action of the sun and moon on the ocean, the function

$$sdx + udy + vdx$$

must not be regarded as an exact differential, since it cannot be integrated even when there is no disturbance in its rotatory motion.

265. Thus a fluid mass or a fluid covering a solid of any form whatever, will rotate about an axis without alteration in the relative position of its particles. This would be the state of the ocean were the earth a solitary body, moving in space; but the attractions of the sun and moon not only trouble the ocean, but also cause commotions in the atmosphere, indicated by the periodic

variations in the heights of the mercury in the barometer. From the vast distance of the sun and moon, their action upon the fluid particles of the ocean and atmosphere, is very small in comparison of that produced by the velocity of the earth's rotation, and by its attraction.

Determination of the Oscillations of a homogeneous Fluid covering a Spheroid, the whole in rotation about an axis; supposing the fluid to be slightly deranged from its state of equilibrium by the action of very small forces.

266. If the earth be supposed to rotate about its axis, uninfluenced by foreign forces, the fluids on its surface would assume a spheroidal form, from the centrifugal force induced by rotation; and a particle in the interior of the fluid would be subject to the action of gravitation and the pressure of the surrounding fluid only. But although the fluids would be moving with great velocity, yet to us they would seem at rest. When in this state the atmosphere and ocean are said to be in equilibrio.

Action of the Sun and Moon.

267. The action of the sun and moon troubles this equilibrium, and occasions tides in both fluids. The whole of this theory is perfectly general, but for the sake of illustration it will be considered with regard to the ocean. If the moon attracted the centre of gravity of the earth and all its particles with equal and parallel forces, the whole system of the earth and the waters that cover it, would yield to these forces with a common motion, and the equilibrium of the seas would remain undisturbed. The difference of the intensity and direction of the forces alone, trouble the equilibrium; for, since the attraction of the moon is inversely as the square of the distance, a molecule at m, under the moon M, is as much more attracted than the centre of gravity of the earth, as the square of EM is greater than the square of mM: hence the particle has a tendency to leave the earth, but is retained by its gravitation, which this tendency diminishes. Twelve hours after, the particle is brought to m' by the rotation of the earth, and is then in

fig. 59.

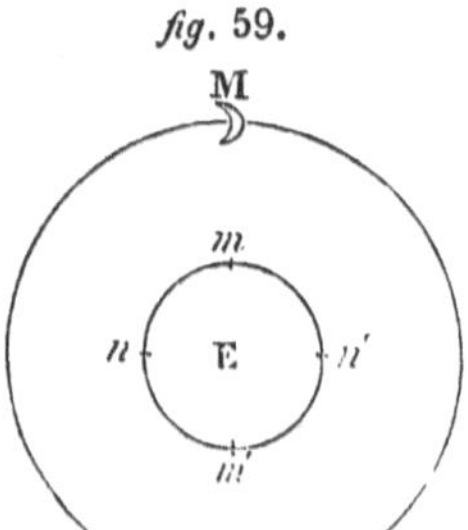

opposition to the moon, which attracts it more feebly than it attracts the centre of the earth, in the ratio of the square of EM to the square of m'M. The surface of the earth has then a tendency to leave the particle, but the gravitation of the particle retains it; and gravitation is also in this case diminished by the action of the moon. Hence, when the particle is at m, the moon draws the particle from the earth; and when it is at m', it draws the earth from the particle: in both instances producing an elevation of the particle above the surface of equilibrium of nearly the same height, for the diminution of the gravitation in each position is almost the same on account of the distance of the moon being great in comparison of the radius of the earth. The action of the moon on a particle at n, 90° distant from m, may be resolved into two forces—one in the direction of the radius nE, and the other tangent to the surface. The latter force alone attracts the particle towards the moon, and makes it slide along the surface; so that there is a depression of the water in n and n', at the same time that it is high water at m and m'. It is evident that, after half a day, the particle, when at n', will be acted on by the same force it experienced at n.

268. Were the earth entirely covered by the sea, the water thus attracted by the moon would assume the form of an oblong spheroid, whose greater axis would point towards the moon; since the column of water under the moon, and the direction diametrically opposite to her, would be rendered lighter in consequence of the diminution of their gravitation: and in order to preserve the equilibrium, the axis 90° distant would be shortened. The elevation, on account of the smaller space to which it is confined, is twice as great as the depression, because the contents of the spheroid always remain the same. If the waters were capable of instantly assuming the form of a spheroid, its summit would always be directed towards the moon, notwithstanding the earth's rotation; but on account of their resistance, the rapid motion of rotation prevents them from assuming at every instant the form which the equilibrium of the forces acting on them requires, so that they are constantly approaching to, and receding from that figure, which is therefore called the *momentary equilibrium* of the fluid. It is evident that the action, and consequently the position of the sun modifies these circumstances, but the action of that body is incomparably less than that of the moon.

Determination of the general Equation of the Oscillations of all parts of the Fluids covering the Earth.

269. Let $pEPQ$, fig. 60, be the terrestrial spheroid, Eo the equatorial radius, Pp the axis of rotation. Suppose the spheroid to be entirely covered with the fluid—the ocean, for example; and let peP, or pEP, represent the bottom of the sea, CD its surface, PD its depth; also let o be the centre of the spheroid and origin of the co-ordinates, and om the radius. Imagine m to be a fluid particle at any point below the surface of the fluid—at the bottom, for example. It is evident that this particle, moved by rotation alone, would be carried to B without changing its distance from the centre of the spheroid, or from the axis of rotation; so that the arcs Pm, PB, are equal to each other, as also the radii om, oB. If peP be assumed as a given meridian, the origin of the time, and γ the first point of Aries, then γPB is the longitude of the particle when arrived at B, and EoB is its latitude.

fig. 60.

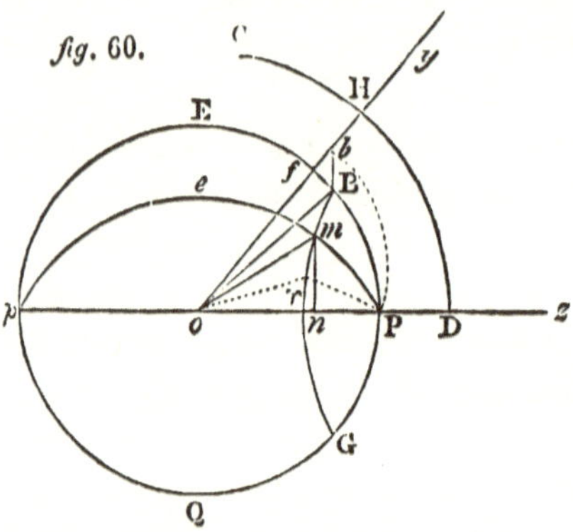

Now, if the disturbing forces were to act on the particle during its rotation from m to B, they would cause it to move to b, some point not far from B. By the disturbing forces alone, the longitude of the particle at B would be increased by the very small angle BPb; the latitude would be diminished by the very small angle Bob, and its distance from the centre of the spheroid increased by fb. The angle γPB is the rotation of the earth, and any may be represented by $nt + \varpi$, since it is proportional to the time, (by Article 213;) but in the time t, the disturbing forces bring the particle to b: therefore the angle $nt + \varpi$ must be increased by BPb or v. Hence

$$\gamma Pb = nt + \varpi + v.$$

Again, if θ be the complement of the latitude EoB, and u, its very small increment, Bob, the angle

$$PoB = \theta + u.$$

In the same manner, if s be the increment of the radius r, then

$$ob = r + s.$$

Hence the co-ordinates of the particle at b are,

$$x = (r + s) \cos (\theta + u),$$
$$y = (r + s) \sin (\theta + u) \cos (nt + \varpi + v),$$
$$z = (r + s) \sin (\theta + u) \sin (nt + \varpi + v).$$

270. v and u very nearly represent the motion of the particle in longitude and latitude estimated from the terrestrial meridian PEp. These are so small, compared with nt the rotatory motion of the earth, that their squares may be omitted. But although the lateral motions v, u of the particle be very small, they are much greater than s, the increase in the length of the radius.

271. If these values of x, y, z, be substituted in (57) the general equation of the motion of fluids; and if to abridge

$$X\delta x + Y\delta y + Z\delta z = \delta V, \text{ then}$$

$$r^2\delta\theta\left\{\left(\frac{d^2u}{dt^2} - 2n \sin\theta \cos\theta\left(\frac{dv}{dt}\right)\right\}$$

$$+ r^2\delta\varpi\left\{\sin^2\theta\left(\frac{d^2v}{dt^2}\right) + 2n \sin\theta \cos\theta\left(\frac{du}{dt}\right) + \frac{2n \sin^2\theta}{r}\left(\frac{ds}{dt}\right)\right\} \quad (70)$$

$$+ \delta r\left\{\left(\frac{d^2s}{dt^2}\right) - 2nr \sin^2\theta\left(\frac{dv}{dt}\right)\right\}$$

$$= \frac{n^2}{2}\,\delta\{(r + s) \sin (\theta + u)\}^2 + \delta V - \frac{\delta p}{\rho},$$

will determine the oscillations of a particle in the interior of the fluid when troubled by the action of the sun and moon. This equation, however, requires modification for a particle at the surface.

Equation at the Surface.

272. If DH, fig. 60, be the surface of the sea undisturbed in its rotation, the particle at B will only be affected by gravitation and the pressure of the surrounding fluid; but when by the action of the sun and moon the tide rises to y, and the particle under consideration is brought to b, the forces which there act upon it will be gravitation, the pressure of the surrounding fluid, the action of the sun and moon, and the pressure of the small column of water between H and y. But the gravitation acting on the particle at b is also different from that which affects it when at B, for b being farther from the centre of gravity of the system of the earth and its fluids, the gravity of the particle at b must be less than at B, consequently the centrifugal force

must be greater: the direction of gravitation is also different at the points B and b.

273. In order to obtain an equation for the motion of a particle at the surface of the fluid, suppose it to be in a state of momentary equilibrium, then as the differentials of v, u, and s express the oscillations of the fluid about that state, they must be zero, which reduces the preceding equation to

$$\frac{n^2}{2}\,\delta\{(r+s)\sin(\theta+u)\}^2 + (\delta V) = 0; \qquad (71)$$

for as the pressure at the surface is zero, $\delta p = 0$, and (δV) represents the value of δV corresponding to that state. Thus in a state of momentary equilibrium, the forces (δV), and the centrifugal force balance each other.

274. Now δV is the sum of all the forces acting on the particle when troubled in its rotation into the elements of their directions, it must therefore be equal to (δV), the same sum suited to a state of momentary equilibrium, together with those forces which urge the particle when it oscillates about that state, into the elements of their directions. But these are evidently the variation in the weight of the little column of water Hy, and a quantity which may be represented by $\delta V'$, depending on the difference in the direction and intensity of gravity at the two points B and b, caused by the change in the situation of the attracting mass in the state of motion, and by the attraction of the sun and moon.

275. The force of gravity at y is so nearly the same with that at the surface of the earth, that the difference may be neglected: and if y be the height of the little column of fluid Hy, its weight will be gy when the sea is in a state of momentary equilibrium; when it oscillates about that state, the variation in its weight will be $g\delta y$, g being the force of gravity; but as the pressure of this small column is directed towards the origin of the co-ordinates and tends to lessen them, it must have a negative sign. Hence in a state of motion,

$$\delta V = (\delta V) + \delta V' - g\delta y,$$

whence

$$(\delta V) = \delta V - \delta V' + g\delta y.$$

276. When the fluid is in momentary equilibrio, the centrifugal force is

$$\frac{n^2}{2}\{(r+s)\sin(\theta+u)\}^2;$$

but it must vary with δy, the elevation of the particle above the surface of momentary equilibrium. The direction Hy does not coincide with that of the terrestrial radius, except at the equator and pole, on account of the spheroidal form of the earth; but as the compression of the earth is very small, these directions may be esteemed the same in the present case without sensible error; therefore $r + s - y$ may be regarded as the value of the radius at y. Consequently

$$- \delta y . rn^2 \sin^2 \theta$$

is the variation of the centrifugal force corresponding to the increased height of the particle; and when compared with $- g\delta y$ the gravity of this little column, it is of the order $\frac{n^2 r}{g}$, the same with the ratio of the centrifugal force to gravity at the equator, or to $\frac{1}{288}$, and therefore may be omitted; hence equation (71) becomes

$$\delta V - \delta V' + g\delta y + \frac{n^2}{2} \delta\{(r + s) \sin(\theta + u)\}^2 = 0.$$

277. As the surface of the sea differs very little from that of a sphere, δr may be omitted; consequently

if $$\frac{n^2}{2} \delta\{(r + \theta) \sin(\theta + u)\}^2$$

be eliminated from equation (70), the result will be

$$r^2\delta\theta \left\{\left(\frac{d^2u}{dt^2}\right) - 2n \sin\theta \cos\theta \left(\frac{dv}{dt}\right)\right\}$$

$$+ r^2\delta\varpi \left\{\sin^2\theta \left(\frac{d^2v}{dt^2}\right) + 2n \sin\theta \cos\theta \left(\frac{du}{dt}\right) + 2n \sin^2\theta \left(\frac{ds}{dt}\right)\right\}$$

$$= - g\delta y + \delta V', \qquad (72)$$

which is the equation of the motion of a particle at the surface of the sea. The variations δy, $\delta V'$ correspond to the two variables θ and ϖ.

278. To complete the theory of the motions of the atmosphere and ocean, the equation of the continuity of the fluid must now be found.

Continuity of Fluids.

Suppose $m'h$, fig. 61, to be an indefinitely small rectangular portion of the fluid mass, situate at B, fig. 60, and suppose the solid to be formed by the imaginary rotation of the area $Bnhh'$ about the axis oz; the centre of gravity of Bnhh' will describe an arc, which on account of the smallness of the solid, may without sensible error be represented by mn, its radius being mA; hence the arc mn is $m\text{A} \times mn$. Now the area Bnhh' multiplied by $m\text{A} \times mn$, is equal to the solid $m'h$, supposing it indefinitely small and rectangular.

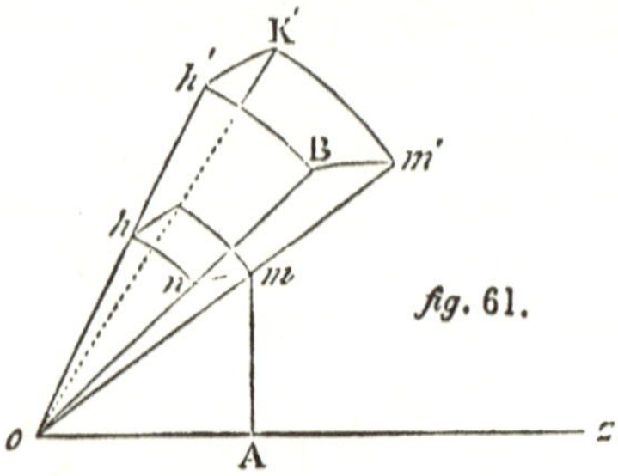

fig. 61.

The colatitude of the point B or Aom $= \theta$, the longitude of B is $nt + \varpi$, then the indefinitely small increments of these angles are $m'o\text{K}' = d\theta$, $m'o\text{B} = d\varpi$, for as the figure is independent of the time, nt is constant. Hence if the radii oB, on, be represented by r' and r, the sectors Boh', noh, are $r'^2 d\theta$ and $r^2 d\theta$; hence

$$\text{the area} \qquad \text{B}nhh' = \frac{(r'^2 - r^2)}{2}\, d\theta = \frac{(r'+r)(r'-r)}{2}\, d\theta.$$

But as the thickness is indefinitely small,

$$r' + r = 2r,\ r' - r = dr;$$

therefore the area $\qquad$ $\text{B}nhh' = rdr.d\theta.$

Again, $\qquad$ $\text{A}m = r \sin\theta,$

consequently, $\qquad$ $\text{A}m.mn = rd\varpi \sin\theta,$

and thus the volume $\qquad$ $m'h = r^2 dr d\theta d\varpi \sin\theta;$

and if ρ be the density,

$$dm = \rho r^2 dr d\theta d\varpi \sin\theta.$$

But in consequence of the disturbing forces, r, θ, and ϖ, become $r + s$, $\theta + u$, $\varpi + v$, after the time t, and dr, $d\theta$, $d\varpi$,

$$\text{become} \qquad dr + \frac{ds}{dr} dr,\ d\theta + \frac{du}{d\theta}.\, d\theta,\ d\varpi + \frac{dv}{d\varpi}.d\varpi;$$

also the density is changed to $\rho + \rho'$. If these values be put in the preceding expression for the solid dm, it becomes after the time t equal to

$$(\rho + \rho')\ (r + s^2)\ \left(1 + \frac{ds}{dr}\right)\left(1 + \frac{du}{d\theta}\right)\left(1 + \frac{dv}{d\varpi}\right) dr d\theta d\varpi \sin(\theta + u),$$

but this must be equal to the original mass; hence

$$(\rho + \rho')\ (r + s^2)\ \left(1 + \frac{ds}{dr}\right)\left(1 + \frac{du}{d\theta}\right)\left(1 + \frac{dv}{d\varpi}\right) \sin(\theta + u) = \rho r^2 \sin\theta.$$

If the squares and products of

$$s, \frac{ds}{dr}, \frac{du}{d\theta}, \frac{dv}{d\varpi}$$

be omitted, and observing that

$$2rs + r^2\frac{ds}{dr} = \frac{d.r^2s}{dr},$$

and $$\sin(\theta + u) = \sin\theta + u\cos\theta;$$

for as u is very small, the arc may be put for the sine, and unity for the cosine, the equation of the continuity of the fluid is

$$0 = r^2\left(\rho' + \rho\left(\left(\frac{du}{d\theta}\right) + \left(\frac{dv}{d\varpi}\right) + \frac{u\cos\theta}{\sin\theta}\right) + \rho\left(\frac{d.r^2s}{dr}\right),\right. \quad (73)$$

expressed in polar co-ordinates.

279. The equations (70), (72), and (73), are perfectly general; and therefore will answer either for the oscillations of the ocean or atmosphere.

Oscillations of the Ocean.

280. The density of the sea is constant, therefore $\rho' = 0$; hence the equation of continuity becomes

$$0 = \left(\frac{d.r^2s}{dr}\right) + r^2\left\{\left(\frac{du}{d\theta}\right) + \left(\frac{dv}{d\varpi}\right) + \frac{u\cos\theta}{\sin\theta}\right\}.$$

In order to find the integral of this equation with regard to r only, it may be assumed, that all the particles that are on any one radius at the origin of the time, will remain on the same radius during the motion; therefore r, v, and u will be nearly the same on the small part of the terrestrial radius between the bottom and surface of the sea; hence, the integral will be

$$0 = r^2s - (r^2s) + r^2\gamma\left\{\left(\frac{du}{d\theta}\right) + \frac{dv}{d\varpi} + \frac{u\cos\theta}{\sin\theta}\right\}$$

(r^2s) is the value of r^2s at the surface of the sea, but the change in the radius of the earth between the bottom and surface of the sea is so small, that $r^2(s)$ may be put for (r^2s); then dividing the whole by r^2, and neglecting the terms $\frac{2\gamma(s)}{r}$, which is the ratio of the depth of the sea to the terrestrial radius, and therefore very small, the mean depth even of the Pacific ocean being only about four miles, whereas the mean radius of the earth is nearly 4000 miles; the preceding equation becomes

$$0 = s - (s) + \gamma\left\{\left(\frac{du}{d\theta}\right) + \left(\frac{dv}{d\varpi}\right) + \frac{u\cos\theta}{\sin\theta}\right\}. \qquad (74)$$

Now $\gamma + s - (s)$ is the whole depth of the sea from the bottom to the highest point to which the tides rise at its surface of momentary equilibrium; and γ varies with the angles ϖ and θ; hence at the surface of equilibrium, it becomes

$$\gamma + u\frac{d\gamma}{d\theta} + v\frac{d\gamma}{d\varpi};$$

and as y is the height of a particle above the surface of equilibrium, it follows that

$$\gamma + s - (s) = -y + \gamma + u\frac{d\gamma}{d\theta} + v\frac{d\gamma}{d\varpi},$$

or

$$s - (s) = -y + u\frac{d\gamma}{d\theta} + v\frac{d\gamma}{d\varpi}.$$

Whence the equation of continuity becomes

$$y = -\frac{d(\gamma u)}{d\theta} - \frac{d(\gamma v)}{d\varpi} - \frac{\gamma u\cos\theta}{\sin\theta}. \qquad (75)$$

281. In order to apply the other equations to the motion of the sea, it must be observed that a fluid particle at the bottom of the sea would in its rotation from m to B always touch the spheroid, which is nearly a sphere; therefore the value of s would be very small for that particle, and would be to v, u, of the order of the eccentricity of the spheroid, to its mean radius taken as unity; consequently at the bottom of the sea, s may be omitted in comparison of u, v. But it appears from equations (74), that $s - (s)$ is a function of u and v independent of r, when the very small quantity $\frac{2\gamma(s)}{r}$ is omitted: hence s is the same throughout every part of the radius r, as it is at the bottom, and may therefore be omitted throughout the whole depth, when compared with u and v, so that equation (72) of the surface of the fluid is reduced to

$$r^2\delta\theta\left\{\left(\frac{d^2u}{dt^2}\right) - 2n\sin\theta\cos\theta\left(\frac{dv}{dt}\right)\right\}, \qquad (76)$$

$$+ r^2\delta\varpi\left\{\sin^2\theta\left(\frac{d^2v}{dt^2}\right) + 2n\sin\theta\cos\theta\left(\frac{du}{dt}\right)\right\} = -g\delta y + \delta V'.$$

282. When the fluid mass is in momentary equilibrium, the equation for the motion of a particle in the interior of the fluid becomes

$$0 = \tfrac{1}{2}n^2\delta\{(r + s)\sin(\theta + u)\}^2 + (\delta V) - \frac{(\delta p)}{\rho},$$

where (δV), (δp), are the values of δV and δp suited to that state. But we may suppose that in a state of motion,

$$\delta V = (\delta V) + \delta V', \text{ and } \delta p = (\delta p) + \delta p';$$

whence $$(\delta V) = \delta V - \delta V', (\delta p) = \delta p - \delta p',$$

and $$\tfrac{1}{2}n^2\delta\{(r + s)\sin(\theta + u)\}^2 = \delta V' - \delta V + \frac{\delta p}{\rho} - \frac{\delta p'}{\rho}.$$

283. If the first member of this expression be eliminated from equation (70), with regard to the independent variation of r alone, it gives

$$\frac{d\left(V' - \dfrac{p'}{\rho}\right)}{dr} = \left(\frac{d^2s}{dt^2}\right) - 2nr\sin^2\theta\left(\frac{dv}{dt}\right). \qquad (77)$$

284. Now $n\left(\dfrac{dv}{dt}\right)$ is of the order y, s, or $\dfrac{7s}{r}$; for if the co-efficients of $\delta\theta$, $\delta\varpi$, be each made zero in equation (76), it will give

$$r^2\left(\frac{d^2u}{dt^2}\right) - 2nr^2\sin\theta\cos\theta\left(\frac{dv}{dt}\right) = -g\left(\frac{dy}{d\theta}\right) + \left(\frac{dV'}{d\theta}\right),$$

$$r^2\sin^2\theta\left(\frac{d^2v}{dt^2}\right) + 2nr^2\sin\theta\cos\theta\left(\frac{du}{dt}\right) = -g\left(\frac{dy}{d\varpi}\right) + \left(\frac{dV'}{d\varpi}\right);$$

add the differential of the last equation relative to t, to the first equation multiplied by $\quad -2n\sin\theta\cos\theta$

and let the second member of this equation be represented by

$$y'.r^2\sin^2\theta,$$

then divide by $\quad r^2\sin^2\theta$, and put $2n\cos\theta = a$,

and there will be found the linear equation

$$\left(\frac{d^3v}{dt^3}\right) + a^2\left(\frac{dv}{dt}\right) = y'.$$

The value of $\dfrac{dv}{dt}$ obtained from the integral of this equation will be a function of y', and as y' is a function of y and V', each of which is of the order s or $\dfrac{7s}{r}$, $\dfrac{dv}{dt}$; consequently,

$$\frac{d\left(V' - \dfrac{p'}{\rho}\right)}{dr}$$

is of the same order. If then equation (77), be multiplied by dr its integral will be

$$V' - \frac{p'}{\rho} = \int dr \left\{\left(\frac{d^2s}{dt^2}\right) - 2nr\sin^2\theta\left(\frac{dv}{dt}\right)\right\} + \lambda.$$

285. Since this equation has been integrated with regard to r only, λ must be a function of θ, ϖ, and t, independent of r, according to the theory of partial equations. And as the function in r is of the order $\frac{\gamma s}{r}$ it may be omitted; and then

$$\delta V' - \frac{\delta p'}{\rho} = \delta\lambda,$$

by which equation (70) becomes

$$r^2\delta\theta\left\{\left(\frac{d^2u}{dt^2}\right) - 2n\sin\theta\cos\theta\left(\frac{dV}{dt}\right)\right\},$$

$$+\, r^2\delta\varpi\left\{\sin^2\theta\left(\frac{d^2v}{dt^2}\right) + 2n\sin\theta\cos\theta\left(\frac{du}{dt}\right)\right\} = \delta\lambda.$$

286. But as $\delta\lambda$ does not contain r, s, or y, it is independent of the depth of the particle; hence this equation is the same for a particle at the surface, or in its neighbourhood, consequently it must coincide with equation (76); and therefore

$$\delta\lambda = \delta V' - g\delta y.$$

287. Thus it appears, that the whole theory of the tides would be determined if integrals of the equations

$$r^2\delta\theta\left\{\left(\frac{d^2u}{dt^2}\right) - 2n\sin\theta\cos\theta\left(\frac{dv}{dt}\right)\right\}$$

$$+\, r^2\delta\varpi\left\{\sin^2\theta\left(\frac{d^2v}{dt^2}\right) + 2n\sin\theta\cos\theta\left(\frac{du}{dt}\right)\right\} = -\,g\delta y + \delta V'$$

$$y = -\frac{d(\gamma u)}{d\theta} - \frac{d(\gamma v)}{d\varpi} - \frac{\gamma u\cos\theta}{\sin.\theta}$$

could be found, for the horizontal flow might be obtained from the first, by making the co-efficients of the independent quantities $\delta\theta$, $\delta\varpi$, separately zero, then the height to which they rise would be found from the second. This has not yet been done, as none of the known methods of analysis have hitherto succeeded.

288. These equations have been formed on the hypothesis of the earth being entirely covered by the sea; hence the integrals, if they

could be found, would be inadequate to determine the oscillations of the ocean retarded or accelerated by the continents, islands, and innumerable other causes, beyond the reach of analysis. No attempt is therefore made to integrate the equations; but the theory of the tides is determined by comparing the general relations which subsist between the observed phenomena and the causes which produce them.

289. In order to integrate the equation of continuity, it was assumed that if the angles P*ob*, *m*P*b*, or rather

$$u, \frac{du}{dt}, \quad v, \frac{dv}{dt},$$

be the same for every particle situate on the same radius throughout the whole depth of the sea at the beginning of the motion, they will always continue to be the same for that set of particles during their motion, therefore all the fluid particles that are at the same instant on any one radius, will continue very nearly on that radius during the oscillations of the fluid. Were this rigorously true, the horizontal flow of the tides would be isochronous, like the oscillations of a pendulum, and their velocity would be inversely as their depth, provided the particles had no motion in latitude; and it may be nearly so in the Pacific, whose mean depth is about four miles, and where the tides only rise to about five feet; but it is very far from being the case in shallow seas, and on the coasts where the tides are high; because the condition of isochronism depends on the omission of quantities of the order of the ratio of the height of the tides to the depth of the sea.

290. The reaction of the sea on the terrestrial spheroid is so small that it is omitted. The common centre of gravity of the spheroid and sea is not changed by this reaction, and therefore the ratio of the action of the sea on the spheroid, is to the reaction of the spheroid on the sea, as the mass of the sea to the solid mass; that is, as the depth of the sea to the radius of the earth, or at most as 1 to 1000, assuming the mean depth of the sea to be four miles. For that reason u, v, express the true velocity of the tides in longitude and latitude, as they were assumed to be.

On the Atmosphere.

291. Experience shows the atmosphere to be an elastic fluid, whose density increases in proportion to the pressure. It is subject to changes of density from the variation of temperature in different latitudes, at different heights, and from various other causes; but in this investigation the temperature is assumed to be constant.

292. Since the air resists compression equally in all directions, the height of the atmosphere must be unlimited if its atoms be infinitely divisible. Some considerations, however, induced Dr. Wollaston to suppose that the earth's atmosphere is of finite extent, limited by the weight of ultimate atoms of definite magnitude, no longer divisible by repulsion of their parts. But whether the particles of the atmosphere be infinitely divisible or not, all phenomena concur in proving its density to be quite insensible at the height of about fifty miles.

Density of the Atmosphere.

293. The law by which the density of the air diminishes as the height above the surface of the sea increases, will appear by considering ρ, ρ' ρ'', to be the densities of three contiguous strata of air, the thickness of each being so small that the density may be assumed uniform throughout each stratum. Let p be the pressure of the superincumbent air on the lowest stratum, p' the pressure on the next, and p'' the pressure on the third; and let m be a coefficient, such that $\rho = \alpha p$. Then, because the densities are as the pressures,

$$\rho' = \alpha p', \qquad \text{and} \qquad \rho'' = \alpha p''.$$

Hence, $\rho - \rho' = \alpha\,(p - p')$ and $\rho - \rho = \alpha\,(p' - p'')$.

But $p - p'$ is equal to the weight of the first of these strata, and $p' - p''$ is equal to that of the second: hence

$$\rho - \rho' : \rho' - \rho'' :: \rho : \rho';$$

consequently $$\rho\rho'' = \rho'^2.$$

The density of the middle stratum is therefore a mean proportional between the densities of the other two; and whatever be the number of equidistant strata, their densities are in continual proportion.

294. If the heights therefore, from the surface of the sea, be taken in an increasing arithmetical progression, the densities of the strata

of air will decrease in geometrical progression, a property that logarithms possess relatively to their numbers.

295. All the circumstances both of the equilibrium and motion of the atmosphere may be determined from equation (70), if the quantities it contains be supposed relative to that compressible fluid instead of to the ocean.

Equilibrium of the Atmosphere.

296. When the atmosphere is in equilibrio v, u, and s are zero, which reduces equation (70) to

$$\frac{n^2}{2} \cdot r^2 . \sin^2 \theta + V - \int \frac{\delta p}{\rho} = \text{constant}.$$

Suppose the atmosphere to be every where of the same density as at the surface of the sea, let h be the height of that atmosphere which is very small, not exceeding $5\frac{1}{2}$ miles, and let g be the force of gravity at the equator; then as the pressure is proportional to the density, $p = h . g . \rho$,

and
$$\int \frac{\delta p}{\rho} = hg . \log. \rho,$$

consequently the preceding equation becomes

$$hg . \log \rho = \text{constant} + V + \frac{n^2}{2} . r^2 . \sin^2 \theta.$$

At the surface of the sea, V is the same for a particle of air, and for the particle of the ocean adjacent to it; but when the sea is in equilibrio
$$V + \frac{n^2}{2} . r^2 . \sin^2 \theta = \text{constant},$$

therefore ρ is constant, and consequently the stratum of air contiguous to the sea is every where of the same density.

297. Since the earth is very nearly spherical, it may be assumed that r the distance of a particle of air from its centre is equal to $R + r'$, R being the terrestrial radius extending to the surface of the sea, and r' the height of the particle above that surface. V, which relates to the surface of the sea, becomes at the height r';

$$V' = V + r' \left(\frac{dV}{dr}\right) + \&c.$$

by Taylor's theorem, consequently the substitution of $R + r'$ for r in the value of hg log. ρ gives

$$hg \,.\, \log \rho = \text{constant} + V + r'\left(\frac{dV}{dr}\right) + \frac{r'^2}{2}\left(\frac{d^2V}{dr^2}\right)$$
$$+ \frac{n^2}{2} \,.\, R^2 \,.\, \sin^2\theta + n^2 \,.\, Rr' \,.\, \sin^2\theta$$

$V, \left(\frac{dV}{dr}\right)$, &c. relate to the surface of the sea where

$$V + \frac{n^2}{2} \,.\, R^2 \sin^2\theta = \text{constant},$$

and as $$-\left(\frac{dV}{dr}\right) - n^2 \,.\, R \,.\, \sin^2\theta,$$

is the effect of gravitation at that surface, it may be represented by g', whence $$hg \,.\, \log \rho = \text{constant} - r'g' + \frac{r'^2}{2} \,.\, \left(\frac{d^2V}{dr^2}\right).$$

298. Since $\left(\frac{d^2V}{dr^2}\right)$ is multiplied by the very small quantity r'^2, it may be integrated in the hypothesis of the earth being a sphere; but in that case $$-\left(\frac{dV}{dr}\right) = g' = \frac{m}{R^2}$$

m being the mass of the earth;

hence $$\left(\frac{d^2V}{dr^2}\right) = -\frac{2m}{R^3} = -\frac{2g'}{R};$$

consequently the preceding equation becomes

$$\log \rho = -\frac{r'}{h} \,.\, \frac{g'}{g}\left(1 + \frac{r'}{R}\right);$$

whence $$\rho = \rho' \,.\, c^{-\frac{r'g'}{hg} \,.\, \left(1 + \frac{r'}{R}\right)};$$

an equation which determines the density of the atmosphere at any given height above the level of the sea; c is the number whose logarithm is unity, and ρ' a constant quantity equal to the density of the atmosphere at the surface of the sea.

299. If g' and g be the force of gravity at the equator and in any other latitude, they will be proportional to l' and l, the lengths of the pendulum beating seconds at the level of the sea in these two places; hence l' and l, which are known by experiment, may be substituted for g' and g, and the formula becomes

$$\rho = \rho' \,.\, c^{-\frac{r'l'}{hl}\left(1 + \frac{r'}{R}\right)} \qquad (78)$$

Whence it appears that strata of the same density are every where very nearly equally elevated above the surface of the sea.

300. By this expression the density of the air at any height may be found, say at fifty-five miles. $\frac{r'}{R}$ is very small and may be neglected; and l may be made equal to l' without sensible error; hence

$$\rho = \rho' c^{-\frac{r'}{h}}.$$

Now the height of an atmosphere of uniform density is only about $h = 5\frac{1}{2}$ English miles;

hence if $\quad r' = 10h = 55, \rho = \rho' c^{-10},$

and as $\quad c = 2.71828, \rho = \frac{\rho'}{22026},$

so that the density at the height of 55 English miles is extremely small, which corresponds with what was said in article 292.

301. From the same formula the height of any place above the level of the sea may be found; for the densities ρ' and ρ, and consequently h, are given by the height of the barometer, l' and l, the lengths of the seconds' pendulum for any latitude are known by experiment; and R, the radius of the earth is also a given quantity; hence r' may be found. But in estimating the heights of mountains by the barometer, the variation of gravity at the height r' above the level of the sea cannot be omitted, therefore $\frac{l'-l}{l'} r'$ must be included in the preceding formula.

Oscillations of the Atmosphere.

302. The atmosphere has the form of an ellipsoid flattened at the poles, in consequence of its rotation with the earth, and its strata by article 299, are everywhere of the same density at the same elevation above the surface of the sea. The attraction of the sun and moon occasions tides in the atmosphere perfectly similar to those of the ocean; however, they are probably affected by the rise and fall of the sea.

303. The motion of the atmosphere is determined by equations (70), (73), which give the tides of the ocean, with the exception of a small change owing to the elasticity of the air; hence the term $\frac{\delta p}{\rho}$, expressing the ratio of the pressure to the density cannot be omitted as it was in the case of the sea.

*

Let $\rho = (\rho) + \rho'$; (ρ) being the density of the stratum in equilibrio, and ρ' the change suited to a state of motion;

hence $$p = hg\,((\rho) + \rho'),$$

and $$\frac{\delta p}{\rho} = hg\,\frac{\delta(\rho)}{(\rho)} + g\,\frac{\delta(h\rho')}{(\rho)}.$$

Let $$\frac{h\rho'}{(\rho)} = y', \text{ then } \frac{\delta p}{\rho} = hg\,\frac{\delta(\rho)}{(\rho)} + g\delta y'.$$

304. The part $hg\dfrac{\delta(\rho)}{(\rho)}$ vanishes, because in equilibrio

$$\frac{n^2}{2}\,\delta\{(r+s)\sin(\theta+u)\}^2 + (\delta V) - hg\,\frac{\delta(\rho)}{(\rho)} = 0,$$

therefore $$\frac{\delta p}{\rho} = g\delta y'.$$

Let ϕ be the elevation of a particle of air above the surface of equilibrium of the atmosphere which corresponds with y, the elevation of a particle of water above the surface of equilibrium of the sea. Now at the sea $\phi = y$, for the adjacent particles of air and water are subject to the same forces; but it is necessary to examine whether the supposition of $\phi = y$, and of y being constant for all the particles of air situate on the same radius are consistent with the equation of continuity (73), which for the atmosphere is

$$0 = r^2\left\{\rho' + (\rho)\left\{\left(\frac{du}{d\theta}\right) + \left(\frac{dv}{d\varpi}\right) + \frac{u\,.\cos\theta}{\sin\theta}\right\}\right\} + (\rho)\,.\left(\frac{d\,.\,r^2 s}{dr}\right).$$

If the value of $\dfrac{\rho'}{(\rho)}$ from this equation be substituted in $h\,\dfrac{\rho'}{(\rho)} = y'$, it becomes

$$y' = -h\,.\left\{\left(\frac{d\,.\,r^2 s}{r^2.\,dr}\right) + \left(\frac{du}{d\theta}\right) + \left(\frac{dv}{d\varpi}\right) + \frac{u\,.\cos\theta}{\sin\theta}\right\}$$

The part of s that depends on the variation of the angles θ and ϖ is so small, that it may be neglected, consequently $s = \phi$; and if $\phi = y$ then $\left(\dfrac{d\phi}{dr}\right) = 0$. Since the value of ϕ is the same for all the particles situate on the same radius. Also y is of the order h or $\dfrac{n^2}{g}$;

consequently $$y' = -h\,.\left\{\left(\frac{du}{d\theta}\right) + \left(\frac{dv}{d\varpi}\right) + \frac{u\,.\cos\theta}{\sin\theta}\right\} \qquad (79)$$

then u and v being the same for all the particles situate primitively on the same radius, the value of y' will be the same for all these particles, and as quantities of the order δs are omitted, equation (70) becomes

$$r^2\delta\theta\left\{\left(\frac{d^2u}{dt^2}\right) - 2n\sin\theta\cos\theta\left(\frac{dv}{dt}\right)\right\}$$
$$+ r^2\delta\varpi\left\{\sin^2\theta\left(\frac{d^2v}{dt^2}\right) + 2n\sin\theta\cos\theta\left(\frac{du}{dt}\right)\right\} \quad (80)$$
$$= \delta V' - g\delta y' - g\delta y.$$

Thus the equations that determine the oscillations of the atmosphere only differ from those that give the tides by the small quantity $g\delta y'$, depending on the elasticity of the air.

305. Finite values of the equations of the motion of the atmosphere cannot be obtained; therefore the ebb and flow of the atmosphere may be determined in the same manner as the tides of the ocean, by estimating the effects of the sun and moon separately. This can only be effected by a comparison of numerous observations.

Oscillations of the Mercury in the Barometer.

306. Oscillations in the atmosphere cause analogous oscillations in the barometer. For suppose a barometer to be fixed at any height above the surface of the sea, the height of the mercury is proportional to the pressure on that part of its surface that is exposed to the action of the air. As the atmosphere rises and falls by the action of the disturbing forces like the waves of the sea, the surface of the mercury is alternately more or less pressed by the variable mass of the atmosphere above it. Hence the density of the air at the surface of the mercury varies for two reasons; first, because it belonged to a stratum which was less elevated in a state of equilibrium by the quantity y, and secondly, because the density of a stratum is augmented when in motion by the quantity $\frac{(\rho)}{h}\,.\,y$. Now if h be the height of the atmosphere in equilibrio when its density is uniform, and equal to (ρ), then

$$h : y :: (\rho) : y\,.\,\frac{(\rho)}{h},$$

the increase of density in a state of motion from the first cause. In the same manner, $y' . \frac{(\rho)}{h}$ is the increase of density from the second cause. Thus the whole increase is

$$(\rho)\,\frac{(y'+y)}{h}.$$

And if H be the height of the mercury in the barometer when the atmosphere is in equilibrio, its oscillations when in motion will be expressed by

$$H\,\frac{(y'+y)}{h}. \qquad (81)$$

The oscillations of the mercury are therefore similar at all heights above the level of the sea, and proportional in their extent to the height of the barometer.

Conclusion.

307. The account of the first book of the *Mècanique Céleste* is thus brought to a conclusion. Arduous as the study of it may seem, the approach in every science, necessarily consisting in elementary principles, must be tedious: but let it not be forgotten, that many important truths, coeval with the existence of matter itself, have already been developed; and that the subsequent application of the principles which have been established, will lead to the contemplation of the most sublime works of the Creator. The general equation of motion has been formed according to the primordial laws of matter; and the universal application of this one equation, to the motion of matter in every form of which it is susceptible, whether solid or fluid, to a single particle, or to a system of bodies, displays the essential nature of analysis, which comprehends every case that can result from a given law. It is not, indeed, surprising that our limited faculties do not enable us to derive general values of the unknown quantities from this equation: it has been accomplished, it is true, in a few cases, but we must be satisfied with approximate values in by much the greater number of instances. Several circumstances in the solar system materially facilitate the approximations; these La Place has selected with profound judgment, and employed with the greatest dexterity.

BOOK II.

CHAPTER I.

PROGRESS OF ASTRONOMY.

308. THE science of astronomy was cultivated very early, and many important observations and discoveries were made, yet no accurate inferences leading to the true system of the world were drawn from them, until a much later period. It is not surprising, that men deceived by appearances, occasioned by the rotation of the earth, should have been slow to believe the diurnal motion of the heavens to be an illusion; but the absurd consequence which the contrary hypothesis involves, convinced minds of a higher order, that the apparent could not be the true system of nature.

Many of the ancients were aware of the double motion of the earth; a system which Copernicus adopted, and confirmed by the comparison of a series of observations, that had been accumulating for ages; from these he inferred that the precession of the equinoxes might be attributed to a motion in the earth's axis. He ascertained the revolution of the planets round the sun, and determined the dimensions of their orbits, till then unknown. Although he proved these truths by evidence which has ultimately dissipated the erroneous theories resulting from the illusions of the senses, and overcame the objections which were opposed to them by ignorance of the laws of mechanics, this great philosopher, constrained by the prejudices of the times, only dared to publish the truths he had discovered, under the less objectionable name of hypotheses.

In the seventeenth century, Galileo, assisted by the discovery of the telescope, was the first who saw the magnificent system of Jupiter's satellites, which furnished a new analogy between the planets and the earth: he discovered the phases of Venus, by which he removed all doubts of the revolution of that planet round the sun. The bright spots which he saw in the moon beyond the line which separates the enlightened from the obscure part, showed the existence and height of its mountains. He observed the spots and rotation of the sun, and the singular appearances exhibited by the rings of Saturn; by which discoveries the rotation of the earth was confirmed: but if the rapid progress of mathematical science had not concurred to establish this essential truth, it would have been overwhelmed and stifled by fanatical zeal. The opinions of Galileo were denounced as heretical by the Inquisition, and he was ordered by the Church of Rome to retract them. At a late period he ventured to promulgate his discoveries, but in a different form, vindicating the system of Copernicus; but such was the force of superstition and prejudice, that he, who was alike an honour to his country, and to the human race, was again subjected to the mortification of being obliged to disavow what his transcendent genius had proved to be true. He died at Arcetri in the year 1642, the year in which Newton was born, carrying with him, says La Place, the regret of Europe, enlightened by his labours, and indignant at the judgment pronounced against him by an odious tribunal.

The truths discovered by Galileo could not fail to mortify the vanity of those who saw the earth, which they conceived to be the centre and primary object of creation, reduced to the rate of but a small planet in a system, which, however vast it may seem, forms but a point in the scale of the universe.

The force of reason by degrees made its way, and persecution ceased to be the consequence of stating physical truths, though many difficulties remained to impede its progress, and no ordinary share of moral courage was required to declare it: 'prejudice,' says an eminent author, 'bars up the gate of knowledge; but he who would learn, must despise the timidity that shrinks from wisdom, he must hate the tyranny of opinion that condemns its pursuit: wisdom is only to be obtained by the bold; prejudices must first be overcome, we must learn to scorn names, defy idle fears, and use the powers of nature to give us the mastery of nature. There are virtues in plants, in metals, even in woods, that to seek alarms the feeble, but to possess constitutes the mighty.'

About the end of the sixteenth, or the beginning of the seventeenth century, Tycho Brahe made a series of correct and numerous observations on the motion of the planets, which laid the foundation of the laws discovered by his pupil and assistant, Kepler.

Tycho Brahe, however, would not admit of the motion of the earth, because he could not conceive how a body detached from it could follow its motion: he was convinced that the earth was at rest, because a heavy body, falling from a great height, falls nearly at the foot of the vertical.

Kepler, one of those extraordinary men, who appear from time to time, to bring to light the great laws of nature, adopted sounder views. A lively imagination, which disposed him eagerly to search for first causes, tempered by a severity of judgment that made him dread being deceived, formed a character peculiarly fitted to investigate the unknown regions of science, and conducted him to the discovery of three of the most important laws in astronomy.

He directed his attention to the motions of Mars, whose orbit is one of the most eccentric in the planetary system, and as it approaches very near the earth in its oppositions, the inequalities of its motions are considerable; circumstances peculiarly favourable for the determination of their laws.

He found the orbit of Mars to be an ellipse, having the sun in one of its foci; and that the motion of the planet is such, that the radius vector drawn from its centre to the centre of the sun, describes equal areas in equal times. He extended these results to all the planets, and in the year 1626, published the Rudolphine Tables, memorable in the annals of astronomy, from being the first that were formed on the true laws of nature.

Kepler imagined that something corresponding to certain mysterious analogies, supposed by the Pythagoreans to exist in the laws of nature, might also be discovered between the mean distances of the planets, and their revolutions round the sun: after sixteen years spent in unavailing attempts, he at length found that the squares of the times of their sidereal revolutions are proportional to the cubes of the greater axes of their orbits; a very important law, which was afterwards found equally applicable to all the systems of the satellites. It was obvious to the comprehensive mind of Kepler, that motions so regular could only arise from some universal principle pervading the whole system. In his work *De Stella Martis*, he observes, that 'two insulated bodies would move towards one another like two magnets, describing spaces reciprocally as their masses. If the earth and moon were not held at the distance that separates them by some force, they would come in contact, the moon describing $\frac{53}{54}$ of the distance, and the earth the remainder, supposing them to be equally dense.' 'If,' he continues, 'the earth ceased to attract the waters of the ocean, they would go to the moon by the attractive force of that body. The attraction of the moon, which extends to the earth, is the cause of the ebb and flow of the sea.' Thus Kepler's work, *De Stella Martis*, contains the first idea of a principle which Newton and his successors have fully developed.

The discoveries of Galileo on falling bodies, those of Huygens on Evolutes, and the centrifugal force, led to the theory of motion in curves. Kepler had determined the curves in which the planets move, and Hook was aware that planetary motion is the result of a force of projection combined with the attractive force of the sun.

Such was the state of astronomy when Newton, by his grand and comprehensive views, combined the whole, and connected the most distant parts of the solar system by one universal principle.

Having observed that the force of gravitation on the summits of the highest mountains is nearly the same as on the surface of the earth, Newton inferred, that its influence extended to the moon, and, combining with her force of projection, causes that satellite to describe an elliptical orbit round the earth. In order to verify this conjecture, it was necessary to know the law of the diminution of gravitation. Newton considered, that if terrestrial gravitation retained the moon in her orbit, the planets must be retained in theirs by their gravitation to the sun; and he proved this to be the case, by showing the areas to be proportional to the times: but it resulted from the constant ratio found by Kepler between the squares of the times of revolutions of the planets, and the cubes of the greater axes of their orbits, that their centrifugal force, and consequently their tendency to the sun, diminishes in the ratio of the squares of their distances from his centre. Thus the law of diminution was proved with regard to the planets, which led Newton to conjecture, that the same law of diminution takes place in terrestrial gravitation.

He extended the laws deduced by Galileo from his experiments on bodies falling at the surface of the earth, to the moon; and on these principles determined the space she would move through in a second of time, in her descent towards the earth, if acted upon by the earth's attraction alone. He had the satisfaction to find that the action of the earth on the moon is inversely as the square of the distance, thus proving the force which causes a stone to fall at the earth's surface, to be identical with that which retains the moon in her orbit.

Kepler having established the point that the planets move in ellipses, having the sun in one of their foci, Newton completed his theory, by showing that a projectile might move in any of the conic sections, if acted on by a force directed to the focus, and inversely as the square of the distance: he determined the conditions requisite to make the trajectory a circle, an ellipse, a parabola, or hyperbola. Hence he also concluded, that comets move round the sun by the same laws as the planets.

A comparison of the magnitude of the orbits of the satellites and the periods of their revolutions, with the same quantities relatively to the planets, made known to him the respective masses and densities of the sun and of planets accompanied by satellites, and the intensity of gravitation at their surfaces. He observed, that the satellites move round their planets nearly as they would have done, had the planets been at rest, whence he concluded that all these bodies obey the same law of gravitation towards the sun: he also concluded, from the equality of action and re-action, that the sun gravitates towards the planets, and the planets towards their satellites; and that the earth is attracted by all bodies which gravitate towards it. He afterwards extended this law to all the particles of matter, thus establishing the general principle, that each particle of matter attracts all other particles directly as its mass, and inversely as the square of its distance.

These splendid discoveries were published by Newton in his *Principia*, a work which has been the admiration of mankind, and which will continue to be so while science is cultivated.

Referring to that stupendous effort of human genius, La Place, who perhaps only yields to Newton in priority of time, thus expresses himself in a letter to the writer of these pages:

'Je publie successivement les divers livres du cinquième volume qui doit terminer mon traité de *Mécanique Céleste*, et dans lequel je donne l'analyse historique des recherches des géomètres sur cette matière. Cela m'a fait relire avec une attention particulière l'ouvrage incomparable des *Principes Mathématiques* de la philosophie naturelle de Newton, qui contient le germe de toutes ces recherches. Plus j'ai étudié cet ouvrage, plus il m'a paru admirable, en me transportant surtout à l'époque où il a été publié. Mais en même tems que j'ai senti l'élégance de la méthode synthétique suivant laquelle Newton a présenté ses découvertes, j'ai reconnu l'indispensable nécessité de l'analyse pour approfondir les questions très difficiles qu'il n'a pu qu'effleurer par la synthèse. Je vois avec un grand plaisir vos mathématiciens se livrer maintenant à l'analyse; et je ne doute point qu'en suivant cette méthode avec la sagacité propre à votre nation, ils ne soient conduits à d'importantes découvertes.'

The reciprocal gravitation of the bodies of the solar system is a cause of great irregularities in their motions; many of which had been explained before the time of La Place, but some of the most important had not been accounted for, and many were not even known to exist. The author of the *Mécanique Céleste* therefore undertook the arduous task of forming a complete system of physical astronomy, in which the various motions in nature should be deduced from the first principles of mechanics. It would have been impossible to accomplish this, had not the improvements in analysis kept pace with the rapid advance in astronomy, a pursuit in which many have acquired immortal fame; that La Place is pre-eminent amongst these, will be most readily acknowledged by those who are best acquainted with his works.

Having endeavoured in the first book to explain the laws by which force acts upon matter, we shall now compare those laws with the actual motions of the heavenly bodies, in order to arrive by analytical reasoning, entirely independent of hypothesis, at the principle of that force which animates the solar system. The laws of mechanics may be traced with greater precision in celestial space than on earth, where the results are so complicated, that it is difficult to unravel, and still more so to subject them to calculation: whereas the bodies of the solar system, separated by vast distances, and acted upon by a force, the effects of which may be readily estimated, are only disturbed in their respective movements by such small forces, that the general equations comprehend all the changes which ages have produced, or may hereafter produce in the system; and in explaining the phenomena it is not necessary to have recourse to vague or imaginary causes, for the law of universal gravitation may be reduced to calculation, the results of which, confirmed by actual observation, afford the most substantial proof of its existence.

It will be seen that this great law of nature represents all the phenomena of the heavens, even to the most minute details; that there is not one of the inequalities which it does not account for; and that it has even anticipated observation, by unfolding the causes of several singular motions, suspected by astronomers, but so complicated in their nature, and so long in their periods, that observation alone could not have determined them but in many ages.

By the law of gravitation, therefore, astronomy is now become a great problem of mechanics, for the solution of which, the figure and masses of the planets, their places, and velocities at any given time, are the only data which observation is required to furnish. We proceed to give such an account of the solution of this problem, as the nature of the subject and the limits of this work admit of.

CHAPTER II.

ON THE LAW OF UNIVERSAL GRAVITATION, DEDUCED FROM OBSERVATION.

309. The three laws of Kepler furnish the data from which the principle of gravitation is established, namely:—

i. That the radii vectores of the planets and comets describe areas proportional to the time.

ii. That the orbits of the planets and comets are conic sections, having the sun in one of their foci.

iii. That the squares of the periodic times of the planets are proportional to the cubes of their mean distances from the sun.

310. It has been shown, that if the law of the force which acts on a moving body be known, the curve in which it moves may be found; or, if the curve in which the body moves be given, the law of the force may be ascertained. In the general equation of the motion of a body in article 144, both the force and the path of the body are indeterminate; therefore in applying that equation to the motion of the planets and comets, it is necessary to know the orbits in which they move, in order to ascertain the nature of the force that acts on them.

311. In the general equation of the motion of a body, the forces acting on it are resolved into three component forces, in the direction of three rectangular axes; but as the paths of the planets, satellites, and comets, are proved by the observations of Kepler to be conic sections, they always move in the same plane: therefore the component force in the direction perpendicular to that plane is zero, and the other two component forces are in the plane of the orbit.

312. Let AmP, fig. 62, be the elliptical orbit of a planet m, having the centre of the sun in the focus S, which is also assumed as the origin of the co-ordinates. The imaginary line Sm joining the centre of the sun and the centre of the planet is the radius vector. Suppose the two

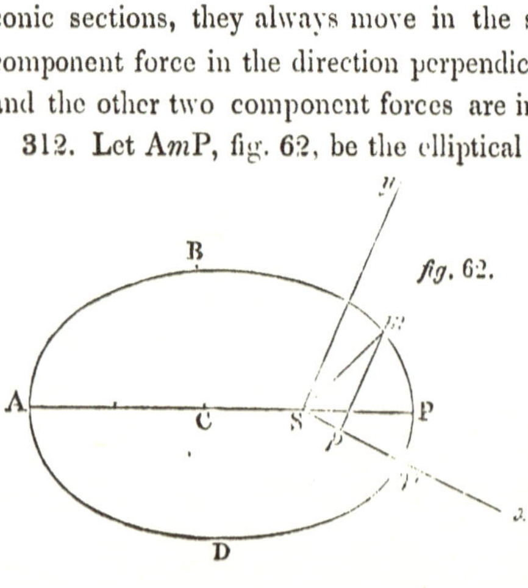

fig. 62.

component forces to be in the direction of the axes Sx, Sy, then the component force Z is zero; and as the body is free to move in every direction, the virtual velocities δx, δy are zero, which divides the general equation of motion in article 144 into

$$\frac{d^2x}{dt^2} = X; \quad \frac{d^2y}{dt^2} = Y;$$

giving a relation between each component force, the space that it causes the body to describe on ox, or oy, and the time. If the first of these two equations be multiplied by $-y$, and added to the second multiplied by x, their sum will be

$$\frac{d(xdy - ydx)}{dt^2} = Yx - Xy.$$

But $xdy - ydx$ is double the area that the radius vector of the planet describes round the sun in the instant dt. According to the first law of Kepler, this area is proportional to the time, so that

$$xdy - ydx = cdt;$$

and as c is a constant quantity,

$$\frac{d(xdy - ydx)}{dt^2} = 0,$$

therefore $$Yx - Xy = 0,$$

whence $$X : Y :: x : y;$$

so that the forces X and Y are in the ratio of x to y, that is as Sp to pm, and thus their resulting force mS passes through S, the centre of the sun. Besides, the curve described by the planet is concave towards the sun, whence the force that causes the planet to describe that curve, tends towards the sun. And thus the law of the areas being proportional to the time, leads to this important result,—that the force which retains the planets and comets in their orbits, is directed towards the centre of the sun.

313. The next step is to ascertain the law by which the force varies at different distances from the sun, which is accomplished by the consideration, that these bodies alternately approach and recede from him at each revolution; the nature of elliptical motion, then, ought to give that law. If the equation

$$\frac{d^2x}{dt^2} = X$$

be multiplied by dx, and $$\frac{d^2y}{dt^2} = Y,$$

by dy, their sum is

$$\frac{dxd^2x + dyd^2y}{dt^2} = \mathrm{X}dx + \mathrm{Y}dy,$$

and its integral is

$$\frac{dx^2 + dy^2}{dt^2} = 2\int(\mathrm{X}dx + \mathrm{Y}dy),$$

the constant quantity being indicated by the integral sign. Now the law of areas gives

$$dt = \frac{xdy - ydx}{c},$$

which changes the preceding equation to

$$\frac{c^2(dx^2 + dy^2)}{(xdy - ydx)^2} = 2\int(\mathrm{X}dx + \mathrm{Y}dy). \quad (82)$$

In order to transform this into a polar equation, let r represent the radius vector Sm, fig. 62, and v the angle $m\mathrm{S}\gamma$,

then S$p = x = r\cos v$; $pm = y = r\sin v$, and $r = \sqrt{x^2 + y^2}$
whence $dx^2 + dy^2 = r^2dv^2 + dr^2$, $xdy - ydx = r^2dv$;
and if the resulting force of X and Y be represented by F, then

$$F : \mathrm{X} :: \mathrm{S}m : \mathrm{S}p :: 1 : \cos v;$$

hence $$\mathrm{X} = -F\cos v;$$

the sign is negative, because the force F in the direction mS, tends to diminish the co-ordinates; in the same manner it is easy to see that

$$\mathrm{Y} = -F\sin v;\ F = \sqrt{\mathrm{X}^2 + \mathrm{Y}^2};\ \text{and } \mathrm{X}dx + \mathrm{Y}dy = -Fdr;$$

so that the equation (82) becomes

$$0 = \frac{c^2\{r^2dv^2 + dr^2\}}{r^4dv^2} + 2\int Fdr. \quad (83)$$

whence $$dv = \frac{cdr}{r\sqrt{-c^2 - 2r^2\int Fdr}}.$$

314. If the force F be known in terms of the distance r, this equation will give the nature of the curve described by the body. But the differential of equation (83) gives

$$F = \frac{c^2}{r^3} - \frac{c^2}{2}\,\frac{d\left\{\frac{dr^2}{r^4dv^2}\right\}}{dr}. \quad (84)$$

Thus a value of the resulting force F is obtained in terms of the variable radius vector Sm, and of the corresponding variable angle $m\mathrm{S}\gamma$; but in order to have a value of the force F in terms of mS alone, it is necessary to know the angle $\gamma\mathrm{S}m$ in terms of Sm.

The planets move in ellipses, having the sun in one of their foci; therefore let ϖ represent the angle γSP, which the greater axis AP makes with the axes of the co-ordinates Sx, and let v be the angle γSm. Then if $\frac{\text{CS}}{\text{CP}}$, the ratio of the eccentricity to the greater axis be e, and the greater axis CP $= a$, the polar equation of conic sections is

$$r = \frac{a(1-e^2)}{1+e\cos(v-\varpi)},$$

which becomes a parabola when $e = 1$, and a infinite; and a hyperbola when e is greater than unity and a negative. This equation gives a value of r in terms of the angle γSm or v, and thence it may be found that

$$\frac{dr^2}{r^4 dv^2} = \frac{2}{ar(1-e^2)} - \frac{1}{r^2} - \frac{1}{a^2(1-e^2)}$$

which substituted in equation (84) gives

$$F = \frac{c^2}{a(1-e^2)} \cdot \frac{1}{r^2}.$$

The coefficient $\frac{c^2}{a(1-e^2)}$ is constant, therefore F varies inversely as the square of r or Sm. Wherefore the orbits of the planets and comets being conic sections, the force varies inversely as the square of the distance of these bodies from the sun.

Now as the force F varies inversely as the square of the distance, it may be represented by $\frac{h}{r^2}$, in which h is a constant coefficient, expressing the intensity of the force. The equation of conic sections will satisfy equation (84) when $\frac{h}{r^2}$ is put for F; whence as

$$h = \frac{c^2}{a(1-e^2)}$$

forms an equation of condition between the constant quantities a and e, the three arbitrary quantities a, e, and ϖ, are reduced to two; and as equation (83) is only of the second order, the finite equation of conic sections is its integral.

315. Thus, if the orbit be a conic section, the force is inversely as the square of the distance; and if the force varies inversely as the square of the distance, the orbit is a conic section. The planets and

comets therefore describe conic sections in virtue of a primitive impulse and an accelerating force directed to the centre of the sun, and varying according to the preceding law, the least deviation from which would cause them to move in curves of a totally different nature.

316. In every orbit the point P, fig. 63, which is nearest the sun, is the perihelion, and in the ellipse the point A farthest from the sun is the aphelion. SP is the perihelion distance of the body from the sun.

317. A body moves in a conic section with a different velocity in every point of its orbit, and with a perpetual tendency to fly off in the direction of the tangent, but this tendency is counteracted by the attraction of the sun. At the perihelion, the velocity of a planet is greatest; therefore its tendency to leave the sun exceeds the force of attraction: but the continued action of the sun diminishes the velocity as the distance increases; at the aphelion the velocity of the planet is least: therefore its tendency to leave the sun is less than the force of attraction which increases the velocity as the distance diminishes, and brings the planet back towards the sun, accelerating its velocity so much as to overcome the force of attraction, and carry the planet again to the perihelion. This alternation is continually repeated.

318. When a planet is in the point B, or D, it is said to be in quadrature, or at its mean distance from the sun. In the ellipse, the mean distance, SB or SD, is equal to CP, half the greater axis; the eccentricity is CS.

319. The periodic time of a planet is the time in which it revolves round the sun, or the time of moving through 360°. The periodic time of a satellite is the time in which it revolves about its primary.

320. From the equation

$$F = \frac{c}{a(1-e^2)} \cdot \frac{1}{r^2},$$

it may be shown, that the force F varies, with regard to different planets, inversely as the square of their respective distances from the sun. The quantity $2a(1-e^2)$ is 2SV, the parameter of the orbit, which is invariable in any one curve, but is different in each conic section. The intensity of the force depends on

$$\frac{c^2}{a(1-e^2)} \text{ or } \frac{c^2}{\text{SV}},$$

which may be found by Kepler's laws. Let T represent the time of the revolution of a planet; the area described by its radius vector in this time is the whole area of the ellipse, or

$$\pi a^2 \,.\, \sqrt{1 - e^2}.$$

where $\pi = 3.14159$ the ratio of the circumference to the diameter. But the area described by the planet during the indefinitely small time dt, is $\frac{1}{2}\, cdt$; hence the law of Kepler gives

$$\tfrac{1}{2}\, cdt : \pi a^2 \sqrt{1 - e^2} :: dt : T;$$

whence
$$c = \frac{2\pi a^2 \sqrt{1 - e^2}}{T}. \qquad (85)$$

But, by Kepler's third law, the squares of the periodic times of the planets are proportional to the cubes of their mean distances from the sun; therefore

$$T^2 = k^2 a^3,$$

k being the same for all the planets.

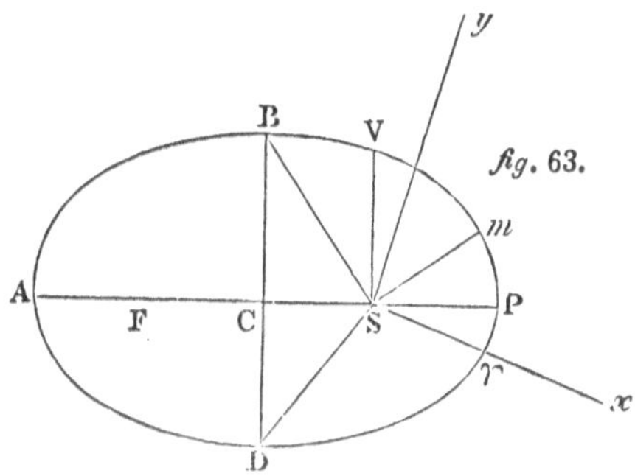

fig. 63.

Hence
$$c = \frac{2\pi \sqrt{a(1 - e^2)}}{k};$$

but $2a(1 - e^2)$ is 2SV, the parameter of the orbit.

Therefore, in different orbits compared together, the values of c are as the areas traced by the radii vectores in equal times; consequently these areas are proportional to the square roots of the parameters of the orbits, either of planets or comets. If this value of c be put in

$$F = \frac{c}{a(1-e^2} \cdot \frac{1}{r^2},$$

it becomes
$$F = \frac{4\pi^2}{k^2} \cdot \frac{1}{r^2}; = h . \frac{1}{r^2}$$

in which $\frac{4\pi^2}{k^2}$ or h, is the same for all the planets and comets; the force, therefore, varies inversely as the square of the distance of each from the centre of the sun: consequently, if all these bodies were placed at equal distances from the sun, and put in motion at the same instant from a state of rest, they would move through equal spaces in equal times; so that all would arrive at the sun at the same instant,—properties first demonstrated geometrically by Newton from the laws of Kepler.

321. That the areas described by comets are proportional to the square roots of the parameters of their orbits, is a result of theory more sensibly verified by observation than any other of its consequences. Comets are only visible for a short time, at most a few months, when they are near their perihelia; but it is difficult to determine in what curve they move, because a very eccentric ellipse, a parabola, and hyperbola of the same perihelion distance coincide through a small space on each side of the perihelion. The periodic time of a comet cannot be known from one appearance. Of more than a hundred comets, whose orbits have been computed, the return of only three has been ascertained. A few have been calculated in very elliptical orbits; but in general it has been found, that the places of comets computed in parabolic orbits agree with observation: on that account it is usual to assume, that comets move in parabolic curves.

322. In a parabola the parameter is equal to twice the perihelion distance, or

$$a(1-e^2) = 2D;$$

hence, for comets, $$c = \frac{2\pi}{k} \sqrt{2D}.$$

For, in this case, $e=1$ and a is infinite; therefore, in different parabolæ, the areas described in equal times are proportional to the square roots of their perihelion distances. This affords the means of ascertaining how near a comet approaches to the sun. Five or six comets seem to have hyperbolic orbits; consequently they could only be once visible, in their transit through the system to which we belong, wandering in the immensity of space, perhaps to visit other suns and other systems.

It is probable that such bodies do exist in the infinite variety of creation, though their appearance is rare. Most of the comets that we have seen, however, are thought to move in extremely eccentric ellipses, returning to our system after very long intervals. Two hundred years have not elapsed since comets were observed with accuracy, a time which is probably greatly exceeded by the enormous periods of the revolutions of some of these bodies.

323. The three laws of Kepler, deduced from the observations of Tycho Brahe, and from his own observations of Mars, form an era of vast importance in the science of astronomy, being the bases on which Newton founded the universal principle of gravitation: they lead us to regard the centre of the sun as the focus of an attractive force, extending to an infinite distance in all directions, decreasing as the squares of the distance increase. Each law discloses a particular property of this force. The areas described by the radius vector of each planet or comet, being proportional to the time employed in describing them, shows that the principal force which urges these bodies, is always directed towards the centre of the sun. The ellipticity of the planetary orbits, and the nearly parabolic motion of the comets, prove that for each planet and comet this force is reciprocally as the square of the distance from the sun; and, lastly, the squares of the periodic times, being proportional to the cubes of the mean distances, proves that the areas described in equal times by the radius vector of each body in the different orbits, are proportional to the square roots of the parameters—a law which is equally applicable to planets and comets.

324. The satellites observe the laws of Kepler in moving round their primaries, and gravitate towards the planets inversely as the square of their distances from their centre; but they must also gravitate towards the sun, in order that their relative motions round their planets may be the same as if the planets were at rest. Hence the satellites must gravitate towards their planets and towards the sun inversely as the squares of the distances. The eccentricity of the orbits of the two first satellites of Jupiter is quite insensible; that of the third inconsiderable; that of the fourth is evident. The great distance of Saturn has hitherto prevented the eccentricity of the orbits of any of its satellites from being perceived, with the exception of the sixth. But the law of the gravitation of the satellites of Jupiter and Saturn is derived most clearly from this ratio,—that, for each system of satellites, the squares of their periodic times are as the cubes of their mean distances from the centres of their respective

fig. 64.

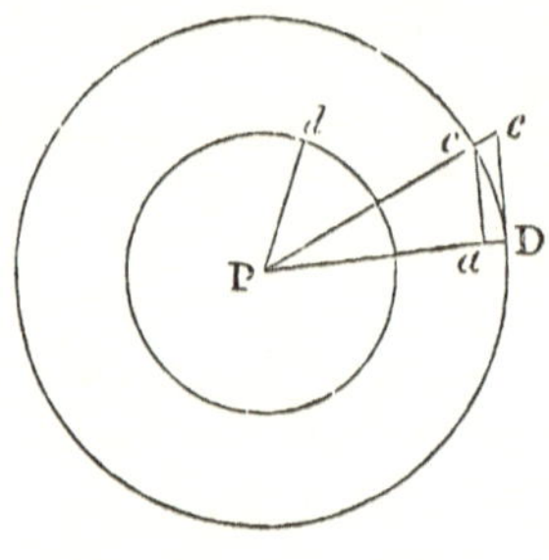

planets. For, imagine a satellite to describe a circular orbit, with a radius PD $= a$, fig. 64, its mean distance from the centre of the planet. Let T be the duration of a sidereal revolution of the satellite, then

$$3.14159 = \pi,$$

being the ratio of the circumference to the diameter, $a \, . \, \frac{2\pi}{T}$ will be the very small arc D*c* that the satellite describes in a second. If the attractive force of the planet were to cease for an instant, the satellite would fly off in the tangent D*e*, and would be farther from the centre of the planet by a quantity equal to *a*D, the versed sine of the arc D*c*. But the value of the versed sine is

$$a \, . \, \frac{2\pi^2}{T^2},$$

which is the distance that the attractive force of the planet causes the satellite to fall through in a second.

Now, if another satellite be considered, whose mean distance is P*d* $= a'$, and $T_{\prime}$ the duration of its sidereal revolution, its deflection will be $a' \frac{2\pi^2}{T_{\prime}^2}$ in a second; but if F and F' be the attractive forces of the planet at the distances PD and P*d*, they will evidently be proportional to the quantities they make the two satellites fall through in a second;

hence $$F : F' :: a \frac{2\pi^2}{T^2} : a' \frac{2\pi^2}{T_{\prime}^2},$$

or $$F : F' :: \frac{a}{T^2} : \frac{a'}{T_{\prime}^2};$$

but the squares of the periodic times are as the cubes of the mean distances; hence

$$T^2 : T_{\prime}^2 :: a^3 : a'^3;$$

whence $$F : F' :: \frac{1}{a^2} : \frac{1}{a'^2}.$$

Thus the satellites gravitate to their primaries inversely as the square of the distance.

325. As the earth has but one satellite, this comparison cannot be made, and therefore the ellipticity of the lunar orbit is the only celestial phenomenon by which we can know the law of the moon's attractive force. If the earth and the moon were the only bodies in the system, the moon would describe a perfect ellipse about the earth; but, in consequence of the action of the sun, the path of the moon is sensibly disturbed, and therefore is not a perfect ellipse; on this account some doubts may arise as to the diminution of the attractive force of the earth as the inverse square of the distance.

The analogy, indeed, which exists between this force and the attractive force of the sun, Jupiter, and Saturn, would lead to the belief that it follows the same law, because the solar attraction acts equally on all bodies placed at the same distance from the sun, in the same manner that terrestrial gravitation causes all bodies in vacuo to fall from equal heights in equal times. A projectile thrown horizontally from a height, falls to the earth after having described a parabola. If the force of projection were greater, it would fall at a greater distance; and if it amounted to 30772.4 feet in a second, and were not resisted by the air, it would revolve like a satellite about the earth, because its centrifugal force would then be equal to its gravitation. This body would move in all respects like the moon, if it were projected with the same force, at the same height.

It may be proved, that the force which causes the descent of heavy bodies at the surface of the earth, diminished in the inverse ratio of the square of the distance, is sufficient to retain the moon in her orbit, but this requires a knowledge of the lunar parallax.

On Parallax.

326. Let m, fig. 65, be a body in its orbit, and C the centre of the earth, assumed to be spherical. A person on the surface of the earth, at E, would see the body m in the direction EmB; but the body would appear, in the direction CmA, to a person in C, the centre of the earth. The angle CmE, which measures the difference of these directions, is the parallax of m. If z be the zenith of

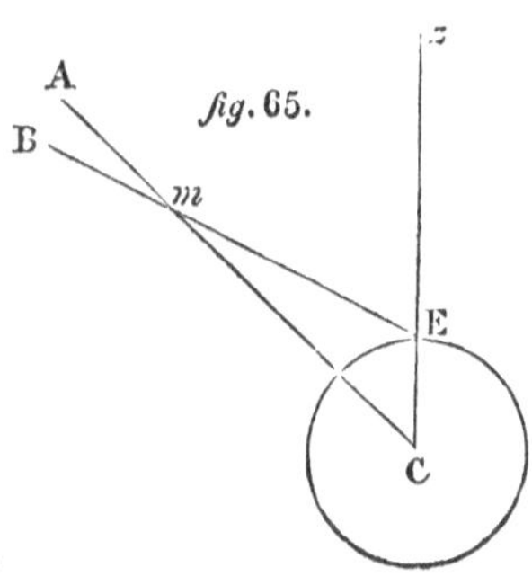

fig. 65.

an observer at E, the angle $z\mathrm{E}m$, called the zenith distance of the body, may be measured; hence $m\mathrm{EC}$ is known, and the difference between $z\mathrm{E}m$ and $z\mathrm{C}m$ is equal to $\mathrm{C}m\mathrm{E}$, the parallax, then if $\mathrm{CE} = R$, $\mathrm{C}m = r$, and $z\mathrm{E}m = z$,

$$\text{sine } \mathrm{C}m\mathrm{E} = \frac{R}{r} \sin z;$$

hence, if CE and $\mathrm{C}m$ remain the same, the sine of the parallax, $\mathrm{C}m\mathrm{E}$, will vary as the sine of the zenith distance $z\mathrm{E}m$; and when $z\mathrm{E}m = 90°$, as in fig. 67,

$$\sin \mathrm{P} = \frac{R}{r};$$

P being the value of the angle $\mathrm{C}m\mathrm{E}$ in this case; then the parallax is a maximum, for $\mathrm{E}m$ is tangent to the earth, and, as the body m is seen in the horizon, it is called the horizontal parallax; hence the sine of the horizontal parallax is equal to the terrestrial radius divided by the distance of the body from the centre of the earth.

327. The length of the mean terrestrial radius is known, the horizontal parallax may be determined by observation, therefore the distance of m from the centre of the earth is known. By this method the dimensions of the solar system have been ascertained with great accuracy. If the distance be very great compared with the diameter of the earth, the parallax will be insensible. If $\mathrm{C}m\mathrm{E}$ were an angle of the fourth of a second, it would be inappreciable; an arc of $1'' = 0.000004848$ of the radius, the fourth of a second is therefore $0.000001212 = \frac{1}{825082}$; and thus, if a body be distant from the earth by 825082 of its semidiameters, or 3265660000 miles, it will be seen in the same position from every point of the earth's surface. The parallax of all the celestial bodies is very small: even that of the moon at its maximum does not much exceed 1°.

328. P being the horizontal parallax, let p be the parallax $\mathrm{E}m\mathrm{C}$, fig. 66, at any height. When P is known, p may be found, and the contrary, for if $\frac{R}{r}$ be eliminated, then $\sin p = \sin P \sin z$, and when P is constant, $\sin p$ varies as $\sin z$.

329. The horizontal parallax is determined as follows: let E and E′, fig. 66, be two places on the same meridian of the earth's surface; that is, which contemporaneously have the same noon. Suppose the latitudes of these two places to be perfectly known; when a body m is on the meridian, let its zenith distances

fig. 66.

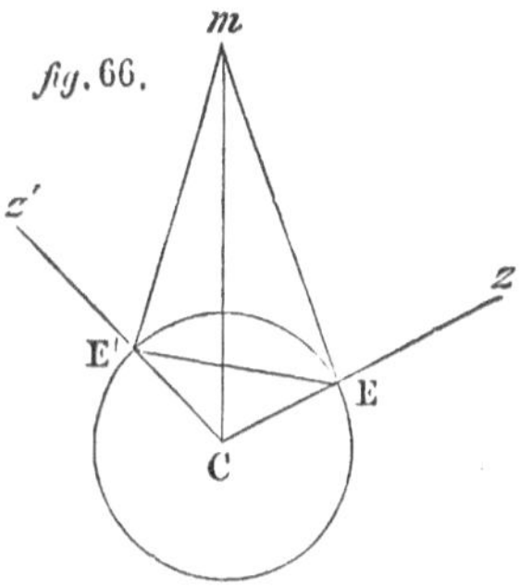

$$zEm = z, \quad z'E'm = z',$$

be measured by two observers in E and E′. Then ECE′, the sum of the latitudes, is known, and also the angles CEm, CE′m; hence EmE′, EmC, and E′mC may be determined; for P is so small, that it may be put for its sine; therefore

$$\sin p = P \sin z, \ \sin p' = P \sin z';$$

and as p and p' are also very small,

$$p + p' = P\{\sin z + \sin z'.\}$$

Now, $p + p'$ is equal to the angle EmE′, under which the chord of the terrestrial arc EE′, which joins the two observers, would be seen from the centre of m, and it is the fourth angle of the quadrilateral CEmE′.

But $$CEm = 180° - z, \ CE'm = 180° - z',$$

and if $$ECm + E'Cm = \phi,$$

then $$180° - z + 180° - z' + p + p' + \phi = 360°;$$

hence $$p + p' = z + z' - \phi;$$

therefore the two values of $p + p'$ give

$$P = \frac{z + z - \phi}{\sin z + \sin z'},$$

which is the horizontal parallax of the body, when the observers are on different sides of Cm; but when they are on the same side,

$$P = \frac{z - z' - \phi}{\sin z - \sin z'}.$$

It requires a small correction, since the earth, being a spheroid, the lines ZE, Z′E′ do not pass through C, the centre of the earth.

The parallax of the moon and of Mars were determined in this manner, from observations made by La Caille at the Cape of Good Hope, in the southern hemisphere; and by Wargesten at Stockholm, which is nearly on the same meridian in the northern hemisphere.

330. The horizontal parallax varies with the distance of the body from the earth ; for it is evident that the greater the distance, the less the parallax. It varies also with the parallels of terrestrial latitude, the earth, being a spheroid, the length of the radius decreases from the equator to the poles. It is on this account that, at the mean distance of the moon, the horizontal parallax observed in different latitudes varies; proving the elliptical figure of the earth. The difference between the mean horizontal parallax at the equator and at the poles, from this cause, is $10''.3$.

331. In order to obtain a value of the moon's horizontal parallax, independent of these inequalities, the horizontal parallax is chosen at the mean distance of the moon from the earth, and on that parallel of terrestrial latitude, the square of whose sine is $\frac{1}{3}$, because the attraction of the earth upon the corresponding points of its surface is nearly equal to the mass of the earth, divided by the square of the mean distance of the moon from the earth. This is called the constant part of the horizontal parallax. The force which retains the moon in her orbit may now be determined.

Force of Gravitation at the Moon.

332. If the force of gravity be assumed to decrease as the inverse square of the distance, it is clear that the force of gravity at E, fig. 67, would be, to the same force at m, the distance of the moon, as the square of Cm to the square of CE; but CE divided by Cm is the sine of the horizontal parallax of the moon, the constant part of which is found by observation to be $57' \, 4''.17$ in the latitude in question; hence the force of gravity, reduced to the distance of the moon, is equal to the force of gravity at E on the earth's surface, multiplied by $\sin^2 57' \, 4''.17$, the square of the sine of the constant part of the horizontal parallax.

fig. 67.

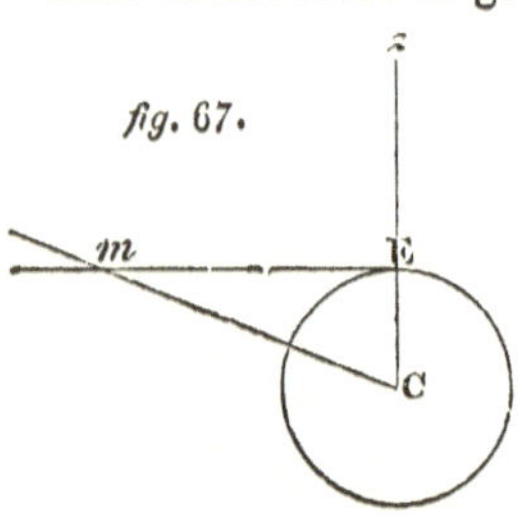

Since the earth is a spheroid, whose equatorial diameter is greater than its polar diameter, the force of gravity increases from the equator to the poles; but it has the same intensity in all points of the earth's surface in the same latitude.

Now the space through which a heavy body would fall during a second in the latitude the square of whose sine is $\frac{1}{3}$, has been ascertained by experiments with the pendulum to be 16.0697 feet; but the effect of the centrifugal force makes this quantity less than it would otherwise be, since that force has a tendency to make bodies fly off from the earth. At the equator it is equal to the 288th part of gravity; but as it decreases from the equator to the poles as the square of the sine of the latitude, the force of gravity in that latitude the square whose sine is $\frac{1}{3}$, is only diminished by two-thirds of $\frac{1}{288}$ or by its 432nd part. But the 432nd part of 16.0697 is 0.0372, and adding it to 16.0697, the whole effect of terrestrial gravity in the latitude in question is 16.1069 feet; and at the distance of the moon it is $16.1069 \cdot \sin^2 57' \, 4''.17$ nearly. But in order to have this quantity more exactly it must be multiplied by $\frac{357}{358}$, because it is found by the theory of the moon's motion, that the action of the sun on the moon diminishes its gravity to the earth by a quantity, the constant part of which is equal to the 358th part of that gravity.

Again, it must be multiplied by $\frac{76}{75}$, because the moon in her relative motion round the earth, is urged by a force equal to the sum of the masses of the earth and moon divided by the square of Cm, their mutual distance. It appears by the theory of the tides that the mass of the moon is only the $\frac{1}{75}$ of that of the earth which is taken as the unit of measure; hence the sum of the masses of the two bodies is

$$1 + \frac{1}{75} = \frac{76}{75}.$$

Then if the terrestrial attraction be really the force that retains the moon in her orbit, she must fall through

$$16.1069 \times \sin^2 57' \, 4''.17 \times \frac{357}{358} \times \frac{76}{75} = 0.00448474$$

of a foot in a second.

333. Let mS, fig. 68, be the small arc which the moon would describe in her orbit in a second, and let C be the centre of the earth. If the attraction of the earth were suddenly to cease, the moon would

fig. 68.

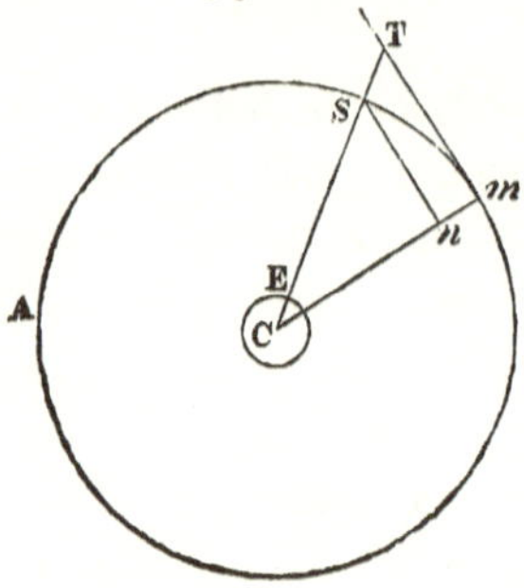

go off in the tangent mT; and at the end of the second she would be in T instead of S; hence the space that the attraction of the earth causes the moon to fall through in a second, is equal to mn the versed sine of the arc Sm.

The arc Sm is found by simple proportion, for the periodic time of the moon is $27^{\text{days}}.32166$, or 2360591″, and since the lunar orbit without sensible error may be assumed equal to the circumference of a circle whose radius is the mean distance of the moon from the earth; it is

$$2\mathrm{C}m\,.\,\pi,\ \text{or if}\ \frac{355}{113}\ \text{be put for}\ \pi,\ 2\mathrm{C}m\,.\,\frac{355}{113},$$

therefore
$$2360591'' : 1'' :: 2\mathrm{C}m\,\frac{355}{113} : \mathrm{S}m$$

and
$$\mathrm{S}m = \frac{2(355)\,.\,\mathrm{C}m\,.\,1^{\text{sec.}}}{113(2360591'')}.$$

The arc Sm is so small that it may be taken for its chord, therefore $(m\mathrm{S})^2 = \mathrm{C}m\,.\,mn$; hence

$$\frac{4(355)^2\,(\mathrm{C}m)^2}{(113)^2\,(2360591'')^2} = 2\mathrm{C}m\,.\,mn;$$

consequently
$$mn = \frac{2(355)^2\,.\,\mathrm{C}m}{(113)^2\,(2360591'')^2}.$$

Again, the radius CE of the earth in the latitude the square of whose sine is $\frac{1}{3}$, is computed to be 20898700 feet from the mensuration of the degrees of the meridian: and since

$$\frac{\mathrm{CE}}{\mathrm{C}m} = \sin 57'\,4''.17,$$

$$\mathrm{C}m = \frac{\mathrm{CE}}{\sin 57'\,4''.17} = \frac{20898700}{\sin 57'\,4''.17},$$

consequently,
$$mn = \frac{2(355)^2\,(20898700)}{(113)^2\,(2360591'')^2\,\sin 57'\,4''.17} = 0.00445983$$

of a foot, which is the measure of the deflecting force at the moon. But the space described by a body in one second from the earth's attraction at the distance of the moon was shown to be 0.00448474 of a foot in a second; the difference is therefore only the 0.00002491 of a foot, a quantity so small, that it may safely be ascribed to errors in observation.

334. Hence it appears, that the principal force that retains the moon in her orbit is terrestrial gravity, diminished in the ratio of the square of the distance. The same law then, which was proved to apply to a system of satellites, by a comparison of the squares of the times of their revolutions, with the cubes of their mean distances, has been demonstrated to apply equally to the moon, by comparing her motion with that of bodies falling at the surface of the earth.

335. In this demonstration, the distances were estimated from the centre of the earth, and since the attractive force of the earth is of the same nature with that of the other celestial bodies, it follows that the centre of gravity of the celestial bodies is the point from whence the distances must be estimated, in computing the effects of their attraction on substances at their surfaces, or on bodies in space.

336. Thus the sun possesses an attracting force, diminishing to infinity inversely as the squares of the distances, which includes all the bodies of the system in its action; and the planets which have satellites exact a similar influence over them.

Analogy would lead us to suppose that the same force exists in all the planets and comets; but that this is really the case will appear, by considering that it is a fixed law of nature that one body cannot act upon another without experiencing an equal and contrary reaction from that body: hence the planets and comets, being attracted towards the sun, must reciprocally attract the sun towards them according to the same law; for the same reason, satellites attract their planets. This property of attraction being common to planets, comets, and satellites, the gravitation of the heavenly bodies towards one another may be considered as a general principle of this universe; even the irregularities in the motions of these bodies are susceptible of being so well explained by this principle, that they concur in proving its existence.

337. Gravitation is proportional to the masses; for supposing the planets and comets to be at the same distance from the sun, and left to the action of gravity, they would fall through equal heights in equal times. The nearly circular orbits of the satellites prove that they gravitate like their planets towards the sun in the ratio of their

masses: the smallest deviation from that ratio would be sensible in their motions, but none depending on that cause has been detected by observation.

338. Thus the planets, comets, and satellites, when at the same distance from the sun, gravitate as their masses; and as reaction is equal and contrary to action, they attract the sun in the same ratio; therefore their action on the sun is proportional to their masses divided by the square of their distances from his centre.

339. The same law obtains on earth; for very correct observations with the pendulum prove, that were it not for the resistance of the air, all bodies would fall towards its centre with the same velocity. Terrestrial bodies then gravitate towards the earth in the ratio of their masses, as the planets gravitate towards the sun, and the satellites towards their planets. This conformity of nature with itself upon the earth, and in the immensity of the heavens, shows, in a striking manner, that the gravitation we observe here on earth is only a particular case of a general law, extending throughout the system.

340. The attraction of the celestial bodies does not belong to their mass alone taken in its totality, but exists in each of their atoms, for if the sun acted on the centre of gravity of the earth without acting on each of its particles separately, the tides would be incomparably greater, and very different from what they now are. Thus the gravitation of the earth towards the sun is the sum of the gravitation of each of its particles; which in their turn attract the sun as their respective masses; besides, everything on earth gravitates towards the centre of the earth proportionally to its mass; the particle then reacts on the earth, and attracts it in the same ratio; were that not the case, and were any part of the earth however small not to attract the other part as it is itself attracted, the centre of gravity of the earth would be moved in space in virtue of this gravitation, which is impossible.

341. It appears then, that the celestial phenomena when compared with the laws of motion, lead to this great principle of nature, that all the particles of matter mutually attract each other as their masses directly, and as the squares of their distances inversely.

342. From the universal principle of gravitation, it may be foreseen, that the comets and planets will disturb each other's motion, so that their orbits will deviate a little from perfect ellipses; and

the areas will no longer be exactly proportional to the time: that the satellites, troubled in their paths by their mutual attraction, and by that of the sun, will sensibly deviate from elliptical motion: that the particles of each celestial body, united by their mutual attraction, must form a mass nearly spherical; and that the resultant of their action at the surface of the body, ought to produce there all the phenomena of gravitation. It appears also, that centrifugal force arising from the rotation of the celestial bodies must alter their spherical form a little by flattening them at their poles; and that the resulting force of their mutual attractions not passing through their centres of gravity, will produce those motions that are observed in their axes of rotation. Lastly, it is clear that the particles of the ocean being unequally attracted by the sun and moon, and with a different intensity from the nucleus of the earth, must produce the ebb and flow of the sea.

343. Having thus proved from Kepler's laws, that the celestial bodies attract each other directly as their masses, and inversely as the square of the distance, La Place inverts the problem, and assuming the law of gravitation to be that of nature, he determines the motions of the planets by the general theorem in article 144, and compares the results with observation.

CHAPTER III.

ON THE DIFFERENTIAL EQUATIONS OF THE MOTION OF A SYSTEM OF BODIES, SUBJECTED TO THEIR MUTUAL ATTRACTIONS.

344. As the earth which we inhabit is a part of the solar system, it is impossible for us to know any thing of its absolute motions; our observations must therefore be limited to its relative motions. In estimating the relative motion of planets, it is usual to refer them to the centre of the sun, and those of satellites to the centres of their primary planets. The sun and planets mutually attract each other; but in estimating the motions of a planet, the sun is supposed to be at rest, and all the motion is referred to the planet, which thus moves in consequence of the difference between its own action, and that of the sun. It is the same with regard to satellites and their primaries.

345. To determine the relative motions of a system of bodies m, m', m'', &c. fig. 69, considered as points revolving about one body S, which is the centre of their motions—

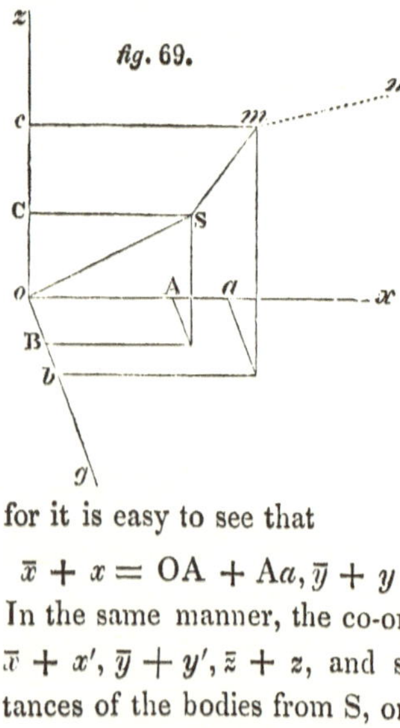

fig. 69.

Let $\bar{x}$, $\bar{y}$, $\bar{z}$, be the co-ordinates of S referred to o as an origin, and x, y, z, x', y', z', &c., the co-ordinates of the bodies m, m', &c. referred to S as their origin. Then the co-ordinates of m when referred to o, are $\bar{x}+x'$, $\bar{y}+y$, $\bar{z}+z$, for it is easy to see that

$$\bar{x} + x = \text{OA} + \text{A}a, \bar{y} + y = \text{OB} + \text{B}b, \bar{z} + z = \text{OC} + \text{C}c.$$

In the same manner, the co-ordinates of m', when referred to o, are $\bar{x} + x'$, $\bar{y} + y'$, $\bar{z} + z$, and so for the other bodies. Let the distances of the bodies from S, or

$$\text{S}m = \sqrt{x^2+y^2+z^2} \quad \text{S}m' = \sqrt{x'^2+y'^2+z'^2}, \text{ \&c.}$$

be represented by r, r', r'', &c. and the masses by m, m', &c. and S. The equations of the motion of m will be first determined.

346. The whole action of the system relative to m consists of three parts:

1. Of the action of S on m.
2. Of the action of all the bodies m', m'', m''', &c. on m.
3. Of the action of all the bodies m, m', m'', &c. on S.

These will be determined separately.

i. The action of S on m is $-\dfrac{S}{r^2}$, that is directly as its mass, and inversely as the square of its distance. It has a negative sign, because the body S draws m towards the origin of the co-ordinates. This force when resolved in the direction ox is $-\dfrac{Sx}{r^3}$; for the force $-\dfrac{S}{r^2}$ is to its component force in ox, as Sm to Aa, that is as r to x.

ii. The distance of m' from m is

$$\sqrt{(x'-x)^2+(y'-y)^2+(z'-z)^2}$$

for x, y, z, x', y', z', being the co-ordinates of m and m' referred to S as their origin, the distance of these bodies from each other is the diagonal of a parallelopiped whose sides are $x'-x$, $y'-y$, $z'-z$. For the same reason, the distance of m'' from m is

$$\sqrt{(x''-x)^2+(y''-y)^2+(z''-z)^2}, \text{ \&c.}$$

In order to abridge, let

$$\lambda = \frac{m.m'}{\sqrt{(x'-x)^2+(y'-y)^2+(z'-z^2)}} + \frac{m.m''}{\sqrt{(x''-x)^2+(y''-y)^2+(z''-z)^2}} + \text{\&c.}$$

it is evident that

$$\frac{1}{m}\left(\frac{d\lambda}{dx}\right) = \frac{m'(x'-x)}{\{(x'-x)^2+(y'-y)^2+(z'-z)\}^{\frac{3}{2}}}$$

$$+\frac{m''(x''-x)}{\{(x''-x)^2+(y''-y)^2+(z''-z^2)\}^{\frac{3}{2}}} + \text{\&c.}$$

is the sum of the actions of all the bodies m', m'', &c. on m when resolved in the direction ox. Hence the whole action of the system on m resolved in the axes ox is

$$\frac{1}{m}\left(\frac{d\lambda}{dx}\right) - \frac{Sx}{r^3};$$

but by the general theorem of motion

$$\frac{1}{m}\left(\frac{d\lambda}{dx}\right) - \frac{Sx}{r^3} = \frac{d^2(\bar{x} + x)}{dt^2}, \tag{86}$$

for $\bar{x} + x$ is the co-ordinate oa, or the distance of m from o in the direction ox

iii. The action of m on S is $\frac{m}{r^2}$, and its component force in ox is $\frac{mx}{r^3}$; likewise the actions of m', m'', &c. on S, when resolved in the same axes, are $\frac{m'x'}{r'^3}$, $\frac{m''x''}{r''^3}$, &c. hence the action of the system on S in the axes ox, may be expressed by $\Sigma\,\frac{mx}{r^3}$; but by the general theorem

$$\Sigma.\,\frac{mx}{r^3} = \frac{d^2\bar{x}}{dt^2},$$

for the co-ordinates of S alone vary by this action. Now, if $\Sigma\frac{mx}{r^3}$ be put for $\frac{d^2\bar{x}}{dt^2}$, in the equation (86) it becomes

$$0 = \frac{d^2x}{dt^2} + \frac{Sx}{r^3} + \Sigma.\,\frac{mx}{r^3} - \frac{1}{m}\left(\frac{d\lambda}{dx}\right),$$

which is the whole action of the system relatively to m, when resolved in the direction ox, and because

$$\Sigma.\,\frac{my}{r^3} = \frac{d^2\bar{y}}{dt^2}.\quad \Sigma.\,\frac{mz}{r^3} = \frac{d^2\bar{z}}{dt^2};$$

the other two component forces are

$$0 = \frac{d^2y}{dt^2} + \frac{Sy}{r^3} + \Sigma\frac{my}{r^3} - \frac{1}{m}\left(\frac{d\lambda}{dy}\right),$$

$$0 = \frac{d^2z}{dt^2} + \frac{Sz}{r^3} + \Sigma\frac{mz}{r^3} - \frac{1}{m}\left(\frac{d\lambda}{dz}\right).$$

The same equations will give the motions of m', m'', &c. round S, if m', x', y', z'; m'', x'', y'', z'', &c. be successively put for m, x, y, z, and *vice versâ*, and the equations

$$\frac{d^2\bar{x}}{dt^2} = \Sigma.\frac{mx}{r^3},\quad \frac{d^2\bar{y}}{dt^2} = \Sigma.\frac{my}{r^3},\quad \frac{d^2\bar{z}}{dt^2} = \Sigma.\frac{mz}{r^3},$$

determine the motion of S.

347. These equations, however, may be put under a more convenient form for

$$\Sigma.\,\frac{mx}{r^3} = \frac{mx}{r^3} + \frac{m'x'}{r'^3} + \text{\&c.}$$

and if $S + m$ the sum of the masses of the sun and of a planet, or of a planet and its satellite, by represented by μ, the equation in x becomes

$$0 = \frac{d^2x}{dt^2} + \frac{\mu x}{r^3} + \frac{m'x'}{r'^3} + \&c. - \frac{1}{m}\left(\frac{d\lambda}{dx}\right).$$

The part $\frac{d^2x}{dt^2} + \frac{\mu x}{r^3}$ relates only to the undisturbed elliptical motion of m round S; it is much greater than the remaining part

$$\frac{m'x'}{r'^3} + \frac{m''x''}{r''^3} + \&c. - \frac{1}{m}\left(\frac{d\lambda}{dx}\right),$$

which contains all the disturbances to which the body m is subject from the action of the other bodies of the system. $-\frac{1}{m}\left(\frac{d\lambda}{dx}\right)$ contains the direct action of the bodies m', m'', &c. on m; but m is also troubled indirectly by the action of these bodies on S, this part is contained in $\frac{m'x'}{r'^3} + \frac{m''x''}{r''^3} + \&c.$

By the latter action S is drawn to or from m; and by the former, m is drawn to or from S; in both cases altering the relative position of S and m. Let

$$R = \frac{m'}{\sqrt{(x'-x)^2+(y'-y)^2+(z'-z)^2}} - \frac{m'(x'x+y'y+z'z)}{r'^3}$$
$$+ \frac{m''}{\sqrt{(x''-x)^2+(y''-y)^2+(z''-z)^2}} - \frac{m''(x''x+y''y+z''z)}{r''^3} + \&c.$$

whence it is easy to see that

$$-\frac{dR}{dx} = \frac{m'x'}{r'^3} + \frac{m''x''}{r''^3} + \&c. - \frac{1}{m}\left(\frac{d\lambda}{dx}\right),$$
$$-\frac{dR}{dy} = \frac{m'y'}{r^3} + \frac{m''y''}{r''^3} + \&c. - \frac{1}{m}\left(\frac{d\lambda}{dy}\right),$$
$$-\frac{dR}{dz} = \frac{m'z'}{r'^3} + \frac{m''z''}{r''^3} + \&c. - \frac{1}{m}\left(\frac{d\lambda}{dz}\right),$$

and therefore the preceding equations become

$$\frac{d^2x}{dt^2} + \frac{\mu x}{r^3} = \left(\frac{dR}{dx}\right),$$
$$\frac{d^2y}{dt^2} + \frac{\mu y}{r^3} = \left(\frac{dR}{dy}\right), \qquad (87)$$
$$\frac{d^2z}{dt^2} + \frac{\mu x}{r^3} = \left(\frac{dR}{dz}\right).$$

The whole motions of the planets and satellites are derived from these equations, for S may either be considered to be the sun, and m, m', &c. planets; or S may be taken for a planet, and m, m', &c. for its satellites.

If one planet only moved round the sun, its orbit would be a perfect ellipse, but by the attraction of the other planets, its elliptical motion is very much altered, and rendered extremely complicated.

348. It appears then, that the problem of planetary motion, in its most general sense, is the determination of the motion of a body when attracted by one body, and disturbed by any number of others. The only results that can be obtained from the preceding equations, which express this general problem, are the principle of areas and living forces; and that the motion of the centre of gravity is uniform, rectilinear, and in no way affected by the mutual action of the bodies. As these properties have been already proved to exist in a system of bodies mutually attracting each other, whatever the law of the force might be, provided that it could be expressed in functions of the distance; it evidently follows, that they must exist in the solar system, where the force is inversely as the square of the distance, which is only a particular case of the more general theorem. As no other results can be obtained from these general equations in the present state of analysis, the effects of one disturbing body is estimated at a time, but as this can be repeated for each body in the system, the disturbing action of all the planets on any one may be found.

349. The problem of planetary motion when so limited is, to determine, at any given time, the place of a body when attracted by one body and disturbed by another, the masses, distances, and positions of the bodies being given. This is the celebrated problem of three bodies; it is extremely complicated, and the most refined and laborious analysis is requisite to select among the infinite number of inequalities to which the planets are liable, those that are perceptible, and to assign their values. Although this problem has employed the greatest mathematicians from Newton to the present day, it can only be solved by approximation.

350. The action of a planet on the sun, or of a satellite on its

primary, shortens its periodic time, if the planet be very large when compared with the sun, or the satellite when compared with its primary; for, as the ratio of the cube of the greater axis of the orbit to the square of the periodic time is proportional to the sum of the masses of the sun and the planet, Kepler's law would vary in the different orbits, according to the masses if they were considerable. But as the law is nearly the same for all the planets, their masses must be very small in comparison to that of the sun; and it is the same with regard to the satellites and their primaries. The volumes of the sun and planets confirm this; if the centre of the sun were to coincide with the centre of the earth, his volume would not only include the orbit of the moon, but would extend as far again, whence we may form some idea of his magnitude; and even Jupiter, the largest planet of the solar system, is incomparably smaller than the sun.

351. Thus any modifications in the periodic times, that could be produced by the action of the planets on the sun, must be insensible. As the masses of the planets are so small, their disturbing forces are very much less than the force of the sun, and therefore their orbits, although not strictly elliptical, are nearly so; and the areas described so nearly proportional to the time, that the action of the disturbing force may at first be neglected; then the body may be estimated to move in a perfect ellipse. Hence the first approximation is, to find the place of a body revolving round the sun in a perfect ellipse at a given time. In the second approximation, the greatest effects of the disturbing forces are found; in the third, the next greatest, and so on progressively, till they become so small, that they may be omitted in computation without sensible error. By these approximations, the place of a body may be found with very great accuracy, and that accuracy is verified by comparing its computed place with its observed place. The same method applies to the satellites.

Fortunately, the formation of the planetary system affords singular facilities for accomplishing these approximations: one of the principal circumstances is the division of the system into partial systems, formed by the planets and their satellites. These systems are such, that the distances of the satellites from their primaries are very much less than the distances of their primaries from the sun. Whence, the action of the sun being very nearly the same on the planet and on its satellites, the satellites move very nearly as if they were only influenced by the attraction of the planet.

Motion of the Centre of Gravity.

352. From this formation it also follows, that the motion of the centre of gravity of a planet and its satellites, is very nearly the same as if all these bodies were united in one mass at that point.

Let C be the centre of gravity of a system of bodies m, m', m'', &c., as, for example, of a planet and its satellites, and let S be any body not belonging to the system, as the sun.

It was shown, in the first book, that the force which urges the centre of gravity of a system of bodies parallel to any straight line, Sx, is equal to the sum of the forces which urge the bodies m, m', &c. parallel to this straight line, multiplied respectively by their masses, the whole being divided by the sum of their masses.

It was also shown, that the mutual action and attraction of bodies united together in any manner whatever, has no effect on the centre of gravity of the system, whether at rest or in motion. It is, therefore, sufficient to determine the action of the body S, not belonging to the system, on its centre of gravity.

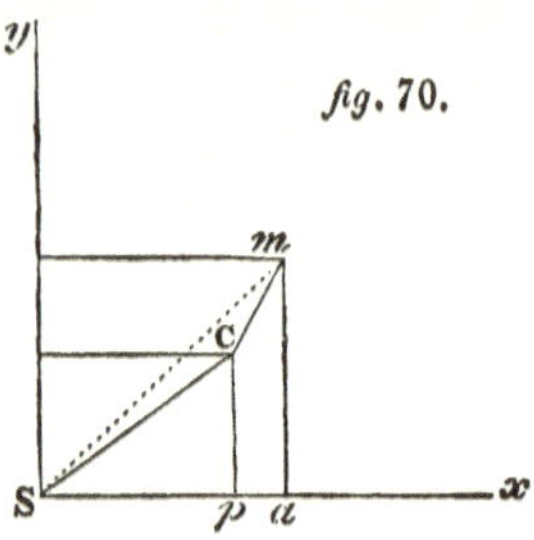

fig. 70.

Let $\bar{x}$, $\bar{y}$, $\bar{z}$, be the co-ordinates of C, fig. 70, the centre of gravity of the system referred to S, the centre of the sun; and let x, y, z, x', y', z', &c., be the co-ordinates of the bodies m, m', m'', &c., referred to C, their common centre of gravity. Imagine also, that the distances Cm, Cm', &c., of the bodies from their centre of gravity, are very small in comparison of SC, the distance of the centre of gravity from the sun. The action of the body m on the sun at S, when resolved in the direction Sx, is

$$\frac{m.(\bar{x} + x)}{r^3},$$

in which m is the mass of the body, and

$$r = \sqrt{(\bar{x} + x)^2 + (\bar{y} + y)^2 + (\bar{z} + z)^2}.$$

But the action of the sun on m is to the action of m on the sun, as

S, the mass of the sun, to m, the mass of the body; hence the action of these two bodies on C, the centre of gravity of the system, is

$$- \text{S} \, . \, \frac{\frac{m(\bar{r} + x)}{r^3}}{m}.$$

The same relation exists for each of the bodies; if we therefore represent the sum of the actions in the axes ox by

$$\Sigma \, . \, \frac{m(\bar{x} + x)}{r^3},$$

and the sum of the masses by $\Sigma . m$, the whole force that acts on the centre of gravity in the direction Sx will be

$$- S \, . \, \frac{\Sigma \, . \, \frac{m(\bar{x} + x)}{r^3}}{\Sigma . m}.$$

Now, $\bar{x} + x$, fig. 70, is equal to Sp + pa, but Sp and pa are the distances of the sun and of the body m from C, estimated on Sx; as pa is incomparably less than Sp, the square of pa may be omitted without sensible error, and also the squares of y and z, together with the products of these small quantities; then if

$$\bar{r} = \text{SC} = \sqrt{\bar{x}^2 + \bar{y}^2 + \bar{z}^2},$$

the quantity $\frac{\bar{x} + x}{r^3}$ becomes

$$\frac{\bar{x} + x}{\{\bar{r}^2 + 2(\bar{x}x + \bar{y}y + \bar{z}z\}^{\frac{3}{2}}}, \text{ or } (\bar{x} + x)\{\bar{r}^2 + 2(\bar{x}x + \bar{y}y + \bar{z}z)\}^{-\frac{3}{2}}.$$

And expanding this by the binomial theorem, it becomes

$$\frac{}{\bar{r}^3} + \frac{x}{\bar{r}^3} - \frac{3\bar{x}\{\bar{x}x + \bar{y}y + \bar{z}z\}}{\bar{r}^5}.$$

Now, the same expression will be found for x', y', z', &c., the coordinates of the other bodies; and as by the nature of the centre of gravity $\Sigma . mx = 0$, $\Sigma . my = 0$, $\Sigma . mz = 0$, the expression

$$- S \frac{\Sigma \, . \, \frac{m(\bar{x} + x)}{r^3}}{\Sigma m} \text{ becomes } - \frac{S . \bar{x}}{r^3},$$

$$\text{or } - \frac{S \, . \, \Sigma \, . \, \frac{mx}{r^3}}{\Sigma m} = - \frac{\text{S} . \bar{r}}{\bar{r}^3};$$

that is, when the squares and products of the small quantities x, y, z, &c., are omitted; hence the centre of gravity of the system is urged by the action of the sun in the direction Sx, as if all the masses were united in C, their common centre of gravity. It is evident that

$$-\frac{S.\bar{y}}{\bar{r}^3}, \qquad -\frac{S.\bar{z}}{\bar{r}^3},$$

are the forces urging the centre of gravity in the other two axes.

353. In considering the relative motion of the centre of gravity of the system round S, it will be found that the action of the system of bodies m, m', m'', &c., on S in the axes ox, oy, oz, are

$$\frac{\bar{x}.\Sigma m}{\bar{r}^3};\ \frac{\bar{y}.\Sigma m}{\bar{r}^3};\ \frac{\bar{z}.\Sigma m}{\bar{r}^3},$$

when the squares and products of the distances of the bodies from their common centre of gravity are omitted. These act in a direction contrary to the origin. Whence the action of the system on S is nearly the same as if all their masses were united in their common centre of gravity; and the centre of gravity is urged in the direction of the axes by the sum of the forces, or by

$$\begin{aligned} &-\{S+\Sigma.m\}\frac{\bar{x}}{\bar{r}^3},\\ &-\{S+\Sigma.m\}\frac{\bar{y}}{\bar{r}^3}, \qquad (88)\\ &-\{S+\Sigma.m\}\frac{\bar{z}}{\bar{r}^3}; \end{aligned}$$

and thus the centre of gravity moves as if all the masses m, m', m'', &c., were united in their common centre of gravity; since the co-ordinates of the bodies m, m', m'', &c., have vanished from all the preceding results, leaving only $\bar{x}$, $\bar{y}$, $\bar{z}$, those of the centre of gravity.

From the preceding investigation, it appears that the system of a planet and its satellites, acts on the other bodies of the system, nearly as if the planet and its satellites were united in their common centre of gravity; and this centre of gravity is attracted by the different bodies of the system, according to the same law, owing to the distance between planets being comparatively so much greater than that of satellites from their primaries.

Attraction of Spheroids.

354. The heavenly bodies consist of an infinite number of particles subject to the law of gravitation; and the magnitude of these bodies

bears so small a proportion to the distances between them, that they act upon one another as if the mass of each were condensed in its centre of gravity. The planets and satellites are therefore considered as heavy points, placed in their respective centres of gravity. This approximation is rendered more exact by their form being nearly spherical: these bodies may be regarded as formed of spherical layers or shells, of a density varying from the centre to the surface, whatever the law may be of that variation. If the attraction of one of these layers, on a point interior or exterior to itself, can be found, the attraction of the whole spheroid may be determined.

Let C, fig. 71, be the centre of a spherical shell of homogeneous matter, and CP $= a$, the distance of the attracted point P from the centre of the shell. As everything is symmetrical round CP, the whole attraction of the spheroid on P must be in the direction of this line. If dm be an element

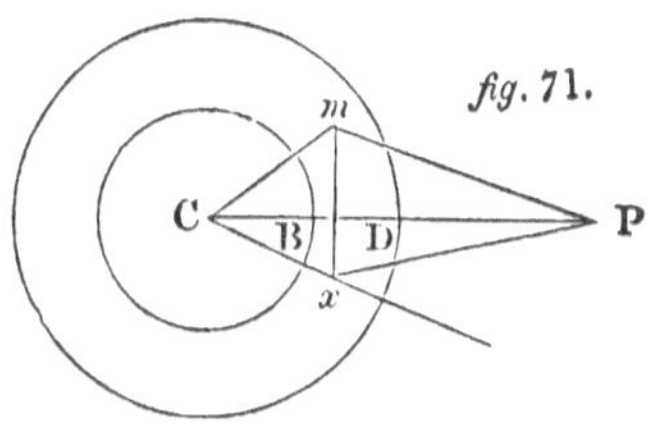

fig. 71.

of the shell at m, and $f = m$P be its distance from the point attracted, then, assuming the action to be in the inverse ratio of the distance, $\frac{dm}{f^2}$ is the attraction of the particle on P; and if C$\text{P}m = \gamma$, this action, resolved in the direction CP, will be $\frac{dm}{f^2} . \cos \gamma$, and the whole attraction A of the shell on P, will be

$$A = \int \frac{dm . \cos \gamma}{f^2}.$$

The position of the element dm, in space, will be determined by the angle $m\text{CP} = \theta$, C$m = r$, and by ω, the inclination of the plane PCm on mCx. But, by article 278, $dm = r^2 \sin \theta \, dr \, d\varpi \, d\theta$; and from the triangle CPm it appears that

$$f^2 = a^2 - 2ar \cos \theta + r^2; \ \cos \gamma = \frac{a - r \cos \theta}{f};$$

hence $$A = \int r^2 \sin \theta . dr d\varpi d\theta \, . \, \frac{a - r \cos \theta}{f^3},$$

is the attraction of the whole shell on P, for the integral must be taken from $r =$ CB to $r =$ CD, and from $\theta = 0$, $\omega = 0$ to $\theta = \pi$, $\omega = 2\pi$, π being the semicircle whose radius is unity. The value

of f, gives
$$\frac{df}{da} = \frac{a - r\cos\theta}{f};$$

hence
$$A = -\int r^2 . \sin\theta \, dr \, d\omega \, d\theta \, . \frac{d\frac{1}{f}}{da};$$

but as r, ω, and θ are independent of a,
$$A = -\frac{d\int\frac{r^2 \sin\theta . dr \, d\omega \, d\theta}{f}}{da}.$$

Thus the whole attraction of the spherical layer on the point P is obtained by taking the differential of
$$\int\frac{r^2 \sin\theta . dr \, d\omega \, d\theta}{f},$$

according to a, and dividing it by da.

Let
$$\int\frac{r^2 \sin\theta . dr \, d\omega \, d\theta}{f} = V.$$

This integral from $\omega = 0$ to $\omega = 2\pi$, is
$$V = 2\pi \int\frac{r^2 dr . d\theta \sin\theta}{f}.$$

But from the value of f, it is easy to find
$$\frac{d\theta \sin\theta}{f} = \frac{1}{ar} df;$$

hence
$$V = \frac{2\pi}{a}\int r dr . df.$$

The integral with regard to θ must be taken from $\theta = 0$ to $\theta = \pi$; but at these limits $f^2 = (a - r)^2$ and $f^2 = (a + r)^2$; and as f must always be positive, when the attracted point is within the spherical layer
$$f = r - a, \text{ and } f = r + a;$$

and when the attracted point P is without the spherical layer
$$f = a - r, \text{ and } f = a + r;$$

hence, in the first case,
$$V = 4\pi \int r dr;$$

and in the second,
$$V = \frac{4\pi}{a}\int r^2 dr.$$

355. But the differential of V, according to a, and divided by da,

when the sign is changed, is the whole attraction of the shell on P. Hence, from the first expression, $\frac{dV}{da} = 0$.

Thus a particle of matter in the interior of a hollow sphere is equally attracted on all sides.

356. The second expression gives

$$-\frac{dV}{da} = \frac{4\pi}{a^2}\int r^2 dr.$$

The integral of this quantity from

$$r = \text{CB} = R' \text{ to } r = \text{CD} = R'',$$

is

$$-\frac{dV}{da} = \frac{4\pi}{3a^2}(R''^3 - R'^3),$$

which is the action of a spherical layer on a point without it.

If M be the mass of the layer whose thickness is $R'' - R'$, it will be equal to the difference of two spheres whose radii are R'' and R'; hence

$$M = \frac{4\pi}{3}(R''^3 - R'^3);$$

and therefore

$$A = \frac{M}{a^2}.$$

Thus the attraction of a spherical layer on a point exterior to it, is the same as if its whole mass were united in its centre.

357. If R', the radius of the interior surface, be zero, the shell will be changed into a sphere whose radius is R''. Hence the attraction of a homogeneous sphere on a point at its surface, or beyond it, is the same as if its mass were united at its centre.

These results would be the same were the attracting solid composed of layers of a density varying, according to any law whatever, from the centre to the surface; for, as they have been proved with regard to each of its layers, they must be true for the whole.

358. The celestial bodies then attract very nearly as if the mass of each was united in its centre of gravity, not only because they are far from one another, but because their forms are nearly spherical.

CHAPTER IV.

ON THE ELLIPTICAL MOTION OF THE PLANETS.

359. The elliptical orbit of the earth is the plane of the ecliptic: the plane of the terrestial equator cuts the plane of the ecliptic in a line passing through the vernal and autumnal equinoxes.

The vernal equinox is assumed as an origin from whence the angular distances of the heavenly bodies are estimated. Astronomers designate that point by the character ♈, the first point of Aries, although these points have not coincided for 2230 years, on account of the precession or retrograde motion of the equinoxes.

360. Angular distance from the vernal equinox, or first point of Aries, estimated on the plane of the ecliptic, is longitude, which is reckoned from west to east, the direction in which the bodies of the solar system revolve round the sun. For example, let EnBN, fig. 72, represent the ecliptic, S the sun, and ♈ the first point of Aries, or vernal equinox. If the earth be in E, its longitude is the angle ♈SE.

361. The earth alone moves in the plane of the ecliptic, the orbits of the other bodies of the system are inclined to it at small angles; so that the planets, in their revolutions, are sometimes seen above that plane, and sometimes below it. The angular distance of a planet above or below the plane of the ecliptic, is its latitude; when the planet is above that plane, it is said to have north latitude, and when below it, south latitude. Latitude is reckoned from zero to 180°.

362. Let EnBN represent the plane of the ecliptic, and let m be a planet moving round the sun S in the direction mPn, the orbit being inclined to the ecliptic at the angle PNE; the part of the orbit NPn is supposed to be above the plane of the ecliptic, and NAn below it. The line NSn, which is the intersection of the plane of the orbit with the plane of the ecliptic, is the line of nodes; it always passes through the centre of the sun. When the

planet is in N, it is in its ascending node; when in n, it is in its descending node. Let mp be a perpendicular from m on the plane of the ecliptic, Sp is the projection of the radius vector Sm, and is the curtate distance of the planet from the sun. ♈SN is the longitude of the ascending node; and it is clear that the longitude of n, the descending node, is 180° greater. The longitude of m is ♈Sm, or ♈Sp, according as it is estimated on the orbit, or on the ecliptic; and mSp, the angular height of m above the plane of the ecliptic, is its latitude. As the position of the first point of Aries is known, it is evident that the place of a planet m in its orbit is found, when the angles ♈Sm, mSp, and Sm, its distance from the sun, are known at any given time, or ♈Sp, pSm, and Sp, which are more generally employed. But in order to ascertain the real place of a body, it is also requisite to know the nature of the orbit in which it moves, and the position of the orbit in space. This depends on six constant quantities, AP, the greater axis of the ellipse; $\frac{CS}{CP}$, the eccentricity; ♈SP, the longitude of P, the perihelion; ♈SN, the longitude of N, the ascending node; ENP, the inclination of the orbit on the plane of the ecliptic; and on the longitude of the epoch, or position of the body at the origin of the time.

fig. 72.

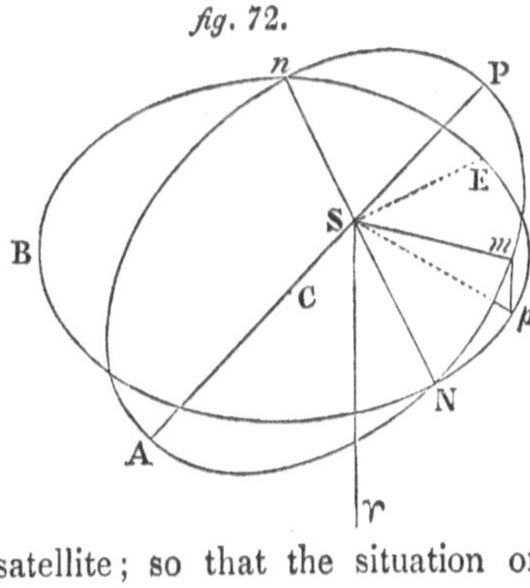

These six quantities, called the elements of the orbit, are determined by observation; therefore the object of analysis is to form equations between the longitude, latitude, and distance from the sun, in values of the time; and from them to compute tables which will give values of these three quantities, corresponding to any assumed time, for a planet or satellite; so that the situation of every body in the system may be ascertained by inspection alone, for any time past, present, or future.

363. The motion of the earth differs from that of any other planet, only in having no latitude, since it moves in the plane of the

ecliptic, which passes through the centre of the sun. In consequence of the mutual attraction of the celestial bodies, the position of the ecliptic is variable to a very minute extent; but as the variation is known, its position can be ascertained.

364. The motions of the celestial bodies, and the positions of their orbits, will be referred to the known position of this plane at some assumed epoch, say 1750, unless the contrary be expressly mentioned. It will therefore be assumed to be the plane of the co-ordinates x and y, and will be called the FIXED PLANE.

Motion of one Body.

365. If the undisturbed elliptical motion of one body round the sun be considered, the equations in article 146 become

$$\frac{d^2x}{dt^2} + \frac{\mu x}{r^3} = 0,$$

$$\frac{d^2y}{dt^2} + \frac{\mu y}{r^3} = 0, \qquad (89)$$

$$\frac{d^2z}{dt^2} + \frac{\mu z}{r^3} = 0,$$

where μ is put for $S + m$, the sum of the masses of the sun and planet, and $r = \sqrt{x^2 + y^2 + z^2}$.

In these three equations, the force is inversely as the square of the distance; they ought therefore to give all the circumstances of elliptical motion. Their finite values will give x, y, z, in values of the time, which may be assumed at pleasure: thus the place of the body in its elliptical orbit will be known at any instant; and as the equations are of the second order, six arbitrary constant quantities will be introduced by their integration, which determine the six elements of the orbit.

366. These give the motion of the planet with regard to the sun; but the equations

$$0 = \frac{d^2x}{dt^2} - \frac{mx}{r^3}; \quad 0 = \frac{d^2y}{dt^2} - \frac{my}{r^3}; \quad 0 = \frac{d^2z}{dt^2} - \frac{mz}{dt^2},$$

of article 346, give values of $\bar{x}$, $\bar{y}$, $\bar{z}$, in terms of the time which will determine the motion of the sun in space; for if the first of them be multiplied by S + m, and added to

$$\frac{d^2x}{dt^2} + \frac{(\mathrm{S}+m)x}{r^3} = 0,$$

multiplied by m, their sum will be

$$(\mathrm{S}+m)\,\frac{d^2\bar{x}}{dt^2} + m\,\frac{d^2x}{dt^2} = 0,$$

the integral of which is

$$\bar{x} = a + bt - \frac{mx}{\mathrm{S}+m};$$

in the same manner,

$$\bar{y} = a' + b't - \frac{my}{\mathrm{S}+m},$$

$$\bar{z} = a'' + b''t - \frac{mz}{\mathrm{S}+m}.$$

These equations give the motion of the sun in space accompanied by m; and as they are the same for each body, if Σm be substituted for m, they will determine the absolute motion of the sun attended by the whole system, when the relative motions of m, m', m'', &c., are known.

367. But in order to ascertain the values of x, y, z, the equations (89) must be integrated. Since these equations are linear and of the second order, their integrals must contain six constant quantities. They are also symmetrical and so connected, that any one of the variable quantities x, y, z, depends on the other two. M. Pontécoulant has determined these integrals with great elegance and simplicity in the following manner.

368. If the first of the equations (89) of elliptical motion multiplied by y, be subtracted from the second multiplied by x, the result will be

$$\frac{xd^2y - yd^2x}{dt^2} = 0;$$

consequently,

$$\frac{xdy - ydx}{dt} = c.$$

*

In the same way it is easy to find that

$$\frac{zdx - xdz}{dt} = c'; \quad \frac{ydz - zdy}{dt} = c'',$$

where c, c', c'', are arbitrary constant quantities introduced by integration. Again, if the first of the same equations be multiplied by $2dx$, the second by $2dy$, and the third by $2dz$, their sum will be

$$\frac{2dxd^2x + 2dyd^2y + 2dzd^2z}{dt^2} + \frac{2\mu\,(xdx + ydy + zdz)}{r^3} = 0.$$

But $\qquad r^2 = x^2 + y^2 + z^2;$

whence $\qquad rdr = xdx + ydy + zdz;$

and the integral of the preceding equation is

$$\frac{dx^2 + dy^2 + dz^2}{dt^2} - \frac{2\mu}{r} + \frac{\mu}{a} = 0, \qquad (90)$$

$\frac{\mu}{a}$ being an arbitrary constant quantity.

If $\qquad \frac{d^2y}{dt^2} = -\frac{\mu y}{r^3}$, multiplied by $c'' = \frac{ydz - zdy}{dt}$,

be subtracted from

$$\frac{d^2x}{dt^2} = -\frac{\mu x}{r^3}, \text{ multiplied by } c' = \frac{zdx - xdz}{dt},$$

the result will be

$$\frac{c'd^2x - c''d^2y}{dt} = \frac{\mu x}{r^3}\,(xdz - zdx) - \frac{\mu y}{r^3}\,(zdy - ydz)$$

$$= \frac{\mu\,(rdz - zdr)}{r^2} = \mu d\,.\,\frac{z}{r}.$$

Whence $\qquad f + \frac{\mu z}{r} = \frac{c'dx - c''dy}{dt};$

and by a similar process values of

$$\mu\,.\,d\,\frac{y}{r}, \text{ and } \mu d\,\frac{x}{r},$$

may be found, the integrals of which are

$$f' + \frac{\mu y}{r} = \frac{c''dz - cdx}{dt}; \quad f'' + \frac{\mu x}{r} = \frac{cdy - c'dz}{dt}.$$

369. Thus the integrals of equations (89) are,

$$c = \frac{xdy - ydx}{dt}; \quad c' = \frac{zdx - xdz}{dt}; \quad c'' = \frac{ydz - zdy}{dt};$$

$$f + \frac{\mu z}{r} = \frac{c'dx - c''dy}{dt},$$

$$f' + \frac{\mu y}{r} = \frac{c''dz - cdx}{dt}, \qquad (91)$$

$$f'' + \frac{\mu x}{r} = \frac{cdy - c'dz}{dt},$$

$$\frac{\mu}{a} - \frac{2\mu}{r} + \frac{dx^2 + dy^2 + dz^2}{dt^2} = 0,$$

containing the seven arbitrary constant quantities c, c', c'', f, f', f'', and a.

370. As two equations of condition exist among the constant quantities, they are reduced to five that are independent, consequently two of the seven integrals are included in the other five. For if the first of these equations be multiplied by z, the second by y, and the third by x, their sum is

$$cz + c'y + c''x = 0. \qquad (92)$$

Again, if the fourth integral multiplied by c, be added to the fifth multiplied by c',

$$fc + f'c' + \mu\frac{cz + c'y}{r} = c'' \cdot \frac{c'dz - cdy}{dt};$$

but $$cz + c'y = -c''x;$$

hence $$-\frac{fc + f'c'}{c''} + \mu\frac{x}{r} = \frac{cdy - c'dz}{dt};$$

but this coincides with the sixth integral, when

$$f'' = -\frac{fc + f'c'}{c''}, \text{ or } f''c'' + f'c' + fc = 0.$$

The six arbitrary quantities being connected by this equation of condition, the sixth integral results from the five preceding.

If the squares of f, f', and f'', from the fourth, fifth, and sixth integrals be added, and $f^2 + f'^2 + f''^2 = l^2$, they give

$$l^2 - \mu^2 = (c^2 + c'^2 + c''^2)\left\{\frac{dx^2 + dy^2 + dz^2}{dt^2} - \frac{2\mu}{r}\right\} - \left\{\frac{cdz + c'dy + c''dx}{dt}\right\}^2$$

but $cz + c'y + c'x = 0$; hence $cdz + c'dy + c''dx = 0$;

consequently, if $c^2 + c'^2 + c''^2 = h^2$,

$$0 = \frac{dx^2 + dy^2 + dz^2}{dt^2} - \frac{2\mu}{r} + \frac{\mu^2 - l^2}{h^2},$$

and comparing this equation with the last of the integrals in article 369, it will appear that

$$\frac{\mu^2 - l^2}{h^2} = \frac{\mu}{a};$$

thus, the last integral is contained in the others; so that the seven integrals and the seven constant quantities are in reality only equal to five distinct integrals and five constant quantities.

371. Although these are insufficient to determine x, y, z, in functions of the time, they give the curve in which the body m moves. For the equation

$$cz + c'y + c''x = 0$$

is that of a plane passing through the origin of the co-ordinates, whose position depends on the constant quantities c, c', c''. Thus the curve in which m moves is in one plane. Again, if the fourth of the integrals in article 269 be multiplied by z, the fifth by y, and the sixth by x, their sum will be

$$fz + f'y + f''x + \frac{\mu(x^2 + y^2 + z^2)}{r} = c'\frac{(zdx - xdz)}{dt} + c''\frac{(ydz - zdy)}{dt}$$
$$+ c\frac{(xdy - ydx)}{dt};$$

but in consequence of the three first integrals in article 369, it becomes $\quad o = \mu r - (c^2 + c'^2 + c''^2) + fz + f'y + f''x,$

or $\quad 0 = \mu r - h^2 + fz + f'y + f''x.$

This equation combined with

$$cz + c'y + c''x = 0, \quad \text{and } r^2 = x^2 + y^2 + z^2,$$

gives the equation of conic sections, the origin of r being in the focus.

fig. 73.

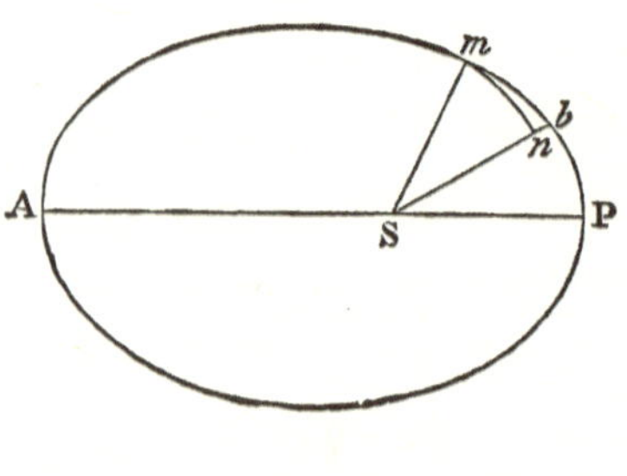

372. Thus the planets and comets move in conic sections having the sun in one of their foci, and their radii vectores describe areas proportional to

the time; for if dv represent the indefinitely small arc mb, fig. 73, contained between

$$Sm = r \text{ and } Sb = r + dr,$$

then $$(mb)^2 = dx^2 + dy^2 + dz^2 = r^2dv^2 + dr^2;$$

but the sum of the squares of the three first of equations (91) is

$$\frac{(x^2+y^2+z^2)(dx^2+dy^2+dz^2)}{dt^2} - \frac{(xdx+ydy+zdz)^2}{dt^2} = h^2,$$

or $$\frac{r^2(dx^2+dy^2+dz^2)}{dt^2} - \frac{r^2dr^2}{dt^2} = h^2;$$

hence $$dv = \frac{hdt}{r^2}. \qquad (93)$$

373. Thus the area $\frac{1}{2}r^2dv$ described by the radius vector r or Sm is proportional to the time dt, consequently the finite area described in a finite time is proportional to the time. It is evident also, that the angular motion of m round S is in each point of the orbit, inversely as the square of the radius vector, and as very small intervals of time may be taken instead of the indefinitely small instants dt, without sensible error, the preceding equation will give the horary motion of the planets and comets in the different points of their orbits.

Determination of the Elements of Elliptical Motion.

374. The elements of the orbit in which the body m moves depend on the constant quantities c, c', c'', f, f', f'', and $\frac{\mu}{a}$. In order to determine them, it must be observed that in the equations (89) the co-ordinates x, y, z are SB, Bp, pm, fig. 74; but if they be referred to γS the line of he equinoxes, so that SD $= x'$, D$p = y'$, $pm = z'$, and if γ SN ENP, the longitude of the node and inclination of the orbit on the fixed plane be represented by θ and ϕ; it is evident, from the method of changing the co-ordinates in article 225, that

fig. 74.

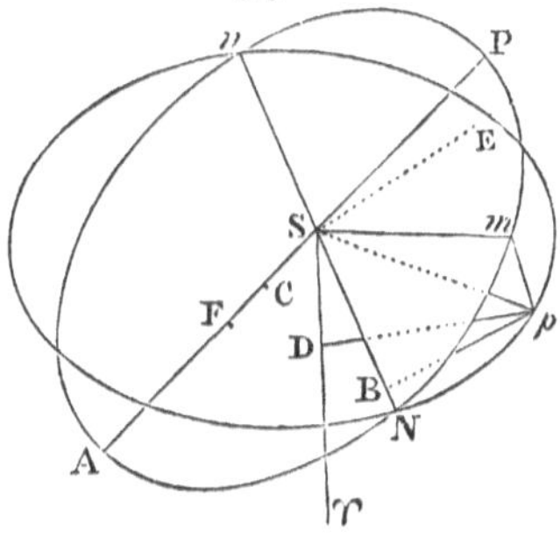

$$x' = x\cos\theta + y\sin\theta,$$
$$y' = y\cos\theta - x\sin\theta,$$
$$z' = y\tan\phi,$$

consequently $\quad z = y\cos\theta\tan\phi - x\sin\theta\tan\phi;$

but if this be compared with

$$0 = c''x + c'y + cz,$$

it will be found that

$$c' = -c\cos\theta\tan\phi,$$
$$c'' = c\sin\theta\tan\phi,$$

whence

$$\begin{cases} \tan\theta = -\dfrac{c''}{c'} \\ \tan\phi = \dfrac{\sqrt{c'^2 + c''^2}}{c} \end{cases} \qquad (94)$$

Thus the position of the nodes and the inclination of the orbit are given in terms of the constant quantities c, c', c''.

375. Now $r^2 = x^2 + y^2 + z^2$, and $rdr = xdx + ydy + zdz$, but at the perihelion the radius vector r is a minimum; hence $dr = 0$, therefore $\quad xdx + ydy + zdz = 0.$

Let $x_{,}$, $y_{,}$, $z_{,}$, be the co-ordinates of the planet when in perihelio, then, substituting the values of c, c', c'', from 269 in the equations in f' and f'' of the same number, and dividing the one by the other, the result in consequence of the preceding relation will be

$$\frac{y_{,}}{x_{,}} = \frac{f'}{f''}.$$

But if $\varpi_{,}$ be the angle ♈SE, the projection of the longitude of the perihelion on the plane Npn, then $\dfrac{y_{,}}{x_{,}} = \tan\varpi_{,}$; hence

$$\tan\varpi_{,} = \frac{f'}{f''};$$

which determines the position of the greater axis of the conic section.

If $\dfrac{dx^2 + dy^2 + dz^2}{dt^2}$ be eliminated from the equation

$$r^2\left(\frac{dx^2 + dy^2 + dz^2}{dt^2}\right) - \frac{r^2dr^2}{dt^2} = h^2,$$

by means of the last of the integrals (91) the result will be

$$2\mu r - \frac{\mu r^2}{a} - \frac{r^2 dr^2}{dt^2} = h^2;$$

but at the extremities of the greater axis $dr = 0$, because the radius vector is either a maximum or minimum at these points, therefore at the aphelion and perihelion

$$0 = r^2 - 2ar + \frac{ah^2}{\mu};$$

whence $$r = a \pm a\sqrt{1 - \frac{h^2}{\mu a}}.$$

The sum of these two values of r is the major axis of the conic section, and their difference is FS or double the eccentricity.

376. Thus a is half of AP, fig. 75, the major axis of the orbit, or it is the mean distance of m from S; and $\sqrt{1 - \frac{h^2}{\mu a}}$, is the ratio of the eccentricity to half the major axis. Let this ratio be represented by e, then as it was shown that $\frac{\mu^2 - l^2}{h^2} = \frac{\mu}{a}$; so also

$$\mu e = l = \sqrt{f^2 + f'^2 + f''^2}.$$

Thus all the elements that determine the nature of the conic section and its position in space are known.

377. The three equations

$r^2 = x^2 + y^2 + z^2$, $\mu r - h^2 + fz + f'y + f''z = 0$, and $c''x + c'y + cz = 0$, give x, y, z, in functions of r; but in order to have values of these co-ordinates in terms of the time, r must be found in terms of the same, which requires another integration. Resume the equation

$$2\mu r - \frac{\mu r^2}{a} - \frac{r^2 dr^2}{dt^2} = h^2,$$

then $$\sqrt{1 - \frac{h^2}{\mu a}} = e,$$

gives $$h^2 = a\mu(1 - e^2)$$

therefore $$dt = \frac{rdr}{\sqrt{\mu}\sqrt{2r - \frac{r^2}{a} - a(1-e^2)}}.$$

To integrate this equation, a value of r must be found from

fig. 75.

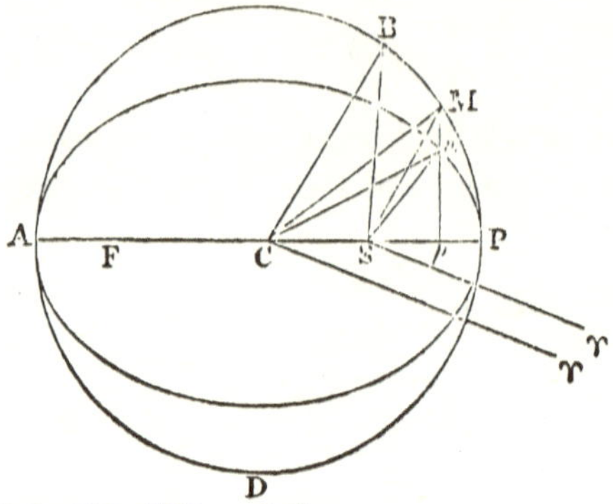

the conic sections. Let AmP, fig. 75, be an ellipse whose major axis is $2a$, its minor axis $2b$, the eccentricity CS $= e'$, and the radius vector S$m = r$.

Let the circle PMA be described on the major axis, draw the perpendicular Mp through m, and join SM, CM, and Cm.

Then $$r^2 = \mathrm{S}p^2 + pm^2, \text{ and if MCP} = u,$$
$$\mathrm{S}p = \mathrm{C}p - \mathrm{CS} = a\cos u - e',$$
or making $$e = \frac{e'}{a}, \ \mathrm{S}p^2 = a^2(\cos u - e)^2.$$
Again, $$pm^2 = b^2 \,.\, \sin^2 u = b^2(1 - \cos^2 u);$$
but $$b^2 = a^2 - e'^2 = a^2(1 - e^2);$$
hence $$r^2 = a^2(1 - e^2)(1 - \cos^2 u) + a^2(\cos u - e)^2,$$
and $$r = a\{1 - e\cos u\}.$$
This value of r and its differential being substituted in the value of dt it becomes $$dt = \frac{a^{\frac{3}{2}}}{\sqrt{\mu}} \,.\, du(1 - e\cos u)$$
the integral of which is $$t + k = \frac{a^{\frac{3}{2}}}{\sqrt{\mu}}\{u - e\sin u\}; \qquad (95)$$
k being an arbitrary constant quantity.

This equation gives u and consequently r in terms of t, and as x, y, z are given in functions of r, the values of these co-ordinates are known at any instant.

When $\mu = 1$ the values of dt and h^2 become
$$\frac{rdr}{\sqrt{2r - \frac{r^2}{a} - a(1 - e^2)}}, \text{ and } a(1 - e^2),$$
and when substituted in $dv = \dfrac{hdt}{r^2}$,

the result is $$dv = \frac{dr \,.\, \sqrt{a(1-e^2)}}{r\sqrt{2r - \frac{1}{a}r^2 - a(1-e^2)}},$$

or
$$dv = -\frac{\frac{a(1-e^2)}{e}d.\frac{1}{r}}{\sqrt{1-\left\{\frac{a(1-e^2)\frac{1}{r}-1}{e}\right\}}};$$

the integral of which is

$$v = \varpi + arc\left\{\cos = \frac{a(1-e^2)\frac{1}{r}-1}{e}\right\},$$

reciprocally
$$r = \frac{a(1-e^2)}{1-e\cos(v-\varpi)},$$

which is the general equation to the conic sections, when the origin of r the radius vector is in the focus; a is half the greater axis, and

$$\cos(v-\varpi) = \cos(♈Sm - ♈SP), \text{ fig. 77.}$$

Elements of the Orbit.

378. Thus the finite values of the equations of elliptical motion are completely determined,

Six arbitrary constant quantities have been introduced, namely,

$2a$, the greater axis of the orbit.

e, the ratio of the eccentricity to half the greater axis.

$\varpi_{,}$, the projection of the longitude of the perihelion.

θ, the longitude of the ascending node.

ϕ, the inclination of the orbit on the plane of the ecliptic, and

ϵ, the longitude of the epoch.

The two first determine the nature of the orbit, the three following its position in space, and the last is relative to the position of the body at a given epoch; or, which is the same thing, it depends on the instant of its passage at the perihelion.

Equations of Elliptical Motion.

379. It now becomes necessary to determine three equations which will give values of the longitude and latitude ♈Sm, mSp, and the distance Sm, fig. 72, in terms of the time from whence tables of the elliptical motions of the planets and satellites may be computed.

380. The motion of a body in an ellipse is not uniform, its velocity is greatest at the perihelion, and least at the aphelion, varying with the angle PSm, which is the true angular motion of the planet; but if the circle PBAD, fig. 75, be described from the centre of the ellipse with the semigreater axis CP, or mean distance from S as radius, the motion of the planet in this circle would be uniform. This is called the mean motion of a body.

381. Were the motion of a planet uniform, the angle PSm described by the planet in any interval of time after leaving perihelion might be found by simple proportion from knowing the periodic time, or time in which it describes 360°; but in order to preserve the equable description of areas, the true place of the planet will be before the mean place in going from perihelion to aphelion; and from aphelion to perihelion the true place will be behind the mean place. These angles are estimated from west to east, the direction in which the bodies of the system move, beginning at the perihelion. If, however, they are estimated from the aphelion, it is only necessary to add 180° to each.

382. The angular distance PCB between the perihelion and the mean place, is the mean anomaly, PSm the angular distance between the true place and the perihelion is the true anomaly; and mSB the angle at the sun, contained between the true and the mean place is called the equation of the centre. If then the mean anomaly be increased or diminished by the equation of the centre, the result will be the true place of the planet in its orbit. The equation of the centre is zero, both at the perihelion and aphelion, for at these points the true and mean places of the planets coincide; it is greatest when the planet is in quadratures, and at its maximum it is equal to an angle measured by twice the eccentricity of the orbit.

383. The mean place of a planet at any given time may be found by simple proportion from its periodic time. The true place of the planet in its orbit, and its distance from the sun, may be found in terms of its mean place by help of the angle PCM, called the eccentric anomaly.

If the time be estimated from the perihelion, $l = 0$, which reduces equation (95) to

$$t = \frac{a^{\frac{3}{2}}}{\sqrt{\mu}}(u - e \,.\, \sin u), \text{ or } nt = u - e \sin u, \text{ if } n = \frac{\sqrt{\mu}}{a^{\frac{3}{2}}}.$$

If the angles u and v be estimated from the perihelion, a comparison of the values of r in article 377, gives

$$1 - e \cos u = \frac{1 - e^2}{1 + e \cos v},$$

whence $$\cos v = \frac{\cos u - e}{1 - e \cos u}; \quad \sin v = \frac{\sin u \,.\, \sqrt{1 - e^2}}{1 - e \cos u},$$

therefore $$\tan \tfrac{1}{2} v = \sqrt{\frac{1 + e}{1 - e}} \,.\, \tan \tfrac{1}{2} u.$$

384. The motions of the celestial bodies in elliptical orbits are therefore obtained from the three equations

$$\begin{aligned} nt &= u - e \sin u, \\ r &= a(1 - e \cos u) \\ \tan \tfrac{1}{2} v &= \sqrt{\frac{1 + e}{1 - e}} \tan \tfrac{1}{2} u. \end{aligned} \qquad (96)$$

Where nt = PCB = mean anomaly, fig. 75,
v = PSm = true anomaly,
u = PCM = eccentric anomaly,
r = Sm = radius vector,
a = CP = mean distance, and
$e = \frac{\text{CS}}{\text{CP}}$ = the ratio of the eccentricity to the mean distance.

385. It appears from these expressions that when u becomes $u + 360°$, r remains the same; and as v is then augmented by 360°, the planet returns to the same point of its orbit, having moved through four right angles, and the time becomes $\text{T} = \frac{a^{\frac{3}{2}}}{\sqrt{\mu}} . 360°$; so that the time of a complete revolution is independent of the eccentricity, and only depends on $2a$, the major axis of the orbit; it is consequently the same as if the planet described a circle at its mean distance from the sun; for in this case $e = 0$, $r = a$, $u = nt$, $v = u$, consequently $v = nt$; the arcs described are therefore proportional to the time, and the planet moves uniformly in the circle whose radius is a. Generally nt represents the arc that a body would

describe in the time t, if it set out from the perihelion at the same instant with a planet m, and moved with a uniform velocity represented by n in a circle described on the major axis of the orbit as diameter. This body would pass the perihelion and aphelion at the same instant with the planet m, but in one half of its revolution the planet would precede the body, and in the other half it would fall behind it. If $a=1$, $\mu=1$, then $n=1$, and $v=t$, the time will therefore be expressed by the arcs described by the planet in the circle whose radius is unity.

Astronomers generally compare the motions of the solar system with those of the earth; they take the mean distance of the sun from the earth as the unit of distance, the sum of the masses of the sun and earth as the unit of mass; and supposing the time to be estimated in mean solar days, the unit of time will be represented by the arc that the earth describes round the sun in one day with its mean motion.

Determination of the Eccentric Anomaly in functions of the Mean Anomaly.

386. If a value of u could be found in terms of nt from the first of these equations, both r and v, and consequently the place of the planet in its orbit at any instant, would be known from the two last.

Now an arc and its sine are incommensurate quantities, so that the one can only be obtained in functions of the other by an infinite series. Therefore a value of u in terms of nt must be found by an infinite series from the first of the preceding equations; but unless the terms of the series decrease rapidly in value u cannot be obtained, for a few of the first terms being computed, the value of the remaining part of the series must be so small that it may be neglected without sensible error. The small eccentricities of the orbits of the planets and satellites afford the means of approximation, for e the ratio of the eccentricity to half the greater axis is still smaller, consequently the powers of such quantities decrease rapidly, and therefore the second part of the equation $u = nt + e \sin u$ may be expanded into a series in functions of the time, and according to the powers of e, which will be sufficiently convergent. This may be accomplished by Maclaurin's *Theorem*, for if u' be the value of u when $e = 0$,

$$u = u' + e \cdot \frac{du'}{de} + \frac{e^2}{1.2} \cdot \frac{d^2u'}{de^2} + \frac{e^3}{1.2.3} \cdot \frac{d^3u'}{de^3} + \&c.$$

But when $e = 0$, $u = nt + e \sin u$, becomes $u' = nt$; and from the same equation

$$\frac{du}{de} = \frac{\sin u}{1 - e\cos u};$$

or when $$e = 0, \frac{du'}{de} = \sin nt.$$

Again, $$\frac{d^2u}{de^2} = \frac{2\cos u \sin u}{(1-e\cos u)^2} - \frac{e\sin^3 u}{(1-e\cos u)^3};$$

or if $$e = 0, \frac{d^2u'}{de^2} = 2\cos nt \sin nt$$

in the same manner, when $e = 0$,

$$\frac{d^3u'}{de^3} = 6\sin nt\cos^2 nt - 3\sin^3 nt,$$

or $$\frac{d^3u'}{de^3} = 6\sin nt - 9\sin^3 nt, \&c. \&c.$$

But $$2\cos nt\sin nt = \sin 2nt,$$

and $6\sin nt - 9\sin^3 nt = -\frac{3}{4}\sin nt + \frac{9}{4}\sin 3nt$; hence

$$\frac{du'}{de} = \sin nt;\ \frac{d^2u'}{de^2} = \tfrac{1}{2}\sin 2nt;\ \frac{d^3u'}{de^3} = \frac{1}{2^2}\{3^2\sin 3nt - 3\sin nt\}\ \&c.$$

consequently,

$$u = nt + e\sin nt + \frac{e^2}{1.2.2} \cdot 2\sin 2nt + \frac{e^3}{2.3.2^2}\{3^2\sin 3nt - 3\sin nt\}$$

$$+ \frac{e^4}{2.3.4.2^3}\{4^3\sin 4nt - 4.2^3\sin 2nt\};$$

$$+ \frac{e^5}{2.3.4.5.2^4} \cdot \{5^4\sin 5nt - 5.3^2\sin 3nt + \frac{5.4}{1.2}\sin nt\}$$

$$+ \ \&c. \quad \&c. \quad \&c.$$

This series converges rapidly in most of the planetary orbits on account of the small value of the fraction which e expresses.

387. Having thus determined u for any instant, corresponding values of v and r may be obtained from the equations $r = a(1 - e\cos u)$

and $$\tan \tfrac{1}{2} v = \sqrt{\frac{1+e}{1-e}} \tan \tfrac{1}{2} u;$$

but it is better to expand these also into series ascending according to the powers of e; and in functions of the sines or cosines of the mean anomaly.

Determination of the Radius Vector in functions of the Mean Anomaly.

Let r' be the value of r when $e = 0$, then

$$\frac{r}{a} = r' + e\frac{dr'}{de} + \frac{e^2}{1.2} \cdot \frac{d^2r'}{de^2} + \&c.$$

but as r is a function of e by the equation $r = a(1 - e\cos u)$; and u is a function of e by $u = nt + e\sin u$, therefore,

$$\frac{dr'}{de} = \frac{dr}{de} + \frac{dr}{du} \cdot \frac{du}{de}.$$

Now when $e = 0$, $\frac{r}{a} = 1$; and $u = nt$ But the differentials of the same equations, when $e = 0$, are

$$\frac{dr}{de} = -\cos nt; \text{ and } \frac{du}{de} = \sin nt;$$

consequently, $$\frac{dr'}{de} = -\cos nt + \sin nt.\frac{dr}{ndt}, \text{ for } du = ndt;$$

or it may be written,

$$\frac{dr'}{de} = -\cos nt + \frac{d\int \sin nt.dr}{ndt}.$$

Again, $$\frac{d^2r'}{de^2} = \frac{d^2\int \sin nt.dr}{ndt.de};$$

but if $\int \sin nt.dr$ be put for r in

$$\frac{dr}{de} = \frac{d\int \sin nt.dr}{ndt},$$

then, $$\frac{d\int \sin nt.dr}{de} = \frac{d\int \sin^2 nt.dr}{ndt}.$$

And if this be substituted in the value of $\frac{d^2r'}{de^2}$ it become

$$\frac{d^2r'}{de^2} = \frac{d^2\int \sin^2 nt \,.\, dr}{(ndt)^2} = \frac{d.\left(\sin^2 nt \,.\, \frac{dr}{ndt}\right)}{ndt}.$$

The differential of the latter expression according to e is

$$\frac{d^3r'}{de^3} = \frac{d^3\int \sin^2 nt \,.\, dr}{(ndt)^2 \,.\, de};$$

and making the same substitution, it becomes

$$\frac{d^3r'}{de^3} = \frac{d^3\int \sin^3 nt \,.\, dr}{(ndt)^3} = \frac{d^2\left(\sin^3 nt \,.\, \frac{dr}{ndt}\right)}{(ndt)^2},$$

and so on. These coefficients being substituted,

$$r = a - ae \cos nt + e \sin nt \,.\, \frac{dr}{ndt} + \frac{e^2}{2} \,.\, \frac{d(\sin^2 nt \,.\, \frac{dr}{ndt})}{ndt} + \&c.$$

But $\qquad r = a\,(1 - e \cos nt)$ gives $\dfrac{dr}{ndt} = ae \,.\, \sin nt$;

hence

$$\frac{r}{a} = 1 - e \cos nt + e^2 \sin^2 nt + \frac{e^3}{2}\,\frac{d \sin^3 nt}{ndt} + \frac{e^4}{2.3} \,.\, \frac{d^2 \,.\, \sin^4 nt}{n^2 dt^2} + \&c.$$

Now $\qquad \sin^2 nt = \frac{1}{2} - \frac{1}{2} \cos 2nt,$

$$\frac{d \sin^3 nt}{ndt} = 3 \sin^2 nt \cos nt = \tfrac{3}{4}\{\cos nt - \cos 3\,nt\}$$

$$\frac{d^2 \sin^4 nt}{(ndt)^2} = 2 \cos 2nt - 2 \cos 4nt, \&c.$$

thus

$$\frac{r}{a} = 1 + \frac{e^2}{2} - e \cos nt - \frac{e^2}{2} \cos 2nt,$$

$$- \frac{e^3}{1.2.2^2} \,.\, \{3 \cos 3nt - 3 \cos nt\},$$

$$- \frac{e^4}{1.2.3.2^3} \,.\, \{4^2 \cos 4nt - 4.2^2 \cos 2nt\},$$

$$- \frac{e^5}{1.2.3.4.2^4} \,.\, \{5^3 \cos 5nt - 5.3^3 \cos 3nt + \frac{5.4}{1.2} \cos nt\}.$$

— &c. &c.

This gives a value of the radius vector in functions of the time.

*

Kepler's Problem. To find a Value of the true Anomaly in functions of the Mean Anomaly.

388. The determination of v in terms of nt is Kepler's problem of finding the true anomaly in terms of the mean anomaly; or, to divide the area of a semicircle in a given ratio by a line drawn from a given point in the diameter—in order to accomplish this, a value of v in functions of u must be obtained from

$$\tan \tfrac{1}{2}v = \sqrt{\frac{1+e}{1-e}} \,.\, \tan \tfrac{1}{2}u;$$

therefore let
$$\lambda = \frac{e}{1+\sqrt{1-e^2}},$$

then
$$e = \frac{2\lambda}{1+\lambda^2}, \text{ and } \sqrt{\frac{1+e}{1-e}} = \frac{1+\lambda}{1-\lambda}.$$

Again, $\sin \tfrac{1}{2}v = c^{v\sqrt{-1}} - 1$, $\cos \tfrac{1}{2}v = c^{v\sqrt{-1}} + 1$,
c being the number whose logarithm is unity; hence the equation in question becomes

$$\frac{c^{v\sqrt{-1}} - 1}{c^{v\sqrt{-1}} + 1} = \frac{1+\lambda}{1-\lambda} \,.\, \frac{c^{u\sqrt{-1}} - 1}{c^{u\sqrt{-1}} + 1},$$

whence
$$c^{v\sqrt{-1}} = \frac{1-\lambda c^{-u\sqrt{-1}}}{1-\lambda c^{u\sqrt{-1}}} \,.\, c^{u\sqrt{-1}};$$

or taking its logarithm,

$$v = u + \frac{\log\{1-\lambda c^{-u\sqrt{-1}}\} - \log\{1 - \lambda c^{u\sqrt{-1}}\}}{\sqrt{-1}}. \text{ Or}$$

$$v = u + \lambda\left\{\frac{c^{u\sqrt{-1}} - c^{-u\sqrt{-1}}}{\sqrt{-1}}\right\} + \frac{\lambda^2}{2}\left\{\frac{c^{2u\sqrt{-1}} - c^{-2u\sqrt{-1}}}{\sqrt{-1}}\right\} + \&c.$$

but
$$\frac{c^{mu\sqrt{-1}} - c^{-mu\sqrt{-1}}}{2\sqrt{-1}} = \sin mu;$$

m being any whole positive number, therefore

$$v = u + 2\lambda \sin u + \frac{2\lambda^2}{2} \sin 2u + \frac{2\lambda^3}{3} \sin 3u + \&c.$$

The true anomaly may now be found in terms of the mean anomaly.

389. In order to have v in terms of the mean anomaly and of the powers of e, values of u, sin u, sin $2u$, must be found in terms of the sines of nt and its multiples; and λ, λ^2, &c. must be developed into series according to the powers of e. Both may be accomplished by *La Grange's* Theorem, for if

$$\phi = \frac{1}{\alpha} = \frac{1}{1+\sqrt{1-e^2}} = \frac{\lambda}{e};$$

when $e = 0$, $\alpha = 2$, $\phi' = \frac{1}{2}$, $\frac{d\phi'}{d\alpha} = -\frac{1}{2^2}$ so that

$$\phi = \frac{\lambda}{e} = \frac{1}{2}\left\{1 + \left(\frac{e}{2}\right)^2 + \frac{4}{2}\left(\frac{e}{2}\right)^4 + \frac{5\,.\,6}{2\,.\,3}\left(\frac{e}{2}\right)^6 + \text{\&c.}\right\}$$

or generally

$$\phi' = \frac{1}{2^i}, \quad \frac{d\phi'}{d\alpha} = -\frac{1}{2^{i+1}},$$

consequently

$$\lambda^i = \frac{e^i}{2^i}\left\{1 + i\left(\frac{e}{2}\right)^2 + \frac{i(i+3)}{2}\left(\frac{e}{2}\right)^4 + \frac{i(i+3)\;(i+5)}{2.3}\left(\frac{e}{2}\right)^6 + \text{\&c.}\right\}$$

If i be successively assumed to be 1, 2, 3, &c., this equation will give all the powers of λ in series, ascending according to the powers of e.

Again. If we assume $\phi = u = nt + e \sin u$, ϕ is a function of u which is a function of e;

hence
$$\frac{d\phi}{de} = \frac{d\phi}{du}\cdot\frac{du}{de};$$

and as $\phi' = nt$, when $e = 0$, so $\frac{du}{de} = \sin nt$. And $\frac{d\phi'}{de} = \sin nt\,\frac{d\phi}{du}$. Whence by the same process it will be found that

$$u = \phi + e \sin nt\,.\,\frac{d\phi}{ndt} + \frac{e^2}{2}\cdot\frac{d\sin^2 nt\,.\,d\phi}{(ndt)^2} + \frac{e^3}{2\,.\,3}\cdot\frac{d^2\sin^3 nt\,.\,d\phi}{(ndt)^3} + \text{\&c. \&c.}$$

Values of u, sin u, sin $2u$, &c., may be determined from this expression by making ϕ successively equal to nt, $e\,.\,\sin nt$, &c. The substitution of these, and of the powers of λ, will complete the development of v, but the same may be effected very easily from the expression $dv = \frac{h\,.\,dt}{r^2}$ of article 372, or rather from

*

$$dv = \sqrt{1-e^2} \cdot \frac{a^2}{r^2} \cdot ndt.$$

390. If $r^i = a^i\,(1 - e\cos nt)^i$ be put for $r = a\,(1 - e\cos nt)$, and $ia^i\,(1 - e\cos nt)^{i-1} \cdot e\sin nt$ for $\frac{dr}{ndt}$ in the development of r in article 387, it becomes

$$\frac{r^i}{a^i} = (1 - e\cos nt)^i + i.e^2.\sin^2.nt\,(1 - e\cos.nt)^{i-1}$$

$$+ \frac{i.e^3.\sin^3.nt\,(1 - e\cos.nt)^{i-1}}{2ndt}$$

$$+ \frac{i.e^4d^2.\sin^4.nt\,(1 - e\cos.nt)^{i-1}}{2.3.n^2dt^2} + \&c.$$

whatever i may be. Let $i = -2$, then

$$\frac{a^2}{r^2} = 1 + 2e\,.\cos.nt + \frac{e^2}{1.2}\,.\,(1 + 5\,.\cos.nt)$$

$$+ \frac{e^3}{1\,.\,2^2}(13\,.\cos.3nt + 3\,.\cos.nt)$$

$$+ \frac{e^4}{1\,.\,2^3\,.\,3}(103.\cos.4nt + 8.\cos.2nt + 9)$$

$$+ \&c.$$

If this quantity be substituted in the preceding expression for dv, when the integration is accomplished, and the approximation only carried to the sixth powers of e, the result will be

$$v = nt + \{2e - \tfrac{1}{4}e^3 + \frac{5}{96}e^5\}\sin nt$$

$$+ \left\{\frac{5}{4}e^2 - \frac{11}{24}e^4 + \frac{17}{192}\,.\,e^6\right\}\sin 2nt$$

$$+ \left\{\frac{13}{12}e^3 - \frac{45}{64}e^5\right\}\sin 3nt$$

$$+ \left\{\frac{103}{96}e^4 - \frac{451}{480}e^6\right\}\sin 4nt,$$

$$+ \quad \&c. \quad \&c.$$

391. The angles v and nt which are the true and mean anomaly, begin at the perihelion; but if they be estimated from the aphelion, it will only be necessary to make e negative in the values of r and v, or to add 180° to each angle. This expression gives $v - nt$ the equation of the centre.

True Longitude and Radius Vector in functions of the Mean Longitude.

392. Instead of fixing the origin of the time at the instant of the planet's passage at the perihelion, let it be fixed at any point whatever, as E, fig. 76, so that nt = ECB, then by adding the constant angle ♈CE represented by ϵ, the whole angle ♈CB $= nt + \epsilon$ is the mean longitude of the planet, ♈ being the equinox of Spring; and if the constant angle ♈CP, which is the longitude of the perihelion, be represented by ϖ, the angle $nt + \epsilon - \varpi$ = PCB must be put for nt, and if v be estimated from ♈, then $v - \varpi$ must be put for v, and the preceding values of v and r become,

fig. 76.

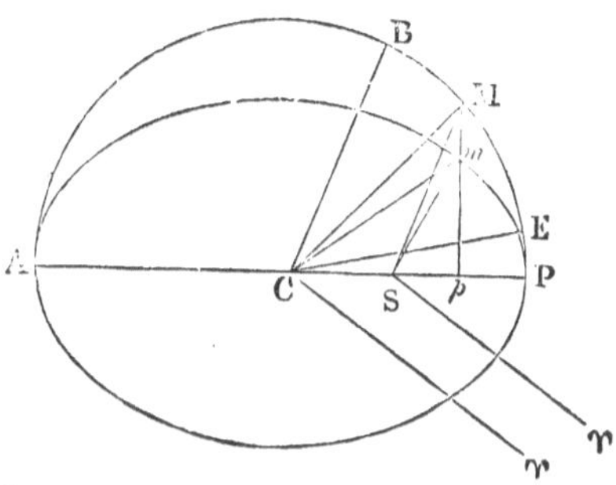

$$v = nt + \epsilon + \{2e - \tfrac{1}{4} e^3\} \sin (nt + \epsilon - \varpi) \qquad (97)$$
$$+ \left\{\frac{5}{4} e^2 - \frac{11}{24} e^4\right\} \sin 2(nt + \epsilon - \varpi),$$
$$+ \qquad \&c. \qquad \&c.$$

$$\frac{r}{a} = 1 + \tfrac{1}{2}e^2 - \{e - \frac{3}{8} e^3\} \cos (nt + \epsilon - \varpi) \qquad (98)$$
$$- \{\tfrac{1}{2} e^2 - \tfrac{1}{3} e^4\} \cos 2 (nt + \epsilon - \varpi),$$
$$- \qquad \&c. \qquad \&c.$$

393. v is the true longitude of the planet and $nt + \varepsilon$ its mean longitude both being estimated on the plane of the orbit. The angle ε = ♈CE is the longitude of the point E, from whence the time is estimated, commonly called the longitude of the epoch.

394. In astronomical series, the quantities which multiply the sines and cosines are the coefficients; and the angles are called the arguments: for example in $(2e - \frac{1}{4} e^2) \sin (nt + \varepsilon - \varpi)$ the part $2e - \frac{1}{4} e^2$ is the coefficient, and $nt + \varepsilon - \varpi$ is the argument.

395. Although the time increases without limit, these series converge: for, as a sine or cosine never can exceed the radius, the values

of the sines and cosines in these series never can be greater than unity, however much the time may increase, and as the powers of e soon become extremely small, they converge rapidly.

396. The values of v and r answer for all the planets and satellites, since they are independent of the masses, for the mass of a planet is so inconsiderable in comparison of that of the sun, that it may be omitted, and as the mass of the sun forms the standard of comparison for the masses of the other bodies of the system, it is assumed to be the unit of measure. The same holds with regard to a planet and its satellites.

Determination of the Position of the Orbit in space.

397. The values of v and r give the place of a body in its orbit, but not its position in space; they however afford the means of ascertaining it. For let NpnG, fig. 77, be the plane of the ecliptic, or fixed plane at the epoch, on which the plane of the orbit PnAN has a very small inclination; then Nn is the line of the nodes; S the sun, and if mp be a perpendicular from the planet on the plane of the ecliptic, it will be the tangent of the latitude mSp. Let ♈SN the longitude of the node be represented by ϵ when estimated on the plane of the orbit, and let θ represent the same angle when projected on the plane of the ecliptic; also let $v_{,} =$ ♈Sp be the true longitude ♈Sm or v, when projected on the plane of the ecliptic.

fig. 77.

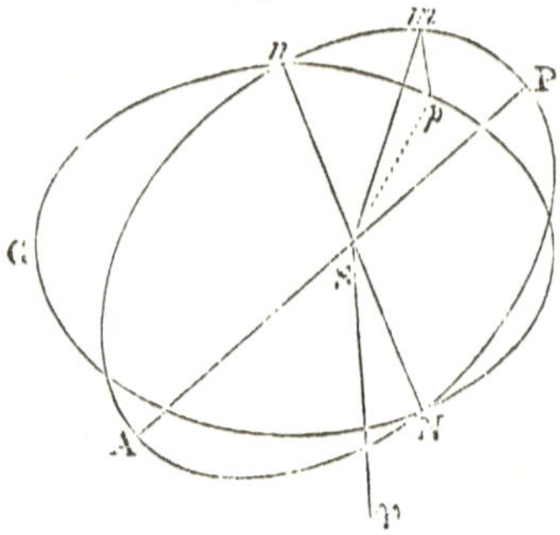

Then $\quad$ N$\mathrm{S}p = v_{,} - \theta$, N$\mathrm{S}m = v - \epsilon_{,}$.

And if ϕ be the inclination of the two planes, it appears from the right angled triangle pNm, that

$$\tan (v_{,} - \theta) = \cos \phi \tan (v - \epsilon). \qquad (99)$$

Projected Longitude in Functions of true Longitude.

398. This gives $v_{,}$ in terms of v, and the contrary. But these two

angles may be obtained in terms of one another in very converging series by means of the expression,

$$\tfrac{1}{2}v = \tfrac{1}{2}u + \lambda \sin u + \frac{\lambda^2}{2} \sin 2u + \frac{\lambda^3}{3} \sin 3u + \&c.$$

which was derived from $\tan \tfrac{1}{2}v = \sqrt{\frac{1+e}{1-e}} \tan \tfrac{1}{2}u$, by making $\lambda = \frac{e}{1+\sqrt{1-e^2}}$. If $v_{,} - \theta$ be put for $\tfrac{1}{2}v$, $v - \text{Є}$ for $\tfrac{1}{2}u$,

and $\cos\phi$ for $\sqrt{\frac{1+e}{1-e}}$;

then $$\lambda = \frac{\cos\phi - 1}{\cos\phi + 1} = -\tan^2 \tfrac{1}{2}\phi,$$

and the series becomes

$$v_{,} - \theta = v - \text{Є} - \tan^2\tfrac{1}{2}\phi . \sin 2(v - \text{Є}) + \tfrac{1}{2}\tan^4\tfrac{1}{2}\phi . \sin 4(v - \text{Є}) - \&c. \quad (100)$$

True Longitude in Functions of projected Longitude.

On the contrary, if $v - \text{Є}$ be put for $\tfrac{1}{2}v$, and $v_{,} - \theta$ for $\tfrac{1}{2}u$, the result will be

$$v - \text{Є} = v_{,} - \theta + \tan^2\tfrac{1}{2}\phi . \sin 2(v_{,} - \theta) + \tfrac{1}{2}\tan^4\tfrac{1}{2}\phi . \sin 4(v_{,} - \text{Є}) + \&c. \quad (101)$$

Projected Longitude in Functions of Mean Longitude.

399. A value of $v_{,} - \theta$, or $\text{NS}p$, may be found in terms of the sines and cosines of nt, and its multiple arcs, from the series

$$v = nt + \epsilon + \{2e - \tfrac{1}{4}e^3\} \sin(nt + \epsilon - \varpi) + \left\{\frac{5}{4}e^2 - \frac{11}{24}e^4\right\} \sin 2(nt + \epsilon - \varpi) + \&c.$$

which may be written

$$v = nt + \epsilon + e\text{Q}.$$

If Є be subtracted from both sides of this equation, and the sines taken in place of the arcs, it becomes

$$\sin(v - \text{Є}) = \sin(nt + \epsilon - \text{Є} + e\text{Q}),$$

which may be expanded into a series, ascending, according to the powers of e, by the method already employed for the development of v and r; if

$$\phi = \sin(v - \text{Є}) = \sin(nt + \epsilon - \text{Є} + e\text{Q}).$$

Whence it may be found that,

$$\sin i\,(v - \varpi) = \sin i\,(nt + \epsilon - \varpi + eQ) = \left\{1 - \frac{i^2e^2Q^2}{1.2} + \frac{i^4e^4Q^4}{1.2.3.4} - \&c.\right\}$$

$$\times \sin i(nt + \epsilon - \varpi) + \left\{ieQ - \frac{i^3e^3Q^3}{1.2.3} + \frac{i^5e^5Q^5}{1.2.3.4.5} - \&c.\right\} \times$$

$$\cos i\,(nt + \epsilon - \varpi) + \&c.$$

Latitude.

400. If mp, the tangent of the latitude, be represented by s, the right-angled triangle mNp gives

$$s = \tan\phi \sin(v_{,} - \theta).$$

Curtate Distances.

401. Let $r_{,}$ be the curtate distance Sp, then Spm, being a right angle,

$$Sp : Sm :: 1 : \sqrt{1 + s^2};$$

hence

$$Sp = \frac{Sm}{\sqrt{1 + s^2}};$$

or

$$r_{,} = r(1 + s^2)^{-\frac{1}{2}} = r\{1 - \tfrac{1}{2}s^2 + \tfrac{3}{8}s^4 - \&c.\} \quad (102)$$

402. Thus $v_{,}$, s, and $r_{,}$, the longitude, latitude, and curtate distance of the planet are determined in convergent series of the sines and cosines of nt and its multiples; if therefore the time be assumed, the place of the body will be known, and the means are thus furnished for computing tables of the motions of the planets and satellites, from which their elliptical places may be ascertained at any instant.

403. A particular period is chosen as an origin from whence the time is estimated, which is called the Epoch of the tables: the elements of the orbits are determined by observation; and the longitude, latitude, and distance of the body from the sun are computed for that period, and for every succeeding day, hour, and minute, if necessary, for any number of years; these are arranged in tables according to the time; so that by inspection alone the corresponding place of the body referred to the fixed plane, or position of the ecliptic at the epoch, may be found.

Fortunately for the facility of astronomical calculations, the orbits of the celestial bodies are either very nearly circular, as in the planets and satellites, or very eccentric, as in the comets. In both circumstances the series which determine the motions of the body may be made to converge rapidly, which would not be the case if the eccentricity bore a mean ratio to the greater axis.

Motion of Comets.

404. If the ratio of the eccentricity to the greater axis be made very nearly equal to unity, instead of a very small fraction, the preceding series will then give the place of a comet in a very eccentric orbit, with this difference, that the terms have the increasing powers of the difference between unity and the ratio of the eccentricity to the greater axis, as coefficients, instead of the powers of that ratio itself. This difference is zero in the parabola; then the value of the radius vector becomes

$$r = \frac{D}{\cos^2 . \frac{1}{2} v},$$

D being the perihelion distance: hence, in the parabola, the distance S*m* is equal to the perihelion distance SP, divided by the square of the cosine of half the true anomaly PS*m*. If, then, the true anomaly were known, the distance of the comet from the sun would be determined from this equation. When the body moves in a parabola, the equation between the mean and true anomaly is reduced to a cubic equation between the time and the tangent of half the true anomaly PS*m*.

Arbitrary Constant Quantities of Elliptical Motion, or Elements of the Orbits.

405. There are six elements in the orbit of each celestial body: four of elliptical motion, namely, the mean distance of the planet from the sun; the eccentricity; the mean longitude of the planet at the epoch; and the longitude of the perihelion at the same epoch. The other two elements relate to the position of the orbit in space, namely, the longitude of the ascending node at the epoch, and the inclination of the orbit on the plane of the ecliptic. The mean values of all these must be determined by observation, before the

motion of the body can be ascertained, or tables computed. Hence there are forty-two elements to be determined for the seven principal planets, and twenty-four more for the four new planets, Ceres, Pallas, Juno, and Vesta, besides those of the moon and satellites. Tables have been computed for most of these bodies; some of the satellites, however, are but little known, and the theory of the four new planets is still imperfect.

The same series that determine the motions of the planets answer equally well for the elliptical motion of the moon and satellites, only the mass of the planet is to be employed in place of that of the sun, omitting the mass of the satellite.

Co-ordinates of a Planet.

406. The simplicity of analytical expressions very much depends on a skilful choice of co-ordinates, which are arbitrary and infinite in number, but so connected, that any one set may be expressed in values of any other. For example, the place of the planet m has been determined by the angles ♈Sm, mSp, and Sm, fig. 77, but these have been changed into ♈Sp, pSm, and Sp, which are the heliocentric longitude, latitude, and curtate distance of m. Again, from the latter, the geocentric longitude, latitude, and distance may be deduced, that is, the place of m as seen from the earth; and, lastly, the right ascension and declination of m, or its place referred to the equator, may be obtained from its geocentric longitude and latitude.

These quantities are given in terms of the mean longitude or time, since the first co-ordinates are given in series of the sines and cosines of that quantity. In the theory of the moon, the series are found to converge more rapidly, if the mean longitude, latitude, and distance are determined in functions of the true longitude. All these co-ordinates are connected by spherical triangles, so that they are easily deduced from one another.

Determination of the Elements of Elliptical Motion.

407. Were the primitive velocity with which the bodies of the solar system projected in space known, the values of the elements

of their orbits might be determined; for if the equation (90) be resumed, and if the first member, which is the square of the velocity, be represented by V^2, then

$$V^2 = \mu \left\{ \frac{2}{r} - \frac{1}{a} \right\}$$

in which r is the radius vector, and a is half the greater axis of the conic section, μ being the masses of the sun and planet. Thus the velocity is independent of the eccentricity of the orbit.

If u be the angular velocity which the planet would have if it described a circle at the distance of unity round the sun, then $r = a = 1$, and the preceding expression gives $u^2 = \mu$; hence

$$V^2 = u^2 \left\{ \frac{2}{r} - \frac{1}{a} \right\},$$

V being the primitive velocity with which the body moved in a conic section. This equation will give a value of a by means of the primitive velocity of m, and its distance from S, fig. 78. a is positive in the ellipse, infinite in the parabola, and negative in the hyperbola; thus the orbit of m is an ellipse, a parabola, or hyperbola, according as V is less, equal to, or greater than $u\sqrt{\frac{2}{r}}$. It is remarkable that the *direction* of the primitive impulse has no influence on the nature of the conic section in which the planet moves; the intensity alone has that effect.

fig. 78.

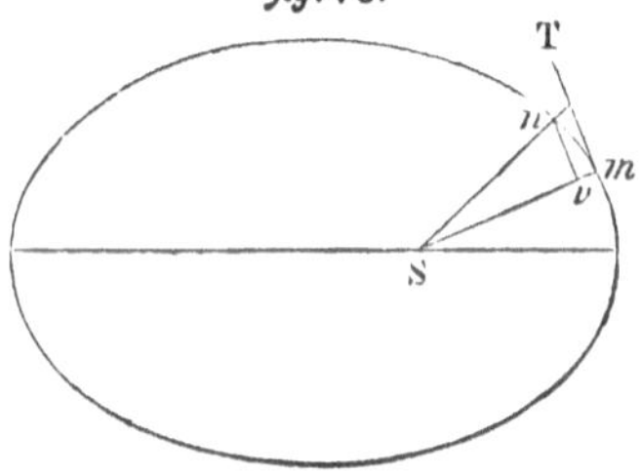

To determine the eccentricity of the orbit, let α be the angle TmS, that the direction of the relative motion of m makes with the radius vector r; then $\quad mn : mv :: ds : dr :: 1 : \cos\alpha$;

then
$$\frac{ds}{dt}\cos\alpha = \frac{dr}{dt}, \text{ but } \frac{ds}{dt} = V,$$

hence

$$V^2\cos^2\alpha = \frac{dr^2}{dt^2}; \text{ or if } \mu\left\{\frac{2}{r} - \frac{1}{a}\right\}$$

be put for V,

$$\frac{dr^2}{dt^2} = \mu\left\{\frac{2}{r} - \frac{1}{a}\right\}\cos^2\alpha;$$

but by article 377,

$$2\mu r - \frac{\mu r^2}{a} - \frac{r^2dr^2}{dt^2} = \mu a\,(1 - e^2);$$

hence $$a\,(1 - e^2)^2 = r^2 \sin^2\alpha\left\{\frac{2}{r} - \frac{1}{a}\right\},$$

which gives the eccentricity of the orbit. The equation of conic sections,

$$r = \frac{a(1 - e^2)}{1 + e\cos v}$$

gives $$\cos v = \frac{a(1 - e^2) - r}{er}.$$

Thus the angle v, that the radius vector makes with the perihelion distance, is found, and, consequently, the position of the perihelion. The equations (96) will then give the angle u, or eccentric anomaly, and, by means of it, the instant of the passage at the perihelion.

In order to have the position of the orbit, with regard to a fixed plane passing through the centre of S, fig. 77, supposed immoveable, let ϕ be the inclination of the two planes, and $\epsilon = m\text{SN}$; also let $mp = z$ be the primitive elevation of the planet above the fixed plane, which s supposed to be known; then

$$r\sin\epsilon\sin\phi = z.$$

So that ϕ, the inclination of the orbit, will be known when ϵ shall be determined. For that purpose, let $\lambda = m\text{R}p$, fig. 79, be the angle made by mR, the primitive direction of the relative motion of m with the plane ENB; then the triangle mSR, in which SmR $= \alpha$, NS$m = \epsilon$, and S$m = r$, gives

fig. 79.

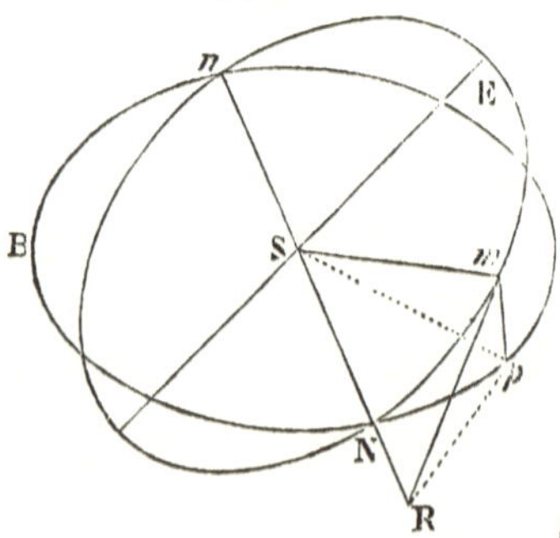

$$m\text{R} = \frac{r\sin\epsilon}{\sin(\epsilon + \alpha)};$$

then $$\frac{z}{m\text{R}} = \sin\lambda,$$

which is given, because λ is supposed to be known; therefore

$$\tan\epsilon = \frac{z\sin\alpha}{r\sin\lambda - z\cos.\alpha}.$$

The elements of the orbit of the planet being determined by these formulæ in terms of r, z, the velocity of the planet, and the direction of its motion, the variations of these elements, corresponding to the supposed variations in the velocity and its direction, may be obtained; and it will be easy, by means of methods that will be hereafter given, to have the differential variations of these elements, arising from the action of the disturbing forces.

Velocity of Bodies moving in Conic Sections.

408. As the actual motions of the bodies of the solar system afford no information with regard to their primitive motions, the elements of their orbits can only be known by observation; but when these are determined, the velocities with which the bodies of the solar system were first projected in space, may be ascertained. If the equation

$$V^2 = u^2\left\{\frac{2}{r} - \frac{1}{a}\right\}$$

be resumed, then in the circle $r = a$, since the eccentricity is zero;

hence $v = u\sqrt{\frac{1}{r}}$; therefore $V : u :: 1 : \sqrt{r}$.

thus the velocities of planets in different circles are as the square roots of their radii.

In the parabola, a is infinite; hence

$$\frac{1}{a} \text{ is zero, and } V = \sqrt{\frac{2}{r}}.$$

Thus the velocities in different points of a parabolic orbit are reciprocally as the square roots of the radii vectores, and the velocity in each point is to the velocity the planet would have if it moved in a circle with a radius equal to r, as $\sqrt{2}$ to 1.

409. When an ellipse is infinitely flattened, it becomes a straight line; hence, in this case, V will express the velocity of m, if it were to descend in a straight line towards the sun; for then Sm, fig. 80, would coincide with SA. If m were to begin to fall from a state of rest at A, its velocity would be zero at that point;

fig. 80.

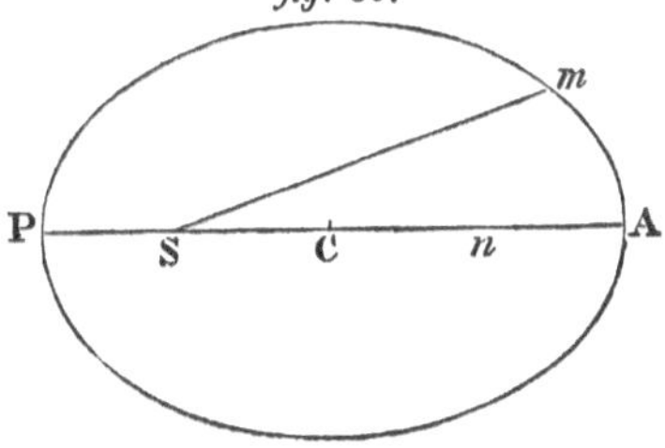

hence $\frac{2}{r} - \frac{1}{a} = 0.$

Now, suppose that, in falling from A to n, the body had acquired the velocity V, then the equation would be

$$V^2 = u^2\left\{\frac{2}{r'} - \frac{1}{a}\right\},$$

and eliminating a, which is common to the two last equations,

$$V = u\sqrt{\frac{2(r - r')}{rr'}},$$

in which $r' = \mathrm{S}n$. This is the relative velocity the body m has acquired in falling from A through $r - r' = \mathrm{A}n$. Imagine the body m to have acquired, by its fall through An, the same velocity with a body moving in a conic section; the velocity of the latter body is

$$V' = u\sqrt{\frac{2}{r} - \frac{1}{a}}.$$

If these two be equated,

$$\mathrm{A}n = (r - r') = \frac{r(2a - r)}{4a - r}.$$

This expression gives the height through which a body moving in a conic section must fall, from the extremity A of the radius vector, in order to acquire the relative velocity which it had at A.

In the circle $a = r$, hence A$n = \frac{1}{3}r$; in the ellipse, An is less than $\frac{1}{2}r$: in the parabola, a is infinite, which gives A$n = \frac{1}{2}r$; and in the hyperbola a is negative, and therefore An is greater than $\frac{1}{2}r$.

CHAPTER V.

THEORY OF THE PERTURBATIONS OF THE PLANETS.

410. THE tables computed on the theory of perfectly elliptical motion, are soon found inadequate to give the true place of a planet, on account of the reciprocal disturbances of the system. It is therefore necessary to investigate what these disturbances are, and to determine their effects.

In the first approximation to the celestial motions, the mutual action of the sun and of one planet was considered: it then appeared that a planet, *m*, moves round the sun in an ellipse N*m*P*n*, fig. 81, inclined to the ecliptic NB*n*, at a very small angle P*np*. Now, if *m* be attracted by another planet *m′*, which is much smaller than the sun, it will no longer go on in its elliptical orbit N*mn*, but will be drawn out of that orbit, and will move in some curved line, *ca*D, which may either be nearer to, or farther from, the plane of the ecliptic, according to the position of the disturbing body. In the first infinitesimal of time, the troubled orbit coincides with the ellipse through an indefinitely small space *ca*; in the second infinitely small interval of time, *am* will be the path of the planet in the ellipse, and *a*D will be its path in its troubled orbit: *am* is described in consequence of the action of the sun alone; *a*D by the combined action of the sun and of the disturbing body: *am* is the second increment of the space; *a*D is the second increment of the space, together with some very small space, FD, introduced by the action of the disturbing force. In consequence of the addition of FD, the longitude of *m* is increased by B*b*; its latitude is changed by the angle DSE, and the radius vector is increased by the difference between SD and S*m*,—these three quantities are the perturbations of the planet in longitude, latitude, and distances.

fig. 81.

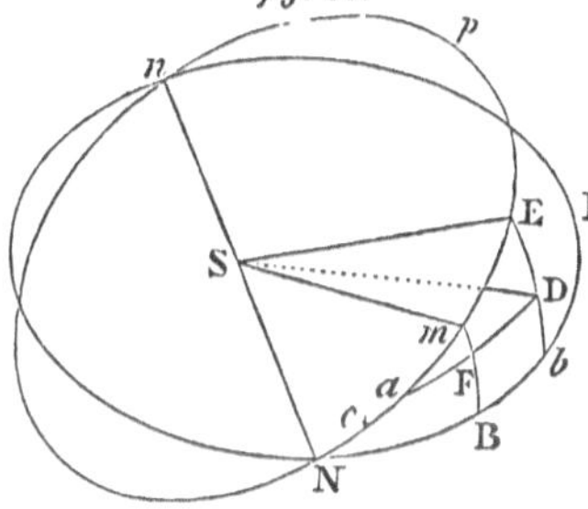

411. It is evident that the perturbations are true variations; and as the longitude, latitude, and radius vector of a planet moving in

an elliptical orbit, have been represented by v, s, and r, the arcs $Bb = \delta v$, $ED = \delta s$ and $SD - Sm = \delta r$, are the variations of these co-ordinates.

412. The perturbations in longitude, latitude, and distance, depend on the configuration of the bodies; that is, on the position of the bodies with regard to each other, to their perihelia and to their nodes. These inequalities, after going through a certain course of increase and decrease, are renewed as often as the bodies return to the same relative positions, and are therefore called Periodic Inequalities.

413. Thus the place of a planet, m, moving in its troubled orbit caD, will be determined by the co-ordinates $v + \delta v$, $s + \delta s$, $r + \delta r$. These, however, are modified by a variation in the elements of the ellipse; for it is evident that, the path of the planet being changed from aE to aD, the elements of the ellipse NmE must vary. The variations of the elements are independent of the configuration or relative position of the bodies, and are only sensible in many revolutions; whereas those depending on the configuration, accomplish their changes in short periods. Thus $v + \delta v$, $s + \delta s$, $r + \delta r$, may be regarded as the co-ordinates of the planet in its true orbit, provided the elements contained in these functions be considered to vary by very slow degrees. This perfectly accords with observation, whence it appears that the perihelia of the orbits of the planets and satellites have a very slow direct motion in space; that the nodes have a slow retrograde motion; and that the eccentricities and inclinations are perpetually varying by very slow degrees. These very slow changes are really periodic, but many ages elapse before they accomplish their revolutions; on that account they are called Secular Inequalities, to distinguish them from the Periodic Inequalities, which pass rapidly from their maxima to their minima. Thus the Periodic Inequalities only depend on the configuration of the bodies, whereas the Secular Inequalities depend on the configuration of the perihelia and nodes alone.

414. La Grange took a new and very elegant view of the subject:—he considered the changes δv, δs, δr, to arise entirely from periodic and secular variations in the elements of elliptical motion, thus referring all the inequalities, to which a planet is liable, to changes in the elements of its orbit alone. In fact, as the curve aD

very nearly coincides with the ellipse, it may be regarded as a portion of a new ellipse, having elements differing from those of the original one by infinitely minute variations. Of *these* a portion will be compensated in a whole revolution, or many revolutions of m, and of the disturbing planet constituting the Periodic Inequalities; but a portion will remain uncompensated, and entirely independent of the position of the bodies with regard to each other. These uncompensated parts increase and diminish with extreme slowness; their effects on the motion of m partake of that character, and constitute what are called Secular Inequalities. Thus, in La Grange's view, the co-ordinates of m in its elliptical orbit are modified, both by periodic and secular variations, in the elements of the ellipse.

415. The secular inequalities depend on the ratio of the disturbing mass to that of the sun, which, by article 350, is a very small fraction. Their arguments are not only different from those of the periodic inequalities, but, though also periodic, their periods are immensely longer.

416. Both periodic and secular inequalities may be represented by supposing a point p to revolve in an ellipse AP, fig. 82, where all the elements are perpetually varying by very slow degrees. Then, suppose a planet m to oscillate round the moveable point p in a curve mab, whose nature depends on the disturbing forces: this oscillating motion will represent the periodic inequalities, and the whole compound motion m represents the real motion of a planet in its troubled orbit.

fig. 82.

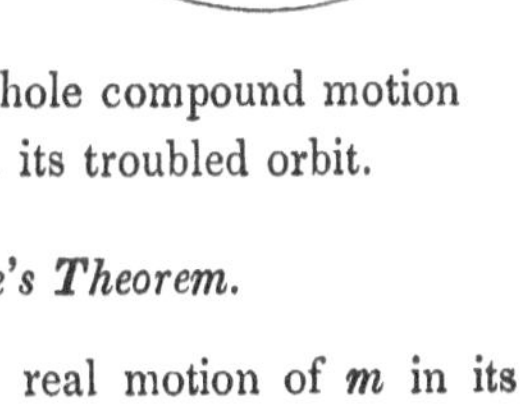

Demonstration of La Grange's Theorem.

417. The equations which determine the real motion of m in its troubled orbit are, by article 347,

$$\frac{d^2x}{dt^2} + \frac{\mu x}{r^3} = \left(\frac{dR}{dx}\right),$$
$$\frac{d^2y}{dt^2} + \frac{\mu y}{r^3} = \left(\frac{dR}{dy}\right), \qquad (87)$$
$$\frac{d^2z}{dt^2} + \frac{\mu z}{r^3} = \left(\frac{dR}{dz}\right).$$

If $R = 0$, these equations would be the same with those in article 365, already integrated. Let a be one of the arbitrary constant quantities, or elements of the orbit of m, introduced by integration. When $R = 0$, then

$$a = \text{Func.}\left(x, y, z\ \frac{dx}{dt}, \frac{dy}{dt}, \frac{dz}{dt}\ .\ t\right)$$

may represent any one whatever of the integrals (91); or, if to abridge $x_{\prime} = \frac{dx}{dt}\ \ y_{\prime} = \frac{dy}{dt}\ \ z_{\prime} = \frac{dz}{dt}$,

$$a = \text{Func.}\ (x, y, z, x_{\prime}, y_{\prime}, z_{\prime}, t). \tag{103}$$

During the instant dt, the ellipse and troubled orbit coincide; therefore $x, y, z, x_{\prime}, y_{\prime}, z_{\prime}$ have the same values in both, and a is constant. But at the end of the instant dt, the velocities $x_{\prime}, y_{\prime}, z_{\prime}$, are respectively augmented, from the action of the disturbing forces, by the indefinitely small quantities

$$\frac{dR}{dx}\,dt, \quad \frac{dR}{dy}, \quad \frac{dR}{dz}\,dt;$$

then a is no longer constant; and when $x_{\prime}, y_{\prime}, z_{\prime}$ are increased by those quantities, the corresponding variation of a is

$$da = \left(\frac{da}{dx_{\prime}}\cdot\ \frac{dR}{dx} + \frac{da}{dy_{\prime}}\cdot\ \frac{dR}{dy} + \frac{da}{dz_{\prime}}\cdot\ \frac{dR}{dz}\right)dt. \tag{104}$$

If equation (103) be regarded as the first integral of the equations (87), when $R = 0$, it will evidently satisfy the same equations when R is not zero, because the values of $x, y, z, x_{\prime}dt, y_{\prime}dt, z_{\prime}dt$, are supposed to be the same in each orbit, since these quantities only differ in the two curves by their second differentials.

Hence, if $(x_{\prime}), (y_{\prime}), (z_{\prime})$ be the values of $x_{\prime}, y_{\prime}, z_{\prime}$, when $R = 0$, then $x_{\prime} = (x_{\prime}),\ y_{\prime} = (y_{\prime}),\ z_{\prime} = (z_{\prime})$,

and $dx_{\prime} = (dx_{\prime}) + \delta x_{\prime},\quad dy_{\prime} = (dy_{\prime}) + \delta y_{\prime},\quad dz_{\prime} = (dz_{\prime}) + \delta z_{\prime}$.

Let func. $(x, y, z, x_{\prime}, y_{\prime}, z_{\prime}, t)$ be the differential of equation (103) when $R = 0$, then will

$$0 = \text{func.}\ (x, y, z, x_{\prime}, y, z_{\prime}, t)$$

and the differential of the same equation, when R is not zero, will be

$$da = \text{func.}\ (x, y, z, x_{\prime}, y_{\prime}, z_{\prime}, t) + \left(\frac{da}{dx_{\prime}}\,\delta x_{\prime} + \frac{da}{dy_{\prime}}\,\delta y_{\prime} + \frac{da}{dz_{\prime}}\,\delta z_{\prime}\right),$$

because, in the latter case, all the quantities vary. If the first differential be subtracted from the second, the result will be

$$da = \left(\frac{da}{dx_{,}}\,\delta x_{,} + \frac{da}{dy_{,}}\,\delta y_{,} + \frac{da}{dz_{,}}\,\delta z_{,}\right). \tag{105}$$

But if $(dx_{,}) + \delta x_{,},\ (dy_{,}) + \delta y_{,},\ (dz_{,}) + \delta z_{,},$
be put, in equations (87), in place of their equals,

$$\frac{d^2x}{dt},\quad \frac{d^2y}{dt},\quad \frac{d^2z}{dt},$$

they become

$$\delta x_{,} = \frac{dR}{dx}\,dt,\ \delta y_{,} = \frac{dR}{dy}\,dt,\ \delta z_{,} = \frac{dR}{dz}\,dt.$$

Since $(dx_{,})$, $(dy_{,})$, $(dz_{,})$, are supposed to satisfy these equations when $R = 0$.

If the preceding values of $\delta x_{,}, \delta y_{,}, \delta z_{,}$, be put in equation (105), it becomes identical with equation (104). Hence the integral (103) satisfies the equations (87), whether the disturbing forces be included or not, the only difference being that, in the first case, a must be regarded as a variable quantity, and in the last it is constant.

The same may be shown of all the first integrals of equations (87), when R is zero.

418. It appears, from what has been said, 1st, that as the motion is performed in the unvaried ellipse during the first element of time, x, y, z, dx, dy, dz, are alike in the varied and unvaried ellipse. 2nd, That as the motion is performed in the variable ellipse during the second element of time, if d^2x, d^2y, d^2z, be considered as belonging to the unvaried ellipse, $d^2x + d\delta x$, $d^2y + d\delta y$, $d^2z + d\delta z$ will belong to the variable orbit of m. Hence the differential equation of the first order, which determines the motion of the body, answers for both orbits during the first instant of the time, the elements of the orbit being constant; in the second increment of time, the equations of elliptical motion have the form

$$\frac{d^2v}{dt^2} + n^2v = 0,$$

the elements of the orbit being constant; but in the troubled orbit they have the form

$$\frac{d^2v}{dt^2} + n^2v + R = 0,$$

where the elements of the orbit are variable, and R is the part containing the disturbing forces.

419. As the elements of the orbit only vary during the second increment of the time, their variation is of the first order; that is, the eccentricity e becomes $e + de$, the inclination ϕ becomes

$$\phi + d\phi, \text{ \&c. \&c.}$$

420. The elegant theory of the variation of the arbitrary constant quantities is due to Euler. La Grange first applied it to the celestial motions.

421. It is proposed, first, to determine the periodic and secular variations of the elements of orbits of any eccentricities and inclinations; in the second place, to find those of the planets and satellites, all of which have nearly circular orbits, slightly inclined to the plane of the ecliptic; and then to determine the periodic inequalities, δv, δs, δr, in longitude, latitude, and distance.

Variation of the Elements, whatever the Eccentricities and Inclinations may be.

422. All the elements of the orbit have been determined from the seven arbitrary constant quantities, c, c', c'', f, f', f'', and a, introduced by the integration of the equations (87) of elliptical motion; but it was shown that the elements of the orbit, as well as the differentials dx, dy, dz, vary during the second element of time by the action of the disturbing forces, and then the differentials of the equations (91) will afford the means of finding the variations of the elements, whatever the eccentricities and inclinations of the orbits may be. Equations (87) give

$$d^2x = dt^2\left(\frac{dR}{dx}\right); \quad d^2y = dt^2\left(\frac{dR}{dy}\right); \quad d^2z = dt\left(\frac{dR}{dz}\right);$$

which are the changes in dx, dy, dz, due to the disturbing forces alone, the elliptical part being omitted. If, therefore, the differentials of equations (91) be taken, considering c, c', c'', f, f', f'', a, dx, dy, dz, alone as variable, when the preceding values of d^2x, d^2y, d^2z are substituted, they become

$$dc = dt\left\{x\left(\frac{dR}{dy}\right) - y\left(\frac{dR}{dx}\right)\right\},$$

$$dc' = dt\left\{z\left(\frac{dR}{dx}\right) - x\left(\frac{dR}{dz}\right)\right\},$$

$$dc'' = dt\left\{y\left(\frac{dR}{dz}\right) - z\left(\frac{dR}{dy}\right)\right\},$$

$$df = dx\left\{z\left(\frac{dR}{dx}\right) - x\left(\frac{dR}{dz}\right)\right\} - dy\left\{y\left(\frac{dR}{dz}\right) - z\left(\frac{dR}{dy}\right)\right\}$$
$$+ c'\,dt\left(\frac{dR}{dx}\right) - c''dt\left(\frac{dR}{dy}\right); \qquad (106)$$

$$df' = dz\left\{y\left(\frac{dR}{dz}\right) - z\left(\frac{dR}{dy}\right)\right\} - dx\left\{x\left(\frac{dR}{dy}\right) - y\left(\frac{dR}{dx}\right)\right\}$$
$$+ c''dt\left(\frac{dR}{dz}\right) - cdt\left(\frac{dR}{dx}\right);$$

$$df'' = dy\left\{x\left(\frac{dR}{dy}\right) - y\left(\frac{dR}{dx}\right)\right\} - dz\left\{z\left(\frac{dR}{dx}\right) - x\left(\frac{dR}{dz}\right)\right\}$$
$$+ cdt\left(\frac{dR}{dy}\right) - c'dt\left(\frac{dR}{dz}\right).$$

$$d\,.\,\frac{\mu}{a} = -\,2dR.$$

423. If values of c, c', c'', f, f', f'', derived from these equations, be substituted instead of their constant values in equations

$$\tan\phi = \frac{\sqrt{c'^2 + c''^2}}{c}, \quad \tan\theta = -\,\frac{c''}{c'},$$

$$h^2 = \mu a\,(1 - e^2) = c^2 + c'^2 + c''^2,$$

$$\tan\varpi_{,} = \frac{f'}{f''}, \text{ and } \mu e = \sqrt{f^2 + f'^2 + f''^2},$$

given in article **374** and those following, they will determine the elements of the disturbed orbit.

The equations
$$c''x + c'y + cz = 0,$$
$$\mu r - h^2 + f''x + f'y + fz = 0;$$
and their differentials
$$c''dx + c'dy + cdz = 0,$$
$$\mu dr + f''dx + f'dy + fdz = 0,$$
will also answer in the disturbed orbit, provided the same substitution be made.

424. The mean distance a gives the mean motion of m, or more correctly that in the disturbed orbit, which corresponds with the mean motion in the elliptical orbit; for

$$n = a^{-\frac{3}{2}}\sqrt{\mu}.$$

*

If ζ be the mean motion of m, then in the elliptical orbit,

$$d\zeta = ndt;$$

but this equation also answers for the disturbed orbit, since the two orbits coincide during the first instant of time. But

$$dd\zeta = dndt,$$

$$dn = \frac{3an}{2\mu} \cdot d\frac{\mu}{a};$$

and as the last of equations (106) is

$$d\frac{\mu}{a} = -2\text{d}R, \text{ so } dn = -\frac{3an}{\mu}\text{d}R;$$

hence
$$dd\zeta = -\frac{3andt \cdot \text{d}R}{\mu};$$

the integral of which is

$$\zeta = -\frac{3}{\mu}\int\int andt \cdot \text{d}R. \tag{107}$$

425. The seven arbitrary constant quantities are only equivalent to five in consequence of the two equations

$$0 = fc + f'c' + f''c'',$$

$$0 = \frac{\mu}{a} + \frac{f^2 + f'^2 + f''^2 - \mu^2}{c^2 + c'^2 + c''^2}.$$

These also exist in the disturbed orbit, when the arbitrary quantities are replaced by their variable values.

426. Since R is given in article 347, all the elements of the disturbed orbit are determined with the exception of ϵ, the longitude of the planet at the epoch. From the equations

$$dv = \frac{hdt}{r^2}, \quad r^2 = \frac{a^2(1-e^2)^2}{(1+e\cos(v-\varpi))^2},$$

it is evident that

$$dv \cdot \frac{(1-e^2)^2}{(1+e\cos(v-\varpi))^2} = \frac{h}{a^2}dt.$$

But
$$h = \sqrt{\mu a(1-e^2)};$$

hence
$$\frac{h}{a^2} = a^{-\frac{3}{2}}\sqrt{\mu}\sqrt{(1-e^2)} = n\sqrt{1-e^2};$$

therefore
$$ndt = dv \cdot \frac{(1-e^2)^{\frac{3}{2}}}{(1+e\cos(v-\varpi))^2}.$$

If
$$\frac{c^{(v-\varpi)\sqrt{-1}} + c^{-(v-\varpi)\sqrt{-1}}}{2}$$
be put for $\cos(v-\varpi)$,
$$\frac{\sqrt{1-e^2}}{1+e\cos(v-\varpi)} = \frac{2\sqrt{1-e^2}}{2+e\{c^{(v-\varpi)\sqrt{-1}} + c^{-(v-\varpi)\sqrt{-1}}\}}.$$

Again, if $\lambda = \dfrac{e}{1+\sqrt{1-e^2}}$; then $e = \dfrac{2\lambda}{1+\lambda^2}$,

which, substituted in the second member of the last equation, gives
$$\frac{1}{1+e\cos(v-\varpi)} = \frac{1}{\sqrt{1-e^2}}\left\{\frac{1-\lambda^2}{1+\lambda^2+\lambda\{c^{(v-\varpi)\sqrt{-1}} + c^{-(v-\varpi)\sqrt{-1}}\}}\right\}.$$

The numerator of the last term is
$$1-\lambda^2 = (1+\lambda c^{-(v-\varpi)\sqrt{-1}}) - \lambda c^{-(v-\varpi)\sqrt{-1}}(1+\lambda c^{(v-\varpi)\sqrt{-1}})$$

And the denominator is equal to
$$(1+\lambda c^{(v-\varpi)\sqrt{-1}})(1+\lambda c^{-(v-\varpi)\sqrt{-1}})$$

hence
$$\frac{1}{1+e\cos(v-\varpi)} = \frac{1}{\sqrt{1-e^2}}\left\{\frac{1}{1+\lambda c^{(v-\varpi)\sqrt{-1}}} - \frac{\lambda c^{-(v-\varpi)\sqrt{-1}}}{1+\lambda c^{-(v-\varpi)\sqrt{-1}}}\right\}.$$

By division,
$$\frac{1}{1+\lambda c^{(v-\varpi)\sqrt{-1}}} = 1 - \lambda c^{(v-\varpi)\sqrt{-1}} + \lambda^2 c^{2(v-\varpi)\sqrt{-1}} - \&c.$$
$$\frac{\lambda c^{-(v-\varpi)\sqrt{-1}}}{1+\lambda c^{-(v-\varpi)\sqrt{-1}}} = \lambda c^{-(v-\varpi)\sqrt{-1}} - \lambda^2 c^{-2(v-\varpi)\sqrt{-1}} + \&c.$$

And the difference of these is
$$\frac{1}{1+e\cos(v-\varpi)} = \frac{1}{\sqrt{1-e^2}}\{1 - \lambda(c^{(v-\varpi)\sqrt{-1}} + c^{-(v-\varpi)\sqrt{-1}})$$
$$+ \lambda^2(c^{2(v-\varpi)\sqrt{-1}} + c^{-2(v-\varpi)\sqrt{-1}}) - \&c.\};$$

but $\quad c^{i(v-\varpi)\sqrt{-1}} + c^{-i(v-\varpi)\sqrt{-1}} = 2\cos i(v-\varpi)$;

hence
$$\frac{1}{1+e\cos(v-\varpi)} = \frac{1}{\sqrt{1-e^2}}\{1 - 2\lambda\cos(v-\varpi)$$
$$+ 2\lambda^2 . \cos 2(v-\varpi) - \&c.\};$$

or
$$\frac{1}{1+e\cos(v-\varpi)} = \frac{1}{\sqrt{1-e^2}} \mp 2\cos i(v-\varpi)\frac{\lambda^i}{\sqrt{1-e^2}},$$

which is the general form of the series, i being any whole positive number.

*

Now, $\frac{1}{de} \cdot d\frac{e}{1+e\cos(v-\varpi)} = \frac{1}{(1+e\cos(v-\varpi))^2} =$

$$\frac{1}{de} \cdot \left\{ d\frac{e}{\sqrt{1-e^2}} \pm 2\cos(v-\varpi) . d\frac{e\lambda^i}{\sqrt{1-e^2}} \right\};$$

but $d . \frac{e}{\sqrt{1-e^2}} = \frac{de}{(1-e^2)^{\frac{3}{2}}}$, and $d\frac{e\lambda^i}{\sqrt{1-e^2}} = \pm \frac{e^i\{1+i\sqrt{1-e^2}\}\,de}{(1-e^2)^{\frac{3}{2}}(1+\sqrt{1-e^2})^i}$

the sign $+$ is used when i is even, and $-$ when it is odd. Hence if to abridge

$$E^{(i)} = \pm\frac{2e^i.\{1+i\sqrt{1-e^2}\}}{(1+\sqrt{1-e^2})^i},$$

the value of ndt becomes,

$$ndt = dv\{1 + E^{(1)}\cos(v-\varpi) + E^{(2)}\cos 2(v-\varpi) + \&c.\}; \quad (108)$$

the integral of which is

$$\int ndt + \epsilon = v + E^{(1)}\sin(v-\varpi) + \tfrac{1}{2}E^{(2)}\sin 2(v-\varpi) + \&c.\},$$

ϵ being arbitrary.

This equation is relative to the invariable ellipse; but in order that it may also suit the real orbit, every quantity in it must vary including e, ϖ, and ϵ; and this differential must coincide with (108) since they are of the first order, and the two orbits coincide during the first element of time. Their difference is

$$d\epsilon = de\left\{\left(\frac{dE^{(1)}}{de}\right)\sin(v-\varpi) + \tfrac{1}{2}\left(\frac{dE^{(2)}}{de}\right)\sin 2(v-\varpi) + \&c.\right\}$$
$$- d\varpi\{E^{(1)}\cos(v-\varpi) + E^{(2)}\cos 2(v-\varpi) + \&c.\}$$

$v-\varpi$ is the true anomaly of m estimated on the orbit, and ϖ is the longitude of the perihelion on the orbit. Now equation (101) is

$$v - \zeta = v_{\prime} - \theta + \tan^2\tfrac{1}{2}\phi\sin 2(v_{\prime}-\theta) + \&c.$$

v being the longitude on the orbit, and $v_{\prime}$ its projection on the fixed plane. If ϖ be put for v and $\varpi_{\prime}$ for $v_{\prime}$; then

$$\varpi - \zeta = \varpi_{\prime} - \theta + \tan^2\tfrac{1}{2}\phi\sin 2(\varpi_{\prime}-\theta) + \&c.$$

Again, if we make v and $v_{\prime}$ zero in equation (101), it becomes

$$\zeta = \theta + \tan^2\tfrac{1}{2}\phi\sin 2\theta + \tfrac{1}{2}\tan^4\tfrac{1}{2}\phi\sin 4\theta + \&c.$$

hence $\quad \varpi = \varpi_{\prime} + \tan^2\tfrac{1}{2}\phi\{\sin 2\theta + \sin 2(\varpi_{\prime}-\theta)\} + \&c.$

therefore $\quad d\varpi = d\varpi_{\prime}\{1 + 2\tan^2\tfrac{1}{2}\phi\cos 2(\varpi_{\prime}-\theta) + \&c.\}$

$$+ 2d\theta\tan^2\tfrac{1}{2}\phi\{\cos 2\theta - \cos 2(\varpi_{\prime}-\theta) + \&c.\}$$
$$+ \frac{d\phi\tan\tfrac{1}{2}\phi}{\cos^2\tfrac{1}{2}\phi}\{\sin 2\theta + \sin 2(\varpi_{\prime}-\theta) + \&c.\}$$

Thus $d\varpi_{\prime}$, $d\theta$, $d\phi$, being determined, we shall have $d\varpi$ from this equation, and from thence $d\epsilon$.

427. It appears from the preceding investigations, that the expressions in series given by the equations in article 392, and those following, of the radius vector, of its projection on the fixed plane, of the longitude, and its projection on the fixed plane, and of the latitude in the invariable orbit will answer for the disturbed orbit, provided nt be changed into $\int ndt$, and all the elements of the variable orbit be determined by the preceding equations; for the finite equations between r, v, s, x, y, z, and $\int ndt$, are the same in both cases, and all the equations in the articles alluded to are determined independently of the constancy or variation of the elements, consequently these expressions will still answer when the elements are variable.

These investigations relate to orbits of any inclination and eccentricity; but the orbits of the planetary system are nearly circular, and very little inclined either to one another, or to the plane of the ecliptic.

Variations of the Elliptical Elements of the Orbits of the Planets.

428. The equation $$n = a^{-\frac{3}{2}}\sqrt{\mu}$$
shows that the mean motions and greater axes of the orbits of the planets are so connected, that one cannot vary independently of the other; and as

$$\frac{\mu}{a} = -2\int \text{d}R,$$

it is clear that the differential of R is taken only with regard to nt the mean motion of m. If the mass of the sun be assumed as the unit, and the mass of the planet omitted in comparison of it, $\mu = 1$, and

$$da = 2a^2\text{d}R;$$

$2a$ being the major axis.

429. The inequalities in the eccentricity and longitude of the perihelion are obtained from

$$\tan\varpi_{,} = \frac{f'}{f''}, \quad \mu e = \sqrt{f^2 + f'^2 + f''^2}$$

$\varpi_{,}$ being the longitude of the perihelion of m when projected on the fixed plane of the ecliptic. If the orbit of the planet m at a given epoch be assumed to be the fixed plane containing the axes x and y, any inclination the orbit may have at a subsequent period being entirely owing to the action of the disturbing forces must be so small, that the true longitude of the perihelion will only differ from its projection on that new fixed plane, by quantities of the order of the squares of the disturbing masses respectively multiplied by the squares of the inclinations of the orbits, therefore without sensible error it may be assumed that $\varpi_{,} = \varpi$; ϖ being the longitude of the perihelion estimated on the orbit; thus

$$\tan \varpi = \frac{f'}{f''},$$

whence $$\sin \varpi = \frac{f'}{\sqrt{f'^2+f''^2}};$$

and $$\cos \varpi = \frac{f''}{\sqrt{f'^2+f''^2}}$$

But by article 370 $f = -\frac{f'c'+f''c''}{c}$. Now c, c', c'' are the areas described by the radius vector of m on its orbit, when projected on the co-ordinate planes; but as the orbit nearly coincides with the fixed plane of the orbit at the epoch containing the axes x and y, the other two co-ordinate planes are nearly at right angles to it; hence c' and c'' are extremely small, and as f is of the same order in consequence of the preceding equation it may be omitted, so that

$$e = \sqrt{f'^2 + f''^2}$$

whence $$f'' = e \cos \varpi; \quad f' = e \sin \varpi,$$

and $$ede = f''df'' + f'df'; \quad e^2 d\varpi = f''df' - f'df'',$$

making $\mu = 1$.

430. Since f is very small df is still smaller, therefore the fourth of the equations (91) may be omitted as well as $c'dt = zdx - xdz$, and $c''dt = ydz - zdy$, on account of the smallness of c' and c''. Also z, the height of the planet above the fixed plane of its orbit, is so small that its square may be neglected; therefore quantities having the factors zdz, or $dz\left(\frac{dR}{dz}\right)$ may be omitted, which reduces the values of the fifth and sixth of equations (106) to

$$df'' = dy\left\{x\left(\frac{dR}{dy}\right) - y\left(\frac{dR}{dx}\right)\right\} + cdt\left(\frac{dR}{dy}\right),$$

$$df' = -dx\left\{x\left(\frac{dR}{dy}\right) - y\left(\frac{dR}{dx}\right)\right\} - cdt\left(\frac{dR}{dx}\right).$$

431. If $r_{,} = \text{S}p$, fig. 77, be the radius vector of m projected on the fixed plane of the orbit of m containing the axes x and y; and if the

angle NSp be represented by $v_,$, and pm the tangent of the latitude of m above the fixed plane of its orbit by s, then

$$x = r_, \cos v_, \; ; \; y = r_, \sin v_, \; ; \; z = r_, s.$$

Since x is a function of $r_,$ and $v_,$,

$$\frac{dR}{dx} = \frac{dR}{dr_,} \cdot \frac{dr_,}{dx},$$

$$\frac{dR}{dx} = \frac{dR}{dv_,} \cdot \frac{dv_,}{dx}.$$

But $$\frac{dr_,}{dx} = \frac{1}{\cos v_,}; \quad \frac{dv_,}{dx} = -\frac{1}{r_, \sin v_,};$$

hence $$\frac{dR}{dx} = \frac{dR}{dr_,} \cdot \frac{1}{\cos v_,}; \quad \frac{dR}{dx} = -\frac{dR}{dv_,} \cdot \frac{1}{r_, \sin v_,}.$$

If the first equation be multiplied by $\cos^2 v_,$, and the second by $\sin^2 v_,$, their sum will be,

$$\frac{dR}{dx} = \left(\frac{dR}{dr_,}\right) \cos v_, - \left(\frac{dR}{dv'}\right) \frac{\sin v_,}{r_,}.$$

In like manner it may be found that

$$\frac{dR}{dy} = \left(\frac{dR}{dr_,}\right) \sin v_, + \left(\frac{dR}{dv_,}\right) \frac{\cos v_,}{r_,};$$

whence $$x\left(\frac{dR}{dy}\right) - y\left(\frac{dR}{dx}\right) = \frac{dR}{dv_,};$$

consequently,

$$df'' = + dy\left(\frac{dR}{dv_,}\right) + cdt \left\{\left(\frac{dR}{dr_,}\right) \sin v_, + \left(\frac{dR}{dv_,}\right) \frac{\cos v_,}{r_,}\right\}$$

$$df' = - dx\left(\frac{dR}{dv_,}\right) - cdt \left\{\left(\frac{dR}{dr_,}\right) \cos v_, - \left(\frac{dR}{dv_,}\right) \frac{\sin v_,}{r_,}\right\};$$

but $$dx = d(r_, \cos v_,) : \quad dy = d(r_, \sin v_,),$$

and $$cdt = xdy - ydx = r_,^2 dv_, ; \text{ so that}$$

$$df'' = + \{dr_, \sin v_, + 2r_, dv_, \cos v_,\} \left(\frac{dR}{dv_,}\right) + r_,^2 dv_, \sin v_, \left(\frac{dR}{dr_,}\right)$$

$$df' = - \{dr_, \cos v_, - 2r_, dv_, \sin v_,\} \left(\frac{dR}{dv_,}\right) - r_,^2 dv_, \cos v_, \left(\frac{dR}{dr_,}\right).$$

432. The values of $r_,$, $dr_,$, $dv_,$ $\left(\frac{dR}{dr_,}\right)$, $\left(\frac{dR}{dv_,}\right)$, are the same from whatever point the longitudes may be estimated; but by diminishing the angle $v_,$ by a right angle, $\sin v_,$ becomes $-\cos v_,$; and $\cos v_,$

becomes $\sin v_{,}$, so that the expression of df'' is changed into that of df', whence it follows, that if the value of df'' be developed into a series of sines and cosines of angles increasing proportionally with the time, and if each of the angles ϵ, ϵ', ϖ, ϖ', θ, θ', be diminished by 90°, the value of df' will be obtained.

433. By articles 398 and 401, the projection of the longitude on the fixed plane of the ecliptic, and the curtate distance are,

$$v_{,} - \theta = v - \zeta - \tan^2\tfrac{1}{2}\phi \sin 2(v - \zeta) + \&c.$$

$$r_{,} = r\{1 - \tfrac{1}{2}s^2 + \&c.\}$$

But when the orbit of m at the epoch is assumed to be the fixed plane, any inclination it may have at a subsequent period, arises entirely from the action of the disturbing forces, and is so very small that the squares of the tangent of that inclination may be neglected, whence, $v_{,} - \theta = v - \zeta$, $r_{,} = r$, $v_{,} = v$, and $\theta = \zeta$.

In the invariable orbit,

$$r = \frac{a(1-e^2)}{1+e\cos(v-\varpi)},\ dr = \frac{r^2dv\,.\,e\sin(v-\varpi)}{a(1-e^2)},$$

$$r^2dv = a^2.n.dt\sqrt{1-e^2},$$

But these equations answer also for the variable orbit, since the two ellipses coincide during the first element of time, and when substitution is made for r, dr, and r^2dv in the last values of df'' and df', they become

$$df'' = \frac{a.ndt}{\sqrt{1-e^2}}\{2\cos v + \tfrac{3}{2}e\cos\varpi + \tfrac{1}{2}e\cos(2v-\varpi)\}\left(\frac{dR}{dv}\right)$$
$$+ a^2.ndt\sqrt{1-e^2}\ \sin v\left(\frac{dR}{dr}\right),$$

$$df' = \frac{a.ndt}{\sqrt{1-e^2}}\{2\sin v + \tfrac{3}{2}e\sin\varpi + \tfrac{1}{2}e\sin(2v-\varpi)\}\left(\frac{dR}{dv}\right)$$
$$- a^2.ndt\sqrt{1-e^2}\cos v\left(\frac{dR}{dr}\right).$$

But $f'' = e\cos\varpi,\ f' = e\sin\varpi$

and by means of these equations the expressions $ede = f''df'' + f'df'$ and $e^2d\varpi = f''df' - f'df''$ in consequence of $\cos(2v - 2\varpi) = 2\cos^2(v-\varpi) - 1$, become

$$de = \frac{a.ndt}{\sqrt{1-e^2}}\{2\cos(v-\varpi) + e + e\cos^2(v-\varpi)\}.\left(\frac{dR}{dv}\right) \quad (109)$$
$$+ a^2.ndt\sqrt{1-e^2}\sin(v-\varpi)\left(\frac{dR}{dr}\right),$$

$$ed\varpi = -\frac{a.ndt}{\sqrt{1-e^2}}\sin(v-\varpi)\left\{2 + e\cos(v-\varpi)\right\}\left(\frac{dR}{dv}\right) \quad (110)$$

$$- a^2ndt\sqrt{1-e^2}\cos(v-\varpi)\}\left(\frac{dR}{dr}\right).$$

The variation of the eccentricity however may be obtained under a more simple form from the equation $c = \sqrt{\mu a(1-e^2)}$ article 422, c' and c'' being zero, for

$$dc = \frac{da\sqrt{a(1-e^2)}}{2a} - \frac{ede\sqrt{a}}{\sqrt{1-e^2}};$$

but $$\frac{dc}{dt} = x\left(\frac{dR}{dy}\right) - y\left(\frac{dR}{dx}\right) = \left(\frac{dR}{dv}\right);$$

hence by comparing the two values of dc, and observing that

$$\frac{da}{2a^2} = \text{d}R,$$

$$ede = -a.ndt\sqrt{1-e^2}\left(\frac{dR}{dv}\right) + a(1-e^2)\,\text{d}R. \quad (111)$$

434. The variation in the longitude of the epoch may be found by the preceding equations (109) and (110). For it was shown in article 392, that if the mean anomaly be estimated from any other point than the perihelion, $nt + \epsilon - \varpi$ may be put for nt, or rather $\int ndt + \epsilon - \varpi$; hence the equations in article 385 are

$$\int ndt + \epsilon - \varpi = u - e\sin u,$$

$$r = a(1 - e\cos u),$$

$$\tan\tfrac{1}{2}(v-\varpi) = \sqrt{\frac{1+e}{1-e}}\tan\tfrac{1}{2}u,$$

and $$r = \frac{a(1-e^2)}{1+e\cos(v-\varpi)}.$$

In the invariable orbit,

$$ndt = du(1 - e\cos u),$$

in which u varies with the time. But if we suppose the time constant, and u to vary only in consequence of the variation of e and ϖ, then in the troubled orbit,

$$d\epsilon - d\varpi = \text{d}u(1 - e\cos u) - de\sin u.$$

From the third of the preceding equations,

$$-\frac{d\varpi}{\cos^2\frac{1}{2}(v-\varpi)} = \frac{du}{\cos^2\frac{1}{2}u}.\sqrt{\frac{1+e}{1-e}} + \frac{2de\tan\frac{1}{2}u}{(1-e)\sqrt{1-e^2}}$$

and substituting for $\cos^2 \frac{1}{2}(v - \varpi)$, its value from the same equation, the result is

$$du = -\frac{d\varpi\,(1 - e\cos u)}{\sqrt{1-e^2}} - \frac{de\sin u}{1-e^2};$$

hence $$d\epsilon - d\varpi = -\frac{d\varpi(1-e\cos u)^2}{\sqrt{1-e^2}} - \frac{de\sin u(2-e^2-e\cos u)}{1-e^2};$$

or $$d\epsilon - d\varpi + d\varpi\sqrt{1-e^2} = \frac{ed\varpi}{\sqrt{1-e^2}}\{2\cos u - e - e\cos^2 u\}$$

$$-\frac{de}{1-e^2}\sin u\,(2 - e^2 - e\cos u).$$

Now $$r = \frac{a(1-e^2)}{1+e\cos(v-\varpi)}, = a(1 - e\cos u),$$

whence $$\cos u = \frac{e+\cos(v-\varpi)}{1+e\cos(v-\varpi)},\ \sin u = \frac{\sqrt{1-e^2}\sin(v-\varpi)}{1+e\cos(v-\varpi)}.$$

And substituting these,

$$d\epsilon - d\varpi(1-\sqrt{1-e^2}) = \sqrt{1-e^2}\frac{\{2\cos(v-\varpi)+e+e\cos^2(v-\varpi)\}}{(1+e\cos(v-\varpi))^2}.ed\varpi$$

$$-\sqrt{1-e^2}\frac{\{2+e\cos(v-\varpi)\}}{\{1+e\cos(v-\varpi)\}^2}de\sin(v-\varpi).$$

If the values of $ed\varpi$ and de, given by equations (109) and 110), be substituted, the result will be

$$d\epsilon = d\varpi(1-\sqrt{1-e^2}) - 2a.ndt.r\left(\frac{dR}{dr}\right);$$

$$\text{but } r\left(\frac{dR}{dr}\right) = a\left(\frac{dR}{da}\right);$$

hence $$d\epsilon = d\varpi\,(1-\sqrt{1-e^2}) - 2a^2\left(\frac{dR}{da}\right).ndt,$$

which is the variation in the epoch.

435. The variations in the inclination of the orbits, and in the longitude of their nodes, are obtained from

$$\tan\varphi = \frac{\sqrt{c'^2+c''^2}}{c},\quad \tan\theta = \frac{-c''}{c'},$$

$$\text{for } \tan\phi\cos\theta = -\frac{c'}{c};\ \tan\phi\sin\theta = \frac{c''}{c};$$

whence $$d.\tan\phi = \frac{1}{c}\{dc''\sin\theta - dc'\cos\theta - dc\tan\phi\},$$

$$d\theta\tan\phi = \frac{1}{c}\{dc''\cos\theta + dc'\sin\theta\}.$$

If substitution be made for $\frac{dc}{dt}$ $\frac{dc'}{dt}$ $\frac{dc''}{dt}$ of their values in article 422, and making

$$x = r\cos v,\ y = r\sin v,$$
$$s = \tan\phi\sin(v - \theta),$$

there will result

$$d.\tan\phi = -\frac{dt\tan\phi\cos(v-\theta)}{c}\{r\left(\frac{dR}{dr}\right)\sin(v-\theta) + \left(\frac{dR}{dv}\right)\cos(v-\theta)\}$$

$$+\frac{(1+s^2)dt}{c}\cos(v-\theta)\left(\frac{dR}{ds}\right) \qquad (112)$$

$$d\theta.\tan\phi = -\frac{dt\tan\phi\sin(v-\theta)}{c}\{r\left(\frac{dR}{dr}\right)\sin(v-\theta) + \left(\frac{dR}{dv}\right)\cos(v-\theta)$$

$$+\frac{(1+s^2)dt}{c}\sin(v-\theta)\left(\frac{dR}{ds}\right).$$

These two equations determine the inclination of the orbit, and motion of the nodes. They give

$$\sin(v-\theta).d\tan\phi - d\theta.\cos(v-\theta)\tan\phi = 0,$$

which may also be obtained from

$$s = \tan\phi\sin(v-\theta).$$

436. If the orbit of m has so small an inclination on the fixed plane, that the squares of s and $\tan\phi$ may be omitted, then

$$d.\tan\phi = \frac{dt}{c}\cos(v-\theta)\left(\frac{dR}{ds}\right),$$

$$d\theta.\tan\phi = \frac{dt}{c}\sin(v-\theta)\left(\frac{dR}{ds}\right);$$

if to abridge

$$p = \tan\phi\sin\theta, \quad q = \tan\phi\cos\theta,$$

and as $$c = \sqrt{a(1-e^2)};\ a = \frac{1}{a^2n^2};$$

$$\frac{1}{c} = \frac{an}{\sqrt{1-e^2}};\ \text{these become}$$

$$dq = \frac{andt}{\sqrt{1-e^2}} \cos v \left(\frac{dR}{ds}\right),$$

$$dp = \frac{andt}{\sqrt{1-e^2}} \sin v \left(\frac{dR}{ds}\right).$$

But $$z = + qy - px;$$

and as the orbit is supposed to have a very small inclination on the fixed plane, $r \cos v$, $r \sin v$, and rs, may be put for x, y, and z, the last equation becomes

$$s = q \sin v - p \cos v, \text{ whence}$$

$$\frac{dR}{ds} = \frac{1}{\sin v} \left(\frac{dR}{dq}\right); \; \frac{dR}{ds} = -\frac{1}{\cos v} \left(\frac{dR}{dp}\right);$$

consequently $$dq = -\frac{andt}{\sqrt{1-e^2}} \left(\frac{dR}{dp}\right)$$

$$dp = \frac{andt}{\sqrt{1-e^2}} \left(\frac{dR}{dq}\right).$$

437. But when the inclination of the orbit is very small,

$$\frac{z}{a} = q \sin (nt + \epsilon) - p \cos (nt + \epsilon)$$

whence $$dq = \frac{andt}{\sqrt{1-e^2}} \cos (nt + \epsilon) \left(\frac{dR}{dz}\right),$$

$$dp = -\frac{andt}{\sqrt{1-e^2}} \sin (nt + \epsilon) \left(\frac{dR}{dz}\right);$$

for $$\frac{dR}{dp} = -\left(\frac{dR}{dz}\right), \cos (nt + \epsilon)$$

$$\frac{dR}{dq} = \left(\frac{dR}{dz}\right); \sin (nt + \epsilon)$$

and $x = a \cos (nt + \epsilon)$, $y = a \sin (nt + \epsilon)$.

438. Since the elliptical and troubled orbits coincide during the first element of the time, the equations of motion are identical for that period, therefore the variation of the elements must be zero; consequently,

$$0 = \left(\frac{dR}{da}\right) da + \left(\frac{dR}{de}\right) de + \left(\frac{dR}{d\varpi}\right) d\varpi + \left(\frac{dR}{d\epsilon}\right) d\epsilon + \left(\frac{dR}{dp}\right) dp + \left(\frac{dR}{dq}\right) dq \qquad (113)$$

Because nt is always accompanied by $-\varpi$, therefore

$$\frac{dR}{dv} = \frac{dR}{ndt} + \frac{dR}{d\varpi},$$

so that the differential de becomes

$$de = -andt\frac{\sqrt{1-e^2}}{e}.\left(\frac{dR}{d\varpi}\right) - a\frac{\sqrt{1-e^2}}{e}(1-\sqrt{1-e^2})\,\mathrm{d}R.$$

If this value of de, and the preceding values of da, $d\epsilon$, dp, dq, be substituted in equation (113), observing that $\frac{\mathrm{d}R}{ndt}$ may be put for $\frac{dR}{d\epsilon}$ and $\frac{dR}{d\varpi}$, it will be reduced to

$$d\varpi = \frac{andt\sqrt{1-e^2}}{e}\left(\frac{dR}{de}\right);$$

whence $d\epsilon = \frac{andt\sqrt{1-e^2}}{e}(1-\sqrt{1-e^2}).\left(\frac{dR}{de}\right) - 2a^2\left(\frac{dR}{da}\right)ndt.$

By article 424, $\qquad d\zeta = -3\int andt.\mathrm{d}R;$

the integral of which is the periodic inequality in the mean motion.

439. The differential equations of the periodic variations of the elements of the orbit of m are therefore

$$da = 2a^2\mathrm{d}R;$$

$$d\zeta = -3\int andt.\mathrm{d}R;$$

$$d\epsilon = \frac{andt\sqrt{1-e^2}}{e}(1-\sqrt{1-e^2})\left(\frac{dR}{de}\right) - 2a^2\left(\frac{dR}{da}\right)ndt;$$

$$de = -\frac{a\sqrt{1-e^2}}{e}(1-\sqrt{1-e^2})\,\mathrm{d}R - \frac{andt\sqrt{1-e^2}}{e}\left(\frac{dR}{d\varpi}\right); \quad (114)$$

$$d\varpi = \frac{andt\sqrt{1-e^2}}{e}\left(\frac{dR}{de}\right);$$

$$dp = \frac{andt}{\sqrt{1-e^2}}\left(\frac{dR}{dq}\right);$$

$$dq = -\frac{andt}{\sqrt{1-e^2}}\left(\frac{dR}{dp}\right).$$

Because ϵ always accompanies nt,

$$\frac{dR}{d\epsilon} = \frac{\mathrm{d}R}{ndt}; \text{ whence } ndt\left(\frac{dR}{d\epsilon}\right) = \mathrm{d}R;$$

so that da may also be expressed by

$$da = 2a^2ndt\left(\frac{dR}{d\epsilon}\right).$$

440. By article 347, R is a given function of $x\,y\,z$, $x'\,y'\,z'$, &c., the co-ordinates of m, m', m'', &c. and is of the first order with regard to the masses; and if the squares and products of the masses be omitted, the elliptical values of $x\,y\,z$, $x'\,y'\,z'$, &c. may be substituted, and then R will be a function of the time, and of the elements of the orbits, and may therefore be developed in a series of sines and cosines containing the time. But the first part of this series is independent of the time, being a function of the elements of the orbits alone, as will be shown immediately, and may be represented by F.

441. As F does not contain the arc nt, its differential with regard to that quantity, is zéro, consequently when F is put for R in the preceding equations they become

$$da = 0\,; \quad d\zeta = 0\,;$$

$$d\epsilon = \frac{andt\sqrt{1-e^2}}{e}(1-\sqrt{1-e^2})\left(\frac{dF}{de}\right) - 2a^2ndt\left(\frac{dF}{da}\right)$$

$$de = -\frac{andt\sqrt{1-e^2}}{e}\left(\frac{dF}{d\varpi}\right);$$

$$d\varpi = \frac{andt\sqrt{1-e^2}}{e}\left(\frac{dF}{de}\right); \qquad (115)$$

$$dp = \frac{andt}{\sqrt{1-e^2}}\left(\frac{dF}{dq}\right);$$

$$dq = -\frac{andt}{\sqrt{1-e^2}}\left(\frac{dF}{dp}\right).$$

The integrals of these equations are the secular variations of the elements of the orbit of m.

442. In the determination of the periodic variations of the elements, all terms of the series R, that do not contain the time, must be omitted; and in the secular variations, all terms of that series that do contain the time must be rejected. Thus the periodic variations in the elements of the planetary orbits depend on the configuration, or relative position of the bodies, and their secular variations do not.

443. These periodic and secular variations, in the elements of elliptical motion, are sufficient for the determination of all the inequalities to which the bodies of the solar system are liable in their revolutions round the sun. On the same principle, the periodic and secular variations in the rotation of the earth and planets may be

found from the variation of the six arbitrary constant quantities introduced by the integration of the equations of rotatory motion. The expressions of these variations are identical in the motions of translation and rotation; and as the perturbations in these two motions arise from the same cause, they are expressed by the same formulæ. The analysis by which La Grange has united the two great problems of the solar system is the most refined and elegant in the science of astronomy.

444. Observation shows the inclinations of the orbits of the planets on the plane of the ecliptic to be very small; hence if EN, fig. 83, be the fixed plane of the ecliptic at a given epoch, PN the orbit of m, P′N′ the orbit of m',

$$\text{ENP} = \phi \qquad \text{EN}'\text{P}' = \phi',$$

the inclination of these orbits on the plane of the ecliptic; and

$$♈\text{SN} = \theta, \; ♈\text{SN}' = \theta',$$

the longitudes of their ascending nodes on the same plane, then if the planet m were moving on the orbit PN, the tangent of its latitude would be

fig. 83.

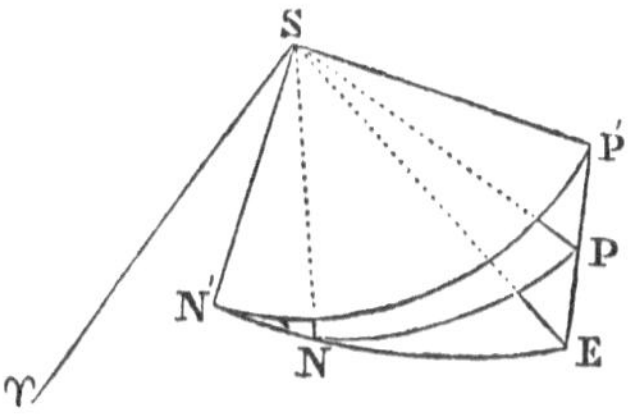

$$z = \text{EP} = \tan\phi \sin(nt + \epsilon - \theta).$$

And if it were moving on the orbit P′N′, the tangent of its latitude would be

$$z' = \text{EP}' = \tan\phi' \sin(nt + \epsilon - \theta').$$

Hence if γ be the tangent of the inclination of the orbit P′N′ on the orbit PN, and Π the longitude of the line of common intersection of these two planes, or of the ascending node of the orbit of m' on that of m, then

$$\tan\phi' \sin(nt + \epsilon - \theta') - \tan\phi \sin(nt + \epsilon - \theta) =$$
$$\gamma \sin(nt + \epsilon - \Pi) = z' - z = \text{PP}' \text{ nearly}.$$

If then as before

$$p = \tan\phi \sin\theta \qquad q = \tan\phi \cos\theta \tag{116}$$
$$p' = \tan\phi' \sin\theta' \qquad q' = \tan\phi' \cos\theta';$$

there will be found

$$\gamma \sin\Pi = p' - p \qquad \gamma \cos\Pi = q' - q \tag{117}$$
$$\gamma^2 = (p' - p)^2 + (q' - q)^2.$$

Now if EN be the primitive orbit of m at the epoch, and PN its orbit at any other period, $z = 0$, $\phi = 0$, and $\gamma = \tan\phi'$; and it is evident that γ, the tangent of the mutual inclination of these two planes, will be of the order of the disturbing forces; and therefore very small, since any inclination the orbit may acquire subsequently to the epoch is owing to the disturbing forces.

445. It is now requisite to develope R into a series of the sines and cosines of the mean angular distances of the bodies.

If the disturbing action of only one body be estimated at a time

$$R = \frac{m}{\sqrt{(x'-x)^2+(y'-y)^2+(z'-z)^2}} - \frac{m'(xx'+yy'+zz')}{(x'^2+y'^2+z'^2)^{\frac{3}{2}}},$$

in which

$$r = \text{S}m = \sqrt{x^2+y^2+z^2}; \quad r' = \text{S}m' = \sqrt{x'^2+y'^2+z'^2},$$

$$mm' = \sqrt{(x'-x)^2+(y'-y)^2+(z'-z)^2}.$$

The orbits of the planets are nearly circular, and their greatest inclination on the plane of the ecliptic does not exceed 7°, R developed according to the powers and products of these quantities must necessarily be very convergent; but as R is independent of the position of the co-ordinate planes, the plane of projection Npn, fig. 84, may be so chosen as to make the inclination still less, consequently z and z' will be very small.

fig. 84.

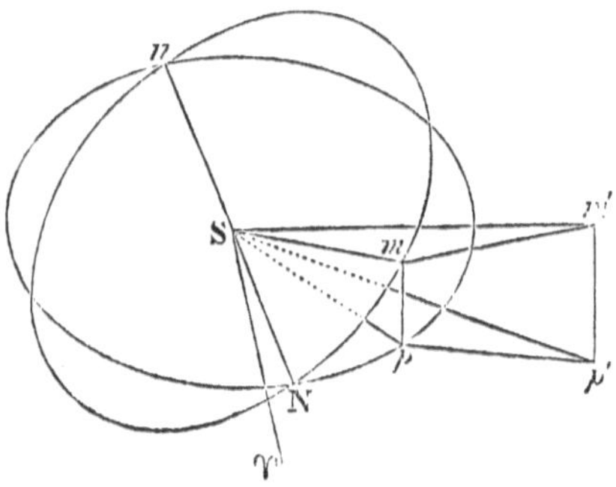

Let $v_{,} = \gamma\text{S}p$, $v'_{,} = \gamma\text{S}p'$, be the projected longitudes of m and m' on the fixed plane, and let

$$r_{,} = \text{S}p \quad r_{,}' = \text{S}p'$$

be their curtate distances; then

$$x = r_{,} . \cos v_{,}; \quad y = r_{,} \sin v_{,};$$

$$x' = r'_{,} . \cos v'_{,}; \quad y' = r'_{,} \sin v'_{,}; \text{ hence}$$

$$R = \frac{m'}{\sqrt{r'^2_{,} - 2r_{,}r'_{,}.\cos(v'_{,}-v_{,}) + r^2_{,} + (z'-z)^2}} - \frac{m'(r_{,}r'_{,}.\cos(v'_{,}-v_{,})+zz')}{(r'^2_{,}+z'^2)^{\frac{3}{2}}}$$

or z and z' being extremely small,

$$R = \frac{m'}{\sqrt{r'^2_{,} - 2r_{,}r'_{,}.\cos(v'_{,}-v_{,}) + r^2_{,}}} - \frac{r_{,}.\cos(v'_{,}-v).m'}{r'^2_{,}}$$

$$-\frac{m'.zz'}{r_{,}'^3} + \frac{3m'r_{,}.z'^2\cos(v'_{,}-v_{,})}{2r'^4_{,}} - \frac{m'(z'-z)^2}{2\{r'^2_{,} - 2r'r'_{,}\cos(v'-v_{,}) + r^2_{,}\}^{\frac{3}{2}}} + \&c.$$

Because the eccentricities and inclinations of the orbits of the planets and satellites are very small, it appears from the values of the radius vector and true longitude in the elliptical orbit developed in article 398, and those following, that

$$r_{,} = a(1+u)\,; \quad r'_{,} = a'\,(1+u')\,;$$
$$v_{,} = nt + \epsilon + v\,; \quad v'_{,} = n't + \epsilon' + v'\,;$$

u, u', v, v', being very small quantities depending on the eccentricities and inclinations, and a, a' the mean distances of m and m', or half the greater axes of their orbits.

If these quantities be substituted in R, and if to abridge

$$n't - nt + \epsilon' - \epsilon = \beta,$$

observing also that,

$$\cos(\beta + v' - v) = \cos\beta \,.\, \cos(v' - v) - \sin\beta \,.\, \sin(v' - v) = \cos\beta - (v' - v)\sin\beta,$$

because $v' - v$ is so small that it may be taken for its sine and unity for its cosine, thus

$$R = -m'\,.\,\frac{a}{a'^2}\,.\,\frac{1+u}{(1+u')^2}\,.\,\cos\beta + m'\,.\,\frac{a}{a'^2}\,.\,\frac{1+u}{(1+u')^2}\,.\,(v'-v)\,.\,\sin\beta$$
$$+ \frac{m'}{\{a^2(1+u)^2 - 2aa'(1+u)\,(1+u')\,.\,\cos\beta + a'^2(1+u')^2\}^{\frac{1}{2}}}$$
$$- \frac{m'\,.\,zz'}{a'^3} + \frac{3m'\,.\,az'^2}{2a'^4}\,.\,\cos\beta$$
$$- \frac{m'(z-z')^2 - 3m'\,.\,az'^2(v'-v)\,.\,\sin\beta}{2\{a^2(1+u)^2 - 2aa'(1+u)\,(1+u')\,.\,\cos\beta + a'^2(1+u')^2\}^{\frac{3}{2}}} + \&c.$$

446. The expansion of this function into a series ascending, according to the powers and products of the very small quantities u, u', v, v', z, and z' is easily accomplished by the theorem for the development of a function of any number of variables, for if R' be the value of R when these small quantities are zero, that is, supposing the orbits to be circular and all in one plane, then

$$R = R' + au\,.\,\frac{dR'}{da} + a'u'\,.\,\frac{dR'}{da'} + (v'-v)\,.\,\frac{dR'}{ndt}$$
$$+ \frac{a^2u^2}{2}\,.\,\frac{d^2R'}{da^2} + \frac{a'^2u'^2}{2}\,.\,\frac{d^2R'}{da'^2} + \&c.$$

because a is the only quantity that varies with u, a' with u', and t with $(v'-v)$.

But $\quad R' = m'\left\{(a^2 - 2aa'\cos\beta + a'^2)^{-\frac{1}{2}} - \frac{a}{a'^2}\cos\beta\right\};$

and if $(a'^2 - 2aa'\cos\beta + a^2)^{-\frac{1}{2}}$ be developed according to the cosines of the multiples of the arc β, it will have the form

$$(a'^2 - 2aa'\cos\beta + a^2)^{-\frac{1}{2}} = \tfrac{1}{2}A_0 + A_1\cos\beta + A_2\cos 2\beta + A_3 \, . \cos 3\beta + \&c.$$

in which A_0, A_1, &c., are functions of a and a' alone; in fact if to abridge $\frac{a}{a'} = \alpha$, the binomial theorem gives

$$A_0 = \frac{2}{a'}\left\{1 + \left(\frac{1}{2}\right)^2\alpha^2 + \left(\frac{1.3}{2.4}\right)^2\alpha^4 . + \left(\frac{1.3.5}{2.4.6}\right)^2 . \alpha^6 + \&c.\right\},$$

the other coefficients are similar functions of the powers of α; but a general method of finding these coefficients in more convergent series will be given afterwards. Thus,

$$R' = m'\left\{\tfrac{1}{2}A_0 + \left(A_1 - \frac{a}{a'^2}\right) . \cos\beta + A_2 . \cos 2\beta + \&c.\right\}$$

and if i represent every whole number either positive or negative including zero, the general term of this series is

$$R' = \frac{m'}{2} . \Sigma . A_i . \cos i\beta,$$

provided that when $i = 1$, $A_1 - \frac{a}{a'^2}$ be put for A_1.

Again, if $(a'^2 - 2aa' . \cos\beta + a^2)^{-\frac{3}{2}} =$

$$\tfrac{1}{2}B_0 + B_1 . \cos\beta + B_2 . \cos 2\beta + B_3 . \cos 3\beta + \&c.$$

its general term is

$$\frac{m'}{2} . \Sigma . B_i . \cos i\beta;$$

and as

$$\frac{dR'}{da} = \frac{m'}{2} . \Sigma . \left(\frac{dA_i}{da}\right) . \cos i\beta; \quad \frac{dR'}{da'} = \frac{m'}{2} . \Sigma . \left(\frac{dA_i}{da'}\right) \cos i\beta;$$

$$\frac{dR'}{ndt} = -\frac{m'}{2} . \Sigma . iA_i . \sin i\beta; \quad \frac{d^2R'}{da^2} = \frac{m'}{2} . \Sigma . \left(\frac{d^2A_i}{da^2}\right) . \cos i\beta;$$

&c. &c.

The development of R is

$$R = \frac{m'}{2} . \Sigma . A_i . \cos i\,(n't - nt + \epsilon' - \epsilon)$$

$$+\frac{m'}{2}.u.\Sigma.a\left(\frac{dA_i}{da}\right).\cos i(n't-nt+\epsilon'-\epsilon),$$

$$+\frac{m'}{2}.u'.\Sigma.a'\left(\frac{dA_i}{da'}\right).\cos i(n't-nt+\epsilon'-\epsilon),$$

$$-\frac{m'}{2}.(v'-v).\Sigma.i.A_i.\sin i(n't-nt+\epsilon'-\epsilon),$$

$$+\frac{m'}{4}.u^2.\Sigma.a^2\left(\frac{d^2A_i}{da^2}\right).\cos i(n't-nt+\epsilon'-\epsilon),$$

$$+\frac{m'}{2}.uu'.\Sigma.aa'\left(\frac{d^2A_i}{da.da'}\right).\cos i(n't-nt+\epsilon'-\epsilon)$$

$$+\frac{m'}{4}.u'^2.\Sigma.a'^2\left(\frac{d^2A_i}{da'^2}\right).\cos i(n't-nt+\epsilon'-\epsilon),$$

$$-\frac{m'}{2}.u.(v'-v).\Sigma.ia\left(\frac{dA_i}{da}\right).\sin i(n't-nt+\epsilon'-\epsilon)$$

$$-\frac{m'}{2}.u'.(v'-v).\Sigma.ia'\left(\frac{dA_i}{da'}\right).\sin i(nt'-nt+\epsilon'-\epsilon),$$

$$-\frac{m'}{4}.(v'-v)^2.\Sigma.i^2A_i.\cos i(n't-nt+\epsilon'-\epsilon),$$

$$-\frac{m'.zz'}{a'^3}+\frac{3m'.a.z'^2}{2.a'^4}.\cos(n't-nt+\epsilon'-\epsilon),$$

$$-\frac{m'(z'-z)^2}{4}.\Sigma.B_i.\cos i(n't-nt+\epsilon'-\epsilon)$$

$$+\frac{3m'.a.z'^2}{4}.(v'-v).\Sigma.B_i.\cos i(n't-nt+\epsilon'-\epsilon),$$

$$+\ \&c. \qquad \&c.$$

a series that may be extended indefinitely.

447. If $v_{\prime}$ be the projection of v, by articles 398 and 401, $v_{\prime}$ and the curtate distance are

$$r_{\prime}=r(1-\tfrac{1}{2}s^2+\tfrac{3}{8}s^4-\&c.),$$

$$v_{\prime}=v-\tan^2\tfrac{1}{2}\phi\{\sin 2v+\tfrac{1}{2}\tan^2\phi.\sin 4v+\}\ \&c.$$

or, if the values of r and v, in article 392, be substituted,

$$r_{\prime}=a(1+\tfrac{1}{2}e^2-e\cos(nt+\epsilon-\varpi)+\&c.).(1-\tfrac{1}{2}s^2+\&c.),$$

$$v_{\prime}=nt+e+2e\sin(nt+\epsilon-\varpi)+\&c.$$

$$-\tan^2\tfrac{1}{2}\phi\{\sin 2v+\tfrac{1}{2}\tan^2\phi\sin 4v+\&c.\}.$$

Where a is half the greater axis of the orbit of m, e the eccentricity, ϖ the longitude of the perihelion, θ the longitude of the ascending node, ϕ the inclination of the orbit of m on the fixed ecliptic at the epoch, and $nt + \epsilon$ the mean longitude of m.

But $\quad r_{,} = a\,(1 + u), \qquad v_{,} = nt + \epsilon + v;$

hence $$u = -e\cos(nt + \epsilon - \varpi) + \tfrac{1}{2}e^2(1 - \cos 2\,(nt + \epsilon - \varpi)) - \tfrac{1}{4}\tan^2\phi\,.\,\sin^2(nt + \epsilon),$$

$$v = 2\,e\,.\sin(nt + \epsilon - \varpi) + \tfrac{5}{4}e^2\,.\sin 2\,(nt + \epsilon - \varpi) - \tan^2\tfrac{1}{2}\phi\,.\,\sin 2\,(nt + \epsilon)$$

when the approximation only extends to the squares and products of the eccentricities and inclinations.

In the same manner,

$$u' = -e'\,.\cos(n't + \epsilon' - \varpi') + \tfrac{1}{2}e'^2\,.(1 - \cos 2\,(n't + \epsilon' - \varpi')) - \tfrac{1}{4}\tan^2\phi'\,.\,\sin^2(n't + \epsilon'),$$

$$v' = 2e'\,.\sin(n't + \epsilon' - \varpi') + \tfrac{5}{4}e'^2\,.\sin 2\,(n't + \epsilon' - \varpi') - \tan^2\tfrac{1}{2}\phi'\,.\,\sin 2\,(n't + \epsilon').$$

448. The substitution of these quantities will give the value of R in series, if the products of the sines and cosines be replaced by the cosines of the sums and differences of the arcs, observing that cosines of the forms

$$\cos\{i\,(n't - nt + \epsilon' - \epsilon) + n't - nt + \epsilon' - \epsilon - \varpi + \varpi'\},$$
$$\cos\{i\,(n't - nt + \epsilon' - \epsilon) + n't + nt + \epsilon' + \epsilon - 2\varpi)$$

become

$$\cos\{i\,(n't - nt + \epsilon' - \epsilon) - \varpi + \varpi'\},$$
$$\cos\{i\,(n't - nt + \epsilon' - \epsilon) + 2nt + 2\epsilon - 2\varpi)$$

by the substitution of $i - 1$ for i, and cosines of the form

$$\cos\{i\,(n't - nt + \epsilon' - \epsilon) - n't + nt - \epsilon' + \epsilon + \varpi' - \varpi\}$$

become

$$\cos\{i\,(n't - nt + \epsilon' - \epsilon) + \varpi' - \varpi\},$$

by the substitution of $i + 1$ for i.

449. Attending to these circumstances, it will be found that

$$R = \frac{m'}{2}\,.\,\Sigma A_i\,.\cos i\,(n't - nt + \epsilon' - \epsilon), \tag{118}$$

$$+ \frac{m'}{2}\,.\,M_0\,.\,e\,.\cos\{i\,(n't - nt + \epsilon' - e) + nt + \epsilon - \varpi\},$$

$$+ \frac{m'}{2}\,.\,M_1\,.\,e'\,.\cos\{i\,(n't - nt + \epsilon' - \epsilon) + nt + \epsilon - \varpi'\},$$

$$+ \frac{m'}{2} \, . \, N_0 \, . \, e^2 \, . \, \cos \{i(n't - nt + \epsilon' - \epsilon) + 2nt + 2\epsilon - 2\varpi\},$$

$$+ \frac{m'}{2} \, . \, N_1 \, . \, ee' \, . \, \cos \{i(n't - nt + \epsilon' - \epsilon) + 2nt + 2\epsilon - \varpi - \varpi'\},$$

$$+ \frac{m'}{2} \, . \, N_2 \, . \, e'^2 \, . \, \cos \{i(n't - nt + \epsilon' - \epsilon) + 2nt + 2\epsilon - 2\varpi'\},$$

$$+ \frac{m'}{2} \, . \, N_3 \, . \, (e^2 + e'^2) \, . \, \cos i \, (n't - nt + \epsilon' - \epsilon),$$

$$+ \frac{m'}{2} \, . \, N_4 \, . \, ee' \, . \, \cos \{i \, (n't - nt + \epsilon' - \epsilon) + \varpi - \varpi'\},$$

$$+ \frac{m'}{2} \, . \, N_5 \, . \, ee' \, . \, \cos \{i \, (n't - nt + \epsilon' - \epsilon) - \varpi + \varpi'\},$$

$$- \frac{m' \, . \, zz'}{a'^3} + \frac{3m' \, . \, a \, . \, z'^2}{2a'^4} \, . \, \cos \, (n't - nt + \epsilon' - \epsilon),$$

$$- \frac{m' \, (z - z')^2}{4} \, . \, \Sigma B_i \, . \, \cos i \, (n't - nt + \epsilon' - \epsilon),$$

$$+ \frac{m'}{4} \, . \, Q_0 \, . \, e'^3 \, . \, \cos \{i \, (n't - nt + \epsilon' - \epsilon) + 3nt + 3\epsilon - 3\varpi'\},$$

$$+ \frac{m'}{4} \, . \, Q_1 \, . \, e'^2 \, . \, e \, . \, \cos \{i \, (n't - nt + \epsilon' - \epsilon) + 3nt + 3\epsilon - 2\varpi' - \varpi\},$$

$$+ \frac{m'}{4} \, . \, Q_2 \, . \, e' \, . \, e^2 \, . \, \cos \{i \, (n't - nt + \epsilon' - \epsilon) + 3nt + 3\epsilon - \varpi' - 2\varpi\},$$

$$+ \frac{m'}{4} \, . \, Q_3 \, . \, e^3 \, . \, \cos \{i \, (n't - nt + \epsilon' - \epsilon) + 3nt + 3\epsilon - 3\varpi\},$$

$$+ \frac{m'}{4} \, . \, z'^2 \, . \, e' \, . \, \Sigma \, . \, B_i \, . \, \cos \{i(n't - nt + \epsilon' - \epsilon) + n't + \epsilon' - \varpi'\},$$

$$+ \frac{m'}{4} \, . \, z'^2 \, . \, e \, . \, \Sigma \, . \, B_i \, . \, \cos \{i(n't - nt + \epsilon' - \epsilon) + nt + \epsilon - \varpi\},$$

$$+ \quad \&c. \qquad\qquad \&c.$$

The coefficients being

$$M_0 = - \left\{a \left(\frac{dA_i}{da}\right) + 2i \, A_i\right\};$$

$$M_1 = - a' \left(\frac{dA_{(i-1)}}{da'}\right) + 2 \, (i - 1) \, A_{(i-1)};$$

$$N_0 = \tfrac{1}{4} \left\{i(4i - 5) \, A_i + 2 \, (2i - 1) \, a \left(\frac{dA_i}{da}\right) + a^2 \left(\frac{d^2 A_i}{da^2}\right)\right\};$$

*

$$N_1 = -\tfrac{1}{2}\{4(i-1)^2 A_{(i-1)} + 2(i-1)a\left(\frac{dA_{(i-1)}}{da}\right)$$

$$-2(i-1)a'\left(\frac{dA_{(i-1)}}{da'}\right) - aa'\left(\frac{d^2A_{(i-1)}}{da\,da'}\right)\};$$

$$N_2 = \tfrac{1}{4}\{(i-2)(4i-3)A_{(i-2)} - 2(2i-3)a'\left(\frac{dA_{(i-2)}}{da'}\right)$$

$$+ a'^2\left(\frac{d^2A_{(i-2)}}{da'^2}\right)\};$$

$$N_3 = -\tfrac{1}{2}\{4i^2A_i - 2a\left(\frac{dA_i}{da}\right) - a^2\left(\frac{d^2A_i}{da^2}\right)\};$$

$$N_4 = \tfrac{1}{2}\{4(i-1)^2A_{(i-1)} - 2(i-1)a\left(\frac{dA_{(i-1)}}{da}\right)$$

$$-2(i-1)a'\left(\frac{dA_{(i-1)}}{da'}\right) + aa'\left(\frac{d^2A_{(i-1)}}{da\,.\,da'}\right)\};$$

$$N_5 = \tfrac{1}{2}\{4(i+1)^2A_{(i+1)} + 2(i+1)a\left(\frac{dA_{(i+1)}}{da}\right)$$

$$+2(i+1)a'\left(\frac{dA_{(i+1)}}{da'}\right) + aa'\left(\frac{d^2A_{(i+1)}}{da\,.\,da'}\right)\}$$

&c. &c.

450. But $z = r_, s = r_, \tan\phi \sin(v_, - \theta)$, by article 435, or substituting the values of $r_,$ and $v_,$ in article 447, and rejecting the product $e \tan\phi$, it becomes

$$z = a\,.\tan\phi\sin(nt + \epsilon - \theta);$$

also
$$z' = a'\,.\tan\phi'\sin(n't + \epsilon' - \theta'),$$

ϕ and ϕ' being the inclinations of the orbits of m and m' on the ecliptic. These values of z and z' are referred to the ecliptic at the epoch; but if the orbit of m at the epoch be assumed to be the fixed plane, $\phi = 0$, $\tan\phi' = \gamma$, the mutual inclination of the orbits of m and m', then Π being the longitude of the ascending node of the orbit of m' on that of m,

$$z = 0,\ z' = a'\gamma\sin(n't + \epsilon' - \Pi),$$

consequently the terms of R depending on z' with regard to γ^2, $e\gamma^2$, and $e'\gamma^2$, become

$$+ \frac{m'}{2} \cdot N_6 \cdot \gamma^2 \cdot \cos\{i(n't - nt + \epsilon' - \epsilon) + 2nt + 2\epsilon - 2\Pi\},$$

$$+ \frac{m'}{2} \cdot N_7 \cdot \gamma^2 \cdot \cos i\,(n't - nt + \epsilon' - \epsilon),$$

$$+ \frac{m'}{4} \cdot Q_4 \cdot \gamma^2 e' \cdot \cos\{i(n't - nt + \epsilon' - \epsilon) + 3nt + 3\epsilon - \varpi' - 2\Pi\},$$

$$+ \frac{m'}{4} \cdot Q_5 \cdot \gamma^2 e \cdot \cos\{i(n't - nt + \epsilon' - \epsilon) + 3nt + 3\epsilon - \varpi - 2\Pi\}.$$

451. It appears from this series that the sum of the terms independent of the eccentricities and inclinations of the orbits, is

$$\frac{m'}{2}\,\Sigma \,.\, A_i \cos i\,(n't - nt + \epsilon' - \epsilon),$$

which is the same as if the orbits were circular and in one plane.

The sum of the terms depending on the first powers of the eccentricities has the form

$$\frac{m'}{2}\,\Sigma \,.\, M \cos\{i\,(n't - nt + \epsilon' - \epsilon) + nt + \epsilon + K\}.$$

Those depending on the squares and products of the eccentricities and inclinations may be expressed by

$$\frac{m'}{2}\,\Sigma\, N \,.\, \cos\{i\,(n't - nt + \epsilon' - \epsilon) + 2nt + 2\epsilon + L\}$$

$$+ \frac{m'}{2}\,\Sigma\, N' \,.\, \cos\{i\,(n't - nt + \epsilon' - \epsilon) + L'\}.$$

Those depending on the third powers and products of these elements are

$$\frac{m'}{4}\,\Sigma\, Q \,.\, \cos\{i\,(n't - nt + \epsilon' - \epsilon) + 3nt + 3\epsilon + U\}$$

$$+ \frac{m'}{4}\,\Sigma \,.\, Q' \,.\, \cos\{i\,(n't - nt + \epsilon' - \epsilon) + nt + \epsilon + U'\},$$

&c. &c.

It may be observed that the coefficient of the sine or cosine of the angle ϖ has always the eccentricity e for factor; the coefficient of the sine or cosine of 2ϖ has e^2 for factor; the sine or cosine of 3ϖ has e^3, and so on: also the coefficient of the sine or cosine of θ has $\tan.\phi$ for factor; the sine or cosine of 2θ has $\tan^2.\phi$ for factor, &c. &c.

Determination of the Coefficients of the Series R.

452. In order to complete the developement of R, the coefficients A_i and B_i, and their differences, must be determined. Let

$$(a'^2 - 2aa' \cos\beta + a^2)^{-s} = \text{A}^{-s} = \tfrac{1}{2}A_0 + A_1 \cos\beta + A_2 \,.\, \cos 2\beta + \&\text{c}.$$

The differential of which is

$$\text{A}^{-s-1}\, 2saa' \sin\beta = A_1 \sin\beta + 2A_2 \sin 2\beta + 3A_3 \sin 3\beta + \&\text{c}.$$

multiplying both sides of this equation by A, and substituting for A^{-s}, it becomes

$$2saa' \sin\beta \left\{\tfrac{1}{2}A_0 + A_1 \cos\beta + A_2 \cos 2\beta + \&\text{c.}\right\}$$
$$= (a'^2 - 2aa' \cos\beta + a^2)\left\{\text{A}_1 \sin\beta + 2A_2 \sin 2\beta + \&\text{c.}\right\}$$

If it be observed that

$$\cos\beta \sin\beta = \tfrac{1}{2}\cos 2\beta, \&\text{c.}$$

when the multiplication is accomplished, and the sines and cosines of the multiple arcs put for the products of the sines and cosines, the comparison of the coefficients of like cosines gives

$$A_2 = \frac{(a^2 + a'^2) A_1 - saa' A_0}{aa' (2 - s)};$$

$$A_3 = \frac{2(a^2+a'^2)A_2 - (1+s)aa'A_1}{aa' (3-s)};$$

and generally

$$A_i = \frac{(i-1)\,(a^2+a'^2)\, A_{(i-1)} - (i+s-2)aa'\, A_{(i-2)}}{(i-s)\; aa'}; \qquad (119)$$

in which i may be any whole number positive or negative, with the exception of 0 and 1. Hence A_i will be known, if A_0, A_1 can be found.

Let $\text{A}^{-s-1} = \tfrac{1}{2}B_0 + B_1 \cos\beta + B_2 \cos 2\beta + \&\text{c}.$ multiplying this by

$$(a^2 - 2aa' \cos\beta + a'^2),$$

and substituting the value of A^{-s} in series

$$\tfrac{1}{2}A_0 + A_1 \cos\beta + A_2 \cos 2\beta + \&\text{c}.$$
$$= (a^2 - 2aa' \cos\beta + a'^2)\,(\tfrac{1}{2}B_0 + B_1 \cos\beta + B_2 \cos 2\beta + \&\text{c.})$$

the comparison of the coefficients of like cosines gives

$$A_i = (a^2 + a'^2) \,.\, B_i - aa' \,.\, B_{(i-1)} - aa'\, B_{(i+1)}.$$

But as relations must exist among the coefficients $B_{(i-1)}$, B_i, $B_{(i+1)}$, similar to those existing among $A_{(i-1)}$, A_i, $A_{(i+1)}$, the equation (119) gives, when $s+1$ and $i+1$ are put for s and i,

$$B_{(i+1)} = \frac{i(a^2+a'^2)\,B_i - (i+s)\,.\,aa'\,B_{(i-1)}}{(i-s)\,aa'}. \qquad (120)$$

If this quantity be put in the preceding value of A_i, it becomes

$$A_i = \frac{2saa'\,B_{(i-1)} - s\,(a^2+a'^2)\,.\,B_i}{i-s}; \qquad (121)$$

or if $i+1$ be put for i,

$$A_{(i+1)} = \frac{2saa'\,B_i - s\,(a^2+a'^2)\,B_{(i+1)}}{i-s+1}; \qquad (122)$$

whence may be obtained, by the substitution of the preceding value of $B_{(i+1)}$,

$$A_{(i+1)} = \frac{s(i+s)\,.\,aa'(a^2+a'^2)\,B_{(i-1)} + s\{2(i-s)a^2a'^2 - i(a^2+a'^2)^2\}B_i}{(i-s)\,(i-s+1)\,.\,aa'}.$$

If $B_{(i-1)}$ be eliminated between this equation and (121), there will result,

$$B_i = \frac{\frac{1}{s}(i+s)\,(a^2+a'^2)\,A_i - \frac{2}{s}(i-s+1)\,.\,aa'\,.\,A_{(i+1)}}{(a'^2-a^2)^2},$$

or substituting for $A_{(i+1)}$ its value given by equation (119),

$$B_i = \frac{\frac{1}{s}\,(s-1)\,(a^2+a'^2)\,.\,A_i + \frac{2}{s}(i+s-1)\,.\,aa'\,A_{(i-1)}}{(a'^2-a^2)^2}.$$

If to abridge $\frac{a}{a'} = \alpha$, the two last equations, as well as equation (119), when both the numerators and the denominators of their several members are divided by a'^2, take the form

$$A_i = \frac{(i-1)\,(1+\alpha^2)\,A_{(i-1)} - (i+s-2)\,.\,\alpha\,.\,A_{(i-2)}}{(i-s)\,\alpha}, \qquad (123)$$

$$B_i = \frac{\frac{1}{s}\,(i+s)\,(1+\alpha^2)\,A_i - \frac{2}{s}(i-s+1)\,\alpha'\,.\,A_{(i+1)}}{(1-\alpha^2)^2\,a'^2}; \qquad (124)$$

$$B_i = \frac{\frac{1}{s}\,(s-i)\,(1+\alpha^2)\,.\,A_i + \frac{2}{s}(i+s-1)\,.\,\alpha'\,.\,A_{(i-1)}}{(1-\alpha^2)^2\,a'^2}, \qquad (125)$$

which is very convenient for computation.

All the coefficients A_2, A_3, &c., B_0, B_1, &c., will be obtained from equations (123) and (125), when A_0, A_1 are known; it only remains, therefore, to determine these two quantities.

454. Because

$$\cos\beta = \frac{c^{\beta\sqrt{-1}} + c^{-\beta\sqrt{-1}}}{2},$$

c being the number whose hyperbolic logarithm is unity; therefore

$$a'^2 - 2aa'\cos\beta + a^2 = \{a' - ac^{\beta\sqrt{-1}}\}\,.\,\{a' - ac^{-\beta\sqrt{-1}}\}$$

consequently,

$$A^{-s} = \{a' - ac^{\beta\sqrt{-1}}\}^{-s}.\,\{a' - ac^{-\beta\sqrt{-1}}\}^{-s}.$$

but

$$(a' - ac^{\beta\sqrt{-1}})^{-s} = \frac{1}{a'^s}\{1 + s\,\alpha c^{\beta\sqrt{-1}} + \frac{s(s+1)}{2}\alpha^2 c^{2\beta\sqrt{-1}} + \&c.\},$$

$$(a' - ac^{-\beta\sqrt{-1}})^{-s} = \frac{1}{a'^s}\{1 + s\,\alpha c^{-\beta\sqrt{-1}} + \frac{s(s+1)}{2}\alpha^2 c^{-2\beta\sqrt{-1}} + \&c.\};$$

the product of which is

$$A^{-s} = \frac{1}{a'^{2s}}\{1 + s^2\alpha^2 + \left(\frac{s(1+s)}{1\,.\,2}\right)^2\alpha^4 + \left(\frac{s(1+s)\,(2+s)}{1\,.\,2\,.\,3}\right)^2\alpha^6 + \&c.\}$$

$$+\frac{2}{a'^{2s}}\{s\alpha + \frac{s^2(1+s)}{1\,.\,2}\alpha^3 + \frac{s(s+1)}{1\,.\,2}\,.\,\frac{s(1+s)\,(2+s)}{1\,.\,2\,.\,3}\alpha^5 + \&c.\}\times$$

$$(c^{\beta\sqrt{-1}} + c^{-\beta\sqrt{-1}}) + \&c.$$

whence it appears that $c^{i\beta\sqrt{-1}}$, and $c^{-i\beta\sqrt{-1}}$ have always the same coefficients; and as $c^{i\beta\sqrt{-1}} + c^{-i\beta\sqrt{-1}} = 2\cos i\beta$, it is easy to see that this series is the same with

$$A^{-s} = (a'^2 - 2aa'\cos\beta + a^2)^{-s} = \tfrac{1}{2}A_0 + A_1\cos\beta + \&c.$$

consequently,

$$A_0 = \frac{2}{a'^{2s}}\{1 + s^2\alpha^2 + \left(\frac{s(1+s)}{1\,.\,2}\right)^2\alpha^4 + \left(\frac{s(s+1)\,(s+2)}{1\,.\,2\,.\,3}\right)^2\alpha^6 + \&c.\},$$

$$A_1 = \frac{2}{a'^{2s}}\{s\alpha + s\,.\,\frac{s(1+s)}{1\,.\,2}\,.\,\alpha^3 + \frac{s\,.\,(s+1)}{1\,.\,2}\,.\,\frac{s\,.\,(s+1)\,(s+2)}{1\,.\,2\,.\,3}\,.\,\alpha^5 + \&c.\}$$

These series do not converge when $s = \frac{1}{2}$; but they converge rapidly when $s = -\frac{1}{2}$; then, however, A_0 and A_1 become the first and second coefficients of the development of

$$(a'^2 - 2aa'\cos\beta + a^2)^{\frac{1}{2}}.$$

Let S and S′ be the values of these two coefficients in this case, then

$$S = a'\{1 + (\tfrac{1}{2})^2\alpha^2 + \left(\frac{1\,.\,1}{2\,.\,4}\right)^2\alpha^4 + \left(\frac{1\,.\,1\,.\,3}{2\,.\,4\,.\,6}\right)^2\alpha^6 + \&c.\}$$

$$S' = -a'\{\alpha - \frac{1.1}{2.4}\alpha^3 - \frac{1.1.1.3}{4.2.4.6}\alpha^5 - \frac{1.3.5.1.1.3.5.7}{4.6.8.2.4.6.8.10}\alpha^7 - \&c.\}$$

and as the values of A_0, $A_{\prime}$ may be obtained in functions of S and S', the two last series form the basis of the whole computation.

Because A_0, $A_{\prime}$ become S and S' when $s = -\frac{1}{2}$, and that B_i becomes A_i; if $s = -\frac{1}{2}$, and $i = 0$, equation (124) gives

$$A_0 = 2 \cdot \frac{(1+\alpha^2)\,S + 3\alpha\,S'}{(1-\alpha^2)^2 \cdot a'^2};$$

and if $s = -\frac{1}{2}$, and $i = 1$, equation (125) gives

$$A_1 = \frac{4\alpha\,S + 3\,(1+\alpha^3)\,S'}{(1-\alpha^2)^2 \cdot a'^2}.$$

If $s = \frac{1}{2}$, and $i = 0$, equation (125) gives

$$B_0 = \frac{(1+\alpha^2)\,A_0 - 2\alpha\,A_1}{a'^2\,(1-\alpha^2)^2};$$

and substituting the preceding values of A_0 and A_1, it becomes

$$B_0 = \frac{2S}{a'^4\,(1-\alpha^2)^2}.$$

In the same manner it will be found that

$$B_1 = \frac{-3S'}{a'^4(1-\alpha^2)^2}.$$

454. It now remains to determine the differences of A_i and B_i with regard to a. Resume

$$\text{A}^{-s} = \tfrac{1}{2}A_0 + A_1\cos\beta + A_2\cos 2\beta + \&c.$$

and take its differential with regard to a, observing that

$$\frac{d\text{A}}{da} = 2(a - a')\cos\beta;$$

then

$$-2s.(a-a')\cos\beta\,.\text{A}^{-s-1} = \tfrac{1}{2}\cdot\frac{dA_0}{da} + \frac{dA_1}{da}\cdot\cos\beta$$

$$+ \frac{dA_2}{da}\cdot\cos 2\beta + \&c.$$

But $$\text{A} = a'^2 - 2aa'\cdot\cos\beta + a^2$$

gives $$a - a'\cos\beta = \frac{\text{A} + a^2 - a'^2}{2a};$$

therefore $A^{-s} + (a^2 - a'^2)\,A^{-s-1} = -\frac{1}{2}\cdot\frac{a}{s}\cdot\frac{dA_0}{da}$

$$-\frac{a}{s}\cdot\frac{dA_1}{da}\cos\beta - \frac{a}{s}\cdot\frac{dA_2}{da}\cos 2\beta - \&c.$$

or, substituting the values of A^{-s} and A^{-s-1} in series

$$\tfrac{1}{2}A_0 + A_1\cos\beta + A_2\cos 2\beta + \&c. + (a^2 - a'^2)\times$$
$$\{\tfrac{1}{2}B_0 + B_1\cos\beta + B_2\cos 2\beta + \&c.\} =$$
$$-\tfrac{1}{2}\cdot\frac{a}{s}\cdot\frac{dA_0}{da} - \frac{a}{s}\cdot\frac{dA_1}{da}\cos\beta - \frac{a}{s}\cdot\frac{dA_2}{da}\cos 2\beta - \&c.$$

and the comparison of like cosines gives the general expression,

$$\frac{dA_i}{da} = \frac{s(a'^2 - a^2)}{a}\cdot B_i - \frac{s}{a}A_i; \qquad (126)$$

or, substituting for B_i its value in (124), it becomes

$$\frac{dA_i}{da} = \left(\frac{ia'^2 + (i+2s)\cdot a^2}{a(a'^2 - a^2)}\right)A_i - \left(\frac{2(i-s+1)a'}{a'^2 - a^2}\right)A_{(i+1)}.$$

If the differentials of this equation be taken with regard to a, and if, in the resulting equations, substitution be made for $\frac{dA_i}{da}$, $\frac{dA_{(i+1)}}{da}$ from the preceding formula, the successive differences of A_i, in functions of $A_{(i+1)}$, $A_{(i+2)}$, will be obtained.

Coefficients of the series R.

455. If $\frac{1}{2}$ be put for s in the preceding equation, and in equation (123), and if it be observed that in the series R, article 446, $\frac{dA_i}{da}$ is always multiplied by a, $\frac{d^2A_i}{da^2}$ by a^2, and so on; then where i is successively made equal to 0, 1, 2, 3, &c. the coefficients and their differences are,

$$A_0 = \frac{2(1+\alpha^2)\,S + 6\alpha S'}{a'^2(1-\alpha^2)^2}$$

$$A_1 = \frac{4\alpha S + 3(1+\alpha^2)S'}{a'^2(1-\alpha^2)^2}$$

$$A_2 = \frac{1}{3\alpha}.\ \{2(1+\alpha^2)A_1 - \alpha A_0\}$$

$$A_3 = \frac{1}{5\alpha}.\ \{4(1+\alpha^2)A_2 - 3\alpha A_1\}$$

$$A_4 = \frac{1}{7\alpha}.\ \{6(1+\alpha^2)\,A_3 - 5\alpha A_2\}$$

$$A_5 = \frac{1}{9\alpha}.\ \{8(1+\alpha^2)\,A_4 - 7\alpha A_3\},$$

&c. &c.

$$a\left(\frac{d.A_0}{da}\right) = \frac{1}{1-\alpha^2}\{\alpha^2 A_0 - \alpha A_1\}$$

$$a\left(\frac{d.A_1}{da}\right) = \frac{1}{1-\alpha^2}\{(1+2\alpha^2)\,A_1 - 3\alpha A_2\}$$

$$a\left(\frac{d.A_2}{da}\right) = \frac{1}{1-\alpha^2}\{(2+3\alpha^2)\,A_2 - 5\alpha A_3\}$$

$$a\left(\frac{d.A_3}{da}\right) = \frac{1}{1-\alpha^2}\{(3+4\alpha^2)\,A_3 - 7\alpha A_4\},$$

&c. &c.

$$a^2\left(\frac{d^2.A_0}{da^2}\right) = \frac{1}{(1-\alpha^2)^2}\{2\alpha^4 A_0 + (\alpha - 3\alpha^3)A_1\},$$

$$a^2\left(\frac{d^2.A_1}{da^2}\right) = \frac{1}{(1-\alpha^2)^2}\{(2-4\alpha^2)A_1 - (\alpha - 3\alpha^3)\,A_0\}$$

$$a^2\left(\frac{d^2.A_2}{da^2}\right) = \frac{1}{(1-\alpha^2)^2}\{A_2\,\{2+3\alpha^2)^2 + 5\alpha^2(1+\alpha^2) - 2(1-\alpha^2)^2\}$$
$$- 5\alpha(5+9\alpha^2)A_3 + 5.7\alpha^2 A_4\}$$

$$a^2\left(\frac{d^2.A_3}{da}\right) = \frac{1}{(1-\alpha^2)^2}\{\{(3+4\alpha^2)^2 + 7\alpha^2(1+\alpha^2) - 3(1-\alpha)^2\}\,A_3$$
$$- 7\alpha\,(7+11\alpha^2)\,A_4 + 7.9.\alpha^2 A_5\}$$

$$a^2\left(\frac{d^2.A_4}{da^2}\right) = \frac{1}{(1-\alpha^2)^2}\{\{(4+5\alpha^2)^2 + 9\alpha^2(1+\alpha^2) - 4(1-\alpha^2)^2\}\,A_4$$
$$- 9\alpha\,(9+13\alpha^2)\,A_5 + 9.11.\alpha^2 A_6\},$$

&c. &c.

456. By the aid of equation (120), it is easy to see that

$$B_0 = \frac{2S}{(a'^2 - a^2)^2}$$

$$B_1 = -\frac{3S'}{(a'^2 - a^2)^2}$$

$$B_2 = \frac{1}{\alpha}\{2(1 + \alpha^2)\, B_1 - 3\alpha\, B_0\}$$

$$B_3 = \frac{1}{3\alpha}\{4(1 + \alpha^2)\, B_2 - 5\alpha\, B_1\}$$

$$B_4 = \frac{1}{5\alpha}\{6(1 + \alpha^2)\, B_3 - 7\alpha\, B_2\}$$

&c. &c.

$$a\left(\frac{dB_0}{da}\right) = \frac{3\alpha^2 B_0 + \alpha B_1}{1 - \alpha^2}$$

$$a\left(\frac{dB_1}{da}\right) = \frac{3\alpha B_0 + (2\alpha^2 - 1)B_1}{1 - \alpha^2},$$

&c. &c.

457. The coefficient A_i and its differences have a very simple form, when expressed in functions of B_i, for equations (121) and (126) give

$$A_0 = (a'^2 + a^2)\, B_0 - 2aa'B_1$$

$$A_1 = 2aa'\, B_0 - (a'^2 + a^2)B_1$$

$$A_2 = \frac{2aa'\, B_1 - (a'^2 + a^2)\, B_2}{3}$$

$$A_3 = \frac{2aa'B_2 - (a'^2 + a^2)B_3}{5},$$

&c. &c.

$$a\left(\frac{dA_0}{da}\right) = a'a\, B_1 - a^2 B_0$$

$$a\left(\frac{dA_1}{da}\right) = a'^2 B_1 - aa'B_0$$

$$a\left(\frac{dA_2}{da}\right) = \tfrac{1}{3}\{(2a'^2 - a^2)\, B_2 - aa'\, B_1\}$$

$$a\left(\frac{dA_3}{da}\right) = \tfrac{1}{5}\{(3a'^2 - 2a^2)\, B_3 - aa'\, B_2\}$$

$$a\left(\frac{dA_4}{da}\right) = \tfrac{1}{7}\{(4a'^2 - 3a^2)\, B_4 - aa'\, B_3\},$$

&c. &c.

$$a^2\left(\frac{d^2A_0}{da^2}\right) = 2a^2 B_0 - a' aB_1$$

$$a^2\left(\frac{d^2A_1}{da^2}\right) = 3aa' B_0 - 2a'^2 B_1,$$

&c. &c.

458. The differences of A_i and B_i with regard to a' are obtained from their differences with regard to a, for A_i being a homogeneous function of a and a' of the dimension -1,

$$a\left(\frac{dA_i}{da}\right) + a'\left(\frac{dA_i}{da'}\right) = -A_i;$$

as readily appears from

$$(a^2 - 2aa' \cos\beta + a'^2)^{-\frac{1}{2}},$$

therefore,

$$a'\left(\frac{dA_i}{da'}\right) = -A_i - a\left(\frac{dA_i}{da}\right)$$

$$a'\left(\frac{d^2A_i}{da.da'}\right) = -2\left(\frac{dA_i}{da}\right) - a\left(\frac{d^2A_i}{da^2}\right)$$

$$a'^2\left(\frac{d^2A_i}{da'^2}\right) = 2A_i + 4a\left(\frac{dA_i}{da}\right) + a^2\left(\frac{d^2A_i}{da^2}\right),$$

&c. &c.

Likewise B_i being a homogeneous function of the dimension -3,

$$a'\left(\frac{dB_i}{da'}\right) + a\left(\frac{dB_i}{da}\right) = -3B_i.$$

459. By means of these, all the differences of A_i, B_i, with regard to a', may be eliminated from the series R, so that the coefficients of article 449 become

$$M_0 = -a\left(\frac{dA_i}{da}\right) - 2iA_i$$

$$M_1 = a\left(\frac{dA_{(i-1)}}{da}\right) + 2(i-1)A_{(i-1)}$$

$$N_0 = \tfrac{1}{4}\left\{i(4i-5)A_i + 2(2i-1)a\left(\frac{dA_i}{da}\right) + a^2\left(\frac{d^2A_i}{da^2}\right)\right\}$$

$$N_1 = -\tfrac{1}{2}\left\{(2i-2)(2i-1)A_{(i-1)} + 2(2i-1)a\left(\frac{dA_{(i-1)}}{da}\right) + a^2\left(\frac{d^2A_{(i-1)}}{da^2}\right)\right\}$$

$$N_2 = \tfrac{1}{4}\left\{(4i^2-7i+2)A_{(i-2)} + 2(2i-1)a\left(\frac{dA_{(i-2)}}{da}\right) + a^2\left(\frac{d^2A_{(i-2)}}{da^2}\right)\right\}$$

*

$$N_3 = -\tfrac{1}{2}\left\{4i^2 A_i - 2a\left(\frac{dA_i}{da}\right) - a^2\left(\frac{dA_i}{da^2}\right)\right\}$$

$$N_4 = \tfrac{1}{2}\left\{(2i-2)(2i-1)A_{(i-1)} - 2a\left(\frac{dA_{(i-1)}}{da}\right) - a^2\left(\frac{d^2A_{(i-1)}}{da^2}\right)\right\}$$

$$N_5 = \tfrac{1}{2}\left\{(2i+2)(2i+1)A_{(i+1)} - 2a\left(\frac{dA_{(i+1)}}{da}\right) - a^2\left(\frac{d^2A_{(i+1)}}{da^2}\right)\right\}$$

$$N_6 = \tfrac{1}{4}aa'\,\Sigma\, B_{(i-1)}$$

$$N_7 = -\tfrac{1}{4}aa'\,(B_{(i-1)} + B_{(i+1)}),$$

&c. &c.

When $i = 1$, $N_6 = \frac{1}{4}aa'\,B_0 - \frac{1}{2}\,\frac{a}{a'^2}$, and $\frac{1}{2}.\frac{a}{a'^2}$ must be added to N_7.

460. The series represented by S and S′ which are the bases of the computation, are numbers given by observation: for if the mean distance of the earth from the sun be assumed as the unit, the mean distances of the other planets determined by observation, may be expressed in functions of that unit, so that $\alpha = \frac{a}{a'}$, the ratio of the mean distance of m to that of m' is a given number, and as the functions are symmetrical with regard to a and a', the denominator of $\frac{a}{a'}$ may always be so chosen as to make α less than unity, therefore if eleven or twelve of the first terms be taken and the rest omitted, the values of S and S′ will be sufficiently exact; or, if their sum be found, considering them as geometrical series whose ratio is $1 - \alpha^2$, the values of S and S′ will be exact to the sixth decimal, which is sufficient for all the planets and satellites. Thus A_i, B_i, their differences, and consequently the coefficients M_0, M_1, N_0, &c. of the series R are known numbers depending on the mean distances of the planets from the sun.

461. All the preceding quantities will answer for the perturbations of m' when troubled by m, with the exception of A_1, which becomes $A_1 - \frac{a'}{a^2}$; and when employed to determine the perturbations of Jupiter's satellites, the equatorial diameter of Jupiter, viewed at his mean distance from the sun, is assumed as the unit of distance, in functions of which the mean distances of the four satellites from the centre of Jupiter are expressed.

CHAPTER VI.

SECULAR INEQUALITIES IN THE ELEMENTS OF THE ORBITS.

Stability of the Solar System, with regard to the Mean Motions of the Planets and the greater axes of their Orbits.

462. WHEN the squares of the disturbing masses are omitted, however far the approximation may be carried with regard to the eccentricities and inclinations, the general form of the series represented by R, in article 449, is

$$m'k \cdot \cos\{i'n't - int + c\} = R,$$

k and c are quantities consisting entirely of the elements of the orbits, k being a function of the mean distances, eccentricities, and inclinations, and c a function of the longitudes of the epochs of the perihelia and nodes. The differential of this expression, with regard to nt the mean motion of m, is

$$\text{d}R = m'kindt \sin\{i'n't - int + c\}.$$

The expression $\text{d}R$ always relates to the mean motion of m alone; when substituted in

$$da = 2a^2\text{d}R,$$

it gives $\quad da = 2a^2m'ik \cdot ndt \cdot \sin\{i'n't - int + c\},$

the integral of which is

$$\delta a = -\frac{2a^2im'nk}{i'n' - in} \cdot \cos\{i'n't - int + c\}.$$

It is evident that if the greater axes of the orbits of the planets be subject to secular inequalities, this value of δa must contain terms independent of the sines and cosines of the angular distances of the bodies from each other. But a must be periodic unless $i'n' - in = 0$; that is, unless the mean motions of the bodies m and m' be commensurable. Now the mean motions of no two bodies in the solar system are exactly commensurable, therefore

$i'n' - in$ is in no case exactly zero; consequently the greater axes of the celestial bodies are not subject to secular inequalities; and on account of the equation $n = a^{-\frac{3}{2}}$, their mean motions are uniform.

Thus, when the squares and products of the masses m, m' are omitted, the differential $\mathrm{d}R$ does not contain any term proportional to the element of the time, however far the approximation may be carried with regard to the eccentricities and inclinations of the orbits, or, which is the same thing, $\frac{\mathrm{d}R}{ndt}$ does not contain a constant term; for if it contained a term of the form $m'k$, then would

$$a = 2fa^2 \,.\, \mathrm{d}R = 2a^2m'knt, \text{ and } \zeta = -3\int\int andt \,.\, \mathrm{d}R$$

would become $\zeta = -3\int\int an^2m'kdt^2 = -3an^2m'kt^2$,

so that the greater axes would increase with the time, and the mean motion would increase with the square of the time, which would ultimately change the form of the orbits of the planets, and the periods of their revolutions. The stability of the system is so important, that it is necessary to inquire whether the greater axes and mean motions be subject to secular inequalities, when the approximation is carried to the squares and products of the masses.

463. The terms depending on the squares and products of the masses are introduced into the series R by the variation of the elements of the orbits, both of the disturbed and disturbing bodies. Hence, if δa, δe, &c. be the integrals of the differential equations of the elements in article 439, the variable elements will be $a + \delta a$, $e + \delta e$, &c. for m, and $a' + \delta a'$, $e' + \delta e'$, &c. for m'; and when these are substituted for a, e, a', e', &c. in the series R, it takes the form

$$R_{,} = R + \delta R + \delta' R;$$

and from what has been said, the greater axis and mean motion of m will not be affected by secular inequalities, unless the differential

$$\mathrm{d}R_{,} = \mathrm{d}R + \mathrm{d}.\delta R + \mathrm{d}.\delta' R$$

contains a term that is not periodic.

$\mathrm{d}R$ is of the first order relatively to the masses, and has been proved in the preceding article not to contain a term that is not periodic. $\mathrm{d}.\delta R$ and $\mathrm{d}.\delta' R$ include the squares and products of the masses; the first is the differential of δR with regard to the elements of the troubled planet m, and $\mathrm{d}.\delta' R$ is a similar function with regard to the disturbing body m'. It is proposed to examine whether either of these contain a term that is not periodic, beginning with $\mathrm{d}.\delta R$.

464. The variation δR regards the elements of m alone, and is

$$\delta R = \frac{dR}{da}\delta a + \frac{dR}{d\epsilon}\delta\epsilon + \frac{d}{de}\,\delta e + \frac{dR}{d\varpi}\delta\varpi + \frac{dR}{dp}\delta p + \frac{dR}{dq}\,\delta q.$$

If the values in article 439, be put for δa, δe, &c. this expression becomes

$$\delta R = 2a^2\left\{\frac{dR}{da}\int\frac{dR}{d\epsilon}\,ndt - \frac{dR}{d\epsilon}\int\frac{dR}{da}\,.\,ndt\right\}$$

$$+\frac{a\sqrt{1-e^2}}{e}\left(1-\sqrt{1-e^2}\right)\left\{\frac{dR}{d\epsilon}\int\frac{dR}{de}\,ndt - \frac{dR}{de}\int\frac{dR}{d\epsilon}\,.\,ndt\right\}$$

$$+\frac{a\sqrt{1-e^2}}{e}\left\{\frac{dR}{d\varpi}\int\frac{dR}{de}\,.\,ndt - \frac{dR}{de}\int\frac{dR}{d\varpi}\,.\,ndt\right\}$$

$$+\frac{a}{\sqrt{1-e^2}}\left\{\frac{dR}{dp}\int\frac{dR}{dq}\,.\,ndt - \frac{dR}{dq}\int\frac{dR}{dp}\,.\,ndt\right\}.$$

And its differential, according to the elements of the orbit of m alone, is obtained by suppressing the signs $\int$ introduced by the integration of the differential equations of the elements in article 439, which reduces this expression to zero; therefore to obtain $d.\delta R$, it is sufficient to take the differential according to nt of those terms in δR that are independent of the sign $\int$.

When the series in article 449 is substituted for R, δR will take the form

$$P\,.\int.\,Qdt - Q\,.\int.\,Pdt.$$

Where P and Q represent a series of terms of the form

$$k.\,{\cos \atop \sin}\,(i'nt - int + c),$$

i' and i being any whole numbers positive or negative.

Let $$k\cos\,(in't - int + c)$$

belong to P, and let $k'\cos\,(i'n't - int + c')$ be the corresponding term of Q, k, k', c, c', being constant quantities.

A term that is not periodic could only arise in

$$\mathrm{d}\delta R = \mathrm{d}\{P\!\int\! Qdt - Q\!\int\! Pdt\},$$

if it contained such an expression as

$$kk'\cos\{i'n't - int + c\}\cos\{i'n't - int + c'\} = \tfrac{1}{2}kk'\cos\,(c - c')$$
$$+\tfrac{1}{2}kk'\cos\{2i'n't - 2int + c + c'\};$$

or a similar product of the sines of the same angles. But when

$k\cos(i'n't - int + c)$ is put for P, and $k'\cos(i'n't - int + c')$ for Q, $\mathrm{d}\delta R$ becomes

$$\mathrm{d}.\delta R = kindt\,.\sin(i'n't - int + c)\,.\int k'dt\,.\cos(i'n't - int + c')$$
$$-k'indt\,.\sin(i'n't - int + c')\,.\int kdt\,.\cos(i'n't - int + c),$$

which is equal to zero when the integrations are accomplished. Whence it may be concluded that $\mathrm{d}.\delta R$ is altogether periodic.

465. It now remains to determine whether the variation of the elements of the orbit of m' produces terms that are not periodic in $\mathrm{d}.\delta' R$. This cannot be demonstrated by the same process, because the function R, not being symmetrical relatively to the co-ordinates of m and m', changes its value in considering the disturbance of m' by m. Let R' be what R becomes with regard to the planet m' troubled by m; then $R' = m\left\{\frac{1}{\sqrt{(x'-x)^2+(y'-y)^2+(z'-z)^2}} - \frac{xx'+yy'+zz'}{r^3}\right\}$

hence $$R = \frac{m'}{m}R' + m'(xx' + yy' + zz')\left(\frac{1}{r^3} - \frac{1}{r'^3}\right);$$

and $$\delta' R = \frac{m'}{m}\delta' R' + m'\delta'\left\{(xx' + yy' + zz')\left(\frac{1}{r^3} - \frac{1}{r'^3}\right)\right\}.$$

If the differential of this equation according to d be periodic, so will $\mathrm{d}.\delta' R$. Now in consequence of the variations of the elements of the orbit of m,

$$\delta' R' = \frac{dR'}{da'}\delta a' + \frac{dR'}{de}\delta e' + \frac{dR'}{d\epsilon'}\delta\epsilon' + \frac{dR'}{d\varpi'}\delta\varpi' + \frac{dR'}{dp'}\delta p' + \frac{dR'}{dq'}\delta q.$$

And as this expression with regard to the planet m' is in all respects similar to that of δR in the preceding article with regard to m, by the same analysis it may be proved, that $\mathrm{d}.\delta R'$ is altogether periodic. Thus the only terms that are not periodic, must arise from the differential of $$m'\delta'\left\{xx' + yy' + zz'\left(\frac{1}{r^3} - \frac{1}{r^3}\right)\right\}.$$

Let $$m'\{xx' + yy' + zz'\}\left(\frac{1}{r^3} - \frac{1}{r^{3\prime}}\right) = L.$$

Then by article 346,

$$\frac{m'x}{r^3} = -\frac{m'}{S}\cdot\frac{d^2x}{dt^2} - \frac{mm'}{S}\cdot\frac{x}{r^3} + \frac{m'}{S}\left(\frac{dR}{dx}\right);$$

likewise $$\frac{m'x'}{r'^3} = -\frac{m'}{S}\cdot\frac{d^2x'}{dt^2} - \frac{m'^2}{S}\cdot\frac{x'}{r'^3} + \frac{m'}{S}\cdot\left(\frac{dR'}{dx'}\right).$$

The co-ordinates $y\ z$, $y'\ z'$, furnish similar equations. Thus,

$$L = \frac{m'}{S}\left\{\frac{d\,(xdx' - x'dx + ydy' - y'dy + zdz' - z'dz)}{dt^2}\right\} + N, \text{ where}$$

$$N = \frac{m'^2}{S}\left(\frac{xx' + yy' + zz'}{r'^3}\right) - \frac{mm'}{S}\left(\frac{xx' + yy' + zz'}{r^3}\right)$$

$$+ \frac{m'}{S}\left\{x'\left(\frac{dR}{dx}\right) - x\left(\frac{dR'}{dx'}\right) + y'\left(\frac{dR}{dy}\right) - y\left(\frac{dR'}{dy'}\right) + z'\left(\frac{dR}{dz}\right)\right.$$

$$\left. - z\left(\frac{dR'}{dz'}\right)\right\}.$$

If N be omitted at first,

$$d.L = -\frac{m'}{S} \cdot d\left\{\frac{d\,(x'dx - xdx' + y'dy - ydy' + z'dz - zdz')}{dt^2}\right\}.$$

466. The elliptical values of the co-ordinates being substituted, every term must be periodic. For example, if

$$x = a \cdot \cos\,(nt + \epsilon - \varpi) \qquad x' = a' \cdot \cos\,(n't + \epsilon' - \varpi')$$

$$\frac{x'dx - xdx'}{dt} = \tfrac{1}{2}aa'(n - n') \cdot \sin\,\{\,n't - nt + \epsilon' - \epsilon - \varpi' + \varpi\,\};$$

a quantity that must be periodic unless $n't - nt = 0$, which never can happen, because the mean motions of no two bodies in the solar system are exactly commensurable; but even if a term that is not periodic were to occur, it would vanish in taking the second differential; and as the same thing may be shown with regard to the other products

$$y'dy - ydy' \quad z'dz - zdz',$$

dL is a periodic function. With regard to the term dL = dN, if the elliptical values of the co-ordinates of m and m' be substituted, it will readily appear that this expression is periodic, for the equations of the elliptical motion of m and m', in article 365, give

$$\frac{xx' + yy' + zz'}{r^3} = -\frac{x'd^2x + y'd^2y + z'd^2z}{(S + m)dt^2},$$

$$\frac{xx' + yy' + zz'}{r'^3} = -\frac{xd^2x' + yd^2y' + zd^2z'}{(S + m')dt^2};$$

so that the function N becomes

$$N = -\frac{m'^2}{S\,(S + m)}\left\{\frac{xd^2x' + yd^2y' + zd^2z'}{dt^2}\right\}$$

$$+ \frac{mm'}{S\,(S + m)}\left(\frac{x'd^2x + y'd^2y + z'd^2z}{dt^2}\right) + \frac{m'}{S}\left\{x'\left(\frac{dR}{dx}\right) - x\left(\frac{dR'}{dx'}\right)\right.$$

$$\left. + y'\left(\frac{dR}{dy}\right) - y\left(\frac{dR'}{dy'}\right) + z'\left(\frac{dR}{dz}\right) - z\left(\frac{dR'}{dz'}\right)\right\}.$$

467. From what has been said, it will readily appear that the terms of this expression, consisting of the products $x'd^2x$, xd^2x', &c. &c., are periodic when the elliptical values are substituted for the co-ordinates, and their differentials.

468. The last term of the value of N is also periodic; for, if the elliptical values of the co-ordinates of m and m' be put in R, it may be developed into a series of cosines of the multiples of the arcs nt and $n't$, and the differential may be found by making R vary with regard to the quantities belonging to m alone; hence this differential may contain the sines and cosines of the multiples of nt, but no sine or cosine of $n't$ alone; and as

$$x' = r' \cos (n't + \epsilon' - \varpi'),$$

the mean motions nt, $n't$, never vanish from $x'\left(\frac{dR}{dx}\right)$, which is consequently periodic; and as the same may be demonstrated for each of the products

$$x\left(\frac{dR'}{dx'}\right), y'\left(\frac{dR}{dy}\right), \text{ \&c. \&c.,}$$

not only N but its differential are periodic, and consequently $d.\delta'R$.

Thus it has been proved that when the approximation is carried to the squares and products of the masses, the expression

$$dR_{,} = dR + d.\delta R + d.\delta' R$$

relatively to the variations of the mean motions of the two planets m and m' is periodic.

469. These results would be the same whatever might be the number of disturbing bodies; for m'' being a second planet disturbing the motion of m, it would add to R the term

$$\frac{m''}{\sqrt{(x''-x)^2+(y''-y)^2+(z''-z)^2}} - \frac{m''(xx''+yy''+zz'')}{r''^3}.$$

The variation of the co-ordinates of m' and m'' resulting from the reciprocal action of these two planets, would produce terms multiplied by mm'' and m''^2 in the variation of R; and by the preceding analysis it follows that all the terms in $d.\delta''R$ are periodic. $\delta''R$ relates to the variation of the elements of the orbit of m''.

The variations of the co-ordinates of m' arising from the action of m'' on m', will cause a variation in the part of R depending on the action of m' on m represented by

$$\frac{m'}{\sqrt{(x'-x)^2+(y'-y)^2+(z'-z)^2}} - \frac{m'(xx'+yy'+zz')}{r'^3}$$

There will arise terms in R, multiplied by $m'm''$, which will be functions of nt, $n't$, $n''t$, when substitution is made of the elliptical values of the co-ordinates; and as the mean motions cannot destroy each other, these terms will only produce periodic terms in $\text{d}R$. Should there be any terms independent of the mean motion nt in the development of R, they will vanish by taking the differential $\text{d}R$. And as terms depending on nt alone will have the form $m'm'' . \text{d}P$, P being a function of the elliptical co-ordinates of m . , there will arise in $\int \text{d} . R$ terms of the form $m'm''\int \text{d}P = m'm'' . P$, since $\text{d}P$ is an exact differential. These terms will then be of the second order after integration, and such terms are omitted in the value of this function.

The variation of the co-ordinates x, y, z, produced by the action of m'' on m only introduce into the preceding part of R terms multiplied by $m'm''$ and functions of the three angles nt, $n't$, $n''t$; and as these three mean motions cannot destroy each other, there can only be periodic terms in $\text{d}R$. The terms depending on nt alone, only produce periodic terms of the order $m'm''$ in $\text{d}R$.

The same may be proved with regard to the part of R depending on the action of m'' on m.

470. Hence whatever may be the number of disturbing bodies, when the approximation includes the squares and products of the masses, the variation of the elliptical elements of the disturbed and disturbing planets only produce periodic terms in $\text{d}R$.

471. Now the variation of

$$\zeta = -3\int\int andt . \text{d}R \text{ is}$$

$$\delta\zeta = -3an\int\int dt . \text{d}.\delta R + 3a^2\int\int (ndt . \text{d}R . \int \text{d}R).$$

It was proved in article 464 that $\text{d}\delta R = 0$ in considering only secular quantities of the order of the squares of the masses. It is easy to see from the form of the series R that $\text{d}R\int \text{d}R = 0$ with regard to these quantities, consequently the variation of the mean motion of a planet cannot contain any secular inequality of the first or second order with regard to the disturbing forces that can become sensible in the course of ages, whatever the number of planets may be that trouble its motion. And as $da = 2a^2\text{d}R$ becomes

$$\delta a = \{ 2a^2 \int dR + 8a^3 \int (dR \int dR) \},$$

by the substitution of $(a + \delta a^2)$ for a^2, δa cannot contain a secular inequality if $\delta \zeta$ does not contain one.

472. It therefore follows, that when periodic inequalities are omitted as well as the quantities of the third order with regard to the disturbing forces, the *mean motions* of the planets, and the *greater axes* of their orbits, are *invariable.*

The whole of this analysis is given in the Supplement to the third volume of the *Mècanique Cèleste;* but that part relating to the second powers of the disturbing forces is due to M. Poisson.

Differential Equations of the Secular Inequalities in the Eccentricities, Inclinations, Longitudes of the Perihelia and Nodes, which are the annual and sidereal variations of these four elements.

473. That part of the series R, in article 449, which is independent of periodic quantities, is found by making $i = 0$, for then

$$\text{Sin}\ 'i\ (n't - nt + \epsilon' - \epsilon) = 0,$$

$$\text{Cos}\ i\ (n't - nt + \epsilon' - \epsilon) = 1;$$

and if the differences of $A_0\ A_1$ with regard to a' be eliminated by their values in article 458, the series R will be reduced to

$$F = \frac{m'}{2} A_0 + \frac{m'}{4} \left\{ a \left(\frac{dA_0}{da} \right) + \tfrac{1}{2} a^2 \left(\frac{d^2 A_0}{da^2} \right) \right\} (e^2 + e'^2)$$

$$+ \frac{m'}{2} \left\{ A_1 - a \left(\frac{dA_1}{da} \right) - \tfrac{1}{2} a^2 \left(\frac{d^2 A_1}{da^2} \right) \right\} ee' \cos (\varpi' - \varpi)$$

$$- \frac{m'}{8} aa' B_1 \gamma^2.$$

But the formulæ in articles 456 and 457 give

$$a \left(\frac{dA_0}{da} \right) + \tfrac{1}{2} a^2 \left(\frac{d^2 A_0}{da^2} \right) = - \frac{3aa' . S'}{2 (a'^2 - a^2)^2},$$

$$A_1 - a \left(\frac{dA_1}{da} \right) - \tfrac{1}{2} a^2 \left(\frac{d^2 A_1}{da^2} \right) = \frac{3(aa'S + (a^2 + a'^2) S')}{(a'^2 - a^2)^2}$$

$$aa' B_1 = - \frac{3aa' . S'}{(a'^2 - a^2)^2};$$

consequently

$$F = \frac{m'}{2} A_0 - \frac{3m' \,.\, aa' \,.\, S'}{2 \,.\, 4 \,.\, (a'^2 - a^2)^2} \,.\, \{e^2 + e'^2 - (p' - p)^2 - (q' - q)^2\}$$
$$+ \frac{3m' (a'a \,.\, S + (a^2 + a'^2) S')}{2 (a'^2 - a^2)^2} \,.\, ee' \,.\, \cos (\varpi' - \varpi);$$

for by article 444

$$\gamma^2 = (p' - p)^2 + (q' - q)^2,$$

whence

$$\frac{dF}{d\varpi} = \frac{3m'(aa'S + (a^2 + a'^2) S')}{2 (a'^2 - a^2)^2} \,.\, ee' \,.\, \sin (\varpi' - \varpi)$$
$$\frac{dF}{de} = - \frac{3m'aa'S'}{4(a'^2 - a^2)^2} \,.\, e$$
$$+ \frac{3 \,.\, m' (aa'S + (a^2 + a'^2) S')}{2 (a'^2 - a^2)^2} \,.\, e' \,.\, \cos (\varpi' - \varpi)$$
$$\frac{dF}{dp} = - \frac{3m' \,.\, aa'S'}{4 (a'^2 - a^2)^2} \,.\, (p' - p)$$
$$\frac{dF}{dq} = - \frac{3m'aa' \,.\, S'}{4 (a'^2 - a^2)^2} \,.\, (q' - q).$$

474. When the squares of the eccentricities are omitted, the differential equations in article 441 become

$$\frac{de}{dt} = - \frac{an}{e} \frac{dF}{d\varpi}; \qquad \frac{d\varpi}{dt} = \frac{an}{e} \frac{dF}{de};$$
$$\frac{dp}{dt} = . \, an \,.\, \frac{dF}{dq}; \qquad \frac{dq}{dt} = - \, an \,.\, \frac{dF}{dp}.$$

If the differentials of F, according to the elements, be substituted in these, and if to abridge

$$- \frac{3m' \,.\, na^2a'S'}{4 (a'^2 - a^2)^2} = (0, 1);$$
$$- \frac{3m' \,.\, an \,.\, (aa'S + (a^2 + a'^2) S')}{2(a'^2 - a^2)^2} = \boxed{0 \,.\, 1};$$

they become

$$\frac{de}{dt} = \boxed{0 \,.\, 1} \, e' \sin (\varpi' - \varpi)$$
$$\frac{d\varpi}{dt} = (0 \,.\, 1) - \boxed{0 \,.\, 1} \, \frac{e'}{e} \cos (\varpi' - \varpi) \qquad (127)$$
$$\frac{dp}{dt} = - (0 \,.\, 1) (q - q')$$
$$\frac{dq}{dt} = (0 \,.\, 1) (p - p').$$

475. But $\tan\phi = \sqrt{p^2 + q^2}$ and $\tan\theta = \frac{p}{q}$, and when the squares of the inclinations are omitted $\cos\phi = 1$, hence

$$d\phi = dp \sin\theta + dq \cos\theta;\ d\theta = \frac{dp\cos\theta - dq\sin\theta}{\tan\phi};$$

and substituting the preceding values of dp, dq, the variations in the inclinations and longitude of the node are,

$$\frac{d\phi}{dt} = (0\,.\,1)\,.\,\tan\phi\,.\,\sin(\theta - \theta')$$

$$\frac{d\theta}{dt} = -(0.1) + (0.1)\,.\,\frac{\tan\phi'}{\tan\phi'}\,.\,\cos(\theta - \theta').$$

476. The preceding quantities are the secular variations in the orbit of m when troubled by m' alone, but all the bodies in the system act simultaneously on the planet m, and whatever effect is produced in the elements of the orbit of m by the disturbing planet m', similar effects will be occasioned by the disturbing bodies m', m'', &c. Hence, as the change produced by m' in the elements of the orbit of m are expressed by the second terms of the preceding equations, it is only necessary to add to them a similar quantity for each disturbing body, in order to have the whole action of the system on m.

The expressions $(0\,.\,1)$, $\boxed{0.1}$ have been employed to represent the coefficients relative to the action of m' on m; for quantities relative to m which has no accent, are represented by 0; and those relating to m' which has one accent, by 1; following the same notation, the coefficients relative to the action of m'' on m will be (0.2), $\boxed{0.2}$; those relating to m''' on m by $(0\,.\,3)$, $\boxed{0.3}$; and so on. Therefore the secular action of m'' in disturbing the elements of the orbit of m will be

$$\boxed{0.2}\; e'' \sin(\varpi'' - \varpi);\ (0\,.\,2) - \boxed{0.2}\;\frac{e''}{e}\cos(\varpi'' - \varpi)$$

$$(0\,.\,2)\tan\phi\sin(\theta - \theta'');\ -(0.2) + (0\,.2)\,\frac{\tan\phi''}{\tan\phi}\cos(\theta - \theta'').$$

477. Therefore the differential equations of the secular inequalities of the elements of the orbit of m, when troubled by the simultaneous action of all the bodies in the system, are

$$\frac{de}{dt} = \boxed{0.1}\, e' \sin(\varpi' - \varpi) + \boxed{0.2}\, e'' \sin(\varpi'' - \varpi)$$

$$+ \boxed{0.3}\, e''' \sin(\varpi''' - \varpi) + \&c.$$

$$\frac{d\varpi}{dt} = (0\,.\,1) + (0\,.\,2) + \&c. - \boxed{0.1}\, \frac{e'}{e} \cos(\varpi' - \varpi)$$

$$- \boxed{0.2}\, \frac{e''}{e} \cos(\varpi'' - \varpi) - \&c. \qquad (128)$$

$$\frac{d\phi}{dt} = (0\,.\,1) \tan\phi' \sin(\theta - \theta') + (0\,.\,2) \tan\phi'' \sin(\theta - \theta'') + \&c.$$

$$\frac{d\theta}{dt} = -\{(0\,.\,1) + (0\,.\,2) + \&c.\} + (0\,.\,1) \frac{\tan\phi'}{\tan\phi} \cos(\theta - \theta')$$

$$+ (0\,.\,2) \frac{\tan\phi''}{\tan\phi} \cos(\theta - \theta'') + \&c.$$

478. All the quantities in these equations are determined by observation for a given epoch assumed as the origin of the time, and when integrated, or (which is the same thing) multiplied by t, they give the annual variation in the elements of the orbit of a planet, on account of the immense periods of the secular inequalities, which admit of one year being regarded as an infinitely short time in which the elements e, ϖ, &c., may be supposed to be constant.

479. It is evident that the secular variations in the elements of the orbits of m', m'', m''', &c., will be obtained from the preceding equations, if every thing relating to m be changed into the corresponding quantities relative to m', and the contrary, and so for the other bodies. Thus the variation in the elements of m', m'', &c., from the action of all the bodies in the system, will be

$$\frac{de'}{dt} = \boxed{1.0}\,.\, e\,.\, \sin(\varpi - \varpi') + \boxed{1.2}\,.\, e''\,.\, \sin(\varpi'' - \varpi') + \&c.$$

$$\frac{de''}{dt} = \boxed{2.0}\,.\, e\,.\, \sin(\varpi - \varpi'') + \boxed{2.1}\,.\, e'\,.\, \sin(\varpi' - \varpi'') + \&c.$$

&c. &c.

$$\frac{d\varpi'}{dt} = (1\,.\,0) + (1\,.\,2) + \&c. - \boxed{1.0}\,.\, \frac{e}{e'}\,.\, \cos(\varpi - \varpi')$$

$$- \boxed{1.2}\,.\, \frac{e''}{e'} \cos(\varpi'' - \varpi')\,.\, - \&c.$$

$$\frac{d\varpi''}{dt} = (2.0) + (2.1) + \&c. - \boxed{2.0} \cdot \frac{e}{e''} \cdot \cos(\varpi - \varpi'')$$

$$- \boxed{2.1} \cdot \frac{e'}{e''} \cdot \cos(\varpi' - \varpi'') - \&c. \qquad (129)$$

&c. &c.

$$\frac{d\phi'}{dt} = (1.0) . \tan\phi . \sin(\theta'-\theta) + (1.2) . \tan\phi'' . \sin(\theta'-\theta'') + \&c.$$

$$\frac{d\phi''}{dt} = (2.0) . \tan\phi . \sin(\theta''-\theta) + (2.1) . \tan\phi' . \sin(\theta''-\theta') + \&c.$$

&c. &c.

$$\frac{d\theta'}{dt} = -\{(1.0) + (1.2) + \&c.\} + (1.0) . \frac{\tan\phi}{\tan\phi'} . \cos(\theta'-\theta)$$

$$+ (1.2) \frac{\tan\phi''}{\tan\phi'} . \cos(\theta' - \theta'') + \&c.$$

$$\frac{d\theta''}{dt} = -\{(2.0) + (2.1) + \&c.\} + (2.0) . \frac{\tan\phi}{\tan\phi''} . \cos(\theta''-\theta)$$

$$+ (2.1) . \frac{\tan\phi}{\tan\phi''} . \cos(\theta'' - \theta') + \&c.$$

&c. &c.

As these quantities do not contain the mean longitude, nor its sines or cosines, they depend on the configuration of the orbits only.

Approximate Values of the Secular Variations in these four Elements in Series, ascending according to the powers of the Time.

480. The annual variations in the elements are readily obtained from these formulæ; but as the secular inequalities vary so slowly that they may be assumed to vary as the time for a great many centuries without sensible error, series may be formed, whence very accurate values of the elements may be computed for at least a thousand years before and after the epoch. Let the eccentricity be taken as an example. With the given values of the masses and mean longitudes of the perihelia determined by observation, let a value of $\frac{de}{dt}$, the variation in the eccentricity, be computed from the preceding

equation for the epoch, say 1750, and another for 1950. If the latter be represented by $\left(\frac{de}{dt}\right)$, and the former by $\frac{d\bar{e}}{dt}$, then

$$\left(\frac{de}{dt}\right) - \frac{d\bar{e}}{dt} = 200 \,.\, \frac{d^2\bar{e}}{dt^2}; \text{ or, } \left(\frac{de}{dt}\right) = \frac{d\bar{e}}{dt} + 200 \,.\, \frac{d^2\bar{e}}{dt^2}$$

the quantities $\frac{d\bar{e}}{dt}$, $\frac{d^2\bar{e}}{dt^2}$, being relative to the year 1750. Hence, $\bar{e}$ being the eccentricity of any orbit at that epoch, the eccentricity e at any other assumed time t, may be found from

$$e = \bar{e} + \frac{d\bar{e}}{dt} \,.\, t + \tfrac{1}{2} \,.\, \frac{d^3\bar{e}}{dt^2}\, t^2 + \&c.$$

with sufficient accuracy for 1000 or 1200 years before and after 1750.

In the same manner all the other elements may be computed from

$$\begin{aligned}
\varpi &= \bar{\varpi} + \frac{d\bar{\varpi}}{dt} \,.\, t + \tfrac{1}{2} \,.\, \frac{d^2\bar{\varpi}}{dt^2} \,.\, t^2 + \&c. \\
\phi &= \bar{\phi} + \frac{d\bar{\phi}}{dt} \,.\, t + \tfrac{1}{2} \,.\, \frac{d^2\bar{\phi}}{dt^2} \,.\, t^2 + \&c. \\
\theta &= \bar{\theta} + \frac{d\bar{\theta}}{dt} \,.\, t + \tfrac{1}{2} \,.\, \frac{d^2\bar{\theta}}{dt^2} \,.\, t^2 + \&c. \qquad (130) \\
\gamma &= \bar{\gamma} + \frac{d\bar{\gamma}}{dt} \,.\, t + \tfrac{1}{2} \,.\, \frac{d^2\bar{\gamma}}{dt^2} \,.\, t^2 + \&c. \\
\Pi &= \bar{\Pi} + \frac{d\bar{\Pi}}{dt} \,.\, t + \tfrac{1}{2} \,.\, \frac{d^2\bar{\Pi}}{dt^2} \,.\, t^2 + \&c.
\end{aligned}$$

For as $\bar{\phi}$ and $\bar{\theta}$ are given by observation, $\bar{\gamma}$ and $\bar{\Pi}$, which are functions of them, may be found. All the quantities in these equations are relative to the epoch.

These expressions are sufficient for astronomical purposes; but as very important results may be deduced from the finite values of the secular variations, the integrals of the preceding differential equations must be determined for any given time.

Finite Values of the Differential Equations relative to the eccentricities and longitudes of the Perihelia.

481. Direct integration is impossible in the present state of analysis, but the differential equations in question may be changed into

linear equations capable of being integrated by the following method of La Grange. Let

$$h = e \sin \varpi \qquad l = e \cos \varpi$$
$$h' = e' \sin \varpi' \qquad l' = e' \cos \varpi',$$
&c. &c.

then
$$\frac{dh}{dt} = \frac{de}{dt} \sin \varpi + \frac{d\varpi}{dt} \cdot e \cos \varpi,$$

$$\frac{dl}{dt} = \frac{de}{dt} \cos \varpi - \frac{d\varpi}{dt} \cdot e \sin \varpi;$$

and substituting the differentials in article 477, the result will be

$$\frac{dh}{dt} = \{(0 \cdot 1) + (0 \cdot 2) + \&c.\} l - \boxed{0.1}\, l' - \boxed{0.2}\, l''$$
$$- \boxed{0.3}\, l''' - \&c.$$
$$\frac{dl}{dt} = -\{(0 \cdot 1) + (0 \cdot 2) + \&c.\} h - \boxed{0.1}\, h' + \boxed{0.2}\, h''$$
$$+ \boxed{0.3}\, h''' + \&c. \qquad (131)$$

likewise

$$\frac{dh'}{dt} = \{(1 \cdot 0) + (1 \cdot 2) + \&c.\} l' - \boxed{1.0}\, l - \boxed{1.2}\, l''$$
$$- \boxed{1.3}\, l''' - \&c.$$
$$\frac{dl'}{dt} = -\{(1 \cdot 0) + (1 \cdot 2) + \&c.\} h' + \boxed{1.0}\, h + \boxed{1.2}\, h''$$
$$+ \boxed{1.3}\, h''' + \&c.$$
&c. &c.

It is obvious that there must be twice as many such equations, and as many terms in each, as there are bodies in the system.

482. The integrals of these equations will be obtained by making

$$h = N \sin (gt + \mathfrak{C}) \qquad l = N \cos (gt + \mathfrak{C})$$
$$h' = N' \sin (gt + \mathfrak{C}) \qquad l' = N' \cos (gt + \mathfrak{C})$$
&c. &c.

It is easy to see why these quantities take this form, for if

$$h' = 0,\ h'' = 0,\ \&c.,\ l = 0;\ l' = 0,\ \&c.,\ \text{then}$$

$$\frac{dh}{dt} = (0 \cdot 1)\, l; \quad \frac{dl}{dt} = -(0 \cdot 1)\, h.$$

Let $$\frac{dh}{dt} = gl \qquad \frac{dl}{dt} = -gh,$$

but $$\frac{d^2h}{dt^2} = g\frac{dl}{dt}, \text{ therefore}$$

$$\frac{d^2h}{dt^2} + g^2h = 0.$$

And by article 214 $h = N \sin(gt + \varepsilon)$, N and ε being arbitrary constant quantities. In the same manner $l = N \cos(gt + \varepsilon)$.

483. If the preceding values of h, h', h'', &c., l, l', l'', &c., and their differentials be substituted in equations (131), the sines and cosines vanish, and there will result a number of equations,

$$Ng = \{(0.1)+(0.2)+(0.3)+\&c.\}N - \boxed{0.1}\,N' - \boxed{0.2}\,N'' - \&c.$$

$$N'g = \{(1.0)+(1.2)+(1.3)+\&c.\}N' - \boxed{1.0}\,N - \boxed{1.2}\,N'' - \&c. \quad (132)$$

&c. &c.

equal to the number of quantities N, N', N'', &c., consequently equal to the number of bodies in the system; hence, if N', N'', N''', &c., be eliminated, N will vanish, and will therefore remain indeterminate, and there will result an equation in g only, the degree of which will be equal to the number of bodies m, m', m'', &c. The roots of this equation may be represented by g, g_1, g_2, &c., which are the mean secular motions of the perihelia of the orbits of m, m', m'', &c., and are functions of the known quantities (0.1), $\boxed{0.1}$, (1.0), $\boxed{1.0}$, &c., only. When successively substituted in equations (132), these equations will only contain the indeterminate quantities N, N', N'', &c.; but it is clear, that for each root of g, N, N', N'', &c., will have different values. Therefore let N, N', N'', &c., be their values corresponding to the root g; N_1, N_1', N_1'', &c., those corresponding to the root g; N_2, N_2', N_2'', &c., those arising from the substitution of g_2, &c. &c.; and as the complete integral of a differential linear equation is the sum of the particular equations, the integrals of (131) are

$$h = N\sin(gt+\varepsilon) + N_1\sin(g_1t+\varepsilon_1) + N_2\sin(g_2t+\varepsilon_2) + \&c.$$

$$h' = N'\sin(gt+\varepsilon) + N_1'\sin(g_1t+\varepsilon_1) + N_2'\sin(g_2t+\varepsilon_2) + \&c. \quad (133)$$

&c. &c.

$$l = N\cos(gt+\varepsilon) + N_1\cos(g_1t+\varepsilon_1) + N_2\cos(g_2t+\varepsilon_2) + \&c.,$$

$$l' = N'\cos(gt + \mathfrak{C}) + N_1'\cos(g_1t + \mathfrak{C}_1) + N_2'\cos(g_2t + \mathfrak{C}_2) + \&c.$$
$$\&c. \qquad\qquad \&c.$$

for each term contains two arbitrary quantities $N, \mathfrak{C}$; $N_1, \mathfrak{C}_1$, &c.

484. Since each term of the equations (132) has one of the quantities N, N', &c., for coefficient, these equations will only give values of the ratios

$$\frac{N'}{N};\ \frac{N''}{N'}\ \&c.,$$

so that for each of the roots g, g_1, g_2, &c., one of the quantities N, N_1, N_2, &c., will remain indeterminate.

To show how these are determined, it must be observed that in the expression

$$\boxed{0.1} = -\frac{3m' . an\,(aa'S + (a^2 + a'^2)\,S')}{2\,(a'^2 - a^2)^2}$$

of article 474, S and S' are the coefficients of the first and second terms of the development of

$$(a^2 - 2aa'\cos\beta + a'^2\}^{\frac{1}{2}},$$

which remains the same when a' is put for a; and the contrary, that is to say, whether the action of m' on m be considered, or that of m on m'. Hence if m, n', and a', be put for m', n, and a,

$$\boxed{1.0} = -\frac{3m . a'n'\,(aa'S + (a'^2 + a^2)\,S')}{2\,(a'^2 - a^2)^2}$$

consequently

$$\boxed{0.1}\ m . n'a' = \boxed{1.0}\ . m' . na.$$

It is also evident that

$$(0 . 1)\ m . n'a' = (1 . 0)\ m' . na.$$

But if the mass of the planet be omitted in comparison of that of the sun considered as the unit,

$$n^2 = \frac{1}{a^3};\ n'^2 = \frac{1}{a'^3},\ \&c.;$$

therefore

$$\boxed{0.1}\ m\sqrt{a} - \boxed{1.0}\ m'\sqrt{a'} = 0,$$

$$\boxed{0.2}\ m\sqrt{a} - \boxed{2.0}\ m''\sqrt{a''} = 0,$$
$$\&c. \qquad\qquad \&c.$$

$$(0 . 1)\ m\sqrt{a} - (1 . 0)\ m'\sqrt{a'} = 0,$$

$$(0 . 2)\ m\sqrt{a} - (2 . 0)\ m''\sqrt{a''} = 0,$$
$$\&c. \qquad\qquad \&c.$$

485. Now let those of equations (131) that give

$$\frac{dh}{dt},\ \frac{dh'}{dt},\ \&c.,$$

be respectively multiplied by

$$Nm\ \sqrt{a},\ N'm'\ \sqrt{a'},\ N''m''\ \sqrt{a''},\ \&c.;$$

then, in consequence of equations (132), and the preceding relations, it will be found that

$$N\frac{dh}{dt}\ m\ \sqrt{a} + N'\frac{dh'}{dt} m'\sqrt{a'} + N''\frac{dh''}{dt}\ m''\ \sqrt{a''} + \&c.$$

$$= g\ \{N\ lm\sqrt{a} + N'l'm'\sqrt{a'} + N''l''m''\ \sqrt{a''} + \&c.\};$$

if the preceding values of h, h', h'', &c., l, l', &c., be put in this, a comparison of the coefficients of like cosines gives

$$0 = NN_1 m\ \sqrt{a} + N'N'_1 m'\ \sqrt{a'} + N''N''_1 m''\ \sqrt{a''} + \&c.$$

$$0 = NN_2 m\ \sqrt{a} + N'N'_2 m'\ \sqrt{a'} + N''N''_2 m'\ \sqrt{a''} + \&c.$$

Again, if the values of h, h', h'', &c., in equations (133) be respectively multiplied by

$$Nm\ \sqrt{a},\ N'm'\ \sqrt{a'},\ \&c.$$

they give

$$Nmh\ \sqrt{a} + N'm'h'\ \sqrt{a'} + N''m''h''\ \sqrt{a''} + \&c. = \quad (134)$$

$$\{\ N^2 m\ \sqrt{a} + N'^2 m'\ \sqrt{a'} + N''^2 m''\ \sqrt{a''} + \&c.\}\sin(gt+\beta),$$

in consequence of the preceding relations.

By the same analysis the values of l, l', l'', &c., give

$$Nml\ \sqrt{a} + N'm'l'\ \sqrt{a'} + N''m''l''\ \sqrt{a''} + \&c. =$$

$$\{\ N^2 m\ \sqrt{a} + N'^2 m'\ \sqrt{a'} + N''^2 m''\ \sqrt{a''} + \&c.\}\cos(gt+\beta).$$

The eccentricities of the orbits of the planets, and the longitudes of their perihelia, are known by observation at the epoch, and if these be represented by $\bar{e}$, $\bar{e}'$, &c. $\bar{\varpi}$. $\bar{\varpi}'$, &c. by article 481,

$$h = \bar{e}\sin\bar{\varpi},\ h' = \bar{e}'\sin\bar{\varpi}',\ \&c.,$$

$$l = \bar{e}\cos\bar{\varpi},\ l' = \bar{e}'\cos\bar{\varpi}',\ \&c.;$$

therefore h, h', &c., l, l', &c., are given at that period. And if it be taken as the origin of the time $t = 0$, and the preceding equations give

$$\operatorname{Tan}\beta = \frac{N.\bar{e}\sin\bar{\varpi}.m\ \sqrt{a} + N'.\bar{e}'\sin\bar{\varpi}'.m'\ \sqrt{a'} + \&c.}{N.\bar{e}\cos\bar{\varpi}.m\sqrt{a} + N'.\bar{e}'\cos\bar{\varpi}'.m'\sqrt{a'} + \&c.}.$$

But, for the root g, the equations (132) give

$$N' = CN, \ N'' = C'N, \ N''' = C''N, \text{ \&c.},$$

C, C', C'' being constant and given quantities; therefore

$$\tan \zeta = \frac{\bar{e} \sin \varpi \,.\, m\sqrt{a} + C.\bar{e}' \sin \varpi'.m'\sqrt{a'} + \text{\&c.}}{\bar{e} \cos \varpi.m\sqrt{a} + C.\bar{e}' \cos \varpi'.m'\sqrt{a'} + \text{\&c.}}.$$

If these values of N', N'', &c., be eliminated from equation (134), it gives

$$N = \frac{\bar{e} \sin \varpi \, m \sqrt{a} + C\bar{e}' \sin \varpi' m' \sqrt{a'} + \text{\&c.}}{\{m\sqrt{a} + C^2 m' \sqrt{a'} + C'^2 m'' \sqrt{a''} + \text{\&c.}\} \sin \zeta}.$$

Thus $\tan \zeta$ and N are determined, and the remaining coefficients N', N'', &c., may be computed from equations (132), for the root g.

In this manner the indeterminate quantities belonging to the other roots g_1, g_2, &c., may be found. Thus the equations (133) are completely determined, whence the eccentricities of the orbits and the longitudes of their perihelia may be found for any instant $\mp t$, before or after the epoch.

486. The roots g, g_1, g_2, &c., express the mean secular motions of the perihelia, in the same manner that n represents the mean motion of a planet.

For example, the periodic time of the earth is about 365¼ days; hence $n = \frac{360°}{365\frac{1}{4}}$, which is the mean motion of the earth for a day, and nt is its mean motion for any time t. The perihelion of the terrestrial orbit moves through 360° in 113270 years nearly; hence,

for the earth, $$g = \frac{360°}{113270} = 19' \, 4''.7$$

in a century; and gt is the mean motion for any time t: so that $nt + \epsilon$ being the mean longitude of a planet, $gt + \zeta$ is the mean longitude of its perihelion at any given time.

487. The equations (133), as well as observation, concur in proving that the perihelia have a motion in space, and that the eccentricities vary slowly. As, however, that variation might in process of time alter the nature of the orbits so much as to destroy the stability of the system, it is of the greatest importance to inquire whether these variations are unlimited, or if limited, what their extent is.

Stability of the Solar System with regard to the Form of the Orbits.

488. Because

$$h = e \sin \varpi, \quad l = e \cos \varpi, \quad e^2 = h^2 + l^2;$$

and in consequence of the values of h and l in equations (133), the square of the eccentricity of the orbit of m becomes

$$e^2 = N^2 + N_1^2 + N_2^2 + \&c. + 2NN_, \cos\{(g_, - g)t + \mathfrak{C}_, - \mathfrak{C}\}$$
$$+ 2NN_2 \cos\{(g_2 - g)t + \mathfrak{C}_2 - \mathfrak{C}\} + \&c. \qquad (135)$$

When the roots g, $g_,$, &c., are all real and unequal, the cosines in this expression will oscillate between fixed limits, and e^2 will always be less than

$$(N + N_, + N_2 + \&c.)^2 = N^2 + N_,^2 + \&c. + 2NN_1 + 2NN_2 + \&c.$$

taken with the same sign, for it could only obtain that maximum if

$$(g_1 - g)t + \mathfrak{C}_1 - \mathfrak{C} = 0, \; (g_2 - g)t + \mathfrak{C}_2 - \mathfrak{C} = 0, \; \&c.,$$

which could never happen unless the time were to vanish; that is, unless $\quad g_1 - g = 0, \;\; g_2 - g = 0, \;\; \&c.;$

thus, if g, g_1, g_2, &c., be real and unequal, the value of e^2 will be limited.

489. If however any of these roots be imaginary or equal, they will introduce circular arcs or exponentials into the values of h, h', &c., l, l', &c.; and as these quantities would then increase indefinitely with the time, the eccentricities would no longer be confined to fixed limits, but would increase till the orbits of the planets, which are now nearly circular, become very eccentric.

The stability of the system therefore depends on the nature of the roots g, g_1, g_2, &c.: however it is easy to prove that they will all be real and unequal, if all the bodies m, m', m'', &c., in the system revolve in the same direction.

490. For that purpose let the equations

$$\frac{de}{dt} = \boxed{0.1}\, e' \sin(\varpi' - \varpi) + \boxed{0.2}\, e'' \sin(\varpi'' - \varpi') + \&c.$$

$$\frac{de'}{dt} = \boxed{1.0}\, e \sin(\varpi - \varpi') + \boxed{1.2}\, e'' \sin(\varpi' - \varpi'') + \&c.$$

&c. &c.

*

be respectively multiplied by

$$me\sqrt{a},\ m'e'\sqrt{a'},\ m''e''\sqrt{a''},\ \&c.,$$

and added; then in consequence of the relations in article 484, and because

$$\sin(\varpi-\varpi') = -\sin(\varpi'-\varpi)$$
$$\sin(\varpi-\varpi'') = -\sin(\varpi''-\varpi),\ \&c.\ \&c.,$$

the sum will be

$$0 = ede.m\sqrt{a} + e'de'.m'\sqrt{a'} + e''de''.m''\sqrt{a''} + \&c.;$$

and as the greater axes of the orbits are constant, its integral is

$$e^2m\sqrt{a} + e'^2m'\sqrt{a'} + e''^2m''\sqrt{a''} + \&c. = C. \qquad (136)$$

491. The radicals $\sqrt{a}$, $\sqrt{a'}$, &c., must all have the same sign if the planets revolve in the same direction; since by Kepler's law they depend on the periodic times; and in analysis motions in one direction have a different sign from those in a contrary direction: but as all the planets and satellites revolve from west to east, the radicals, and consequently all the terms of the preceding equations must have positive signs; therefore each term is less than the constant quantity C.

But observation shows that the orbits of the planets and satellites are nearly circular, hence each of the quantities

$$e^2m\sqrt{a},\ e'^2m'\sqrt{a'},\ \&c.$$

is very small; and C being a very small constant quantity given by observation, the first number of equation (136) is very small.

As C never could have changed since the system was constituted as it now is, so it never can change while the system remains the same; therefore equation (136) cannot contain any quantity that increases indefinitely with the time; so that none of the roots g, g_1, g_2, &c., are either equal or imaginary.

492. Since the greater axes and masses are invariable, and the eccentricities are perpetually changing, they have the singular property of compensating each other's variation, so that the sum of their squares, respectively multiplied by the coefficients

$$m\sqrt{a},\ m'\sqrt{a'},\ \&c.,$$

remains constant and very small.

493. To remove all doubts on a point so important, suppose some of the roots, g, g_1, g_2, &c., to be imaginary, then some of the cosines or sines will be changed into exponentials; and, by article 215, the general value of h in (133) would contain the term $Cc^{\alpha t}$, c being the number whose hyperbolic logarithm is unity. If $Dc^{\alpha t}$, $C'c^{\alpha t}$, $D'e^{\alpha t}$, &c., be the corresponding terms introduced by these imaginary roots in h, h', l', &c., then e^2 would contain a term $(C^2 + D^2)\,c^{2\alpha t}$, e'^2 would contain $(C'^2 + D'^2)\,c^{2\alpha t}$, and so on; hence the first number of equation (136) would contain

$$C^{2\alpha t}\{m\sqrt{a}\,(C^2 + D^2) + m'\sqrt{a'}\,(C'^2 + D'^2) + \&c.\},$$

a quantity that increases indefinitely with the time.

If $C^{\alpha t}$ be the greatest exponential that h, l, h', l', &c., contain, $C^{2\alpha t}$ will be the greatest in the first member of equation (136); therefore the preceding term cannot be destroyed by any other term in that equation. In order, therefore, that its first member may be reduced to a constant quantity, the coefficient of $C^{2\alpha t}$ must itself be zero; hence

$$m\sqrt{a}\,(C^2 + D^2) + m'\sqrt{a'}\,(C'^2 + D'^2) + \&c. = 0.$$

But if the radicals $\sqrt{a}$, $\sqrt{a'}$, &c., have the same sign, that is, if all the bodies m, m', &c., move in the same direction, this coefficient can only be zero when each of the quantities C, D, C', D', &c., is zero separately; thus, h, l, h', l', &c., do not contain exponentials, and therefore the roots of g, $g_{\prime}$, &c., are all real. If the roots g and $g_{\prime}$ be equal, then the preceding integral becomes

$$h = (b + b')\,C^{\alpha t} = (b + b')\,(1 + \frac{\alpha t}{2} + \frac{\alpha^2 t^2}{1.2} + \&c.$$

Thus the general value of h will contain a finite number of terms of the form Ct^r, which increases indefinitely with the time; the same roots would introduce the terms Dt^r, $C't^r$, $D't^r$, &c., in the general value of l, h', l', &c.; therefore the first member of equation (136) would contain the term

$$t^2\,\{m\sqrt{a}\,(C^2 + D^2) + m'\sqrt{a'}\,(C'^2 + D'^2) + \&c.\};$$

and if t^r be the highest power of t in h, l, h', l', &c., t^{2r} will be the highest power of t in equation (136); consequently its first member can only be constant when

$$m\sqrt{a}\,(C^2 + D^2) + m'\sqrt{a'}\,(C'^2 + D'^2) + \&c. = 0,$$

which cannot happen when all the planets revolve in the same direction, unless

$$C = 0, D = 0, C' = 0, D' = 0, \text{\&c.}$$

Thus, h, l, h', l', &c., neither contain exponentials nor circular arcs, when the bodies of the solar system revolve in the same direction, and as they really do so, the roots g, g_1, g_2, &c., are all real and unequal.

494. Because the equation (135) does not contain any quantity that increases with the time, on account of the roots $g, g_{\prime}$, &c., being real and unequal, and that the eccentricities themselves and their variations are extremely small, the eccentricities increase and decrease with the cosines, between fixed but very narrow limits, in immense periods: for, considering only the mutual disturbances of Jupiter and Saturn, the eccentricities of their orbits would take no less than 70,414 years to accomplish their changes; but if more than two planets be taken, and compound periods established, they would evidently extend to millions of years.

495. The positions of the perihelia now remain to be considered.

$$e \sin \varpi = h, \; e \cos \varpi = l \text{ give } \tan \varpi = \frac{h}{l},$$

and substituting the values of h and l in article 483,

$$\tan \varpi = \frac{N \sin (gt + \mathfrak{C}) + N_{\prime} \sin (g_{\prime}t + \mathfrak{C}_{\prime}) + \text{\&c.}}{N \cos (gt + \mathfrak{C}) + N_{\prime} \cos (g_{\prime}t + \mathfrak{C}_{\prime}) + \text{\&c.}};$$

or, if $gt + \mathfrak{C}$ be subtracted from ϖ,

$$\tan (\varpi - gt - \mathfrak{C}) = \frac{\tan \varpi - \tan (gt + \mathfrak{C})}{1 + \tan \varpi \tan (gt + \mathfrak{C})};$$

and when substitution is made for $\tan \varpi$,

$$\tan (\varpi - gt - \mathfrak{C})$$

$$= \frac{N' \sin \{(g_{\prime} - g)t + \mathfrak{C}_{\prime} - \mathfrak{C}\} + N_2 \sin \{g_2 - g)t + \mathfrak{C}_2 - \mathfrak{C}\}}{N + N_{\prime} \cos \{(g_{\prime} - g)t + \mathfrak{C}_{\prime} - \mathfrak{C}\} + N_2 \cos \{(g_2 - g)t + \mathfrak{C}_2 - \mathfrak{C}) + \text{\&c.}\}}.$$

This tangent never can be infinite, if the sum $N + N_{\prime} + N_2 + $ &c., of the coefficients in the denominator be less than N with a positive sign; for in this case the denominator never can be zero; so that the angle $\varpi - gt - \mathfrak{C}$ never can attain to a quadrant, but will oscillate between $+ 90°$ and $- 90°$; hence the true motion of the perihelion is $gt + \mathfrak{C}$.

From this equation it appears that the motions of the perihelia are not uniform, and that they may experience variations in the course of ages, to which no limits can be assigned, though observation shows that the variations are very slow.

496. Because the equations which give the secular variations in the eccentricities and longitudes of the perihelia do not contain the mean longitudes nor the inclinations of the orbits, they are independent of the configuration of the planets, and would be the same if all the bodies revolved in one plane, at least when the approximation does not extend to the higher powers of the eccentricities, inclinations, or masses. These secular inequalities depend on the angular distances of the perihelia of all the planets taken two and two, that is, on the configuration of the orbits.

497. It may be concluded from the preceding analysis, that when periodic inequalities are omitted, the mean motions of the planets are uniform; and that the system is stable with regard to the species of the orbits, which, retaining the greater axis invariable, deviate but little from the circular form; the eccentricities being subject to the condition expressed by equations (136)—that the sum of their squares, multiplied by the masses of the bodies, and the square roots of the greater axes of their orbits is invariably the same. The perihelia alone are subject to unlimited variations.

Secular Variations in the Inclinations of the Orbits and Longitudes of their Nodes.

498. In order to determine the secular inequalities in the inclinations of the orbits and longitudes of the nodes, let the equations in article 474 be resumed

$$\frac{dp}{dt} = (0.1)\,(q' - q)$$

and

$$\frac{dq}{dt} = -\,(0.1)\,(p' - p),$$

which express the variations in the position of the orbit of m, when troubled by m' alone. But as all the bodies in the system act simultaneously on m, each of them will produce a variation in the inclination of its orbit, and in the longitude of its nodes, similar to those caused by the action of m'; hence

$$\frac{dp}{dt} = (0.1)(q' - q) + (0.2)(q'' - q) + \&c.$$

$$\frac{dq}{dt} = -(0.1)(p' - p) - (0.2)(p'' - p) - \&c.$$

will express the whole action of the system on the position of the orbit of m. Similar equations must exist for every body in the system: there will consequently be the following series of equations,

$$\frac{dp}{dt} = -\{(0.1)+(0.2)+\&c.\}q+(0.1)q'+(0.2)q''+\&c.$$

$$\frac{dq}{dt} = \{(0.1)+(0.2)+\&c.\}p-(0.1)p'-(0.2)p''-\&c. \quad (137)$$

$$\frac{dp'}{dt} = -\{(1.0)+(1.2)+\&c.\}q'+(1.0)q+(1.2)q''+\&c.$$

$$\frac{dq'}{dt} = \{(1.0)+(1.2)+\&c.\}p'-(1.0)p-(1.2)p''-\&c.$$

&c. &c.

These equations are perfectly similar to those in article 481, and may be integrated on the same principle; whence

$$\begin{aligned} p &= N \sin(gt + \mathfrak{C}) + N_{\prime} \sin(g_{\prime}t + \mathfrak{C}_{\prime}) + \&c. \\ q &= N \cos(gt + \mathfrak{C}) + N_{\prime} \cos(g_{\prime}t + \mathfrak{C}_{\prime}) + \&c. \qquad (138) \\ p' &= N' \sin(gt + \mathfrak{C}) + N_{\prime}' \sin(g_{\prime}t + \mathfrak{C}_{\prime}) + \&c. \\ q' &= N' \cos(gt + \mathfrak{C}) + N_{\prime}' \cos(g_{\prime}t + \mathfrak{C}_{\prime}) + \&c. \end{aligned}$$

Stability of the Solar System with regard to the Inclination of the Orbits.

499. The equation in g resulting from these, has g, g_1, g_2, &c. for its roots, and the constant quantities $N, N_{\prime}$, &c. and $\mathfrak{C}$, $\mathfrak{C}_{\prime}$, &c. are determined in a similar manner to what was employed for the eccentricities. For since $\overline{\phi}$, $\overline{\theta}$, &c. are the values of ϕ, θ, &c. when $t = 0$,

$$p = \tan\overline{\phi}\cos\overline{\theta} \quad q = \tan\overline{\phi}\sin\overline{\theta},$$

$$p' = \tan\overline{\phi}'\cos\overline{\theta}' \quad q' = \tan\overline{\phi}'\sin\overline{\theta}',$$

&c. &c.

hence, if all the inclinations of the orbits of the planets, and the longitudes of their nodes be known by observation at any given epoch, when $t = 0$, there will be a sufficient number of equations to determine all the quantities N, $N_{,}$, &c. and $\mathfrak{C}$, $\mathfrak{C}_{,}$, &c.

500. Also the roots g, $g_{,}$, &c. are real and unequal, for if the equations (137) be respectively multiplied by

$$m\sqrt{a}\,.\,p\,;\; m\sqrt{a}\,.\,q\,;\quad m'\sqrt{a'}\,.\,p'\,;\; m'\sqrt{a'}\,.\,q, \text{ \&c.}$$

and added, the integral of their sum will be

$$(p^2 + q^2)\,m\sqrt{a} + (p'^2 + q'^2)\,m'\sqrt{a'} + \text{\&c.} = C \qquad (139)$$

in consequence of the relations

$$(0.1)m\sqrt{a} = (1.0)m'\sqrt{a'}$$
$$(0.2)m\sqrt{a} = (2.0)m''\sqrt{a''},$$
$$\text{\&c.} \qquad\qquad \text{\&c.}$$

Whence we may be assured by the same reasoning employed with regard to the eccentricities, that this equation neither contains arcs of circles nor exponentials, when the bodies all revolve in the same direction, so that all the roots are real and unequal.

501. Now $\tan\phi = \sqrt{p^2 + q^2}$,

and if the values of p and q be substituted

$$\tan\phi = \sqrt{p^2+q^2} =$$
$$\sqrt{\{N^2 + N_{,}^2 + \text{\&c.} + 2NN_{,}\cos\{(g_{,} - g)t + \mathfrak{C}_{,} - \mathfrak{C}\} + 2NN_2}$$
$$\cos\{(g_2 - g)t + \mathfrak{C}_2 - \mathfrak{C}\} + \text{\&c.}\}.$$

The expression $\sqrt{p^2+q^2}$ is less than $N + N_1 + N_2 +$ &c., on account of these coefficients being multiplied by cosines which diminish their values. The maximum of $\tan\phi$ would be $N + N_{,} +$ &c., which it never can attain, since the differences of the roots $g_1 - g$, $g_2 - g$ are never zero; and as the inclinations of the orbits of the planets on the plane of the ecliptic are very small, the coefficients N, N_1, &c., which depend on the inclinations, are very small also, and will always remain so. And the inclinations of the orbits will oscillate between very narrow limits in periods depending on the roots g, g_1, &c.

502. The plane of the ecliptic in which the earth moves, changes its position in space from the action of the planets, each producing a retrograde motion in the intersection of the plane of the ecliptic, and that of its own orbit; whence it appears, that if EN be the orbit of

fig. 85.

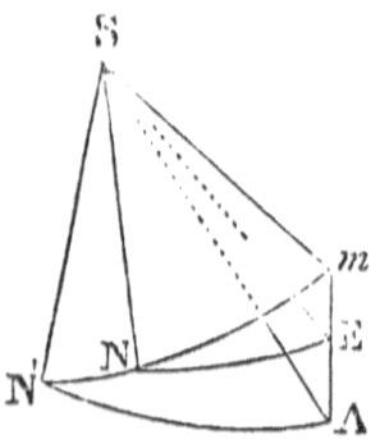

the earth at a given epoch, AN′ will be its position at a subsequent period, and so on. The secular inequality in the position of the terrestrial orbit changes the obliquity of the ecliptic; but as it is determined from equations (138) it oscillates between narrow limits, never exceeding 3°, therefore the equator never has coincided, and never will coincide with the ecliptic, supposing the system constituted as it is at present, so that there never was, and there never will be eternal spring.

503. Since $p^2 + q^2 = \tan^2\phi$, $p'^2 + q'^2 = \tan^2\phi'$, &c. equation (139) becomes

$$m\sqrt{a}\tan^2\phi + m'\sqrt{a'}\tan^2\phi' + \&c. = C. \qquad (140)$$

Whence it may be concluded that the sum of the masses of all the bodies in the system multiplied by the square roots of half the greater axes of their orbits, and by the squares of the tangents of their inclinations on a fixed plane, will always be the same. If this sum be very small at any one period, and if all the radicals have the same sign, that is, if all the bodies revolve in the same direction, it will always remain so; and as in nature, the inclinations of all the orbits on the plane of the ecliptic are very small, and the bodies revolve in the same direction, the variations of the inclinations compensate each other, so that this expression will remain for ever constant, and very small.

504. Other two integrals may be obtained from the equations (137). For if the first be multiplied by $m\sqrt{a}$, the third by $m'\sqrt{a'}$, the fifth by $m''\sqrt{a''}$, &c. &c. their sum will be

$$m\sqrt{a}\,\frac{dp}{dt} + m'\sqrt{a'}\,\frac{dp'}{dt} + m''\sqrt{a''}\,\frac{dp''}{dt} + \&c. = 0,$$

in consequence of the relations in article 484, the integral of which is

$$m\sqrt{a}\,.\,p + m'\sqrt{a'}\,.\,p' + m''\sqrt{a''}\,p'' + \&c. = \text{constant}.$$

In a similar manner the differential equations in q, q', give

$$m\sqrt{a}\,.\,q + m'\sqrt{a'}\,q' + m''\sqrt{a''}\,q'' + \&c. = \text{constant}.$$

505. With regard to the nodes $\tan\theta = \frac{p}{q}$, and substituting for p and q,

$$\tan\theta = \frac{N\sin(gt+\epsilon)+N_{\prime}\sin(g_1t+\epsilon_{\prime})+\&c.}{N\cos(gt+\epsilon)+N_{\prime}\cos(g_1t+\epsilon_1)+\&c.};$$

or subtracting $gt+\epsilon$ from θ,

$$\tan(\theta-gt-\epsilon)=\frac{N_1\sin\{(g_{\prime}-g)t+\epsilon_{\prime}-\epsilon\}+N_2\sin\{(g_2-g)t+\epsilon_2-\epsilon\}+\&c.}{N+N_1\cos\{(g_1-g)t+\epsilon_1-\epsilon\}+N_2\cos\{(g_2-g)t+\epsilon_2-\epsilon\}+\&c.}$$

If the sum of the coefficients $N+N_1+N_2+$ &c. of the cosines in the denominator taken positively be less than N, $\tan(\theta-gt-\epsilon)$ never can be infinite; hence the angle $\theta-gt-\epsilon$ will oscillate between $+90°$ and $-90°$, so that $gt+\epsilon$ is the true motion of the nodes of the orbit of m, and $g=\dfrac{360°}{\text{period of ☊ of } m.}$ As in general the periods of the motions of the nodes are great, the inequalities increase very slowly. From these equations it may be seen, that the motion of the nodes is indefinite and variable.

The method of computing the constant quantities will be given in the theory of Jupiter, whence the laws, periods, and limits of the secular variations in the elements of his orbit, will be determined.

506. The equations which give p, q, p' q' may be expressed by a diagram. Let An be the orbit of the planet m at any assigned time, as the beginning of January, 1750, which is the epoch of many of the French tables. After a certain time, the action of the disturbing body m' alone on the planet m, changes the inclination of its orbit, and brings it to the position Bn. But m'' acting simultaneously with m' brings the orbit into the position Cn : m'' acting along with the preceding bodies changes it to Dn'', and so on. It is evident that the last orbit will be that in which m moves. So the whole inclination of the orbit of m on the plane An, after a certain time, will be the sum of the finite and simultaneous changes. Hence if N be the inclination of the circle Bn on the fixed plane An, and $\gamma Sn = gt+\epsilon$ the longitude of its ascending node; N′ the inclination of the circle Cn' on Bn, and $\gamma Sn' = g't+\epsilon'$ the longitude of the node n'; N″ the inclination of the circle Dn'' on Cn', and $\gamma Sn'' = g_2t+\epsilon_2$ the longitude of the node n''; and so on for each disturbing body, the last circle will be the orbit of m.

fig. 86.

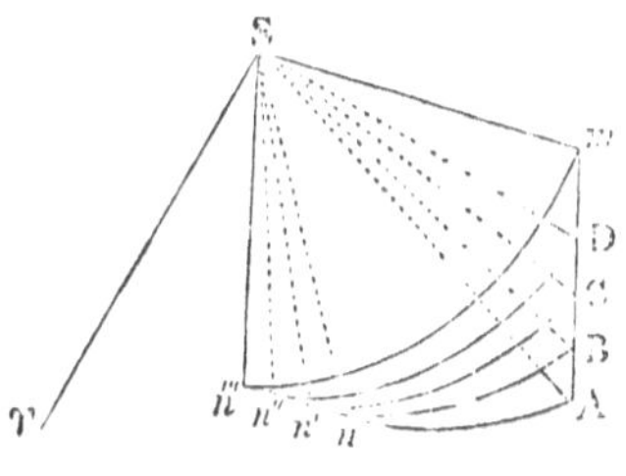

507. Applying the same construction to h and l (133), it will be found that the tangent of the inclination of the last circle on the fixed plane is equal to the eccentricity of the orbit of m; and that the longitude of the intersection of this circle with the same plane is equal to that of the perihelion of the orbit of m.

508. The values of p and q in equations (138) may be determined by another construction; for let C, fig. 90, be the centre of a circle whose radius is N; draw any diameter Da, and take the arc

fig. 87.

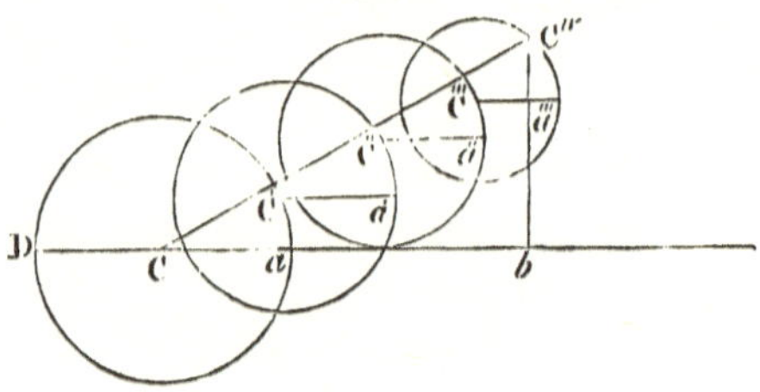

$$aC' = gt + \zeta;$$

on C′ as a centre with radius equal to $N_{,}$, describe a circle, and having drawn C′a' parallel to Ca, take $a'C' = g_{,}t + \zeta_{,}$: on C″ as a centre with radius equal to N_{2}, describe a circle, and having drawn C″a'' parallel to Ca, take the a'' C″ $= g_2 t + \zeta_2$, and so on. Let a^{iv} C^{iv} be the arc in the last circle, then if $C^{iv}b$ be perpendicular to Ca produced, it is evident that

$$C^{iv}b = p, \; Cb = q,$$

and if CC^{iv} be joined,

$$\tan\phi = \sqrt{p^2+q^2}, \tan\theta = \frac{p}{q},$$

θ being the angle $C^{iv}Cb$.

509. The equations which determine the secular variations in the inclinations and motions of the nodes being independent of the eccentricities, are the same as if the orbits of the planets were circular.

Annual and Sidereal Variations in the Elements of the Orbits, with regard to the variable Plane of the Ecliptic.

510. Equations (128) give the annual variations in the inclinations and longitudes of the nodes with regard to a fixed plane, but astronomers refer the celestial motions to the moveable orbit of the earth whence observations are made; its motion occasioned by the action of the planets is indeed extremely minute, but it is important to know the secular variations in the position of the orbits with

fig. 88.

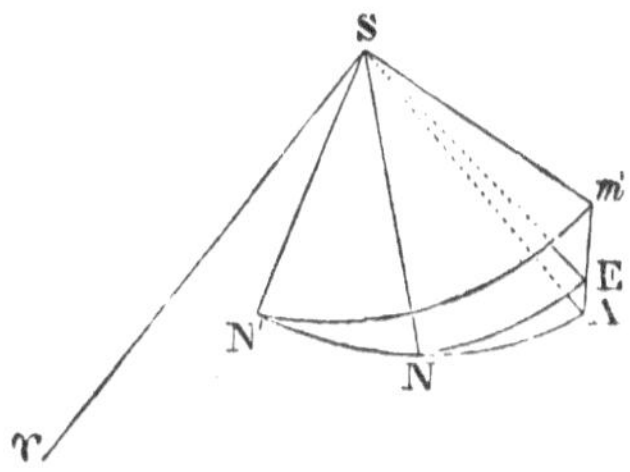

regard to it. Suppose AN fig. 88, to be the plane of the ecliptic or orbit of the earth, EN the variable plane of the ecliptic in which the earth is moving at a subsequent period, and m'N$'$ the orbit of a planet m', whose position with regard to EN is to be determined.

By article 444,

$$\text{EA} = q \sin (n't + \epsilon') - p \cos (n't + \epsilon')$$

is the latitude of m above AN; and the latitude of m' above AN$'$ is

$$\text{A}m' = q' \,.\, \sin (n't + \epsilon') - p' \cos (n't + \epsilon').$$

As the inclinations are supposed to be very small, the difference of these two, or m'A $-$ EA is very nearly equal to m'E the latitude of m' above the variable plane of the ecliptic EN.

If ϕ' be the inclination of m'N$'$ the orbit of m' to EN the variable ecliptic, and θ' the longitude of its ascending node, then will

$$\tan \phi' .\, \sin \theta' = p' - p; \quad \tan \phi' \cos \theta' = q' - q.$$

Whence $\tan \phi' = \sqrt{(p'-p)^2 + (q'-q)^2} \quad \tan \theta' = \dfrac{p'-p}{q'-q}.$

If EN be assumed to be the fixed plane at a given epoch, then $p = 0$, $q = 0$, but neither dp nor dq are zero; hence

$$d\phi' = (dp' - dp) \,.\, \sin \theta' + (dq' - dq) \cos \theta',$$

$$d\theta' = \frac{(dp' - dp) \,.\, \cos \theta' - (dq' - dq) \sin \theta'}{\tan \phi',}$$

and substituting the values in article 498 in place of the differentials dp, dq, &c. there will result

$$\frac{d\phi'}{dt} = \{(1.2) - (0.2)\} \tan \phi'' \sin (\theta' - \theta'') + \{(1.3) - (0.3)\}$$

$$\times \tan \phi''' \sin (\theta' - \theta''') + \&c.$$

$$\frac{d\theta'}{dt} = - \{(1.0) + (1.2) + (1.3) + \&c.\} - (0.1) \qquad (141)$$

$$+ \{(1.2)-(0.2)\} .\, \frac{\tan \phi''}{\tan \phi'} \,.\, \cos (\theta' - \theta'')$$

$$+ \{(1.3)-(0.3)\} .\, \frac{\tan \phi'''}{\tan \phi'} \,.\, \cos (\theta' - \theta'''),$$

$$+ \&c.$$

*

Motion of the Orbits of two Planets.

511. Imagine two planets m and m' revolving round the sun so remotely from the rest of the system, that they are not sensibly disturbed by the other bodies.

Let $\gamma = \sqrt{(p'-p)^2 + (q'-q)^2}$ be the mutual inclination of the two orbits supposed to be very small. If the orbit of m at the epoch be assumed as the fixed plane

$$\phi = 0, \quad \gamma = \phi', \quad p = 0, \quad q = 0,$$

and

$$\tan^2\phi' = \tan^2\gamma = p'^2 + q'^2.$$

In this case, equations (140) and (128) become

$$m'\sqrt{a'}\tan^2\phi' = C, \qquad \frac{d\theta'}{dt} = -(1.0).$$

Since the greater axes of the orbits are constant, the first shows that the inclination is constant, and the second proves the motion of the node of the orbit of m' on that of m to be uniform and retrograde, and the motion of the intersection of the two orbits on the orbit of m, in consequence of their mutual attraction, will be $-(1.0)t$.

Secular Variations in the Longitude of the Epoch.

512. The mean place of a planet in its orbit at a given instant, assumed to be the origin of the time, is the longitude of the epoch. It is one of the most important elements of the planetary orbits, being the origin whence the antecedent and subsequent longitudes are estimated. If the mean place of the planet at the origin of the time should vary from the action of the disturbing forces, the longitudes estimated from that point would be affected by it; to ascertain the secular inequalities of that element is therefore of the greatest consequence.

The differential equation of the longitude of the epoch in article 441, is

$$d\epsilon = \frac{an\sqrt{1-e^2}}{e}.(1 - \sqrt{1-e^2}).\frac{dF}{de}dt - 2a^2n\frac{dF}{da}dt.$$

By article 473,

$$\frac{dF}{da} = -\frac{m'}{2} \cdot \left(\frac{3a'S' + 2aS}{(a'^2 - a^2)^2}\right)$$
$$-\frac{m'}{4} \cdot \frac{a'^2}{a} \cdot \left(\frac{3S'(2a'^2 - 3a^2) + 6aa'S}{(a'^2 - a^2)^3}\right) ee' \cos(\varpi' - \varpi)$$
$$+\frac{m'}{2.4} aa' \cdot \left(\frac{6a'S - 3aS'}{(a'^2 - a^2)^3}\right) \{e^2 + e'^2 - (p'-p)^2 - (q'-q)^2\}$$
$$\frac{dF}{de} = -\frac{3m'aa'.S'}{4(a'^2 - a^2)^2} \cdot e$$
$$+\frac{3m'}{2(a'^2 - a^2)^2} \{(a'^2 + a^2)S' + aa'S\} e' \cos(\varpi' - \varpi).$$

If these be put in the value of $d\epsilon$, rejecting the powers of e above the second, and if to abridge

$$C = \frac{m'.na^2.(2aS + 3a'S')}{(a'^2 - a^2)^2},$$
$$C_1 = -\frac{3m'.na^2a'(4aa'S - (3a^2 - a'^2)S')}{2.4.(a'^2 - a^2)^3},$$
$$C_2 = -\frac{3m'.na.\{(a^2 - 5a'^2)aa'S + (a^4 + 6a^2a'^2 - 5a'^4)S'\}}{4.(a'^2 - a^2)^3},$$
$$C_3 = \frac{3m'.na^3a'(2a'S - aS')}{4(a'^2 - a^2)^3},$$

$d\epsilon$ becomes

$$\frac{d\epsilon}{dt} = C + C_1e^2 + C_2ee' \cos(\varpi' - \varpi)$$
$$+ C_3\{(p' - p)^2 + (q' - q)^2 - e'^2\}.$$

But
$$h = e \sin \varpi \quad l = e \cos \varpi,$$
$$h' = e' \sin \varpi' \quad l' = e' \cos \varpi';$$

hence
$$\frac{d\epsilon}{dt} = C + C_1(h^2 + l^2) + C_2(hh' + ll')$$
$$+ C_3\{(p' - p)^2 + (q' - q)^2 - h'^2 - l'^2\}.$$

513. This equation only expresses the variation in the epoch of m when troubled by m'; but, in order to have the effect of the whole system in disturbing the epoch of m, a similar set of terms must be added for each of the planets; but if the two planets m and m' alone be considered, their mutual inclination will be constant by article 511, hence $\gamma^2 = (p' - p)^2 + (q' - q)^2 = M^2$, a constant quantity.

Again by article 483,

$$h^2 + l^2 = N^2 + N_{\prime}^2 + 2NN_{\prime} \cos\{(g_{\prime} - g)t + \mathcal{C}_{\prime} - \mathcal{C}\}$$
$$h'^2 + l'^2 = N'^2 + N_{\prime}'^2 + 2N'N_{\prime}' \cos\{(g_{\prime} - g)t + \mathcal{C}_{\prime} - \mathcal{C}\}$$
$$hh' + ll' = NN' + N_{\prime}N_{\prime}' + (NN'_{\prime} + N'N) \cos\{g_{\prime} - g)t + \mathcal{C}_{\prime} - \mathcal{C}\}.$$

*

Substituting these in $d\epsilon$, and to abridge, making

$$A'n = C + C_{,}(N^2 + N_{,}^2) + C^2(NN' + N'N_{,}') + C_3(M^2 - N'^2 - N_{,}'^2),$$

$$B' = 2C_{,}NN' - 2C_3N'N_{,}' + C_2(NN_{,}' + N_{,}N'),$$

it becomes

$$d\epsilon = A' \cdot ndt + B' \cos\{(g_{,} - g)t + \mathfrak{C}_{,} - \mathfrak{C}\} dt.$$

The integral of which is

$$\delta\epsilon = A'nt + \frac{B'}{g_{,} - g} \sin\{(g_{,} - g)t + \mathfrak{C}_{,} - \mathfrak{C}\}$$

514. The term $A'nt$ only augments the mean primitive motion of the planet m in the ratio of 1 to $1 + A'$. so that the mean motion which should result from observation would be $(1 + A')nt$, corresponding to the mean distance $\frac{a}{(1+A')^{\frac{2}{3}}}$.

Knowing this distance, which is given by a comparison of the periodic times, the primitive distance a may be determined; but as A' is an infinitely small fraction of the order of the masses m and m', this correction in the mean distance is insensible. The term $A'nt$ may therefore be omitted, so that the secular variation in the epoch is

$$\delta\epsilon = \frac{B'}{g_{,} - g} \sin\{(g_{,} - g)t + \mathfrak{C}' - \mathfrak{C}.\} \qquad (142)$$

The variation in the epoch, like the other secular inequalities in article 480, may be expressed in series ascending according to the powers of the time; but as the term depending on its first power is insensible, it will have the form

$$\delta\epsilon = H t^2 + \&c.$$

This inequality is insensible for the planets; its greatest effect is produced in the theory of Jupiter and Saturn: but even then it is only $\delta\epsilon = - 0''.0000006501 . t^2$ for Jupiter, and for Saturn $\delta\epsilon' = + 0''.0000015114 . t^2$, t being any number of Julian years from 1750. This inequality is not the 60th part of a sexagesimal second in a century, a quantity altogether insensible. Like all other inequalities it is periodic; but its period, which depends on $g_{,} - g$, is for Jupiter and Saturn no less than 70414 years. The variation $\delta\epsilon$, though of the order of disturbing forces, may, in the course of many centuries, become sensible, on account of the small divisor $g_{,} - g$ introduced by integration; but although it is insensible with regard to the planets, it is of much importance in the theories of the Moon and of Jupiter's Satellites.

Stability of the System, whatever may be the powers of the Disturbing Masses.

515. The stability of the system has been proved with regard to the greater axes of the orbits, even when the approximation extends to the squares of the disturbing forces, and to all powers of the eccentricities and inclinations. Its invariability with regard to the other elements has only been proved on the hypothesis of the orbits being nearly circular, and very little inclined to each other and to the plane of the ecliptic; but as the same results may be derived from the general equations of the motion of a system of bodies, they equally exist whatever the eccentricities and inclinations may be, and when the approximation includes the squares of the disturbing forces, and they remain the same whatever changes the secular inequalities may introduce in the lapse of ages.

516. If the equations of the motion of a system of bodies in article 346 be resumed, and the equations in x, x', &c., multiplied respectively by

$$my - m \cdot \frac{\Sigma m \cdot y}{S + \Sigma m}; \qquad m'y' - m' \cdot \frac{\Sigma m \cdot y}{S + \Sigma m}; \text{ \&c.}$$

and those in y, y', &c., by

$$-mx + m \cdot \frac{\Sigma m \cdot x}{S + \Sigma m}; \qquad -m'x' + m' \cdot \frac{\Sigma m \cdot x}{S + \Sigma m}; \text{ \&c.}$$

their sum will be

$$\Sigma m \cdot \left(\frac{xd^2y - yd^2x}{dt^2}\right) + \frac{\Sigma \cdot my}{S + \Sigma m} \cdot \Sigma m \cdot \frac{d^2x}{dt^2}$$
$$- \frac{\Sigma \cdot mx}{S + \Sigma m} \cdot \Sigma m \cdot \frac{d^2y}{dt^2} = 0;$$

for the nature of the function λ gives

$$y \cdot \frac{d\lambda}{dx} + y' \cdot \frac{d\lambda}{dx'} + \text{\&c} = 0; \; -x\frac{d\lambda}{dy} - x' \cdot \frac{d\lambda}{dy'} - \text{\&c.} = 0,$$
$$\frac{d\lambda}{dx} + \frac{d\lambda}{dx'} + \text{\&c.} = 0; \; \frac{d\lambda}{dy} + \frac{d\lambda}{dy'} + \text{\&c.} = 0;$$

as may be seen by trial. The integral of the preceding equation is

$$\Sigma m \cdot \left(\frac{xdy - ydx}{dt}\right) + \frac{\Sigma \cdot my}{S + \Sigma m} \cdot \Sigma m \cdot \frac{dx}{dt}$$
$$- \frac{\Sigma mx}{S + \Sigma m} \cdot \Sigma m \cdot \frac{dy}{dt} = C.$$

A similar equation may be found in x, z, and y, z; and when $S+m'=1$, it will be found that

$$\Sigma m \,.\, \frac{ydx-xdy}{dt}+\Sigma mm'\left(\frac{xdy'-y'dx+x'dy-ydx'}{dt}\right)=C$$

$$\Sigma m \,.\, \frac{xdz-zdx}{dt}+\Sigma mm'\left(\frac{zdx'-x'dz+z'dx-xdz'}{dt}\right)=C' \quad (143)$$

$$\Sigma m \,.\, \frac{zdy-ydz}{dt}+\Sigma mm'\left(\frac{y'dz-zdy'+y'dz-zdy'}{dt}\right)=C''$$

C, C', C'', being constant quantities. Now $\dfrac{ydx-xdy}{dt}$ is double the area described in the time dt by the projection of the radius vector of m on the plane xy. This area on the orbit is $\sqrt{a(1-e^2)}$; and if ϕ be the inclination of the orbit on the plane xy, $\cos\phi\sqrt{a(1-e^2)}$ is its projection. In the same manner

$$\frac{y'dx'-x'dy'}{dt}=\cos\phi'\sqrt{a'(1-e'^2)}$$

is the area described by the projection of the radius vector of m' on the same plane, and so on. In consequence of these the first of the preceding equations becomes

$$m\sqrt{a(1-e^2)}\cos\phi+m'\sqrt{a'(1-e'^2)}\cos\phi'+\&c.$$
$$=mm'\left(\frac{ydx'-x'dy+y'dx-xdy'}{dt}\right)+\&c.+C.$$

If the elliptical values of x, y, x', y', be substituted, the first term of the second member of this equation must always be periodic; for, in consequence of the observations in article 466, the arcs nt, $n't$, never destroy one another in the expressions ydx', $x'dy$, &c. Hence, if periodic quantities and those of the fourth order be neglected, the last number of the equation is constant. If the products ydx', $x'dy$, &c., contained constant terms, they would be of the first order with regard to the masses; and as they are functions of the elliptical elements, their variation is of the second order; consequently, the variation of the terms $mm'\,.\,y'dx$, &c., is of the fourth order. If the periodic part of the values of the elliptical elements be substituted in the first member of the preceding equation, any terms resulting from that substitution that are not periodic will be of the third order, and may be regarded as constant. The second member of the equation in question may therefore be esteemed constant. Hence,

$$m\sqrt{a(1-e^2)}\cos\phi+m'\sqrt{a'(1-e'^2)}\cos\phi'+\&c.=C. \quad (144)$$

517. Again, $\frac{xdz-zdx}{dt}$ and $\frac{zdy-ydz}{dt}$

are the areas described by the radius vector of m in the time dt, projected on the co-ordinate planes $x\,z$, and $y\,z$. But it is easy to see by trigonometry that the cosines of the inclination of the orbit on these planes are $\sin\phi\cos\theta$, and $\sin\phi\sin\theta$;

hence, $$\frac{xdz-zdx}{dt}=\sqrt{a(1-e^2)}\sin\phi\cos\theta,$$

and $$\frac{zdy-ydz}{dt}=\sqrt{a(1-e^2)}\sin\phi\sin\theta.$$

Similar expressions exist for all the bodies; and as the same reasoning applies to the two last equations (143), as to the first, they give

$$m\sqrt{a(1-e^2)}\sin\phi\cos\theta+m'\sqrt{a'(1-e'^2)}\sin\phi'\cos\theta'+\&c.=C',$$
$$m\sqrt{a(1-e^2)}\sin\phi\sin\theta+m'\sqrt{a'(1-e'^2)}\sin\phi'\sin\theta'+=\&c.C''. \quad (145)$$

518. These relations exist whatever the eccentricities and inclinations may be, and whatever may be the changes that they undergo in the course of ages from their secular inequalities, the approximation extending to the third order inclusively, and even to the squares of the disturbing forces.

519. A variety of results may be derived from them. Because

$$\cos\phi=\frac{1}{\sqrt{1+\tan^2\phi}}, \text{ equation (144) gives}$$

$$m\sqrt{\frac{a(1-e^2)}{1+\tan^2\phi}}+m'\sqrt{\frac{a'(1-e'^2)}{1+\tan^2\phi'}}+\&c.=C.$$

If e^4 and $e^2\phi^2$ be omitted,

$$m\sqrt{\frac{a(1-e^2)}{1+\tan^2\phi}}=m\sqrt{a(1-e^2)}\,(1+\tan^2\phi)^{-\frac{1}{2}}=m\sqrt{a}$$
$$-\tfrac{1}{2}m\sqrt{a}\,(e^2+\tan^2\phi),$$

consequently

$$\tfrac{1}{2}m\sqrt{a}\,(e^2+\tan^2\phi)+\tfrac{1}{2}m'\sqrt{a'}\,(e'^2+\tan^2\phi')+\&c.=2m\sqrt{a}$$
$$+2m'\sqrt{a'}+\&c.+2C.$$

But the last member is altogether constant: hence

$$m\sqrt{a}\,(e^2+\tan^2\phi)+m'\sqrt{a'}\,(e'^2+\tan^2\phi')+\&c.=\text{constant}.$$

It was shown that when the squares and products of the eccentricities and inclinations are omitted, the variations in the eccentricities are the same as if all the planets moved in one plane; and that the variations in the inclinations are the same as if the orbits were circular, as these quantities vary independently of one another, e, e', &c., and ϕ, ϕ', &c., may be alternately zero in the last equation, consequently,

$$m\sqrt{a}\,.\,e^2 + m'\sqrt{a'}\,.\,e'^2 + m''\sqrt{a''}\,.\,e''^2 + \text{\&c.} = \text{constant};$$

$$m\sqrt{a}\tan^2\phi + m'\sqrt{a'}\tan^2\phi' + m''\sqrt{a''}\tan^2\phi'' = \text{constant};$$

results that are the same with equations (136) and (140).

If quantities of the order of the squares of the eccentricities and inclinations be omitted, the tangents of the very small quantities ϕ, ϕ', may be taken in place of their sines, so that by the substitution of

$$\begin{aligned} p &= \tan\phi\sin\theta, & q &= \tan\phi\cos\theta, \\ p' &= \tan\phi'\sin\theta', & q &= \tan\phi'\cos\theta', \\ &\text{\&c.} & &\text{\&c.} \end{aligned}$$

in equations (145) they become

$$m\sqrt{a}\,.\,q + m'\sqrt{a'}\,.\,q' + m''\sqrt{a''}\,q'' + \text{\&c.} = \text{constant},$$

$$m\sqrt{a}\,.\,p + m'\sqrt{a'}\,.\,p' + m''\sqrt{a''}\,p'' + \text{\&c.} = \text{constant}.$$

520. Since the eccentricities and inclinations of all the orbits in the solar system are very small, the constant quantities in all the preceding equations of condition must be very small, provided the radicals $\sqrt{a}$, $\sqrt{a'}$, &c., have the same signs, that is, if the bodies all move in one direction, which is the case in nature; it may therefore be concluded that the elements vary within very narrow limits.

521. Let there be only two bodies m and m', the mutual inclination of their orbits being

$$\cos\gamma = \cos\phi\cos\phi' + \sin\phi\sin\phi'\cos(\theta' - \theta);$$

then if the squares of the equations (144) and (145) be added, the result will be

$$m^2a(1 - e^2) + m'a'(1 - e'^2) + 2mm'\sqrt{a(1 - e^2)}\,.\,\sqrt{a'(1 - e'^2)} \times \cos\gamma = \text{constant}. \qquad (146)$$

Neglecting quantities of the fourth order, and putting all the constant quantities in the second member, it becomes

$$m\sqrt{a}\,.\,e^2 + m'\sqrt{a'}\,e'^2 + \frac{4mm'\sqrt{aa'}\sin^2\frac{1}{2}\gamma}{m\sqrt{a} + m'\sqrt{a'}} = \text{constant},$$

for $\cos\gamma = 1 - 2\sin^2\frac{1}{2}\gamma$.

The constant in the second part of this equation is equal to the first member at a given epoch, for at that epoch all the elements are supposed to be known by observation; it ought, therefore, to be independent of the variation of the elements e, e', and γ: its variation will be

$$m\sqrt{a}\,.\,e\delta e + m'\sqrt{a'}\,e'\delta e' + \frac{2mm'\sqrt{aa'}\,.\,\gamma\delta\gamma}{m\sqrt{a}+m'\sqrt{a'}} = 0, \qquad (147)$$

for a and a' are constant. This relation must always exist among the secular variations of the eccentricities of the two orbits and their mutual inclination.

If the constant part of equation (146) be included in the second member it becomes

$m^2ae^2 + m'^2a'e'^2 - 2mm'\,a^2a'^2nn'\sqrt{1-e^2}\sqrt{1-e'^2}\cos\gamma =$ constant, by the substitution of a^2n and a'^2n' for $\sqrt{a}$ and $\sqrt{a'}$; and if it be observed that

$$\sqrt{1-e^2} = 1 - \frac{e^2}{1+\sqrt{1-e^2}}, \quad \sqrt{1-e'^2} = 1 - \frac{e'^2}{1+\sqrt{1-e'^2}},$$

$$\cos\gamma = 1 - \frac{\sin^2\gamma}{1+\cos\gamma},$$

then will

$$\sqrt{1-e^2}\sqrt{1-e'^2}\cos\gamma = 1 - \frac{e^2\sqrt{1-e'^2}\cos\gamma}{1+\sqrt{1-e^2}} - \frac{e'^2\cos\gamma}{1+\sqrt{1-e'^2}} - \frac{\sin^2\gamma}{1+\cos\gamma}.$$

If this value be put in the preceding equation, and all constant quantities included in the second member, it becomes

$$m^2\,.\,ae^2 + m'^2\,.\,a'e'^2 + 2mm'\,.\,a^2a'^2nn'\,.\,\frac{e^2\sqrt{1-e'^2}\,.\,\cos\gamma}{1+\sqrt{1-e^2}}$$

$$+2mm'\,.\,a^2a'^2nn'\,.\,\frac{e'^2\cos\gamma}{1+\sqrt{1-e^2}} + 2mm'a'^2a^2nn'\frac{\sin^2\gamma}{1+\cos\gamma} = C_{,};$$

$C_{,}$ being an arbitrary constant quantity.

$C_{,}$ is a very small quantity with regard to the squares and products of m and m', since they are multiplied by e^2, e'^2, $\sin\gamma^2$; and that the mutual inclination of the two planes and their eccentricities are supposed to be very small, as is really the case in nature. Each term of the first member of this equation will therefore remain very small with regard to the squares and products of m and m'; if all the terms have the same sign, each term will then be less than $C_{,}$. But because all the planets revolve in the same direction round the sun, nt, $n't$, will have the same sign. Hence all the terms in the first member will be positive as long as γ is less than 90°. But if $\gamma = 90°$, then $\sin\gamma = 1$; $\cos\gamma = 0$, which reduces the equation to

$$m^2ae^2 + m'^2a'e'^2 + 2mm'a^2a'^2nn' = C,$$

and the last term is no longer very small with regard to mm', which is impossible, since C, is very small with regard to the product of m and m', and that the other terms of the first member are positive. Thus, because the angle γ never can attain to 90°, it follows that γ, the inclination, and the eccentricities e, e', of the two orbits, will always be small; for, as $\cos\gamma$ never can become negative, every term in the first member of the equation under discussion will be positive, and will remain very small with regard to the squares and products of the masses m and m'. That is to say, the coefficients e^2 e'^2 $\sin^2\gamma$ will always remain very small, because they are small at present.

522. This reasoning would be the same whatever might be the number of planets, since each of them would only add terms to the first member of the equation under consideration, similar to those that compose it.

523. Thus it may be concluded that the planetary system is stable with regard to the eccentricities, the inclinations, and greater axes of the orbits, however far the approximation may be carried with regard to the elements of the orbits, even including the second powers of the disturbing forces.

524. La Place and Poisson have proved the stability of the solar system when the approximation extends to the first and second powers of the disturbing force, on the hypothesis that all the planets revolve in nearly circular orbits, little inclined to each other; but in a very able paper read before the Royal Society on the 29th April, 1830, Mr. Lubbock has shown that these conditions are not necessary in a system subject to the law of gravitation. He has obtained expressions for the variations of the elliptical constants, which are rigorously true, whatever the power of the disturbing force may be, whence it appears, that, however far the approximation may be carried, the eccentricities, the major axes, and the inclinations of the orbits to a fixed plane, contain no term that varies with the time, and that their secular variations oscillate between fixed limits in very long periods.

The Invariable Plane.

525. It has been already mentioned that in the motion of a system of bodies there exists an invariable plane, which, always retaining a parallel position, is easily found, because the sum of the masses of the bodies of the system respectively multiplied by the projections of the areas described by their radii vectores in a given time, is a maximum on that plane, and the sum of the projections on any other planes at right angles to it is zero. It is principally in the solar system that this plane is of importance, on account of the proper motions of the stars, and of the plane of the ecliptic, which render it difficult to determine the celestial motions with precision, this difficulty indeed is already perceptible, and will increase when very accurate observations, separated by very long intervals of time, must be compared with each other.

If I be the inclination of the invariable plane on the fixed plane which contains the co-ordinates x and y, and if Ω be the longitude of its ascending node, by article 166

$$\text{Tan } I \sin \Omega = \frac{C''}{C}; \ \tan I \cos \Omega = \frac{C'}{C};$$

and substituting the values of C, C', C'', given by equations (144) and (145),

$$\text{Tan } I \sin \Omega = \frac{m\sqrt{a(1-e^2)}\sin\phi\sin\theta + m'\sqrt{a'(1-e'^2)}\sin\phi'\sin\theta' + \&c.}{m\sqrt{a(1-e^2)}\cos\phi + m'\sqrt{a'(1-e'^2)}\cos\phi' + \&c.}$$

$$\text{Tan } I \cos \Omega = \frac{m\sqrt{a(1-e^2)}\sin\phi\cos\theta + m'\sqrt{a'(1-e'^2)}\sin\phi'\cos\theta' + \&c.}{m\sqrt{a(1-e^2)}\cos\phi + m'\sqrt{a'(1-e'^2)}\cos\phi' + \&c.}.$$

The second members of these two equations have been proved to be invariable, even in carrying the approximation to the squares and products of the masses, whatever changes the secular variations may induce in the course of ages; and, by what Mr. Lubbock has shown, they must be constant, whatever the power of the disturbing force may be: hence it follows, that the invariable plane retains its position, notwithstanding the secular variations in the elliptical elements of the planetary system.

526. The determination of this plane requires a knowledge of the

masses of all the bodies in the system, and of the elements of their orbits. Approximate values of these are only known with regard to the planets, but of the masses of the comets we are in total ignorance; however, as the mutual gravitation of the planets is sufficient to represent all their inequalities, it shows that, hitherto at least, the action of the comets on the planetary system is insensible. Besides, the comet of 1770 approached so near to the earth that its periodic time was increased by 2.046 days; and if its mass had been equal to that of the earth, it would have increased the length of the sidereal year by nearly one hour fifty-six minutes, according to the computation of La Place; but he adds, that if an increase of only two seconds had taken place in the length of the year, it would have been detected by Delambre, when he computed his astronomical tables from the observations of Dr. Maskelyne; whence the mass of the comet must have been less than the $\frac{1}{5000}$ part of the mass of the earth. The same comet passed through the Satellites of Jupiter in the years 1767 and 1779, without producing the smallest effect. Thus, though comets are greatly disturbed by the action of the planets, they do not appear to produce any sensible effects by their reaction.

527. If the position of the ecliptic in the beginning of 1750 be assumed as the fixed plane of the co-ordinates x and y, and if the line of the equinoxes be taken as the origin of the longitudes, it is found that at the epoch 1750 the longitude of the ascending node of the invariable plane was $\Omega = 102^\circ\ 57'\ 30''$, and its inclination on the ecliptic $I = 1^\circ 35'\ 31''$; and if the values of the elements for 1950 be substituted in the preceding formulæ, it will appear that in 1950

$$\Omega = 102^\circ\ 57'\ 15''; \quad I = 1^\circ\ 35'\ 31'';$$

which differ but little from the first.

528. The position of this plane is really approximate, since it has been determined in the hypothesis of the solar system being an assemblage of dense points mutually acting on one another, whereas the celestial bodies are neither homogeneous nor spherical; but as the quantities omitted have hitherto been insensible, the position of the plane as it is here given, will enable future astronomers to ascertain the real changes that may have taken place in the forms and positions of the planetary orbits.

CHAPTER VII.

PERIODIC VARIATIONS IN THE ELEMENTS OF THE PLANETARY ORBITS.

Variations depending on the first Powers of the Eccentricities and Inclinations.

529. THE differential dR relates to the arc nt alone, consequently the differential equation $da = 2a^2 \,.\, \text{d}R$ in article 439 becomes

$$da = + m'a^2 \,.\, in \,.\, \Sigma A_i \sin i\,(n't - nt + \epsilon' - \epsilon)$$
$$+ m'a^2 en\,(i-1) \,.\, M_0 \sin\{i(n't - nt + \epsilon' - \epsilon) + nt + \epsilon - \varpi\}$$
$$+ m'a^2 e'n\,(i-1) \,.\, M_1' \sin\{i(n't - nt + \epsilon' - \epsilon) + nt + \epsilon - \varpi'\}.$$

The integral of this equation is the periodic variation in the mean distance, and if represented by δa, then

$$\delta a = - m'a^2 \frac{n}{n'-n} \,.\, \Sigma A_i \cos i\,(n't - nt + \epsilon' - \epsilon)$$
$$- m'a^2 e \frac{(i-1)n}{i(n'-n)+n} M_0 \cos\{i(n't - nt + \epsilon' - \epsilon) + nt + \epsilon - \varpi\}$$
$$- m'a^2 e' \frac{(i-1)n}{i(n'-n)+n} M_1 \cos\{i(n't - nt + \epsilon' - \epsilon) + nt + \epsilon - \varpi'\}.$$

In a similar manner it may be found that the periodic variation in the mean motion $d\zeta = - 3\int andt\, \text{d}R$ is,

$$\delta\zeta = \tfrac{3}{2} \,.\, m'a \,.\, \frac{n^2}{i\,(n'-n)^2} \,.\, A_i \sin i\,(n't - nt + \epsilon' - \epsilon)$$
$$+ \tfrac{3}{2} m'ae \frac{(i-1)n^2}{\{i(n'-n)+n\}^2} M_0 \sin\{i(n't - nt + \epsilon' - \epsilon) + nt + \epsilon - \varpi\}$$
$$+ \tfrac{3}{2} m'ae' \frac{(i-1)n^2}{\{i(n'-n)+n)\}^2} \,.\, M_1 \sin\{i(n't - nt + \epsilon' - \epsilon) + nt + \epsilon - \varpi'\}.$$

From the other differential equations in article 439 it may also be found that the periodic variation in the eccentricity is

$$\delta e = \tfrac{1}{2} m'a \frac{n}{i(n'-n)+n} M_0 \cos\{i(n't - nt + \epsilon' - \epsilon) + nt + \epsilon - \varpi\}$$

$$+ \tfrac{1}{4} m'ae \frac{n}{n'-n} A_i \cos i(n't - nt + \epsilon' - \epsilon)$$

$$+ m'ae' \frac{n}{i(n'-n)+2n} N_0 \cos\{i(n't-nt+\epsilon'-\epsilon)+2nt+2\epsilon-2\varpi\}$$

$$+ \tfrac{1}{2} m'a.e' \frac{n}{i(n'-n)+2n} N_1 \cos\{i(n't-nt+\epsilon'-\epsilon)+2nt+2\epsilon-\varpi-\varpi'\}$$

$$- \tfrac{1}{2} m'ae' \frac{n}{i(n'-n)} N_4 \cos\{i(n't-nt)+\epsilon'-\epsilon)+\varpi-\varpi'\}$$

$$+ \tfrac{1}{2} m'ae' \frac{n}{i(n'-n)} N_5 \cos\{i(n't-nt+\epsilon'-\epsilon)-\varpi+\varpi'\}.$$

The variation of the epoch

$$\delta\epsilon = - ma' \frac{n}{i(n'-n)} a\left(\frac{dA_i}{da}\right) \sin i(n't - nt + \epsilon' - \epsilon)$$

$$+ \tfrac{1}{4} m'ae \frac{n}{i(n'-n)+n} M_0 \sin\{i(n't-nt+\epsilon'-\epsilon)+nt+\epsilon-\varpi\}$$

$$- m'a^2e \frac{n}{i(n'-n)+n} \cdot \frac{dM_0}{da} \sin\{i(n't-nt+\epsilon'-\epsilon)+nt+\epsilon-\varpi\}$$

$$- m'a^2e' \frac{n}{i(n'-n)+n} \cdot \frac{dM_1}{da} \sin\{i(n't-nt+\epsilon'-\epsilon)+nt+\epsilon-\varpi'\}.$$

The variation in the longitude of the perihelion

$$e\delta\varpi = \tfrac{1}{2} m'a \frac{n}{i(n'-n)+n} M_0 \sin\{i(n't-nt+\epsilon'-\epsilon)+nt+\epsilon-\varpi\}$$

$$+ m'ae \frac{n}{i(n'-n)+2n} N_0 \sin\{i(n't-nt+\epsilon'-\epsilon)+2nt+2\epsilon-2\varpi\}$$

$$+ mae \cdot \frac{n}{i(n'-n)} N_3 \sin i(n't - nt + \epsilon' - \epsilon)$$

$$+ \tfrac{1}{2} m'ae' \frac{n}{i(n'-n)+2n} N_1 \sin\{i(n't-nt+\epsilon'-\epsilon)+2nt+2\epsilon-\varpi-\varpi'\}$$

$$+ \tfrac{1}{2} m'ae' \frac{n}{i(n'-n)} N_4 \sin\{i(n't-nt+\epsilon'-\epsilon)+\varpi-\varpi'\}$$

$$+ \tfrac{1}{2} m'ae' \frac{n}{i(n'-n)} \cdot N_5 \sin\{i(n't-nt+\epsilon'-\epsilon)-\varpi+\varpi'\}.$$

When e^2, $e\gamma$, $e'\gamma$, are omitted, the differentials of p and q in article 437 become

$$dp = a^2ndt \sin(nt+\epsilon) \frac{dR}{dz}$$

$$dq = a^2ndt \cos(nt+\epsilon) \frac{dR}{dz}.$$

When the orbit of m at the epoch is assumed to be the fixed plane,

$$z = 0, \text{ and } z' = a'\gamma \sin (n't + \epsilon' - \Pi).$$

the products of the inclination by the eccentricities being omitted.

Now although z be zero, its differential is not, therefore $\frac{dR}{dz}$ must be determined from

$$R = \frac{m'zz'}{a'^3} + \frac{m'(z'-z)^2}{4} \Sigma B_i \cos i(n't - nt + \epsilon' - \epsilon);$$

whence

$$\frac{dR}{dz} = \frac{m'z'}{a'^3} - \frac{m'z'}{2} \Sigma B_i \cos i(n't - nt + \epsilon' - \epsilon),$$

and

$$\frac{dR}{dz} = \frac{-m'}{a'^2} \gamma \sin \{ (n't + \epsilon' - \Pi)$$

$$+ \frac{m'}{2} a'\Sigma B_{(i-1)} \gamma \sin \{ i(n't - nt + \epsilon' - \epsilon) + nt + \epsilon - \Pi \}$$

where i may be any whole number, positive or negative, except zero. When this quantity is substituted in dp, dq, their integrals are

$$\delta p = -\frac{m'}{2} \cdot \frac{a^2 n}{a'^2} \gamma \left\{ \frac{1}{n'-n} \sin (n't - nt + \epsilon' - \epsilon - \Pi) - \frac{1}{n'+n} \times \right.$$

$$\sin (n't + n't + \epsilon' + \epsilon - \Pi) \}\},$$

$$+ \frac{m'}{4} a^2 a' n \Sigma B_{(i-1)} \gamma \left\{ \frac{1}{i(n'-n)} \sin (i(n't - nt + \epsilon' - \epsilon) - \Pi) \right.$$

$$\left. - \frac{1}{i(n'-n)+2n} \sin \{ i(n't - nt + \epsilon' - \epsilon) + 2nt + 2\epsilon - \Pi \} \right\}$$

$$\delta q = \frac{m'}{2} \cdot \frac{a^2 n}{a'^2} \gamma \left\{ \frac{1}{n'+n} \cos (n't + nt + \epsilon' + \epsilon - \Pi) \right.$$

$$\left. + \cdot \frac{1}{n'-n} \cos \{ n't - nt + \epsilon' - \epsilon) - \Pi \} \right\}$$

$$- \frac{m'}{4} a^2 a' n \Sigma B_{(i-1)} \gamma \left\{ \frac{1}{i(n'-n} \cos (i(n't - nt + \epsilon' - \epsilon) + \Pi) \right.$$

$$\left. + \frac{1}{i(n'-n)+2n} \cos \{ i(n't - nt + \epsilon' - \epsilon) + 2nt + 2\epsilon - \Pi \} \right\}.$$

530. The equations which determine the variations in the greater axes and mean motion show that these two elements are subject to very considerable periodic variations, depending on the con-

figurations of the bodies, when the divisor $i(n' - n) + n$ or $i'n' - in$ is very small.

There is no instance of the mean motions of any two of the celestial bodies being so exactly commensurable as to have $i'n' - in = 0$, therefore the greater axes and mean motions have no secular inequalities, but in several instances this divisor is a very small fraction, and as a quantity is increased in value when divided by a fraction, the divisor $i'n' - in$, and still more its square, increases the values of these periodic variations very much. For this reason the periodic variation in the mean motion is much greater than that in the greater axis, evidently arising from the double integration in the former.

531. It is unnecessary to add constant quantities to the preceding integrals, for they may be included in the elements of elliptical motion, which then become

$$a + a_{,}, \ e + e_{,}, \ \varpi + \varpi_{,}, \ \epsilon + \epsilon_{,}, \ p + p_{,}, \ q + q_{,};$$

and in the troubled orbit they are

$$a + a_{,} + \delta a ; \ e + e_{,} + \delta e ; \ \varpi + \varpi_{,} + \delta\varpi ;$$
$$\epsilon + \epsilon_{,} + \delta\epsilon ; \ p + p_{,} + \delta p ; \ q + q_{,} + \delta q.$$

Since $a_{,}$, $e_{,}$, &c., $\delta a_{,}$, δe, &c., are very small quantities of the order m', $a + a_{,}$, $e + e_{,}$, &c., may be substituted in the latter quantities instead of a, e, &c., they will then be functions of the time and of the six constant quantities $a + a_{,}$, $e + e_{,}$, &c.: so that the formulæ of troubled motion in reality contain but six arbitrary constant quantities, as they ought to do. In order to determine $a_{,}$, $e_{,}$, &c., suppose the perturbations of the planet m were required during a given interval of time. The quantities a, e, &c., are given by observation at the epoch when $t = 0$ in the elliptical orbit, that is, assuming the disturbing force to be zero; but as $a_{,} + \delta a$, $e_{,} + \delta e$, &c., arise entirely from the disturbing force, they must also be zero at the epoch; therefore, values of the arbitrary constant quantities $a_{,}$, $e_{,}$, &c., are obtained from the equations

$$a_{,} + \delta\bar{a} = 0, \ e_{,} + \delta\bar{e} = 0, \ \varpi_{,} + \delta\bar{\varpi} = 0, \text{ \&c.},$$

$\delta\bar{a}$, $\delta\bar{e}$, &c., being the values of δa, δe, &c., at the epoch.

The effect of the disturbing forces upon each of the elliptical elements will be completely expressed by $a_{,} + \delta\bar{a}$; $e_{,} + \delta\bar{e}$, &c. during the time under consideration. Thus both the periodic and secular variations of the elements of the orbits are determined.

CHAPTER VIII.

PERTURBATIONS OF THE PLANETS IN LONGITUDE, LATITUDE, AND DISTANCE.

532. THE position of a planet in space is fixed when its curtate distance S*p*, fig. 77, its projected longitude γS*p*, and its latitude *pm*, are known. The determination of these three co-ordinates in functions of the time is the principal object of Physical Astronomy; these quantities in series ascending according to the powers of the eccentricities and inclinations are given in article 399, and those following, supposing the planet to move in a perfect ellipse; but if values of the elements of the orbits corrected by their periodic and secular variations be substituted instead of their elliptical elements, the same series will determine the motion of the planet in its real perturbed orbit.

533. The projected longitude and curtate distance only differ from the true longitude and distance on the orbit by quantities of the second order with regard to the inclinations; and when the orbit at the epoch is assumed to be the fixed plane, these quantities as well as those of the latitude that depend on the product of the inclination by the eccentricity are so small that they are insensible, as will readily appear if it be considered that any inclination the orbit may have acquired subsequently to the epoch, can only have arisen from the small secular variation in the elements; besides the epoch may be chosen to make it so, being arbitrary. Hence the perturbations in the longitude and radius vector may be determined as if the orbits were in the same plane, and the latitude may be found in the hypothesis of the orbits being circular, provided the orbit at the epoch be taken as the fixed plane: circumstances which greatly facilitate the determination of the perturbations.

The following very elegant method of finding the perturbations, by considering the troubled orbit as an ellipse whose elements are varying every instant, was employed by La Grange; but La Place's method, which will be explained afterwards, has the advantage of greater simplicity, especially in the higher approximations.

534. In the elliptical hypothesis the radius vector and true longitude are expressed, in article 392, by

$$r = \text{functions} \,.\, (a, \zeta, e, \epsilon, \varpi),$$
$$v = \text{functions} \,.\, (\zeta, e, \epsilon, \varpi),$$

but in the true orbit these quantities become

$$a + \delta a, \quad \zeta + \delta\zeta, \quad e + \delta e, \quad \epsilon + \delta\epsilon, \quad \varpi + \delta\varpi;$$

therefore

$$\delta r = \frac{dr}{da}.\delta a + \frac{dr}{d\zeta}.\delta\zeta + \frac{dr}{de}.\delta e + \frac{dr}{d\epsilon}.\delta\epsilon + \frac{dr}{d\varpi}.\delta\varpi,$$

$$dv = \frac{dv}{d\zeta}.\delta\zeta + \frac{dv}{de}.\delta e + \frac{dv}{d\epsilon} \,.\, \delta\epsilon + \frac{dv}{d\varpi}.\delta\varpi;$$

and if the values of the periodic variations in the elements in article 529 be substituted instead of δa, $\delta\zeta$, &c., the perturbations in the radius vector and true longitude will be obtained; the approximation extending to the first powers of the eccentricities and inclinations inclusively.

535. The perturbations in longitude may be expressed under a more simple form; for by article 372,

$$dv = \frac{\sqrt{a(1-e^2)}}{r^2} \,.\, dt,$$

an equation belonging both to the elliptical and to the real orbit, since it is a differential of the first order; on that account it ought not to change its form when the elements vary; hence

$$d \,.\, \delta v = \tfrac{1}{2}\sqrt{\frac{1-e^2}{a}}.\frac{\delta a}{r^2}.dt - \sqrt{\frac{a}{1-e^2}}.\frac{e\delta e}{r^2}.dt - 2\sqrt{a(1-e^2)}\frac{\delta r}{r^3}.dt;$$

and neglecting the squares of the disturbing forces, the integral is

$$\delta v = \frac{1}{2a}.\int \delta a.dv - \frac{e}{1-e^2}.\int de \,.\, dv - 2.\int \frac{\delta r}{r}.dv.$$

But $\qquad h = \sqrt{a.(1-e^2)}$, then $\dfrac{\delta h}{h} = \dfrac{1}{2a}.\delta a - \dfrac{e}{1-e^2}\delta e;$

therefore

$$\delta v = \int\left(\frac{\delta h}{h} - \frac{2\delta r}{r}\right).dv \qquad (148)$$

will give the perturbations in longitude when those in the radius vector are known.

Perturbations in the Radius Vector.

536. By article 392,

$$r = a(1 + \tfrac{1}{2}e^2 - e\cos(nt + \epsilon - \varpi) - \tfrac{1}{2}e^2\cos 2(nt + \epsilon - \varpi));$$

whence

$$\delta r = (\delta a - a\delta e\cos(nt+\epsilon-\varpi) - ae\delta\varpi\sin(nt+\epsilon-\varpi))(1+2e\cos(nt+\epsilon-\varpi))$$
$$- 3e\delta a\cos(nt+\epsilon-\varpi) + 2ae\delta e + ae(\delta\zeta+\delta\epsilon)\sin(nt+\epsilon-\varpi).$$

If the values of δa, δe, $\delta\varpi$, $\delta\zeta$, $\delta\epsilon$, from article 529, be substituted in this expression, after the reduction of the products of the sines and cosines to the cosines of multiple arcs, and substitution for M_0, M_1, N_3, N_4 N_5, from article 459, it becomes

$$\frac{\delta r}{a} = \frac{m'}{2}.\Sigma.C_i.\cos i(n't - nt + \epsilon' - \epsilon), \qquad (149)$$
$$+ m'.e.\Sigma.D_i.\cos\{i(n't - nt + \epsilon' - \epsilon) + nt + \epsilon - \varpi\},$$
$$+ m'.e'.\Sigma.E_i.\cos\{i(n't - nt + \epsilon' - \epsilon) + nt + \epsilon - \varpi'\},$$

where

$$C_i = \frac{n^2}{n^2 - i^2(n'-n)^2}\left\{\frac{2n}{n-n'}aA_i + a^2\frac{dA_i}{da}\right\},$$

$$D_i = \frac{n^2}{\{i(n'-n)+n\}^2 - n^2}\left\{\frac{3n}{n'-n}aA_i - \frac{i^2(n'-n)\{i(n'-n)-n\}-3n^2}{n^2}C_i\right.$$
$$\left. + \tfrac{1}{2}a^3\left(\frac{d^2A_i}{da^2}\right)\right\},$$

$$E_i = \frac{n^2}{\{i(n'-n)+n\}^2 - n^2}\left\{\frac{(i-1)(2i-1)n}{i(n'-n)+n}.aA_{(i-1)}\right.$$
$$\left. - \frac{i^2(n'-n)+n}{i(n'-n)+n}a^2\frac{dA_{(i-1)}}{da} - \tfrac{1}{2}a^3\frac{d^2A_{(i-1)}}{da^2}\right\}.$$

The Perturbations in Longitude.

537. Having thus determined the perturbations in the radius vector, the term $\frac{2\delta r}{r}$ is known; and if substitution be made for δa and δe, from article 529, $\frac{\delta h}{h}$ will be obtained, and the integral of equation (148) will give

$$\delta v = \frac{m'}{2}\Sigma.F_i.\sin i(n't - nt + \epsilon' - \epsilon)$$
$$+ m'e.\Sigma.G_i.\sin\{i(n't - nt + \epsilon' - \epsilon) + nt + \epsilon - \varpi\}$$
$$+ m'e'.\Sigma.H_i.\sin\{i(n't - nt + \epsilon' - \epsilon) + nt + \epsilon - \varpi'\}.$$

Where

$$F_i = \frac{n}{i(n-n')}\left\{-\frac{n}{n-n'} \cdot aA_i + 2C_i\right\}$$

$$G_i = \frac{n}{i(n'-n)+n}\left\{\frac{(i-1)n}{n'-n} \cdot aA_i - 2D_i - \frac{i\{i(n'-n)-n\}+6n}{2n} \cdot C_i\right\}$$

$$H_i = \frac{n}{i(n'-n)+n} \times \left\{\frac{-(i-1)(2i-1)naA_{(i-1)} - (i-1)na^2\dfrac{dA_{(i-1)}}{da}}{2(i(n'-n)+n)} - 2E_i\right\}$$

538. In these values of δr and δv, i includes all whole numbers, either positive or negative, zero excepted: δr and δv will now be determined in the latter case, which is very important, because it gives the part of the perturbations that is not periodic.

539. If $i = 0$ in the series R in article 449, the only constant term introduced by this value into δr will be

$$\frac{m}{2} a^3 \left(\frac{dA_0}{da}\right).$$

Again, in finding the integral δa the arbitrary constant a_i that ought to have been added, would produce a constant term in δr. In order to find it, let the origin of the time be at the instant of the conjunction of the two bodies m and m',

when $$n't - nt + \epsilon' - \epsilon = 0;$$

whence $\cos 0 = 1$, and the first term of δa in article 529 becomes

$$\delta a = -2m'a^2 \frac{n}{n'-n} \Sigma A_i,$$

whence $$\delta r = \frac{m'}{2} a^3 \frac{dA_0}{da} - 2m'a^2 \frac{n}{n'-n} \Sigma A_i;$$

where Σ extends to all positive values of i from $i = 1$ to $i = \infty$.

540. If these values of δr and δa be put in equation (148), the result will be

$$\delta v = m'a\left\{\frac{3n}{n'-n} \Sigma A_i - a\left(\frac{dA_0}{da}\right)\right\} \cdot nt.$$

And as by article 392 the elliptical parts of r and v that are not periodic, or that do not depend on sines and cosines, are $r = a$, and $v = nt + \epsilon$: those parts of the radius vector and true longitude that are not periodic are expressed by

$$r + \delta r = a - 2m'a^2 \frac{n}{n'-n} . \Sigma A_i + \tfrac{1}{2}m'a^3 \left(\frac{dA_0}{da}\right)$$

$$v + \delta v = nt + \epsilon + m'a \left\{\frac{3n}{n'-n} \Sigma A_i - a\left(\frac{dA_0}{da}\right)\right\} nt \qquad (150)$$

in the real orbit.

Thus the perturbations in longitude seem to contain a term that increases indefinitely with the time; were that really the case, the stability of the solar system would soon be at an end. This term however is only introduced by integration, since the differential equations of the perturbations contain no such terms; it is therefore foreign to their nature, and may be made to vanish by a suitable determination of the arbitrary constant quantities. In fact the true longitude of a planet in its disturbed orbit consists of three parts,—of the mean motion, of the equation of the centre, and of the perturbations. The mean motion of the planet is the only quantity in the problem of three bodies that increases with the time: the equation of the centre is a periodic correction which is zero in the apsides and at its maximum in quadratures; and the perturbations being functions of the sines of the mean longitudes of the disturbed and disturbing bodies are consequently periodic, and are applied as corrections to the equation of the centre. All the coefficients of these quantities are functions of the elements of the orbits, which vary periodically but in immensely long periods. The arbitrary constant quantities introduced by integration, must therefore be determined so that the mean motion of the troubled planet may be entirely contained in that part of the longitude represented by v.

541. The values of a, n, e, ϵ, and ϖ, in the preceding equations, are for the epoch $t = 0$, and would be the elliptical values of the elements of the orbit of m, if at that instant the disturbing forces were to cease. Let $n_{\prime}t$ be the mean motion of m given by observation, then the second of the equations under consideration gives

$$n_{\prime} = n\left\{1 + m'a\left(\frac{3n}{n'-n} . \Sigma A_i - a\left(\frac{dA_0}{da}\right)\right)\right\},$$

*

and let $a_{,}$ be the mean distance corresponding to $n_{,}$ resulting from the equation,

$$n_{,}^2 = \frac{S+m}{a_{,}^3}.$$

If in this last expression $n + n_{,} - n$, and $a + a_{,} - a$, be put for $n_{,}$ and $a_{,}$, and if $(n_{,} - n)^2$, $(a_{,} - a)^2$, which are very small be omitted, then

$$2n\,(n_{,} - n) = -\frac{3n^2}{a}\,(a_{,} - a);$$

and substituting for $n_{,}$ it becomes

$$a - a_{,} = \frac{2m'a^2}{3}\left\{\frac{3n}{n'-n}\;\Sigma A_i - a\left(\frac{dA_0}{da}\right)\right\};$$

and as a may be put for $a_{,}$ in the terms multiplied by m', the equations (150) become

$$r + \delta r = a_{,} - \tfrac{1}{6}\,m'a_{,}^3\left(\frac{dA_0}{da_{,}}\right)$$

$$v + \delta v = n_{,}t + \epsilon.$$

Thus δv no longer contains a term proportional to the time, and the mean motion of the disturbed planet is altogether included in the part of the longitude expressed by v, in consequence of the introduction of the arbitrary constant quantities $n_{,}$ and $a_{,}$, instead of n and a.

The part of δr depending on the first powers of the eccentricities may be found by making $i = 0$ in the values of δa, δe, &c., in article 529; after which their substitution in δr of article 536, will give

$$\delta r = -\frac{m'}{4}\,ae\left\{3a\left(\frac{dA_0}{da}\right) + \tfrac{1}{2}a^2\left(\frac{d^2A_0}{da^2}\right)\right\}\cos\,(nt + \epsilon - \varpi)$$

$$-\frac{m'}{4}\,ae'\left\{3A_1 - 3a\left(\frac{dA_{,}}{da}\right) - \tfrac{1}{2}a^2\left(\frac{d^2A_1}{da^2}\right)\right\}\cos\,(nt + \epsilon - \varpi').$$

The corresponding part of δv from article 535 is

$$\delta v = \frac{m'}{2}\,ae\left\{3a\left(\frac{dA_0}{da}\right) + \tfrac{1}{2}a^2\left(\frac{d^2A_0}{da^2}\right)\right\}\sin\,(nt + \epsilon - \varpi)$$

$$+\frac{m'}{2}\,ae'\left\{2A_1 - 2a\left(\frac{dA_1}{da}\right) - \tfrac{1}{2}a^2\left(\frac{d^2A_1}{da^2}\right)\right\}\sin\,(nt + \epsilon - \varpi').$$

542. If the different parts of the value of δr and δv be added, and if

$$f = \tfrac{1}{4}\left\{3a^2\left(\frac{dA_0}{da}\right) + \tfrac{1}{2}a^3\left(\frac{d^2A_0}{da^2}\right)\right\}$$

$$f' = \tfrac{1}{4}\left\{3aA_1 - 3a^2\left(\frac{dA_1}{da}\right) - \tfrac{1}{2}a^3\left(\frac{d^2A_1}{da^2}\right)\right\}$$

$$f'' = \tfrac{1}{4}\left\{2A_1 - 2a\left(\frac{dA_1}{da}\right) - \tfrac{1}{2}a^2\left(\frac{d^2A_1}{da^2}\right)\right\}$$

The periodic inequalities in the radius vector and true longitude of m when troubled by m', are

$$\frac{\delta r}{a} = -\frac{m'}{6}a_{,}^{3}\left(\frac{dA_0}{da_{,}}\right) + \frac{m'}{2}.\Sigma.\,C_i.\,\cos i\,(n't - nt + \epsilon' - \epsilon)$$
$$- m'.e.f.\,\cos(nt + \epsilon - \varpi) - m'e'f'\cos(nt + \epsilon - \varpi')$$
$$+ m'.e.\Sigma.D_i.\cos\{i\,(n't - nt + \epsilon' - \epsilon) + nt + \epsilon - \varpi\}$$
$$+ m'.e'.\Sigma.E_i.\,\cos\{i\,(n't - nt + \epsilon' - \epsilon) + nt + \epsilon - \varpi'\},$$

$$\delta v = \frac{m'}{2}.\Sigma.F_i.\,\sin i\,(n't - nt + \epsilon' - \epsilon)$$
$$+ 2m'.e.f.\,\sin(nt + \epsilon - \varpi) + 2m'.e'.f''.\,\sin(nt + \epsilon - \varpi')$$
$$+ m'.e.\Sigma.G_i.\,\sin\{i\,(n't - nt + \epsilon' - \epsilon) + nt + \epsilon - \varpi\}$$
$$+ m'.e'.\Sigma.H_i.\,\sin\{i\,(n't - nt + \epsilon' - \epsilon) + nt + \epsilon - \varpi'\}.$$

The action of each disturbing body will produce a similar effect on the radius vector and longitude of m, and the sum of all will be perturbations in these two co-ordinates arising from the disturbing action of the whole system on the planet m.

543. It has been already observed that each of the periodic variations δa, δe, &c., ought to contain an arbitrary constant quantity $a_{,}$, $e_{,}$, $\varpi_{,}$, &c., introduced by their integrations, so that their true values are
$$a_{,} + \delta a\,;\ e_{,} + \delta e\,;\ \varpi_{,} + \delta\varpi\,;\ \&c.\ \&c.$$
Now, if the values of δr, δv, are to express the effects of the disturbing forces on the radius vector and longitude during a given time, these constant quantities must be so determined, that when $t = 0$, they must give
$$e_{,} + \delta e = 0,\ \varpi_{,} + \delta\varpi = 0,\ \&c.\ \&c.,$$
as was done with δa.

Substituting these values in place of δe, $\delta\varpi$, &c., in equation (149), the resulting values will complete those of δr and δv, which will no longer contain any arbitrary quantity, but will express the whole change in the longitude and distance arising from the action of the disturbing forces. Hence, if (r) (v) be the elliptical values of r and v, given in article 392, but corrected for the secular variation of the elements, the radius vector and longitude of m in its troubled orbit will be determined by

$$r = (r) + \delta r, \quad v = (v) + \delta v.$$

Perturbations in Latitude.

544. If the second powers of the masses be omitted as well as the squares of the eccentricities, and the products of the eccentricities by the inclination, the orbit at the epoch being the fixed plane, then by article 437

$$\delta s = \frac{\delta z}{a}, \qquad \delta z = y\delta q - x\delta p,$$

and in this case $y = a \sin(nt + \epsilon)$, $x = a\cos(nt + \epsilon)$,

then $$\frac{\delta z}{a} = \delta q . \sin(nt + \epsilon) - \delta p \cos(nt + \epsilon),$$

and substituting the values of δq, δp, from article 529,

$$\delta s = \frac{m' . n^2}{n'^2 - n^2} \cdot \frac{a^2}{a'^2} \gamma \sin(n't + \epsilon' - \Pi)$$

$$+ \frac{m'.n^2.a^2a'}{2} \gamma \Sigma \frac{B_{(i-1)}}{n^2 - (n + i(n' - n))^2} \sin\{i(n't - nt + \epsilon' - \epsilon) + nt + \epsilon - \Pi).$$

Now if a plane very little inclined to the orbit of m be assumed for the fixed plane instead of that of the orbit at the epoch, and if ϕ, ϕ', θ, θ', be the inclinations and longitudes of the nodes of the orbits of m and m' on this new plane; then as γ is the tangent of the mutual inclination of the two orbits, and Π the longitude of their mutual intersection, by article 444,

$$\gamma \sin \Pi = p' - p; \quad \gamma \cos \Pi = q' - q.$$

If these values be substituted in δs, and if $(s) + \delta s = s$ be the whole latitude of m in its troubled orbit above the fixed plane, then will

$$s = q \sin(nt + \epsilon) - p\cos(nt + \epsilon)$$

$$+ \frac{m' . n^2}{n'^2 - n^2} \cdot \frac{a^2}{a'^2} \{(q' - q)\sin(n't + \epsilon') - (p' - p)\cos(n't + \epsilon')\}$$

$$- \frac{m'n^2.a^2.a'}{2}(q' - q)\Sigma \frac{B_{(i-1)}}{\{i(n' - n) + n\}^2 - n^2} \times$$

$$\sin(i(n't - nt + \epsilon - \epsilon) + nt + \epsilon)$$

$$+ \frac{m'n^2.a^2.a'^2}{2}(p' - p)\Sigma \frac{B_{(i-1)}}{\{i(n' - n)^2 + n\}^2 - n^2} \times$$

$$\sin(i(nt - nt + \epsilon - \epsilon) + nt + \epsilon).$$

The two terms independent of m' are the latitude of m above the fixed plane when m remains on the plane of its primitive orbit. If the exact latitude of m be substituted for these two terms, this expression will be more correct.

Each disturbing planet will add an expression to s similar to δs; the sum of the whole will be the true latitude of m when troubled by all the bodies in the system.

545. By a similar process, the perturbations depending on the other powers and products of the eccentricities may be obtained, but it would lead to long and intricate reductions, from which La Place's method, deduced directly from the equations (87), is exempt.

CHAPTER IX.

SECOND METHOD OF FINDING THE PERTURBATIONS OF A PLANET IN LONGITUDE, LATITUDE, AND DISTANCE.

Determination of the general Equations.

546. To determine the perturbations δv, δr, δs from the three general equations,

$$\frac{d^2x}{dt^2} + \frac{\mu x}{r^3} = \frac{dR}{dx}$$

$$\frac{d^2y}{dt^2} + \frac{\mu y}{r^3} = \frac{dR}{dx}$$

$$\frac{d^2z}{dt} + \frac{\mu z}{r^3} = \frac{dR}{dz}.$$

The sum of these equations respectively multiplied by dx, dy, dz is

$$\frac{dxd^2x + dyd^2y + dzd^2z}{dt^2} + \frac{\mu(xdx + ydy + zdz)}{r^3} \quad (151)$$

$$= dx\left(\frac{dR}{dx}\right) + dy\left(\frac{dR}{dy}\right) + dz\left(\frac{dR}{dz}\right).$$

The integral of which is evidently

$$\frac{dx^2 + dy^2 + dz^2}{dt^2} - \frac{2\mu}{r} + \frac{\mu}{a} = 2\int \mathrm{d}R. \quad (152)$$

The differential of R is only relative to the co-ordinates of m, because the motions of that body alone are under consideration; a is an arbitrary constant quantity introduced by integration; it is half the greater axis of the orbit of m when R is zero. Again, the same equations, respectively multiplied by x, y, and z, and added to the preceding integral, give

$$\frac{xd^2x + yd^2y + zd^2z}{dt^2} + \frac{dx^2 + dy^2 + dz^2}{dt^2} + \frac{\mu(x^2 + y^2 + z^2)}{r^3}$$

$$- \frac{2\mu}{r} + \frac{\mu}{a} = x\left(\frac{dR}{dx}\right) + y\left(\frac{dR}{dy}\right) + z\left(\frac{dR}{dz}\right) + 2\int \mathrm{d}R.$$

The two first members of this equation are equal to $\frac{1}{2}\frac{d^2r^2}{dt^2}$, the third is $\frac{\mu}{r}$, and if to abridge rR' be put for

$$x\left(\frac{dR}{dx}\right)+y\left(\frac{dR}{dy}\right)+z\left(\frac{dR}{dz}\right),$$

the equation becomes

$$\frac{1}{2}\frac{d^2r^2}{dt^2}-\frac{\mu}{r}+\frac{\mu}{a}=2\int d.R+rR'.$$

Let dv be the indefinitely small angle mSh, fig. 89, contained between

$$Sm=r, \text{ and } Sh=r+\delta r, \text{ then } mh^2=ma^2+ah^2;$$

but $$ma=rdv, \text{ and } ah=dr,$$

hence $$mh^2=dr^2+r^2dv^2=dx^2+dy^2+dz^2.$$

But $$xd^2x+yd^2y+zd^2z=d(xdx+ydy+zdz)$$
$$-(dx^2+dy^2+dz^2)=rd^2r-r^2dv^2;$$

so that the equation in question becomes,

$$\frac{rd^2r-r^2.dv^2}{dt^2}+\frac{\mu}{r}=rR',$$

whence $$\frac{dv^2}{dt^2}-\frac{d^2r}{rdt_2}-\frac{\mu}{r^3}=-\frac{1}{r}R'.$$

547. In solving equations by approximation, a value of the unknown quantity is found by omitting some of the smaller terms, then the value so found is substituted in the equation, and a new value is sought, including the terms that were at first omitted.

Now values of r and v have been determined in the elliptical orbit by omitting the parts containing the disturbing forces, but if $r+\delta r$ $v+\delta v$ be put for r and v in the preceding equations, the parts containing the elliptical motion may be omitted, and the remaining terms will give the perturbations. It is evident, however, that this substitution must not be made in R, since it contains the first powers of the disturbing action already. Consequently, the equations in question become

$$\frac{d^2.r\delta r}{dt^2}+\frac{\mu r\delta r}{r^3}=2\int dR+rR'$$

$$\frac{2r^2dv.d\delta v}{dt^2}=\frac{rd^2\delta r-\delta r.d^2r}{dt^2}-\frac{3\mu r\delta r}{r^3}-rR'$$

and eliminating $\frac{\mu r\delta r}{r^3}$ from the second by means of the first

$$d\delta v = \frac{d\{dr\,.\,\delta r + 2r\,.\,d\delta r\} - dt^2(3\int dR + 2rR')}{r^2 dv}. \qquad (153)$$

548. The integral of the angle $dv = mSh$ does not lie all in one plane, but it is easy to obtain a value of that indefinitely small angle in functions of its projection $pSB = dv'$. For let pB be the projection of the arc mh on the fixed plane of the ecliptic, then $Sm = r$, $mp = s$, the tangent of the latitude, and the curtate distance $Sp = \frac{r}{\sqrt{1+s^2}}$. Draw mE at right angles to Bh; and describe the arc pc with the radius Sp. Now the arc mh being indefinitely small, its projection pB is indefinitely small, therefore both may be taken in place of their sines; also

fig. 89.

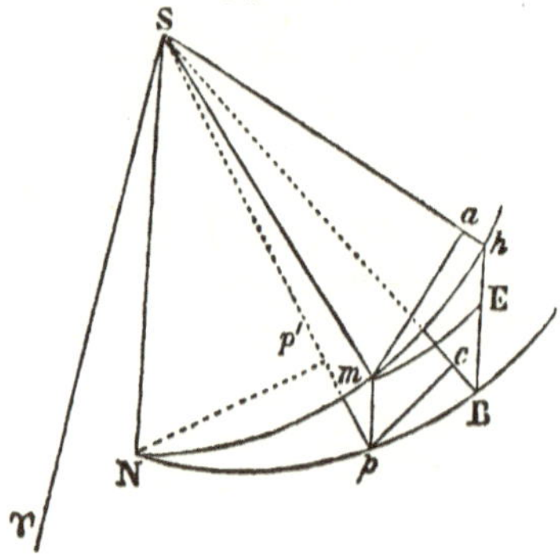

$$mh^2 = m\mathrm{E}^2 + \mathrm{E}h^2$$

and as $m\mathrm{E} = p\mathrm{B}$,

therefore $$pc^2 + c\mathrm{B}^2 + \mathrm{E}h^2 = mh^2.$$

But $pc = \frac{rdv'}{\sqrt{1+s^2}}$; $c\mathrm{B} = d.\frac{r}{\sqrt{1+s^2}} = \frac{dr(1+s^2) - rsds}{(1+s^2)^{\frac{3}{2}}}$.

Again $$mp = \frac{sr}{\sqrt{1+s^2}}.$$

And $$h\mathrm{E} = .dmp = \frac{rds + sdr(1+s^2)}{(1+s^2)^{\frac{3}{2}}};$$

hence $$mh^2 = \frac{r^2dv'^2}{1+s^2} + dr^2 + \frac{r^2ds^2}{(1+s^2)^2};$$

but $$mh^2 = r^2dv + dr^2,$$

and lastly, $$dv' = dv.\frac{\sqrt{(1+s^2)^2 - \frac{ds^2}{dv^2}}}{\sqrt{1+s^2}}. \qquad (154)$$

Thus dv' is known when ds is determined; however, if the latitude be estimated from a known position of the orbit of the planet itself at any given epoch, instead of from the fixed plane of the ecliptic, it will be zero at that instant; and any latitude the planet may have at a subsequent period, can only arise from the disturbing forces, and must on that account be very small.

549. Assuming therefore NpB, fig. 89, to have been the orbit of the planet m at any given time, s and ds will be so small, that their squares may be omitted, and then $dv = dv'$. Also the radius vector r, only differs from the curtate distance $\frac{r}{\sqrt{1+s^2}}$ by the extremely small quantity $\frac{1}{2}rs^2$ which may be omitted, and then the true longitude v may be estimated on the plane NpB without sensible error; so that $SN = x = r\cos v$; $Np' = y = r\sin v$;

and as $z\left(\frac{dR}{dz}\right)$ is so small that it may be omitted,

$$rR' = x\left(\frac{dR}{dx}\right) + y\left(\frac{dR}{dy}\right) = r\left(\frac{dR}{dr}\right);$$

hence, the equation which determines the perturbations in the radius vector becomes

$$\frac{d^2.r\delta r}{dt^2} + \frac{\mu r\delta r}{r^3} = 2\int \mathrm{d}R + r\left(\frac{dR}{dr}\right) \qquad (155)$$

550. It was shown in article 372, that r^2dv is the area described by the body in the indefinitely small time dt,

therefore $\quad r^2dv = \sqrt{\mu a(1-e^2)}\, dt = na^2\sqrt{1-e^2}\,.\,dt$

hence the value of $d\delta v$ becomes

$$d\delta v = \frac{1}{\sqrt{1-e^2}}\left\{\frac{d(2rd\,.\,\delta r + dr\,.\,\delta r)}{a^2ndt^2} - \frac{an}{\mu}\left(3\int \mathrm{d}R + 2r\left(\frac{dR}{dr}\right)\right)\right\}$$

and its integral is

$$\delta v = \frac{1}{\sqrt{1-e^2}}\left\{\frac{2rd\,.\,\delta r + dr\,.\,\delta r}{a^2ndt} - \frac{an}{\mu}\left(3\iint dt\,.\,\mathrm{d}R + 2\int r\left(\frac{dR}{dr}\right)dt\right)\right\} (156)$$

which determines the perturbations of m in longitude.

551. Since the orbit of m at the epoch is taken as the fixed plane, the only latitude the planet will have at a subsequent period must arise from the perturbations, and may therefore be represented by δs; hence $\quad z = r\delta s.$

And substituting this value of z in the third of the equations of motion in article 546, it becomes

$$\frac{d^2.r\delta s}{dt^2} + \frac{\mu\,.\,r\delta s}{r^3} - \frac{dR}{dz} = 0. \qquad (157)$$

A value of δs may easily be found from this; and if it be then desired to refer the position of the planet to a plane which is but little inclined to that of its primitive orbit, it will only be necessary to add to this value of δs the latitude of the planet, supposing it not to quit the plane of its primitive orbit.

Perturbations in the Radius Vector.

552. These are obtained by successive approximations from the equation $$\frac{d^2 r dr}{dt^2} + \frac{\mu r \delta r}{r^3} = 2\int \mathrm{d}R + r\left(\frac{dR}{dr}\right).$$

A value of δr is first determined by omitting the eccentricities; that value is substituted in the same equation, and a new value of δr is found, including the first powers of the eccentricities; that is again substituted, and a third value of δr is obtained, containing the squares and products of the eccentricities and inclinations, and so on, till the remaining or rejected quantities are less than the errors of observation.

553. Supposing the orbits to be circular, then $r^{-3} = a^{-3}$; and by article 383, $\frac{\mu}{a^3} = n^2$. And if the mass of the planet be omitted when compared with that of the sun taken as the unit, the preceding equation, after these substitutions, becomes

$$\frac{d^2 . r\delta r}{dt^2} + n^2 r dr = 2\int \mathrm{d}R + r\left(\frac{dR}{dr}\right).$$

But $$r\left(\frac{dR}{dr}\right) = a\left(\frac{dR}{du}\right);$$

and in this case $R = \frac{m'}{2} \Sigma A_i \cos i(n't - nt + \epsilon' - \epsilon)$.

When $i = 0$, $\cos i(n't - nt + \epsilon' - \epsilon) = 1$,

$$R = \frac{m}{2} A_0 + \frac{m}{2} . \Sigma . A_i \cos i(n't - nt + \epsilon' - \epsilon),$$

and as $\mathrm{d}R$ is the differential of R with regard to nt alone,

therefore $$2\int \mathrm{d}R + r\left(\frac{dR}{dr}\right) =$$

$$2m'g + \frac{m'}{2} a\left(\frac{dA_0}{da}\right) + \frac{m'}{2}\Sigma\left\{\frac{2n}{n-n'} A_i + a\left(\frac{dA_i}{da}\right)\right\} \cos i(n't - nt + \epsilon' - \epsilon);$$

whence $$\frac{d^2 r\delta r}{dt^2} + n^2 r\delta r = 2m'g + \frac{m'}{2} a . \left(\frac{dA_0}{da}\right)$$

$$+ \frac{m'}{2}\Sigma\left\{\frac{2n}{n-n'} A_i + a\left(\frac{dA_i}{da}\right)\right\} \cos i(n't - nt + \epsilon' - \epsilon).$$

The integral of this equation is

$$\frac{r\delta r}{a^2} = B + B' \,.\, \cos i\,(n't - nt + \epsilon' - \epsilon),$$

B and B' being indeterminate coefficients, then

$$\frac{d^2 r\delta r}{dt^2} + n^2 r\delta r = Bn^2a^2$$

$$+ B'a^2\,(n^2 - i^2\,(n' - n)^2)\cos i\,(n't - nt + \epsilon' - \epsilon).$$

And comparing the coefficients of like cosines,

$$B = 2m'ag + \frac{m'}{2}\,a^2\left(\frac{dA_0}{da}\right),$$

$$B' = \frac{m'}{2}\,\frac{an^2}{n^2 - i^2\,(n-n')^2}\,.\,\Sigma.\left\{\frac{2n}{n-n'}\,A_i + a\left(\frac{dA_i}{da}\right)\right\}, \text{ and so}$$

$$\frac{r\delta r}{a^2} = 2m'ag + \frac{m'}{2}\,a^2\left(\frac{dA_0}{da}\right) + \frac{m'n^2}{2}\,.\,\Sigma\,.\,\frac{\left\{\frac{2n}{n-n'}\,aA_i + a^2\left(\frac{dA_i}{da}\right)\right\}}{n^2 - i^2\,(n - n')^2}$$

$$\times \cos i\,(n't - nt + \epsilon' - \epsilon)\,;$$

or if a be put for r in the first member, and because by article 536,

$$C_i = \frac{n^2\left\{\frac{2n}{n-n'}\,aA_i + a^2\left(\frac{dA_i}{da}\right)\right\}}{n^2 - i^2\,(n'-n)^2}$$

$$\frac{\delta r}{a} = 2m'ag + \frac{m'}{2}\,a^2\left(\frac{dA_0}{da}\right) + \frac{m'}{2}\,\Sigma C_i \cos i\,(n't - nt + \epsilon' - \epsilon)\,;$$

which is the first approximation.

554. When the first powers of the eccentricities are included,

$$r^{-3} = \frac{1}{a^3}\,\{1 + 3e\cos\,(nt + \epsilon - \varpi)\}\,;$$

and therefore

$$\frac{d^2 r\delta r}{dt^2} + n^2 r\delta r + 3n^2 a\,.\,\delta r\,.\,e\cos\,(nt + \epsilon - \varpi) = 2\int dR + r\left(\frac{dR}{dr}\right),$$

but $\quad R = \frac{m'}{2}\,\Sigma M_0\, e\cos\{i\,(n't - nt + \epsilon' - \epsilon) + nt + \epsilon - \varpi\}$

$$+ \frac{m'}{2}\,\Sigma M_1\, e'\cos\{i\,(n't - nt + \epsilon' - \epsilon) + nt + \epsilon - \varpi'\}$$

therefore $\quad 2\int dR + r\left(\frac{dR}{dr}\right) =$

$$\frac{m'}{2}\,\Sigma\left\{\frac{2\,(i-1)\,n}{i\,(n-n') - n}\,M_0 + a\left(\frac{dM_0}{da}\right)\right\} e\cos\{i(n't - nt + \epsilon' - \epsilon) + nt + \epsilon - \varpi)$$

$$+ \frac{m'}{2}\,\Sigma\left\{\frac{2(i-1)\,n}{i\,(n-n') - n}\,M_1 + a\left(\frac{dM_1}{da}\right)\right\}\,.\,e'.\cos\{i(n't - nt + \epsilon' - \epsilon)$$

$$+ nt + \epsilon - \varpi'\}.$$

*

By the substitution of this quantity, and of the preceding value of $\frac{\delta r}{a}$,

$$\frac{d^2 r\delta r}{dt^2} + n^2 r\delta r = -\frac{m'}{2}\,\Sigma\left\{3a^2n^2\,C_i - \frac{2\,(i-1)n}{i(n-n')-n}\,M_0 - a\,\frac{dM_0}{da}\right\}$$

$$\times\ .\ e\cos\{i\,(n't - nt + \epsilon' - \epsilon) + nt + \epsilon - \varpi\}$$

$$+\frac{m'}{2}\,\Sigma\left\{\frac{2\,(i-1)n}{i(n-n')-n}\,M_1 + a\left(\frac{dM_1}{da}\right)\right\}\times$$

$$e'\,.\,\cos\{i\,(n't - nt + \epsilon' - \epsilon) + nt + \epsilon - \varpi'\}.$$

Let
$$\frac{r\delta r}{a^2} = \frac{m'}{2}\,Be\cos\{i\,(n't - nt + \epsilon' - \epsilon) + nt + \epsilon - \varpi\}$$

$$+\frac{m'}{2}\,B'e'\cos\{i\,(n't - nt + \epsilon' - \epsilon) + nt + \epsilon - \varpi'\}$$

B and B' being indeterminate coefficients,

then
$$\frac{d^2 r\delta r}{dt^2} + n^2 r\delta r =$$

$$\frac{m'}{2}\,.\,Ba^2\{n^2 - (i(n'-n)+n)^2\}.e.\cos\{i(n't-nt+\epsilon'-\epsilon)+nt+\epsilon-\varpi\}$$

$$+\frac{m'}{2}.\,B'a^2\{n^2 - (i(n'-n)+n)^2\}\,e'\cos\{i(n't-nt+\epsilon'-\epsilon)+nt+\epsilon-\varpi'\}$$

If to abridge

$$K_i = 3C_i - \frac{2(i-1)n}{i\,(n-n')-n}\,a\,M_0 - a^2\,.\,\frac{dM_0}{da}$$

$$L_i = -\frac{2(i-1)n}{i\,(n-n')-n}\,a\,M_1 - a^2\,\frac{dM_1}{da},$$

$$B = \frac{-n^2.K_i}{n^2-\{i(n-n')+n\}^2};\quad B' = \frac{-n^2.L_i}{n^2-\{i(n-n')+n\}^2};$$

and because $a^3n^2 = 1$,

$$\frac{r\delta r}{a^2} = \Sigma\,\frac{m'n^2}{2\{i(n'-n)+n\}^2-2n^2}\left\{\begin{array}{l}+K_i.e\cos\{i(n't-nt+\epsilon'-\epsilon)+nt+\epsilon-\varpi\}\\ +L_i.e'\cos\{i(n't-nt+\epsilon'-\epsilon)+nt+\epsilon-\varpi'\}\end{array}\right\};$$

where i may have any whole value, positive or negative, except zero. But in order to have the complete value of $\frac{r\delta r}{a^2}$, according to the theory of linear equations the integral of

$$\frac{d^2\,.\,r\delta r}{dt^2} + n^2 r\delta r = 0$$

must be added. The true integral of this equation is

$$\frac{rdr}{a^2} = m'fe\,.\,\cos\,(nt(1+c)+\epsilon-\varpi) + m'f'e'.\cos\,(nt(1+c')+\epsilon-\varpi')$$

where c and c' are given functions of the elements; but if it be assumed as is generally done, that the elliptical elements have already been corrected by their secular variations c and c' may be omitted, and then

$$\frac{r\delta r}{a^2} = m'fe \cos(nt + \epsilon - \varpi) + m'f'e' \cos(nt + \epsilon - \varpi').$$

If all the parts of $\frac{r\delta r}{a^2}$ that have been determined in this and in the first approximation be collected, and if $a(1 - \cos(nt + \epsilon - \varpi)$ be put for r, then will

$$\frac{\delta r}{a} = 2m'ag + \frac{m'}{2} a^2 \left(\frac{dA_0}{da}\right)$$

$$+ \frac{m'}{2} \cdot \frac{n^2}{(n^2 - i^2(n'-n)^2)} \cdot \Sigma \left\{ \frac{2n}{n-n'} aA_i + a^2 \left(\frac{dA_i}{da}\right) \right\} \cos\ (n't - nt + \epsilon' - \epsilon)$$

$$+ m'f \,.\, e \,.\, \cos(nt + \epsilon - \varpi) + m'f' \,.\, e' \,.\, \cos(nt + \epsilon - \varpi')$$

$$+ \frac{m'}{2} \cdot \Sigma \cdot \left\{ C_i + \frac{n^2 K_i}{\{i(n'-n)+n\}^2 - n^2} \right\} . e . \cos\{i(n't - nt + \epsilon' - \epsilon) + nt + \epsilon - \varpi\}$$

$$+ \frac{m'}{2} \cdot \Sigma \cdot \left\{ \frac{n L_i}{\{i(n'-n)+n\}^2 - n^2} \right\} . e' . \cos\{i(n't - nt + \epsilon' - \epsilon) + nt + \epsilon - \varpi'\}.$$

If substitution be made for K_i and L_i, it will be found that the coefficients in this expression are identical with those in article 536, so that

$$\frac{n^2 \Sigma \left\{ \frac{2n}{n-n'} aA_i + a^2 \left(\frac{dA_i}{da}\right) \right\}}{n^2 - i^2(n'-n)^2} = C_i,$$

$$C_i + \frac{n^2 K_i}{\{i(n'-n)+n\} - n^2} = D_i;$$

$$\frac{n^2 L_i}{\{i(n'-n)+n\}^2 - n^2} = F_i,$$

consequently

$$\frac{\delta r}{a} = 2m'ag + \frac{m'}{2} \cdot a^2 \left(\frac{dA_0}{da}\right) + \frac{m'}{2} \Sigma\, C_i \cos i\,(n't - nt + \epsilon' - \epsilon)$$

$$+ m' \,.\, fe \,.\, \cos(nt + \epsilon - \varpi) + m' \,.\, f' \,.\, e' \,.\, \cos(nt + \epsilon - \varpi)$$

$$+ m' \,.\, \Sigma \,.\, D_i \,.\, e \,.\, \cos\{i(n't - nt + \epsilon' - \epsilon) + nt + \epsilon - \varpi\}$$

$$+ m' \,.\, \Sigma \,.\, F_i \,.\, e' \,.\, \cos\{i(n't - nt + \epsilon' - \epsilon) + nt + \epsilon - \varpi'\}.$$

Perturbations in Longitude.

555. The perturbations in longitude may now be found from equation (156); which becomes, when e^2 is omitted, and $\mu = a^3n^2 = 1$,

$$\delta v = \frac{2rd \,.\, \delta r + dr \,.\, \delta r}{a^2 \,.\, ndt} - 3an\iint dt\, dR - 2an\int r\left(\frac{dR}{dr}\right) dt.$$

By the substitution of the preceding values of R and δr it will be found that the perturbations in longitude are

$$\delta v = -m'a\left(3g + a\left(\frac{dA_0}{da}\right)\right) . nt$$

$$+\frac{m'}{2}\Sigma\left\{-\frac{n^2}{i(n-n')^2}aA_i + \frac{2n^3\left\{\frac{2n}{n-n'}aA_i + a^2\left(\frac{dA_i}{da}\right)\right\}}{i(n-n')\{n^2-i^2(n'-n)^2\}}\right\}$$

$$\sin i(n't - nt + \epsilon' - \epsilon)$$

$$+ m'f_i e \sin(nt + \epsilon - \varpi) + m'f_i'e' . \sin(nt + \epsilon - \varpi')$$

$$+ m'e\,\Sigma\,G_i \sin\{i(nt - nt + \epsilon - \varpi) + nt + \epsilon - \varpi\}$$

$$+ m'e'\,\Sigma\,H_i \sin\{i(nt - nt + \epsilon - \varpi) + nt + \epsilon - \varpi'\} + C,$$

where $$f_i = 3a^2\frac{dA_0}{da} + a^3\frac{d^2A_0}{da^2} + 2ag - 2f$$

$$f_i' = \frac{3}{2}aA_1 - \frac{3}{2}a^2\frac{dA_1}{da} - a^3\frac{d^2A_1}{da^2} - 2f'.$$

556. If all the periodic terms be omitted in the expressions $r + \delta r$ and $v + \delta v$, they become

$$r + \delta r = a + 2m'ag + \tfrac{1}{2}m'a^2\left(\frac{dA_0}{da}\right)$$

$$v + \delta v = nt + \epsilon - m'\left(3ag + a^2\left(\frac{dA_0}{da}\right)\right) . nt;$$

$v + \delta v$ is the mean longitude of the planet at the end of the time t; and if it be asumed that this longitude is the same as the elliptical orbit of the planet, and in the orbit it really describes, this condition will determine g.

Whence $$g = -\tfrac{1}{3}a\left(\frac{dA_0}{da}\right);$$

and, as before, $$r + \delta r = a - \frac{m'}{6}a^3\left(\frac{dA_0}{da}\right),$$

which is the constant part of the radius vector in the troubled orbit.

Thus a is not the mean distance of the planet from the sun in the troubled orbit, as it is in the elliptical orbit. In the latter case a is deduced from the mean motion by the equation

$$n^2 = \frac{1}{a^3},$$

whereas in the troubled orbit it is

$$a - \frac{m'}{6}a^3\left(\frac{dA_0}{da}\right).$$

Therefore the mean motion and periodic time are different from what they would have been had there been no disturbance; but when once they are produced they are permanent, and unchangeable in their quantity by the subsequent action of the other bodies of the system.

The perturbations in the co-ordinates of a planet depend on the angular distances of the disturbed and disturbing bodies, that is, on the differences of their mean longitudes; but terms of the form

$$f_{\prime}e \sin (nt + \epsilon - \varpi),\ f_{\prime}'e' \sin (nt + \epsilon - \varpi')$$

belong to elliptical motion; they form a part of the equation of the centre, but they will vanish from δv if $f_{\prime}$ and $f_{\prime}'$, which are perfectly arbitrary, be made zero; in that case

$$f = \tfrac{1}{2}\left(\tfrac{7}{3}\, a^2 \frac{dA_0}{da} + a^3 \frac{d^2A_0}{da^2}\right);$$

$$f' = \tfrac{1}{2}\left\{\tfrac{3}{2}\, aA_1 - \tfrac{3}{2}\, a^2 \frac{dA_1}{da} - a^3 \frac{d^2A_1}{da^2}\right\}.$$

and as the arbitrary constant quantity C may be made zero, the perturbations in the radius vector and longitude of m are

$$\delta r = -\frac{m'}{6} a^2 \left(\frac{dA_0}{da}\right) + \frac{m'}{2} \Sigma\, C_i \cos i\,(n't - nt + \epsilon' - \epsilon) \quad (158)$$

$$+ m'fe \cos (nt + \epsilon - \varpi) + m'f'e' \cos (nt + \epsilon - \varpi')$$
$$+ m'e\,\Sigma\, D_i \cos \{ i\,(n't - nt + \epsilon' - \epsilon) + nt + \epsilon - \varpi\}$$
$$+ m'e'\,\Sigma\, E_i \cos \{ i\,(n't - nt + \epsilon' - \epsilon) + nt + \epsilon - \varpi'\};$$

$$\delta v = \frac{m'}{2} \Sigma\, F_i \sin i\,(n't - nt + \epsilon' - \epsilon) \quad (159)$$

$$+ m'e\,\Sigma\, G_i \sin \{ i\,(n't - nt + \epsilon' - \epsilon) + nt + \epsilon - \varpi\}$$
$$+ m'e'\,\Sigma\, H_i \sin \{ i\,(n't - nt + \epsilon' - \epsilon) + nt + \epsilon - \varpi'\}.$$

The coefficients being the same as in article 537, and i may have any whole value, except zero.

557. The integral

$$\frac{r\delta r}{a^2} = m'.f.c.\cos (nt(1+c) + \epsilon - \varpi) + m'.f'.e'.\cos (nt(1+c') + \epsilon - \varpi'),$$

by the resolution of the cosines becomes

$$\frac{r\delta r}{a^2} = m'.f.e.\cos (nt + \epsilon - \varpi) + m'.f'.e'.\cos (nt + \epsilon - \varpi')$$

$$- m'.f.e.cnt.\sin (nt + \epsilon - \varpi) - m'.f'.e'.c'nt.\sin (nt + \epsilon - \varpi');$$

and as it is given under this form by direct integration, it was very very embarrassing to mathematicians, because the terms containing the arcs cnt, $c'nt$, as coefficients, increase indefinitely with the time, and if such inequalities really had existence in our system, its stability would soon be at end. The expression (98) for the radius vector does not contain a term that increases with the time, neither does the series R; consequently the arc nt could not be introduced into the differential equation (155), unless R contained terms of the form $A \cdot {\sin \atop \cos}\left(\alpha + \beta t + \gamma t^2 + \&c.\right)$, the differential of which would produce them.

Now, the powers and products of sines and cosines introduce the sines and cosines of multiple arcs, but never the sines or cosines of the powers of arcs; consequently R does not contain terms of the preceding form, and therefore the differential equation (155) does not contain any term that increases with the time. Terms in the finite equations, that have the arc nt as coefficient, really arise from the imperfection of analysis, by which, in the course of integration, periodic terms, such as $A \cdot \cos(nt + \epsilon - \varpi)$, are introduced under their developed form $\alpha + \beta t + \gamma t^2 + \&c.$;
and, as Mr. Herschel observes, that is not done at once, but by degrees; a first approximation giving only α, the next βt, and so on. In stopping here, it is obvious that we should mistake the nature of this inequality, and that a really periodical function, from the effect of an imperfect approximation, would appear under the form of one not periodical, and would lead to erroneous conclusions as to the stability of the system and the general laws of its perturbations.

When, by this manner of integration, terms that increase with the time are introduced, the method of reducing the integrals to the periodic form will be found in a Memoir by La Place, in the *Mem. Acad. Sci.*, 1772, and in the fifth chapter, second book, of the *Mécanique Céleste*.

Perturbations in Latitude.

558. These are found by substituting

$$\frac{dR}{dz} = -\frac{m'}{a'^2}\gamma \sin(n't + \epsilon' - \Pi)$$

$$+ \frac{m'}{2} a' \Sigma B_{(i-1)} \gamma \sin\{i(n't - nt + \epsilon' - \epsilon) + nt + \epsilon - \Pi\} \text{ in}$$

$$\frac{d^2 r\delta s}{dt^2} + n^2 . r\delta s = \frac{dR}{dz};$$

where the primitive orbit of m is assumed to be the fixed plane, and the product of the eccentricity by the inclination neglected. Making $a^3n^2 = 1$, and integrating, the result is

$$\frac{\delta s}{a} = \frac{m'n^2}{n'^2 - n^2} \cdot \frac{a^2}{a'^2} \gamma \sin (n't + \epsilon' - \Pi) \qquad (160)$$

$$+ \frac{m'}{2} \cdot \frac{a'}{a} \gamma \frac{\Sigma B_{(i-1)} \cdot}{n^2 - (i(n'-n)+n)^2} \sin(i(n't-nt+\epsilon'-\epsilon)+nt+\epsilon-\Pi).$$

This expression is the same with that in article 544. No constant quantities are added, because, being arbitrary, they are assumed to be zero, which does not interfere with the generality of the problem, and is more convenient for use: i may have any whole value, positive or negative, zero excepted.

Perturbations, including the Squares of the Eccentricities and Inclinations.

559. When the approximation extends to the squares and products of the eccentricities and inclinations

$$r^{-3} = \frac{1}{a^3} \{1 + 3e \cos (nt + \epsilon - \varpi) + 3e^2 \cos 2 (nt + \epsilon - \varpi)\};$$

and by article 451,

$$R = \frac{m'}{2} . \Sigma N . \cos \{ i (n't - nt + \epsilon' - \epsilon) + 2nt + 2\epsilon + L \}$$

$$+ \frac{m'}{2} . \Sigma N' . \cos \{ i (n't - nt + \epsilon' - \epsilon) + L' \};$$

whence

$$2\int dR + r \left(\frac{dR}{dr}\right) =$$

$$\frac{m'}{2} . \Sigma \left\{a \frac{dN}{da} + N. \frac{2(2-i)n}{i(n'-n)+2n}\right\} \cos\{i(n't-nt+\epsilon'-\epsilon)+2nt+2\epsilon+L\}$$

$$+ \frac{m'}{2} . \Sigma \left\{a \frac{dN'}{da} - N' . \frac{2n}{(n'-n)}\right\} \cos \{ i (n't-nt+\epsilon'-\epsilon)+L' \}$$

hence

$$\frac{d^2 r\delta r}{dt^2} + n^2 r\delta r + 3n^2 a . \delta r . \{e \cos (nt+\epsilon-\varpi) + e^2 \cos 2 (nt+\epsilon-\varpi)\}$$

$$= \frac{m'}{2} \cdot \Sigma \left\{ a\frac{dN}{da} + N\frac{2(2-i)n}{i(n'-n)+2n} \right\} . \cos\{i(n't-nt+\epsilon'-\epsilon)+2n+L\}$$

$$+ \frac{m'}{2} . \Sigma \left\{ a\frac{dN'}{da} - N'\frac{2n}{i(n-n)} \right\} . \cos\{i(n't-nt+\epsilon'-\epsilon)+L'\}.$$

The value of $\frac{\delta r}{a}$, given in article 558, must be substituted in the last term of the first member; but as all terms are rejected that do not contain the squares or products of the eccentricities and inclinations, the only part of $\frac{\delta r}{a}$ that is requisite is

$$\frac{\delta r}{a} = \frac{m'}{2} \Sigma . C_i \cos i\,(n't - nt + \epsilon' - \epsilon)$$
$$+ m'e\,\Sigma\, D_i \cos\{i\,(n't - nt + \epsilon' - \epsilon) + nt + \epsilon - \varpi\}$$
$$+ m'e'\,\Sigma\, E_i \cos\{i\,(n't - nt + \epsilon' - \epsilon) + nt + \epsilon - \varpi'\};$$

where i may have every value, positive or negative, zero excepted. But if i be made negative in the two last terms, and if D'_i E'_i be the two coefficients in this case, then

$$\frac{\delta r}{a} = \frac{m'}{2} \Sigma\, C_i \cos i\,(n't - nt + \epsilon' - \epsilon) \qquad (161)$$
$$+ m'e\,\Sigma\, D_i \cos\{i\,(n't - nt + \epsilon' - \epsilon) + nt + \epsilon - \varpi\}$$
$$+ m'e\,\Sigma\, D'_i \cos\{-i\,(n't - nt + \epsilon' - \epsilon) + nt + \epsilon - \varpi\}$$
$$+ m'e'\,\Sigma\, E_i \cos\{i\,(n't - nt + \epsilon' - \epsilon) + nt + \epsilon - \varpi'\}$$
$$+ m'e'\,\Sigma\, E'_i \cos\{-i\,(n't - nt + \epsilon' - \epsilon) + nt + \epsilon - \varpi'\};$$

but now i can only be a *positive* whole number.

When this quantity is substituted in the last term of the first member for δr, and terms of the second order alone retained, the differential equation becomes, when integrated by the method of indeterminate coefficients, or otherwise,

$$\frac{r\delta r}{a^2} = \frac{m' . n^2}{\{i\,(n'-n)+3n\} . \{i\,(n'-n)+n\}} \times \qquad (162)$$

$$\left\{ \begin{array}{l} \frac{3}{2}e^2 . \Sigma . \{\frac{1}{2}C_i + D_i\} . \cos\{i\,(n't-nt+\epsilon'-\epsilon)+2nt+2\epsilon-2\varpi\} \\ + \frac{3}{2}ee' . \Sigma . E_i . \cos\{i\,(n't-nt+\epsilon'-\epsilon)+2nt+2\epsilon-\varpi-\varpi'\} \\ -\frac{1}{2}\Sigma\left\{\frac{2(2-i)n}{in'+(2-i)n}aN + a^2\frac{dN}{da}\right\} . \cos\{i(n't-nt+\epsilon'-\epsilon)+2nt+L\} \end{array} \right\}$$

$$+ \frac{m' \,.\, n^2}{\{i(n'-n)-n\} \,.\, \{i(n'-n)+n\}} \times$$

$$\left\{\begin{array}{l} \frac{3}{2}\, ee' \,.\, \Sigma \,.\, E_i \,.\, \cos\{i(n't-nt+\epsilon'-\epsilon)-\varpi'+\varpi\} \\ +\frac{3}{2}\, ee' \,.\, \Sigma \,.\, E'_i \,.\, \cos\{i(n't-nt+\epsilon'-\epsilon)+\varpi'-\varpi\} \\ +\frac{3}{2}\,\Sigma\{D_i+D'_i\} \,.\, e^2 \,.\, \cos i(n't-nt+\epsilon'-\epsilon) \\ -\frac{1}{2}\,\Sigma\{a^2\frac{dN'}{da}-\frac{2n}{n'-n}aN'\} \,.\, \cos\{i(n't-nt+\epsilon'-\epsilon+L'\}. \end{array}\right\}$$

Now in order to obtain the value of $\frac{\delta r}{a}$ from this expression it must be observed that

$$\frac{r\delta r}{a^2} = \frac{r}{a} \,.\, \frac{\delta r}{a};$$

and when the elliptical value of r is substituted it becomes

$$\frac{r\delta r}{a^2} = \frac{\delta r}{a}\{1+\tfrac{1}{2}e^2-e \,.\, \cos(nt+\epsilon-\varpi)-\tfrac{1}{2}e^2\cos 2(nt+\epsilon-\varpi)\}$$

whence

$$\frac{\delta r}{a} = \frac{r\delta r}{a^2} - \frac{\delta r}{a}\{\tfrac{1}{2}e^2-e\cos(nt+\epsilon-\varpi)-\tfrac{1}{2}e^2\cos 2(nt+\epsilon-\varpi)\}.$$

If the value of $\frac{\delta r}{a}$ from equation (161) be substituted in the second member, it will be found, after the reduction of the products of the cosines, that the perturbations in the radius vector depending on the second powers of the eccentricities and inclinations are expressed by

$$\begin{aligned} \frac{\delta r}{a} = \frac{r\delta r}{a^2} &+ \frac{m'}{4} \,.\, \Sigma \,.\, \{\tfrac{1}{2}C_i+2D_i\} \times \\ & e^2 \,.\, \cos\{i(n't-nt+\epsilon'-\epsilon)+2nt+2\epsilon-2\varpi\} \quad (163) \\ &+ \frac{m'}{2} \,.\, \Sigma \,.\, E_i ee' \,.\, \cos\{i(n't-nt+\epsilon'-\epsilon)+2nt+2\epsilon-\varpi-\varpi'\} \\ &+ \frac{m'}{2} \,.\, \Sigma\{D_i+D'_i-\tfrac{1}{2}C_i\} \,.\, e^2\cos i(n't-nt+\epsilon'-\epsilon) \\ &+ \frac{m'}{2} \,.\, \Sigma \,.\, E_i ee' \,.\, \cos\{i(n't-nt+\epsilon'-\epsilon)+\varpi-\varpi'\} \\ &+ \frac{m'}{2} \,.\, \Sigma \,.\, E_i \,.\, ee' \,.\, \cos\{i(n't-nt+\epsilon'-\epsilon)-\varpi+\varpi'\}; \end{aligned}$$

where $\frac{r\delta r}{a^2}$ represents equation (162).

560. With the values of $\frac{\delta r}{a}$ in (163) and (161), together with those terms of R that depend on the second powers and products of the eccentricities and inclinations, equation (156) gives the perturbations in longitude equal to

$$\delta v = \frac{1}{\sqrt{1-e^2}}\left\{\frac{2d\,(r\delta r)}{a^2.ndt} + \frac{m'}{2}.\Sigma.\{D_i - D'_i\}e^2.\sin i\,(n't - nt + \epsilon' - \epsilon)\right.$$

$$+ \frac{m'}{2}\,.\,\Sigma\,.\,E_i\,.\,ee'\,.\,\sin\{i\,(n't - nt + \epsilon' - \epsilon) + \varpi - \varpi'\}$$

$$- \frac{m'}{2}\,.\,\Sigma\,.\,E'_i\,ee'\,.\,\sin\{i\,(n't - nt + \epsilon' - \epsilon) - \varpi + \varpi'\} \quad (164)$$

$$- \frac{m'}{2}\,.\,\Sigma\,.\,\{\tfrac{1}{2}C_i + D_i\}\,e^2.\sin\{i\,(n't - nt + \epsilon' - \epsilon) + 2nt + 2\epsilon - 2\varpi\}$$

$$- \frac{m'}{2}\,.\,\Sigma E_i\,ee'\,.\,\sin\{i\,(n't - nt + \epsilon' - \epsilon) + 2nt + 2\epsilon - \varpi - \varpi'\}$$

$$- \frac{m'}{2}\Sigma\left\{\frac{(6-3i)\,n^2}{i(n'-n)+2n)^2}\,.\,aN + a^2\,.\,\frac{dN}{da}\,.\,\frac{2n}{i(n'-n)+2n}\right\}.\times$$
$$\sin\{i\,(n't - nt + \epsilon - \epsilon) + 2nt + L\},$$

$$- \frac{m'}{2}\Sigma\left\{\frac{2n}{i\,(n'-n)}\,.\,a^2\,.\,\frac{dN'}{da} - \frac{3n^2}{i\,(n'-n)^2}\,.\,aN'\right\}\times$$
$$\sin\{i\,(n't - nt + \epsilon' - \epsilon) + L'\}\}.$$

561. The inequalities of this order are very numerous, it is therefore necessary to select those that have the greatest values and to reject the rest, which can only be done in each particular case from the values of the divisors

$$i\,(n'-n)+3n,\quad i\,(n'-n)+2n,\quad i\,(n'-n)+n,\quad \text{and } i\,(n'-n).$$

For if the mean motions of the bodies m and m' be so nearly commensurable as to make any of these a small fraction, the inequality to which it is divisor will in general be of sufficient magnitude to be computed.

562. The inequalities in latitude will be determined afterwards.

Perturbations depending on the Cubes and Products of three Dimensions of the Eccentricities and Inclinations.

563. These perturbations are only sensible when the divisor $i\,(n'-n) + 3n$, is a very small fraction, that is, when the mean motions of the two bodies are nearly commensurable; but as this divisor arises from the angle $i\,(n't - nt + \epsilon' - \epsilon) + 3nt + 3\epsilon$ alone, the only part of the series R that is requisite by article 451, is

$$R = \frac{m'}{4}\,Q_0 e'^3 \cos\{i\,(n't - nt + \epsilon' - \epsilon) + 3nt + 3\epsilon - 3\varpi'\}$$

$$+ \frac{m'}{4} Q_1 e'^2 e \cos\{i(n't - nt + \epsilon' - \epsilon) + 3nt + 3\epsilon - 2\varpi' - \varpi\}$$

$$+ \frac{m'}{4} Q_2 e' e^2 \cos\{i(n't - nt + \epsilon' - \epsilon) + 3nt + 3\epsilon - \varpi' - 2\varpi\}$$

$$+ \frac{m'}{4} Q_3 e^3 \cos\{i(n't - nt + \epsilon' - \epsilon) + 3nt + 3\epsilon - 3\varpi\}$$

$$+ \frac{m'}{4} Q_4 e'\gamma^2 \cos\{i(n't - nt + \epsilon' - \epsilon) + 3nt + 3\epsilon - \varpi - 2\Pi\}$$

$$+ \frac{m'}{4} Q_5 e\gamma^2 \cos\{i(n't - nt + \epsilon' - \epsilon) + 3nt + 3\epsilon - \varpi - 2\Pi\}$$

But $\cos\{i(n't - nt + \epsilon' - \epsilon) + 3nt + 3\epsilon - 3\varpi\}$

$$= \cos 3\varpi . \cos\{i(n't - nt + \epsilon' - \epsilon) + 3nt + 3\epsilon)$$
$$+ \sin 3\varpi . \sin\{i(n't - nt + \epsilon' - \epsilon) + 3nt + 3\epsilon\};$$

each cosine may be resolved in the same manner; and if

$$P = \tfrac{1}{4}\{Q_0 \,.\, e'^3 . \sin 3\varpi' + Q_1 e'^2 e \sin(2\varpi' + \varpi) + Q_2 ee'^2 \sin(\varpi' + 2\varpi) + Q_3 e^3 \sin 3\varpi) + Q_4 e'\gamma^2 \sin(2\Pi + \varpi') + Q_5 e\gamma^2 \sin(2\Pi + \varpi)\}. \quad (165)$$

$$P' = \tfrac{1}{4}\{Q_0 e'^3 \cos 3\varpi' + Q_1 e'^2 e \cos(2\varpi' + \varpi) + Q_2 ee'^2 \cos(\varpi' + 2\varpi) + Q_3 e^3 \cos 3\varpi) + Q_4 e'\gamma^2 \cos(2\Pi + \varpi') + Q_5 e\gamma^2 \cos(2\Pi + \varpi)\}. \quad (166)$$

This part of R becomes

$$R = m' P \,.\, \sin\{i(n't - nt + \epsilon' - \epsilon) + 3nt + 3\epsilon\} + m'P' \,.\, \cos\{i(n't - nt + \epsilon' - \epsilon) + 3nt + 3\epsilon\}. \quad (167)$$

564. Let $\frac{r\delta r}{a^2} = m'K \cos\{i(n't - n't + \epsilon - \epsilon) + 2nt + 2\epsilon + B\}$

be the part of the equation (162) that has the divisor

$$i(n' - n) + 3n;$$

by the substitution of this, and of the preceding value of R, equation (155) gives, when integrated,

$$\frac{r\delta r}{a^2} = -\frac{2(i-3)m'n}{i(n' - n) + 3n}\begin{cases} aP\sin\{i(n't - nt + \epsilon' - \epsilon) + 3nt + 3\epsilon\} \\ + aP'\cos\{i(n't - nt + \epsilon' - \epsilon) + 3nt + 3\epsilon\} \end{cases}$$
$$- \tfrac{3}{2} m'eK \cos\{i(n't - nt + \epsilon' - \epsilon) + 3nt + 2\epsilon + B - \varpi\}$$
$$+ \tfrac{1}{2} m'eK \cos\{i(n't - nt + \epsilon' - \epsilon) + nt + \epsilon + B + \varpi\};$$

and because $\frac{r\delta r}{a^2} = \frac{r}{a} . \frac{\delta r}{a}$

$$= \frac{r}{a} m'K \cos\{i(n't - nt + \epsilon' - \epsilon) + 2nt + 2\epsilon + B\}$$

*

the whole perturbations in the radius vector having the divisor

$$i(n'-n)+3n, \text{ are}$$

$$\frac{\delta r}{a} = m'K\cos\{i(n't-nt+\epsilon'-\epsilon)+2nt+2\epsilon+B\} \quad (168)$$

$$-m'Ke\cos\{i(n't-nt+\epsilon'-\epsilon)+3nt+3\epsilon-\varpi+B\}$$

$$+m'Ke\cos\{i(n't-nt+\epsilon'-\epsilon)+nt+\epsilon+\varpi+B\}$$

$$-\frac{2(i-3)nm'}{i(n'-n)+3n}\left\{\begin{array}{l} aP\,.\,\sin\{i(n't-nt+\epsilon'-\epsilon)+3nt+3\epsilon\} \\ +aP'.\cos\{i(n't-nt+\epsilon'-\epsilon)+3nt+3\epsilon\}\end{array}\right\}.$$

565. If this quantity and the preceding value of R be substituted in equation (156) the result will be,

$$\delta v = -\frac{3(3-i)m'n^2}{\{i(n'-n)+3n\}^2}\left\{\begin{array}{l} aP'\sin\{i(n't-nt+\epsilon'-\epsilon)+3nt+3\epsilon\} \\ -aP\cos\{i(n't-nt+\epsilon'-\epsilon)+3nt+3\epsilon\}\end{array}\right\} \quad (169)$$

$$+\frac{2m'n}{i(n'-n)+3n}\left\{\begin{array}{l} a^2\left(\dfrac{dP}{da}\right)\cos\{i(n't-nt+\epsilon'-\epsilon)+3nt+3\epsilon\} \\ -a^2\left(\dfrac{dP'}{da}\right)\sin\{i(n't-nt+\epsilon'-\epsilon)+3nt+3\epsilon\}\end{array}\right\}$$

$$-\frac{m'e}{2}K\sin\{i(n't-nt+\epsilon'-\epsilon)+3nt+3\epsilon-\varpi+B\}$$

$$+\frac{5}{2}m'eK\sin\{i(n't-nt+\epsilon'-\epsilon)+nt+\epsilon+\varpi+B\}.$$

And as that part of δv article 560, having the divisor $i(n-n)+3n$ is nearly

$$\delta\zeta = m'He\sin\{i(n't-nt+\epsilon'-\epsilon)+2nt+2\epsilon+B\}$$

the terms $\frac{5}{4}m'eK.\sin\{i(n't-nt+\epsilon'-\epsilon)+nt+\epsilon+\varpi+B\}$ must be added to the preceding value of δv, which will then be the whole perturbations in longitude having the divisor in question.

Secular Variation of the Elliptical Elements during the periods of the Inequalities.

566. An inequality

$$\frac{C}{\{5n'-2n\}}\begin{matrix}\sin \\ \cos\end{matrix}\{(5n'-2n)t+B\}$$

is at its maximum when the sine or cosine is unity; and if $5n'-2n$ be a small fraction, the coefficient

$$\frac{C}{\{5n'-2n\}^2}$$

will be very great. The period of an inequality is the time the argument or angle $(5n'-2n)\,t+B$ takes to increase from zero to 360°; it is evident that the period will be the greater, the less the difference $5n'-2n$.

Thus, the perturbations in longitude expressed by

$$\delta v = -\frac{3\,(3-i)\,m'n^2}{\{i(n'-n)+3n\}^2}\cdot\left\{\begin{array}{l} aP'\sin\{i(n't-nt+\epsilon'-\epsilon)+3nt+3\epsilon\} \\ -aP\,.\cos\{i(n't-nt+\epsilon'-\epsilon)+3nt+3\epsilon\}\end{array}\right\}$$

are very great, and of long periods, when $i\,(n'-n)+3n$ is a small fraction.

567. The square of the divisor could only be introduced by a double integration, consequently the preceding value of δv is the integral of the part

$$\delta v = -\,3a\textstyle\int\!\int ndt\,.\,\mathrm{d}R$$

of equation (156), which is the periodic inequality in the mean motion of m, when troubled by m', in article 439. Thus, when the mean motions are nearly commensurable, all terms having the small divisors in question, must be applied as corrections to the mean motion of the troubled planet.

568. In some cases the periods of these inequalities extend to many centuries; in so long a time the secular variations of the elements of the orbits have a very sensible influence on these perturbations; and in order to include this effect, the expression

$$\delta v = -\,3\textstyle\int\!\int andt\,.\,\mathrm{d}R$$

must be integrated *by parts* in the hypothesis of P and P' being variable functions of the elements. Now

$$R = m'P\sin\{i\,(n't-nt+\epsilon'-\epsilon)+3nt+3\epsilon\}$$
$$+\,m'P'\cos\{i\,(n't-nt+\epsilon'-\epsilon)+3nt+3\epsilon\};$$

whence

$$\mathrm{d}R = +\,m'P\,.\,(3-i)\,ndt\,.\cos\{i\,(n't-nt+\epsilon'-\epsilon)+3nt+3\epsilon\}$$
$$-\,m'P'\,.(3-i)\,ndt\,.\sin\{i\,(n't-nt+\epsilon'-\epsilon)+3nt+3\epsilon\}$$

and $$-\,3a\textstyle\int\!\int ndt\,.\,\mathrm{d}R =$$

$$3a\,(3-i)\,.\,m'\textstyle\int\!\int P'\,.\,n^2dt^2\,.\sin\{i\,(n't-nt+\epsilon'-\epsilon)+3nt+3\epsilon\}$$
$$-\,3a\,(3-i)\,m'\textstyle\int\!\int P\,.\,n^2dt^2\,.\cos\{i\,(n't-nt+\epsilon'-\epsilon)+3nt+3\epsilon\}.$$

From the integration of this equation it will be found that the periodic inequality in the mean motion, depending on the third dimensions of the eccentricities and inclinations, and affected by the secular variations during its period, is

$$\delta v = \delta \zeta = \tag{170}$$

$$\frac{3(3-i)m'n^2}{\{i(n'-n)+3n\}^2}.\left\{aP-\frac{2a.dP'}{\{i(n'-n)+3n\}dt}-\frac{3a.d^2P}{\{i(n'-n)+3n\}^2dt^2}-\&c.\right\}\times$$

$$\cos\{i(n't-nt+\epsilon'-\epsilon)+3nt+3\epsilon\}$$

$$-\frac{3(3-i)m'n^2}{\{i(n'-n)+3n\}^2}.\left\{aP'+\frac{2a.dP}{\{i(n'-n)+3n\}dt}-\frac{3a.d^2P'}{\{i(n'-n)+3n\}^2dt^2}+\&c.\right\}\times$$

$$\sin\{i(n't-nt+\epsilon'-\epsilon)+3nt+3\epsilon\}.$$

This correction must be applied to the mean motions in the elliptical part of such planets as have their motions nearly commensurable.

569. The same method of integration may be employed for the term in equation (164), that has the divisor

$$n^2 - \{i(n'-n)+2n\}^2$$

when the quantity $\quad i(n'-n)+3n$

is a small fraction, and in general to all inequalities of long periods having small divisors.

The variation of the elements during the periods of the inequalities may be estimated by the following approximate method, which will answer for several centuries before and after the epoch. By the method employed in article 563 the sum of the terms in equations (164) depending on the angle

$$i(n't-nt+\epsilon'-\epsilon)+2nt+2\epsilon$$

may be put under the form

$$\delta v = m'\bar{P}\sin\{i(n't-nt+\epsilon'-\epsilon)+2nt+2\epsilon\}$$
$$+\, m'\bar{P}'\cos\{i(n't-nt+\epsilon'-\epsilon)+2nt+2\epsilon\};$$

$\bar{P}$ and $\bar{P}'$ being functions of the elements of the orbits of m and m' determined by observation for a given epoch, say 1750. Since $\bar{P}$ and $\bar{P}'$ are known quantities, let

$$\frac{\bar{P}'}{\bar{P}} = \tan\bar{E}, \text{ and } \sqrt{\bar{P}^2+\bar{P}'^2} = \bar{F}$$

sin $\bar{E}$ having the same sign with $\bar{P}'$ and cos $\bar{E}$ with $\bar{P}$; hence

$$\delta v = m'\bar{F}\sin\{i(n't-nt+\epsilon'-\epsilon)+2nt+2\epsilon+\bar{E}\}$$

are the perturbations in question for the epoch 1750. Now if the time t be made equal to 500 in the expressions for the elements in article 480, values of P and P' will be found for the year 2250, with which new values of F and E may be computed for that era. Again, values of P and P' may be obtained from the same formulæ for the year 2750, and by the method employed in article 480, the series

$$F = \bar{F} + \frac{d\bar{F}}{dt}t + \tfrac{1}{2}\frac{d^2\bar{F}}{dt^2}t^2 + \&c.$$

$$E = \bar{E} + \frac{d\bar{E}}{dt}t + \tfrac{1}{2}\frac{d^2\bar{E}}{dt^2}t^2 + \&c.$$

will give values of the variable coefficients for any time t during many centuries, consequently

$$\delta v = m'\left\{\bar{F} + \frac{d\bar{F}}{dt}t + \&c.\right\}\sin\left\{i(n't - nt + \epsilon' - \epsilon) + 2nt + 2\epsilon + \bar{E} + \frac{d\bar{E}}{dt}t^2 + \&c.\right\} \qquad (171)$$

will give the perturbations, including the secular variations in the elements of the orbits during their periods, $\bar{F}$, $\bar{E}$ and their differences being relative to the epoch 1750.

570. The formulæ that have been obtained will give the places of all the planets at any instant with great accuracy, except those of Jupiter and Saturn, which are so remote from the rest, as to be almost beyond the sphere of their disturbing influence; but their proximity to one another, and their immense magnitude, render their mutual disturbances greater than those of any of the other planets. They may be regarded as forming with the sun a system by themselves; and as there are some circumstances in their motions peculiar to them alone, their theory will form a separate subject of consideration.

CHAPTER X.

THE THEORY OF JUPITER AND SATURN.

571. By comparing ancient with modern observations, Halley discovered that the mean motion of Jupiter had been accelerated, and that of Saturn retarded. Halley, Euler, La Grange, La Place, and other eminent mathematicians, were led by their researches to the certain conclusion that these inequalities do not depend on the configuration of the orbits; and as La Place proved that they are not occasioned by the action of comets, or bodies foreign to the system, he could only suppose them to belong to the class of periodic inequalities.

Observation had already shown that five times the mean motion of Saturn is so nearly equal to twice the mean motion of Jupiter, that the difference of these two quantities, or $5n' - 2n$, is an extremely small fraction, being about the 74th part of the mean motion of Jupiter. La Place perceived that the square of this minute quantity is divisor to some of the perturbations in the longitude of Jupiter and Saturn, which led him to conjecture that the nearly commensurable ratio in the mean motions might be the cause of this anomaly in the theory of these two planets; a conjecture which computation amply confirmed, showing that a great inequality of $48' \, 2''.207$ at its maximum exists in the theory of Saturn, which at the present time increases the mean motion of the planet, and accomplishes its changes in about 929 years; and that the mean motion of Jupiter is also affected by a corresponding and contrary inequality of nearly the same period, only amounting to $19' \, 46''.62$ at its maximum, which diminishes the mean motion of Jupiter.

These two inequalities attained their maximum in the year 1560; from that period, the apparent mean motion of the two planets approached to their true motions, and became equal to them in 1790, which accounts for Halley finding the mean motion of Saturn slower, and that of Jupiter faster, by a comparison of ancient with modern observations, than modern observations alone showed them to be: whilst on the other hand, modern observations indicated to Lambert an acceleration in Saturn's motion, and a retardation in that of Jupiter; and the quantities of the inequalities found by these astronomers are nearly the same with those determined by La Place.

Recorded observations of these mean motions at very remote periods enable us to ascertain the chronology of the nations in which science had made early advances. Thus the Indians determined the mean motions of Jupiter and Saturn, when the mean motion of Jupiter was at its maximum of acceleration, and that of Saturn at its greatest retardation; the two periods at which that was the case, were 3102 years before the Christian era, and 1491 years after it.

The formulæ of the motions of Jupiter and Saturn determined by La Place, agree with their oppositions, the error not amounting to 12″.96, when it is to be recollected that only twenty years ago the errors in the best tables exceeded 1296″. These formulæ also represent with great precision the observations of Flamstead, of the Arabian astronomers, and of Ptolemy, leaving no grounds to doubt that La Place has succeeded in solving this difficulty, by assigning the true cause of these inequalities, which had for so many ages baffled the acuteness of astronomers; so that anomalies which seemed at variance with the law of gravitation, do in fact furnish the strongest corroboration of the universal influence it exerts throughout the solar system. Such, says La Place, has been the fate of that brilliant discovery of Newton, that every difficulty which has been raised against it, has formed a new subject of triumph, the sure characteristic of a law of nature.

The precision with which these two greatest planets of our system have obeyed the laws of mutual gravitation from the earliest periods at which we have records of their motions, proves the stability of the system, since Saturn has experienced no sensible action of foreign bodies from the time of Hipparchus, although the sun's attraction on Saturn is about a hundred times less than that exerted on the earth.

Periodic Variations in the Elements of the Orbits of Jupiter and Saturn, depending on the First Powers of the Disturbing Forces.

572. If i be made equal to 5 in equation (169), the great inequality of Jupiter, including the secular variations of the elements of both orbits during its period of 929 years, is

$$\delta v = \delta \zeta = \tag{172}$$

$$\frac{6m'n^2}{(5n'-2n)^2}\left\{aP' + \frac{2a\,.\,dP}{(5n'-2n)dt} - \&c.\right\}.\sin(5n't - 2nt + 5\epsilon' - 2\epsilon)$$

$$-\frac{6m'n^2}{(5n'-2n)^2}\left\{aP - \frac{2a.dP'}{(5n'-2n)dt} - \&c.\right\}.\cos(5n't - 2nt + 5\epsilon' - 2\epsilon)$$

$$+\frac{2m'n}{5n'-2n}\left\{\begin{array}{l} a^2\,.\,\dfrac{dP}{da}\,.\cos(5n't - 2nt + 5\epsilon' - 2\epsilon) \\ -a^2.\dfrac{dP'}{da}.\sin(5n't - 2nt + 5\epsilon' - 2\epsilon)\end{array}\right\}$$

$$-\frac{m'}{2}\,.\,eK\,.\,\sin(5n't - 2nt + 5\epsilon' - 2\epsilon - \varpi + \beta)$$

$$+\frac{5m'}{2}.\,eK\,.\,\sin(5n't - 2nt + 5\epsilon' - 2\epsilon + \varpi + \beta)$$

$$+\,m'.He\,.\,\sin(5n't - 2nt + 5\epsilon' - 2\epsilon + \varpi + \beta),$$

which must be applied as a correction to the mean motion of Jupiter.

573. Because of the equality and opposition of action and reaction, the great inequality in the mean motion of Saturn may be determined when that of Jupiter is known, and *vice versâ;* for by article 546,

$$\frac{dx^2 + dy^2 + dz^2}{dt^2}. - 2\,\frac{(S+m)}{r} = 2\int dR$$

may be assumed to belong to Jupiter, and

$$\frac{dx'^2 + dy'^2 + dz'^2}{dt^2} - 2\left(\frac{S+m'}{r'}\right) = 2\int dR'$$

to Saturn, dR and dR' relate to the co-ordinates of m and m'. Their sum, when the first equation is multiplied by m, and the second by m', is

$$2m\int dR + 2m'\int dR' = -\,2m\,\frac{(S+m)}{r} + m\,\frac{dx^2 + dy^2 + \delta z^2}{dt^2}$$

$$-\,2m'\,\frac{(S+m')}{r'} + m'\,\frac{dx'^2 + dy'^2 + dz'^2}{dt^2}.$$

The second member of this equation does not contain any term of the order of the squares of the disturbing masses having the divisor $5n' - 2n$, which can only arise from the integration of the sines or cosines of the angle $5n't - 2nt$; because, when the elliptical values are substituted instead of x, y, z, the part

$$-\,\frac{2m(s+m)}{r} + m\,\frac{dx^2 + dy^2 + dz^2}{dt^2},$$

will only contain the sines or cosines of the angle nt, and the remaining part of the second member is a function of $n't$ only; and as such terms as have the square of the divisor $5n' - 2n$ are alone under consideration, the second member may be omitted,

then $$m\int dR + m'\int dR' = 0. \qquad (173)$$

574. When $S + m = \mu$ is restored, which has hitherto been assumed equal to unity, the general expression for the periodic inequality in the mean motion of Jupiter is

$$\delta\zeta = -\,3\int\int.\,\frac{andt\,.\,dR}{S+m}.$$

The corresponding inequality in the mean motion of Saturn is

$$\delta\zeta' = -\,3\int\int.\,\frac{a'n'dt.dR'}{S+m'}.$$

From these two it is easy to find

$$m(S + m)\,.\,a'n'.\,\delta\zeta + m'(S + m')\,.\,an\,.\,\delta\zeta +$$
$$3m\,.\,a'n'\int\int andt\,.\,dR + 3m'.\,an\,.\int\int a'n'dt\,.\,dR' = 0.$$

And in consequence of equation (173)

$$m(S + m)\,.\,a'n'\,.\,\delta\zeta + m'(S + m')\,.\,an\,.\,\delta\zeta = 0.$$

But $$n = \frac{\sqrt{S + m}}{a^{\frac{3}{2}}} \qquad n' = \frac{\sqrt{S + m'}}{a'^{\frac{3}{2}}};$$

and if the masses m and m' be omitted in $(S + m)$, $(S + m')$; in comparison of the mass of the sun taken as the unit, the preceding equation becomes

$$m\sqrt{a}\,.\,\delta\zeta = -\,m'\sqrt{a'}\,.\,\delta\zeta'.$$

Thus the periodic inequality in the mean motion of Jupiter is contrary to that in the mean motion of Saturn when n and n' have the same signs, which must always be the case, because both planets revolve about the sun in the same direction, so that one body is accelerated when the other is retarded, which corresponds with observation. These inequalities are in the ratio of $m\sqrt{a}$ to $m'\sqrt{a'}$; hence, if the inequality in the mean motion of Jupiter be known, that in the mean motion of Saturn will be found from

$$\delta\zeta' = -\frac{m\sqrt{a}}{m'\sqrt{a'}}\,\delta\zeta. \qquad (174)$$

575. As the whole of the following analyses depend on the angle

$$5n't - 2nt + 5\epsilon' - 2\epsilon,$$

it will be represented by λ for the sake of abridgment. If i be made equal to 5 in equation (167), it becomes

$$R = m'P \,.\, \sin\lambda + m'P' .\cos\lambda.$$

From this, values of dR, $\frac{dR}{de}$, $\frac{dR}{d\varpi}$ may be found; but equations (165) and (166), show that

$$\left(\frac{dP}{d\varpi}\right) = e\left(\frac{dP'}{de}\right); \quad \left(\frac{dP'}{d\varpi}\right) = -e\left(\frac{dP}{de}\right);$$

consequently, by the substitution of dR, $\frac{dR}{de}$, $\frac{dR}{d\varpi}$ in equations (114), the periodic variations in the eccentricity, longitude of the perihelion, and semigreater axis of Jupiter's orbit, depending on the third powers of the eccentricities and inclinations, are easily found to be

$$\delta e_{,} = +\frac{m'.an}{5n'-2n}\left\{\frac{dP}{de}\,.\,\sin\lambda + \frac{dP'}{de}\,.\,\cos\lambda\right\} \qquad (175)$$

$$e\delta\varpi_{,} = -\frac{m'.an}{5n'-2n}\left\{\frac{dP}{de}\,.\,\cos\lambda - \frac{dP'}{de}\,.\,\sin\lambda\right\}. \qquad (176)$$

576. The periodic inequalities in γ and Π, the mutual inclination of the orbits of Jupiter and Saturn, and the longitude of the ascending node of the orbit of Saturn on that of Jupiter, are obtained from

$$R = \frac{m'}{4}\,.\,Q_4\,e'\gamma^2.\cos(\lambda - 2\Pi - \varpi')$$

$$+\frac{m'}{4}\,.\,Q_5\,e\gamma^2.\cos(\lambda - 2\Pi - \varpi); \text{ or}$$

$$R = \frac{m'}{4} \cdot \gamma^2 \cos 2\Pi \, \{Q_4 \, . \, e' \cos(\lambda - \varpi') + Q_5 \, . \, e \cos(\lambda - \varpi)\}$$

$$+ \frac{m'}{4} \cdot \gamma^2 \sin 2\Pi \, \{Q_4 \, . \, e' \sin(\lambda - \varpi') + Q_5 \, . \, e \sin(\lambda - \varpi)\};$$

or to abridge

$$R = \frac{m'}{4} \cdot \gamma^2 \cos 2\Pi \, . \, A + \frac{m'}{4} \cdot \gamma^2 \sin 2\Pi \, . \, B.$$

But from article 444 it appears that

$$\gamma^2 \, . \cos 2\Pi = (q'-q)^2 - (p'-p)^2; \quad \gamma^2 \sin 2\Pi = 2(q'-q)(p'-p);$$

whence

$$R = \frac{m'}{4}\{(q'-q)^2 - (p'-p)^2\} \, . \, A + \frac{m'}{4} \, . \, 2(q'-q)(p'-p) \, . \, B,$$

and $$\frac{dR}{dp} = \frac{m'}{2}(p'-p) \, . \, A - \frac{m'}{2}(q'-q) \, . \, B,$$

or $$\frac{dR}{dp} = \frac{m'}{2} \, . \, \gamma \sin \Pi \, . \, A - \frac{m'}{2} \, . \, \gamma \cos \Pi \, . \, B;$$

restoring the values of A and B, and reducing the products of the sines and cosines,

$$\frac{dR}{dp} = -\frac{m'}{2} \, . \, Q_4 \, . \, e'\gamma \, . \sin(\lambda - \varpi' - \Pi) - \frac{m'}{2} \, . \, Q_5 \, . \, e\gamma \, . \sin(\lambda - \varpi - \Pi).$$

But $\sin(\lambda - \varpi' - \Pi) =$

$$\sin(\lambda + \Pi) \, . \, \cos(\varpi + 2\Pi) - \cos(\lambda + \Pi) \, . \, \sin(\varpi + 2\Pi),$$

hence

$$\frac{dR}{dp} = \frac{m'}{2}\{Q_4 \, . \, e'\gamma \, . \, \sin(\varpi' + 2\Pi) + Q_5 \, . \, e\gamma \, . \, \sin(\varpi + 2\Pi)\} \, . \cos(\lambda + \Pi)$$

$$+ \frac{m'}{2}\{Q_4 \, . \, e'\gamma \, . \cos(\varpi' + 2\Pi) + Q_5 \, . \, e\gamma \, . \cos(\varpi + 2\Pi)\} \, . \sin(\lambda + \Pi);$$

and from equations (165) and (166) it is clear that

$$\frac{dR}{dp} = m' \frac{dP}{d\gamma} \, . \, \cos(\lambda + \Pi) - m' \, . \, \frac{dP'}{d\gamma} \, . \, \sin(\lambda + \Pi).$$

In the same manner it may be found that

$$\frac{dR}{dq} = -m' \, . \, \frac{dP}{d\gamma} \, . \, \sin(\lambda + \Pi) - m' \frac{dP'}{d\gamma} \, . \, \cos(\lambda + \Pi);$$

with these values the two last of equations (114) become, when integrated,

$$\delta p = \frac{m' \, . \, an}{5n' - 2n} \, . \, \left\{\frac{dP}{d\gamma} \, . \cos(\lambda + \Pi) - \frac{dP'}{d\gamma} \, . \, \sin(\lambda + \Pi)\right\};$$

$$\delta q = -\frac{m' \, . \, an}{5n' - 2n} \, . \, \left\{\frac{dP}{d\gamma} \, . \, \sin(\lambda + \Pi) + \frac{dP'}{d\gamma} \, . \cos(\lambda + \Pi)\right\}.$$

If s be the latitude of Jupiter, by article 436

$$s = q \sin v - p \cos v;$$

hence
$$\delta s = \delta q \,.\, \sin v - \delta p \,.\, \cos v,$$

and substituting for δp, δq,

$$\delta s = -\frac{m'.an}{5n' - 2n}\left\{\frac{dP}{d\gamma} \,.\, \cos(\lambda - v + \Pi) - \frac{dP'}{d\gamma} \,.\, \sin(\lambda - v + \Pi)\right\} \tag{177}$$

which is the only sensible inequality in the latitude of Jupiter in this approximation.

The latitude of Jupiter above the primitive orbit of Saturn is

$$s = -\gamma \sin(v - \Pi)$$

whence
$$-\delta s = \delta\gamma \sin(v - \Pi) - \gamma\delta\Pi \cos(v - \Pi)$$

and a comparison of the two values of δs, gives

$$\delta\gamma' = \frac{m' \,.\, an}{5n' - 2n}\left\{\frac{dP'}{d\gamma} \cos\lambda + \frac{dP}{d\gamma} \sin\lambda\right\}$$

$$\gamma\delta\Pi' = -\frac{m'.an}{5n' - 2n}\left\{\frac{dP}{d\gamma} \,.\, \cos\lambda - \frac{dP'}{d\gamma} \sin\lambda\right\}.$$

These are the variations occasioned by the action of Saturn in the mutual inclination of the two orbits, and in the ascending node of their common intersection; but Jupiter produces a corresponding effect in these two quantities, and if it be expressed by $\delta\gamma''$, $\gamma\delta\Pi''$, then the whole variations will be

$$\delta\gamma = \delta\gamma' + \delta\gamma'', \qquad \delta\Pi = \delta\Pi' + \delta\Pi'';$$

but by article

$$\delta\gamma'' = \frac{m.a'n'}{m'.an} \,.\, \delta\gamma'; \quad \delta\Pi'' = \frac{m.a'n'}{m'.an} \,.\, \delta\Pi';$$

or, substituting for n and n', the whole variations in the two elements in question are

$$\delta\gamma = \frac{m'.an}{5n' - 2n} \cdot \frac{m\sqrt{a} + m'\sqrt{a'}}{m'\sqrt{a'}} \cdot \left\{\frac{dP'}{d\gamma} \cos\lambda + \frac{dP}{d\gamma} \sin\lambda\right\}$$

$$\gamma\delta\Pi = +\frac{m' \,.\, an}{5n' - 2n} \cdot \frac{m\sqrt{a} + m'\sqrt{a'}}{m'\sqrt{a'}} \cdot \left\{\frac{dP'}{\delta\gamma} \sin\lambda - \frac{dP}{d\gamma} \cos\lambda\right\}. \tag{178}$$

577. The corresponding periodic inequalities in the latitude and elements of the orbit of Saturn are

$$\delta s = - \frac{2a'n' \cdot m}{(5n'-2n)m'} \left\{ \begin{array}{l} \frac{dP}{d\gamma} \sin\{4n't - 2nt + 4\epsilon' - 2\epsilon - v + \Pi\} \\ - \frac{dP'}{d\gamma} \cos\{4n't - 2nt + 4\epsilon' - 2\epsilon - v + \Pi\} \end{array} \right\}$$

$$\delta\zeta' = - \frac{m\sqrt{a}}{m'\sqrt{a'}} \cdot \delta\zeta,$$

$$\delta e' = \frac{m \cdot a'n'}{5n' - 2n} \left\{ \frac{dP'}{de'} \cos\lambda + \frac{dP}{de'} \sin\lambda \right\}, \qquad (179)$$

$$e'\delta\varpi' = - \frac{m \cdot a'n'}{5n' - 2n} \left\{ \frac{dP}{de'} \cos\lambda - \frac{dP'}{de'} \sin\lambda \right\}.$$

It is evident that the variations in the mean motions are by much the greatest, on account of the divisor $(5n' - 2n)^2$.

Periodic Variations in the Elements of the Orbits of Jupiter and Saturn, depending on the Squares of the Disturbing Forces.

578. The equations in the preceding articles, which determine the periodic inequalities in the elements of the orbits of Jupiter and Saturn, are functions of the sines and cosines of their mean motions; and when the mean motions are corrected by the application of their great inequalities, the equations in question give secular as well periodic inequalities in the elements of both orbits, depending on the squares and products of the disturbing masses.

The great inequalities may be put under a convenient form for this analysis, if the value of R, in article 563, be expressed by

$$R = m' \cdot \Sigma \cdot Q \cdot \cos\{5n't - 2nt + 5\epsilon' - 2\epsilon - \beta\},$$

where β is a function of the longitudes of the perihelia and node of the common intersection of the two orbits. The substitution of this in

$$\delta\zeta = - 3\iint \cdot andt \cdot \text{d}R,$$

gives

$$\delta\zeta = - 6m'\iint \cdot an^2dt^2 \cdot \Sigma Q \cdot \sin(5n't - 2nt + 5\epsilon' - 2\epsilon - \beta\}. \qquad (180)$$

Since $\delta\zeta$ and $\delta\zeta'$ represent the great inequalities of Jupiter and Saturn, their corrected mean motions are

$$nt + \delta\zeta, \text{ and } n't + \delta\zeta';$$

and, by the substitution of these in the preceding equation, it becomes

$$(\delta\zeta) = -6m'\iint . an^2dt^2 . \Sigma Q . \sin\{5n't - 2nt + 5\epsilon' - 2\epsilon - \beta + 5\delta\zeta' - 2\delta\zeta\} \quad (181)$$

$(\delta\zeta)$ being the great inequality of Jupiter when the mean motions are corrected. In order to abridge, let

$$5n't - 2nt + 5\epsilon' - 2\epsilon = \lambda,$$

then $\sin(\lambda - \beta + 5\delta\zeta' - 2\delta\zeta) =$

$$\sin(\lambda - \beta)\cos(5\delta\zeta' - 2\delta\zeta) + \cos(\lambda - \beta)\sin(5\delta\zeta' - 2\delta\zeta).$$

But $5\delta\zeta' - 2\delta\zeta$ is so small, that it may be taken for its sine, and unity for its cosine; and as quantities of the order of the square of the disturbing forces are alone to be retained, $\sin(\lambda - \beta)$ may be omitted; hence

$$\sin(\lambda - \beta + 5\delta\zeta' - 2\delta\zeta) = \{5\delta\zeta' - 2\delta\zeta\}\cos(\lambda - \beta);$$

or, as

$$\delta\zeta' = -\frac{m\sqrt{a}}{m'\sqrt{a'}} . \delta\zeta$$

therefore

$$\sin(\lambda - \beta + 5\delta\zeta' - 2\delta\zeta) = -\left\{\frac{5m\sqrt{a} + 2m'\sqrt{a'}}{m'\sqrt{a'}}\right\} . \delta\zeta . \cos(\lambda - \beta)$$

but the integral of equation (180) is

$$\delta\zeta = \frac{6m' . an^2 . \Sigma . Q}{(5n' - 2n)^2} . \sin(\lambda - \beta),$$

consequently

$$\sin(\lambda - \beta + 5\delta\zeta' - 2\delta\zeta) =$$

$$-\frac{(3m' . an^2 . \Sigma . Q)^2}{(5n' - 2n)^2} . \left\{\frac{5m\sqrt{a} + 2m'\sqrt{a'}}{m'\sqrt{a'}}\right\} . \sin(2\lambda - 2\beta).$$

When this quantity is substituted in equation (181), instead of the sine, its integral

$$(\delta\zeta) = -\frac{(3m'.an^2.\Sigma Q)^2}{2(5n' - 2n)^4} . \left\{\frac{5m\sqrt{a} + 2m'\sqrt{a'}}{m'\sqrt{a'}}\right\} . \sin 2(5n't - 2nt + 5\epsilon' - 2\epsilon - \beta) \quad (182)$$

is the variation in the mean motion of Jupiter, and on account of the relation in article 574, the corresponding inequality in the mean motion of Saturn is

$$(\delta\zeta') = \frac{(3m \,.\, an^2 . \Sigma Q)^2}{2(5n' - 2n)^4} \cdot \frac{5m'\sqrt{a'} + 2m\sqrt{a}}{m'\sqrt{a'}} \cdot \frac{m\sqrt{a}}{m'\sqrt{a'}} \times$$

$$\sin 2\{5n't - 2nt + 5\epsilon' - 2\epsilon\}. \qquad (183)$$

These inequalities have a sensible effect, on account of the minute divisor $(5n' - 2n)^4$.

579. The great inequalities in the mean motions also occasion variations in the eccentricities and longitudes of the perihelia, depending on the squares of the disturbing forces.

The principal term of the great inequality is sufficient for this purpose; and if the secular variations in the elements of the orbits during the period of the inequalities be omitted, the first term of the great inequality in the mean motion of Jupiter (172), when λ is put for $5n't - 2nt + 5\epsilon' - 2\epsilon$, is,

$$-\frac{6m' \,.\, an^2}{(5n' - 2n)^2}\{P \cos\lambda - P' \sin\lambda\}.$$

The corresponding inequality in the mean motion of Saturn is

$$+\frac{6m' \,.\, an^2}{(5n' - 2n)^2} \cdot \frac{m\sqrt{a}}{m'\sqrt{a'}}\{P \cos\lambda - P' \sin\lambda\}.$$

If these be applied as corrections to nt and $n't$, in the differential of equation (175), or

$$d\delta e = +\, m' \,.\, andt \,.\left\{\frac{dP}{de} \,.\, \cos\lambda - \frac{dP'}{de}\sin\lambda\right\},$$

it will be found, by the same analysis that was employed in the last article, that

$$d.\delta e = +\, m' \,.\, andt\left\{\frac{dP}{de} \,.\, \cos\lambda - \frac{dP'}{de}\sin\lambda\right\}$$

$$-\, m' \,.\, andt \,.\, \frac{dP}{de}\left\{\frac{6m' \,.\, an^2}{(5n' - 2n)^2} \cdot \frac{5m'\sqrt{a'} + 2m\sqrt{a}}{m'\sqrt{a'}}\right\} . \times$$

$$\{P \,.\, \cos\lambda \sin\lambda - P' \sin^2\lambda)\} \qquad (184)$$

$$-\, m' \,.\, andt \,.\, \frac{dP'}{de}\left\{\frac{6m' \,.\, an^2}{(5n' - 2n)^2} \cdot \frac{5m\sqrt{a'} + 2m\sqrt{a}}{m'\sqrt{a'}}\right\} . \times$$

$$\{P \,.\, \cos^2\lambda - P' \cos\lambda \sin\lambda)\}.$$

But $P\cos\lambda\sin\lambda - P'\sin^2\lambda = \frac{1}{2}P\sin 2\lambda + \frac{1}{2}P'\cos 2\lambda - \frac{1}{2}P'$

$$P\cos^2\lambda - P'\cos\lambda\sin\lambda = \tfrac{1}{2}P\cos 2\lambda - \tfrac{1}{2}P'\sin 2\lambda + \tfrac{1}{2}P;$$

and, as terms depending on the first powers of the masses are to be rejected, the periodic part of the preceding equation is

$$\delta e_2 = -\frac{3m'^2.a^2n^3}{2(5n'-2n)^3}\cdot\frac{5m\sqrt{a}+2m'\sqrt{a'}}{m'\sqrt{a'}}\cdot\left\{P'.\frac{dP}{de}+P.\frac{dP'}{de}\right\}\times$$

$$\sin 2(5n't-2nt+5\epsilon'-2\epsilon) \qquad (185)$$

$$-\frac{3m'^2.a^2n^3}{2(5n'-2n)^3}\cdot\frac{5m\sqrt{a}+2m'\sqrt{a'}}{m'\sqrt{a'}}\cdot\left\{P'.\frac{dP'}{de}-P\frac{dP}{de}\right\}\times$$

$$\cos 2(5n't-2nt+5\epsilon'-2\epsilon).$$

By the same process it may be found that the periodic variations of nt, and $n't$, produce the periodic variation

$$\delta\varpi_2 = \frac{3m'^2.a^2n^3}{2e(5n'-2n)^3}\cdot\frac{5m\sqrt{a}+2m'\sqrt{a'}}{m'\sqrt{a'}}\cdot\left\{P.\frac{dP}{de}-P'\frac{dP'}{de}\right\}\times$$

$$\sin 2(5n't-2nt+5\epsilon'-2\epsilon) \qquad (186)$$

$$+\frac{3m'^2.a^2n^3}{2e(5n'-2n)^3}\cdot\frac{5m\sqrt{a}+2m'\sqrt{a'}}{m'\sqrt{a'}}\cdot\left\{P'.\frac{dP}{de}+P.\frac{dP'}{de}\right\}\times$$

$$\cos 2(5n't-2nt+5\epsilon'-2\epsilon),$$

in the longitude of the perihelion of Jupiter. These are the only sensible periodic inequalities in the elements of Jupiter's orbit of this order. Corresponding variations obtain in those of the orbit of Saturn.

Secular Variations in the Elements of the Orbits of Jupiter and Saturn, depending on the Squares of the Disturbing Forces.

580. The secular variations in the elements of the orbits of Jupiter and Saturn depending on the first powers of the disturbing forces, are determined by the formulæ (130), in common with the other planets; but to these must be added their variations depending on the squares of the masses, quantities only sensible in the motions of Jupiter and Saturn.

The secular part of equation (184), arising from the corrected values of nt, $n't$, is

$$(\delta e) = -\frac{3m'^2 . a^2 n^3}{(5n'-2n)^3} . t . \frac{5m\sqrt{a} + 2m'\sqrt{a'}}{m'\sqrt{a'}} . \times$$

$$\left\{P . \frac{dP'}{de} - P' . \frac{dP}{de}\right\}. \qquad (187)$$

and the corresponding variation in the longitude of the perihelion of Jupiter's orbit, depending on the squares of the disturbing forces, is

$$(\delta\varpi) = \frac{3m'^2 a^2 n^3}{e(5n'-2n)^2} . t . \frac{5m\sqrt{a} + 2m'\sqrt{a'}}{m'\sqrt{a'}} . \times$$

$$\left\{P . \frac{dP}{de} + P' . \frac{dP'}{de}\right\}. \qquad (188)$$

The corresponding inequalities for Saturn are,

$$(\delta e') = -\frac{3m^2 . a^3 n^3 . t}{a'(5n'-2n)^2} . \frac{5m\sqrt{a} + 2m'\sqrt{a'}}{m\sqrt{a}} . \times$$

$$\left\{P . \frac{dP'}{de'} - P' . \frac{dP}{de'}\right\} \qquad (189)$$

$$(\delta\varpi') = \frac{3m^2 . a^3 n^3 t}{a'e'(5n'-2n)} . \frac{5m\sqrt{a} + 2m'\sqrt{a'}}{m\sqrt{a}} . \times$$

$$\left\{P . \frac{dP}{de'} + P' . \frac{dP'}{de'}\right\}.$$

581. Thus the periodic inequalities in the mean motions cause both periodic and secular variations in the elements of the two orbits of the order of the squares of the disturbing forces; but the periodic variations in the other elements have the same effect; for, making

$$5n't - 2nt + 5\epsilon' - 2\epsilon = \lambda,$$

the differential of equation (175) is

$$d\epsilon = + m' . andt \left\{\frac{dP}{de} \cos\lambda - \frac{dP'}{de} \sin\lambda\right\};$$

and when all the elements are variable except the mean motion, the effects of which have already been determined,

$$\delta . de = m' . andt . \left\{-\delta e \left(\frac{d^2P'}{de^2} . \sin\lambda - \frac{d^2P}{de^2} . \cos\lambda\right)\right.$$

$$- \delta\varpi \left(\frac{d^2P'}{de\, d\varpi} \cdot \sin\lambda - \frac{d^2P}{de\, d\varpi} \cdot \cos\lambda\right)$$

$$- \delta e' \left(\frac{d^2P'}{de\, de'} \cdot \sin\lambda - \frac{d^2P}{de\, de'} \cdot \cos\lambda\right)$$

$$- \delta\varpi' \left(\frac{d^2P'}{de.d\varpi'} \cdot \sin\lambda - \frac{d^2P}{de.d\varpi'} \cdot \cos\lambda\right)$$

$$- \delta\gamma \left(\frac{d^2P'}{de.d\gamma} \cdot \sin\lambda - \frac{d^2P}{de.d\gamma} \cdot \cos\lambda\right)$$

$$- \delta\Pi \left(\frac{d^2P'}{de.d\Pi} \cdot \sin\lambda - \frac{d^2P}{de.d\Pi} \cdot \cos\lambda\right)\Big\}$$

If the values of $\delta\varpi$, δe, $\delta e'$, $\delta\varpi'$, $\delta\gamma$, and $\delta\Pi$, from article 575, and those that follow, be substituted, observing at the same time, that equations (165) and (166) give

$$\frac{d^2P}{de.d\varpi} = e \cdot \frac{d^2P'}{de^2}; \qquad \frac{d^2P'}{de.d\varpi} = -e \cdot \frac{d^2P}{de^2};$$

$$\frac{d^2P}{de.d\varpi'} = e' \cdot \frac{d^2P'}{de.de'}; \qquad \frac{d^2P'}{de.d\varpi'} = -e' \cdot \frac{d^2P}{de.de'}; \quad (190)$$

$$\frac{d^2P}{de.d\Pi} = \gamma\, \frac{d^2P'}{de.d\gamma}; \qquad \frac{d^2P'}{de.d\Pi} = -\gamma\, \frac{d^2P}{de.d\gamma};$$

it will be found, when the periodic terms are omitted, and equation (187) added, that the whole secular variation in the eccentricity of Jupiter's orbit, depending on the squares of the disturbing forces, is

$$(\delta e) = -\frac{3m'^2.a^2n^3.t}{(5n'-2n)^2}\left\{\frac{2m'\sqrt{a'}+5m\sqrt{a}}{m'\sqrt{a'}}\right\}.\left\{P.\left(\frac{dP'}{de}\right) - P'\left(\frac{dP}{de}\right)\right\}$$

$$+ \frac{m'^2.a^2n^2.t}{5n'-2n} \times \quad (191)$$

$$\left\{\left(\frac{dP'}{de}\right) \cdot \left(\frac{d^2P}{de^2}\right) - \left(\frac{dP}{de}\right) \cdot \left(\frac{d^2P'}{de^2}\right) + \left(\frac{dP'}{d\gamma}\right) \cdot \left(\frac{d^2P}{de\, d\gamma}\right)\right.$$

$$\left. - \left(\frac{dP}{d\gamma}\right) \cdot \left(\frac{d^2P'}{de\, d\gamma}\right)\right\} + \frac{mm'.aa'.nn't}{5n'-2n} \times$$

$$\left\{\left(\frac{dP'}{de'}\right) \cdot \left(\frac{d^2P}{de\, de'}\right) - \left(\frac{dP}{de'}\right) \cdot \left(\frac{d^2P'}{de\, de'}\right) + \left(\frac{dP'}{d\gamma}\right) \cdot \left(\frac{d^2P}{de.d\gamma}\right)\right.$$

$$\left. - \left(\frac{dP}{d\gamma}\right) \cdot \left(\frac{d^2P'}{de\, d\gamma}\right)\right\}.$$

By the same process it may be found, that where the periodic terms which are quite insensible are omitted, the secular variation in the longitude of the perihelion of Jupiter's orbit, depending on the squares of the disturbing force, including the equation (188), is

$$(\delta\varpi)=\frac{3m'^2.a^2n^3.t}{e(5n'-2n)^2}\cdot\frac{5m\sqrt{a}+2m'\sqrt{a'}}{m'\sqrt{a'}}\cdot\left\{P\left(\frac{dP}{de}+P'\frac{dP'}{de}\right)\right\}$$

$$+\frac{m'^2.a^2n^2.t}{e(5n'-2n)}\times$$

$$\left\{\left(\frac{dP}{de}\right)\cdot\left(\frac{d^2P}{de^2}\right)+\left(\frac{dP'}{de}\right)\cdot\left(\frac{d^2P'}{de^2}\right)+\left(\frac{dP}{d\gamma}\right)\cdot\left(\frac{d^2P}{de.d\gamma}\right)\right.$$

$$\left.+\left(\frac{dP'}{d\gamma}\right)\cdot\left(\frac{d^2P'}{de.d\gamma}\right)\right\}+\frac{mm'.aa'.nn'.t}{e(5n'-2n)}\times \qquad (192)$$

$$\left\{\left(\frac{dP}{de'}\right)\cdot\left(\frac{d^2P}{de.de'}\right)+\left(\frac{dP'}{de'}\right)\cdot\left(\frac{d^2P'}{dede'}\right)+\left(\frac{dP}{d\gamma}\right)\cdot\left(\frac{d^2P}{ded\gamma}\right)\right.$$

$$\left.+\left(\frac{dP'}{d\gamma}\right)\cdot\left(\frac{d^2P}{ded\gamma}\right)\right\}.$$

582. The corresponding variations for Saturn, including equations (189), are,

$$(\delta e')=-\frac{3m^2.a^3n^3.t}{a'(5n'-2n)^2}\cdot\frac{5m\sqrt{a}+2m'\sqrt{a'}}{m\sqrt{a}}.\left\{P.\left(\frac{dP'}{de'}\right)-P'\left(\frac{dP}{de'}\right)\right\}$$

$$+\frac{m^2.a'^2n'^2}{5n'-2n}.t\times \qquad (193)$$

$$\left\{\left(\frac{dP'}{de'}\right)\cdot\left(\frac{d^2P}{de'^2}\right)-\left(\frac{dP}{de'}\right)\cdot\left(\frac{d^2P'}{de'^2}\right)+\left(\frac{dP'}{d\gamma}\right)\cdot\left(\frac{d^2P}{de'd\gamma}\right)\right.$$

$$\left.-\left(\frac{dP}{d\gamma}\right)\cdot\left(\frac{d^2P'}{de'd\gamma}\right)\right\}+\frac{mm'.aa'.nn'}{5n'-2n}t\times$$

$$\left\{\left(\frac{dP'}{de}\right)\cdot\left(\frac{d^2P}{dede'}\right)-\left(\frac{dP'}{de}\right)\cdot\left(\frac{d^2P}{dede'}\right)-\left(\frac{dP'}{d\gamma}\right)\cdot\left(\frac{d^2P}{de'd\gamma}\right)\right.$$

$$\left.-\left(\frac{dP}{d\gamma}\right)\cdot\left(\frac{d^2P'}{de'd\gamma}\right)\right\};$$

$$(\delta\varpi')=\frac{3m^2.a^3n^3.t}{a'e'(5n'-2n)^2}\cdot\frac{5m\sqrt{a}+2m'\sqrt{a'}}{m\sqrt{a}}.\left\{P.\left(\frac{dP}{de'}\right)+P'\left(\frac{dP'}{de'}\right)\right\}$$

(194)

$$+\frac{m^2.a'^2.n'^2.t}{e'(5n'-2n)}\Big\{\Big(\frac{dP}{de'}\Big).\Big(\frac{d^2P}{de'^2}\Big)+\Big(\frac{dP'}{de'}\Big).\Big(\frac{d^2P'}{de'^2}\Big)+\Big(\frac{dP}{d\gamma}\Big).\Big(\frac{d^2P}{de'd\gamma}\Big)$$

$$+\Big(\frac{dP'}{d\gamma}\Big).\Big(\frac{d^2P'}{de'd\gamma}\Big)\Big\}+\frac{mm'.aa'.nn'}{e'(5n'-2n)}t\times$$

$$\Big\{\Big(\frac{dP}{de}\Big).\Big(\frac{d^2P}{dede'}\Big)+\Big(\frac{dP'}{de}\Big).\Big(\frac{d^2P'}{dede'}\Big)+\Big(\frac{dP}{d\gamma}\Big).\Big(\frac{d^2P}{de'd\gamma}\Big)$$

$$+\Big(\frac{dP}{d\gamma}\Big).\Big(\frac{d^2P'}{de'd\gamma}\Big)\Big\}.$$

583. Secular variations, depending on the squares of the disturbing forces, arise from the same cause in the mutual inclination of the orbits, and in the longitude of the ascending node of the orbit of Saturn on that of Jupiter. These are obtained from equations (178), considering the elements to be variable; then the substitution of their periodic variations will give, in consequence of

$$\Big(\frac{dP'}{d\gamma}\Big).\Big(\frac{d^2P}{d\gamma^2}\Big)-\Big(\frac{dP}{d\gamma}\Big).\Big(\frac{d^2P'}{d\gamma^2}\Big)=0.$$

$$(\delta\gamma)=-\frac{3m'^2.a^2n^3}{(5n'-2n)^2}t.\frac{m\sqrt{a}+m'\sqrt{a'}}{m'\sqrt{a'}}.\frac{5m\sqrt{a}+2m'\sqrt{a'}}{m'\sqrt{a'}}.$$

$$\times\Big\{P\Big(\frac{dP'}{d\gamma}\Big)-P'\Big(\frac{dP}{d\gamma}\Big)\Big\} \qquad (195)$$

$$+\frac{m'^2.a^2n^2}{5n'-2n}.t.\frac{m\sqrt{a}+m'\sqrt{a'}}{m'\sqrt{a'}}.\times$$

$$\Big\{\Big(\frac{dP'}{de}\Big).\Big(\frac{d^2P}{de.d\gamma}\Big)-\Big(\frac{dP}{de}\Big).\Big(\frac{d^2P'}{de.d\gamma}\Big)\Big\}$$

$$+\frac{mm'.aa'.nn'}{5n'-2n}.t.\frac{m\sqrt{a}+m'\sqrt{a'}}{m'\sqrt{a'}}.\times$$

$$\Big\{\Big(\frac{dP'}{de'}\Big).\Big(\frac{d^2P}{de'd\gamma}\Big)-\Big(\frac{dP}{de'}\Big).\Big(\frac{d^2P'}{de'.d\gamma}\Big)\Big\};$$

$$(\delta\Pi)=\frac{3m'^2.a^2n^3}{\gamma(5n'-2n)^2}.t.\frac{m\sqrt{a}+m'\sqrt{a'}}{m'\sqrt{a'}}.\frac{5m\sqrt{a}+2m'\sqrt{a'}}{m'\sqrt{a'}}. \qquad (196)$$

$$\Big\{P\Big(\frac{dP}{d\gamma}\Big)+P'\Big(\frac{dP'}{d\gamma}\Big)\Big\}$$

$$+\frac{m'^2.a^2n^2}{\gamma(5n'-2n)}.t.\frac{m\sqrt{a}+m'\sqrt{a'}}{m'\sqrt{a'}}.\Big\{\Big(\frac{dP}{de}\Big).\Big(\frac{d^2P}{ded\gamma}\Big)$$

$$+\Big(\frac{dP'}{de}\Big).\Big(\frac{d^2P'}{de.d\gamma}\Big)+\Big(\frac{dP}{d\gamma}\Big).\Big(\frac{d^2P}{d\gamma^2}\Big)+\Big(\frac{dP'}{d\gamma}\Big).\Big(\frac{d^2P'}{d\gamma^2}\Big)\Big\}$$

$$+\frac{mm'\,.\,aa'\,.\,nn'}{\gamma(5n'-2n)}.t.\frac{m\sqrt{a}+m'\sqrt{a'}}{m'\sqrt{a'}}.\left\{\left(\frac{dP}{de'}\right).\left(\frac{d^2P}{de'.d\gamma}\right)\right.$$

$$\left.+\left(\frac{dP'}{de'}\right).\left(\frac{d^2P'}{de'd\gamma}\right)+\left(\frac{dP}{d\gamma}\right).\left(\frac{d^2P}{d\gamma^2}\right)+\left(\frac{dP'}{d\gamma}\right).\left(\frac{d^2P'}{d\gamma^2}\right)\right\}.$$

584. These are the variations with regard to the plane of Jupiter's orbit at a given time, but the variations in the position of the orbits of Jupiter and Saturn with regard to the ecliptic may easily be found, for ϕ, ϕ', being the inclinations of the orbits of m and m' on the fixed ecliptic at the epoch, and θ, θ' the longitudes of the ascending nodes estimated on that plane, by article 444,

$$p'-p=\gamma\sin\Pi;\quad q'-q=\gamma\cos\Pi;$$

or

$$\phi'\sin\theta'-\phi\sin\theta=\gamma\sin\Pi,$$

$$\phi'\cos\theta'-\phi\cos\theta=\gamma\cos\Pi.$$

and on account of the action and reaction of Jupiter and Saturn,

$$\delta(\phi'.\sin\theta')=-\frac{m\sqrt{a}}{m'\sqrt{a'}}.\ \delta(\phi\sin\theta),$$

$$\delta(\phi'.\cos\theta')=-\frac{m\sqrt{a}}{m'\sqrt{a'}}.\ \delta(\phi.\cos\theta).$$

And from these four equations, it will readily be found, that

$$(\delta\phi)\ =-\frac{m'\sqrt{a'}}{m\sqrt{a}+m'\sqrt{a'}}\{\delta\gamma.\cos(\Pi-\theta)-\gamma\delta\Pi.\sin(\Pi-\theta)\}$$

$$(\phi\delta\theta)\ =-\frac{m'\sqrt{a'}}{m\sqrt{a}+m'\sqrt{a'}}\{\delta\gamma.\sin(\Pi-\theta)+\gamma\delta\Pi.\cos(\Pi-\theta)\}$$

$$(\delta\phi')\ =\frac{m\sqrt{a}}{m\sqrt{a}+m'\sqrt{a'}}\{\delta\gamma.\cos(\Pi-\theta')-\gamma\delta\Pi\sin(\Pi-\theta')\}\quad(197)$$

$$(\phi'\delta\theta')=\frac{m\sqrt{a}}{m\sqrt{a}+m'\sqrt{a'}}\{\delta\gamma.\sin(\Pi-\theta')+\gamma\delta\Pi\cos(\Pi-\theta')\}.$$

Thus when $\delta\gamma$ and $\gamma\delta\Pi$ are computed, the variations in the inclinations and longitude of the nodes when referred to the fixed plane of the ecliptic may be found.

585. The periodic variations in the eccentricities, inclinations, longitudes of the perihelia, and nodes, do not affect the mean motion

with any sensible inequalities depending on the squares and product of the masses ; for if the variation of

$$\delta\zeta = + \frac{6m' . an^2}{(5n'-2n)^2} \{P' . \sin\lambda - P\cos\lambda\}$$

be taken, considering all the elements as variable, the substitution of their periodic variations will make the whole vanish in consequence of the relations between the partial differences.

586. The longitude of the epoch is not affected by any variations of this order that are sensible in the planets, but they are of much importance in the theories of the moon and Jupiter's satellites.

587. The variations in the elements depending on the squares of the disturbing forces, are insensible in the theories of all the planets, except those of Jupiter and Saturn ; they are only perceptible in the motions of these two planets, on account of the nearly commensurable ratio in their mean motions introducing the minute divisor $5n' - 2n$; therefore, if

$$(\delta\bar{e}), \quad (\delta\bar{\varpi}), \quad (\delta\bar{\gamma},) \quad (\delta\bar{\Pi}), \quad (\delta\bar{\phi}), \quad (\delta\bar{\theta}),$$

be the secular variations in the elements depending on the second powers of the disturbing forces, and computed for the epoch from the equations in articles 580, and the two following, the equations (130) become, with regard to Jupiter and Saturn only,

$$e = \bar{e} + \left\{\frac{d\bar{e}}{dt} + (\delta\bar{e})\right\}t + \&c.$$

$$\varpi = \bar{\varpi} + \left\{\frac{d\bar{\varpi}}{dt} + (\delta\bar{\varpi})\right\}t + \&c.$$

$$\gamma = \bar{\gamma} + \left\{\frac{d\bar{\gamma}}{dt} + (\delta\bar{\gamma})\right\}t + \&c. \qquad (198)$$

$$\pi = \bar{\pi} + \left\{\frac{d\bar{\Pi}}{dt} + (\delta\bar{\Pi})\right\}t + \&c.$$

$$\phi = \phi + \left\{\frac{d\bar{\phi}}{dt} + (\delta\bar{\phi})\right\}t + \&c.$$

$$\theta = \bar{\theta} + \left\{\frac{d\bar{\theta}}{dt} + (\delta\bar{\theta})\right\}t + \&c.$$

Whence the elements of the orbits of these two planets may be determined with great accuracy for 1000 or 1200 years before and after the time assumed as the epoch.

Periodic Perturbations in Jupiter's Longitude depending on the Squares of the disturbing Forces.

588. Where e^2 is omitted, equation (97) becomes

$$v = 2e \sin (nt + \epsilon - \varpi).$$

The eccentricity and longitude of the perihelion, when corrected for their periodic inequalities (175), and (185) (176) and (186), become,

$$e + \delta e_1 + \delta e_2 \quad \text{and } \varpi + \delta\varpi_1 + \delta\varpi_2,$$

and the longitude of the epoch when corrected by its periodic variation, is $\epsilon + \delta\epsilon_1$; by the substitution of these v becomes

$$\delta v = (2e + 2\delta e_1 + 2\delta e_2) \sin \{nt + \epsilon - \varpi + \delta\epsilon_1 - \delta\varpi_1 - \delta\varpi_2\}:$$

when the quantities that do not contain the squares of the disturbing forces are rejected, the developement of this expression is

$$\delta v = \{2\delta e_2 + 2e\delta\varpi_1 . \delta\epsilon_1 - e\,\delta\varpi^2\} \sin (nt + \epsilon - \varpi)$$
$$- \{2e\delta\varpi_1 + 2e\delta e_1 . \delta\varpi_1 - 2\delta e . \delta\epsilon\} \cos (nt + \epsilon - \varpi);$$

when the values of the periodic variations are substituted, the result will be the inequality

$$\delta v = - \frac{3m'^2 . a^2 n^3}{(5n' - 2n)^3} . \frac{5m\sqrt{a} + 4m'\sqrt{a'}}{m'\sqrt{a'}} . \left\{P\left(\frac{dP'}{de}\right) + P'\left(\frac{dP}{de}\right)\right\} \times$$
$$\cos \{5n't - 10nt + 5\epsilon' - 10\epsilon - \varpi\}$$
$$- \frac{3m'^2 . a^2 n^3}{(5n' - 2n)^3} . \frac{5m\sqrt{a} + 4m'\sqrt{a'}}{m'\sqrt{a'}} . \left\{P'\left(\frac{dP'}{de}\right) - P\left(\frac{dP}{de}\right)\right\} \times$$
$$\sin \{5n't - 10nt + 5\epsilon' - 10\epsilon - \varpi\}. \quad (199)$$

The corresponding inequality for Saturn is found from

$$v' = 2e' \sin (n't + \epsilon - \varpi').$$

589. The radii vectores and true longitudes of m and m' in their elliptical orbits have been represented by r, r', v, v',

but as δr, $\delta r'$, δv, $\delta v'$

are the periodic perturbations of these quantities, these two co-ordinates of m and m' in their troubled orbits, are

$$r + \delta r, r' + \delta r', v + \delta v, v' + \delta v'.$$

When these quantities are substituted in

$$R = \frac{m'(rr' \cos (v' - v)) + zz'}{(r'^2 + z'^2)^{\frac{3}{2}}} - \frac{m'}{\sqrt[3]{r^2 - 2rr' \cos (v' - v) + r'^2}},$$

R becomes a function of the squares and products of the masses, it consequently produces terms of that order in the mean motion

$$\zeta = - 3\iint . andt . dR$$

having the factor $(5n'-2n)^2$; they therefore form a part of the great inequalities in the mean motions of Jupiter and Saturn. A mistake has been observed in La Place's determination of these inequalities, which has been, and still is, a subject of controversy between three of the greatest mathematicians of the present age, MM. Plana, Poisson, and Pontécoulant, to whose very learned papers the reader is referred for a full investigation of this difficult subject.

590. The numerical values of the perturbations of Jupiter in longitude are computed from equations (159), (164), (172), (182), and (199), together with some terms depending on the fifth powers of the eccentricities and inclinations which may be determined by the same process as in the other approximations; his perturbations in latitude are computed from equations (160) and (177), and those in his radius vector from (158) and (163).

591. Hitherto the mass of the planet has been omitted when compared with that of the sun taken as the unit; so that half the greater axes has been determined by the equation $a^3 = \frac{1}{n^2}$, whereas its real value is found from

$$\frac{1+m}{a^3} = n^2, \text{ or } a = n^{-\frac{2}{3}}(1 + \tfrac{1}{3}m);$$

the semigreater axes of the orbits of Jupiter and Saturn ought therefore to be augmented by $\frac{1}{3}ma$, $\frac{1}{3}m'a'$, quantities that are only sensible in these two planets.

CHAPTER XI.

INEQUALITIES OCCASIONED BY THE ELLIPTICITY OF THE SUN.

592. As the sun has hitherto been considered a sphere, his action was assumed to be the same as if his mass were united in his centre of gravity; but from his rotatory motion, his form must be spheroidal on account of his centrifugal force, therefore the excess of matter at his equator may have an influence on the motions of the planets.

In the theory of spheroids it is found that the attraction of the redundant matter at the equator is expressed by

$$(\rho - \tfrac{1}{2}\psi) \cdot \frac{R'^2}{r^3} \cdot (\eta^2 - \tfrac{1}{3}).$$

Where ρ is the ellipticity of the sun, ψ the ratio of the centrifugal force to gravity at the solar equator, η the declination of a planet m relative to this equator, R' the semidiameter of the sun, his mass being unity. Therefore, the attraction of the elliptical part of the sun's mass adds the term

$$(\rho - \tfrac{1}{2}\psi) \cdot \frac{R'^2}{r^3} \cdot (\eta^2 - \tfrac{1}{3})$$

to the disturbing action expressed by the series R in article 449. If this disturbing action of the sun's spheroidal form be alone considered, omitting η^2, and substituting

$$\frac{1}{a^3}(1 - \tfrac{3}{2}e^2), \text{ for } r^{-3},$$

it gives, with regard to secular quantities alone,

$$F = -\tfrac{1}{3}(\rho - \tfrac{1}{2}\psi) \cdot \frac{R'^2}{a^3}(1 - \tfrac{3}{2}e^2),$$

and
$$\frac{dF}{de} = e(\rho - \tfrac{1}{2}\psi) \cdot \frac{R'^2}{a^3}.$$

*

The substitution of which in

$$d\varpi = \frac{andt}{e} \cdot \frac{dF}{de},$$

gives by integration,

$$\delta\varpi = (\rho - \tfrac{1}{2}\psi) \cdot \frac{R'^2}{a^2} \cdot nt.$$

Thus the action of the excess of matter at the sun's equator produces a direct motion in the perihelia of the planetary orbits.

593. The effect of the sun's ellipticity on the position of the orbits may be ascertained from the last of equations (115),

or
$$dp = andt \cdot \frac{dF}{dq}.$$

Since η is the declination of the planet m on the plane of the sun's equator, if the equator be taken as the fixed plane, then will

$$\eta^2 = \frac{z^2}{r^2}.$$

And if the eccentricity be omitted,

$$F = (\rho - \tfrac{1}{2}\psi) \cdot \frac{R'^2}{a^5}(z^2 - a^2),$$

therefore
$$\frac{dF}{dz} = 2 \cdot (\rho - \tfrac{1}{2}\psi) \cdot \frac{R'^2}{a^5} \cdot z.$$

But
$$\frac{dF}{dq} = \frac{dF}{dz} \cdot \frac{dz}{dq} = \frac{dF}{dz} \cdot a \sin(nt + \epsilon)$$

on account of equation,

$$\frac{z}{a} = q \cdot \sin(nt + \epsilon) - p \cos(nt + \epsilon)$$

consequently,

$$\frac{dF}{dq} = 2 \cdot (\rho - \tfrac{1}{2}\psi) \cdot \frac{R'^2}{a^5} \cdot z \cdot \sin(nt + \epsilon)$$

or substituting $a \cdot \tan\phi \cdot \sin(nt + \epsilon - \theta)$ for z,

$$\frac{dF}{dq} = -(\rho - \tfrac{1}{2}\psi) \cdot \frac{R'^2}{a^3} \cdot \cos\theta \cdot \tan\phi,$$

whence
$$dp = -ndt \cdot (\rho - \tfrac{1}{2}\psi) \cdot \frac{R'^2}{a^2} \cdot \cos\theta \cdot \tan\phi.$$

But
$$p = \tan\phi \cdot \sin\theta;$$

whence
$$dp = d\theta \cdot \tan\phi \cdot \cos\theta.$$

therefore $$d\theta = -ndt.(\rho - \tfrac{1}{2}\psi).\frac{R'^2}{a^2},$$

and $$\delta\theta = -nt.(\rho - \tfrac{1}{2}\psi).\frac{R'^2}{a^2}.$$

Thus the nodes of the planetary orbits have a retrograde motion on the plane of the solar equator equal to the direct motion of their perihelia on the same plane, both so small that they are scarcely perceptible even in Mercury. As neither the eccentricities nor the inclinations are affected by this disturbance, it has no influence on the stability of the system.

CHAPTER XII.

PERTURBATIONS IN THE MOTIONS OF THE PLANETS OCCASIONED BY THE ACTION OF THEIR SATELLITES.

594. THE common centre of gravity of a planet and its satellites very nearly describes an ellipse round the sun. If that orbit be considered to be the orbit of the planet itself, the respective positions of the satellites with regard to each other, and to the sun, will give that of the planet with regard to their common centre of gravity, and consequently the perturbations produced by the satellites on their primary.

Let G, fig. 90, be the common centre of gravity of a planet, and of its satellites, S the sun, γ the equinoctial point, and $\bar{x}$, $\bar{y}$, $\bar{z}$, the co-ordinates of G, so that $SG = \bar{x}$, and $\bar{z}$ perpendicular to the plane of the orbit. Then if x, y, z, be the co-ordinates of a satellite m, and $v = \gamma SG$,

fig. 90.

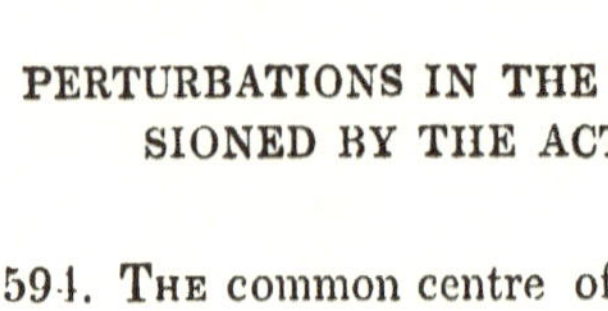

$U = \gamma Gm$, the longitudes of G and m; it is evident that $Gp = x - \bar{x}$, and r being the radius Gm,

$$Gp = x - \bar{x} = r \,.\, \cos (U - v)\,;$$

hence, if Σm be the sum of the masses of the satellites, and P that of their primary,

$$\Sigma m \,.\, x = \bar{x} \,.\, P + \Sigma m \,.\, r \cos (U - v),$$

or,

$$\Sigma mx = \bar{x} \,.\, P + mr \,.\, \cos (U - v) + m'r' \,.\, \cos (U - v') + \&c.$$

In the same manner

$$\Sigma my = \bar{y} \,.\, P + mr \,.\, \sin (U - v) + m'r' \,.\, \sin (U - v') + \&c.$$

$$\Sigma mz = \bar{z}P + m \,.\, rs + m' \,.\, r's' + \&c.$$

s, s', s'', &c., being the latitudes of the satellites above the orbit of their common centre of gravity. But by the property of the centre of gravity,

$$\Sigma m \,.\, x = 0, \quad \Sigma m \,.\, y = 0, \quad \Sigma m \,.\, z = 0;$$

consequently,

$$0 = \bar{x} \,.\, P + mr \,.\, \cos(U - v) + \&c.$$
$$0 = \bar{y} \,.\, P + mr \,.\, \sin(U - v) + \&c.$$
$$0 = \bar{z} \,.\, P + mr \,.\, s + m'r's' \,.\, + \&c.$$

By article 353 the centre of gravity is urged in a direction parallel to the co-ordinates, by the forces

$$-(P + \Sigma m)\,\bar{x}; \quad -\frac{(P + \Sigma m)\,\bar{y}}{\bar{r}}; \quad -\frac{(P + \Sigma m)\,\bar{z}}{\bar{r}}.$$

$\bar{r} = SG$, the radius vector of the centre of gravity. These forces vary very nearly as $\bar{x}$, $\frac{\bar{y}}{\bar{r}}$, and $\frac{\bar{z}}{\bar{r}}$;

therefore the perturbations in the radius vector SG are very nearly proportional to $\bar{x}$, that is, to

$$-\frac{m}{P} \,.\, r\cos(U - v) - \frac{m'}{P} \,.\, r'\cos(U - v') - \&c.$$

The perturbations in longitude are nearly proportional to

$$-\frac{m}{P} \,.\, \frac{r}{\bar{r}}\sin(U - v) - \frac{m'}{P} \,.\, \frac{r'}{\bar{r}}\sin(U - v') - \&c.;$$

and those in latitude to

$$-\frac{m}{P} \,.\, \frac{rs}{\bar{r}} - \frac{m'r's'}{\bar{r}} - \&c.$$

The masses of Jupiter's satellites compared with the mass of that planet are so small, and their elongations seen from the sun subtend so small an angle, that the perturbations produced by them in Jupiter's motions are insensible; and there is reason to believe this to be the case also with regard to Saturn and Uranus.

495. But the Earth is sensibly troubled in its motions by the Moon, her action produces the inequalities

$$\delta r = -\frac{m}{E} \,.\, r\cos(U - v)$$

$$\delta v = -\frac{m}{E} \,.\, \frac{r}{\bar{r}}\sin(U - v)$$

$$\delta s = -\frac{m}{E} \,.\, \frac{r}{r} \,.\, s;$$

or, more correctly,

$$\delta r = - \frac{m}{E+m} \cdot r \cos(U - v)$$

$$\delta v = - \frac{m}{E+m} \cdot \frac{r}{\bar{r}} \cdot \sin(U - v) \qquad (200)$$

$$\delta s = - \frac{m}{E+m} \cdot \frac{r}{\bar{r}} \cdot s;$$

in the radius vector, longitude and latitude of the Earth, E and m being the masses of the Earth and Moon.

CHAPTER XIII.

DATA FOR COMPUTING THE CELESTIAL MOTIONS.

596. The data requisite for computing the motions of the planets determined by observation for any instant arbitrarily assumed as the epoch or origin of the time, are

The masses of the planets ;
Their mean sidereal motions for a Julian year of 365.25 days ;
The mean distances of the planets from the sun ;
The ratios of the eccentricities to the mean distances ;
The inclinations of the orbits on the plane of the ecliptic ;
The longitudes of the perihelia ;
The longitudes of the ascending nodes on the ecliptic ;
The longitudes of the planets.

Masses of the Planets.

597. Satellites afford the means of ascertaining the masses of their primaries ; the masses of such planets as have no satellites are found from a comparison of their inequalities determined by analysis, with values of the same obtained from numerous observations. The secular inequalities will give the most accurate values of the masses, but till they are perfectly known the periodic variations must be employed. On this account there is still some uncertainty as to the masses of several bodies. It is only necessary to know the ratio of the mass of each planet to that of the sun taken as the unit; the masses are consequently expressed by very small fractions.

598. If T be the time of a sidereal revolution of a planet m, whose mean distance from the sun is a, π the ratio of the circumference to the diameter, and $\mu = \sqrt{m + S}$ the sum of the masses of the sun and planet, by article 383,

$$T = \frac{2\pi \cdot a^{\frac{3}{2}}}{\sqrt{\mu}}.$$

From this expression the masses of such planets as have satellites may be obtained.

Suppose this equation relative to the earth, and that the mass of the earth is omitted when compared with that of the sun, it then becomes

$$T = \frac{2\pi \cdot a^{\frac{3}{2}}}{\sqrt{S}}.$$

Again, let $\mu = m + m'$ the sum of the masses of a planet and of its satellite m', T' being the time of a sidereal revolution of the planet at the mean distance a' from the sun, then

$$T' = \frac{2\pi \cdot a'^{\frac{3}{2}}}{\sqrt{m+m'}};$$

and dividing the one by the other the result is,

$$\frac{m+m'}{S} = \frac{a'^3}{a^3} \cdot \frac{T^2}{T'^2}.$$

If the values of T, T', a and a', determined from observation, be substituted in this expression, the ratio of the sum of the masses of the planet and of its satellite to the mass of the sun will be obtained; and if the mass of the satellite be neglected when compared with that of its primary, or if the ratio of these masses be known, the preceding equation will give the ratio of the mass of the planet to that of the sun. For example,

599. Let m be the mass of Jupiter, that of his satellite being omitted, and let the mass of the sun be taken as the unit, then

fig. 91.

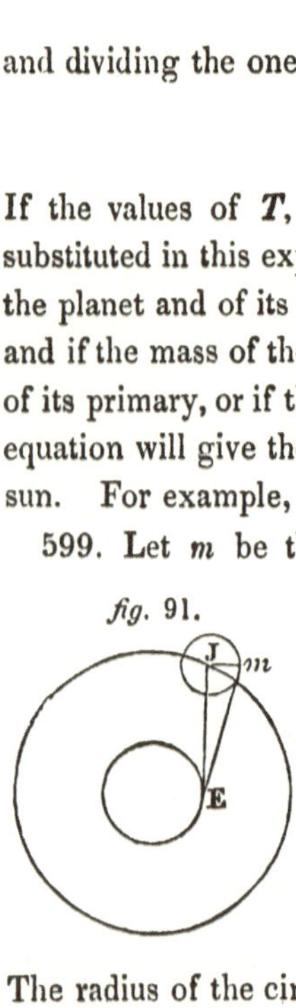

$$m = \frac{a'^3}{a^3} \cdot \frac{T^2}{T'^2}.$$

Jm the mean radius of the orbit of the fourth satellite at the mean distance of the earth from the sun taken as the unit, is seen under the angle

$$JEm = 2580''.579$$

The radius of the circle reduced to seconds is 206264″.8; hence the mean radii of the orbit of the fourth satellite and of the terrestrial orbit are in the ratio of these two numbers. The time of a sidereal revolution of the fourth satellite is 16.6890 days, and the sidereal year is 365.2564 days, hence

$$a = 206264.8$$
$$a' = 2580.58$$
$$T = 365.2564$$
$$T' = 16.6890.$$

With these data it is easy to find that the mass of Jupiter is

$$m = \frac{1}{1066.09}.$$

The sixth satellite of Saturn accomplishes a sidereal revolution in 15.9453 days; the mean radius of its orbit, at the mean distance of the planet, is seen from the sun under an angle of 179″; whence the mass of Saturn is $\frac{1}{3359.40}$.

By the observations of Sir William Herschel the sidereal revolutions of the fourth satellite of Uranus are performed in 13.4559 days, and the mean radius of its orbit seen from the sun at the mean distance of the planet is 44″.23. With these data the mass of Uranus is found to be $\frac{1}{19504}$.

600. This method is not sufficiently accurate for finding the mass of the Earth, on account of the numerous inequalities of the Moon. It has already been observed, that the attraction of the Earth on bodies at its surface in the parallel where the square of the sine of the latitude is $\frac{1}{3}$, is nearly the same as if its mass were united at its centre of gravity. If R be the radius of the terrestrial spheroid drawn to that parallel, and m its mass, this attraction will be

$$g = \frac{m}{R^2}; \text{ whence } m = g \, . \, R^2.$$

Then, if a be the mean distance of the Sun from the Earth, T the duration of the sidereal year,

$$T = \frac{2\pi \, . \, a^{\frac{3}{2}}}{\sqrt{S}};$$

and, by division,

$$\frac{m}{S} = \frac{g \, . \, R^2 \, . \, T^2}{4\pi^2 \, . \, a^3}.$$

R, g, T, and a, are known by observation, therefore the ratio of the mass of the Earth to that of the Sun may be found from this expression.

The sine of the solar parallax at the mean distance of the sun from the earth, and in the latitude in question, is

$$\sin P = \frac{R}{a} = \sin 8''.75\,;$$

the attraction of the Earth, and the terrestrial radius in the same parallel, are

$$g = 2\,.\,16.1069 = 32.2138$$
$$R = 2089870,$$

and the sidereal year is

$$T = 31558152''.9$$

with these data the mass of the earth is computed to be

$$\frac{1}{337103},$$

the mass of the sun being unity. This value varies as the cube of the solar parallax compared with that adopted.

601. The compression of the three larger planets, and the ring of Saturn, probably affect the values of the masses computed from the elongations of their satellites; but the comparison of numerous well chosen observations, with the disturbances determined from theory, will ultimately give the masses of all the planets with great accuracy.

The action of each disturbing body adds a term of the form $m'\delta v'$ to the longitude, so that the longitude of m at any given instant in its troubled orbit, is

$$v + m'\delta v' + m''\delta v'' + \&c. \qquad v_{,,}\ \delta v',\ \delta v'',\ \&c.$$

are susceptible of computation from theory; and as they are given by the Tables of the Motions of the Planets, the true longitude of m is
$$v + m'\delta v' + m''\delta v'' + \&c. = L.$$

When this formula is composed with a great number of observations, a series of equations,

$$m'\delta v' + m''\delta v'' + \&c. = L - v,$$
$$m'\delta v_2' + m''\delta v_2'' + \&c. = L' - v_2,$$
$$\&c. = \&c.$$

are obtained, where m', m'', &c., are unknown quantities, and by the resolution of these the masses of the planets may be estimated by the perturbations they produce.

602. As there are ten planets, ten equations would be sufficient to

give their masses, were the observed longitudes and the computed quantities v, $\delta v'$, $\delta v''$, &c., mathematically exact; but as that is far from being the case, many hundreds of observations made on all the planets must be employed to compensate the errors. The method of combining a series of equations more numerous than the unknown quantities they contain, so as to determine these quantities with all possible accuracy, depends on the theory of probabilities, which will be explained afterwards. The powerful energy exercised by Jupiter on the four new planets in his immediate vicinity occasions very great inequalities in the motions of these small bodies, whence that highly distinguished mathematician, M. Gauss, has obtained a value for the mass of Jupiter, differing considerably from that deduced from the elongation of his satellites, it cannot however be regarded as conclusive till the perturbations of these small planets are perfectly known.

603. The mass of Venus is obtained from the secular diminution in the obliquity of the Ecliptic. The plane of the terrestrial equator is inclined to the plane of the ecliptic at an angle of 23° 28′ 47″ nearly, but this angle varies in consequence of the action of the planets. A series of tolerably correct observations of the Sun's altitude at the solstices chiefly by the Chinese and Arabs, have been handed down to us from the year 1100 before Christ, to the year 1473 of the Christian era; by a comparison of these, it appears that the obliquity was then diminishing, and it is still decreasing at the rate of 50″.2 in a century. From numerous observations on the obliquity of the ecliptic made by Bradley about a hundred years ago, and from later observations by Dr. Maskelyne, Delambre determined the maximum of the inequalities produced by the action of Venus, Mars, and the Moon, on the Earth, and by comparing these observations with the analytical formulæ, he obtained nearly the same value of the mass of Venus, whether he deduced it from the joint observations of Bradley and Maskelyne, or from the observations of each separately. From this correspondence in the values of the mass of Venus, obtained from these different sets of observations, there can be little doubt that the secular diminution in the obliquity of the ecliptic is very nearly 50″.2, and the probability of accuracy is greater as it agrees with the observations made by the Chinese and Arabs so

many centuries ago. Notwithstanding doubts still exist as to the mass of Venus.

604. The mass of Mars has been determined by the same method, though with less precision than that of Venus, because its action occasions less disturbance in the Earth's motions, for it is evident that the masses of those bodies that cause the greatest disturbance will be best known. The action of the new planets is insensible, and that of Mercury has a very small influence on the motions of the rest. An ingenious method of finding the mass of that planet has been adopted by La Place, although liable to error.

605. Because mass is proportional to the product of the density and the volume, if m, m', be the masses of any two planets of which ρ, ρ', are the densities, and V, V', the volumes, then

$$m : m' :: \rho \,.\, V : \rho' \,.\, V'.$$

But as the planets differ very little from spheres, their volumes may be assumed proportional to the cubes of their diameters; hence if D, D', be the diameters of m, and m',

$$m : m' :: \rho \,.\, D^3 : \rho' \,.\, D'^3;$$

whence

$$\frac{\rho}{\rho'} = \frac{D'^3}{D^3} \cdot \frac{m}{m'} \qquad (201).$$

The apparent diameters of the planets have been measured so that D and D' are known; this equation will therefore give the densities if the masses be known, and *vice versâ*.

By comparing the masses of the Earth, Jupiter, and Saturn, with their volumes, La Place found that the densities of these three planets are nearly in the inverse ratio of their mean distances from the sun, and adopting the same hypothesis with regard to Mercury, Mars, and Jupiter, he obtained the preceding values of the masses of Mars and Mercury, which are found nearly to agree with those determined from other data. Irradiation, or the spreading of the light round the disc of a planet, and other difficulties in measuring the apparent diameters, together with the uncertainty of the hypothesis of the law of the densities, makes the values of the masses obtained in this way the more uncertain, as the hypothesis does not give a true result for the masses of Venus and Saturn. Fortunately the influence of Mercury on the solar system is very small.

606. The mass of the Sun being unity, the masses of the planets are,

Mercury	$\frac{1}{2025810}$
Venus	$\frac{1}{405871}$
The Earth	$\frac{1}{354936}$
Mars	$\frac{1}{2546320}$
Jupiter	$\frac{1}{1070.5}$
Saturn	$\frac{1}{3512}$
Uranus	$\frac{1}{17918}$

Densities of the Planets.

607. The densities of bodies are proportional to the masses divided by the volumes, and when the masses are spherical, their volumes are as the cubes of their radii; as the sun and planets are nearly spherical, their densities are therefore as their masses divided by the cubes of their radii; but the radii must be taken in those parallels of latitude, the squares of whose sines are $\frac{1}{3}$.

The mean apparent semidiameters of the Sun and Earth at their mean distance are,

Sun	$961''$.
The Earth	$8''.6$

The radius of Jupiter's spheroid in the latitude in question, when viewed at the mean distance of the earth from the sun, is $94''.341$; and the corresponding radius of Saturn at his mean distance from the sun is $8''.1$. Whence the densities are,

Sun	1
The Earth	3.9326
Jupiter	0.99239
Saturn	0.59496

Thus the densities decrease with the distance from the sun; however that of Uranus does not follow this law, being greater than that of Saturn, but the uncertainty of the value of its apparent diameter may possibly account for this deviation.

Intensity of Gravitation at the Surfaces of the Sun and Planets.

608. Let g and g' represent the force of gravity at the surfaces of two bodies m and m', whose apparent diameters are D and D'. If the bodies be spherical and without rotation, the force of gravity at their equators will be as their masses divided by the squares of their diameters;

hence $$g = g' \cdot \frac{m}{m'} \cdot \frac{D'^2}{D^2}.$$

Because the masses, apparent diameters, and the intensity of gravity at the terrestrial equator are known, g, the intensity of the gravitating force at the equator of any other body may be found; and as the rotation of the sun and planets is determined by observation, their centrifugal forces, and consequently the intensity of gravitation at their surfaces may be computed. With the preceding values of the masses and apparent diameters it will be found, that if the weight of a body at the terrestrial equator be the unit, the same body transported to the equator of Jupiter, would weigh 2.716; but this would be diminished by about a ninth, on account of the centrifugal force. The same body would weigh 27.9 at the sun's equator, and a body at the sun's equator would fall through 448.39 feet in the first second of its descent, that would only fall through 16.0436 feet at the earth's equator.

To determine the fall of bodies at the surfaces of the sun and planets was hopeless till Newton's immortal discovery connected us with remote worlds.

609. The mean sidereal motions of the planets in a Julian year of 365.25 days are the second data.

When the sun is in the tropics his declination is a maximum, and equal to the obliquity of the ecliptic; the time at which that happens is found by observing his declination at noon for several days before and after the instant of a solstice, so that an equation can be formed between the time and the declination, which is sufficiently exact for a few days. If the differential of the declination in this equation be made zero, the instant of the solstice and the obliquity of the ecliptic will be obtained. The instant of the equinoxes is determined in the same manner, only that in the equation between the time and the declination, the declination is made zero, for in these points the sun is in the plane of the ecliptic. The length of the year is determined by comparing together the time of the sun's being in either equinox, or in either tropic, with the time of his being in the same point for another year distant from the former by a long period; the interval reckoned in days and parts of a day, divided by the number of years elapsed, will give the true length of the year; and the greater the interval, the more correct will it be. The length of the year however, like all astronomical data, was determined by successive approximations, but it was very early known to be 365.25 days.

The Julian year being known, if the synodic revolutions of the planets be known, their mean motions for any given interval may be found.

610. The longitude of an inferior planet in conjunction, or of a superior planet in opposition, is the same as if viewed from the centre of the sun. The synodical revolution of the planet, which is the interval between two conjunctions, or two oppositions, may be ascertained by observation, and from thence its periodic time. Let T be the synodic revolution of a planet, P its periodic time, then

$$P : 365.25 :: 360^\circ : 360^\circ \pm \alpha,$$

the angle described by the planet in 365.25 days. If it be an inferior planet, its angular motion will be greater than that of the earth; hence the angle described in 365.25 days is equal to 360° plus the angle gained by the planet on the earth, or $360^\circ + \alpha$. But if it be a superior planet, its angular velocity being less than that of the earth, the angle described in a Julian year is $360^\circ - \alpha$. But these angles are as the times in which they are described, therefore

$$360^\circ : 360^\circ \pm \alpha :: T : 365.25 \pm T;$$

hence $$P : 365.25 :: T : 365.25 \pm T,$$

and $$P = \frac{365.25 \times T}{365.25 \pm T}.$$

As the synodic revolutions are known, the sidereal revolutions of the planets are as follow:

	Days.
Mercury	87.9705
Venus	224.7
The Earth	365.2564
Mars	686.99
Vesta	1592.69
Juno	1331.
Ceres	1681.42
Pallas	1686.56
Jupiter	4332.65
Saturn	10759.4
Uranus	30687.5

Whence it will be found by simple proportion that the mean sidereal motions of the planets in a Julian year of 365.2564 days, or the values of n, n', &c., are

Mercury	5381034″.99
Venus	2106644″.82
The Earth	1295977″.74
Mars	689051″.63
Vesta	355681″.17
Juno	297216″.21
Ceres	281531″.00
Pallas	280672″.32
Jupiter	109256″.78
Saturn	43996″.13
Uranus	15425″.64

These have been determined by approximation, continually corrected by a long series of observations on the oppositions and conjunctions of the planets.

Mean Distances of the Planets, or Values of a, a', a'', &c.

611. The mean distances are obtained from the mean motions of the planets: for, assuming the mean distance of the earth from the sun as the unit, Kepler's law of the squares of the periodic times being as the cubes of the mean distances, gives the following values of the mean distances of the planets from the sun.

Mercury	0.3870981
Venus	0.7233316
The Earth	1.0000000
Mars	1.5236923
Vesta	2.3678700
Juno	2.6690090
Ceres	2.7672450
Pallas	2.7728860
Jupiter	5.2011524
Saturn	9.5379564
Uranus	19.1823927

Ratio of the Eccentricities to the Mean Distances, or Values of e, e′, &c., for 1801.

612. The eccentricity of an orbit is found by ascertaining that heliocentric longitude of the planet at which it is moving with its mean angular velocity, for there the increments of the true and mean anomaly are equal to one another, and the equation of the centre, or difference between the mean and true anomaly is a maximum, and equal to half the eccentricity. By repeating this process for a series of years, the effects of the secular variations will become sensible, and may be determined; and when they are known, the eccentricity may be determined for any given period. The values of e, e', e'', &c., for 1801, are

	Mercury	0.20551494
	Venus	0.00686074
	The Earth	0.01685318
	Mars	0.09330700
1820	Vesta	0.08913000
1820	Juno	0.25784800
1820	Ceres	0.07843900
1820	Pallas	0.24164800
	Jupiter	0.04816210
	Saturn	0.05615050
	Uranus	0.04661080

*

Inclinations of the Orbits on the Plane of the Ecliptic, in 1801.

613. When the earth is in the line of a planet's nodes, if the planet's elongation from the sun and its geocentric latitude be observed, the inclination of the orbit may be found; for the sine of the elongation is to the radius, as the tangent of the geocentric latitude to the tangent of the inclination. If the planet be 90° distant from the sun, the latitude observed is just equal to the inclination. By this method Kepler determined the inclination of the orbit of Mars. The secular inequalities become sensible after a course of years. The values of ϕ, ϕ', ϕ'', &c. were in 1801

		°	′	″
	Mercury	7	0	9.1
	Venus	3	23	28.5
	Mars	1	51	6.2
1820	Vesta	7	8	9.0
1820	Juno	13	4	9.7
1820	Ceres	10	37	26.2
1820	Pallas	34	34	55.0
	Jupiter	1	18	51.3
	Saturn	2	29	35.7
	Uranus	0	46	28.4

Longitudes of the Perihelia.

614. The angular velocity of a body is least in aphelion, and greatest in perihelion; consequently, if its longitude be observed when the increments of the angular velocity are greatest or least, these points will be in the extremities of the major axis: if these be really the two observed longitudes, the interval between them will be exactly half the time of a revolution, a property belonging to no other diameter in the ellipse. As it is very improbable that the observations should differ by 180°, they require a small correction to reduce them to the true times and longitudes. On this principle the longitudes of the perihelia may be determined, and if the observations be continued for a series of years, their secular motions will be obtained, whence their places may be computed for any epoch. The longitude of the perihelion is the distance of the perihelion from the ascending node estimated on the orbit, plus the longitude of the node. In the beginning of 1801, the values of ϖ, ϖ', ϖ'', &c., were,

		°	′	″
	Mercury	74	21	46.8
	Venus	128	43	53.0
	The Earth	99	30	4.8
	Mars	332	23	56.4
1820	Vesta	249	33	24.2
1820	Juno	53	33	46.0
1820	Ceres	147	7	31.1
1820	Pallas	121	7	4.3
	Jupiter	11	8	34.4
	Saturn	89	9	29.5
	Uranus	167	30	23.7

Longitudes of the Ascending Nodes.

615. When a planet is in its nodes, it is in the plane of the ecliptic; its longitude is then the same with the longitude of its node, and its latitude is zero. The place of the nodes may therefore be found by a series of observations, and if they be continued long enough, their secular motions will be obtained; whence their positions at any time may be computed. In the beginning of 1801 the values of θ, θ', θ'', &c., were,

		°	′	″
	Mercury	45	57	30.9
	Venus	74	54	12.9
	Mars	48	0	3.5
1820	Vesta	103	13	18.2
1820	Juno	171	7	40.4
1820	Ceres	80	41	24.0
1820	Pallas	172	39	26.8
	Jupiter	98	26	18.9
	Saturn	111	56	37.3
	Uranus	72	59	35.4

616. Mean longitudes of the planets on the 1st January, 1801, at midnight, or values of ϵ, ϵ', ϵ'', &c.

	°	′	″
Mercury	163	56	26.9
Venus	10	44	21.6
The Earth	100	9	12.9

		°	′	″
	Mars	64	6	59.9
1820 1st Jan. at noon.	Vesta	278	30	0.4
	Juno	200	16	19.1
	Ceres	123	16	11.9
	Pallas	108	24	57.9
	Jupiter	112	12	51.3
	Saturn	135	19	5.5
	Uranus	177	48	1.1

All the longitudes are estimated from the mean equinox of spring, the epoch being the 1st January, 1801.

617. With these data the motions of the planets are computed; they are, however, only approximate, since each element is determined independently of the rest; whereas they are so connected, that their values ought to be determined simultaneously by equations of condition formed from thousands of observations.

618. Elements of the orbits of the three comets belonging to the solar system.

Halley's Comet of 1682.

Period of revolution 76 years, nearly. Instant of passage at perihelion 1835, October 31st, 2.

Half the greater axis	17.98355
Eccentricity	0.967453
Longitude of perihelion on orbit	304° 34′ 19″
Longitude of ascending node	55 6 59
Inclination	17 46 50

Motion retrograde.

Enke's Comet of 1819.

Period of revolution 1203.687 days. Passage at perihelion 1829, January 10th, 573.

Mean diurnal motion	1069″.557
Half the greater axis	2.224346
Eccentricity	0.8446862
Longitude of perihelion	157° 18′ 35″
Longitude of ascending node	334 24 15
Inclination	13 22 34

Claussen and Gambart's Comet of 1825.

Period of revolution 6, 7 years. Passage at perihelion 1832, November 27th, 4808.

Half the greater axis	3.53683
Eccentricity	0.7517481
Longitude of perihelion	109° 56′ 45″
Longitude of ascending node . . .	248 12 24
Inclination	13 13 13

The computation, in the next Chapter, of the perturbations of Jupiter and Saturn will be sufficient to show the method of finding their numerical values, especially as there are many peculiar to these two planets.

CHAPTER XIV.

NUMERICAL VALUES OF THE PERTURBATIONS OF JUPITER.

619. THE epoch assumed for this computation is that of the French Tables, namely, the 31st of December, at midnight, 1749, mean time at Paris. The data for that epoch are as follow:—

Values of e, e', &c.

Mercury	0.20551320
Venus	0.00688405
The Earth	0.01681395
Mars	0.09308767
Jupiter	0.04807670
Saturn	0.05622460
Uranus	0.04669950

Values of ϖ, ϖ', ϖ'', &c.

Mercury	73°.5661
Venus	127.9117
The Earth	98.6211
Mars	331.473
Jupiter	10.3511
Saturn	88.1519
Uranus	166.614

Values of ϕ, ϕ', ϕ'', &c.

Mercury	7°
Venus	3.3931
Mars	1.8499
Jupiter	1.3172
Saturn	2.4986
Uranus	0.7736

Values of θ, θ', θ'', &c.

Mercury	45°.3452
Venus	74.4384
Mars	47.6438
Jupiter	97.906
Saturn	111.5064
Uranus	72.6314

The longitudes are estimated from the mean equinox of spring.

620. The series represented by S and S' in article 453 form the basis of the whole computation, but twelve or fourteen of the first terms of each will be sufficiently correct for all the planets.

The numerical values of the coefficients, A_0, A_1, &c. B_0, B_1, &c., and their differences, for Jupiter and Saturn, are obtained from the formulæ in article 455, and those that follow. The mean distances of these two planets are, according to La Place,

$$a = 5.20116636, \qquad a' = 9.5378709,$$

whence

$$\alpha = 0.54531725.$$

$$S = 10.2612 \qquad S' = -4.99987,$$

$$A_0 = 0.228576 \qquad A_1 = 0.065071 \qquad A_2 = 0.027012$$

$$A_3 = 0.012369 \qquad A_4 = 0.005929 \qquad A_5 = 0.002918$$

$$A_6 = 0.001458 \qquad A_7 = 0.000738 \qquad A_8 = 0.000376$$

$$A_9 = 0.000189 \qquad A_{10} = 0.000091 \qquad A_{11} = 0.000034$$

$$\frac{dA_0}{da} = 0.008891 \qquad \frac{dA_1}{da} = 0.016305 \qquad \frac{dA_2}{da} = 0.012149$$

$$\frac{dA_3}{da} = 0.007987 \qquad \frac{dA_4}{da} = 0.004983 \qquad \frac{dA_5}{da} = 0.00302$$

$$\frac{dA_6}{da} = 0.001798 \qquad \frac{dA_7}{da} = 0.001056 \qquad \frac{dA_8}{da} = 0.000617$$

$$\frac{dA_9}{da} = 0.000364 \qquad \frac{dA_{10}}{da} = 0.000223.$$

$$\frac{d^2A_0}{da^2} = 0.003314 \qquad \frac{d^2A_1}{da^2} = 0.002942 \qquad \frac{d^2A_2}{da^2} = 0.004058$$

$$\frac{d^2A_3}{da^2} = 0.004070 \qquad \frac{d^2A_4}{da^2} = 0.003453 \qquad \frac{d^2A_5}{da^2} = 0.002654$$

$\frac{d^2A_6}{da^2} = 0.001919$ $\frac{d^2A_7}{da^2} = 0.001319$ $\frac{d^2A_8}{da^2} = 0.000877$

$\frac{d^2A_9}{da^2} = 0.000559.$

$\frac{d^3A_0}{da^3} = 0.001466$ $\frac{d^3A_1}{da^3} = 0.001556$ $\frac{d^3A_2}{da^3} = 0.001551$

$\frac{d^3A_3}{da^3} = 0.001868$ $\frac{d^3A_4}{da^3} = 0.002061$ $\frac{d^3A_5}{da^3} = 0.002013$

$\frac{d^3A_6}{da^3} = 0.001808$ $\frac{d^3A_7}{da^3} = 0.001478$ $\frac{d^3A_8}{da^3} = 0.001156$

$\frac{d^4A_0}{da^4} = 0.001069$ $\frac{d^4A_1}{da^4} = 0.001064$ $\frac{d^4A_2}{da^4} = 0.001107$

$\frac{d^4A_3}{da^4} = 0.001138$ $\frac{d^4A_4}{da^4} = 0.001284$ $\frac{d^4A_5}{da^4} = 0.001308$

$\frac{d^4A_6}{da^4} = 0.001503$ $\frac{d^4A_7}{da^4} = 0.001469$

$\frac{d^5A_0}{da^5} = 0.000993$ $\frac{d^5A_1}{da^5} = 0.001001$ $\frac{d^5A_2}{da^5} = 0.001011$

$\frac{d^5A_3}{da^5} = 0.001044$ $\frac{d^5A_4}{da^5} = 0.001088$ $\frac{d^5A_5}{da^5} = 0.001175$

$\frac{d^5A_6}{da^5} = 0.001212.$

$B_0 = 0.005026$ $B_1 = 0.003674$ $B_2 = 0.0024$

$B_3 = 0.001493$ $B_4 = 0.000904$ $B_5 = 0.000537$

$B_6 = 0.000315$ $B_7 = 0.000183$ $B_8 = 0.000107$

$B_9 = 0.000062.$

$\frac{dB_0}{da} = 0.001774$ $\frac{dB_1}{da} = 0.000184$ $\frac{dB_2}{da} = 0.000162$

$\frac{dB_3}{da} = 0.000128$ $\frac{dB_4}{da} = 0.000943$ $\frac{dB_5}{da} = 0.000661$

$\frac{dB_6}{da} = 0.000448$ $\frac{dB_7}{da} = 0.000448$ $\frac{dB_8}{da} = 0.000293$

$\frac{dB_9}{da} = 0.000189.$

$\frac{d^2B_0}{da^2} = 0.001225$ $\frac{d^2B_1}{da^2} = 0.001203$ $\frac{d^2B_2}{da^2} = 0.001181$

$$\frac{d^2B_3}{da^2} = 0.001101 \qquad \frac{d^2B_4}{da^2} = 0.000951 \qquad \frac{d^2B_5}{da^2} = 0.000774$$

$$\frac{d^2B_6}{da^2} = 0.000602 \qquad \frac{d^2B_7}{da^2} = 0.000453.$$

$$\frac{d^3B_0}{da^3} = 0.001102 \qquad \frac{d^3B_1}{da^3} = 0.001102 \qquad \frac{d^3B_2}{da^3} = 0.001076$$

$$\frac{d^3B_3}{da^3} = 0.001043 \qquad \frac{d^3B_4}{da^3} = 0.000984 \qquad \frac{d^3B_5}{da^3} = 0.000885$$

$$\frac{d^3B_6}{da^3} = 0.000764.$$

Jupiter and Mercury.

$$a' = 0.38709812 \qquad a = 5.20116636$$

$$\alpha = 0.0744256$$

$$S = 5.20887 \qquad S' = -0.38683.$$

Jupiter and Venus.

$$a' = 0.7233323 \qquad \alpha = 0.13907116$$

$$S = 5.22634 \qquad S' = -0.721579.$$

Jupiter and the Earth.

$$a' = 1. \qquad \alpha = 0.19226461$$

$$S = 5.24933 \qquad S' = -0.995358.$$

Jupiter and Mars.

$$a' = 1.52369352 \qquad \alpha = 0.29295212$$

$$S = 5.31338 \qquad S' = -1.50717.$$

Jupiter and Uranus.

$$a' = 19.183305 \qquad \alpha = 0.2711298$$

$$S = 19.5375 \qquad S' = -5.1528.$$

Secular Variations of Jupiter and Saturn.

621. These are given by the numerical values of equations (198), which are computed from the formulæ

$$\boxed{4.0} = -\frac{3m'.an\{aa'S+(a^2+a'^2)S'\}}{2(a'^2-a^2)^2}$$

$$(4.0) = -\frac{3m'.a^2a'n.S'}{4(a'^2-a^2)^2},$$

as the numerical values of all the quantities in these expressions are given, it is easy to find by their substitution, that

$$
\begin{array}{llll}
(4.0) = 0''.000226 & \boxed{4.0} = 0.000021, & & (202) \\
(4.1) = 0.004291 & \boxed{4.1} = 0.00744, & & \\
(4.2) = 0.009862 & \boxed{4.2} = 0.002359, & & \\
(4.3) = 0.00451 & \boxed{4.3} = 0.001633, & & \\
(4.5) = 7.702 & \boxed{4.5} = 5.0342, & & \\
(4.6) = 0.09665 & \boxed{4.6} = 0.03247, & &
\end{array}
$$

where the digits 0, 1, 2, 3, &c. refer to Mercury, Venus, the Earth, Mars, Jupiter, Saturn, and Uranus.

622. By the substitution of the preceding data, equations (128) and (141), give the following results, when multiplied by the radius reduced to seconds, or, by 206264''.8, where $\frac{d\varpi}{dt}$ is the sidereal motion of the perihelion of Jupiter in longitude at the epoch 1750, during a period of 365¼ days: $2\frac{de}{dt}$ is the annual variation of the equation of the centre: $\frac{d\phi}{dt}$ is the annual variation of the orbit of Jupiter on the fixed ecliptic of 1750; $\frac{d\phi'}{dt}$ is the annual variation of the inclination on the true ecliptic: $\frac{d\theta}{dt}$ is the annual and sidereal motion of the ascending node of the orbit of Jupiter on the fixed ecliptic of 1750; and $\frac{d\theta'}{dt}$ is the same variation with regard to the true ecliptic.

$$
\begin{array}{ll}
\frac{d\bar{\varpi}}{dt} = 6''.5998 & \frac{d\bar{e}}{dt} = 0''.27721 \\
\frac{d\bar{\phi}}{dt} = -0''.07814 & \frac{d\bar{\phi}'}{dt} = -0''.223178 \\
\frac{d\bar{\theta}}{dt} = 6''.4562, & \frac{d\bar{\theta}'}{dt} = -14''.6634.
\end{array}
$$

By article 484,

$$(4.0) = \frac{m\sqrt{a}}{m'\sqrt{a'}}(0.4)\,; \qquad \boxed{4.0} = \frac{m\sqrt{a}}{m'\sqrt{a'}}\boxed{0.4}\,;$$

if then, the quantities (202) relating to Jupiter, be multiplied by $\frac{m\sqrt{a}}{m'\sqrt{a'}}$, those corresponding to Saturn will be found, and the formulæ (128) give for Saturn

$$\frac{d\bar{\varpi}'}{dt} = 16''.1127 \qquad \frac{d\bar{\epsilon}'}{dt} = 0''.54021$$

$$\frac{d\bar{\phi}'}{dt} = 0''.099741 \qquad \frac{d\bar{\theta}'}{dt} = -\,9''.0053.$$

By article 444,

$$\phi' \sin\theta' - \phi\sin\theta = \gamma\sin\bar{\Pi},$$

$$\phi' \cos\theta' - \phi\cos\theta = \gamma\cos\bar{\Pi}\,;$$

and by the substitution of the numerical values of article 613 and 615, it will readily be found, that in 1750

$$\bar{\gamma} = 1^\circ\,15'\,30'' \qquad \bar{\Pi} = 125^\circ\,44'\,34'',$$

γ being the mutual inclination of the orbits of Jupiter and Saturn, and Π the longitude of the ascending node of the orbit of Saturn on that of Jupiter. If the differential of these equations be taken and the numerical values of

$$\frac{d\theta'}{dt},\ \frac{d\theta}{dt},\qquad \frac{d\phi}{dt},\ \frac{d\phi'}{dt}$$

substituted, it will be found, that

$$\frac{d\bar{\gamma}}{dt} = -\,0''.000105, \qquad \frac{d\bar{\Pi}}{dt} = -\,26''.094.$$

623. The variations in the elements that depend on the squares of the disturbing forces must now be computed, and for that purpose the numerical values of P, P', and their differences, must be found from equations (165) and (166).

The coefficients Q_0, Q_1, &c., are given by the expansion of R, article 446; so that

$$Q_0 = -\tfrac{1}{12}\left\{389A_2 + 201a\,.\,\frac{dA_2}{da} + 27a^2.\,\frac{d^2A_2}{da^2} + a^3\,.\,\frac{d^3A_2}{da^3}\right\},$$

$$Q_1 = \tfrac{1}{4}\left\{402A_3 + 193a\,.\,\frac{dA_3}{da} + 26a^2.\,\frac{d^2A_3}{da^2} + a^3.\,\frac{d^3A_3}{da^3}\right\},$$

$$Q_2 = -\tfrac{1}{4}\left\{396A_4 + 184a.\frac{dA_4}{da} + 25a^2.\frac{d^2A_4}{da^2} + a^3.\frac{d^3A_4}{da^3}\right\},$$

$$Q_3 = \tfrac{1}{12}\left\{380A_5 + 174a.\frac{dA_5}{da} + 24a^2.\frac{d^2A_5}{da^2} + a^3\frac{d^3A_5}{da^3}\right\},$$

$$Q_4 = -\tfrac{1}{4}aa'\left\{10B_3 + a\frac{dB_3}{da}\right\},$$

$$Q_5 = \tfrac{1}{4}aa'\left(7B_4 + a\frac{dB_4}{da}\right\}.$$

If the values of A_2, A_3, &c., and their differences in article 620, be substituted, then will

$$\begin{aligned} Q_0 &= -\ 2.199192 \\ Q_1 &= \quad 4.0292, \\ Q_2 &= -\ 2.43538, \\ Q_3 &= \quad 0.487332, \\ Q_4 &= -\ 0.267808, \\ Q_5 &= \quad 0.139264. \end{aligned}$$

With the preceding data, equations (165) and (166) give

$$P = 0.0000114596, \qquad P' = -\ 0.000107267;$$

and as the differences of these equations are

$$\frac{dP}{de} = \tfrac{1}{4}\{Q_1e'^2 \sin(2\varpi' + \varpi) + 2Q_2ee' \sin(\varpi' + 2\varpi)$$

$$+ 3Q_3e^2 \sin 3\varpi + Q_5\gamma^2 \sin(2\Pi + \varpi)\},$$

$$\frac{dP'}{de} = \tfrac{1}{4}\{Q_1e'^2 \cos(2\varpi' + \varpi) + 2Q_2ee' \cos(\varpi' + 2\varpi)$$

$$+ 3Q_3e^2 \cos 3\varpi + Q_5\gamma^2 \cos(2\Pi + \varpi)\}$$

&c. &c.

all the quantities in equations (191), (192), (195), and (196), are known; whence for Jupiter at the epoch,

$$(\delta\bar{\varpi}) = 0''.352941, \qquad (\delta\bar{e}) = 0''.052278,$$

$$(\delta\bar{\gamma}) = 0''.000184, \qquad (\delta\bar{\Pi}) = -\ 0''.007631,$$

and from equations (193) and (194), the corresponding variations in the elements of the orbit of Saturn are

$$(\delta\bar{\varpi}') = 3''.242722, \qquad (\delta\bar{e}') = -\ 0''.102763;$$

so that

$$\frac{d\bar{\varpi}}{dt} = 6''.95281; \quad \frac{d\bar{\varpi}'}{dt} = 19''.355448; \quad \frac{d\bar{e}}{dt} = 0''.329487;$$

$$\frac{d\bar{e}'}{dt} = -0''.642968; \quad \frac{d\bar{\gamma}}{dt} = -0''.000079; \quad \frac{d\bar{\Pi}}{dt} = -26''.10163.$$

But at the epoch,

$$\bar{e} = 9916''.53; \quad \bar{e}' = 11597''.1; \quad \bar{\varpi} = 10^\circ.35108,$$

$$\bar{\varpi}' = 88^\circ.15194; \quad \bar{\gamma} = 1^\circ.25838; \quad \bar{\Pi} = 125^\circ.74278.$$

Consequently the elements of the two orbits at any time t are

$$\begin{aligned} e &= 9916''.53 + 0''.329487.t, \\ \varpi &= 10^\circ.35108 + 6''.95281.t, \\ e' &= 11597''.1 - 0''.642968.t, \\ \varpi' &= 88^\circ.15194 + 19''.3555448.t, \\ \gamma &= 1^\circ.25838 + 0''.000079.t, \\ \Pi &= 125^\circ.74278 - 26''.10163.t. \end{aligned} \qquad (203)$$

If $t = 0$ these expressions will give the elements in 1950, and if the computation be repeated with them it will be found that in 1950

$$\frac{d\varpi}{dt} = 7''.053178; \quad \frac{d\varpi'}{dt} = 19''.424739; \quad \frac{de}{dt} = 0''.326172;$$

$$\frac{de'}{dt} = -0''.648499; \quad \frac{d\gamma}{dt} = -0''.001487; \quad \frac{d\Pi}{dt} = -26''.402056.$$

The differences between these and their values for 1750, divided by 200, will be their second differences, therefore the formulæ (198), with regard to Jupiter and Saturn, are

$$\begin{aligned} e &= 9916''.53 + 0''.329487.t - 0''.0000082871.t^2, \\ \varpi &= 10^\circ\,21'\,4'' + 6''.952808.t + 0''.0002509259.t^2, \\ e' &= 11597''.1 - 0''.642968.t - 0''.0000138275.t^2, \\ \varpi' &= 88^\circ\,9'\,6''.4 + 19''.355448.t + 0''.0001732274.t^2, \\ \gamma &= 1^\circ\,15'\,30''.2 + 0''.000078.t - 0''.0000391311.t^2, \\ \Pi &= 125^\circ\,44'\,33'' - 26''.1028\,.t - 0''.0007507307.t^2, \end{aligned} \qquad (204)$$

which will give the elements of the orbits of these two planets for 1000 or 1200 years before or after 1750.

Periodic Inequalities of Jupiter.

624. The inequalities in the radius vector and longitude, which are independent of the eccentricities and inclinations, are computed from

$$\frac{\delta r}{a} = -\frac{m'}{6}\,a^2.\,\frac{dA_0}{da} + \frac{m'}{2}\,.\,\Sigma.\,C_i\,.\,\cos i(n't - nt + \epsilon' - \epsilon),$$

$$\delta v = \frac{m'}{2}\,.\,\Sigma\,.\,F_i\,.\,\sin i(n't - nt + \epsilon' - \epsilon);$$

If $i = 1$, then by articles 536 and 537

$$C_1 = \frac{n^2}{n'(2n-n')}\left\{\frac{2n}{n-n'}.aA_1 + a^2.\frac{dA_1}{da}\right\}$$

$$F_1 = \frac{n}{n-n'}\left\{-\frac{n}{n-n'}.\,aA_1 + 2C_1\right\}.$$

But $n = 109256''$; $n' = 43996''.7$; $a = 5.20116636$;

$$m' = \frac{1}{3359.4};\quad A_1 = 0.0078973;\quad \frac{dA_1}{da} = 0.00531108.$$

$$\frac{2n}{n-n'}.\,aA_1 = 0.1375352;\quad a^2.\frac{dA_1}{da} = 0.143676;$$

$$\frac{2n}{n-n'}\,aA_1 + a^2\frac{dA_1}{da} = 0.281209;$$

$$\log 0.281209 = 9.4490293$$

$$\log\frac{2n^2}{n'(2n-n')} = 0.4926697$$

$$\log 2C_1 = 9.9416990 = \log 0.874378$$

hence

$$-\frac{n}{n-n'}\,aA_1 + 2C_1 = 0.8056104.$$

$$\log 0.8056104 = 9.9061248$$

$$\log\frac{n}{n-n'} = 0.2238068$$

$$\text{log of radius in seconds} = 5.3144256$$

the sum is 5.4443572

$$\log 3359.4 = 3.5262617$$

$$\log 82''.812 = 1.9180955$$

Consequently, when $i = 1$, $\delta v = 82''.821.\ \sin(n't - nt + \epsilon' - \epsilon)$. Hence if i be made successively equal to all the positive numbers from 1 to 9, and the corresponding quantities substituted in the preceding formulæ, it will be found that the inequalities of this order in the longitude and radius vector of Jupiter arising from the action of Saturn, are

$$\delta v = \left\{\begin{array}{ll} 82''.811711 & \sin\ (n't - nt + \epsilon' - \epsilon) \\ -204''.406384 & \sin 2(n't - nt + \epsilon' - \epsilon) \\ -\ 17''.071564 & \sin 3(n't - nt + \epsilon' - \epsilon) \\ -\ \ 3''.926319 & \sin 4(n't - nt + \epsilon' - \epsilon) \\ -\ \ 1''.210573 & \sin 5(n't - nt + \epsilon' - \epsilon) \\ -\ \ 0''.42843 & \sin 6(n't - nt + \epsilon' - \epsilon) \\ -\ \ 0''.170923 & \sin 7(n't - nt + \epsilon' - \epsilon) \\ -\ \ 0''.076086 & \sin 8(n't - nt + \epsilon' - \epsilon) \\ -\ \ 0''.041273 & \sin 9(n't - nt + \epsilon' - \epsilon) \end{array}\right.$$

$$\delta r = \begin{cases} - 0.0000620586 \\ + 0.000676876 \cos (n't - nt + \epsilon' - \epsilon) \\ - 0.00289662 \cos 2(n't - nt + \epsilon' - \epsilon) \\ - 0.0003021367 \cos 3(n't - nt + \epsilon' - \epsilon) \\ - 0.0000782514 \cos 4(n't - nt + \epsilon' - \epsilon) \\ - 0.0000258952 \cos 5(n't - nt + \epsilon' - \epsilon) \\ - 0.0000094779 \cos 6(n't - nt + \epsilon' - \epsilon) \\ - 0.000003756 \cos 7(n't - nt + \epsilon' - \epsilon) \\ - 0.0000014781 \cos 8(n't - nt + \epsilon' - \epsilon) \\ - 0.0000004799 \cos 9(n't - nt + \epsilon' - \epsilon). \end{cases}$$

625. The inequalities depending on the first powers of the eccentricities are obtained from

$$\delta r = m'fe \cos (nt + \epsilon - \varpi) + m'f'e'. \cos (nt + \epsilon - \varpi')$$
$$+ m'e . \Sigma . D_i . \cos \{i (n't - nt + \epsilon' - \epsilon) + nt + \epsilon - \varpi\}$$
$$+ m'e' . \Sigma . E_i . \cos \{i (n't - nt + \epsilon' - \epsilon) + nt + \epsilon - \varpi'\},$$
$$\delta v = m'e . \Sigma . G_i . \sin \{i (n't - nt + \epsilon' - \epsilon) + nt + \epsilon - \varpi\}$$
$$+ m'e' . \Sigma . H_i . \sin \{i (n't - nt + \epsilon' - \epsilon) + nt + \epsilon - \varpi'\}.$$

by making i successively equal to the whole positive numbers, from 1 to 7, and to the whole negative numbers, from -1 to -5, and substituting the numerical data corresponding to each in the coefficients D_i, E_i, &c., which are given in articles 536 and 537. The values of e and e' at the epoch are sufficiently exact for all the terms of this order, except those having the arguments $2n't - nt + 2\epsilon' - \epsilon$, and $3n' - 2nt + 3\epsilon' - 2\epsilon$, whose periods are so long, that $9916''.53 + 0''.329487 . t$, and $11597''.1 - 0''.642968 . t$ must be employed instead of e and e'. It will then be found that the perturbations of Jupiter are

$$\delta r = \begin{cases} 0.0000206111 \cos (nt + \epsilon - \varpi) \\ - 0.0000795246 \cos (n't + \epsilon' - \varpi) \\ + 0.0000492096 \cos (n't + \epsilon' - \varpi') \\ - 0.000292213 \cos \{2n't - nt + 2\epsilon' - \epsilon - \varpi\} \\ + 0.0001688085 \cos \{2n't - nt + 2\epsilon' - \epsilon - \varpi'\} \\ - 0.0004584483 \cos \{3n't - 2nt + 3\epsilon' - 2\epsilon - \varpi\} \\ + 0.0009047822 \cos \{3n't - 2nt + 3\epsilon' - 2\epsilon - \varpi'\} \\ + 0.0001259429 \cos \{4n't - 3nt + 4\epsilon' - 2\epsilon - \varpi\} \\ - 0.0002424413 \cos \{4n't - 3nt + 4\epsilon' - 3\epsilon - \varpi'\} \\ + 0.0000268383 \cos \{5n't - 4nt + 5\epsilon' - 4\epsilon - \varpi\} \\ - 0.0000516048 \cos \{5n't - 4nt + 5\epsilon' - 4\epsilon - \varpi'\} \\ + 0.0000579151 \cos \{2nt - n't + 2\epsilon - \epsilon' - \varpi\} \\ - 0.000134653 \cos \{3nt - 2n't + 3\epsilon - 2\epsilon' - \varpi). \end{cases}$$

$$
\delta v = \left\{ \begin{array}{l}
\quad 8''.608489 \sin (n't + \epsilon' - \varpi) \\
- \; 9''.692386 \sin (n't + \epsilon' - \varpi') \\
- \{138''.373337 + t \,.\, 0''.0045985\} \sin \{2n't - nt + 2\epsilon' \\
\qquad\qquad\qquad\qquad\qquad\qquad - \epsilon - \varpi\} \\
+ \{56''.634099 - t \,.\, 0''.0031398\} \sin \{2n't - nt + 2\epsilon' \\
\qquad\qquad\qquad\qquad\qquad\qquad - \epsilon - \varpi'\} \\
- \{44''.460822 + t \,.\, 0''.0014775\} \sin \{3n't - 2nt + 2\epsilon' \\
\qquad\qquad\qquad\qquad\qquad\qquad - 2\epsilon - \varpi\} \\
+ \{84''.942569 - t \,.\, 0''.004794 \} \sin \{3n't - 2nt + 3\epsilon' \\
\qquad\qquad\qquad\qquad\qquad\qquad - 2\epsilon - \varpi'\} \\
+ \;\; 7''.925312 \sin \{4n't - 3nt + 4\epsilon' - 3\epsilon - \varpi\} \\
- \; 15''.629621 \sin \{4n't - 3nt + 4\epsilon' - 3\epsilon - \varpi'\} \\
+ \;\; 1''.047717 \sin \{5n't - 4nt + 5\epsilon - 4\epsilon - \varpi\} \\
- \;\; 2''.781664 \sin \{5n't - 4nt + 5\epsilon - 4\epsilon - \varpi'\} \\
+ \;\; 0''.407251 \sin \{6n't - 5nt + 6\epsilon - 5\epsilon - \varpi\} \\
- \;\; 0''.913302 \sin \{6n't - 5nt + 6\epsilon' - 5\epsilon - \varpi'\} \\
+ \;\; 0''.149277 \sin \{7n't - 6nt + 7\epsilon' - 6\epsilon - \varpi\} \\
- \;\; 0''.325592 \sin \{7n't - 6nt + 7\epsilon' - 6\epsilon - \varpi'\} \\
- \;\; 5''.208122 \sin \{2nt - n't + 2\epsilon - \epsilon' - \varpi\} \\
- \;\; 0''.569738 \sin \{2nt - n't + 2\epsilon - \epsilon' - \varpi'\} \\
+ \; 12''.87665 \;\; \sin \{3nt - 2n't + 3\epsilon' - 2\epsilon' - \varpi\} \\
- \;\; 0''.352399 \sin \{3nt - 2n't + 3\epsilon - 2\epsilon' - \varpi'\} \\
+ \;\; 1''.287482 \sin \{4nt - 3n't + 4\epsilon - 3\epsilon' - \varpi\} \\
- \;\; 0''.172892 \sin \{4nt - 3n't + 4\epsilon - 3\epsilon' - \varpi'\} \\
+ \;\; 0''.356627 \sin \{5nt - 4n't + 5\epsilon - 4\epsilon' - \varpi\} \\
- \;\; 0''.083189 \sin \{5nt - 4n't + 5\epsilon - 4\epsilon' - \varpi'\}.
\end{array} \right.
$$

Inequalities depending on the Squares of the Eccentricities and Inclinations.

626. These are computed by making i successively equal to 1, 2, 3, &c. in formulæ (163) and (164).

If $i = 1$, that part of the perturbations in longitude, depending on the argument $n't + nt + \epsilon' + \epsilon$, is

$$
\delta v = \frac{1}{\sqrt{1-e^2}} \left\{ \frac{2d \,.\, (r\delta r)}{a^2 \,.\, ndt} - \frac{m'}{2} \left\{ \begin{array}{l} (\tfrac{1}{2}C_1 + D_1)e^2 \,.\, \sin(n't + nt + \epsilon' + \epsilon - 2\varpi) \\ + E_1 ee' \,.\, \sin(n't + nt + \epsilon' + \epsilon - \varpi - \varpi') \end{array} \right. \right.
$$

$$
- \frac{m}{2} \left\{ \frac{3n^2}{(n'+n)^2} \,.\, \Sigma \,.\, aN + \Sigma a^2 \,.\, \frac{dN}{da} \,.\, \frac{2n}{n'+n} \right\} \sin(n't + nt + \epsilon' + \epsilon + L);
$$

where

$$\frac{2d(r\delta r)}{a^2.ndt} = -\frac{m'.n^2}{n'^2+2nn'}\left\{\begin{array}{l}3\left(\tfrac{1}{2}C_1+D_1\right)e^2.\sin(n't+nt+\epsilon'+\epsilon-2\varpi)\\+3E_1.ee'.\sin(n't+nt+\epsilon'+\epsilon-\varpi-\varpi')\end{array}\right.$$

$$+\left\{\frac{2n}{n'+n}.\Sigma.dN+\Sigma.a^2\frac{dN}{da}\right\}.\sin(n't+nt+\epsilon'+\epsilon+L)\Big\}.$$

$$C_1 = \frac{n^2}{n^2-(n'-n)^2}.\left\{\frac{2n}{n-n'}.aA_1+a^2\frac{dA_1}{da}\right\}$$

$$D_1 = \frac{n^2}{n'^2-n^2}\left\{\frac{3n}{n'-n}.aA_1-\frac{(n'-n)(n'-2n)-3n^2}{n^2}.C_1+\tfrac{1}{2}a^3\frac{d^2A_1}{da^2}\right\}$$

$$E_1 = -\frac{n^2}{n'^2-n^2}.\left\{a^2\frac{dA_0}{da}+\tfrac{1}{2}a^3\frac{d^2A_0}{da^2}\right\}.$$

$$\begin{aligned}\Sigma.N.\sin(n't+nt+\epsilon'+\epsilon-L=&\\ N_0.e^2.\sin(n't+nt+\epsilon'+\epsilon-2\varpi)&\\ +N_1.ee'.\sin(n't+nt+\epsilon'+\epsilon-\varpi-\varpi')&\\ +N_2.e'^2.\sin(n't+nt+\epsilon'+\epsilon-2\varpi')&\\ +N_6.\gamma^2.\sin(n't+nt+\epsilon'+\epsilon-2\Pi).&\end{aligned}$$

The coefficients N_0, N_1, &c., are given in article 459, and if the numerical values of A_0, A_1, their differences, and also $n = 109256''$, $n'' = 43996''.6$, be substituted, it will be found that δv takes the form

$$\begin{aligned}\delta v = b\ .e^2.\sin(n't+nt+\epsilon'+\epsilon-2\varpi)&\\ +b_1.ee'.\sin(n't+nt+\epsilon'+\epsilon-\varpi-\varpi')&\\ +b_2.e'^2.\sin(n't+nt+\epsilon'+\epsilon-2\varpi')&\\ +b_3.\gamma^2.\sin(n't+nt+\epsilon'+\epsilon-2\Pi),&\end{aligned}$$

where b, b_1, b_2 and b_3 are given numbers. But δv may be expressed by

$$\begin{aligned}\delta v = P.\sin(n't+nt+\epsilon'+\epsilon)&\\ -P'.\cos(n't+nt+\epsilon'+\epsilon),&\end{aligned}$$

Where,

$$\begin{aligned}P' &= be^2.\sin 2\varpi+b_1ee'.\sin(\varpi+\varpi')\\ &\quad+b_2.e'^2\sin 2\varpi'+b_3.\gamma^2\sin 2\Pi\\ P &= be^2.\cos 2\varpi+b_1ee'\cos(\varpi+\varpi')\\ &\quad+b_2.e'^2\cos 2\varpi'+b_3.\gamma^2\cos 2\Pi;\end{aligned}$$

substituting the values of the elements given in article 619, it will be found by the method in article 569, that

$$\sqrt{P^2+P'^2} = 1''.004\quad \frac{P'}{P} = -\tan 45^\circ.4894. = -\frac{\sin 45^\circ.4894}{\cos 45^\circ.4894}.$$

*

Consequently the inequality depending on $i = 1$ becomes

$$\delta v = 1''.004 \,.\, \sin(n't + nt + \epsilon' + \epsilon + 45^\circ.4894).$$

627. It will be found by this method of computation that all the sensible inequalities in longitude and in the radius vector depending on the squares and products of the eccentricities and inclinations, are included in the following expressions; observing that the inequality having the argument $3n't - 5nt + 3\epsilon' - 5\epsilon$, must be computed with the formulæ (204), on account of the great length of its period.

$$\delta v = \begin{cases} \quad 1''.004 \,.\, \sin(n't+nt+\epsilon'+\epsilon+45^\circ.4894) \\ - \; 5''.57871 \,.\, \sin(2n't + 2\epsilon' + 15^\circ.93999) \\ + \; 11''.72425 \,.\, \sin(3n't-nt+3\epsilon'-\epsilon+79^\circ.6633) \\ - \; 18''.07528 \,.\, \sin(4n't-2nt+4\epsilon'-2\epsilon-57^\circ.2072) \\ +\{169''.2659 - t.0''.004277\} \,.\, \sin(3n't - 5nt + 3\epsilon' - 5\epsilon + \\ \qquad 55^\circ.6802 + t \,.\, 50''.5084) \\ + \; 1''.64714 \,.\, \sin(6n't-4nt+6\epsilon'-4\epsilon-54^\circ.43) \\ + \; 2''.4764 \,.\, \sin(n't-nt+\epsilon-\epsilon'+43^\circ.2836) \\ - \; 5''.288 \,.\, \sin(2n't-2nt+2\epsilon'-2\epsilon+42^\circ.6789) \end{cases}$$

$$\delta v = \begin{cases} \quad 0.000082242 \,.\, \cos(2n't+2\epsilon+11^\circ.0153) \\ +0.000022625 \,.\, \cos(3n't - nt + 3\epsilon' - 2\epsilon \\ \qquad - 21^\circ.7884) \\ -0.0001010533 \,.\, \cos(4n't - 2nt + 4\epsilon' - 2\epsilon \\ \qquad - 51^\circ.0677) \\ -\{0.00211145 - t.0.00000005323\} \,.\, \cos(3n't-5nt+3\epsilon'-5\epsilon \\ \qquad + 55^\circ.597 \,.\, + 50''.4144 \,.\, t) \\ -0.0000652204 \,.\, \cos(2n't - 2nt + 2\epsilon' - 3\epsilon \\ \qquad + 54^\circ.1477). \end{cases}$$

Perturbations depending on the Third Powers and Products of the Eccentricities and Inclinations.

628. These are contained in equation (172). But, in order to find the numerical value of the principal term, the differences of P and P' must be computed. By article 623,

$$P = 0.0000114596, \qquad P' = -0.000107267$$

are the values of these quantities in 1750; but their values in the years 2250, and 2750, will be obtained by making t successively equal to 500 and 1000, in equations (204); whence the elements of the orbits of Jupiter and Saturn at these two periods will be known; and if the same computation that was employed for the determination of P and P' be repeated with them, the results in 2250, and 2750, will be

$$P = -\ 0.000008407$$
$$P' = -\ 0.00010552$$
$$P = -\ 0.000027365$$
$$P' = -\ 0.00010009;$$

and, by the method of article 480

$$\frac{dP}{dt} = -\ 0.000000040645;$$

$$\frac{dP'}{dt} = -\ 0.0000000002249;$$

$$\frac{dP}{dt} = -\ 0.00000000003642;$$

$$\frac{dP'}{dt} = -\ 0.000000000014865;$$

with these data the principal term of the great inequality put under the form of equation (171) becomes

$$\begin{aligned}\delta v = &\{1263''.79967 - 0''.008418\,.\,t - 0''.00001925\,.\,t^2\}\\ &\sin(5n't - 2nt + 5\epsilon' - 2\epsilon)\\ &+ \{119''.52695 - 0''.473686\,.\,t - 0''.000078562\,.\,t^2\}\\ &\cos(5n't - 2nt + 5\epsilon' - 2\epsilon).\end{aligned}$$

In order to compute the inequality

$$\delta v = -\frac{2m'n}{5n' - 2n}\left\{\begin{array}{l} a^2\ .\ \dfrac{dP}{da}\ .\ \cos(5n't - 2nt + 5\epsilon - 2\epsilon \\ -\ a^2\ \dfrac{dP'}{da}\ .\ \sin(5n't - 2nt + 5\epsilon' - 2\epsilon \end{array}\right\}$$

equation (165), gives

$$\frac{dP}{da} = \frac{dQ_0}{da}\ .\ e'^3 \sin 3\varpi' + \frac{dQ_1}{da}\ .\ e'^2\ .\ e\ .\ \sin(\varpi' + \varpi)$$

$$+ \frac{dQ_2}{da} . e'e^2 \sin (\varpi + 2\varpi') + \frac{dQ_3}{da} . e^3 \sin 3\varpi$$

$$+ \frac{dQ_4}{da} . e'\gamma^2 \sin (2\Pi + \varpi') + \frac{dQ_5}{da} . e\gamma . \sin (2\Pi + \varpi).$$

The quantities $\frac{dQ_0}{da}$, &c. are obtained from the values of Q_0, Q_1, &c. in article 623,

With which and the numerical values of the elements at the epoch 1750, the preceding value of $\frac{dP}{da}$ gives

$$- \frac{2m'. n}{5n' - 2n} a^2 . \frac{dP}{da} = - 17''.22886;$$

and, by changing the sines into cosines, the same expression gives

$$\frac{2m'n}{5n' - 2n} a^2 . \frac{dP'}{da} = 5''.360016.$$

If t be made equal to 200 in the equations (204), and the computation repeated with the resulting values of the elements, it will be found that in 1950

$$- \frac{2m'n}{5n' - 2n} . a^2 \frac{dP}{da} = - 16''.836801$$

$$\frac{2m'n}{5n' - 2n} . a^2 \frac{dP'}{da} = 6''.449839;$$

but $$\frac{-17''.22886 + 16.83680}{200} = - 0''.0019603,$$

and $$\frac{6''.449839 - 5\cdot360016}{200} = 0''.0054491;$$

hence

$$\delta v = - \{17''.228862 - 0''.0019603 . t\} . \sin (5n't - 2nt + 5\epsilon' - 2\epsilon)$$
$$+ \{5''.360016 + 0''.0054491\ t .\} . \cos (5n't - 2nt + 5\epsilon' - 2\epsilon).$$

The only remaining inequalities of this order are,

$$- m'Ke . \sin (5n't - 2nt + 5\epsilon' - 2\epsilon - \varpi + B)$$
$$+ \frac{5m'}{2} . Ke . \sin (5n't - 2nt + 5\epsilon - 2\epsilon + \varpi - B)$$
$$+ m'He . \sin (5n't - 2nt + 5\epsilon' - 2\epsilon + \varpi + B),$$

the numerical values of which may easily be found equal to

$$\delta v = (0''.8203 - 0''.00059324 \,.\, t) \,.\, \sin(5n't - 2nt + 5\epsilon' - 2\epsilon)$$
$$- (1''.83796 - 0''.00000149 \,.\, t) \cos(5n't - 2nt + 5\epsilon' - 2\epsilon)$$
$$+ 10''.0847 \,.\, \sin(4nt - 5n't + 4\epsilon' - 5\epsilon - 45^\circ.36225).$$

The great inequality of Jupiter also contains the terms

$$\delta v = (12''.5365 - 0''.001755 \,.\, t) \,.\, \sin(5n't - 2nt + 5\epsilon' - 2\epsilon)$$
$$- (8''.1211 + 0''.004885 \,.\, t) \,.\, \cos(5n't - 2nt + 5\epsilon' - 2\epsilon);$$

depending on the fifth powers and products of the eccentricities and inclinations, the computation of these is exactly the same with the examples given, but very tedious on account of the form of the coefficients of the series R. If all the terms depending on the argument $5n't - 2nt + 5e' - 2\epsilon$ be collected, it will be found that the great inequality of Jupiter is

$$\delta v = \left\{ \begin{array}{l} \{1261''.56 - 0''.013495 \,.\, t - 0''.00001925 \,.\, t^2\} \,. \\ \qquad \sin(5n't - 2nt + 5\epsilon' - 2\epsilon) \\ + \{96''.4661 - 0''.47466 \,.\, t + 0''.00007856 \,.\, t^2\} \,. \\ \qquad \cos(5n't - 2nt + 5\epsilon' - 2\epsilon) \end{array} \right.$$

Inequalities depending on the Squares of the Disturbing Force

629. These are given by equations (182) and (199): their numerical values are

$$\delta v = 4''.0248 \,.\, \sin(5nt - 10n't + 5\epsilon - 10\epsilon' + 51^\circ.3653)$$

$- 13''.2389$ sin (twice the argument of the great inequality of Jupiter).

The inequality mentioned in article 589, according to Pontécoulant, is

$$\delta\zeta = 2''.16304 \,.\, \sin(5n't - 2nt + 5\epsilon' - 2\epsilon) + 16''.9712 \times \cos(5'nt - 2nt + 5\epsilon' - 2\epsilon) \text{ for Jupiter};$$

and

$$\delta\zeta' = 3''.4645 \,.\, \sin(5n't - 2nt + 5\epsilon' - 2\epsilon) - 40''.3437 \times \cos(5n't - 2nt + 5\epsilon' - 2\epsilon), \text{ for Saturn.}$$

Periodic Inequalities in the Radius Vector, depending on the Third Powers and Products of the Eccentricities and Inclinations.

630. These are occasioned by Saturn, and are easily found from equation (168) to be

$$\delta r = \left\{ \begin{array}{l} -\ 0.0003042733 \,.\, \cos(5n't - 2nt + 5\epsilon' - 2\epsilon - 12^{\circ}.14694) \\ +\ 0.0001001860 \,.\, \cos(5n't - 2nt + 5\epsilon' - 2\epsilon + 45^{\circ}.27972) \end{array} \right\}.$$

Periodic Inequalities in Latitude.

631. These are obtained from equations (160) and (177).

$$\phi = 1^{\circ}.3172,$$

is the inclination of Jupiter's orbit on the fixed ecliptic of 1750,

$$\frac{d\phi}{dt} = -\ 0''.07821 \text{ is its secular variation,}$$

and
$$\frac{d\phi'}{dt} = -\ 0''.22325,$$

is the same, with regard to the variable ecliptic;

also
$$\theta = 97^{\circ}.906,$$

is the longitude of the ascending node of Jupiter's orbit on the fixed ecliptic; $\frac{d\theta}{dt} = 6''.4571$, is its secular variation with regard to that plane, and $\frac{d\theta'}{dt} = -\ 14''.6626$ is its secular variation with regard to the variable ecliptic. Equations (197) give

$$(\delta\phi) = -\ 0.0000726, \quad \text{and } (\delta\theta) = 0.0008113,$$

for the variations depending on the squares of the disturbing forces;

hence
$$\frac{d\phi}{dt} = -\ 0''.078283, \quad \frac{d\theta}{dt} = 6''.457,$$

with regard to the fixed ecliptic, and

$$\frac{d\phi'}{dt} = -\ 0''.22325, \quad \frac{d\theta'}{dt} = -\ 14''.6626.$$

With these it will be found that

$$\delta s = \left\{ \begin{array}{l} 0''.564458 \,.\, \sin(n't + \epsilon' - \Pi) \\ +\ 0''.663927 \,.\, \sin(2n't - nt + 2\epsilon' - \epsilon - \Pi) \\ +\ 1''.119782 \,.\, \sin(3n't - 2nt + 3\epsilon' - 2\epsilon - \Pi) \\ -\ 0''.279382 \,.\, \sin(4n't - 3nt + 4\epsilon' - 3\epsilon - \Pi) \\ -\ 0''.26913 \,.\, \sin(2nt - n't + 2\epsilon - \epsilon' - \Pi) \end{array} \right\}$$

$$+\ 3''.94168 \,.\, \sin(3nt - 5n't + 3\epsilon - 5\epsilon' + 59^{\circ}.5097;$$

which are the only sensible inequalities in the latitude of Jupiter.

632. The action of the earth occasions the inequalities

$$\delta v = \left\{ \begin{array}{l} 0''.120833 \,.\, \sin(n't - nt + \epsilon' - \epsilon) \\ -\, 0''.000086 \,.\, \sin 2\,(n't - nt + \epsilon' - \epsilon) \end{array} \right\}$$

in the longitude of Jupiter, n' being the mean motion of the earth, and the action of Uranus is the cause of the following perturbations in the longitude of Jupiter,

$$\delta v = \left\{ \begin{array}{ll} 0''.051737 & .\, \sin(n't - nt + \epsilon' - \epsilon) \\ -\, 0''.427296 & .\, \sin 2\,(n't - nt + \epsilon' - \epsilon) \\ -\, 0''.044085 & .\, \sin 3\,(n't - nt + \epsilon' - \epsilon) \\ -\, 0''.005977 & .\, \sin 4\,(n't - nt + \epsilon' - \epsilon) \\ +\, 0''.123506 & .\, \sin(nt + \epsilon - \varpi) \\ -\, 0''.23524 & .\, \sin(nt + \epsilon - \varpi') \\ -\, 0''.53308 & .\, \sin(2n't - nt + 2\epsilon' - \epsilon - \varpi) \\ +\, 0''.102673 & .\, \sin(2n't - nt + 2\epsilon' - \epsilon - \varpi') \\ -\, 0''.127963 & .\, \sin(3n't - 2nt + 3\epsilon' - \epsilon - \varpi') \end{array} \right.$$

where n' is the mean motion of Uranus.

These are all the inequalities that are sensible in the motions of Jupiter; those of Saturn may be computed in the same manner.

On the Laws, Periods, and Limits of the Variations in the Orbits of Jupiter and Saturn.

633. When the values of p, p', q, q', are substituted in equations (137) they give

$$gN = (4.5)\,(N' - N)\,; \qquad gN' = (5.4)\,(N - N')\,;$$

and as

$$(5.4) = (4.5)\,\frac{m\sqrt{a}}{m'\sqrt{a'}},$$

$$g^2 + g\left\{\frac{m'\sqrt{a'} + m\sqrt{a}}{m'\sqrt{a'}}\right\}(4.5) = 0.$$

The roots of which are,

$$g_1 = 0; \quad g = -\,\frac{m'\sqrt{a'} + m\sqrt{a}}{m'\sqrt{a'}}\,(4.5)$$

so that equations (138) become

$$p = N \,.\, \sin(gt + \mathfrak{C}) + N_{,} \,.\, \sin \mathfrak{C}_{,}$$

$$q = N . \cos(gt + \epsilon) + N_{,} . \cos \epsilon_{,} \qquad (205)$$
$$p' = N' . \sin(gt + \epsilon) + N_{,} . \sin \epsilon_{,}$$
$$q' = N' . \cos(gt + \epsilon) + N_{,} . \cos \epsilon_{,}.$$

Whence,

$$p'-p = (N' - N) \sin(gt + \epsilon); \quad q'-q = (N' - N) \cos(gt + \epsilon),$$

and at the epoch when $t = 0$

$$\tan \epsilon = \frac{p'-p}{q'-q}.$$

But as $$N' = -\frac{m\sqrt{a}}{m'\sqrt{a'}} . N;$$

and $$p' - p = (N' - N) \sin \epsilon,$$

so $$N = -\frac{m'\sqrt{a'}\,(p' - p)}{(m\sqrt{a} + m'\sqrt{a'}) \sin \epsilon}.$$

Again, by article 504,

$$m\sqrt{a} . p + m'\sqrt{a'} . p' = \text{constant},$$
$$m\sqrt{a} . q + m'\sqrt{a'} . q' = \text{constant};$$

or in consequence of $Nm\sqrt{a} + N'm'\sqrt{a'} = 0$

$$(m\sqrt{a} + m'\sqrt{a'})\, N_{,} . \sin \epsilon_{,} = \text{constant},$$
$$(m\sqrt{a} + m'\sqrt{a'})\, N_{,} . \cos \epsilon_{,} = \text{constant}.$$

whence $$\tan \epsilon_{,} = \frac{m\sqrt{a}.p + m'\sqrt{a'}.p'}{m\sqrt{a}.q + m'\sqrt{a'}.q'},$$

and $$N_{,} = \frac{m\sqrt{a}.p + m'\sqrt{a'}.p'}{(m\sqrt{a} + m'\sqrt{a'}) \sin \epsilon_{,}}$$

and as at the epoch

$$p = \tan \bar{\phi} . \sin \bar{\theta} \qquad q = \tan \bar{\phi} . \cos \bar{\theta}$$
$$p' = \tan \bar{\phi}' . \sin \bar{\theta}' \qquad q' = \tan \bar{\phi}' . \cos \bar{\theta}'$$

are given, all the constant quantities g, $g_{,}$, ϵ, $\epsilon_{,}$, N, N', and $N_{,}$, are obtained from the preceding equations.

The variations in the inclinations are at their maxima and minima when $gt + \epsilon - \epsilon_{,}$ is either zero or 180°; hence if $\epsilon_{,}$ be substituted for $gt + \epsilon$, equations (205) give

$$\tan\phi = N + N_{,}\,; \qquad \tan\phi' = N' + N_{,}$$

for the maxima of the inclinations; and when $\epsilon_{,} + 180°$ is put for $gt + \epsilon$, they give for the minima,

$$\tan\phi = N - N'; \quad \tan\phi' = N' - N_{,}.$$

The maxima and minima of the longitude of the nodes are given by the equations $d\theta = 0$, $d\theta' = 0$, or $d.\tan\theta = 0$,
whence

$$q\,\frac{dp}{dt} - p\,\frac{dq}{dt} = 0,$$

and therefore $pp' + qq' = p^2 + q^2$, and by the substitution of the quantities in equations (205), it becomes

$$N + N_{,}\,.\,\cos(gt + \epsilon - \epsilon_{,}) = 0,$$

or
$$\cos(gt + \epsilon - \epsilon_{,}) = -\frac{N}{N_{,}}.$$

If $N_{,}$ be greater than N independently of the signs, the nodes will have a libratory motion; but if $N_{,}$ be less than N, they will circulate in one direction.

$\text{Tan}\,\phi = \sqrt{N_{,}^2 - N^2}$ corresponds to the preceding value of

$$\cos(gt + \epsilon - \epsilon_{,});$$

it gives the inclination corresponding to the stationary points of the node.

These points are attained when

$$\cos(gt + \epsilon - \epsilon_{,}) = -\frac{N}{N_{,}},$$

whereas the maxima and minima of the inclinations happen when

$$\cos(gt + \epsilon - \epsilon_{,}) = \pm 1.$$

The stationary positions of the nodes therefore do not correspond either to the maxima or minima of the inclination, or to the semi-intervals between them.

In 1700, by Halley's Tables,

$$\phi = 1°\ 19'\ 10' \qquad \theta = 97°\ 34'\ 9''$$

$$\phi' = 2°\ 30'\ 10'' \qquad \theta' = 101°\ \ 5'\ 6''$$

hence at that time,

$$p = 0.02283 \qquad q = -0.00303$$
$$p' = 0.04078 \qquad q' = -0.01573,$$

with these values, Mr. Herschel found

$$N_{,} = 0.02905 \qquad N' = 0.01537 \qquad N = -0.00661$$
$$\complement = 125^\circ\ 15'\ 40'' \qquad \complement_{,} = 103^\circ\ 38'\ 40'' \qquad g = -25''.5756,$$

consequently for Jupiter

$$\tan\phi = 0\,.\,02980\,.\,\sqrt{1-0.43290\,.\,\cos\{21^\circ\ 37' - t\times 25''.5756\}}$$

and for Saturn,

$$\tan\phi' = 0.03287\,.\,\sqrt{1+0.82665\,.\,\cos\{21^\circ\ 37' - t\times 25''.5756\}}.$$

Also $\quad N_{,} + N' = 0.04442 \quad N_{,} - N' = 0.01368$;

so that the maxima and minima of the inclinations of Saturn's orbit are 2° 32′ 40″ and 0° 47′, and its greatest deviation from its mean state does not exceed 52′ 50″. In Jupiter's orbit, the maximum is 2° 2′ 30″, and the minimum 1° 17′ 10″, and the greatest deviation from a mean state is 0° 22′ 40″.

The longitude of the node θ has a maximum and minimum in both orbits, because $N_{,} > N'$. The extent of its librations in Jupiter's orbit will be 13° 9′ 40″, and in Saturn's 31° 56′ 20″, on either side of its mean station on the plane of the ecliptic supposed immoveable. The period in which the inclinations vary from their greatest to their least values, and the nodes from their greatest to their least longitudes, is by article 486

$$= \frac{360^\circ}{g} = \frac{360^\circ}{25''.5756} = 50673 \text{ Julian years.}$$

634. The limits and periods of the variations in the eccentricities and longitudes of the perihelia are obtained by a similar process, from equations (133), and those in article 485. The quantities

$$h = e\sin\varpi, \quad l = e\cos\varpi, \quad h' = e'\sin\varpi', \quad l' = e'\cos\varpi',$$

are known at the epoch, and equations (132) give

$$g^2 - g\left\{\frac{m'\sqrt{a'}+m\sqrt{a}}{m'\sqrt{a'}}\right\}(4.5) = \frac{m\sqrt{a}}{m'\sqrt{a'}}\left\{\,\overline{|4.5}^{\,2} - (4.5)^2\right\};$$

$$\text{whence } g = 3''.5851 \qquad g_{,} = 21''.9905,$$

$$N = -0.01715\,;\quad N_{\prime} = 0.04321\,;\quad N' = 0\cdot 04877\,;$$
$$N_{\prime}' = 0.03532\,;\quad \mathfrak{C}_{\prime} = 210^{\circ}\,16'\,40''\,;\quad \mathfrak{C} = 306^{\circ}\,34'\,40''\,;$$

and equation (135) gives

$$e = \sqrt{h^2 + l^2},\text{ or}$$

$$e = 0.04649\,\sqrt{1 + 0.68592\cos(83^{\circ}\,42' - t.18''.4054)}$$

for the eccentricity of Jupiter's orbit;

and
$$e' = \sqrt{h'^2 + l'^2},\text{ or}$$

$$e' = 0.06021\,\sqrt{1 - 0.95009\cos(83^{\circ}\,42' - t.18''.4054)}$$

for that of Saturn for any number t of Julian years after the epoch.

The longitudes of the perihelia are found from the value of tan ϖ in article 495. The gréatest deviation of these from their mean place will happen when

$$\cos\{(g - g_{\prime})\,t + \mathfrak{C} - \mathfrak{C}_{\prime}\} = -\frac{gN'^2 + g_{\prime}N_{\prime}'^2}{(g + g_{\prime})\,N'.N_{\prime}'}.$$

If this fraction be less than unity, the perihelia will librate like the nodes about a mean position, if not, they will move continually in one direction. In the case of Jupiter and Saturn $gN'^2 + g_{\prime}\,N_{\prime}'^2$ is greater than
$$(g + g_{\prime})\,N'.N_{\prime}'\,;$$
so that the perihelia go on for ever in one direction.

The period in which the eccentricities accomplish their changes is

$$\frac{360^{\circ}}{g - g_{\prime}} = \frac{360^{\circ}}{18''.4054} = 70414 \text{ Julian years.}$$

The greatest and least values of the eccentricities are expressed by

$$N' \pm N_{\prime}' \text{ and } N \pm N_{\prime}.$$

For Saturn these are

$$0.08409 \text{ and } 0.01345,$$

and for Jupiter

$$0.06036 \text{ and } 0.02606\,;$$

the maximum of one planet corresponding to the minimum of the other.

The numerical values of the perturbations of the other planets will be found in the *Mécanique Céleste;* it is therefore only necessary to observe the circumstances that are peculiar to each planet.

Mercury.

635. The motions of Mercury are less disturbed than those of any other body, on account of his proximity to the sun, his greatest elongation not exceeding 28°.8. His periodic inequalities are caused by Venus, the Earth, Jupiter, and Saturn, those from Saturn are very small, and Mars only affects the elements of his orbit.

The secular variations in the elements of Mercury's orbit were in the beginning of the year 1801, in the eccentricity

$$0.000003867;$$

secular and sidereal variation in the longitude of the perihelion,

$$9' \, 43''.5;$$

secular and sidereal variation in the longitude of the node,

$$-\,13' \, 2'';$$

secular variation of the inclination of the orbit on the true ecliptic,

$$19''.8.$$

636. Mercury sometimes appears as a morning and sometimes as an evening star, and exhibits phases like the moon. He occasionally is seen to pass over the disc of the sun like a black spot: these transits are true annular eclipses of the sun, proving that Mercury is an opaque body shining only by reflected light. The recurrence of the transits of Mercury depends on his periodic time being nearly equal to four times that of the earth. This ratio can be expressed by several pairs of small whole numbers, so that if the planet be in conjunction with the sun while in one of his nodes, he will be in conjunction again at the same node, after the Earth and he have completed a certain number of revolutions. The periodic revolutions of the earth have the following ratios to those of Mercury:

Periods of the Earth,	7	=	29	periods of Mercury.
	13	=	54	
	33	=	137	
	&c.		&c.	

Consequently transits of Mercury will happen at intervals of 7, 13, 33, &c. years.

Had the orbit of Mercury coincided with the plane of the ecliptic, there would have been a transit at each revolution; but in consequence of the inclination of his orbit, transits do not happen often; for when a transit takes place, the latitude of Mercury must be less than the apparent semi-diameter of the sun. The return of the transits are also irregular from the great eccentricity of the orbit, which makes the motion of Mercury very unequal; the retrograde motion of the nodes also prevents the planet from returning to the same latitude when it returns to the same conjunction. A transit of Mercury took place at the descending node in 1799, the next that will happen at that node will be in 1832.

Transits happened at the ascending node in the years 1802, 1815, and 1822.

The mean apparent diameter of Mercury is $6''.9$.

Venus.

637. 'The Morning Star' is the only planet mentioned in the sacred writings, and has been the theme of the poet's song, from Hesiod and Homer, to the days of Milton.

Venus is next to Mercury, and exhibits similar phenomena. Like him she is alternately an evening and a morning star, has phases, and when in her nodes, occasionally appears to pass over the sun's disc, though her transits are not so frequent as those of Mercury. The returns of the transits of Venus depend on five times the mean motion of the earth being nearly equal to three times that of Venus: this however cannot be expressed by pairs of small whole numbers as in the case of Mercury; therefore the transits of Venus do not happen so often. It appears from the ratio of the periodic time of Venus to that of the earth, that eight periods of the earth's revolution are nearly equal to thirteen periods of the revolution of Venus, and 235 periods of the earth are nearly equal to 382 of Venus; hence a transit of Venus may happen at the same node after an interval of eight years, but if it does not happen, it cannot take place again at the same node for 235 years. At present, the heliocentric longitude of Venus's ascending node is something less than 75°, and that of her descending node is about 164°. The earth, as seen from the sun, has nearly the former longitude in the beginning of December, and the latter in the beginning of June; hence the transits of Venus for ages to come will happen in December and June. Those of Mercury will take place in May and November.

Table of the Transits of Venus.

Year.		
1631	6th December, ascending node.	
1639	4th ,,	same.
1761	5th June, descending node.	
1769	3d ,,	same.
1874	8th December, ascending node.	
1882	6th ,,	same.
2004	7th June, descending node.	

The transits of Venus afford the most accurate method of finding the sun's parallax, and consequently his distance from the earth, from whence the true magnitude of the whole system is determined; for unless the actual distance of the sun were known, only the ratios of the magnitudes could have been ascertained.

638. The sun's parallax EmE′, fig. 65, which is the angle subtended at the sun by the earth's radius, can be found, if another angle EmE′, fig. 66, subtended by a chord EE′ lying between two known places on the earth's surface be known; that is, if the sun's parallax at any one altitude be known, his horizontal parallax may be determined, as it has been shown in article 329. However, the method employed in that number is not sufficiently accurate when applied to the sun, because in measuring the zenith distances, an error of three or four seconds might happen, which is immaterial in the case of the moon, whose parallax is nearly a degree, but an error of that magnitude in the parallax of the sun, which is less than nine seconds, would render

the results useless; hence, astronomers have endeavoured to compute the angle EmE′ instead of measuring it. Let AB, fig. 92, represent the equator, S and V the discs of the sun and Venus perpendicular to it: suppose them both to be moving in the equator, the motion of Venus retrograde, that of the sun direct. To a person at A, the internal contact, or total ingress of Venus on the sun commences, when to a spectator at B, the edge of Venus's disc is distant from the sun by the angle VBS. The difference between the times of total ingress as seen from B and A is the time of describing VBS by the approach of the sun and Venus to each other. Hence from the difference of the times, and the rate at which Venus and the sun approach each other, the angle VBS may be found, because the motions of both the sun and Venus are known. And sine VBS is to sine VSB, as Venus's distance from the sun to Venus's distance from the earth. But the ratio of Venus's distance from the sun to her distance from the earth is known, therefore the angle ASB is found, and CSB, the parallax of the sun may be computed, and from that his horizontal parallax; whence the distance of the sun from the earth may be determined in multiples of the terrestrial radius, or even in miles since the length of the radius is known. The computation of the transit is complicated chiefly on account of the inclination of Venus's orbit to the ecliptic, and the situations of the places of observation A and B being always at different distances from the equator. The investigation of this problem, and the computation of the parallax, will be found in Biot's and Woodhouse's *Astronomy*.

fig. 92.

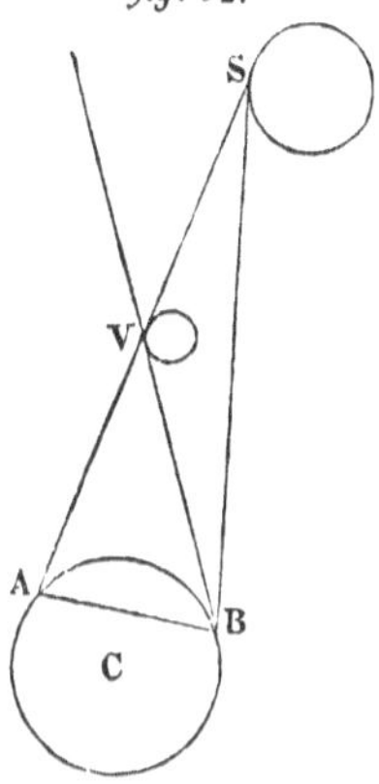

The times of internal contact can be observed with much greater accuracy than any angular distance can be measured, and on this depends the superiority of the preceding method of finding the parallax.

At inferior conjunction, the sun and Venus approach each other at the rate of 4″ in a minute; hence, if the time of contact be erroneous at each place of observation 4″ of time, the angle VBS, fig. 92, may be erroneous $\frac{4 \times 8}{60} = \frac{8}{15}$ of a second, therefore the limit of the error in ASB is about $\frac{1}{15}$ of a second, and thus by the transit of Venus, an angle only $\frac{4}{15}$ of a second can be measured, a less quantity than can be determined by any other method.

639. The preceding method requires the difference of longitudes of the two places A and B to be accurately known, in order to compare the actual times of contact. In 1761 a transit of Venus was observed at the Cape of Good Hope, and at many places in Europe, the longitudes of all being well known: by comparing the observations the mean result determined the parallax to be 8″.47; this is only an approximate value, but it was useful in obtaining the true value from the transit of 1769, which was observed at Wardhus in Lapland, and at Otaheite in the southern hemisphere; but as the longitude of the latter was unknown, astronomers avoided the difficulty by changing their method of calculation. In place of observing the ingress only, they observed the duration of the transit, and from the difference of duration at different places, they deduced the parallax.

Let P be Venus, E the earth, W Wardhus towards the north pole; O Otaheite towards the south; and VA the disc of the sun: then the true line of transit seen from E, the centre of the earth, would be VA, at W the transit would appear to be in the line *va*, and from O it would be seen in *v′a′*.

fig. 93.

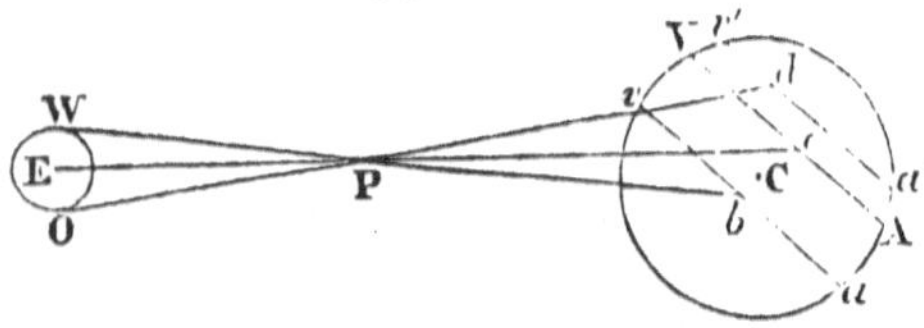

If T be the true duration of the transit, or the time of describing VA, then the time of describing *va* nearer to the sun's centre, and therefore greater than VA, would be T + *t*; whilst that of describing *v′a′*, which is farther from the centre, and therefore less than VA, would be T — *t′*. The difference of the durations of the transits seen from O and W is $T + t - (T - t') = t + t'$, which is entirely the effect of parallax. With an approximate value of the parallax, *t* and *t′*, the differences in the durations at W and O from what they would have been if observed at C, the centre of the earth may be computed; then comparing the computed value of $t + t'$ with its observed value, the error in the assumed parallax will be found. With the parallax 8″.83 it has been calculated that at Wardhus the duration was lengthened by . . 11′.16″.9

And diminished at Otaheite by . .	12. 10
Sum $t + t'$	23′.26″.9
But by observation	23. 10
Difference	16″.9

Consequently the parallax 8″.45 is less than that assumed; therefore to make the observed and computed differences of durations agree, the parallax must be 8″.72. This does not differ much from what is given by the lunar theory 8″.6, but an error recently detected by M. Bessel, reduces it to 8″.575. The transit commenced at Otaheite at half past nine in the morning, and ended at half-past three in the afternoon.

640. Venus is by far the most brilliant and beautiful of the planets, but her splendour is variable. Her phases increase with her distance from the earth, and therefore she ought to become brighter as her disc enlarges; but the increase of the distance diminishes her lustre, since the intensity of light decreases proportionally to the square of the distance: there is, however, a mean position in which Venus is more brilliant than in any other; the interval of her returns to that position is about eight years, depending on the ratio of her periodic time to that of the earth. She is then visible to the naked eye during the day, but she is also visible in daylight every eighteen months though less distinctly.

The variations in the apparent diameter of Venus are very great; she is nearest the earth in her transit; her apparent diameter is then 61″.236. M. Arago has found its mean value to be 16″.904.

Shrŏeter, by observing the horns of Venus, determined her rotation about an axis, considerably inclined to the plane of the ecliptic, to be performed in 23^h 21′; he discovered also very high mountains on her surface.

641. Venus is too near the sun to be very irregular in her motions, her greatest elongation not exceeding 47° 7′. In 1801, the secular variation in the eccentricity of her orbit was 0.000062711.

In the longitude of the perihelion,	4′ 28″.
In the longitude of the ascending node,	− 31′ 10″
In the inclination on the true ecliptic,	4″.5.

The Earth.

642. Uranus is too distant to have a sensible influence on the earth. Besides the disturbances occasioned by the other planets, there are some inequalities produced by the moon which are to be found in article 498.

It will be shown in the theory of the moon, that if $U - \Omega$ be her distance from her ascending node, the greatest inequality in her latitude is $$18542''.8 \cdot \sin(U - \Omega),$$
and if $S = 18542''.8$, the inequality (195) in the earth's latitude is

$$-\frac{m}{E} \cdot \frac{r}{\bar{r}} \cdot 18542''.8 \cdot \sin(U - \Omega).$$

In order to compute the inequalities occasioned by the moon, it is requisite to know the ratio of the mass of the moon to that of the earth. The theory of the tides shows that the action of the moon in raising the waters of the ocean is 2.35333 times greater than that of the sun. The action of the moon on the earth, resolved in the direction r, is $\frac{E+m}{r^3}$; and the action of the sun, according to his radius vector $\bar{r}$, is $\frac{S}{\bar{r}^3}$: S and m being the masses of the sun and moon;

hence $$\frac{E+m}{r^3} = 2.35333 \cdot \frac{S}{r^3}.$$

By the theory of central forces,

$$\frac{E+m}{r^3} = n_{,}^2, \text{ and } \frac{S}{\bar{r}^3} = n^2;$$

n and $n_{,}$ being the mean motions of the earth and moon; whence

$$\frac{m}{E+m} = 2.35333. \frac{n^2}{n_{,}^2}.$$

By observation, $$\frac{n}{n_{,}} = 0.0748301;$$

hence $$\frac{m}{E+m} = \frac{1}{75.928};$$

and if the mass of the earth be taken as the unit, the mass of the moon is $$\frac{m}{E} = m = \tfrac{1}{75} \text{ nearly.}$$

Again, the ratio of the earth's distance from the sun to its distance from the moon is equal to the horizontal parallax of the sun, divided by the mean horizontal parallax of the moon, as will appear by considering that, as the parallax of both the sun and moon is very small, the arc may be taken for the sine, and the mean horizontal parallax of the moon is then the mean terrestrial radius divided by the mean distance of the moon from the earth; and the solar parallax is equal to the same terrestrial radius divided by the mean distance of the earth from the sun. The parallax of the sun is known, by observation, to be 8″.575, that of the moon is 3454″.16; hence the ratio of the distances is $\frac{8''.575}{3454''.16}$.

With these data, the coefficients are

$$\delta v = -\ 6''.8274 \ .\ \sin (U - v),$$
$$\delta r = -\ 0.0000331 \ .\ \cos (U - v),$$
$$\delta s = -\ 0''.61377 \ .\ \sin (U - ☊).$$

643. The inequality caused by the moon in the earth's radius vector is small; the mass of the moon being only $\frac{1}{75}$ part of that of the earth, the distance of the common centre of gravity of the earth and moon from the centre of the former must be less than the semidiameter, that is, it must be within the mass of the earth, and therefore the inequality in the earth's place must be less than 8″.575, the sun's horizontal parallax.

644. The inequality produced by the moon in the earth's longitude is the lunar equation of the tables of the sun; it is of much importance for correcting the value of the mass of the moon. Its coefficient being computed with a value of the mass of the moon determined from the theory of the tides, compared with the coefficient of the same inequality determined by observation, will give the error in the mass of the moon, supposing the parallax of the sun and moon to be correct.

645. The irregularities communicated to the earth by the moon and planets are referred to the sun by observers on the earth's surface; therefore the sun appears to have a motion in longitude, by which he alternately advances before, and falls behind the point that describes the elliptical orbit in the heavens. In like manner he seems alternately to ascend above the plane of the ecliptic, and to

descend below it by the disturbance in latitude. The perturbations in latitude, by the action of the planets, are computed from (160), and are

$$\delta s'' = \begin{cases} 0''.991803 \sin (2n''t - n't + 2\epsilon'' - \epsilon' - \theta') \\ 0''.234256 \sin (4n''t - 3n't + 4\epsilon'' - 3\epsilon' - \theta') \end{cases}$$

$$+ (0''.164703 \sin (2n^{iv}t - n''t + 2\epsilon^{iv} - \epsilon'' - \theta^{iv});$$

this, added to $- 0.61377 \sin (U - ☊)$,

is the whole periodic disturbance in the earth's motion in latitude, taken with a different sign. It affects the obliquity of the ecliptic, determined from the observations of the altitude of the sun in the solstices; it also has an influence on the time of the equinoxes, determined from observations of the sun at that period, as well as on the right ascensions and declinations of the fixed stars, determined by comparison with the sun; for it is clear that any inequalities in the motion of the earth will be referred to the observations made at its surface.

Considering the great accuracy of modern observations, these circumstances must be attended to. It is easy to see that this variation in the sun's latitude will increase his apparent declination by

$$-\frac{\delta s''. \cos \{\text{obliquity of ecliptic}\}}{\cos \{\text{declination of sun}\}};$$

and his apparent right ascension by

$$\frac{\delta s'' . \sin \{\text{obliquity of ecliptic}\} . \cos \{\text{sun's R.A.}\}}{\cos \{\text{declination of} \odot\}}$$

The observed right ascensions and declinations of the sun must therefore be diminished by these quantities, in order to have those that would be observed if the sun never left the plane of the ecliptic.

Secular Inequalities in the Terrestrial Orbit.

646. The eccentricity and place of the perihelion of the terrestrial orbit may be determined with sufficient accuracy for 1000 or 1200 years before and after the epoch 1750, from

$$e = 2\bar{e} - 0''.187638\, t - 0''.000006721\, t^2,$$

and $$\varpi = \bar{\varpi} + 11''.949588\, t + 0''.000079522\, t^2,$$

$\bar{e}$ and $\bar{\varpi}$ are the eccentricity and longitude of the perihelion at the epoch.

The secular diminution of the eccentricity is 18″.79, about 3914 miles, in reality an exceedingly small fraction in astronomy, though it appears so great in terrestrial measures. Were the diminution uniform, which there is no reason to believe, the earth's orbit would become a circle in 36300 years; its variation has a great influence on the motions of the moon.

The longitude of the perihelion increases annually at the rate of 11″.9496, so that it accomplishes a sidereal revolution in 109758 years.

647. A remarkable period in astronomy was that in which the greater axis of the terrestrial orbit coincided with the line of the equinoxes, then the true equinox coincided with the mean. This occurred 4084 years before the epoch in which chronologists place the creation of man; at that time the solar perigee coincided with the equinox of spring. This however is but an approximate value, on account of the masses of the planets and the doubts as to the exact value of precession; the error may therefore be 80 years, which is not much in such a quantity.

Another remarkable astronomical period was, when the greater axis of the terrestrial orbit was perpendicular to the line of equinoxes; it was then that the true and mean solstice were united; this coincidence took place in the year 1248 of the Christian era. It is evident that these two periods depend on the direct motion of the perihelion and precession of the equinoxes conjointly.

648. The position of the ecliptic is changed by the reciprocal action of the planets on one another, and on the earth, each of them producing a retrograde motion in the intersection of the plane of its own orbit with the plane of the ecliptic. This action also changes the position of the plane of the ecliptic, with regard to itself, a change that may be determined from the values of p and q by formulæ (138), or rather from

$$p = \quad 0''.0767209\ t + 0''.000021555\ .\ t^2,$$
$$q = -\ 0''.5009545\ t + 0''.000067473\ .\ t^2.$$

These will give the variation of the ecliptic, with regard to its fixed position in 1750, for 1000 or 1200 years, before and after that epoch.

This change in the ecliptic alters its position with regard to the earth's equator; but as the formulæ in article 498 are periodic, these two planes never have and never will coincide. It occasions also a small motion in the equinoxes of about 0″.0846 annually. Both of these variations are entirely independent of the form of the earth, and would be the same were it a sphere. However, the action of the sun and moon on the protuberant matter at the earth's equator is the cause of the precession of the equinoxes, or of that slow angular motion by which the intersection of the equator and ecliptic goes backward at the rate of 50″.34 annually, so that the pole of the equator describes a circle round the pole of the ecliptic in the space of 25748 years. This motion is diminished by the very small secular inequality 0″.0846, arising from the action of the planets on the ecliptic. The formulæ for computing the obliquity of the ecliptic and precession of the equinoxes depend on the rotation of the earth.

Mars.

649. Mars is troubled by all the planets except Mercury. Jupiter alone affects the latitude of Mars. The secular variations in the elements of his orbit were, in 1801, as follow:

In the eccentricity	0.000090176
In the longitude of the perihelion . .	26″.22
In the inclination on the true ecliptic . .	1″.5
In the longitude of the ascending node .	—38′ 48″

The eccentricity is diminishing.

The greatest elongation of Mars is 126.°8. By spots on his surface it appears that he rotates in one day about an axis that is inclined to the plane of the ecliptic at an angle of 59°.697. His equatorial is to his polar diameter in the ratio of 194 to 189 ; his apparent diameter subtends an angle of 6″.29, at his mean distance, and of 18″.28 at his greatest distance, when his parallax is nearly twice that of the sun. The disc of Mars is occasionally gibbous. Spots near his poles that augment or diminish according as they are exposed to the sun, give the idea of masses of ice.

The New Planets.

650. The orbits of Vesta, Juno, Ceres and Pallas are situate between those of Mars and Jupiter. Ceres was discovered by Piazzi, at Palermo, on the first day of the present century ; Pallas was discovered by Olbers, in 1802 ; Juno in 1803, by Harding; and Vesta

in 1807, by Olbers. These bodies are nearly at equal distances from the sun, their periodic times are therefore nearly the same. The eccentricities of the orbits of Juno and Vesta, and the position of their nodes are nearly the same.

These small planets are much disturbed by the proximity and vast magnitude of Jupiter and Saturn, and the series which determine their perturbations converge slowly, on account of the greatness of the eccentricities and inclinations of their orbits. The inclination of the old planets is so small, that they are all contained within the zodiac, which extends 8° on each side of the ecliptic, but those of the new planets very much exceed these limits. They are invisible to the naked eye, and so minute that their apparent diameters have not yet been measured. Sir William Herschel estimated that they cannot amount to the fourth of a second, which would make the real diameter less than 65 miles. However, Juno, the largest of these asteroids, is supposed to have a real diameter of about 200 miles.

Jupiter.

651. Jupiter is the largest planet in the system, and with his four moons exhibits one of the most splendid spectacles in the heavens. His form is that of an oblate spheroid whose polar diameter is 35″.65, and his equatorial = 38″.44; he rotates in 9 hours 56 minutes about an axis nearly perpendicular to the plane of the ecliptic. The circumference of Jupiter's equator is about eleven times greater than that of the earth, and as the time of his rotation is to that of the earth as 1 to 0.414, it follows that during the time a point of the terrestrial equator describes 1°, a point in the equator of Jupiter moves through 2°.41; but these degrees are longer than the terrestrial degrees in the ratio of 11 to 1, consequently each point in Jupiter's equator moves 26 times faster than a point in the equator of the earth. In the beginning of 1801 the secular variations of his orbit were,

In the eccentricity . . .	0.00015935
In the longitude of the perihelion .	11′ 4″
In the longitude of the ascending node .	– 26′ 17″
In the inclination on the true ecliptic .	23″

Saturn.

652. Viewed through a telescope Saturn is even more interesting than Jupiter: he is surrounded by a ring concentric with himself, and of the same or even greater brilliancy; the ring exhibits a variety of appearances according to the position of the planet with regard to the sun and earth, but is generally of an elliptical form: at times it is invisible to common observation, and can only be seen with superior instruments; this happens when the plane of the ring either passes through the centre of the sun or of the earth, for its edge, which is very thin, is then directed to the eye. On the 29th September, 1832, the plane of the ring will pass through the centre of the earth, and will be seen with a very high magnifying power like a line across the disc of the planet. On the 1st December of the same year, the plane of the ring will pass through the sun. Professor Struve has discovered that the rings are not concentric with the planet. The interval between the outer edge of the globe and the outer edge of the ring on one side is $11''.037$, and on the other side the interval is $11''.288$, consequently there is an eccentricity of the globe in the ring of $0''.215$. In 1825 the ring of Saturn attained its greatest ellipticity; the proportion of the major to the minor axis was then as 1000 to 498, the minor being nearly half the major. Stars have been observed between the planet and his ring. It is divided into two parts by a dark concentric band, so that there are really two rings, perhaps more. These revolve about the planet on an axis perpendicular to their plane in about $10^h\ 29^m\ 17^s$, the same time with the planet.

The form of Saturn is very peculiar. He has four points of greatest curvature, the diameters passing through these are the greatest; the equatorial diameter is the next in size, and the polar the least; these are in the ratio of 36, 35, and 32. Besides the rings, Saturn is attended by seven satellites which reciprocally reflect the sun's rays on each other and on the planet. The rings and moons illuminate the nights of Saturn; the moons and Saturn enlighten the rings, and the planet and rings reflect the sun's beams on the satellites when they are deprived of them in their conjunctions. The rings reflect more light than the planet. Sir William Herschel observed, that with a magnifying power of 570, the colour of Saturn was yellowish, whilst that of the rings was pure white. Saturn has several belts parallel to his equator: changes have been observed in the colour of these and in the brightness of the poles, according as they are turned to or from the sun, probably occasioned by the melting of the snows. Saturn's motions are disturbed by Jupiter and Uranus alone; the secular variations in the elements of his orbit were as follows, in the beginning of 1801.

Eccentricity	0.000312402
Longitude of perihelion . . .	32′ 17″
Longitude of ascending node . .	− 37′ 54″
Inclination on true ecliptic . .	15′ 5″

Uranus, or the Georgium Sidus.

653. This planet was discovered by Sir William Herschel, in 1781. The period of his sidereal revolution is 30687 days. If we judge of the distance of the planet by the slowness of its motion, it must be on the very confines of the solar system; its greatest elongation is 103°.5, and its apparent diameter 4″: it is accompanied by six satellites, only visible with the best telescopes. The only sensible perturbations in the motions of this planet arise from the action of Jupiter and Saturn; the secular variations in the elements of its orbit were, in 1801, as follow:

Eccentricity	0.000025072
Longitude of perihelion	4′
Longitude of ascending node .	— 59′.57″
Inclination on true ecliptic . . .	3″.7

The rotation of Saturn has not been determined.

654. It is remarkable that the rotation of the celestial bodies is from west to east, like their revolutions; and that Mercury, Venus, the Earth, and Mars, accomplish their rotations in about twenty-four hours, while Jupiter and Saturn perform theirs in $\frac{4}{10}$ of a day.

On the Atmosphere of the Planets.

655. Spots and belts are observed on the discs of some of the planets varying irregularly in their position, which shows that they are surrounded by an atmosphere; these spots appear like clouds driven by the winds, especially in Jupiter. The existence of an atmosphere round Venus is indicated by the progressive diffusion of

the sun's rays over her disc. Schroëter measured the extension of light beyond the semicircle when she appeared like a thin crescent, and found the zone that was illuminated by twilight to be at least four degrees in breadth, whence he inferred that her atmosphere must be much more dense than that of the earth. A small star hid by Mars was observed to become fainter before its appulse to the body of the planet, which must have been occasioned by his atmosphere. Saturn and his rings are surrounded by a dense atmosphere, the refraction of which may account for the irregularity apparent in his form: his seventh satellite has been observed to hang on his disc more than 20′ before its occultation, giving by computation a refraction of two seconds, a result confirmed by observation of the other satellites. An atmosphere so dense must have the effect of preventing the radiation of the heat from the surface of the planet, and consequently of mitigating the intensity of cold that would otherwise prevail, owing to his vast distance from the sun. Schroëter observed a small twilight in the moon, such as would be occasioned by an atmosphere capable of reflecting the sun's rays at the height of about a mile. Had a dense atmosphere surrounded that satellite, it would have been discovered by the duration of the occultations of the fixed stars being less than it ought to be, because its refraction would have rendered the stars visible for a short time after they were actually behind the moon, in the same manner as the refraction of the earth's atmosphere enables us to see celestial objects for some minutes after they have sunk below our horizon, and after they have risen above it, or distant objects hid by the curvature of the earth. A friend of the author's was astonished one day on the plain of Hindostan, to behold the chain of the Himala mountains suddenly start into view, after a heavy shower of rain in hot weather.

The Bishop of Cloyne says, that the duration of the occultations of stars by the moon is never lessened by 8″ of time, so that the horizontal refraction at the moon must be less than 2″: if therefore a lunar atmosphere exists, it must be 1000 times rarer than the atmosphere at the surface of the earth, where the horizontal refraction is nearly 2000″. Possibly the moon's atmosphere may have been withdrawn from it by the attraction of the earth. The radiation of the heat occasioned by the sun's rays must be rapid and constant, and must cause intense cold and sterility in that cheerless satellite.

The Sun.

656. The sun viewed with a telescope, presents the appearance of an enormous globe of fire, frequently in a state of violent agitation or ebullition; black spots of irregular form rarely visible to the naked eye sometimes pass over his disc, moving from east to west, in the space of nearly fourteen days: one was measured by Sir W. Herschel in the year 1779, of the breadth of 30,000 miles. A spot is surrounded by a penumbra, and that by a margin of light, more brilliant than that of the sun. A spot when first seen on the eastern edge, appears like a line, progressively extending in breadth till it reaches the middle, when it begins to contract, and ultimately disappears at the western edge: in some rare instances, spots re-appear on the east side; and are even permanent for two or three revolutions, but they generally change their aspect in a few days, and disappear: sometimes several small spots unite into a large one, as a large one separates into smaller ones which soon vanish.

The paths of the spots are observed to be rectilinear in the beginning of June and December, and to cut the ecliptic at an angle of 7° 20′. Between the first and second of these periods, the lines described by the spots are convex towards the north, and acquire their maximum curvature about the middle of that time. In the other half year the paths of the spots are convex towards the south, and go through the same changes. From these appearances it has been concluded, that the spots are opaque bodies attached to the surface of the sun, and that the sun rotates about an axis, inclined at an angle of 7° 20′ to the axis of the ecliptic. The apparent revolution of a spot is accomplished in twenty-seven days; but during that time, the spot has done more, having gone through a revolution, together with an arc equal to that described by the sun in his orbit in the same time, which reduces the time of the sun's rotation to $25^d\ 9^m\ 36^s$.

These phenomena induced Sir W. Herschel to suppose the sun to be a solid dark nucleus, surrounded by a vast atmosphere, almost always filled with luminous clouds, occasionally opening and discovering the dark mass within. The speculations of La Place were different: he imagined the solar orb to be a mass of fire, and that the violent effervescences and explosions seen on its surface are occasioned by the eruption of elastic fluids formed in its interior, and that the spots are enormous caverns, like the craters of our volcanoes.

Light is more intense in the centre of the sun's disc than at the edges, although, from his spheroidal form, the edges exhibit a greater surface under the same angle than the centre does, and therefore might be expected to be more luminous. The fact may be accounted for, by supposing the existence of a dense atmosphere absorbing the rays which have to penetrate a greater extent of it at the edges than at the centre; and accordingly, it appears by Bouguer's observations on the moon, which has little or no atmosphere, that it is more brilliant at the edges than in the centre.

657. A phenomenon denominated the zodiacal light, from its being seen only in that zone, is somehow connected with the rotation of the sun. It is observed before sunrise and after sunset, and is a luminous appearance, in some degree similar to the milky way, though not so bright, in the form of an inverted cone with the base towards the sun, its axis inclined to the horizon, and only inclined to the plane of the ecliptic at an angle of 7°; so that it is perpendicular to the axis of the sun's rotation. Its length from the sun to its vertex varies from 45° to 120°. It is seen under the most favourable circumstances after sunset in the beginning of March: its apex extends towards Aldebaran, making an angle of 64° with the horizon. The zodiacal light varies in brilliancy in different years.

It was discovered by Cassini in 1682, but had probably been seen before that time. It was observed in great splendour at Paris on the 16th of February, 1769.

658. The elliptical motion of the planets is occasioned by the action of the sun; but by the law of reaction, the planets must disturb the sun, for the invariable point to which they gravitate is not the centre of the sun, but the centre of gravity of the system; the quantity of motion in the sun in one direction must therefore be equal to that of all the planets in a contrary direction. The sun thus describes an orbit about the centre of gravity of the system, which is a very complicated curve, because it results from the action of a system of bodies, perpetually changing their relative positions; it is such however as to furnish a centrifugal force with regard to each planet, sufficient to counteract the gravitation towards it.

Newton has shown that the diameter of the sun is nearly equal to 0.009 of the radius of the earth's orbit. If all the great planets of the system were in a straight line with the sun, and on the same side of him, the centre of the sun would be nearly the farthest possible from the common centre of gravity of the whole; yet it is found by computation, that the distance is not more than 0.0085 of the radius vector of the earth; so that the centre of the sun is never distant from the centre of gravity of the system by as much as his own diameter.

Influence of the Fixed Stars in disturbing the Solar System.

659. It is impossible to estimate the effects of comets in disturbing the solar system, on account of our ignorance of the elements of their orbits, and even of the existence of such as have a great perihelion distance, which nevertheless may trouble the planetary motions; but there is every reason to believe that their masses are too small to produce a sensible influence; the effect of the fixed stars may, however, be determined.

Let m' be the mass of a fixed star, x', y', z', its co-ordinates referred to the centre of gravity of the sun, and r' its distance from that point. Also let x, y, z, be the co-ordinates of a planet m, and r its radius vector; then the disturbing influence of the star is

$$R = \frac{m'}{\sqrt{(x'-x)^2+(y'-y)^2+(z'-z)^2}} - \frac{m'(xx+yy+zz)}{r'^3};$$

or $$R = +\frac{m'}{r'} - \frac{m'r^2}{2r'^3} + \tfrac{3}{2}\, m'\, \frac{(xx'+yy'+zz') - \frac{1}{2}r^2)^2}{r'^5} + \&c.$$

when developed according to the powers of r'. The fixed plane being the orbit of m at the epoch, then

$$x = r\cos v, \qquad y = r\sin v, \qquad z = rs,$$

let l be the latitude of the fixed star, and u its longitude, then

$$x' = r'.\cos l.\cos u, \quad y' = r'.\cos l.\sin u, \quad z' = r'.\sin l;$$

and if all the powers of r' above the cube be omitted, it will be found that

$$R = +\frac{m'}{r'} - \frac{m'r^2}{4r'^3}\{2 - 3\cos^2 l$$

$$- 3\cos^2 l.\cos(2v-2u) - 6s.\sin 2l.\cos(v-u)\}.$$

But neglecting s, the substitution of this in equation (155) gives

$$\frac{\delta r}{a} = -\frac{m'a^3nt}{r'^3}.\{(1-\tfrac{3}{2}\cos^2 l)e\sin(v-\varpi)-\tfrac{3}{4}\cos^2 l.e.\sin(v+\varpi-2u)\}.$$

But $$r = a(1+e\cos(v-\varpi));$$

whence $$\frac{\delta r}{a} = \delta e\cos(v-\varpi) + e\delta\varpi.\sin(v-\varpi);$$

and comparing the two values of $\frac{\delta r}{a}$, there will be found

$$\delta\varpi = -\frac{m'a^3}{r'^3e}.nt\{1-\tfrac{3}{2}\cos^2 l - \tfrac{3}{4}\cos^2 l.\cos(2\varpi-2u)\}$$

$$\delta e = \frac{3m'a^3}{4.r'^3}.\cos^2 l\, nt.e.\sin(2\varpi-2u).$$

Whence it appears, that the star occasions secular variations in the eccentricity and longitude of the perihelion of m, but these variations are incomparably less than those caused by the planets. For if m be the earth, the distance of the star from the centre of the sun cannot be less than 100,000 times the mean distance of the earth from the sun, because the annual parallax of the nearest fixed star is less than 1″; therefore assuming $r' = 100,000.a$ the coefficient $\frac{m'a^3}{r'^3}nt$ does not exceed $0''.0000000013.m't$, t being any number of Julian years. This quantity is incomparably less than the corresponding variation in the eccentricity of the earth's orbit, arising from the action of the planets, which is

$$-0''.0938191.t,$$

unless the mass m' of the fixed stars be much greater than what is probable. Whence it may be concluded that the attraction of the fixed stars has no sensible influence on the form of the planetary orbits; and it may be easily proved, that the positions of the orbits are also uninfluenced.

Disturbing Effect of the Fixed Stars on the Mean Motions of the Planets.

660. The part of equation (156) that depends on R, when $\mu = 1$, is

$$d.\delta\zeta = -3a\int ndt.dR - 2a.ndt.r\left(\frac{dR}{dr}\right).$$

The preceding value of R gives

$$d.\delta\zeta = \frac{m'a^3}{r'^3}\,ndt\,(2 - 3\cos^2 l) - \frac{6m'a^3}{r'^3}\,.\,s.\sin 2l.\,\cos(v-u)$$

$$-\ \frac{9}{2}.\,m'.a^3.ndt\int d.\,\frac{s.\sin 2l}{r'^3}.\,\cos(v-u),$$

which is the whole variation in the mean motion of m from the action of the fixed stars. The parts will be examined separately.

Let r'' and l' be the distance and latitude of the star at the epoch 1750, and let it be assumed, that these quantities diminish annually by α and β, then t being any indefinite time, r' and l become

$$r' = r''(1 - \alpha t),\ \ l = l'(1 - \beta t)$$

whence the first term of $d.\delta\zeta$ becomes

$$d.\delta\zeta = \frac{3m'.a^3}{r''^3}\,(1 - \tfrac{3}{2}\cos^2 l')\,\alpha n t^2 - \frac{3m'a^3}{2r''^3}\,.\,\sin 2l'\,.\,\beta\,.\,nt^2.$$

We know nothing of the changes in the distance of the fixed stars; but with regard to the earth, they may be assumed to vary 0''.324 annually in latitude;

hence $\beta = 0''.324,\quad r'' = 100.000a,$

so that $\frac{m'a^3}{r'^3}\,.\,\beta\,.\,nt^2$ becomes

$$\frac{m't^2.\,2''.0357}{10^{15}}.$$

a quantity inappreciable from the earliest observations.

With regard to the terms in s,

$$s = t\,.\,\frac{dp}{dt}\sin v - t.\,\frac{dq}{dt}\cos v\,;$$

consequently, rejecting the periodic part,

$$d.\frac{s.\sin 2l}{r'^3}.\,\cos(v-u) = \frac{\sin 2l}{2r'^3}\left\{\frac{dp}{dt}.\,\sin u - \frac{dq}{dt}.\cos u\right\},$$

so that

$$d.\delta\zeta = -\,\frac{21}{4}\,.\,\frac{m'a^3}{r'^3}.\,n\,.\,tdt\,.\,\sin 2l\left\{\frac{dp}{dt}\sin u - \frac{dq}{dt}\,.\,\cos u\right\};$$

the integral of which is

$$\delta\zeta = -\,\frac{21}{8}\,.\,\frac{m'a^3}{r'^3}.\,nt^2.\,\sin 2l\left\{\frac{dp}{dt}\sin u - \frac{dq}{dt}\,.\cos u\right\}.$$

But with regard to the earth

$$p = \quad 0''.076721.t. + 0''.000021555.t^2$$
$$q = -0''.50096 . t \quad + 0''.0000067474 . t^2.$$

If these quantities be substituted, it will be found that the secular inequalities in the mean motion of the earth are quite insensible; the earliest records also prove them to be so. The same results will be obtained for the most distant planets, whence it may be concluded that the fixed stars are too remote to affect the solar system.

Construction of Astronomical Tables.

661. The motion of a planet in longitude consists of three parts, of the mean or circular motion; of a correction depending on the eccentricity, which is the equation of the centre; and of the periodic inequalities.

In the construction of tables, the mean longitude of the body, and the mean longitude of the aphelion, or perihelion, are determined in degrees, minutes, seconds, and tenths, at the instant assumed as the origin of the tables. These initial values are generally computed for the beginning of each year, and are called the epoch of the tables; from them subsequent values are deduced at convenient intervals, by adding the daily increments. These intervals are longer or shorter according to the motion of the body, or its importance, and the intermediate values are found by simple proportion, or by tables of proportional parts. The mean anomaly is given by the tables, since it is the difference between the mean longitudes of the body and of the aphelion.

The tables of the equation of the centre, and of the mean longitude of the aphelion, give these quantities for each degree of mean anomaly. To these are added tables of the periodic inequalities in longitude, and of the secular inequalities in the eccentricity and longitude of the aphelion. From these tables the true longitude of the body may be known at any instant, by applying the corrections to the mean longitude.

The radius vector consists of three parts,—of a mean value, which is equal to half the greater axis of the orbit; of the elliptical variations, and of its periodic inequalities. The two latter are given in the tables for every degree of mean anomaly. The latitude is computed in terms of the mean anomaly at stated intervals: besides these, the mean longitude of the ascending node and the inclination of the orbit at the beginning of each year, and the secular inequalities of these two quantities are given. Thus the mean motions are given, and the true motions are found by applying the inequalities, the numerical values of which are called equations: for, in astronomy, an equation signifies the quantities that must be added or taken from the mean results, to make them equal to the true results.

The mean motion and equation of the centre are computed from Kepler's problem; the motions of the nodes and perihelia, the secular inequalities of the elements, and the periodic inequalities, are computed from the formulæ determined by the problem of three bodies.

Method of correcting Errors in the Tables.

662. As astronomical tables are computed from analytical formulæ, determined on the principles of universal gravitation, no error can arise from that source; but the elements of the orbit, though determined with great accuracy by numerous observations, will lead to errors, because each element is found separately; whereas these quantities are so connected with each other, that a perfectly correct value of one, cannot be determined independently of the others. For example, the expressions in page 261 show, that the eccentricity depends on the longitudes of the perihelia, and the longitude of the perihelion is given in terms of the eccentricities. A reciprocal connexion exists also between the inclination of the orbit and the longitude of the nodes. Hence, in an accurate determination of the elements, it is necessary to attend to this reciprocal connexion.

The tables are computed with the observed values of the elements; an error in one of the elements will affect every part of the tables, and will be perceived in the comparison of the place of the body derived from them, with its place determined by observation. Were the observation exact, the difference would be the true error of the tables; but as no observation is perfectly accurate, the comparison is made with 1000, or even many thousands of observations, so that their errors are compensated by their numbers.

The simultaneous correction is accomplished, by comparing a longitude of the body derived from observation, with the longitude corresponding to the same instant in the tables.

Suppose the tables of the sun to require correction, and let E represent the error of the tables, or the difference between the longitude of the tables and that deduced from observation, at that point of the orbit where his mean anomaly is 198°. There are three sources from whence this error may arise, namely, the mean longitude of the perigee, the greatest equation of the centre, and the epoch of the tables; for, if an error has been made in computing the initial longitude, it will affect every subsequent longitude. Now, as we do not know to which of these quantities to attribute the discrepancy, part of it is assumed to arise from each. Let P be the unknown error in the longitude of the perigee, e that in the greatest equation of the centre, and ϵ that in the epoch. In order to determine these three errors, let us ascertain what effect would be produced on the place of the sun, where his mean anomaly is 198°, by an error of 60″ in the longitude of the perigee. As the mean anomaly is estimated from perigee, a minute of change in the perigee will produce the change of one minute in the mean anomaly corresponding to each longitude; but the table of the equation of the centre shows that the change of 60″ in the mean anomaly at that part of the orbit which corresponds to 198° produces an increment of 1″.88 in the equation of the centre; and as that quantity is subtractive at that part of the orbit, the true longitude of the sun is diminished by 1″.88; hence if 60″ produce a change of 1″.88 in the true longitude, the error P will produce a change of

$$\frac{1''.88}{60''} P = 0''.3133\, P.$$

Again, if we suppose the greatest equation of the centre to be augmented by any arbitrary quantity as 17″.18, it is easy to see by the tables that the equation of the centre at that point of the orbit where the mean anomaly is 198° is increased by 5″.1; whence the true longitude is diminished by 5″.1. Thus, if 17″.18 produce a change of 5″.1 in the true longitude, the error e will produce the change

$$-\frac{5''.1}{17''.18} e = -0''.2969\, e.$$

Hence the sum of the three errors is equal to E, the error of the tables

$$\epsilon + 0''.3133\, P - 0''.2969\, e = E.$$

This is called an equation of condition between the errors, because it expresses the condition that the sum of the errors must fulfil.

As there are three unknown quantities, three equations would be sufficient for their determination, if the observations were accurate; but as that is not the case, a great number of equations of condition must be formed from an equal number of observed longitudes, and they must be so combined by addition or subtraction, as to form others that are as favourable as possible for the determination of each element. For example, in finding the value of P before the other two, the numerous equations must be so combined, as to render the coefficient of P as great as possible; and the coefficients of e and ϵ as small as may be; this may always be accomplished by changing the signs of all the equations, so as to have the terms containing P positive, and then adding them; for some of the other terms will be positive, and some negative, as they may chance to be; therefore the sum of their coefficients will be less than that of P.

Having determined this equation, in which P has the greatest coefficient possible, two others must be formed on the same principle, in which the coefficients of the other two errors must be respectively as great as possible, and from these three equations values of the three errors will be easily obtained, and their accuracy will be in proportion to the number of observations employed. These values are referred to the mean interval between the first and last observations, supposing them not to be separated by any great length of time, and that the mean motion is perfectly known. Were it not, as might happen in the case of the new planets, an additional error may be assumed to arise from this source, which may be determined in the same manner as the others. This method of correcting errors in astronomical tables was employed by Mayer, in computing tables of the moon, and is applicable to a variety of subjects.

663. The numerous equations of condition of the form

$$E = \epsilon + 0''.3133\,P + 0''.2969\,e,$$

may be combined in a different manner, used by Legendre, called the principle of the least squares.

If the position of a point in space, is to be determined, and if a series of observations had given it the positions n, n', n'', &c., not differing much from each other, a mean place M must be found, which differs as little as possible from the observed positions n, n', n'',

&c.: hence it must be so chosen that the sum of the squares of its distances from the points n, n', n'', &c., may be a minimum; that is,

$$(Mn)^2 + (Mn')^2 + (Mn'')^2 + \&c. = \text{minimum}.$$

A demonstration of this is given in Biot's Astronomy, vol. ii.; but the rule for forming the equation of the minimum, with regard to one of the unknown errors, as P, is to multiply every term of all the equations of conditions by $0''.3133$, the coefficient of P, taken with its sign, and to add the products into one sum, which will be the equation required. If a similar equation be formed for each of the other errors, there will be as many equations of the first degree as errors; whence their numerical values may be found by elimination.

It is demonstrated by the Theory of Probabilities, that the greatest possible chance of correctness is to be obtained from the method of least squares; on that account it is to be preferred to the method of combination employed by Mayer, though it has the disadvantage of requiring more laborious computations.

The principle of least squares is a corollary that follows from a proposition of the Loci Plani, that the sum of the squares of the distances of any number of points from their centre of gravity is a minimum.

664. Three centuries have not elapsed since Copernicus introduced the motions of the planets round the sun, into astronomical tables: about a century later Kepler introduced the laws of elliptical motion, deduced from the observations of Tycho Brahe, which led Newton to the theory of universal gravitation. Since these brilliant discoveries, analytical science has enabled us to calculate the numerous inequalities of the planets, arising from their mutual attraction, and to construct tables with a degree of precision till then unknown. Errors existed formerly, amounting to many minutes; which are now reduced to a few seconds, a quantity so small, that a considerable part of it may perhaps be ascribed to inaccuracy in observation.

BOOK III.

CHAPTER I.

LUNAR THEORY.

665. There is no object within the scope of astronomical observation which affords greater variety of interesting investigation to the inhabitant of the earth, than the various motions of the moon: from these we ascertain the form of the earth, the vicissitudes of the tides, the distance of the sun, and consequently the magnitude of the solar system. These motions which are so obvious, served as a measure of time to all nations, until the advancement of science taught them the advantages of solar time; to these motions the navigator owes that precision of knowledge which guides him with well-grounded confidence through the deep.

Phases of the Moon.

666. The phases of the moon depend upon her synodic motion, that is to say, on the excess of her motion above that of the sun. The moon moves round the earth from west to east; in conjunction she is between the sun and the earth; but as her motion is more rapid than that of the sun, she soon separates from him, and is first seen in the evening like a faint crescent, which increases with her distance till in quadrature, or 90° from him, when half of her disc is enlightened: as her elongation increases, her enlightened disc augments till she is in opposition, when it is full moon, the earth being between her and the sun. In describing the other half of her orbit, she decreases by the same degrees, till she comes into conjunction with the sun again. Though the moon receives no light from the sun when in conjunction, she is visible for a few days before and after it, on account of the light reflected from the earth.

The law of the variation of the phases of the moon proves her form to be spherical, since they vary as the versed sine of her angular distance from the sun.

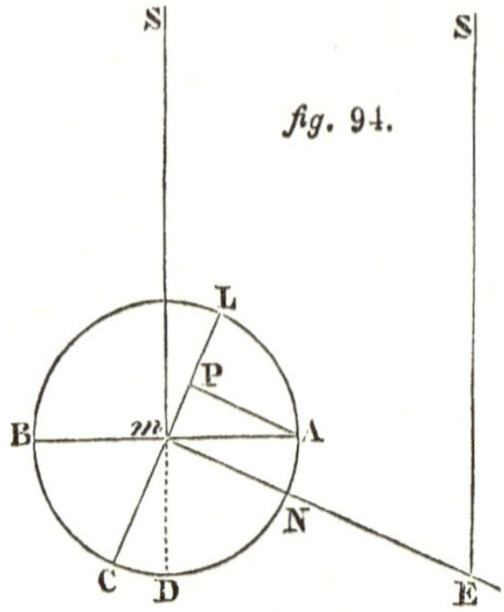

fig. 94.

If E be the earth, fig. 94, *m* the centre of the moon, supposed to be spherical, and S*m*, SE parallel rays from the sun. Then, if AB be at right angles to the ray *m*S, BLA is the part of the disc that is enlightened by the sun; and CL, being at right angles to *m*E, the part of the moon that is turned to the earth will be CNL; hence the only part of the enlightened disc seen from the earth is LA; or, if it be projected on CL, it is PL, the versed sine of AL. But A*m*L is complement to A*m*N, and is therefore equal to D*m*N, or to *m*ES, the elongation or angular distance of the moon from the sun. When the moon is in quadrature, that is, either 90° or 180° from the sun, a little more than half her disc is enlightened; for when the exact half is visible, the moon is a little nearer to the sun than 90°; at that instant, which is known by the division between the light and the dark half being a straight line; the lunar radius E*m*, fig. 95, is perpendicular to *m*S, the line joining the centres of the sun and moon; hence, in the right-angled triangle E*m*S, the angle E, at the observer, may be measured, and therefore we can determine SE, the distance of the sun from the earth, by the solution of a right-angled triangle, when the moon's distance from the earth is known. The difficulty of ascertaining the exact time at which the moon is bisected, renders this method of ascertaining the distance of the sun incorrect. It was employed by Aristarchus of Samos at Alexandria, about two hundred and eighty years before the Christian era,

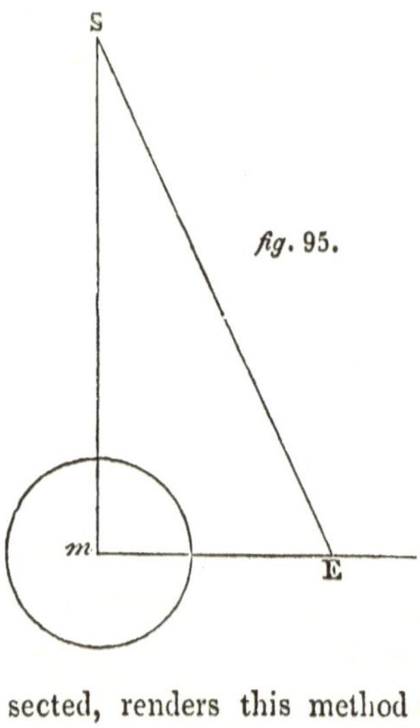

fig. 95.

and was the first circumstance that gave any notion of the vast distance and magnitude of the sun.

Mean or Circular Motion of the Moon.

667. The mean motion of the moon may be determined by comparing ancient with modern observations. The moon when eclipsed is in opposition, and her place is known from the sun's place, which can be accurately computed back to the earliest ages of antiquity. Three eclipses of the moon observed at Babylon in the years 720 and 719 before the Christian era, are the oldest observations recorded with sufficient precision to be relied on. By comparing these with modern observations, it is found that the mean arc described by the moon in one hundred Julian years, or the difference of the mean longitudes of the moon in a century, was 481267°.8793 in the year 1800; it is called the moon's tropical motion, which, omitting 1336 entire circumferences, is 307°.8793; and dividing it by 365.25, the number of days in the Julian year, her diurnal tropical motion is 13°.17636, about thirteen times greater than that of the sun.

668. From the tropical motion of the moon, her periodic revolution, or the time she employs in returning to the same longitude, may be found by simple proportion; for

$$481267°.8793 : 360° :: 365.25 : 27.321582,$$

the periodic revolution of the moon, or a periodic lunar month.

669. By subtracting 5010″, or the precession of the equinoxes for a century, from the secular tropical motion of the moon, her sidereal motion in a century is 481266°.48763; or, omitting the whole circumferences, it is 306°.48763; whence, by simple proportion, her sidereal revolution is $27^{d}\ 7^{h}\ 43'\ 11''.5$. These two motions of the moon only differ by the precession of the equinoxes: her sidereal daily motion is, therefore, 13° 10′ 35″.034.

670. The synodic revolution of the moon is her mean motion from conjunction to conjunction, or from opposition to opposition. The mean motion of the moon in a century being 481267°.8793, and that of the sun being 36000°.7625, their difference, 445267°.1168, is the excess of the moon's motion above the sun's in one hundred Julian years; hence her motion through 360° is accomplished in

$29^d\ 12^h\ 44'\ 2''.8$, a lunar month. The lunar month is to the tropical as 19 to 235 nearly, so that 19 solar years are equal to 235 lunar months. The mean motion of the moon is variable, which affects all the preceding results.

671. The apparent diameter of the moon is either measured by a micrometer, or computed from the duration of the occultations of the fixed stars. Its greatest value is thus found to be $2011''.1$, and the least $1761''.91$. The analogous values in the apparent diameter of the sun are $1955''.6$ and $1890''.96$; whence the variations in the moon's distance from the earth are much greater than those of the sun; consequently the eccentricity of the lunar orbit is much greater than that of the terrestrial orbit.

672. It appears from observation, that the horizontal parallax of the moon takes all possible values between the limits $1^\circ.0248$ and $0^\circ.8975$ which give 55.9164 and 63.8419 for the least and greatest distances of the moon from the earth; consequently, her mean distance is nearly sixty times the terrestrial radius. The solar parallax shows, that the sun is immensely more distant. Because the lunar parallax is equal to the radius of the terrestrial spheroid divided by the moon's distance from the earth, it is evident that, at the same distance of the moon, the parallax varies with the terrestrial radii; consequently, the variations in the parallax not only prove that the moon moves in an ellipse, having the earth in one of its foci, but that the earth is a spheroid.

Elliptical Motion of the Moon.

673. The greatest inequality in the moon's motion is the equation of the centre, which was discovered at a very early period: it is by this quantity alone that the undisturbed elliptical motion of a body differs from its mean or circular motion; it therefore arises entirely from the eccentricity of the orbit, being zero in the apsides, where the elliptical motion is the same with the mean motion, and greatest at the mean distance, or in quadratures, where the two motions differ most. Its maximum is found, by observation, to be $6^\circ\ 17'\ 28''$. This quantity which appears to be invariable, is equal to twice the eccentricity; and if the radius be unity, an arc of

$$3^\circ\ 8'\ 44'' = 0.0549003 = e,$$

the eccentricity of the lunar orbit when the mean distance of the moon from the earth is one.

674. In consequence of the action of the sun, the perigee of the lunar orbit has a direct motion in space. Its mean motion in one hundred Julian years, deduced from a comparison of ancient with modern observations, was 4069°.0395 in 1800, with regard to the equinoxes, which by simple proportion gives $3231^d.4751$ for its tropical revolution, and $3232^d.5807$, or a little more than nine years for its sidereal revolution; hence its daily mean motion is 6′ 41″. These motions change on account of the secular variation in the motion of the perigee.

675. The anomalastic revolution of the moon is her revolution with regard to her apsides, because the moon moves in the same direction with her perigee; after separating from that point, she only comes to it again by the excess of her velocity. That excess is 477198°.69184 in one hundred Julian years; therefore by simple proportion, the moon's anomalastic year is $27^d.5546$.

676. The nodes of the lunar orbit have a retrograde motion, which may be computed from observation, in the same manner with the motion of the perigee. The mean tropical motion of the nodes in 1800 was 1936°.940733, which gives $6788^d.54019$ for their tropical revolution, and $6793^d.42118$ for their sidereal revolution, or 3′ 10″.64 in a day; hence the moon's daily motion, with regard to her node, is 13° 13′ 45″.534. The motion of the perigee and nodes arises from the disturbing action of the sun, and depends on the ratio of his mass to that of the earth; this being very great, is the reason why the greater axis and nodes of the lunar orbit move so much more rapidly than those of any other body in the system.

Lunar Inequalities.

677. The moon is troubled in her motion by the sun; by her own action on the earth, which changes the relative positions of the bodies, and thus affects her motions; by the direct action of the planets; by their disturbing action on the earth, and by the form of the terrestrial spheroid.

678. Previous to the analytical investigations, it may perhaps be

of use to give some idea of the action of the sun, which is the principal cause of the lunar inequalities.

The moon is attracted by the sun and by the earth at the same time, but her elliptical motion is only troubled by the difference of the actions of the sun on the earth and on herself. Were the sun at an infinite distance, he would act equally and in parallel straight lines, on the earth and moon, and their relative motions would not be troubled by an action common to both; but the distance of the sun although very great, is not infinite. The moon is alternately nearer to the sun and farther from him than the earth; and the straight line Sm, fig. 96, which joins the centres of the sun and moon, makes angles more or less acute with SE, the radius vector of the earth. Thus the sun acts unequally, and in different directions, on the earth and moon; whence inequalities result in the lunar motions, depending on mES, the elongation of the sun and moon, on their distances and the moon's latitude.

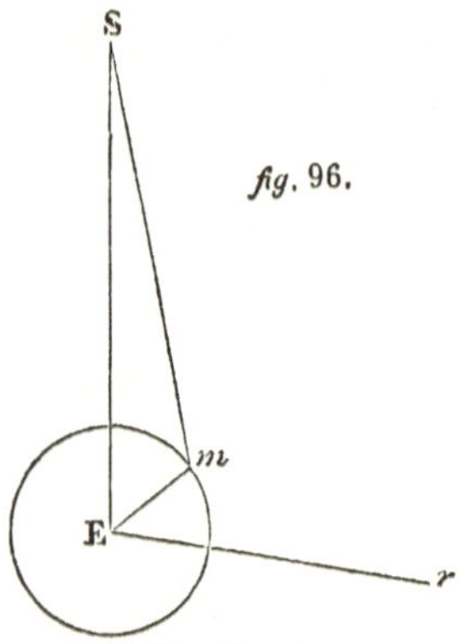

fig. 96.

When the moon is in conjunction at m, fig. 97, she is nearer the sun than the earth is; his action is therefore greater on the moon than it is on the earth; the difference of their actions tends to diminish the moon's gravitation to the earth. In opposition at m', the earth is nearer to the sun than the moon is, and therefore the sun attracts the earth more powerfully than he attracts the moon. The difference of these actions tends also to diminish the moon's gravitation to the earth. In quadratures, at Q and q, the action of the sun on the moon resolved in the direction of the radius vector QE, tends to augment the gravitation of the moon to the earth; but this increment of gravitation in quadratures is only half of the diminution of gravitation in syzigies; and thus, from the whole action of the sun on the moon in the course of a synodic revolution, there results a mean force directed according to the radius

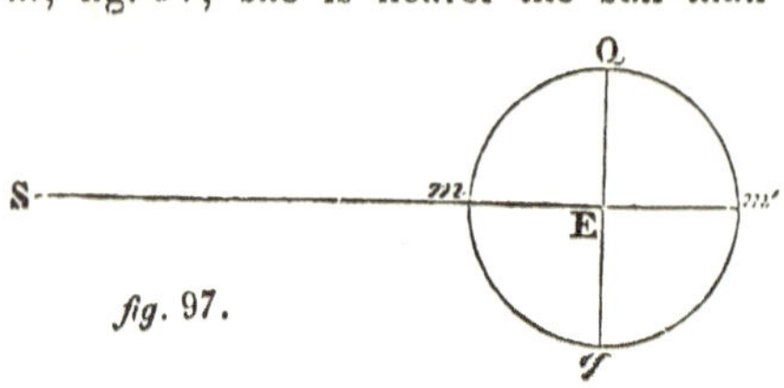

fig. 97.

vector of the moon, which diminishes her gravity to the earth, and may be determined as follows:—

679. Let M, fig. 98, be the moon in her nearly circular orbit nMN; E and S the earth and sun in the plane of the ecliptic; nmN the moon's orbit projected on the same. Then Mm is the tangent of the moon's latitude, and Em her curtate distance. Let SE, Em, be represented by r' and r, and the angle AEm by x, m' being the mass of the sun.

The attraction of the sun on the moon at M is $\frac{m'}{(\text{SM})^2}$. This force may be resolved into three; one in the direction Mm, which troubles the moon in latitude; another in mE, which, being directed towards the centre E, increases the gravity of the moon to the earth, and

fig. 98. *fig.* 99.

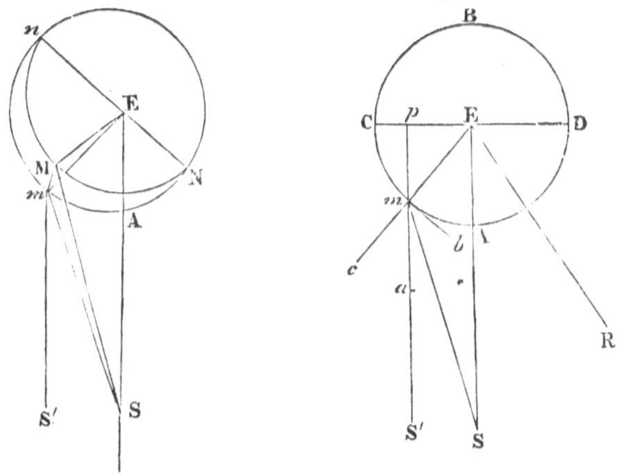

does not disturb the equable description of areas; and into a third in the direction mS′, the excess of which above that by which the sun attracts the earth disturbs the relative position of the moon and earth. The inclination of the lunar orbit is so small that it may be omitted at first, and then the force $\frac{m'}{(\text{S}m)^2}$, fig. 99, is resolved into two, one in the direction mE, which only increases the gravity of the moon, and the other in mS′, which disturbs her motion. Let ma represent this last force, and suppose it resolved into mb and mc. The force mb accelerates the moon in the quadrants CA and DB, and retards her in the other two; the force mc lessens the gravity of the moon.

680. The analytical expression of these forces is readily found. For the action of the sun on the moon in the direction Sm, is $\frac{m'}{(\mathrm{S}m)^2}$, but on account of the great distance of the sun,

$$\mathrm{S}m = \mathrm{SE} - mp = r' - r\cos x, \text{ nearly,}$$

hence the action of the sun on the moon in Sm is

$$\frac{m'}{(r' - r\cos x)^2},$$

which, resolved in the direction SE, is

$$\frac{m'}{r'^2} + \frac{m'}{r'^3}\, 3r\cos x, \text{ nearly.}$$

But the action of the sun on the earth is $\frac{m'}{r'^2}$, and their difference

$\frac{3m'}{r'^3} \cdot r\cos x$ is the force ma.

Now $\frac{m}{r'^3} \cdot 3r\cos^2 x$, is the force ma resolved in mc, and

$$\frac{m'}{r'^3} \cdot 3r\sin x\cos x = \frac{m'}{r'^3} \cdot \tfrac{3}{2}r\sin 2x,$$

is the same resolved in mb. But the force in mE which increases the moon's gravity to the earth, is evidently $\frac{m'r}{r'^3}$; hence the whole force by which the sun increases or diminishes the gravity of the moon to the earth is,

$$\text{force in } m\mathrm{E} - \text{force in } mc, \text{ or } \frac{m'r}{r'^3}(1 - 3\cos^2 . x).$$

In syzigy $x = 0°$, or $180°$, and $\cos^2 x = +1$; thus the action of the sun in conjunction and opposition is $-\frac{2m'r}{r'^3}$. In quadratures $x = 90°$, or $270°$; hence $\cos x = 0$, and the sun's action at these points is $\frac{m'r}{r'^3}$. The mean value of the force $\frac{m'r}{r'^3}(1 - 3\cos^2 x)$ for an entire revolution, is the integral of

$$\frac{m'r}{r'^3}(1 - 3\cos^2 x)\,dx = \frac{m'r}{r'^3}(1 - \tfrac{3}{2} - \tfrac{3}{2}\cos 2x)\,dx,$$

or

$$-\frac{m'r}{r'^3}(\tfrac{1}{2}x + \tfrac{3}{4}\sin 2x);$$

and when $x = 360°$, it becomes $-\frac{m'r}{2r'^3}$, which is the mean disturbing force acting on the moon in the direction of the radius vector.

681. In order to have the ratio of this mean force to the gravity of the moon, we must observe that if E and m be the masses of the earth and moon, $\frac{E+m}{r^2}$ is the force that retains the moon in her orbit, and $\frac{m'}{r'^2}$ is the force that retains the earth in its orbit. But these forces are as

$$\frac{r}{(27.321661)^2} \text{ to } \frac{r'}{(365.25)^2},$$

which are the radii vectores of the moon and earth divided by the squares of their periodic times, whence

$$\frac{m'r}{r'^3} = \frac{1}{179} \cdot \frac{m+E}{r^2};$$

and thus it appears that the mean action of the sun diminishes the gravity of the moon to the earth by its 358th part, for

$$\frac{m'r}{2r'^3} = \frac{1}{358} \cdot \frac{m+E}{r^2}.$$

682. In consequence of this diminution of the moon's gravity by its 385th part, she describes her orbit at a greater distance from the earth with a less angular velocity, and in a longer time than if she were urged to the earth by her gravity alone; but as the force is in the direction of the radius vector, the areas are not affected by it; hence, if her radius vector be increased by its 358th part, and her angular velocity diminished by its 179th part, the areas described will be the same as they would have been without that action. The

force in the tangent mb disturbs the equable description of areas, and that in mM troubles the moon in latitude. The true investigation of these forces can only be conducted by an analytical process, which will now be given, without carrying the approximation so far as may be necessary, referring for the complete developement of the series, to Damoiseau's profound analysis in the *Memoirs* of the *French Institute* for 1827.

683. The peculiar disturbances to which the moon is liable, and the variety of inferences that may be drawn from them, render her motions better adapted to prove the universal prevalence of the law of gravitation, than those of any other body. The perfect coincidence of theory with observation, shows that analytical formulæ not only express all the observed phenomena, but that they may be employed as a means of discovery not less certain than observation itself.

684. Although the motions of the moon be similar to those of a planet, they cannot be determined by the same analysis, on account of the great eccentricity of the lunar orbit, and the immense magnitude of the sun, which make it necessary to carry the approximation at least to the fourth powers of the eccentricities, and to the square of the disturbing force; and although the smallness of the mass of the moon compared with that of the earth, enables us to obtain her perturbations by successive approximations, yet the series converge slowly when the disturbing action of the sun is expressed in functions of the mean longitudes of the sun and moon; and as the facility of analytical investigations, and the fitness of formulæ for computation, depend on a skilful choice of co-ordinates, the motions of the moon are first determined in functions of the true longitudes, and then her co-ordinates are obtained by reversion of series in functions of the mean longitudes of the two bodies.

685. The successive approximations are determined by the magnitude of the coefficients. Those terms belong to the first approximation which have for coefficients, either the ratio of the mean motion of the sun to that of the moon, or the eccentricities of the earth and moon, or the inclination of the lunar orbit on the ecliptic. Those terms belong to the second approximation, which have the squares of these quantities as coefficients; those which have their cubes belong to the third, and so on.

The terms having the constant ratio $\frac{a}{a'} = \frac{1}{400}$ of the parallax of the sun to that of the moon for coefficients, are included in the second approximation, and also those depending on the disturbing force of the sun, which is of the order

$$\frac{m'a^3}{a'^3}, \text{ or } m^2;$$

for it has been observed that a permanent change is produced by the disturbing forces in the mean distance: hence if

$$a', a, n', n, m', m,$$

be the mean distances, mean motions and masses of the sun and moon, and $\bar{a}$ the value of a in the troubled orbit, so that $a = \bar{a}$ when there is no disturbing force, then will

$$\frac{a^2}{\sqrt{\bar{a}}} = \frac{1}{n}, \text{ and as } \frac{m'}{a'^3} = n'^2,$$

therefore $$\frac{m'a^3}{a'^3} \cdot \frac{a_{\prime}}{\bar{a}} = \frac{n'^2}{n^2} = \left(\frac{1}{13.368}\right)^2 = 0.005595;$$

but the mass of the moon is $m = 0.0748013$, consequently

$$m^2 = 0.005595,$$

so that $$\frac{m'a^3}{a'^3} \cdot \frac{a}{\bar{a}} = m^2; \text{ or if } \frac{m'a^3}{a'^3} = \bar{m}^2,$$

then $$\frac{\bar{m}^2}{\bar{a}} = \frac{m^2}{a}.$$

686. By arranging the series according to the magnitudes of their terms, each approximation may be had separately by taking a certain part and rejecting the rest. This process must be continued till the value of the remainder is so small as to be insensible to observation; but even then it is necessary to ascertain not only that it is so at present, but that it will remain so after the lapse of ages. Besides selecting from the innumerable terms of the series those that have considerable coefficients, it is requisite to examine what values the different terms acquire in the determination of the finite values of the perturbations from their indefinitely small changes, for it has been shown that by integration some of the terms acquire divisors, which increase their values so much that great errors would ensue from omitting them.

Analytical Investigation of the Lunar Inequalities.

687. Suppose the motion of the earth to be referred to the sun, and that both sun and moon revolve round the earth assumed to be at rest in E, fig. 100. Let M be the moon in her orbit, m her place projected on the plane of the ecliptic, so that Em is her curtate distance; and let Ep, pm, Mm, or x, y, z, be the co-ordinates of the moon, and x', y', z', those of the sun in S, both referred to the centre of the earth, and to the fixed ecliptic at a given epoch.

fig. 100.

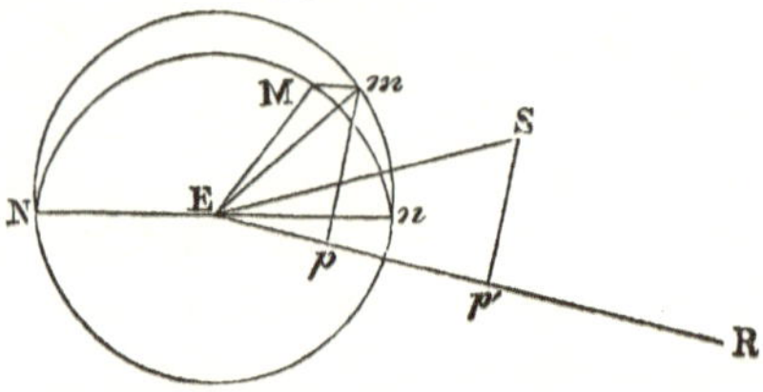

If m', E, m, be the masses of the sun, the earth, and the moon, the equations of article 347 are

$$0 = \frac{d^2x}{dt^2} + \frac{E+m}{r^3} - \frac{1}{m}\left(\frac{d\lambda}{dx}\right);$$

$$0 = \frac{d^2y}{dt^2} + \frac{E+m}{r^3} - \frac{1}{m}\left(\frac{d\lambda}{dy}\right);$$

$$0 = \frac{d^2z}{dt^2} + \frac{E+m}{r^3} - \frac{1}{m}\left(\frac{d\lambda}{dz}\right).$$

In which $r = \sqrt{x^2 + y^2 + z^2}$ is the radius vector of the moon,

$$\lambda = \frac{mm'}{\sqrt{(x'-x)^2+(y'-y)^2+(z'-z)^2}};$$

and the element of the time is assumed to be constant in taking the differentials; but if that element be variable, and if

$$R = \frac{(E+m)}{r} - \frac{m'(xx+yy+zz)}{(x'^2+y'^2+z'^2)^{\frac{3}{2}}} + \frac{m'}{\sqrt{(x'-x)^2+(y'-y)^2+(z'-z)^2}}$$

the relative motion of the moon and earth will be determined by the following equations,

$$\frac{d^2x}{dt^2} - \frac{dxd^2t}{dt^3} = \left(\frac{dR}{dx}\right)$$

$$\frac{d^2y}{dt^2} - \frac{dyd^2t}{dt^3} = \left(\frac{dR}{dy}\right)$$

$$\frac{d^2z}{dt^2} - \frac{dzd^2t}{dt^3} = \left(\frac{dR}{dz}\right).$$

688. In very small angles the arc may be taken for its sine; hence the lunar parallax is the radius of the terrestrial spheroid divided by the moon's distance from the earth, and thus the parallax varies inversely as the radius vector. Then if R' be the radius of the earth, and r the radius vector of the moon, the lunar parallax will be $\frac{R'}{r}$, which thus becomes the third co-ordinate of the moon. But if the earth be assumed to be spherical, its radius may be taken equal to unity, and then the lunar parallax will be $\frac{1}{r}$. Therefore let $u = \frac{1}{r}$, RE$m = v$; and $mM = s$, the tangent of the moon's latitude;

then
$$r = \sqrt{x^2+y^2+z^2} = \frac{\sqrt{1+ss}}{u},$$

$$x = \frac{\cos v}{u}, \quad y = \frac{\sin v}{u}, \quad \text{and } z = \frac{s}{u}.$$

But in taking the differentials of these, dv must be constant, since dt is assumed to be variable.

689. Let the first of the preceding equations multiplied by $-\sin v$ be added to the second multiplied by $\cos v$; and let the first multiplied by $\cos v$ be added to the second multiplied by $\sin v$; then, if the foregoing values of x, y, z, be substituted, and if to abridge

$$\left(\frac{dR}{dx}\right)\sin v - \left(\frac{dR}{dy}\right)\cos v = \pi$$

$$\left(\frac{dR}{dx}\right)\cos v + \left(\frac{dR}{dy}\right)\sin v = \Pi,$$

the result will be

$$\frac{d^2v}{dt^2} - \frac{2dvdu}{udt^2} - \frac{dvd^2t}{dt^3} = -\pi u;$$

$$\frac{d^2u}{u.dt^2} + \frac{dv^2}{dt^2} - \frac{2du^2}{u^2dt^2} - \frac{dud^2t}{udt^3} = -\Pi u; \qquad (206)$$

$$\frac{d^2s}{dt^2} + \frac{sdv^2}{dt^2} - \frac{dsd^2v}{dvdt^2} = \pi u\frac{ds}{dv} - \Pi su + u\left(\frac{dR}{dz}\right).$$

The first of these equations multiplied by $\frac{2dv}{u^3}$, and integrated, is

$$\left(\frac{dv}{u^2dt}\right)^2 = h^2 - \int 2\pi \,.\, \frac{dv}{u^3},$$

h^2 being a constant quantity;

whence
$$dt = \frac{dv}{u^2\sqrt{h^2 - 2\int\frac{\pi dv}{u^3}}}.$$

The elimination of d^2t between the first and second of equations (206), gives

$$\frac{dud^2v}{u^2dvdt^2} - \frac{d^2u}{u^2dt^2} - \frac{dv^2}{udt^2} = \Pi - \frac{\pi du}{udv};$$

and if dv be assumed to be constant, and substituting for dt its preceding value, it becomes

$$\frac{d^2u}{dv^2} + u = -\frac{\Pi - \pi\dfrac{du}{udv}}{u^2\left(h^2 - 2\int\dfrac{\pi dv}{u^3}\right)}.$$

In the same manner the third of equations (206) gives

$$\frac{d^2s}{dv^2} + s = \frac{\left(\dfrac{dR}{dz}\right) - \Pi s + \dfrac{\pi ds}{dv}}{u^3\left(h^2 - 2\int\dfrac{\pi dv}{u^3}\right)}.$$

Now
$$dR = dx\left(\frac{dR}{dx}\right) + dy\left(\frac{dR}{dy}\right) + dz\left(\frac{dR}{dz}\right),$$

and when substitution is made for dx, dy, dz,

$$dR = -\frac{du}{u^2}\left\{\left(\frac{dR}{dx}\right)\cos v + \left(\frac{dR}{dy}\right)\sin v + \left(\frac{dR}{dz}\right)s\right\}$$
$$-\frac{dv}{u}\left\{\left(\frac{dR}{dx}\right)\sin v - \left(\frac{dR}{dy}\right)\cos v\right\} + \frac{ds}{u}\left(\frac{dR}{dz}\right).$$

But
$$dR = du\left(\frac{dR}{du}\right) + dv\left(\frac{dR}{dv}\right) + ds\left(\frac{dR}{ds}\right);$$

hence, by comparison,

$$\frac{dR}{du} = -\frac{1}{u^2}\left\{\left(\frac{dR}{dx}\right)\cos v + \left(\frac{dR}{dy}\right)\sin v + \left(\frac{dR}{ds}\right)s\right\}$$
$$\frac{dR}{dv} = -\frac{1}{u}\left\{\left(\frac{dR}{dx}\right)\sin v - \left(\frac{dR}{dy}\right)\cos v\right\}$$
$$\frac{dR}{ds} = \frac{1}{u}\left(\frac{dR}{dz}\right).$$

Whence $$\Pi = -u^2\left(\frac{dR}{du}\right) - su\left(\frac{dR}{ds}\right)$$

$$\pi = -u\left(\frac{dR}{dv}\right), \text{ and } \frac{dR}{dz} = u\left(\frac{dR}{ds}\right).$$

690. Thus the differential equations which determine the motions of the moon become

$$dt = \frac{dv}{hu^2\left\{1 + \frac{2}{h^2}\int\left(\frac{dR}{dv}\right)\cdot\frac{dv}{u^2}\right\}^{\frac{1}{2}}}$$

$$0 = \left(\frac{d^2u}{dv^2} + u\right)\left\{1 + \frac{2}{h^2}\int\left(\frac{dR}{dv}\right)\cdot\frac{dv}{u^2}\right\} + \frac{du}{h^2u^2dv}\left(\frac{dR}{dv}\right)$$
$$- \frac{1}{h^2}\left(\frac{dR}{du}\right) - \frac{s}{h^2u}\left(\frac{dR}{ds}\right); \qquad (207)$$

$$0 = \left(\frac{d^2s}{dv^2} + s\right)\left\{1 + \frac{2}{h^2}\int\left(\frac{dR}{dv}\right)\cdot\frac{dv}{u^2}\right\} + \frac{1}{h^2u^2}\cdot\frac{ds}{dv}\left(\frac{dR}{dv}\right)$$
$$- \frac{s}{h^2u}\left(\frac{dR}{du}\right) - \frac{(1+s^2)}{h^2u^2}\left(\frac{dR}{ds}\right).$$

In the determination of these equations no quantities have been neglected, therefore the influence of such terms as may be omitted in the final result can be fully appreciated.

691. In order to develope the disturbing forces represented by R, the action of the sun alone will be first considered, assuming the masses of the three bodies to be spherical, and $m + \mathrm{E} = 1$. If x', y', z', be the co-ordinates of the sun, and r' its radius vector, then

$$\frac{1}{\sqrt{(x'-x)^2 + (y'-y)^2 + (z'-z)^2}} = \frac{1}{\sqrt{r^2 + r'^2 - 2xx' - 2yy' - 2zz'}}$$

and the second member developed according to the powers of $\frac{1}{r'}$ is

$$\frac{1}{r'} + \frac{xx'+yy'+zz'-\frac{1}{2}r^2}{r'^3} + \frac{3}{2}\,\frac{(xx'+yy'+zz'-\frac{1}{2}r^2)^2}{r'^5} + \&c.$$

692. Since the earth is assumed to be a sphere, its radius may be taken equal to unity, and therefore the solar parallax will be $\frac{1}{r'}$;

and if $u' = \frac{1}{r'}$, then $r' = \frac{\sqrt{1+s'^2}}{u'}$. But the ecliptic may be taken for the plane of projection, although it be not fixed; for in its secular motion it carries the orbit of the moon with it, as will be shown afterwards, so that the mean inclination of this orbit on the ecliptic remains constant, and the phenomena depending on the relative inclination of the two planes are always the same; hence $s' = 0$, and therefore $z' = 0$, and the co-ordinates of the sun are

$$x' = \frac{\cos v'}{u'} \qquad y' = \frac{\sin v'}{u'}.$$

693. Now the distance of the sun from the earth being nearly 400 times greater than that of the moon, $u' = \frac{1}{r'}$ is very small in comparison of $u = \frac{1}{r}$, consequently in the lunar theory u'^5 may be omitted; and if the preceding values of x, y, z; $x', y', \frac{1}{r}$, and $\frac{1}{r'}$, be substituted in R, it becomes

$$R = \frac{u}{\sqrt{1+s^2}} + m'u' + \frac{m'u'^3}{4u^2}\{1+3\cos(2v-2v')-2s^2\} \quad (208)$$
$$+ \frac{m'u'^4}{8u^3}\{3(1-4s^2)\cos(v-v') + 5\cos(3v-3v')\}.$$

694. But the quantities u, u', v', and s, in the elliptical hypothesis, become
$$u+\delta u, \quad u'+\delta u', \quad v'+\delta v', \quad s+\delta s,$$
in the troubled orbit; and as the mass of the sun is so great that the second powers of the disturbing forces must be taken into account, the co-ordinates of the moon must not only contain R but δR.

695. Now

$$\frac{dR}{du} = \frac{1}{\sqrt{1+s^2}} - \frac{m'u'^3}{2u^3}\{1+3\cos(2v-2v')-2s^2\}$$
$$-\frac{3m'u'^4}{8u^4}\{3(1-4s^2)\cos(v-v')+5\cos(3v-3v')\}$$

$$\frac{dR}{dv} = -\frac{3m'u'^3}{2u^2}\sin(2v-2v') - \frac{3m'u'^4}{8u^3}\{(1-4s^2)\sin(v-v')$$
$$+5\sin(3v-3v')\}$$

$$\frac{dR}{ds} = -\frac{su}{(1+s^2)^{\frac{3}{2}}} - \frac{m'u'^3}{u^2}\cdot s - \frac{3m'u'^4}{u^3}\cdot s\cdot\cos(v-v');$$

and if the approximation be only carried to terms of the third order inclusively, the co-ordinates of the moon in her troubled orbit will be

$$\frac{d^2u}{dv^2} + u = \frac{1}{h^2(1+s^2)^{\frac{3}{2}}} - \frac{m'u'^3}{2h^2u^3}$$

$$- \frac{3m'u'^3}{2h^2u^3} \cdot \cos(2v - 2v')$$

$$+ \frac{3m'u'^3}{2h^2u^4} \cdot \frac{du}{dv} \sin(2v - 2v') \qquad (209)$$

$$- \frac{9}{8}\frac{m'u'^4}{h^2u^4} \cos(v - v')$$

$$+ \left(\frac{d^2u}{dv^2} + u\right)\frac{3m'}{h^2}\int\frac{u'^3dv}{u^4} \sin(2v - 2v')$$

$$+ \left(\frac{d^2u}{dv^2} + u\right)\frac{3m'}{4h^2}\int\frac{u'^4dv}{u^5} \sin(v - v')$$

$$+ \delta\left\{\frac{1}{h^2(1+s^2)^{\frac{3}{2}}} - \frac{m'u'^3}{2h^2u^3} - \frac{3m'u'^3}{2h^2u^3}\cos(2v - 2v')\right.$$

$$+ \frac{3m'u'^3}{2h^2u^4} \cdot \frac{du}{dv} \sin(2v - 2v')$$

$$- \frac{3m'u'^4}{8h^2u^4}\{3\cos(v - v') + 5\cos(3v - 3v')\}$$

$$+ \frac{m'u'^4}{8h^2u^5} \cdot \frac{du}{dv}\{3\sin(v - v') + 15\sin(3v - 3v')\}$$

$$+ \frac{3m'}{h^2}\left(\frac{d^2u}{dv^2} + u\right)\left\{\int\frac{u'^3}{u^4}dv\sin(2v-2v') + \frac{1}{4}\int\frac{u'^4}{u^5}dv\sin(v-v')\right\}$$

$$\left.+ \frac{3m'}{h^2}\left\{\frac{d^2u}{dv^2} + u\right\}\left\{\int\frac{u'^4}{4u^5}dv \cdot 5\sin(3v - 3v')\right\}\right\}.$$

$$\frac{d^2s}{dv^2} + s = -\frac{3m'u'^3}{2h^2u^4}s - \frac{3m'u'^3}{2h^2u^4}s \cdot \cos(2v - 2v')$$

$$+ \frac{3m'u'^3}{2h^2u^4}\frac{ds}{dv} \cdot \sin(2v - 2v')$$

$$+ \frac{3m'u'^4}{2h^2u^5}\left\{11us\cos(v - v') + \frac{1}{4}\frac{ds}{dv}\sin(v - v')\right\} \qquad (210)$$

$$+ \frac{3m'}{h^2}\left(\frac{d^2s}{dv^2} + s\right) \cdot \int\frac{u'^3dv}{u^4}\sin(2v - 2v')$$

$$- \delta \left\{ \frac{3m'u'^3}{2h^2u^4} s + \frac{3m'u'^3}{2h^2u^4} \cos (2v - 2v') \right.$$

$$- \frac{3m'u'^3}{2h^2u^4} \cdot \frac{ds}{dv} \cdot \sin (2v - 2v')$$

$$\left. - \frac{3m'}{h^2} \left(\frac{d^2s}{dv^2} + s \right) \int \frac{u'^3 dv}{u^4} \sin (2v - 2v') \right\}$$

$$dt = \frac{dv}{h^2(u+\delta u)^2 \sqrt{1 - \frac{3m'}{h^2} \int \frac{u'^3 dv}{u^4} \{ \sin (2v - 2v') + \frac{u'}{4u} \sin (v - v') \}}}. \qquad (211)$$

696. These equations contain five unknown quantities, u, v, u', v', and s; but u' and v' may be eliminated by their functions in v by integrating the equations (207) when $R = 0$, that is, assuming the moon to move without disturbance. By the method already employed, the two last of these equations give

$$s = \gamma \sin (v - \theta)$$

$$u = \frac{1}{h^2(1 + \gamma^2)} \{ \sqrt{1 + s^2} + e \cos (v - \varpi) \}$$

γ being the tangent of the inclination of the lunar orbit on the ecliptic, θ the longitude of the ascending node, e the eccentricity, and ϖ the longitude of the perigee.

697. In these equations the lunar orbit is assumed to be immoveable, but observation shows that the nodes and perigee have a rapid motion in space from the action of the sun; the latter accomplish a revolution in a little more than nine years, so that the lunar ellipse revolves in its own plane in the same direction with the moon's motion; hence if c be such that $1 : 1 - c :: v$, the moon's motion in longitude, is to the motion of the apsides, then $v\ (1 - c)$ will be the angle described by the apsis, while the moon describes v. Assuming the instant when the apsis coincided with the axis of x as the origin of the time, then $v - v\ (1 - c) = cv$ will be the moon's true anomaly. In the same manner $(g - 1)\ v$ will represent the retrograde motion of the node, while the moon moves through v. Hence if gv and cv be put for v in the preceding values of s and u, they become

$$s = \gamma \cdot \sin (gv - \theta) \qquad (212)$$

$$u = \frac{1}{h^2(1 + \gamma^2)} \{ 1 + \tfrac{1}{4} \gamma^2 + e \cos (cv - \varpi) - \tfrac{1}{4} \gamma^2 \cos 2 (gv - \theta) \}$$

which are the latitude and parallax of the moon in her orbit considered as a revolving ellipse.

This value of u put in $dt = \dfrac{dv}{h^2u^2}$, which is the first of equations (207), when $R = 0$, gives

$$dt = h^3dt\left\{\begin{array}{l} 1 + \frac{3}{2}(e^2+\gamma^2) - 2e\,(1+\frac{3}{2}e^2+\frac{5}{4}\gamma^2)\cos(cv-\varpi) \\ +\frac{3}{2}e^2\cos(2cv-2\varpi) - e^3\cos(3cv-3\varpi) + \frac{1}{2}\gamma^2\cos(2gv-2\theta) \\ -\frac{3}{4}e\gamma^2\{\cos(2gv+cv-\varpi-2\theta)+\cos(2gv-cv+\varpi-2\theta)\} \end{array}\right\}$$

its integral is

$$t = \text{constant} + h^3\Big\{v\,(1+\tfrac{3}{2}e^2+\tfrac{3}{2}\gamma^2) - \frac{2e}{c}(1+\tfrac{3}{2}e^2+\tfrac{5}{4}\gamma^2)\sin(cv-\varpi)$$

$$+\frac{3e^2}{4c}\sin(2cv-2\varpi) - \frac{e^3}{3c}\sin(3cv-3\varpi) + \frac{\gamma^2}{4g}\sin(2gv-2\theta)$$

$$-\frac{3e\gamma^2}{4(2g+c)}\sin(2gv+cv-\varpi-2\theta) - \frac{3e\gamma^2}{4(2g-c)}\times$$

$$\sin(2gv-cv+\varpi-2\theta)\Big\}.$$

698. The coefficients are somewhat modified by the action of the sun. In elliptical motion the coefficient of dv is $a^{\frac{3}{2}}$; a being half the greater axis of the lunar orbit, hence

$$h^3\,(1+\tfrac{3}{2}e^2+\tfrac{3}{2}\gamma^2) = a^{\frac{3}{2}}.$$

699. Again, because $m = 0.0748013$

$c = 1 - \frac{3}{2}m^2 = 0.991548$, $\quad g = 1 + \frac{3}{4}m^2 = 1.00402175$, nearly, therefore c and g may be taken equal to unity in the coefficients of the preceding integral, which becomes, when quantities of the order e^3 are rejected and n put for $a^{-\frac{3}{2}}$,

$$\begin{aligned} nt + \epsilon = v &- 2e\sin(cv-\varpi) \\ &+ \tfrac{3}{4}e^2\sin(2cv-2\varpi) \\ &+ \tfrac{1}{4}\gamma^2\sin(2gv-2\theta) \\ &- \tfrac{3}{4}e\gamma^2\sin(2gv-cv-2\theta+\varpi), \end{aligned} \tag{213}$$

ϵ being the arbitrary constant quantity.

700. Now, when quantities of the order γ^4 are omitted, the coefficient of the second of equations (212) becomes

$$\frac{1}{h^2(1+\gamma^2)} = h^{-2}(1-\gamma^2);$$

but $$h^{-2} = \frac{1}{a}(1+e^2+\gamma^2+\complement),$$

$\complement$ being the remaining part of the developement of h^{-2}, and therefore of the fourth order in e and γ, consequently

$$\frac{1}{h^2(1+\gamma^2)} = \frac{1}{a}(1 + e^2 + \complement),$$

and the parallax becomes

$$u = \frac{1}{a}\{1+e^2+\tfrac{1}{4}\gamma^2+\complement+e(1+e^2)\cos(cv-\varpi)-\tfrac{1}{4}\gamma^2\cos(2gv-2\theta)\}$$

the constant part of which is

$$\frac{1}{a}(1 + e^2 + \tfrac{1}{4}\gamma^2 + \complement);$$

but as this is modified by the action of the sun, it will be expressed by

$$\frac{1}{\bar{a}}(1 + e^2 + \tfrac{1}{4}\gamma^2 + \complement'),$$

so that without that action $\frac{1}{a} = \frac{1}{\bar{a}}$;

and when quantities of the fourth order are omitted,

$$u = \frac{1}{\bar{a}}\{1+e^2+\tfrac{1}{4}\gamma^2+e(1+e^2)\cos(cv-\varpi)-\tfrac{1}{4}\gamma^2\cos(2gv-2\theta)\}. \quad (214)$$

701. If accented letters be employed for the sun, his parallax and mean longitude will be,

$$u' = \frac{1}{a'}\{1+e'^2+e'(1+e'^2)\cos(c'v'-\varpi')\} \quad (215)$$

$$n't + \epsilon' = v' - 2e'\sin(c'v' - \varpi') \quad (216)$$
$$+ \tfrac{3}{4}e'^2\sin(2c'v' - 2\varpi').$$

For $\gamma' = 0$ since the sun moves in the plane of the ecliptic, and $g' = 1$, $c' = 1$ without error in the coefficients.

In order to abridge, let $n't + \epsilon' = v' + \phi'$, and for the same reason, equation (213) may be expressed by $nt + \epsilon = v + \phi$. For the elimination of v', suppose the sun and moon to have the same epoch; hence $\epsilon = 0$ $\epsilon' = 0$, and comparing their mean motions

$$v' = m(v + \phi) - \phi', \text{ since } \frac{n'}{n} = m.$$

By the substitution of this in

$$\phi' = -2e'\sin(c'v' - \varpi') + \tfrac{3}{4}e'^2\sin(2c'v' - \varpi'),$$

it becomes

$$\phi' = -2e' \sin\{c'mv - \varpi' + (c'm\phi - c'\phi')\} + \tfrac{3}{4}e'^2 \sin\{2c'mv - 2\varpi' + 2(c'm\phi - c'\phi')\};$$

or

$$\begin{aligned}\phi' = &- 2e' \sin(c'mv - \varpi' + c'm\phi) \cos c'\phi' \\ &+ 2e' \cos(c'mv - \varpi' + c'm\phi) \sin c'\phi' \\ &+ \tfrac{3}{4}e'^2 \sin(2'cmv - 2\varpi' + 2c'm\phi) \cos 2c'\phi' \\ &- \tfrac{3}{4}e'^2 \cos(2c'mv - 2\varpi' + 2c'm\phi) \sin 2c'\phi'.\end{aligned}$$

But if $c' = 1$ and $\cos c'\phi' = 1 - \tfrac{1}{2}\phi'^2 + \&c.$

$$\sin c'\phi' = \phi' - \tfrac{1}{6}\phi'^3 + \&c.,$$

then omitting ϕ'^3 the result will be,

$$\begin{aligned}\phi' = &- 2e' \sin(c'mv - \varpi' + c'm\phi) \\ &+ 2e'\phi' \cos(c'mv - \varpi' + c'm\phi) \\ &+ e'\phi'^2 \sin(c'mv - \varpi + c'm\phi) \\ &+ \tfrac{3}{4}e'^2 \sin(2c'mv - 2\varpi' + 2c'm\phi) \\ &- \tfrac{3}{2}e'^2\phi' \cos(2c'mv - 2\varpi' + 2c'm\phi) \\ &\quad \&c. \qquad \&c.\end{aligned}$$

If substitution be again made for ϕ', and the same process repeated, it will be found, that

$$\begin{aligned}\phi' = &- e'(2 - \tfrac{1}{4}e'^2) \sin(c'mv - \varpi') - e'(2 - \tfrac{1}{4}e'^2)m\phi \cos(c'mv - \varpi') \\ &- \tfrac{5}{4}e'^2 \sin(2c'mv - 2\varpi') - \tfrac{5}{2}m'e'^2\phi \cos(2c'mv - 2\varpi').\end{aligned}$$

If this value of ϕ' be put in $v' = m(v + \phi) - \phi'$ the value of ϕ restored, and the products of the sines and cosines reduced to the sines of the sums and differences of the arcs, when e^3 is rejected, the result will be

$$\begin{aligned}v' = mv &- 2me \sin(cv - \varpi) \\ &+ \tfrac{3}{4}e^2m \sin(2cv - 2\varpi) \\ &+ \tfrac{1}{4}m\gamma^2 \sin(2gv - 2\theta) \\ &- \tfrac{3}{4}me\gamma^2 \sin(2gv - cv + \varpi - 2\theta) \\ &+ 2e'(1 - \tfrac{1}{8}e'^2) \sin(c'mv - \varpi') \\ &- 2mee' \sin(cv + c'mv - \varpi - \varpi') \\ &- 2mee' \sin(cv - c'mv - \varpi + \varpi') \\ &+ \tfrac{5}{4}e'^2 \sin(2c'mv - 2\varpi').\end{aligned} \qquad (217)$$

702. If this value of v' be expressed by $v' = mv + \psi$, and substituted in equation (215) it becomes

$$u' = \frac{1}{a'}\{1 + e'^2 + e'(1 + e'^2) \cos(c'mv - \varpi' + c'\psi)\}.$$

It will readily appear by the same process, when all powers of the eccentricities above the second are rejected, that

$$u' = \frac{1}{a'}\left\{\begin{array}{l}1+e'(1-\frac{1}{8}e'^2)\cos(c'mv-\varpi')+e'^2\cos(2c'mv-2\varpi') \\ +mee'\cos(cv-c'mv-\varpi+\varpi')-mee'\cos(cv+cmv-\varpi-\varpi')\end{array}\right\} \quad (218)$$

703. By the same substitution,

$$\cos(v-v') = \cos(v-m'v)\cos\psi + \sin(v-mv)\sin\psi;$$

but $\cos\psi = 1 - \frac{1}{2}\psi^2 + \&c.\ \sin\psi = \psi - \frac{1}{6}\psi^3 + \&c.$

hence
$$\begin{aligned}\cos(v-v') = {} & \cos(v-mv) \\ & + \psi\sin(v-mv) \\ & - \tfrac{1}{2}\psi^2\cos(v-mv) \\ & - \tfrac{1}{6}\psi^3\sin(v-mv) \\ & + \&c. \qquad \&c.;\end{aligned}$$

and
$$\begin{aligned}\cos(v-v') = {} & \cos(v-mv) \qquad (219)\\ & - me\cos(v-mv-cv+\varpi) \\ & + me\cos(v-mv+cv-\varpi) \\ & + \tfrac{3}{8}me^2\cos(2cv-v+mv-2\varpi) \\ & - \tfrac{3}{8}me^2\cos(2cv+v-mv-2\varpi) \\ & + \tfrac{1}{8}m\gamma^2\cos(2gv-v+mv-2\theta) \\ & - \tfrac{1}{8}m\gamma^2\cos(2gv+v-mv-2\theta) \\ & - \tfrac{3}{8}me\gamma^2\cos(v-mv-2gv+cv-\varpi+2\theta) \\ & + \tfrac{3}{8}me\gamma^2\cos(v-mv+2gv-cv+\varpi-2\theta) \\ & + e'(1-\tfrac{1}{8}e'^2)\cos(v-mv-c'mv+\varpi') \\ & - e'(1-\tfrac{1}{8}e'^2)\cos(v-mv+c'mv-\varpi') \\ & + \qquad \&c. \qquad \&c.\end{aligned}$$

Thus the series expressing $\cos(v-v')$ may extend to any powers of the disturbing force and eccentricities.

704. Now
$$\begin{aligned}\cos(2v-2v') = {} & \cos(2v-2mv) \\ & + 2\psi\sin(2v-2mv) \\ & - 2\psi^2\cos(2v-2mv) \\ & - \tfrac{4}{3}\psi^3\sin(2v-2mv) \\ & \&c. \qquad \&c.\end{aligned}$$

which shows that $\cos(2v-2v')$ may be readily obtained from the developement of $\cos(v-v')$ by putting $2v$ for v, and 2ψ for ψ; and the same for any cosine, as $\cos i(v-v')$.

705. Again, if $90^\circ + v$ be put for v, $\cos(v-mv)$ becomes

$$\cos\{(v+90^\circ)(1-m)\} = -\sin(v-mv);$$

hence also the expansion of sin $(v - v')$ may be obtained from the expression (219), and generally the developement of sin $i(v - v')$ may be derived from that of cos $i(v - v')$.

Thus all the quantities in the equations of the moon's motions in article 695 are determined, except the variation δu, $\delta u'$, $\delta v'$, and δs.

706. It is evident from the value of $\frac{d^2u}{dv^2} + u$ in equation (209), that u is a function of the cosines of all the angles contained in the products of the developements of u, u', cos $(v - v')$ cos $(2v - 2v')$ &c.; and δu, being the part of u arising from the disturbing action of the sun, must be a function of the same quantities: hence if A_0, A_1, A_2, &c. be indeterminate coefficients, it may be assumed, that

$$\begin{aligned}
a\delta u = {} & A_0 \cos(2v-2mv) \qquad (220)\\
& + A_1\, e.\cos(2v-2mv-cv+\varpi)\\
& + A_2\, e.\cos(2v-2mv+cv-\varpi)\\
& + A_3\, e'.\cos(2v-2mv+c'mv-\varpi')\\
& + A_4\, e'.\cos(2v-2mv-c'mv+\varpi')\\
& + A_5\, e'.\cos(c'mv-\varpi')\\
& + A_6\, ee'.\cos(2v-2mv-cv+c'mv+\varpi-\varpi')\\
& + A_7\, ee' \cos(2v-2mv-cv-c'mv+\varpi+\varpi')\\
& + A_8\, ee'.\cos(cv+c'mv-\varpi-\varpi')\\
& + A_9\, ee'.\cos(cv-c'mv-\varpi+\varpi')\\
& + A_{10}\, e^2.\cos(2cv-2\varpi)\\
& + A_{11}\, e^2.\cos(2cv-2v+2mv-2\varpi)\\
& + A_{12}\, \gamma^2.\cos(2gv-2\theta)\\
& + A_{13}\, \gamma^2 \cos(2gv-2v+2mv-2\theta)\\
& + A_{14}\, e'^2 \cos(2c'mv-2\varpi')\\
& + A_{15}\, e\gamma^2 \cos(2gv-cv-2\theta+\varpi)\\
& + A_{16}\, e\gamma^2 \cos(2v-2mv-2gv+cv+2\theta-\varpi)\\
& + A_{17}\, \frac{a}{a'} \cos(v-mv)\\
& + A_{18}\, \frac{a}{a'}\, e'.\cos(v-mv+c'mv-\varpi')\\
& + A_{19}\, \frac{a}{a'}\, e'.\cos(v-mv-c'mv+\varpi')\\
& + A_{20}\, \frac{a}{a'} \cos(3v-3mv).
\end{aligned}$$

The term depending on cos $(cv - \varpi)$ which arises from the disturbing action of the sun is omitted, because it has already been included in the value of u.

707. It is evident from equation (210) that δs, the variation of the tangent of the latitude, can only have the form

$$\begin{aligned}
\delta s = & B_0 \ \gamma \sin (2v-2mv-gv+\theta) \qquad (221)\\
& + B_1 \ \gamma \sin (2v-2mv+gv-\theta)\\
& + B_2 \ e\gamma \sin (gv+cv-\theta-\varpi)\\
& + B_3 \ e\gamma \sin (gv-cv-\theta+\varpi)\\
& + B_4 \ e\gamma \sin (2v-2mv-gv+cv+\theta-\varpi)\\
& + B_5 \ e\gamma \sin (2v-2mv+gv-cv-\theta+\varpi)\\
& + B_6 \ e\gamma \sin (2v-2mv-gv-cv+\theta+\varpi)\\
& + B_7 \ e'\gamma \sin (gv+c'mv-\theta-\varpi')\\
& + B_8 \ e'\gamma \sin (gv-c'mv-\theta+\varpi')\\
& + B_9 \ e'\gamma \sin (2v-2mv-gv+c'mv+\theta-\varpi')\\
& + B_{10} \ e'\gamma \sin (2v-2mv+gv-c'mv-\theta+\varpi')\\
& + B_{11} \ e^2\gamma \sin (2cv-gv-2\varpi+\theta)\\
& + B_{12} \ e^2\gamma \sin (2v-2mv-2cv+gv+2\varpi-\theta)\\
& + B_{13} \ e^2\gamma \sin (2cv+gv-2v+2mv-2\varpi-\theta)\\
& + B_{14} \frac{a}{a'} \gamma \sin (gv-v+mv-\theta)\\
& + B_{15} \frac{a}{a'} \gamma \sin (gv+v-mv-\theta),
\end{aligned}$$

B_0, B_1, &c. being indeterminate coefficients.

708. The variation in the longitude of the earth from the action of the planets troubles the motion of the moon. Equation (216), when $\delta(nt+\epsilon)$ is put for δv, gives

$$\delta v' = m\delta(nt+\epsilon)\{1+2e' \cos (c'mv-\varpi') - \tfrac{5}{2}e'^2 \cos (2c'mv-2\varpi')\} \qquad (222)$$

But δv or $\delta(nt+\epsilon)$, arising from the disturbing force, is entirely independent of equation (213), which belongs to the elliptical motion only; and from equation (211) it appears that if C_6, C_9, &c. be indeterminate coefficients,

$$\begin{aligned}
\delta(nt+\epsilon) = & C_6 \sin (2v-2mv) \qquad (223)\\
& + C_9 \ e' \sin (2v-2mv+c'mv-\varpi')\\
& + C_{10} \ e' \sin (2v-2mv-c'mv+\varpi')\\
& + \quad \&c. \qquad \&c.
\end{aligned}$$

By this value, equation (222) becomes

$$\delta v' = m\{C_6 + C_9\, e'^2 + C_{10}\, e'\} \sin(2v - 2mv) \qquad (224)$$
$$+ \qquad \&c. \qquad \&c.$$

709. But the longitude of the earth is troubled by the action of the moon as well as by that of the planets, and thus the moon indirectly troubles her own motions. In the theory of the earth it is found that the action of the moon occasions the inequality

$$\delta v' = \mu\, \frac{r}{r'} \sin(v - v')$$

in the earth's longitude, and thus the whole variation of v' is

$$\delta v' = m\{C_6 + C_9\, e'^2 + C_{10}\, e'^2\} \sin(2v - 2mv) \qquad (225)$$
$$+ \mu\, \frac{u'}{u} \sin(v - v');$$

where μ is the ratio of the mass of the moon to the sum of the masses of the earth and moon.

710. The parallax of the moon is troubled by both these causes, but that arising from the action of the planets may be omitted at present. The moon's attraction produces the inequality

$$\delta r' = \mu r \cos(v - v')$$

in the radius vector of the earth, and consequently the variation

$$\delta u' = -\frac{\mu u'^2}{u} \cos(v - v') \qquad (226)$$

in the solar parallax.

711. Lastly, $\frac{du}{dv}$ is obtained from equation (214).

712. Thus every quantity in the equation of article 695 are determined, and by their substitution, the co-ordinates of the moon will be obtained in her troubled orbit in functions of her true longitude.

The Parallax.

713. The substitution of the given quantities in the differential equation (209) of the parallax is extremely simple, though tedious. The first term

$$-\frac{1}{h^2(1+s^2)^{\frac{3}{2}}} = -\frac{1}{h^2}(1 - \tfrac{3}{2}s^2)$$

when the higher powers of s^2 are omitted; putting

$$\frac{1}{\bar{a}}(1 + e^2 + \gamma^2 + \zeta) \text{ for } h^{-2}$$

and $$\tfrac{1}{2}\gamma^2 - \tfrac{1}{2}\gamma^2 \cos(2gv - 2\theta)$$

for s^2 becomes

$$-\frac{1}{h^2(1+s^2)^{\frac{3}{2}}} = -\frac{1}{\bar{a}}\{1+e^2+\tfrac{1}{4}\gamma^2+\zeta+\tfrac{3}{4}\gamma^2(1+e^2-\tfrac{1}{4}\gamma^2)\cos(2gv-2\theta)\}.$$

Again, $$u'^3 = \frac{1}{a'^3}\{1 + \tfrac{3}{2}e'^2 + 3e'\cos(c'mv - \varpi') + \&c.\}$$

$$u^{-3} = a^3\{1 - \tfrac{3}{4}\gamma^2 - 3e\cos(cv - \varpi) + \&c.\};$$

and as by article 685, $$\frac{m'a^3}{a'^3} = \bar{m}^2$$

$$\frac{m'u'^3}{2h^2u^3} = \frac{\bar{m}^2}{2\bar{a}}\{1 + e^2 + \tfrac{1}{4}\gamma^2 + \tfrac{3}{2}e'^2$$
$$- 3e(1 + \tfrac{1}{2}e^2 + \tfrac{3}{2}e'^2)\cos(cv - \varpi) + \&c.\}$$

In this and all the other terms, ζ is omitted, being of the fourth order in e and γ.

714. Terms of the form $\frac{9m'u'^4}{8h^2u^4}\cos(v - v')$ become

$$\frac{9\bar{m}^2}{8\bar{a}}\cdot\frac{a}{a'}(1 + 2e^2 + 2e'^2)\cos(v - mv)$$
$$+ \frac{9\bar{m}^2}{8\bar{a}}\cdot\frac{a}{a'}e'\cos(v - mv + c'mv - \varpi')$$
$$+ \frac{27\bar{m}^2}{8\bar{a}}\cdot\frac{a}{a'}e'\cos(v - mv - c'mv + \varpi');$$

and, by comparing their coefficients with observation, serve for the determination of $\frac{a}{a'}$, the ratio of the parallax of the sun to that of the moon; but as it is a very small quantity, any error would be sensible, and on that account the approximation must extend to quantities of the fifth order inclusively with regard to the angle $v - v'$; but in every other case, it will only be carried to quantities of the third order.

715. Attending to these circumstances, and observing that in the variation of $\frac{1}{h^2(1+s^2)^{\frac{3}{2}}}$ the square of δs must be included, so that

$$\delta\frac{1}{h^2(1+s^2)^{\frac{3}{2}}} = -\frac{3s\delta s}{h^2} + \frac{3}{2\bar{a}}\delta s^2$$

and as $\frac{\overline{m}^2}{\bar{a}} = \frac{m^2}{a}$,

it will readily be found, that

$$
\begin{aligned}
0 = \frac{d^2u}{dv^2} &+ u \qquad\qquad\qquad\qquad\qquad\qquad\qquad\qquad\qquad\qquad\qquad\qquad (227) \\
&- b_0 \\
&- b_1\ e \cos(cv - \varpi) \\
&+ b_2 \cos(2v - 2mv) \\
&+ b_3\ e \cos(2v - 2mv - cv + \varpi) \\
&- b_4\ e \cos(2v - 2mv + cv - \varpi) \\
&- b_5\ e' \cos(2v - 2mv + c'mv - \varpi') \\
&+ b_6\ e' \cos(2v - 2mv - c'mv + \varpi') \\
&+ b_7\ e' \cos(c'mv - \varpi') \\
&+ b_8\ ee' \cos(2v - 2mv - cv + c'mv + \varpi - \varpi') \\
&- b_9\ ee' \cos(2v - 2mv - cv - c'mv + \varpi + \varpi') \\
&- b_{10}\ ee' \cos(cv + c'mv - \varpi - \varpi') \\
&- b_{11}\ ee' \cos(cv - c'mv - \varpi + \varpi') \\
&+ b_{12}\ e^2 \cos(2cv - 2\varpi) \\
&+ b_{13}\ e^2 \cos(2cv - 2v + 2mv - 2\varpi) \\
&- b_{14}\ \gamma^2 \cos(2gv - 2\theta) \\
&+ b_{15}\ \gamma^2 \cos(2gv - 2v + 2mv - 2\theta) \\
&+ b_{16}\ e'^2 \cos(2c'mv - 2\varpi') \\
&- b_{17}\ e\gamma^2 \cos(2gv - cv - 2\theta + \varpi) \\
&- b_{18}\ e\gamma^2 \cos(2v - 2mv - 2gv + cv + 2\theta - \varpi) \\
&+ b_{19} \frac{a}{a'} \cos(v - mv) \\
&+ b_{20}\ e' \frac{a}{a'} \cos(v - mv + c'mv - \varpi') \\
&+ b_{21}\ e' \frac{a}{a'} \cos(v - mv - c'mv + \varpi').
\end{aligned}
$$

716. The coefficients being

$$
\begin{aligned}
b_0 = \frac{1}{\bar{a}} \{1 + e^2 + \tfrac{1}{4}\gamma^2 + \text{C}\} &- \frac{\overline{m}^2}{2\bar{a}} \{1 + e^2 + \tfrac{1}{4}\gamma^2 + \tfrac{3}{2}e'^2\} \\
&+ \frac{3\overline{m}^2}{4\bar{a}} (4 - 3m - m^2)\, A_0 . (1 - \tfrac{3}{2}e'^2) - \frac{3}{4\bar{a}} B_0^2 . \gamma^2
\end{aligned}
$$

$$b_1 = \frac{3m^2}{4a}\left\{\begin{array}{l} 2+e^2+3e'^2-2(B_2+B_3)\dfrac{\gamma^2}{m^2}+(1+2m-c)A_0(1-\frac{5}{2}e'^2) \\ -4\left\{1+2m+(4(1-m)^2-1)\left(\dfrac{1+m}{2-2m-c}+\dfrac{1-m}{2-2m+c}\right)\right\} \\ \qquad\times A_0(1-\frac{5}{2}e'^2) \\ +\dfrac{1}{1-m}\{(1+6m+c)(1-m)+7+(2-2m-c)^2\}A_1(1-\frac{5}{2}e'^2) \\ -\frac{1}{2}(9+m+c)A_6 . e'^2+\frac{7}{2}(9+3m+c)A_7e'^2 \\ +3(A_8+A_9) . e'^2 \end{array}\right\}$$

$$b_2 = \frac{3m^2}{2a}\left\{\begin{array}{l} 1+(1+2m)e^2+\frac{1}{4}\gamma^2-\frac{5}{2}e'^2+\dfrac{1}{1-m}(1+3e^2+\frac{1}{4}\gamma^2-\frac{5}{2}e'^2) \\ -A_0-(B_0-B_1)\dfrac{\gamma^2}{m^2} \end{array}\right\}$$

$$b_3 = \frac{3m^2}{a}\left\{\begin{array}{l} \frac{1}{4}c\{1+\dfrac{e^2}{4}(2-19m)-\frac{5}{2}e'^2\} \\ -\frac{1}{4}(3+4m)(1+\frac{1}{2}e^2-\frac{5}{2}e'^2)+\dfrac{1-c^2}{4(1-m)} \\ -\dfrac{2(1+m)}{2-2m-c}(1+\frac{7}{4}e^2-\frac{5}{2}e'^2) \\ -\frac{1}{2}(A_1-2A_0)+\frac{1}{2}(B_5-B_6)\dfrac{\gamma^2}{m^2} \end{array}\right\}$$

$$b_4 = \frac{3m^2}{4a}\left\{3+c-4m+\frac{8(1-m)}{2-2m+c}+2A_2\right\}$$

$$b_5 = \frac{3m^2}{4a}\left\{\frac{4-m}{2-m}+2B_9\frac{\gamma^2}{m^2}+2A_3\right\}$$

$$b_6 = \frac{3m^2}{4a}\left\{\frac{7(4-3m)}{2-3m}-2B_{10}\frac{\gamma^2}{m^2}-2A_4\right\}$$

$$b_7 = \frac{3m^2}{2a}\left\{\begin{array}{l} 1+e^2+\frac{1}{4}\gamma^2+\frac{9}{8}e'^2+(B_7+B_8)\dfrac{\gamma^2}{m^2}-\frac{3}{2}(1+2m)A_0 \\ -\dfrac{2(1-2m)(3-2m)(3-m)}{(2-3m)(2-m)}A_0-2A_3-(2-3m)A_4 \\ +(B_9+B_{10})B_0\dfrac{\gamma^2}{m^2}-A_5-11C_6-2C_9+2C_{10} \end{array}\right\}$$

$$+\frac{6m'}{a}\{4A_0+A_3-A_4-10A_1e^2+\tfrac{5}{2}(A_7-A_6)e^2\}$$

$$b_8 = \frac{3m^2}{2a}\left\{\frac{3+2m-c}{4} + \frac{(2+m)}{2-m-c} - \tfrac{3}{2}A_1 - A_6 - \left(\frac{3+m-c}{2} + \frac{4}{2-m-c}\right) A_9\right\}$$

$$b_9 = \frac{3m^2}{2a}\left\{\frac{7(3+6m-c)}{4} + \frac{7(2+3m)}{2-3m-c} + \tfrac{3}{2}A_1 + A_7 + \left(\frac{3-m-c}{2} + \frac{4}{2-3m-c}\right) A_8\right\}$$

$$b_{10} = \frac{3m^2}{2a}\left\{\frac{3+2m}{2} - \left(\frac{1+2m+c}{4} + \frac{2}{c+m}\right) A_1 + A_8 + \left(\frac{1+3m+c}{2} + \frac{4}{c+m}\right) A_7\right\}$$

$$b_{11} = \frac{3m^2}{2a}\left\{\frac{3-2m}{2} + A_9 + 7\left(\frac{1+2m+c}{4} + \frac{2}{c-m}\right) A_1 + \left(\frac{1+m+c}{2} + \frac{4}{c-m}\right) A_6\right\}$$

$$b_{12} = \frac{3m^2}{2a}\left\{1 - B_{11} \cdot \frac{\gamma^2}{m^2} - A_{10}\right\}$$

$$b_{13} = \frac{3m^2}{4a}\left\{\begin{array}{l} \dfrac{2+11m+8m^2}{2} - \dfrac{(10+19m+8m^2)}{2c-2+2m} + 4A_1 \\ + \dfrac{\{8A_{10}+10A_1^2\}}{2c-2+2m} - 2A_{11} \end{array}\right\}$$

$$b_{14} = \frac{3}{4a}\left\{1 + e^2 - \tfrac{1}{4}\gamma^2 - \tfrac{1}{2}m^2 + 2m^2 A_{12}\right\}$$

$$b_{15} = \frac{3m^2}{4a}\left\{\begin{array}{l} \dfrac{1+2m-2g}{4} + \dfrac{(4g^2-1)}{4(1-m)} - \dfrac{(2+m)}{2g-2+2m} \\ + \dfrac{2B_0}{m^2} - 2A_{13} + \dfrac{8A_{12}}{2g-2+2m} \end{array}\right\}$$

$$b_{16} = \frac{3m^2}{2a}\left(\tfrac{3}{2} - A_{14}\right)$$

$$b_{17} = \frac{3m^2}{2a}\left\{\begin{array}{l} \tfrac{1}{2} + \dfrac{B_3}{m^2} + \dfrac{(1+c-2g-10m)}{4} A_1 - (10+5m) A_{13} \\ + (5+m) A_{16} - \dfrac{1}{m^2} B_0 \cdot B_5 + A_{15} \end{array}\right\}$$

$$b_{18} = \frac{3m^2}{4a}\left\{1 + 2m + \frac{(5+m)}{1-2m} - \frac{3(1-m)}{3-2m} + 2A_{16} - \frac{2}{m^2} B_4 + \frac{10}{1-2m} A_{15}\right\}$$

$$b_{19} = \frac{m^2}{a}\left\{\begin{array}{l} \{\tfrac{9}{8}(1-2\mu)(1+2e^2+2e'^2) + \dfrac{3(1-2\mu)(1+\frac{9}{2}e^2+2e'^2)}{4(1-m)} \\ \quad + \dfrac{3(1+m)}{2(1-m)} \times A_{18} e'^2 \\ - \dfrac{(36+21m-15m^2)}{4(1-m)} A_{17} - \dfrac{(57-38m)}{4(1-m)} A_0 + \tfrac{3}{2}(B_{14}+B_{15})\dfrac{\gamma^2}{m^2} \end{array}\right\}$$

$$b_{20} = \frac{3m^2}{2a}\left\{\frac{5(1-2\mu)}{4} - A_{16} + \frac{(4+m)}{4}A_{17} - (5+m)A_{19}\right\}$$

$$b_{21} = \frac{3m^2}{2a(1-2m)}\left\{\frac{15-18m}{4}(1-2\mu) - \frac{(76-33m)}{4}A_{17} - 5A_{18} - (1-2m)A_{19}\right\}.$$

717. The integral of the preceding equation is evidently

$$u = \frac{1}{a}\{1 + e^2 + \tfrac{1}{4}\gamma^2 + \mathfrak{C} + e(1+e^2)\cos(cv - \varpi) \quad (228)$$

$$- \tfrac{1}{4}\gamma^2(1 + e^2 - \tfrac{1}{4}\gamma^2)\cos(2gv - 2\theta)\} + \delta u.$$

Where δu is given by equation (220.)

718. In order to find values of the indeterminate coefficients A_0, A_1, &c., this value of u must be substituted in equation (227); but to determine the unknown quantity c, both e and ϖ must vary in the term $e(1+e^2)\cos(cv - \varpi)$, which expresses the motion of the perigee.

Hence, when $\frac{d\frac{e}{a}}{dv^2}$ is omitted,

a comparison of the coefficients of corresponding sines and cosines gives

$$0 = 1 + e^2 + \tfrac{1}{4}\gamma^2 + \mathfrak{C} - ab_0 \quad (229)$$

$$0 = 1 - \left(c - \frac{d\varpi}{dv}\right)^2 - \frac{ab_1}{1+e^2}$$

$$0 = \frac{e(1+e^2)}{a}\cdot\frac{d^2\varpi}{dv^2} - 2\left(c - \frac{d\varpi}{dv}\right)\frac{d\cdot e\frac{(1+e^2)}{a}}{dv}$$

$$0 = A_0(1 - 4(1-m)^2) + ab_2$$

$$0 = A_1(1 - (2-2m-c)^2) + ab_3$$

$$0 = A_2(1 - (2-2m+c)^2) - ab_4$$

$$0 = A_3(1 - (2-m)^2) - ab_5$$

$$0 = A_4(1 - (2-3m)^2) + ab_6$$

$$0 = A_5(1 - m^2) + ab_7$$

$$0 = A_6(1 - (2-m-c)^2) + ab_8$$

$$0 = A_7(1 - (2-3m-c)^2) - ab_9$$

$$0 = A_8(1 - (c+m)^2) - ab_{10}$$

$$0 = A_9(1 - (c-m)^2) - ab_{11}$$

$$0 = A_{10}(1 - 4c^2) + ab_{12}$$

$$0 = A_{11}(1 - (2c-2+2m)^2) + ab_{13}$$

$$0 = A_{12}(1 - 4g^2) + ab_{14}$$

$$0 = A_{13}(1 - (2g - 2 + 2m)^2) + ab_{15}$$
$$0 = A_{14}(1 - 4m^2) + ab_{16}$$
$$0 = A_{15}(1 - (2g - c)^2) - ab_{17}$$
$$0 = A_{16}(1 - (2 - 2m - 2g + c)^2) - ab_{18}$$
$$0 = A_{17}(1 - (1 - m)^2) + ab_{19}$$
$$0 = ab_{20}$$
$$0 = A_{19}(1 - (1 - 2m)^2) + ab_{21}$$
$$0 = A_{20}(1 - (3 - 3m)^2).$$

719. The secular inequalities in the form of the lunar orbit are derived from the three first of these equations; from the rest are obtained values of the indeterminate coefficients A_0, A_1, &c. &c. It is evident that these coefficients will be more correct, the farther the approximation is carried in the developement of equation (209).

Secular Inequalities in the Lunar Orbit.

720. When the action of the sun is omitted, by article 685, $\dfrac{1}{a} = \dfrac{1}{\bar{a}}$; and $\mathfrak{C}$, being of the fourth order, may be omitted: hence $1 + e^2 + \frac{1}{4}\gamma^2 - ab_0 = 0$ becomes

$$\frac{1}{a} = \frac{1}{\bar{a}} - \frac{\overline{m}^2}{2\bar{a}}(1 + \tfrac{3}{2}e'^2) + \frac{3\overline{m}^2}{4\bar{a}}(1 - \tfrac{5}{2}e'^2)(4 - 3m - m^2)A_0$$
$$- \frac{3}{4\bar{a}}B_0^2\gamma^2. \qquad (230)$$

Since a is the mean distance of the moon from the earth, or half the greater axis of the lunar orbit, the constant part of the moon's parallax is proportional to $\dfrac{1}{a}$. But the action of the planets produces a secular variation in e', the eccentricity of the terrestrial orbit, without affecting $2a'$, the greater axis. The preceding value of $\dfrac{1}{a}$ must therefore be subject to a secular inequality, in consequence of the variation of the term $-\dfrac{3\overline{m}^2}{4\bar{a}}e'^2$; but this variation will always be insensible.

721. The motion of the perigee may be obtained from the second of equations (229), put under the form

$$1 - \left(c - \frac{d\varpi}{dv}\right)^2 - p - p'e'^2 = 0;$$

for since b_1 is a function of e'^2, the quantity $\frac{ab_1}{1+e^2}$ may be expressed by $p + p'e'^2$.

If $\frac{d\varpi}{dv}$ be omitted, $c = \sqrt{1 - p - p'.e'^2}$, so that c varies in consequence of e'^2.

Now,
$$\frac{d\varpi}{dv} = c - \sqrt{1-p} + \frac{p'e'^2}{2\sqrt{1-p}},$$
the integral of which is
$$\varpi = cv - v\sqrt{1-p} + \frac{p'}{2\sqrt{1-p}}\int e'^2 dv + \epsilon;$$
for e'^2 is variable, and p, p' may be regarded as constant, without sensible error, as appears from the value of b_1, and ϵ is a constant quantity, introduced by integration; hence
$$\cos(cv - \varpi) = \cos\left\{v\sqrt{1-p} - \frac{p'}{2\sqrt{1-p}}\int e'^2 dv - \epsilon\right\}. \quad (231)$$

722. Thus, from theory, we learn that the perigee has a motion equal to
$$(1 - \sqrt{1-p})v + \frac{p'}{2\sqrt{1-p}}\int e'^2 dv,$$
which is confirmed by observation; but this motion is subject to a secular inequality, expressed by
$$\frac{p'}{2\sqrt{1-p}}\int e'^2 dv, \quad (232)$$
on account of the variation in e'^2, the eccentricity of the earth's orbit.

In consequence of the preceding value of c, ϖ is equal to the constant quantity ϵ, together with the secular equation of the motion of the perigee.

723. The eccentricity of the moon's orbit is affected by a secular variation similar to that in the parallax, and proportional to $\frac{d\varpi}{dv}$, but as the variations of the latter are only sensible in the integral $\int \frac{d\varpi}{dv} dv$, the eccentricity of the lunar orbit may be regarded as constant.

Latitude of the Moon.

724. The developement of the parallax will greatly assist in that of the latitude, as most of the terms differ only in being multiplied by s, its variation, or its differentials; and by substitution of the requisite quantities in equation (210), it will readily be found, when all the powers of the eccentricities and inclination above the cubes are omitted, that

$$0 = \frac{d^2s}{dv^2} + s \tag{233}$$

$$+ a_0\,\gamma \sin(gv - \theta)$$
$$- a_1\,\gamma \sin(2v - 2mv - gv + \theta)$$
$$+ a_2\,\gamma \sin(2v - 2mv + gv - \theta)$$
$$+ a_3\,e\gamma \sin(gv + cv - \theta - \varpi)$$
$$+ a_4\,e\gamma \sin(gv - cv - \theta + \varpi)$$
$$+ a_5\,e\gamma \sin(2v - 2mv - gv + cv + \theta - \varpi)$$
$$+ a_6\,e\gamma \sin(2v - 2mv + gv - cv - \theta + \varpi)$$
$$+ a_7\,e\gamma \sin(2v - 2mv - gv - cv + \theta + \varpi)$$
$$+ a_8\,e'\gamma \sin(gv + c'mv - \theta - \varpi')$$
$$+ a_9\,e'\gamma \sin(gv - c'mv - \theta + \varpi')$$
$$+ a_{10}\,e'\gamma \sin(2v - 2mv - gv + c'mv + \theta - \varpi')$$
$$+ a_{11}\,e'\gamma \sin(2v - 2mv - gv - c'mv + \theta + \varpi')$$
$$+ a_{12}\,e^2\gamma \sin(2cv - gv - 2\theta + \varpi)$$
$$+ a_{13}\,e^2\gamma \sin(2v - 2mv - 2cv + gv + 2\varpi - \theta)$$
$$+ a_{14}\,e^2\gamma \sin(2cv + gv - 2v + 2mv - 2\varpi - \theta)$$
$$+ a_{15}\,\frac{a}{a'}\,\gamma \sin(gv - v + mv - \theta)$$
$$+ a_{16}\,\frac{a}{a'}\,\gamma \sin(gv + v - mv - \theta).$$

725. The coefficients in consequence of $\frac{\bar{m}^2}{\bar{a}} = \frac{m^2}{a}$ being

$$a_0 = \frac{3m^2}{2}\left\{1 + 2e^2 - \tfrac{1}{4}\gamma^2 + \tfrac{3}{2}e'^2\right.$$
$$- \tfrac{1}{2}(1 - \tfrac{5}{2}e'^2)\left(\frac{(3 - 2m - g)(g + m)}{1 - m}B_0 + 4A_0\right)$$
$$- \tfrac{7}{2}(3 - 3m - g)B_{10}\,e'^2 + \tfrac{1}{4}(3 - m - g)B_9\,e'^2$$
$$\left. + \tfrac{3}{2}(B_7 + B_8)\,e^2\right\}$$

$$a_1 = \frac{3m^2}{4} \ \{(1+g)(1+2e^2 - \tfrac{1}{4}(2+m)\gamma^2 - \tfrac{5}{2}e'^2)$$

$$+ \frac{(1-g^2)}{1-m} - 4A_0 + 10A_1e^2 - 2B_0\}$$

$$a_2 = \frac{3m^2}{2} \cdot \left\{\frac{1-g}{2} + B_1\right\}$$

$$a_3 = \frac{3m^2}{2} \cdot \{B_2 - 2 + (1-m)(3-2m-g)B_0\}$$

$$a_4 = \frac{3m^2}{2} \cdot \{B_3 - 2 - 2A_1 + (1+m)(3-2m-g)B_0\}$$

$$a_5 = \frac{3m^2}{2} \cdot \{(1+g)(1-m) - 2B_0 + B_4\}$$

$$a_6 = \frac{3m^2}{2} \cdot \{(g-1)(1+m) + B_5 - 2A_1\}$$

$$a_7 = \frac{3m^2}{2} \cdot \{(1+g)(1+m) + B_6 + 2A_1 - 2B_0\}$$

$$a_8 = \frac{3m^2}{4} \cdot \{3 + 2B_7 + \tfrac{1}{2}(3-2m-g)B_0 - (3-3m-g)B_{10}\}$$

$$a_9 = \frac{3m^2}{4} \ \{3 + 2B_8 - \tfrac{7}{2}(3-2m-g)B_0 - (3-m-g)B_9\}$$

$$a_{10} = \frac{3m^2}{4} \ \left\{\frac{1+g}{2} + 2B_9 + 3B_0 - (1+g-m)B_8\right\}$$

$$a_{11} = \frac{3m^2}{4} \ \{2B_{10} - \tfrac{7}{2}(1+g) + 3B_0 - (1+g+m)B_7$$

$$a_{12} = \frac{3m^2}{4} \ \{2B_{11} - 5 - 10A_1 + 4A_{11} - (3-2m-2c+2g)B_{12}$$

$$+ (10+19m+8m^2)B_0\left(\frac{3-3m-g}{4} + \frac{(2-2m-g)^2-1}{2(2c+2m-2)}\right)\Bigg\}$$

$$a_{13} = \frac{3m^2}{4} \ \{2B_{12} + (1-g) \cdot \tfrac{1}{4}(10+19m+8m^2) + 10A_1$$

$$- 4A_{11} - 2B_{11}\}$$

$$a_{14} = \frac{3m^2}{4} \ \{\tfrac{1}{2}(10+19m+8m^2) + 2B_{13} + 10A_1 - 4A_{11} - 5B_0\}$$

$$a_{15} = \frac{3m^2}{4} \ \{3 + 2B_{14}\}$$

$$a_{16} = \frac{3m^2}{4} \ \{\tfrac{5}{2} + 2B_{15}\}.$$

726. The integral of the differential equation of the latitude is

$$s = \gamma \sin(gv - \theta) + \delta s; \ \delta s \text{ is given in } (221). \qquad (234)$$

If this quantity be substituted in equation (233) instead of s, a comparison of the coefficients of like sines and cosines will furnish a sufficient number of equations; whence the indeterminate coefficients B_0, B_1, &c. will be known, but in order to find a value of the unknown quantity g, both θ and γ must vary in the terms $\gamma \sin(gv-\theta)$ in taking the differentials of s. Attending to these circumstances, it will readily be found that,

$$2\frac{d\gamma}{dv}\left(g - \frac{d\theta}{dv}\right) - \gamma\frac{d^2\theta}{dv^2} = 0 \qquad (234)$$

$$\gamma - \gamma\left(g - \frac{d\theta}{dv}\right)^2 + \frac{d^2\gamma}{dv^2} + \gamma a_0 = 0$$

$$B_0\,(1 - (2 - 2m - g)^2) - a_1 = 0$$
$$B_1\,(1 - (2 - 2m + g)^2) + a_2 = 0$$
$$B_2\,(1 - (g + c)^2) + a_3 = 0$$
$$B_3\,(1 - (g - c)^2) + a_4 = 0$$
$$B_4\,(1 - (2 - 2m + c - g)^2) + a_5 = 0$$
$$B_5\,(1 - (2 - 2m - c + g)^2) + a_6 = 0$$
$$B_6\,(1 - (2 - 2m - c - g)^2) + a_7 = 0$$
$$B_7\,(1 - (g + m)^2) + a_8 = 0$$
$$B_0\,(1 - (g - m)^2) + a_9 = 0$$
$$B_9\,(1 - (2 - m - g)^2) + a_{10} = 0$$
$$B_{10}\,(1 - (2 - 3m - g)^2) + a_{11} = 0$$
$$B_{11}\,(1 - (2c - g)^2) + a_{12} = 0$$
$$B_{12}\,(1 - (2 - 2m - 2c + g)^2) + a_{13} = 0$$
$$B_{13}\,(1 - (2c + g - 2 + 2m)^2) + a_{14} = 0$$
$$B_{14}\,(1 - (g - 1 + m)^2) + a_{15} = 0$$
$$B_{15}\,(1 - (g + 1 - m)^2) + a_{16} = 0.$$

The two first of these equations will give the secular variations in the nodes and inclination of the orbit, the rest serve for the determination of the coefficients B_0, B_1, &c.

Secular Inequalities in the Position of the Lunar Orbit.

727. The coefficient a_0 may be represented by $q + q'e'^2$, then the second of the equations in the last article becomes

$$\frac{d^2\gamma}{dv^2} + \gamma\left(1 - \left(g - \frac{d\theta}{dv}\right)^2\right) + \gamma\,(q + q'e'^2) = 0;$$

q' is a function of A_0 and B_0; and as these are functions of $1 - \frac{3}{2}e'^2$, therefore $q'e'^2$ may be omitted, as well as $\frac{d^2\gamma}{dv^2}$, which is insensible, and neglecting $\frac{d\theta}{dv}$ in the first instance,

$$g = \sqrt{1 - q - q'e'^2},$$

so that g varies with e'^2.

But $$\frac{d\theta}{dv} = g - \sqrt{1+q} - \frac{q'}{2\sqrt{1+q}}\,e'^2;$$

and as q and q' may be regarded as constant, the integral is

$$\theta = gv - v\sqrt{1+q} - \frac{q'}{2\sqrt{1+q}}\int e'^2 dv + \alpha,$$

α being a constant quantity introduced by integration; hence

$$\sin(gv - \theta) = \sin\left\{v\sqrt{1+q} + \frac{q'}{2\sqrt{1+q}}\int e'^2 dv - \alpha\right\}, \quad (235)$$

which shows the nodes of the lunar orbit to have a retrograde motion on the true ecliptic equal to

$$(\sqrt{1+q} - 1)\,v + \frac{q'}{2\sqrt{1+q}}\int e'^2 dv,$$

which accords with observation. This motion is not uniform, but is affected by a secular inequality expressed by

$$\frac{q'}{2\sqrt{1+q}}\int e'^2 dv, \quad (236)$$

corresponding to the secular variation of e', the eccentricity of the terrestrial orbit.

728. The first of the equations (234) determines the inclination of the lunar orbit on the plane of the ecliptic. Its integral is

$$\gamma = \left\{H\left(g - \frac{d\theta}{dv}\right)\right\}^{-\frac{1}{2}}.$$

H being an arbitrary constant quantity.

Hence it appears that the inclination is subject to a secular inequality; but as it is quite insensible, the inclination γ may be regarded as constant, which is the reason why the most ancient observations do not indicate any change in the inclination of the lunar orbit on the plane of the ecliptic, although the position of the ecliptic has varied sensibly during that interval.

The Mean Longitude of the Moon.

729. When the square root is extracted, equation (211) becomes

$$dt = \frac{dv}{h^2(u^2 + 2u\delta u + \delta u^2)}\{1 - \frac{3m}{h^2}\int\frac{u'^3dv}{u^4}\sin(2v - 2v')$$
$$+ \frac{3m}{2h^4}\left(\int\frac{u^3dv}{u^4}\sin(2v - 2v')\right)^2 - \&c.\};$$

and, making the necessary substitutions, there will result

$$\begin{aligned}
dt = \frac{a^2dv}{\sqrt{\bar{a}}}\{&x_0 + x_1 e\cos(cv - \varpi)\\
&+ x_2 e^2\cos(2cv - 2\varpi) \qquad\qquad (237)\\
&+ x_3 e^3\cos(3cv - 3\varpi)\\
&+ x_4 \gamma^2\cos(2gv - 2\theta)\\
&+ x_5 e\gamma^2\cos(2gv - cv - 2\theta + \varpi)\\
&+ x_6 e\gamma^2\cos(2gv + cv - 2\theta - \varpi)\\
&+ x_7 \cos(2v - 2mv)\\
&+ x_8 e\cos(2v - 2mv - cv + \varpi)\\
&+ x_9 e\cos(2v - 2mv + cv - \varpi)\\
&+ x_{10} e'\cos(2v - 2mv + c'mv - \varpi')\\
&+ x_{11} e'\cos(2v - 2mv - c'mv + \varpi')\\
&+ x_{12} e'\cos(c'mv - \varpi')\\
&+ x_{13} ee'\cos(2v - 2mv - cv + c'mv + \varpi - \varpi')\\
&+ x_{14} ee'\cos(2v - 2mv - cv - cmv + \varpi + \varpi')\\
&+ x_{15} ee'\cos(cv + c'mv - \varpi - \varpi')\\
&+ x_{16} ee'\cos(cv - c'mv - \varpi + \varpi')\\
&+ x_{17} e^2\cos(2cv - 2v + 2mv - 2\varpi)\\
&+ x_{18} \gamma^2\cos(2gv - 2v + 2mv - 2\theta)\\
&+ x_{19} e'^2\cos(2c'mv - 2\varpi')
\end{aligned}$$

$$+ x_{20} \frac{a}{a'} \cos(v - mv)$$

$$+ x_{21} \frac{a}{a'} e' \cos(v - mv + c'mv - \varpi').\}$$

730. The coefficients of which are

$$x_0 = 1 + \frac{27m^4}{64(1-m)^2} + \frac{3m^2 \,.\, A_0}{4(1-m)} + \tfrac{3}{2}\{A_2^2 + A_1^2 e^2\}$$

$$x_1 = -2(1 - \tfrac{1}{4}\gamma^2) + \frac{15m^2}{4(1-m)} A_1 + 3A_0 \,.\, A_1$$

$$x_2 = \tfrac{3}{2} + \tfrac{1}{4} e^2 - \tfrac{3}{2}\gamma^2 - 2A_{10}$$

$$x_3 = -1$$

$$x_4 = \tfrac{1}{2}(1 + \tfrac{3}{2}e^2 - \tfrac{1}{2}\gamma^2 - 2A_{12} + 3A_{15} e^2$$

$$x_5 = -\tfrac{3}{4} - 2A_{15}$$

$$x_6 = -\tfrac{3}{4}$$

$$x_7 = -\frac{3m^2(1+2e^2+\frac{5}{2}e'^2)}{4(1-m)} - 3m^2e^2\left\{\frac{1+m}{2-2m-c} + \frac{1-m}{2-2m+c}\right\}$$

$$- 2A_0(1 + \tfrac{1}{2}e^2 - \tfrac{1}{4}\gamma^2) + 3e^2 A_1 + 3e^2 A_2$$

$$x_8 = \frac{3m^2(1+2e^2-\frac{1}{4}\gamma^2-\frac{5}{2}e'^2)}{4(1-m)} + \frac{3m^2(1+m)(1+\frac{3}{4}e^2-\frac{1}{4}\gamma^2-\frac{5}{2}e'^2)}{2-2m-c}$$

$$- \frac{3m^2e^2(10+19m+8m^2)}{8(2c-2+2m)} - 2A_1(1+\tfrac{1}{2}e^2-\tfrac{1}{4}\gamma^2) + 3A_0 + 3e^2A_{11}$$

$$x_9 = \frac{3m^2}{4(1-m)} + \frac{3m^2(1-m)}{2-2m+c} - 2A_2 + 3A_0 - 3A_1 e^2$$

$$x_{10} = \frac{3m^2}{4(2-m)} - 2A_3 + 3A_6 e^2$$

$$x_{11} = -\frac{21m^2}{4(2-3m)} - 2A_4 + 3A_7 e^2$$

$$x_{12} = -3m\{4A_0 + A_3 - A_4 - 10A_1 e^2 + \tfrac{5}{2}(A_7 - A_6)e^2\}$$

$$+ \left\{\frac{3m^2A_0}{4} + \frac{27m^4}{32(1-m)}\right\}\left\{\frac{7}{2-3m} - \frac{1}{2-m}\right\}$$

$$+ \left\{\frac{3m^2}{4(1-m)} + 3A_0\right\}\{A_3 + A_4\} - 2A_5(1+\tfrac{1}{2}e^2-\tfrac{1}{4}\gamma^2)$$

$$+ 3(A_8 + A_9)e^2 + 3A_1(A_6 + A_7)e^2$$

$$+ \frac{3m^2}{4}\{11C_6 + 2C_9 - 2C_{10}\}$$

$$x_{13} = -\frac{3m^2(2+m)}{4(2-m-c)} - \frac{3m^2}{4(2-m)} - 2A_6 + 3A_3$$

$$x_{14} = \frac{21m^2(2+3m)}{4(2-3m-c)} + \frac{21m^2}{4(2-3m)} - 2A_7 + 3A_4$$

$$x_{15} = -2A_8 + 3A_5$$

$$x_{16} = -2A_9 + 3A_5$$

$$x_{17} = \frac{3m^2(10+19m+8m^2)}{8(2c-2+2m)} - \frac{3m^2(1+m)}{2-2m-c} - \frac{9m^2}{16(1-m)}$$
$$- 3A_0 + 3A_1 - 2A_{11} - \frac{3m^2A_{10} + \frac{15}{4}m^2A_1{}^2}{2c-2+2m}$$

$$x_{18} = \frac{3m^2(2+m)}{8(2g-2+2m)} - \frac{3m^2}{16(1-m)} - 2A_{13} - \tfrac{3}{4}A_0 - \frac{3m^2A_{12}}{2g-2+2m}$$

$$x_{19} = -A_{14}$$

$$x_{20} = -\frac{3m^2}{8(1-m)} + \frac{3m^2(5+3m)}{4(1-m)}A_{17} - 2A_{17}(1+\tfrac{1}{2}e^2-\tfrac{1}{4}\gamma^2)$$
$$+ 3A_0 \,.\, A_{17}$$

$$x_{21} = -2A_{18}.$$

731. Now if quantities of the order m^4 be omitted,

$$\frac{a^2dv}{\sqrt{\bar{a}}}\, x_0 \text{ becomes } \frac{a^2dv}{\sqrt{\bar{a}}};$$

but in this case equation (230) is reduced to

$$\frac{1}{a} = \frac{1}{\bar{a}}\left\{1 - \frac{m^2}{2} - \frac{3m^2}{4}e'^2\right\},$$

because m^2 differs very little from $\bar{m}^2$,

whence
$$\left(\frac{a}{\bar{a}}\right)^2 = 1 + m^2 + \tfrac{3}{2}m^2e'^2,$$

and
$$\frac{a^2dv}{\sqrt{\bar{a}}} = (\bar{a})^{\frac{3}{2}}\left\{(1+m^2) + \tfrac{3}{2}m^2e'^2\right\}dv, \qquad (238)$$

so that $\frac{a^2dv}{\sqrt{\bar{a}}}$ varies with e', the eccentricity of the terrestrial orbit; but if that variation be omitted, the part that is not periodic of

$$\frac{a^2dv}{\sqrt{\bar{a}}} = (\bar{a})^{\frac{3}{2}}(1+m^2)\,dv.$$

If the action of the sun be omitted $a = \bar{a}$, and if $\frac{1}{n}$ be put for $a^{\frac{3}{2}}$, then the part that is not periodic becomes

$$\frac{a^2dv}{\sqrt{\bar{a}}} = \frac{dv}{n} = a^{\frac{3}{2}}(1+m^2)\,.\,dv,$$

and equation (238) is transformed to

$$\frac{a^2dv}{\sqrt{\bar{a}}} = \frac{dv}{n} + \frac{3m^2}{2n}e'^2dv,$$

and
$$dt = \frac{a^2dv}{\sqrt{\bar{a}}}x_0$$

becomes
$$ndt = dv + \tfrac{3}{2}m^2e'^2dv,$$

the integral of which is

$$nt + \epsilon = v + \tfrac{3}{2}m^2\int(e'^2 - \bar{e}^2)\,dv,$$

$\bar{e}^2$ being a constant quantity equal to the eccentricity of the earth's orbit at the epoch.

732. Thus the mean longitude of the moon is affected by a secular inequality, occasioned by the variation of the eccentricity of the earth's orbit, and the true longitude of the moon in functions of her mean longitude contains the secular inequality

$$-\tfrac{3}{2}m^2\int(e'^2 - \bar{e}^2)\,dv, \text{ or } -\tfrac{3}{2}m^2\int(e'^2 - \bar{e}^2)\,ndt,$$

called the acceleration; hence the secular inequalities in the mean longitude of the moon, in the longitude of her perigee and nodes, are as the three quantities

$$3\bar{m}^2, \quad -\frac{p'}{\sqrt{1-p}}, \quad \frac{q'}{\sqrt{1+q}}.$$

It is true that the terms depending on the squares of the disturbing force alter the value of the secular equations in the mean longitude a little; but the terms of this order that have a considerable influence on the secular equation of the perigee have but little effect on that of the mean motion.

733. Thus the integral of equation (237) is

$$\begin{aligned} nt + \epsilon = v &+ \tfrac{3}{2}m^2\int(e'^2 - \bar{e}^2)\,dv \qquad (239)\\ &+ C_0\, e\sin(cv - \varpi)\\ &+ C_1\, e^2\sin(2cv - 2\varpi)\\ &+ C_2\, e^3\sin(3cv - 3\varpi)\\ &+ C_3\, \gamma^2\sin(2gv - 2\theta)\\ &+ C_4\, e\gamma^2\sin(2gv - cv - 2\theta + \varpi)\\ &+ C_5\, e\gamma^2\sin(2gv + cv - 2\theta - \varpi)\\ &+ C_6\sin(2v - 2mv)\\ &+ C_7\, e\sin(2v - 2mv - cv + \varpi)\\ &+ C_8\, e\sin(2v - 2mv + cv - \varpi)\\ &+ C_9\, e'\sin(2v - 2mv + c'mv + \varpi')\\ &+ C_{10}\, e'\sin(2v - 2mv - c'mv + \varpi') \end{aligned}$$

$$+ C_{11} e' \sin (c'mv - \varpi')$$
$$+ C_{12} ee' \sin (2v - 2mv - cv + c'mv + \varpi - \varpi')$$
$$+ C_{13} ee' \sin (2v - 2mv - cv - c'mv + \varpi + \varpi')$$
$$+ C_{14} ce' \sin (cv + cmv - \varpi - \varpi')$$
$$+ C_{15} ce' \sin (cv - cmv - \varpi + \varpi')$$
$$+ C_{16} e^2 \sin (2cv - 2v + 2mv - 2\varpi)$$
$$+ C_{17} \gamma^2 \sin (2gv - 2v + 2mv - 2\theta)$$
$$+ C_{18} e'^2 \sin (2c'mv - 2\varpi')$$
$$+ C_{19} \frac{a}{a'} \sin (v - mv)$$
$$+ C_{20} \frac{a}{a'} e' \sin (v - mv + c'mv - \varpi').$$

734. If the differential of this equation be compared with equation (237), the following values will be obtained for the indeterminate coefficients—

$$C_0 = \frac{x_1}{c} \qquad C_{10} = \frac{x_{11}}{2-3m}$$
$$C_1 = \frac{x_2}{2c} \qquad C_{11} = \frac{x_{12}}{m}$$
$$C_2 = \frac{x_3}{3c} \qquad C_{12} = \frac{x_{13}}{2-m-c}$$
$$C_3 = \frac{x_4}{2g} \qquad C_{13} = \frac{x_{14}}{2-3m-c}$$
$$C_4 = \frac{x_5}{2g-c} \qquad C_{14} = \frac{x_{15}}{c+m}$$
$$C_5 = \frac{x_6}{2g+c} \qquad C_{15} = \frac{x_{16}}{c-m}$$
$$C_6 = \frac{x_7}{2-2m} \qquad C_{16} = \frac{x_{17}}{2c-2+2m}$$
$$C_7 = \frac{x_8}{2-2m-c} \qquad C_{17} = \frac{x_{18}}{2g-2+2m}$$
$$C_8 = \frac{x_9}{2-2m+c} \qquad C_{18} = \frac{x_{19}}{m}$$
$$C_9 = \frac{x_{10}}{2-m} \qquad C_{19} = \frac{x_{20}}{1-m}$$
$$C_{20} = -2A_{18}.$$

CHAPTER II.

NUMERICAL VALUES OF THE COEFFICIENTS.

735. THE following data are obtained by observation.

$$m = 0.0748013$$
$$e = 0.05486281$$
$$\gamma = 0.0900807$$
$$c = 0.99154801$$
$$g = 1.00402175$$
$$e' = 0.016814, \text{ at the epoch } 1750,$$
$$\mu = \frac{1}{75}.$$

e and γ result from the comparison of the coefficients of the sines of the angles $cv - \varpi$ and $gv - \theta$, computed from observation with those from theory. With these data equation (230) gives

$$\frac{1}{a} = \frac{1}{a_1} . 0.9973020 ; \quad \frac{a^2}{\sqrt{a_1}} = 1.0003084 = \frac{1}{n} ;$$

whence

$$\frac{1}{a} = \sqrt[3]{\frac{n^2 (1.0003084)^2}{0.9973020}}.$$

With these the formulæ of articles 718 and 726 and 734 give

$$A_0 = 0.00709262 \qquad A_{11} = 0.349187$$
$$A_1 = 0.201816 \qquad A_{12} = 0.0026507$$
$$A_2 = -0.00372953 \qquad A_{13} = 0.0077734$$
$$A_3 = -0.00300427 \qquad A_{14} = -0.012989$$
$$A_4 = 0.0284957 \qquad A_{15} = -0.742373$$
$$A_5 = -0.00571628 \qquad A_{16} = -0.041378$$
$$A_6 = -0.0698493 \qquad A_{17} = -0.113197$$
$$A_7 = 0.516751 \qquad A_{18} = 1.08469$$
$$A_8 = -0.20751 \qquad A_{19} = 0.001601$$
$$A_9 = 0.274122 \qquad B_0 = 0.0282636$$
$$A_{10} = 0.0008107 \qquad B_1 = -0.0000024$$

$B_2 = -0.0055075$ $C_4 = 0.722823$
$B_3 = 0.019553$ $C_5 = -0.250034$
$B_4 = 0.0063661$ $C_6 = -0.0091988$
$B_5 = -0.0013668$ $C_7 = -0.414046$
$B_6 = -0.021272$ $C_8 = 0.0129865$
$B_7 = 0.07824$ $C_9 = 0.0039255$
$B_8 = -0.0833684$ $C_{10} = -0.0387853$
$B_9 = -0.0327678$ $C_{11} = 0.196755$
$B_{10} = 0.0720448$ $C_{12} = 0.12765$
$B_{11} = 0.491954$ $C_{13} = -1.081734$
$B_{12} = 0.0061023$ $C_{14} = 0.373115$
$B_{13} = 0.0920621$ $C_{15} = -0.616738$
$B_{14} = -0.0125619$ $C_{16} = 0.272377$
$B_{15} = 0.0038663$ $C_{17} = 0.033825$
$C_0 = -2.003974$ $C_{18} = 0.173647$
$C_1 = 0.752886$ $C_{19} = -0.236616$
$C_2 = -0.336175$ $C_{20} = -2.16938$
$C_3 = 0.243118$

736. If these coefficients be reduced to sexagesimal seconds, the mean longitude of the moon will become

$$
\begin{aligned}
nt + \epsilon = v &+ \tfrac{3}{2} m^2 \int (e'^2 - \bar{e}^2)\, dv \\
&- 22677''.5 \,.\, \sin (cv - \varpi) \\
&+ 467.42 \,.\, \sin 2 (cv - \varpi) \\
&- 11.45 \,.\, \sin 3 (cv - \varpi) \\
&+ 406.92 \,.\, \sin (2gv - 2\theta) \\
&+ 66.37 \,.\, \sin (2gv - cv + \varpi - 2\theta) \\
&- 22.96 \,.\, \sin (2gv + cv - \varpi - 2\theta) \\
&- 1906.93 \,.\, \sin (2v - 2mv) \qquad (240) \\
&- 4685.46 \,.\, \sin (2v - 2mv - cv + \varpi) \\
&+ 147.68 \,.\, \sin (2v - 2mv + cv - \varpi) \\
&+ 13.61 \,.\, \sin (2v - 2mv + cmv - \varpi') \\
&- 134.51 \,.\, \sin (2v - 2mv - cmv + \varpi') \\
&+ 682.37 \,.\, \sin (c'mv - \varpi') \\
&+ 24.29 \,.\, \sin (2v - 2mv - cv + c'mv + \varpi - \varpi') \\
&- 205.82 \,.\, \sin (2v - 2mv - cv - c'mv + \varpi + \varpi') \\
&+ 70.99 \,.\, \sin (cv + c'mv - \varpi - \varpi') \\
&- 117.35 \,.\, \sin (cv - c'mv - \varpi + \varpi')
\end{aligned}
$$

$$+\ 169.09 \ . \sin(2cv - 2v + 2mv - 2\varpi)$$
$$+\ 56.62 \ . \sin(2gv - 2v + 2mv - 2\theta)$$
$$+\ 10.13 \ . \sin(2c'mv - 2\varpi')$$
$$+\ 122.014 \ . (1+i)\sin(v - mv)$$
$$-\ 18.81 \ . (1+i)\sin(v - mv + c'mv - \varpi').$$

737. The two last terms have been determined in supposing

$$\frac{a}{a'} = \frac{(1+i)}{400}.$$

This fraction is the ratio of the parallax of the sun to that of the moon; it differs very little from $\frac{1}{400}$, but for greater generality it is multiplied by the indeterminate coefficient $1 + i$; and by comparing the coefficient of $\sin(v - mv)$ with the result of observations the solar parallax is obtained, as will be shown afterwards.

738. It has been shown that the action of the moon produces the inequality

$$\mu \ . \frac{a}{a'} \sin(v - mv)$$

in the earth's longitude. This action of the moon changes the earth's place, and, consequently, the moon's place with regard to the sun, so that the moon indirectly troubles her own motion, producing in her mean longitude the inequality

$$0.54139 \ . \mu \ . \frac{a}{a'} . \sin(v - nv).$$

Thus the direct action of the moon is weakened by reflection in the ratio of 0.54139 to unity.

739. Equation (233) gives the tangent of the latitude, but the expression of the arc by the tangent s is

$$s - \tfrac{1}{3}s^3 + \tfrac{1}{5}s^5 - \&c.$$

Thus the latitude is nearly

$$\gamma\,(1 - \tfrac{1}{4}\gamma^2)\sin(gv - \theta) + \delta s \times$$
$$\{1 - \tfrac{1}{2}\gamma^2 + \tfrac{1}{2}\gamma^2\cos(2gv - 2\theta) + \tfrac{1}{12}\gamma^3\sin(3gv - 3\theta)\}.$$

And from the preceding data the latitude of the moon is easily found to be

$$s = 18542''.0 \ . \sin(gv - \theta) \qquad (241)$$
$$+\ 12.57 \ . \sin(3gv - 3\theta)$$
$$+525.23 \ . \sin(2v - 2mv - gv + \theta)$$

$$
\begin{aligned}
&+ \quad 1''.14 \,.\, \sin(2v - 2mv + gv - \theta)\\
&- \quad 5.53 \,.\, \sin(gv + cv - \theta - \varpi)\\
&+ \; 19.85 \,.\, \sin(gv - cv - \theta + \varpi)\\
&+ \quad 6.46 \,.\, \sin(2v - 2mv - gv + cv + \theta - \varpi)\\
&- \quad 1.39 \,.\, \sin(2v - 2mv + gv - cv - \theta + \varpi)\\
&- \; 21.6 \;\;\,.\, \sin(2v - 2mv - gv - cv + \theta + \varpi)\\
&+ \; 24.34 \,.\, \sin(gv + c'mv - \theta - \varpi')\\
&- \; 25.94 \,.\, \sin(gv - c'mv - \theta + \varpi')\\
&- \; 10.2 \;\;\,.\, \sin(2v - 2mv - gv + c'mv + \theta - \varpi')\\
&+ \; 22.42 \,.\, \sin(2v - 2mv - gv - cmv + \theta + \varpi')\\
&+ \; 27.41 \,.\, \sin(2cv - gv - 2\varpi + \theta)\\
&+ \quad 5.29 \,.\, \sin(2cv + gv - 2v + 2mv - 2\varpi - \theta).
\end{aligned}
$$

740. The sine of the horizontal parallax of the moon is

$$\frac{R'}{r} = \frac{R'u}{\sqrt{1 + ss}},$$

R' being the terrestrial radius, but as this arc is extremely small, it may be taken for its sine ; hence, if

$$\frac{1}{a}\{1+e^2+\tfrac{1}{4}\gamma^2+e(1+e^2)\cos(cv-\varpi)-\tfrac{1}{4}\gamma^2\cos(2gv-2\theta)\}+\delta u$$

be put for u, and quantities of the order $\frac{R'}{a}e^4$ rejected, the parallax

will be
$$\frac{R'}{r} = \frac{R'u}{\sqrt{1+s^2}} =$$

$$\frac{R'}{a}(1+e^2)\{1+e[1-\tfrac{1}{4}\gamma^2+\tfrac{1}{4}\gamma^2\cos(2gv-2\theta)]\cos(cv-\varpi)+a\delta u-s\delta s\}.$$

In the untroubled orbit of the moon the radius vector, and, consequently, the parallax, varies according to a fixed law through every point of the ellipse. Its mean value, or the constant part of the horizontal parallax, is $\frac{R'}{a}$, to which the rest of the series is applied as corrections arising both from the ellipticity of the orbit and the periodic inequalities to which it is subject.

741. In order to compute the constant part of the parallax, let σ be the space described by falling bodies in a second in the latitude, the square of whose sine is $\frac{1}{3}$, l and R' the corresponding lengths of the pendulum and terrestrial radius, π the ratio of the semicircumference

to the radius, E and m the masses of the earth and moon; then, supposing

$$E + m = 1,$$

$$\frac{E}{(E+m)R'^2} = 2\sigma = \pi^2 l, \text{ also } n = \frac{2\pi}{T},$$

T being the number of seconds in a sidereal revolution of the moon; and by article 735

$$\frac{1}{a} = \sqrt[3]{\frac{n^2\,(1.0003084)^2}{0.9973020}},$$

therefore

$$\frac{R'}{a} = \sqrt[3]{\frac{E}{E+m} \cdot \frac{R'}{l} \cdot \frac{4(1.0003084)^2}{T^2\,0.9973020}}.$$

Now the length of the pendulum, independent of the centrifugal force, is

$$l = 32.648 \text{ feet},$$

also

$$R' = 20898500 \text{ feet},$$

$$T = 2360591''.8;$$

and if

$$m = \frac{E}{58.6}$$

it will be found that

$$\frac{R'}{a} = 0.01655101, \text{ and therefore } \frac{R'}{a}(1+e^2) = 3424''.16;$$

this value augmented by 3''.74, to reduce it to the equator, is 3427''.9; hence the equatorial parallax of the moon in functions of its true longitude is

$$\begin{aligned}
\frac{1}{r} = 3427''.9 & \\
& + 187.48 \cos(cv - \varpi) \\
& + 24.68 \cos(2v - 2mv) \\
& + 47.92 \cos(2v - 2mv - cv + \varpi) \\
& - 0.7 \cos(2v - 2mv + cv - \varpi) \\
& - 0.17 \cos(2v - 2mv + c'mv - \varpi') \qquad (242) \\
& + 1.64 \cos(2v - 2mv - c'mv + \varpi') \\
& - 0.33 \cos(c'mv - \varpi') \\
& - 0.22 \cos(2v - 2mv - cv + c'mv + \varpi - \varpi') \\
& + 1.63 \cos(2v - 2mv - cv - cmv + \varpi + \varpi')
\end{aligned}$$

$$
\begin{aligned}
&- 0''.45 \cos (cv + c'mv - \varpi - \varpi') \\
&+ 0.86 \cos (cv - c'mv - \varpi + \varpi') \\
&+ 0.01 \cos (2cv - 2\varpi) \\
&+ 3.6 \cos (2cv - 2v + 2mv - 2\varpi) \\
&+ 0.07 \cos (2gv - 2\theta) \\
&- 0.18 \cos (2gv - 2v + 2mv - 2\theta) \\
&- 0.01 \cos (2c'mv - 2\varpi') \\
&- 0.95 \cos (2gv - cv - 2\theta + \varpi) \\
&- 0.06 \cos (2v - 2mv - 2gv + cv + 2\theta - \varpi) \\
&- 0.97 (1 + i) \cos (v - mv) \\
&+ 0.16 (1 + i) \cos (v - mv + c'mv - \varpi') \\
&- 0.04 \cos (2v - 2mv + cv - c'mv - \varpi + \varpi') \\
&- 0.15 \cos (4v - 4mv - cv + \varpi) \\
&+ 0.05 \cos (4v - 4mv - 2cv + 2\varpi) \\
&+ 0.13 \cos (2cv - 2v + 2mv + c'mv - 2\varpi - \varpi') \\
&+ 0.02 \cos (2cv + 2v - 2mv - 2\varpi) \\
&- 0.12 (1 + i) \cos (cv - v + mv - \varpi).
\end{aligned}
$$

The greatest value of the parallax is $1° \, 1' \, 29''.32$, which happens when the moon is in perigee and opposition; the least, $58' \, 29''.93$, happens when the moon is in apogee and conjunction.

742. With $m = \frac{E}{74}$, Mr. Damoiseau finds the constant part of the equatorial parallax equal to $3431''.73$.

743. The lunar parallax being known, that of the sun may be determined by comparing the coefficients of the inequality

$$122''.014 \, (i + 1) \sin (v - mv)$$

in the moon's mean longitude with the same derived from observation. In the tables of Burg, reduced from the true to the mean longitude, this coefficient is $122''.378$; hence

$$i + 1 = \frac{122''.378}{122''.014} = 1''.00298, \text{ and } \frac{a}{a_{\prime}} = \frac{1''.00298}{400}.$$

But the solar parallax is

$$\frac{R'}{a'} = \frac{R'}{a}\,\frac{a}{a'} = \frac{R'}{a}\,.\,\frac{1''.00298}{400},$$

but

$$\frac{R'}{a} = 0.01655101,$$

hence

$$\frac{R'}{a'} = \frac{1''.00298 \times 0.01655101}{400} = 8''.5602,$$

which is the mean parallax of the sun in the parallel of latitude, the square of whose sine is $\frac{1}{3}$.

Burckhardt's tables give 122″.97 for the value of the coefficent, whence the solar parallax is 8″.637, differing very little from the value deduced from the transit of Venus. This remarkable coincidence proves that the action of the sun upon the moon is very nearly equal to his action on the earth, not differing more than the three millionth part.

744. The constant part of the lunar parallax is 3432″.04, by the observations of Dr. Maskelyne, consequently the equation

$$3432''.04 = \sqrt[3]{\frac{E}{E+m}\,.\,\frac{R'}{l}\,.\,\frac{4\,(1.0003084)^2}{T^2(0.9973020)}}$$

gives the mass of the moon equal to

$$\frac{1}{74.2}$$

of that of the earth.

Since by article 646, $\frac{R'}{a} = 0.01655101$, in the latitude the square of whose sine is $\frac{1}{3}$; if R', the mean radius of the earth, be assumed as unity, the mean distance of the moon from the earth is 60.4193 terrestrial radii, or about 247583 English miles.

745. As theory combined with observations with the pendulum, and the mensuration of the degrees of the meridian, give a value of the lunar parallax nearly corresponding with that derived from astronomical observations, we may reciprocally determine the magnitude of the earth from these observations; for if the radius of the

earth be assumed as the unknown quantity in the expression in article 646, it will give its value equal to 20897500 English feet.

'Thus,' says La Place, 'an astronomer, without going out of his observatory, can now determine with precision the magnitude and distance of the earth from the sun and moon, by a comparison of observations with analysis alone; which in former times it required long voyages in both hemispheres to accomplish.'

746. The apparent diameter of the moon varies with its parallax, for if P be the horizontal parallax, R' the terrestrial radius, r the radius vector of the moon, D her real, and A her apparent diameters; then

$$P = \frac{R'}{r}, \quad A = \frac{D}{r}; \quad \text{whence } \frac{P}{A} = \frac{R'}{D}$$

a ratio that is constant if the earth be a sphere. It is also constant at the same point of the earth's surface, whatever the figure of the earth may be.

If $\qquad P = 57'4''.168 \quad \text{and } \tfrac{1}{2}A = 31'\,7''.73;$

then $\qquad \dfrac{A}{2P} = 0.27293 = \tfrac{3}{11}$ nearly;

thus if $\frac{11}{3}$ be multiplied by the moon's apparent semidiameter, the corresponding horizontal parallax will be obtained.

Secular Inequalities in the Moon's Motions.

747. It has been shown, that the action of the planets is the cause of a secular variation in the eccentricity of the earth's orbit, which variation produces analogous inequalities in the mean motion of the moon, in the motion of her perigee and in that of her nodes.

The Acceleration.

748. The secular variation in the mean motion of the moon denominated the Acceleration, was discovered by Halley; but La Place first showed that it was occasioned by the variation in the eccentricity in the earth's orbit. The acceleration in the mean mo-

tion of the moon is ascertained by comparing ancient with modern observations; for if the ancient observations be assumed as observed longitudes of the moon, a calculation of her place for the same epoch from the lunar tables will render the acceleration manifest, since these tables may be regarded as data derived from modern observations.

An eclipse of the moon observed by the Chaldeans at Babylon, on the 19th of March, 721 years before the Christian era, which began about an hour after the rising of the moon, as recorded by Ptolemy, has been employed. As an eclipse can only happen when the moon is in opposition, the instant of opposition may be computed from the solar tables, which will give the true longitude of the moon at the time, and the mean longitude may be ascertained from the tables. Now, if we compare this result with another mean longitude of the moon computed from modern observations, the difference of the longitudes augmented by the requisite number of circumferences will give the arc described by the moon parallel to the ecliptic during the interval between the observations, and the mean motion of the moon during 100 Julian years may be ascertained by dividing this arc by the number of centuries elapsed. But the mean motion thus computed by Delambre, Bouvard, and Burg, is more than 200″ less than that which is derived from a comparison of modern observations with one another. The same results are obtained from two eclipses observed by the Chaldeans in the years 719 and 720 before the Christian era. This acceleration was confirmed by comparing less ancient eclipses with those that happened recently; for the epoch of intermediate observations being nearer modern times, the differences of the mean longitudes ought to be less than in the first case, which is perfectly confirmed, by the eclipses observed by Ibn-Junis, an Arabian astronomer of the eleventh century. It is therefore proved beyond a doubt, that the mean motion of the moon is accelerated, and her periodic time consequently diminished from the time of the Chaldeans.

Were the eccentricity of the terrestrial orbit constant, the term

$$\tfrac{3}{2}m^2\int(e'^2 - \bar{e}^2)\,dv$$

would be united with the mean angular velocity of the moon;

but the variation of the eccentricity, though small, has in the course of time a very great influence on the lunar motions. The mean motion of the moon is accelerated, when the eccentricity of the earth's orbit diminishes, which it has continued to do from the most ancient observations down to our times; and it will continue to be accelerated until the eccentricity begins to increase, when it will be retarded. In the interval between 1750 and 1850, the square of the eccentricity of the terrestrial orbit has diminished by 0.00000140595. The corresponding increment in the angular velocity of the moon is the 0.0000000117821th part of this velocity. As this increment takes place gradually and proportionally to the time, its effect on the motion of the moon is less by one half than if it had been uniformly the same in the whole course of the century as at the end of it. In order, therefore, to determine the secular equation of the moon at the end of a century estimated from 1801, we must multiply the secular motion of the moon by half the very small increment of the angular velocity; but in a century the motion of the moon is 1732559351″.544, which gives 10″.2065508 for her secular equation. Assuming that for 2000 years before and after the epoch 1750, the square of the eccentricity of the earth's orbit diminishes as the time, the secular equation of the mean motion will increase as the square of the time: it is sufficient then during that period to multiply 10″.2065508 by the square of the number of centuries elapsed between the time for which we compute and the beginning of the nineteenth century; but in computing back to the time of the Chaldeans, it is necessary to carry the approximation to the cube of the time. The numerical formula for the acceleration is easily found, for since

$$\tfrac{3}{2}m\int(e'^2 - \bar{e}^2)dv$$

is the acceleration in the mean longitude of the moon, the true longitude of the moon in functions of her mean longitude will contain the term

$$-\tfrac{3}{2}m^2\int(e'^2 - \bar{e}^2)ndt,$$

$\bar{e}$ being the eccentricity of the terrestrial orbit at the epoch 1750. If then, t be any number of Julian years from 1750, by article 480,

$$2e' = 2\bar{e} - 0''.171793t - 0''.0000681941t^2$$

is the eccentricity of the earth's orbit at any time t, whence the acceleration is

$$10.''1816213 \,.\, T^2 + 0''.018538444 \,.\, T^3,$$

T being any number of centuries before or after 1801.

In consequence of the acceleration, the mean motion of the moon is 7′ 30″ greater in a century now than it was 2548 years ago.

Motion of the Moon's Perigee.

749. In the first determination of the motion of the lunar perigee, the approximation had not been carried far enough, by which the motion deduced from theory was only one half of that obtained by observation; this led Clairaut to suppose that the law of gravitation was more complicated than the inverse ratio of the squares of the distance; but Buffon opposed him on the principle that, the primordial laws of nature being the most simple, could only depend on one principle, and therefore their expression could only consist of one term. Although such reasoning is not always conclusive, Buffon was right in this instance, for, upon carrying the approximation to the squares of the disturbing force, the law of gravitation gives the motion of the lunar perigee exactly conformable to observation, for $\bar{e}'$ being the eccentricity of the terrestrial orbit at the epoch, the equation $c = \sqrt{1-p-p'\bar{e}^2}$ when reduced to numbers is $c = 0.991567$, consequently $(1 - c)v$ the motion of the lunar perigee is $0.008433 \,.\, v$; and with the value of c in article 735 given by observation, it is $0.008452 \,.\, v$, which only differs from the preceding by 0.000019. In Damoiseau's theory it is $0.008453.v$, which does not differ much from that of La Place. The terms depending on the squares of the disturbing force have a very great influence on the secular variation in the motion of the lunar perigee; they make its value three times as great as that of the acceleration: for the secular inequality in the lunar perigee is

$$\frac{p'}{2\sqrt{1+p}}\int(e'^2 - \bar{e}^2)ndt,$$

or, when the coefficient is computed, it is

$$3.00052\, \tfrac{3}{2}\, m^2 \int(e'^2 - \bar{e}^2)ndt,$$

and has a contrary sign to the secular equation in the mean motion.

The motion of the perigee becomes slower from century to century, and is now 8′.2 slower than in the time of Hipparchus.

Motion of the Nodes of the Lunar Orbit.

750. The sidereal motion of the node on the true ecliptic as determined by theory, does not differ from that given by observation by a 350th part; for the expression in article 727 gives the retrograde motion of the node equal to $0.0040105v$, and by observation

$$(g-1)v = 0.00402175v,$$

the difference being 0.00001125. Mr. Damoiseau makes it

$$g-1 = 0.0040215.$$

The secular inequality in the motion of the node depends on the variation in the eccentricity of the terrestrial orbit, and has a contrary sign to the acceleration. Its analytical expression gives

$$\frac{q'}{2\sqrt{1+q}}\int(e'^2-\bar{e}^2)\,dv = 0.735452\,\tfrac{3}{2}m^3\int(e'^2-\bar{e}^2)dv.$$

As the motion of the nodes is retrograde, this inequality tends to augment their longitudes posterior to the epoch.

751. It appears from the signs of these three secular inequalities, as well as from observation, that the motion of the perigee and nodes become slower, whilst that of the moon is accelerated; and that their inequalities are always in the ratio of the numbers 0.735452, 3.00052, and 1.

752. The mean longitude of the moon estimated from the first point of Aries is only affected by its own secular inequality; but the mean anomaly estimated from the perigee is affected both by the secular variation of the mean longitude, and by that of the perigee; it is therefore subject to the secular inequality $-4.00052\,\tfrac{3}{2}m^2\int(e'^2-\bar{e}^2)\,dv$ more than four times that of the mean longitude. From the preceding values it is evident that the secular motion of the moon with regard to the sun, her nodes, and her perigee, are as the numbers 1; 0.265; and 4; nearly.

753. At some future time, these inequalities will produce variations equal to a fortieth part of the circumference in the secular motion of the moon; and in the motion of the perigee, they will amount to

no less than a thirteenth part of the circumference. They will not always increase: depending on the variation of the eccentricity of the terrestrial orbit they are periodic, but they will not run through their periods for millions of years. In process of time, they will alter all those periods which depend on the position of the moon with regard to the sun, to her perigee, and nodes; hence the tropical, synodic, and sidereal revolutions of the moon will differ in different centuries, which renders it vain to attempt to attain correct values of them for any length of time.

Imperfect as the early observations of the moon may be, they serve to confirm the results that have been detailed, which is surprising, when it is considered that the variation of the eccentricity of the earth's orbit is still in some degree uncertain, because the values of the masses of Venus and Mars are not ascertained with precision; and it is worthy of remark, that in process of time the developement of the secular inequalities of the moon will furnish the most accurate data for the determination of the masses of these two planets.

754. The diminution of the eccentricity of the earth's orbit has a greater effect on the moon's motions than on those of the earth. This diminution, which has not altered the equation of the centre of the sun by more than 8′.1 from the time of the most ancient eclipse on record, has produced a variation of 1° 8′ in the longitude of the moon, and of 7°.2 in her mean anomaly.

Thus the action of the sun, by transmitting to the moon the inequalities produced by the planets on the earth's orbit, renders this indirect action of the planets on the moon more considerable than their direct action.

755. The mean action of the sun on the moon contains the inclination of the lunar orbit on the plane of the ecliptic; and as the position of the ecliptic is subject to a secular variation, from the action of the planets, it might be expected to produce a secular variation in the inclination of the moon's orbit. This, however, is not the case, for the action of the sun retains the lunar orbit at the same inclination on the orbit of the earth; and thus in the secular motion of the ecliptic, the orbit of the earth carries the orbit of the moon along with it, as it will be demonstrated, the change in the ecliptic affecting only the declination of the moon. No perceptible

change has been observed in the inclination of the lunar orbit since the time of Ptolemy, which confirms the result of theory.

756. Although the inclination of the orbit does not vary from the change in the plane of the ecliptic; yet, as the expressions which determine the inclination and eccentricity of the lunar orbit, the parallax of the moon, and generally the coefficients of all the moon's inequalities, contain the eccentricity of the terrestrial orbit, they are all subject to secular inequalities corresponding to the secular variation of that quantity. Hitherto they have been insensible, but in the course of time will increase to an estimable quantity. Even now, it is necessary to include the effects of this variation in the inequality called the annual equation, when computing ancient eclipses.

757. The three co-ordinates of the moon have been determined in functions of the true longitudes, because the series converge better, but these quantities may be found in functions of the mean longitudes by reversion of series. For if nt, ϖ, θ, and ϵ, represent the mean motion of the moon, the longitudes of her perigee, ascending node and epoch, at the origin of the time, together with their secular equations for any time t, equation (240) becomes

$$v - (nt + \epsilon) = -\{C_0 . e . \sin(cv - \varpi) + C_1 . e^2 \sin 2(cv - \varpi) \\ + C_2 . e^3 . \sin 3(cv - \varpi) + \&c.\}$$

or to abridge $$v - (nt + \epsilon) = S.$$

The general term of the series is

$$Q . \sin(\zeta v + \psi).$$

And if Q' be the sum of the coefficients arising from the square of the series S, and depending on the angle $\zeta v + \psi'$; Q'' the sum of the coefficients arising from the cube of S, and depending on the angle $\zeta v + \psi$, &c. &c., the general term of the new series, which gives the true longitude of the moon in functions of her mean longitude, is

$$-\{Q + \tfrac{1}{2}\zeta . Q' - \tfrac{1}{6}\zeta^2 . Q'' - \tfrac{1}{24}\zeta^3 . Q''' + \&c.\} . \sin(\zeta(nt + \epsilon) + \psi)$$

La Place does not give this transformation, but Damoiseau has computed the coefficients for the epoch of January 1st, 1801, and has found that the true longitude of the moon in functions of its mean longitude $nt + \epsilon = \lambda$ is

$$
\begin{aligned}
v = nt + \epsilon &+ 22639''.7 \,.\, \sin\{c\lambda - \varpi\} \\
&+ 768''.72 \sin(2c\lambda - 2\varpi) \\
&+ 36''.94 \sin(3c\lambda - 3\varpi) \\
&- 411''.67 \sin(2g\lambda - 2\theta) \\
&+ 39''.51 \sin(c\lambda - 2g\lambda - \varpi + 2\theta) \\
&- 45''.12 \sin(c\lambda + 2g\lambda - \varpi - 2\theta) \\
&+ 2370''.00 \sin(2\lambda - 2m\lambda) \\
&+ 4589''.61 \sin(2\lambda - 2m\lambda - c\lambda + \varpi) \\
&+ 192''.22 \sin(2\lambda - 2m\lambda + c\lambda - \varpi) \\
&- 24''.82 \sin(2\lambda - 2m\lambda + c'm\lambda - \varpi') \\
&+ 165''.56 \sin(2\lambda - 2m\lambda - c'm\lambda + \varpi') \\
&- 673''.70 \sin(c'm\lambda - \varpi') \\
&- 28''.67 \sin(2\lambda - 2m\lambda - c\lambda + c'm\lambda + \varpi - \varpi') \\
&+ 207''.09 \sin(2\lambda - 2m\lambda - c\lambda - c'm\lambda + \varpi + \varpi') \\
&- 109''.27 \sin(c\lambda + cm\lambda - \varpi - \varpi') \\
&+ 147''.74 \sin(c\lambda - cm\lambda - \varpi + \varpi') \\
&+ 211''.57 \sin(2\lambda - 2m\lambda - 2c\lambda + 2\varpi) \\
&+ 54''.83 \sin(2\lambda - 2m\lambda - 2g\lambda + 2\theta) \\
&- 7''.34 \sin(2c'm\lambda - 2\varpi') \\
&- 122''.48 \sin(\lambda - m\lambda) \\
&- 17''.56 \sin(\lambda - m\lambda + c'm\lambda - \varpi').
\end{aligned}
$$

This is only the transformation of La Place's equation (240), but Damoiseau carries the approximation much farther.

758. The first term of this series is the mean longitude of the moon, including its secular variation.

The second term

$$22639''.7 \sin\{c\lambda - \varpi)$$

is the equation of the centre, which is a maximum when

$$\sin(c\lambda - \varpi) = \pm 1,$$

that is, when the mean anomaly of the moon is either 90° or 270°. Thus, when the moon is in quadrature, the equation of the centre is $\pm 6^\circ\ 17'\ 19''.7$. double the eccentricity of the orbit. In syzigies it is zero.

759. The most remarkable of the periodic inequalities next to the equation of the centre, is the evection

$$4589''.61 \sin\{2\lambda - 2m\lambda - c\lambda + \varpi),$$

which is at its maximum and $= \pm 4589''.61$, when $2\lambda - 2m\lambda - c\lambda + \varpi$ is either 90° or 270°, and it is zero when that angle is either 0° or 180°. Its period is found by computing the value of its argument in a given time, and then finding by proportion the time required to describe 360°, or a whole circumference. The synodic motion of the moon in 100 Julian years is

$$445267°.1167992 = \lambda - m\lambda$$

and $$890534°.2335984 = 2\{\lambda - m\lambda\}$$

is double the distance of the sun from the moon in 100 Julian years. If 477198°.839799 the anomalistic motion of the moon in the same period be subtracted, the difference 413335°.3937994 will be the angle $2\lambda - 2m\lambda - c\lambda + \varpi$, or the argument of the evection in 100 Julian years: whence

$$413335°.3937994 : 360° :: 365^d.25 : 31^d.811939 =$$

the period of the evection. If t be any time elapsed from a given period, as for example, when the evection is zero, the evection may be represented for a short time by

$$4589''.61 \,.\, \sin \left\{\frac{360° \,.\, t}{31.811939}\right\}.$$

This inequality is a variation in the equation of the centre, depending on the position of the apsides of the lunar orbit. When the apsides are in syzigies, as in figure 101, the action of the sun increases the eccentricity of the moon's orbit or the equation of the centre. For if the moon be in conjunction at m, the sun draws her from the earth; and if she be in opposition in m', the sun draws the earth from her; in both cases increasing the moon's distance from the earth, and thereby the eccentricity or equation of the centre. When the moon is in any other point of her orbit, the action of the sun may be resolved into two, one in the direction of the tangent, and the other according to the radius vector. The latter increases the moon's gravitation to the earth, and is at its maximum when the moon is in quadratures; as it tends to diminish the distance QE, it makes the

fig. 101.

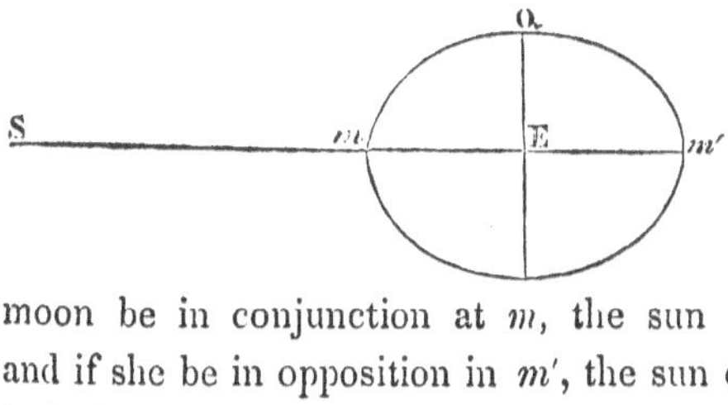

ellipse still more eccentric, which increases the equation of the centre. This increase is the evection. Again, if the line of apsides be at right angles to SE, the line joining the centres of the sun and the earth, the action of the sun on the moon at m or m', figure 102, by increasing the distance from the earth, augments the breadth of the orbit, thereby making it approach the circular form, which diminishes the eccentricity. If the moon be in quadratures, the increase in the moon's gravitation diminishes her distance from the earth, which also diminishes the eccentricity, and consequently the equation of the centre. This diminution is the evection. Were the changes in the evection always the same, it would depend on the angular distances of the sun and moon, but its true value varies with the distance of the moon from the perigee of her orbit. The evection was discovered by Ptolemy, in the first century after Christ, but Newton showed on what it depends.

fig. 102.

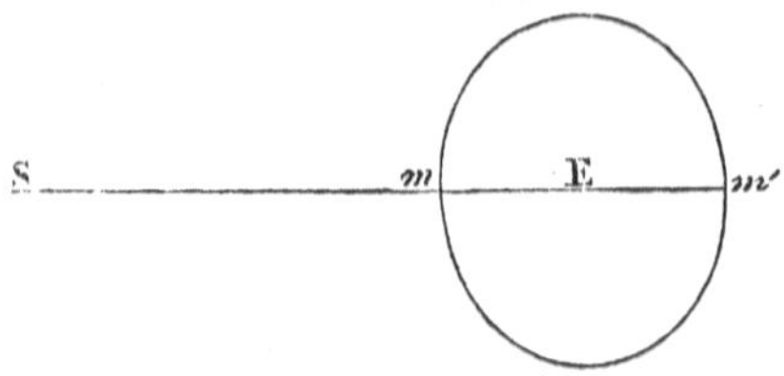

760. The variation is an inequality in the moon's longitude, which increases her velocity before conjunction, and retards her velocity after it. For the sun's force, acting on the moon according to Sm, fig. 103, may be resolved into two other forces, one in the direction of mE, which produces the evection, and the other in the direction of mT, tangent to the lunar orbit. The latter produces the variation which is expressed by

fig. 103.

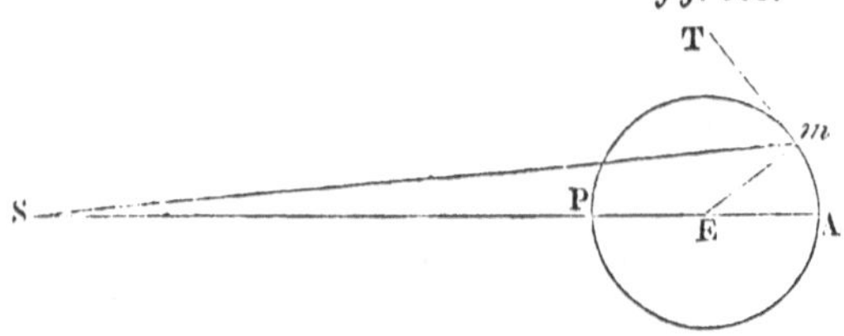

$$2370'' \sin 2\{\lambda - m\lambda\}.$$

This inequality depends on the angular distance of the sun from the moon, and as she runs through her period whilst that distance increases 90°, it must be proportional to the sine of twice the angular distance. Its maximum happens in the octants when $\lambda - m\lambda = 45°$, it is zero when the angular distance of the moon from the sun is either zero, or when the moon is in quadratures. Thus the variation vanishes in syzigies and quadratures, and is a maximum in the octants.

The angular distance of the moon from the sun depends on its synodic motion: it varies $\frac{360^\circ}{29^d.530588}$ daily, and

$$2\,(\lambda - m\lambda) = \frac{2\,.\,360^\circ}{29^d.530588},$$

hence its period is

$$\frac{29^d.530588}{2} = 14^d.765294.$$

Thus the period of the variation is equal to half the moon's synodic revolution. The variation was discovered by Tycho Brahe, and was first determined by Newton.

761. The annual equation

$$673''.70 \sin \{ c'm\lambda - \varpi' \}$$

is another remarkable periodic inequality in the moon's longitude. The action of the sun which produces this inequality is similar to that which causes the accéleration of the moon's mean motion. The annual equation is occasioned by a variation in the sun's distance from the earth, it consequently arises from the eccentricity of the terrestrial orbit. When the sun is in perigee his action is greatest, and he dilates the lunar orbit, so that the angular motion of the moon is diminished; but as the sun approaches the apogee the orbit contracts, and the moon's angular motion is accelerated. This change in the moon's angular velocity is the annual equation. It is a periodic inequality similar to the equation of the centre in the sun's orbit, which retards the motion of the moon when that of the sun increases, and accelerates the motion of the moon when the motion of the sun diminishes, so that the two inequalities have contrary signs.

The period of the annual equation is an anomalistic year. It was discovered by Tycho Brahe by computing the places of the moon for various seasons of the year, and comparing them with observation. He found the observed motion to be slower than the mean motion in the six months employed by the sun in going from perigee to apogee, and the contrary in the other six months. It is evident that as the action of the sun on the moon varies with his distance, and therefore depends on the eccentricity of the earth's orbit, whatever affects the eccentricity must influence all the motions of the moon.

762. The variation has been ascribed to the effect of that part of the sun's force that acts in the direction of the tangent; and the evection to the effect of the part which acts in the direction of the radius vector, and alters the ratio of the perigean and apogean gravities of the moon from that of the inverse squares of the distance. The annual equation does not arise from the direct effect of either, but from an alteration in the mean effect of the sun's disturbing force in the direction of the radius vector which lessens the gravity of the moon to the earth.

763. Although the causes of the lesser inequalities are not so easily traced as those of the four that have been analysed, yet some idea of the sources from whence they arise may be formed by considering that when the moon is in her nodes, she is in the plane of the ecliptic, and the action of the sun being in that plane is resolved into two forces only; one in the direction of the moon's radius vector, and the other in that of the tangent to her orbit. When the moon is in any other part of her orbit, she is either above or below the plane of the ecliptic, and the line joining the sun and moon, which is the direction of the sun's disturbing force, being out of that plane, the sun's force s resolved into three component forces; one in the direction of the moon's radius vector, another in the tangent to her orbit, and the third perpendicular to the plane of her orbit, which affects her latitude. If then the absolute action of the sun be the same in these two positions of the moon, the component forces in the radius vector and tangent must be less than when the moon is in her nodes by the whole action in latitude. Hence any inequality like the evection, whose argument does not depend on the place of the nodes, will be different in these two positions of the moon, and will require a correction, the argument of which should depend on the position of the nodes. This circumstance introduces the inequality

$$54''.83 \cdot \sin(2g\lambda - 2\lambda + 2m\lambda - 2\theta)$$

in the moon's longitude. The same cause introduces other inequalities in the moon's longitude, which are the corrections of the variation and annual equation. But the annual equation requires a correction from another cause which will introduce other terms in the perturbations of the moon in longitude; for since it arises from a change in the mean effect of the sun's disturbing force, which diminishes the moon's gravity, its coefficient is computed for a certain value of the moon's gravity, consequently for a given distance of the moon from the earth; hence, when she has a different distance, the annual equation must be corrected to suit that distance.

764. In general, the numerical coefficients of the principal inequalities are computed for particular values of the sun's disturbing force, and of the moon's gravitation; as these are perpetually changing, new inequalities are introduced, which are corrections to the inequalities computed in the first hypothesis. Thus the perturbations are a series of corrections. How far that system is to be carried, depends on the perfection of astronomical instruments, since it is needless to compute quantities that fall within the limits of the errors of observation.

765. When La Place had determined all the inequalities in the moon's longitude of any magnitude arising from every source of disturbance, he was surprised to find that the mean longitude computed from the tables in Lalande's astronomy for different epochs did not correspond with the mean longitudes computed for the same epochs from the tables of Lahere and Bradley, the difference being as follows:—

Epochs.	Errors.
1766	$-3''$
1779	$9''.3$
1789	$17''.6$
1801	$28''\ 5$

Whence it was to be presumed that some inequality of a very long period affected the moon's mean motion, which induced him to revise the whole theory of the moon. At last he found that the series which determines the mean longitude contains the term

$$\gamma^2 e'^2 e \cdot \frac{a}{a'} \cdot \frac{\sin\{3v-3mv+3c'mv-2gv-cv+2\theta+\varpi-3\varpi'\}}{\{3-3m+3c'm-2g-c\}^2}$$

$$= \gamma^2 e'^3 \cdot e \cdot \frac{a}{a'} \cdot \frac{\sin\{2\theta+\varpi-3\varpi'\}}{\{3-3m+3c'm-2g-c\}^2}$$

depending on the disturbing action of the sun, that appeared to be the cause of these errors.

The coefficient of this inequality is so small that its effect only becomes sensible in consequence of the divisor

$$\{3 - 3m + 3c'm - 2g - c\}^2$$

acquired from the double integration. Its maximum, deduced from the observations of more than a century, is 15″.4. Its argument is twice the longitude of the ascending node of the lunar orbit, plus the longitude of the perigee, minus three times the longitude of the sun's perigee, whence its period may be found to be about 184 years.

The discovery of this inequality made it necessary to correct the whole lunar tables.

766. By reversion of series the moon's latitude in functions of her mean motion is found to be

$$\begin{aligned} s = {} & 18539''.8 \sin\{g\lambda - \theta\} \\ & + 12''.6 \sin\{3g\lambda - 3\theta\} \\ & + 527''.7 \sin\{2\lambda - 2m\lambda - g\lambda + \theta) \\ & + \quad 1''.0 \sin\{2\lambda - 2m\lambda + g\lambda - \theta) \\ & - \quad 1''.3 \sin\{g\lambda + c\lambda - \varpi - \theta) \\ & - 14''.4 . \sin\{c\lambda - g\lambda - \varpi + \theta) \\ & + \quad 1''.8 \sin\{2\lambda - 2m\lambda - g\lambda + c\lambda - \varpi + \theta) \\ & - \quad 0''.3 \sin\{2\lambda - 2m\lambda + g\lambda - c\lambda + \varpi - \theta) \\ & - 15''.8 \sin\{2\lambda - 2m\lambda - g\lambda - c\lambda + \varpi + \theta) \\ & + 23''.8 \sin\{g\lambda + c'm\lambda - \varpi' - \theta) \\ & - 25''.1 \sin\{g\lambda - c'm\lambda + \varpi' - \theta) \\ & - 10''.3 \sin\{2\lambda - 2m\lambda - g\lambda + c'm\lambda - \varpi' + \theta) \\ & + 22''.0 \sin\{2\lambda - 2m\lambda - g\lambda - cm\lambda + \varpi' + \theta) \\ & + 25''.7 \sin\{2c\lambda - g\lambda - 2\varpi + \theta) \\ & - \quad 5''.4 . \sin\{2\lambda - 2m\lambda - 2c\lambda - g\lambda + 2\varpi + \theta) \end{aligned}$$

767. The only inequality in the moon's latitude that was discovered by observation is

$$527''.7 \sin(2\lambda - 2m\lambda - g\lambda + \theta).$$

Tycho Brahe observed, in comparing the greatest latitude of the moon in different positions with regard to her nodes, that it was not always the same, but oscillated about its mean value of 5° 9′, and as the greatest latitude is the measure of the inclination of the orbit, it was evident that the inclination varied periodically. Its period is a semi-revolution of the sun with regard to the moon's nodes.

768. By reversion of series it will be found that the lunar parallax at the equator in terms of the mean motions is

$$\frac{1}{r} = 3420''.89$$

$$\begin{aligned}
&+186''.48 \cos\{c\lambda - \varpi\} \\
&+\ 28''.54 \cos\{2\lambda - 2m\lambda\} \\
&+\ 34''.43 \cos\{2\lambda - 2m\lambda - c\lambda + \varpi) \\
&+\ \ 3''.05 \cos\{2\lambda - 2m\lambda + c\lambda - \varpi) \\
&-\ \ 0''.26 \cos\{2\lambda - 2m\lambda + c'm\lambda - \varpi'\} \\
&+\ \ 1''.92 \cos\{2\lambda - 2m\lambda - c'm\lambda + \varpi'\} \\
&-\ \ 0''.32 \cos\{c'm\lambda - \varpi'\} \\
&-\ \ 0''.24 \cos\{2\lambda - 2m\lambda - c\lambda + c'm\lambda + \varpi - \varpi'\} \\
&+\ \ 1''.45 \cos\{2\lambda - 2m\lambda - c\lambda - c'm\lambda + \varpi + \varpi'\} \\
&+\ \ 1''.20 \cos\{c\lambda - c'm\lambda - \varpi + \varpi'\} \\
&-\ \ 0''.92 \cos\{c\lambda + c'm\lambda - \varpi - \varpi'\} \\
&+\ 10''.24 \cos\{2c\lambda - 2\varpi\} \\
&-\ \ 0''.41 \cos\{2c\lambda - 2\lambda + 2m\lambda - 2\varpi\} \\
&+\ \ 0''.03 \cos\{2g\lambda - 2\theta\} \\
&-\ \ 0''.15.\cos\{2\lambda - 2m\lambda - 2g\lambda + 2\theta\} \\
&-\ \ 0''.70.\cos\{c\lambda - 2g\lambda - \varpi + 2\theta\} \\
&-\ \ 0''.06 \cos\{2\lambda - 2m\lambda - 2g\lambda + c\lambda + 2\theta - \varpi\} \\
&-\ \ 0''.98 \cos\{\lambda - m\lambda\} \\
&+\ \ 0''.14 \cos\{\lambda - m\lambda + c'm\lambda - \varpi'\} \\
&+\ \ 0''.18 \cos\{2\lambda - 2m\lambda + c\lambda - c'm\lambda - \varpi + \varpi'\} \\
&+\ \ 0''.57 \cos\{4\lambda - 4m\lambda - c\lambda + \varpi\} \\
&+\ \ 0''.4\ \ \cos\{4\lambda - 4m\lambda - 2c\lambda + 2\varpi\} \\
&-\ \ 0''.03.\cos\{2\lambda - 2m\lambda - 2c\lambda - c'm\lambda + 2\varpi + \varpi'\} \\
&+\ \ 0''.14 \cos\{2c\lambda + 2\lambda - 2m\lambda - 2\varpi\}.
\end{aligned}$$

769. The planets are at so great a distance from the sun, and from one another, that their form has no perceptible effect on their mutual motions; and, considered as spheres, their action is the same as if their mass were united in their centre of gravity: but the satellites are so near their respective planets that the ellipticity of the latter has a considerable influence on the motions of the former. This is particularly evident in the moon, whose motions are troubled by the spheroidal form of the earth.

CHAPTER III.

INEQUALITIES FROM THE FORM OF THE EARTH.

770. THE attraction of the disturbing matter is equal to the sum of all the molecules in the excess of the terrestrial spheroid above a sphere whose radius is half the axis of rotation, each molecule being divided by its distance from the moon; and the finite values of this action, after it has been resolved in the direction of the three co-ordinates of the moon, are the perturbations in longitude, latitude, and distance, caused by the non-sphericity of the earth. In the determination of these inequalities, therefore, results must be anticipated that can only be obtained from the theory of the attraction of spheroids. By that theory it is found that if ρ be the ellipticity of the earth, R' its mean radius, ϕ the ratio of the centrifugal force at the equator to gravity, and ν the sine of the moon's declination, the attraction of the redundant matter at the terrestrial equator is

$$(\tfrac{1}{2}\phi - \rho)\frac{R'}{r^3}(\nu - \tfrac{1}{3})$$

the sum of the masses of the earth and moon being equal to unity. Hence the quantity R which expresses the disturbing forces of the moon in equation (208) must be augmented by the preceding expression.

771. By spherical trigonometry ν, the sine of the moon's declination in functions of her latitude and longitude, is

$$\nu = \sin\omega\sqrt{1 - s^2}\sin fv + s\cos\omega,$$

in which ω is the obliquity of the ecliptic, s the tangent of the moon's latitude, and fv her true longitude, estimated from the equinox of spring. The part of the disturbing force R that depends on the action of the sun, has the form Qr^2 when the terms depending on the solar parallax are rejected. Hence

$$R = Qr^2 - (\rho - \tfrac{1}{2}\phi) \,.\, \frac{R'^2}{r^3}(\sin^2\omega \,.\, \sin^2 fv + 2s \sin\omega \,.\, \cos\omega \,.\, \sin fv)$$

very nearly; but $s = \gamma \sin(gv - \theta)$ by article 696, and if

$\frac{1}{a^3}$ be put for $\frac{1}{r^3}$

$$R = Qr^2 - (\rho - \tfrac{1}{2}\phi)\frac{R'^2}{a^3} \sin\omega \,.\, \cos\omega \,.\, \gamma\cos(gv - fv - \theta);$$

when all terms are rejected except those depending on the angle $gv - fv - \theta$, which alone have a sensible effect in troubling the motion of the moon.

772. If this force be resolved in the direction of the three co-ordinates of the moon, and the resulting values of

$$\frac{dR}{du} \quad \frac{dR}{dv} \quad \frac{dR}{ds}$$

substituted in the equations in article 695, they will determine the effect which the form of the earth has in troubling the motions of that body. But the same inequalities are obtained directly and with more simplicity from the differential of the periodic variation of the epoch in article 439, which, in neglecting the eccentricity of the lunar orbit, becomes

$$d\epsilon = -2a^2\left(\frac{dR}{da}\right)ndt.$$

Now

$$2a^2\left(\frac{dR}{da}\right) = 4ar^2Q + 6(\rho - \tfrac{1}{2}\phi) \,.\, \frac{R'^2}{a^3} \,.\, \sin\omega \,.\, \cos\omega \,.\, \gamma\cos(gv - fv - \theta).$$

But by article 438 the variation of dR is zero, consequently the co-efficient of $\cos(gv - fv - \theta)$ must be zero in R. Then if $\delta \,.\, r^2Q$ be the part of r^2Q that depends on the compression of the earth,

$$0 = \delta \,.\, r^2Q - (\alpha\rho - \tfrac{1}{2}\alpha\phi) \,.\, \frac{R'^2}{a^3} \,.\, \sin\omega \,.\, \cos\omega \,.\, \gamma\cos(gv - fv - \theta),$$

and eliminating δr^2Q,

$$2\delta a^2\left(\frac{dR}{da}\right) = 10(\rho - \tfrac{1}{2}\phi)\frac{R'^2}{a^3} \,.\, \sin\omega \,.\, \cos\omega \,.\, \gamma\cos(gv - fv - \theta).$$

And if dv be put for ndt,

$$d\epsilon = -10(\rho - \tfrac{1}{2}\phi) \,.\, \frac{R'^2}{a^2} \,.\, \sin\omega \,.\, \cos\omega \,.\, \gamma dv \,.\, \cos(gv - fv - \theta).$$

This equation is referred to the plane of the lunar orbit, but in order to reduce it to the plane of the ecliptic the equation (154) must be resumed, which is

$$dv_{,} = dv\left(1 + \tfrac{1}{2}s^2 - \tfrac{1}{2}\frac{ds^2}{dv^2}\right)$$

$dv_{,}$ being the arc dv projected on the plane of the ecliptic, or fixed plane. By article 436

$$s = q \sin fv - p \cos fv,$$

whence

$$\frac{ds}{dv} = \left(fq - \frac{dp}{dv}\right)\cos fv + \left(fp + \frac{dq}{dv}\right)\sin fv + \&c.$$

and neglecting periodic quantities depending on fv,

$$dv_{,} = dv + \frac{qdp - pdq}{2}, \text{ very nearly.}$$

Hence, in order to have $d\,.\,\delta v_{,}$ it is only necessary to add

$$\frac{qdp-pdq}{2} \text{ to } d\,.\,\delta v.$$

For the same reason the angle $d\epsilon$ will be projected on the plane of the ecliptic if $\frac{qdp-pdq}{2}$ be added to it, so that

$$d\epsilon_{,} = d\epsilon + \frac{qdp-pdq}{2}.$$

Now $s = \gamma \sin(gv - \theta)$ may be put under the form

$$s = \gamma\cos(gv - fv - \theta)\sin fv + \gamma \sin(gv - fv - \theta)\cos fv,$$

and comparing it with

$$s = q \sin fv - p \cos fv,$$

the result is

$$p = -\gamma\sin(gv - fv - \theta) \qquad q = \gamma\cos(gv - fv - \theta),$$

whence

$$dp = -(g - f)\,qdv$$
$$dq = (g - f)\,pdv$$

$$R = r^2 Q - (\rho - \tfrac{1}{2}\phi)\frac{R'^2}{a^2}\sin\omega\cos\omega\,.\,q,$$

and

$$\frac{dR}{dq} = -(\rho - \tfrac{1}{2}\phi)\frac{R'^2}{a^2}\sin\omega\cos\omega;$$

in consequence of this the values of dp, dq, in article 439, become

$$dp = -(g - f)\,qdv + (\rho - \tfrac{1}{2}\phi)\frac{R'^2}{a^2}\sin\omega\cos\omega\,.\,dv$$
$$dq = (g - f)\,pdv;$$

therefore dp contains the term

$$\frac{(\rho - \frac{1}{2}\phi)}{g - f} \cdot \frac{R'^2}{a^2} \sin \omega \cos \omega . dv;$$

and as $ds = dq . \sin fv - dp \cos fv$, the latitude of the moon is subject to the inequality

$$-\frac{(\rho - \frac{1}{2}\phi)}{g - f} \cdot \frac{R'^2}{a^2} \sin \omega \cos \omega \, \gamma \sin fv. \qquad (243)$$

773. The constant part of q produces in $\frac{qdp - pdq}{2}$ the term

$$\tfrac{1}{2} . (\rho - \tfrac{1}{2}\phi) \frac{R^2}{a^2} . \sin \omega \cos \omega \, \gamma \cos (gv - fv - \theta);$$

whence $d\epsilon_{,}$, which is the value of $d\epsilon$ when referred to the plane of the ecliptic, becomes

$$d\epsilon = -\frac{19}{2} . (\rho - \tfrac{1}{2}\phi) \frac{R'^2}{a^2} \sin \omega \cos \omega . \gamma \cos (gv - fv - \theta) . dv,$$

which gives in ϵ, and consequently in the true longitude of the moon the inequality

$$-\frac{19}{2} \frac{(\rho - \frac{1}{2}\phi)}{g - f} \cdot \frac{R'^2}{a^2} . \sin \omega \cos \omega \, \gamma \sin (gv - fv - \theta). \quad (244)$$

774. Now, $\frac{R'}{a} = 0.0165695$, $\omega = 23° \; 28'$, &c.

$$\phi = \frac{1}{289}, \quad \gamma = 0.0900684, \quad g = 1.00402175,$$

and the argument $gv - fv - \theta$ is the mean longitude of the moon. Thus every quantity is known, except ρ, the compression, which may therefore be determined by comparing the coefficient, computed with these data, with the coefficient of the same inequality given by observation. By Burg's Tables, it is $-6''.8$, and by Burckhardt's, $-7''.0$. The mean of these $-6''.9$ gives the compression

$$\rho = \frac{1}{303.22}.$$

By the theory of the rotation of spheroids, it is found that if the earth be homogeneous, the compression is $\frac{1}{230}$. Consequently the earth is of variable density.

That the inequalities of the moon should disclose the interior structure of the earth, is a singular instance of the power of analysis.

775. The inequality in the moon's latitude, depending on the same cause, confirms these results. Its coefficient, determined by Burg and Burckhardt from the combined observations of Maskelyne and Bradley, is $-8''.0$, which, compared with the coefficient of

$$-\frac{(\rho-\frac{1}{2}\phi)}{g-f}\cdot\frac{R^2}{a^2}\sin\omega\cos\omega\,\gamma\,\sin fv,$$

computed with the preceding data, gives $\frac{1}{305.26}$ for the compression, which also proves that the earth is not homogeneous.

776. Since the coefficients of both inequalities are greater in supposing the earth to be homogeneous, it affords another proof that the gravitation of the moon to the earth is composed of the attraction of all its particles. Thus the eclipses of the moon in the early ages of astronomy showed the earth to be spherical, and her motions, when perfectly known, determine its deviation from that figure. The ellipticity of the earth, obtained from the motions of the moon, being independent of the irregularities of its form, has an advantage over that deduced from observations with the pendulum, and from the arcs of the meridian.

777. The inequality in the moon's latitude, arising from the ellipticity of the earth, may be represented by supposing that the orbit of the moon, in place of moving with the earth on the plane of the ecliptic, and preserving the same inclination of 5° 9′ to that plane, moves with a constant inclination of 8″ on a plane NMn passing between the ecliptic and the equator, and through nN, the line of the equinoxes. The inequality in question diminishes the inclination of the lunar orbit to the ecliptic, when its ascending node coincides with the equinox of spring; it augments it when this node coincides with the autumnal equinox.

fig. 104.

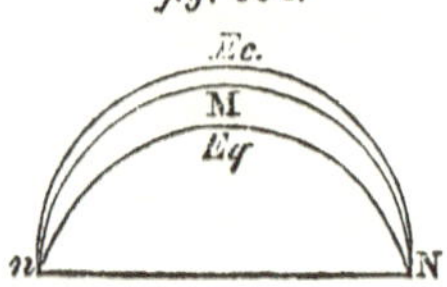

778. This inequality is the re-action of the nutation in the terrestrial axis, discovered by Bradley; hence there would be equilibrium round the centre of gravity of the earth, in consequence of the forces which produce the terrestrial nutation and this inequality in the moon's latitude, if all the molecules of the earth and moon were fixedly united by means of a lever; the moon compensating the smallness of the force which acts on her by the length of the lever to which she is attached, for the distance of the common centre of gravity of

the earth and moon from the centre of the earth is less than the earth's semidiameter.

The proof of this depends on the rotation of the earth; but some idea may be formed of this re-action from the annexed diagram. Let EC be the plane of the ecliptic, seen edgewise; Q the earth's equator; E its centre, and M the moon. Then QEC is the obliquity of the ecliptic, and MEC the latitude of the moon.

fig. 105.

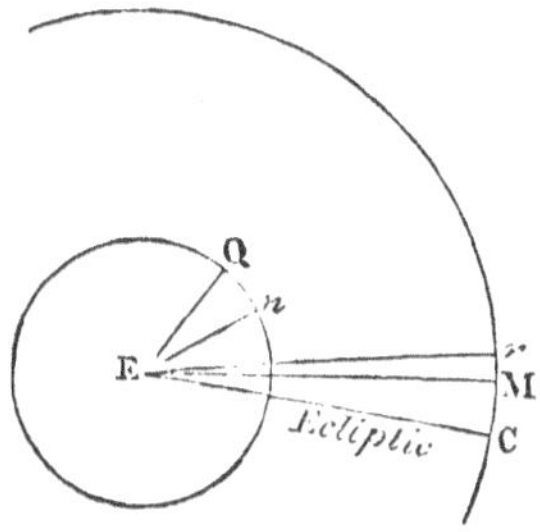

The moon, by her action on the redundant matter at Q, draws the equator to some point n nearer to the ecliptic, producing the nutation QEn; but as re-action is equal and contrary to action, the matter at Q draws the moon from M to some point r, thereby producing the inequality MEr in her latitude, that has been determined. La Place finds the analytical expressions of the areas MEr and QEn, and thence their moments; the one from the preceding inequality in the moon's latitude, the other from the formulæ of nutation in the axis of the earth's rotation from the direct action of the moon. These two expressions are identical, but with contrary signs, proving them, as he supposed, to be the effects of the direct and reflected action of the moon.

779. The form of the earth increases the motion of the lunar nodes and perigee by 0.00000008548$4v$, an insensible quantity. The ellipticity of the lunar spheroid has no perceptible effect on her motion.

CHAPTER IV.

INEQUALITIES FROM THE ACTION OF THE PLANETS.

780. THE action of the planets produces three different kinds of inequalities in the motions of the moon. The first, and by far the greatest, is that arising from their influence on the eccentricity of the earth's orbit, which is the cause of the secular inequalities in the mean motion, in the perigee, and nodes of the lunar orbit. The other two are periodic inequalities in the moon's longitude; one from the direct action of the planets on the moon, the other from the perturbations they occasion in the longitude and radius vector of the earth, which are reflected back to the moon by means of the sun.

fig. 106.

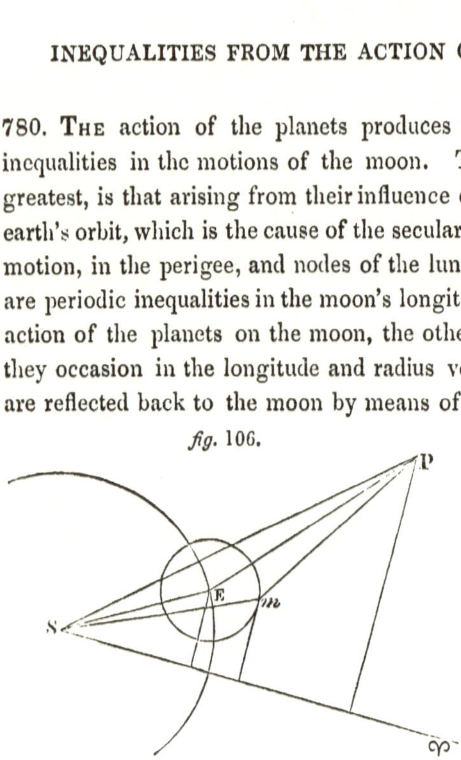

For, let S be the sun, E and m the earth and moon, P a planet, and γ the first point of Aries: then, if P be the mass of the planet, its direct action on the moon is $\frac{P}{(Pm)^2}$, which alters the position of the moon with regard to the earth. Again, the disturbing action of the planet on the earth is $\frac{P}{(PE)^2}$, which changes the position of the earth with regard to the moon, in each case producing inequalities of the same order. The latter become sensible from the very small divisors they acquire by integration.

The direct action will be determined first.

If X', Y', Z', x, y, z, be the co-ordinates of the planet and moon, referred to the centre of the earth, and f the distance of the planet from this centre, then

$$f = \sqrt{(X - x)^2 + (Y - y)^2 + (Z - z)^2}.$$

But if X', Y', Z', x', y', z', be the co-ordinates of the planet and the earth referred to the centre of the sun,

$$X = X' - x', \quad Y = Y' - y', \quad Z = Z' - z';$$

and $f = \sqrt{(X' - (x'+x))^2 + (Y' - (y'+y))^2 + (Z' - (z'+z))^2}$;

and the attraction of the planet on the moon is

$$\frac{P}{f} - \frac{\frac{1}{2}Pr^2}{f^3} + \tfrac{3}{2}P\frac{(X'x + Y'y + Z'z - xx' - yy' - zz')^2}{f^5} + \&c.$$

The ecliptic being the fixed plane,

$$z' = 0, \; r' = \frac{1}{u'}, \; v' = \gamma SE.$$

Then, if $R_{,} = SP$, $U = \gamma SP$, and S, be the radius vector, longitude, and heliocentric latitude of the planet, it is evident that

$$x' = \frac{\cos v'}{u'}, \; y' = \frac{\sin v'}{u'}, \; r' = \frac{\sqrt{1 + s^2}}{u'},$$

$$X' = R_{,} \cos U, \quad Y' = R_{,} \sin U, \quad Z' = R_{,}S;$$

hence $\qquad f = \sqrt{R^2(1 + S^2) + r'^2 - 2Rr' \cos (U - v')}$;

therefore the action of the planet on the moon is

$$\frac{P}{f} - \frac{\frac{1}{2}P(1 + s^2)}{u^2 f^3} + \tfrac{3}{2}P\frac{(R_{,}\cos (v - U) - r\cos (v - v') + R_{,}sS)^2}{u^2 f^5} + \&c.$$

or, omitting S^2, it is

$$\frac{P}{f} + \frac{P(1 - 2s^2)}{4u^2 f^3} + \&c. \&c.$$

The first term does not contain the co-ordinates of the moon, and therefore does not affect her motion; and the only term of the remainder of the series that has a sensible influence is $\frac{P}{4u^2 f^3}$, which, therefore, forms a part of R in (208); and, with regard to the action of the planets alone, $R = \frac{P}{4u^2 f^3}$. But, by article 446, the developement of f is

$$\tfrac{1}{2}A^{(0)} + A^{(1)} \cos (U - v') + A^{(2)} \cos 2(U - v') + \&c.$$

If i be the ratio of the mean motion of the planet to that of the moon, by equation (212)

$$U = iv - 2ie \sin (cv - \varpi) + \&c.$$

Hence, if iv be put for U, and mv for v', it is evident that

$$R = \frac{P}{4u^2}\{\tfrac{1}{2}A^{(0)} + A^{(1)}\cos(i-m)v + A^{(2)}\cos 2(i-m)v + \&c.\}$$

The only term of the parallax in which this value of R is sensible is

$$-\frac{1}{h^2}\left(\frac{dR}{du}\right)$$

which becomes

$$\frac{PA^{(0)}}{4h^2u^3} + \frac{P}{2h^2u^3}\{A^{(1)}\cos(i-m)v + A^{(2)}\cos 2(i-m)v + \&c.\};$$

or, if e^2 and γ^2 be neglected, $u^{-3} = a^3$, and the periodic part of $-\frac{1}{h^2}\left(\frac{dR}{du}\right)$ is

$$\frac{Pa^3}{2h^2}\{A^{(1)}\cos(i-m)v + A^{(2)}\cos 2(i-m)v + \&c.\}$$

But, by the second of equations (209), $-\frac{1}{h^2}\left(\frac{dR}{du}\right)$ contains the variation of $\frac{m'u'^3}{2h^2u^3}$ which is

$$-\frac{3m'u'^3}{2h^2u^4}\,\delta u = -\frac{3m^2}{2}\,.\,\delta u.$$

Let

$$\delta u = G_1\cos(i-m)v + G_2\cos 2(i-m)v + G_3\cos 3(i-m)v + \&c.$$

Therefore the direct disturbance of the planets gives

$$\frac{d^2u}{dv^2} + u =$$

$$-\frac{Pa^3}{2}\{A_1\cos(i-m)v + A_2\cos 2(i-m)v + \&c.\} +$$

$$\frac{3m^2}{2}\{G_1\cos(i-m)v + G_2\cos 2(i-m)v + \&c.\} =$$

$$G_1(1-(i-m)^2)\cos(i-m)v + G_2(1-4(i-m)^2)\cos 2(i-m)v + \&c.$$

And comparing similar cosines,

$$G_1 = -\frac{\tfrac{1}{2}P.A_1.a^3}{1-\tfrac{3}{2}m^2-(i-m)^2}$$

$$G^2 = -\frac{\tfrac{1}{2}P.A_2.a^3}{1-\tfrac{3}{2}m^2-4(i-m)^2}$$

$$G_3 = -\frac{\tfrac{1}{2}P.A_3.a^3}{1-\tfrac{3}{2}m^2-9(i-m)^2}$$

&c. &c.

and thus the integral u or (228) acquires the term

$$-\tfrac{1}{2}Pa^3\left\{\frac{A_1\cos(i-m)v}{1-\frac{3}{2}m^2-(i-m)^2}+\frac{A_2\cos 2(i-m)v}{1-\frac{3}{2}m^2-4(i-m)^2}+\&c.\right\}$$

consequently, the mean longitude $nt+\epsilon$ contains the term

$$\frac{Pa^3}{i-m}\left\{\frac{A_1\sin(i-m)v}{1-\frac{3}{2}m^2-(i-m)^2}+\frac{\frac{1}{2}A_2\sin 2(i-m)v}{1-\frac{3}{2}m^2-4(i-m)^2}+\&c.\right\}$$

or if a^3 be eliminated by $\dfrac{m'a^3}{a'^3}=m^2$

$$\frac{\frac{P}{m'}m^2a'^3}{i-m}\left\{\frac{A_1\sin(i-m)v}{1-\frac{3}{2}m^2-(i-m)^2}+\frac{\frac{1}{2}A_2\sin 2(i-m)v}{1-\frac{3}{2}m^2-4(i-m)^2}+\&c.\right\}\quad(245)$$

m' being the mass of the sun.

If B_1, B_2, &c., be put for A_1, A_2, &c., it becomes

$$\frac{\frac{P}{m'}m^2a^3}{i-m}\left\{\frac{B_1\sin(i-m)v}{1-\frac{3}{2}m^2-(i-m)^2}+\frac{\frac{1}{2}B_2\sin 2(i-m)v}{1-\frac{3}{2}m^2-4(i-m)^2}+\&c.\right\}\quad(246)$$

which is the inequality in the moon's mean longitude, arising from the action of a planet inferior to the earth.

And if α be the ratio of the mean distance of the planet from the sun to that of the sun from the earth, the substitution of α^3B_1, α^3B_2, &c., for A_1, A_2, &c., in equation (245), gives

$$\frac{\frac{P}{m'}m^2\,.\,a'^3\alpha^3}{i-m}\left\{\frac{B_1\sin(i-m)v}{1-\frac{3}{2}m^2-(i-m)^2}+\frac{\frac{1}{2}B_2\sin(i-m)v}{1-\frac{3}{2}m^2-4(i-m)^2}+\&c.\right\}\quad(247)$$

for the action of a superior planet on the mean longitude of the moon.

781. Besides these disturbances, which are occasioned by the direct action of the planets on the moon, there are others of the same order caused by the perturbations in the radius vector of the earth. The variation of u' was omitted in the developement of the co-ordinates of the moon, but

$$\delta.\frac{m'u'^3}{2h^2u^3}=\frac{3m'u'^2}{2h^2u^3}\delta u'$$

and when the eccentricities are omitted,

$$h^2=a,\text{ and}\qquad \frac{m'a^3}{a'^2}=a'm^2.$$

So
$$\delta \frac{m'u'^3}{2h^2u^3} = \frac{3a'm^2}{2a}\delta u';$$

since $\delta u' = \frac{\delta r'}{a'}$ are the periodic inequalities in the radius vector of the earth produced by the action of a planet, they are given in (158), and may be represented by

$$a'\delta u' = -\frac{P}{m'}\{K_1 \cos(i-m)v + K_2 \cos 2(i-m)v + \&c.\}$$

where the coefficients K_1 K_2, &c. are known, and $(i-m)v$ is the mean longitude of the planet minus that of the earth. Thus

$$\frac{3a'm^2}{2a}\delta u' = -\frac{3m^2}{2a}\cdot\frac{P}{m'}\{K_1 \cos(i-m)v + K_2 \cos 2(i-m)v + \&c.\}$$

By the method of indeterminate coefficients, it will be found that $a\delta u$ contains the function

$$\frac{3m^2}{2}\cdot\frac{P}{m'}\left\{\frac{K_1 \cos(i-m)v}{1-\frac{3}{2}m^2-(1-m)^2} + \frac{K_2 \cos 2(i-m)v}{1-\frac{3}{2}m^2-4(i-m)^2} + \&c.\right\}$$

and the mean longitude of the moon is subject to the inequality

$$-\frac{3m^2}{i-m}\cdot\frac{P}{m'}\left\{\frac{K_1 \sin(i-m)v}{1-\frac{3}{2}m^2-(i-m)^2} + \frac{K_2 \sin 2(i-m)v}{1-\frac{3}{2}m^2-4(i-m)^2} + \&c.\right\} \quad (248)$$

Numerical Values of the Lunar Inequalities occasioned by the Action of the Planets.

782. With regard to the action of Venus, the data in articles 611 and 610 give $a = 0.7233325$; $i - m = 0.04679$

and $\frac{P}{m'} = \frac{1}{356632}$; hence because

$$a'^3B_1 = 8.872894,$$
$$a'^3B_2 = 7.386580,$$
$$a'^3B_3 = 5.953940,$$

function (246) becomes

$$+0''.62015 \sin(i-m)v + 0''.25990 \sin 2(i-m)v + 0''.14125 \sin 3(i-m)v$$

which is the direct action of Venus on the moon. Now $\delta r' = -\frac{\delta u'}{u'^2}$, and when the eccentricity is omitted, $u'^2 = \frac{1}{a'^2}$; hence $\frac{\delta r'}{a'} = -a'\delta u'$.

But if the action of Venus on the radius vector of the earth be computed by the formula (158), it will be found that

$$\begin{aligned}a'\delta u' = & \ 0.0000064475 \cos (i-m)v \\ & - 0.0000184164 \ \cos 2(i-m)v \\ & + 0.000002908 \ \cos 3(i-m)v.\end{aligned}$$

This gives the numerical values of the coefficients K^0, K^1, &c.; hence formula (248) becomes

$$\begin{aligned}& + 0''.482200 \sin (i-m)v \\ & - 0''.69336 \,.\, \sin 2(i-m)v \\ & - 0''.07380 \,.\, \sin 3(i-m)v,\end{aligned}$$

which is the indirect action of the planets on the moon's longitude. Added to the preceding the sum is

$$\begin{aligned}& + 1''.10235 \,.\, \sin (i-m)v \\ & - 0''.43336 \,.\, \sin 2(i-m)v \\ & + 0''.06745 \,.\, \sin 3(i-m)v,\end{aligned}$$

the whole action of Venus on the moon's mean longitude.

783. Relative to Mars:

$$\begin{aligned}\alpha &= 0.65630030 \\ a'^3B_1 &= 5.727893 \\ a'^3B_2 &= 4.404530 \,.\, i - m = -0.0350306 \\ a'^3B_3 &= 3.255964 \\ \frac{P}{m} &= \frac{1}{1846082}.\end{aligned}$$

and by formula (158) with regard to Mars,

$$\begin{aligned}a'\delta u' = & + 0''.00000017778 \cos (i-m)v \\ & + 0 \,.\, 0000026121 \ \cos 2(i-m)v \\ & + 0 \,.\, 000000111 \ \cos 3(i-m)v;\end{aligned}$$

whence the action of Mars on the moon's mean longitude, both direct and indirect, is

$$\begin{aligned}& + 0''.025583 \sin (i-m)v \\ & + 0''.389283 \sin 2(i-m)v \\ & - 0''.027337 \sin 3(i-m)v.\end{aligned}$$

784. With regard to Jupiter,

$$\begin{aligned}\alpha &= 0.192205 \\ a'^3B_1 &= 0.618817 \\ a'^3B_2 &= 0.147980 \\ a'^3B_3 &= 0.0331045 \\ i - m &= -0.06849523 \\ \frac{P}{m} &= \frac{1}{1067.09}\end{aligned}$$

And formula (158) gives for the action of Jupiter on the radius vector of the earth,

$$
\begin{aligned}
a'\delta u' = &- 0.0000159055 \cos (i-m)v \\
&- 0.0000090791 \cos 2(i-m)v \\
&- 0.00000064764 \cos 3(i-m)v.
\end{aligned}
$$

Whence it is easy to see that the whole action of Jupiter on the mean longitude of the moon, both direct and indirect, is

$$
\begin{aligned}
&0''.74435 \sin (i-m)v \\
&- 0''.24440 \sin 2(i-m)v \\
&- 0''.01282 \sin 3(i-m)v.
\end{aligned}
$$

If all these inequalities, resulting from the action of the planets on the moon, be taken with a contrary sign, we shall have the inequalities that this action produces in the expression of the true longitude of the moon, $(i-m)v$ being supposed equal to the mean motion of the planet minus that of the earth.

785. The secular action of the planets on the moon, and the elements of her orbit, may be determined from the term $\frac{PA^0}{4h^2u^3}$; but as it is insensible, the investigation may be omitted.

CHAPTER V.

EFFECTS OF THE SECULAR VARIATION IN THE PLANE OF THE ECLIPTIC.

786. HAVING developed all the inequalities to which the moon is subject, we shall now show that the secular variation in the plane of the ecliptic has no effect on the inclination of the lunar orbit.

The latitude of the earth s', being extremely small, was omitted in the values of R, No. (208) : it can only arise from disturbances either secular or periodic ; both oscillate between fixed limits ; but we shall suppose s' to relate only to the secular variations in the plane of the ecliptic, and according to equations (138) shall only assume it to be equal to a series of terms of the form,

$$\Sigma K . \sin (v' + it + \epsilon),$$

i being a very small coefficient. Then omitting quantities of the order s^3, the tangent of the moon's latitude is

$$s = \gamma \sin (gv - \theta) + \Sigma K \sin (v + it + \epsilon) + \delta s;$$

equation (205), which determines the latitude, is

$$0 = \frac{d^2 s}{dv^2} + s + \frac{3m'u'^3 s}{2h^2u^4} - \frac{3mu'^3 s}{2h^2u^4} - \frac{3mu'^3 s}{h^2u^4} \cos(v - v') + \frac{3mu'^3 s}{2h^2u^4} \delta s.$$

Now $$\frac{3m'u'^3 s}{2h^2u^4} = \frac{3m^2}{2} . \frac{a'}{a} \Sigma K \sin (v + it + \epsilon).$$

The following term gives the same quantity with a contrary sign. And if $$\delta s = \Sigma bK \sin (v + it + \epsilon),$$
the last term gives

$$\frac{3m^2}{2} \frac{a'}{a} \Sigma bK \sin (v + it + \epsilon),$$

so that the differential equation of the moon's latitude becomes

$$0 = \frac{d^2 s}{dv^2} + s + \frac{3m^2}{2} . \frac{a'}{a} \Sigma bK \sin (v + it + \epsilon)$$

and if $\Sigma bK \sin (v + it + \epsilon)$ be put for $\frac{d^2 s}{dv^2} + s$, the equation becomes $0 = \Sigma (1 + b) K \{1 + (1 + i)^2\} \sin (v + iv + \epsilon)$

$$+ \frac{3m^2}{2} \frac{a'}{a} \Sigma bK \sin(v + iv + \epsilon)$$

for iv may be put for it, whence

$$b = -\frac{1-(1+i)^2}{1-(1+i)^2 + \frac{3m^2}{2} \cdot \frac{a'}{a}} = \frac{2i+i^2}{\frac{3m^2}{2} \cdot \frac{a'}{a} - 2i - i^2}.$$

Hence the variation of s, the moon's latitude, with regard to the secular motion of the ecliptic is

$$\frac{\Sigma (2i + i^2) K \sin(v + iv + \epsilon)}{\frac{3m^2}{2} \cdot \frac{a'}{a} - 2i - i2}.$$

This quantity is insensible, for iv is only about 16″ a year, and

$$\frac{3m^2}{2} \cdot \frac{a'}{a}$$

being nearly 40° 37′, the value of the factor

$$\frac{2i + i^2}{\frac{3m^2}{2} \cdot \frac{a'}{a} - 2i - i^2}$$

is only 0″.00022.

So that the ecliptic in its motion carries the orbit of the moon along with it.

787. The coincidence of theory with observation, in explaining the inequalities in the motions of the moon, affords the most conclusive proof of the universality of the law of gravitation. Having deduced all these inequalities from that one cause, La Place established the correctness of the results obtained by analysis by comparing them with the lunar tables computed by Mason from 1137 observations made by Bradley between the years 1750 and 1760, and corrected by Burg by means of upwards of 3000 observations made by Maskelyne between the years 1765 and 1793. He had the satisfaction to find that the greatest difference did not exceed 8″ in the longitude, while the difference in latitude was only 1″.94, a degree of accuracy sufficient to warrant the tables of latitude being regarded as equivalent to the result of theory: the approximations in latitude, indeed, are more simple and convergent than those in longitude. The inequalities in the lunar parallax are so small, that theory will determine them more correctly than observation. Accurate as these results are, it is still possible that the motions of the moon may be affected by the resistance of an ethereal medium surrounding the sun.

CHAPTER VI.

EFFECTS OF AN ETHEREAL MEDIUM ON THE MOTIONS OF THE MOON.

788. In order to determine its effects in the hypothesis of its existence, let x, y, z be the co-ordinates of the moon referred to the centre of gravity of the earth, and x', y', z' those of the earth referred to the centre of the sun. The absolute velocity of the moon round the sun will be

$$\frac{\sqrt{(dx + dx')^2 + (dy + dy')^2 + (dz + dz')^2}}{dt}.$$

If K be a coefficient depending on the density of the ether, on the surface of the moon, and on her density; and if the resistance of the ether be assumed proportional to the square of the velocity, it will be

$$\frac{K\{(dx + dx')^2 + (dy + dy')^2 + (dz + dz'^2)\}}{dt^2}.$$

In the same manner

$$\frac{K'(dx'^2 + dy'^2 + dz'^2}{dt^2}$$

is the resistance the earth experiences from the ether, K' being a coefficient for the earth similar to, but different from K. In the theory of the moon the earth is assumed to be at rest, therefore this resistance must be in a contrary direction from that acting on the moon, consequently the whole action of the ether in disturbing the moon will be the difference of these forces: so with regard to the action of the ether alone, (208) becomes

$$R = \frac{K'(dx'^2 + dy'^2 + dz'^2)}{dt^2} - \frac{K\{(dx+dx')^2 + (dy+dy')^2 + (dz+dz')^2\}}{dt^2}$$

and because the resistance is in the plane of the orbit, its component forces are parallel to the axes x and y only; hence

$$\frac{dR}{dx} = K'\frac{dx'}{dt^2} \cdot \sqrt{dx'^2 + dy'^2 + dz'^2}$$

$$-K\frac{(dx+dx')}{dt^3}\cdot\sqrt{(dx+dx')^2+(dy+dy')^2+(dz+dz')^2}$$

$$\frac{dR}{dy}=K'\frac{dy'}{dt^2}\cdot\sqrt{dx'^2+dy'^2+dz'^2}$$

$$-K\frac{(dy+dy')}{dt^2}\cdot\sqrt{(dx+dx')^2+(dy+dy')^2+(dz+dz')^2}$$

But in the theory of the moon

$$x=\frac{\cos v}{u},\quad y=\frac{\sin v}{u},\quad z=\frac{s}{u},\quad x'=\frac{\cos v'}{u'},\quad y'=\frac{\sin v'}{u'},$$

and if the ecliptic of 1750 be assumed as the fixed plane $z'=0$: v' is the heliocentric longitude of the earth.

Let $\sqrt{dx'^2+dy'^2+dz'^2}$, the little arc described by the earth in the time dt be represented by $r'ds'$. This arc is to that described by the moon in her relative motion round the earth as $\frac{a'm}{a}$ to unity, consequently at least thirty times as great. If the eccentricity of the terrestrial orbit be omitted, $ds'=mdt$. If these quantities be substituted for the co-ordinates

$$\sqrt{(dx+dx')^2+(dy+dy')^2+(dz+dz')^2}=$$
$$ma'dt-dx\,.\sin v'+dy\,.\cos v';$$

and if quantities depending on the arc $2v'$ be rejected,

$$\frac{dR}{dx}=\frac{(K-K')m^2}{u'^2}\,.\sin v'-\frac{3Km}{2u'}\cdot\frac{dx}{dt}\qquad(249)$$

$$\frac{dR}{dy}=\frac{(K'-K)m^2}{u'^2}\,.\cos v'-\frac{3Km}{2u'}\cdot\frac{dy}{dt}.$$

But $$d\frac{1}{a}=-2dR=-2dx\left(\frac{dR}{dx}\right)-2dy\left(\frac{dR}{dy}\right).\qquad(250)$$

and $$d\frac{1}{a}=-\frac{2Km^2}{u'^2}\cdot\{dx\,.\sin v'-dy\,.\cos v'\}\qquad(251)$$

$$+\frac{3Km^2}{u'}\cdot\left\{\frac{dx^2+dy^2}{dt}\right\}.$$

The different quantities contained in this equation must now be determined.

789. The distance of the moon from the earth is $Em=\frac{1}{u}$, that of

the earth from the sun is $ES = \frac{1}{u'}$, and that of the moon from the sun is

$$mS = u' \sqrt{1 + \frac{u'^2}{u^2} - 2\frac{u'}{u}\cos(v - v')}$$

but $\frac{u'^2}{u^2}$ is a very small fraction that may be omitted; consequently, when the square root is extracted, the distance of the moon from the sun is

$$mS = u' - \frac{u'^2}{u} \cdot \cos(v - v').$$

If we assume the density of the ether to be proportional to a function of the distance from the sun, and represent that function by $\phi(u')$, with regard to the moon, it will be

$$\phi(u') - \frac{u'^2}{u} \cdot \phi'(u') \cdot \cos(v - v')$$

$\phi'(u')$ being the differential of $\phi(u')$ divided by du'. As K is a quantity depending on the density of the ether it is variable, hence it may be assumed that

$$K = H \cdot \phi(u') - \frac{Hu'^2}{u} \cdot \phi'(u') \cdot \cos(v - v').$$

But as $x = \frac{\cos v}{u}$, $y = \frac{\sin v}{u}$, $u = \frac{1}{a}(1 + e\cos(cv - \varpi))$, therefore

$$dx = -a^2(udv \cdot \sin v + du \cdot \cos v) \cdot (1 - 2e\cos(cv - \varpi)),$$

$$dy = a^2(udv \cdot \cos v - du \cdot \sin v) \cdot (1 - 2e\cos(cv - \varpi)),$$

also $$dt = dv(1 - 2e \cdot \cos(cv - \varpi)).$$

790. By the substitution of these quantities in equation (241) it will be found, after rejecting periodic quantities, and integrating, that

$$\frac{1}{a} = -Hma^2\left\{\frac{3\phi(u')}{u'} - m \cdot \phi'(u')\right\} \cdot v$$

$$+ Hma^2\left\{\frac{6\phi(u')}{u'} - \frac{9}{2}m.\phi'(u')\right\}.e\sin(cv - \varpi),$$

which is the secular variation in the mean parallax of the moon in consequence of the resistance of the ether.

In order to abridge, let

$$\alpha = Hma^2\left\{\frac{3\phi(u')}{u'} - m\phi'(u')\right\}$$

$$\text{G} = Hma^2 \left\{\frac{6\phi\,(u')}{u'} - \frac{9}{2}\,m\,.\,\phi'(u')\right\},$$

then
$$\frac{1}{a} = -\alpha v + \text{G}\,.\,e\sin(cv - \varpi).$$

The value of $\frac{1}{a}$ in equation (225) will be augmented by αv, therefore a will be diminished by αv.

Since
$$d\frac{1}{a} = -2dR,$$

therefore
$$dR = \frac{\alpha}{2\bar{a}}\,dv - \frac{\text{G}}{2\bar{a}}\,dv\,.\,e\,.\cos(cv-\varpi).$$

Consequently, when periodic quantities are omitted, $\zeta = -3\int adv\,.\,dR$ gives
$$\zeta = -\frac{3\alpha}{4\bar{a}}\,av^2$$

or, omitting the action of the sun,
$$\zeta = -\tfrac{3}{4}\,\alpha v^2.$$

Thus the mean motion is affected by a secular variation from the resistance of the ethereal medium; but it may easily be shown, from the value of R in article 788, that this medium has no effect whatever on the motion of the lunar nodes or perigee. However, in consequence of that action the second of equations (224), which is the coefficient of $\sin(cv - \varpi)$, ought to be augmented by $\text{G}\,.\,e$; hence, rejecting c^2, $d\varpi$, and making $c = 1$ it gives

$$\text{G}\,.\,edv = 2\,.\,d\frac{e}{a},$$

or
$$\frac{e}{a} = \text{constant}\,(1 + \tfrac{1}{2}\,\text{G}v);$$

but as $\frac{1}{a}$ must be augmented by αv, if the square of v be omitted,

$$e = \text{constant}\,(1 - (\alpha - \tfrac{1}{2}\text{G})\,v).$$

Thus the eccentricity of the lunar orbit is affected by a secular inequality from the resistance of ether, but it is insensible when compared with the corresponding inequality in the mean motion.

It appears then that the mean motion of the moon is subject to a secular variation in consequence of the resistance of ether, which neither affects the motion of the perigee nor the position of the orbit; and, as the secular inequalities of the moon deduced theoretically from the variation of the eccentricity of the earth's orbit are perfectly confirmed by the concurrence of ancient and modern observations, they cannot be ascribed to the resistance of an ethereal medium.

791. The action of the ether on the motions of the earth may be found by the preceding formulæ to be

$$\frac{dR}{dx} = K'm^2a'^2 \,.\, \sin v'$$

$$\frac{dR}{dy} = -\,K'm^2a'^2 \,.\, \cos v'\,;$$

when the eccentricity of the earth's orbit is omitted, so that

$$u' = \frac{1}{a'}.$$

Consequently the general equation (250) gives

$$\text{d}R = -\,K'.a'^3.m^3.dt, \text{ and therefore}$$

$$\delta v = -\,\frac{3a'}{m'}\,.\iint dt.\text{d}R = \tfrac{3}{2}\,\frac{K'.a'^4m^4.t^2}{m'},$$

m' being the mass of the sun.

If $\phi\,(u')$ be a function of the distance of the earth from the moon, then must $K' = H'\,.\,\phi\,(u')$, H' being a constant quantity depending on the mass and surface of the earth. Whence it may be found by the same method with that employed, that the resistance of ether in the mean motion of the earth would be

$$\zeta = \tfrac{3}{2}\,\frac{H'a'^4m^4t^2.\phi\,(u')}{m'}.$$

Whence it appears that the acceleration in the mean motion of the moon is to that in the mean motion of the earth as unity to

$$\frac{2H'\,.\,m\,.\,\phi\,(u')}{H\left\{3\phi\,(u') - \dfrac{m}{a'}\,\phi'\,(u')\right\}},$$

or as unity to $\quad \tfrac{2}{3}\,m\,.\,\dfrac{H'}{H}$, if $-\dfrac{m}{a'}\,\phi'\,(u')$

be omitted, and because $\dfrac{m'a^3}{a^3} = m^2.$

Now H' and H depend on the masses and surfaces of the earth and moon; and as the resistance is directly as the surface, and inversely as the mass, therefore

$$H = \frac{\text{surface}}{\text{mass}}.$$

But by article 652, if the radius of the earth be unity, the moon's true diameter =

$$\frac{\frac{1}{2}\text{ moon's apparent diameter}}{\text{moon's horizontal parallax}};$$

hence surface of moon =

$$\frac{(\text{apparent diameter})^2}{(\text{lunar parallax})^2}$$

and $$H = \frac{\{\frac{1}{2}\text{ apparent diameter of moon}\}^2}{\text{mass of moon }\{\text{ lunar parallax }\}^2}.$$

But as the terrestrial radius is assumed = 1, the earth's surface is unity; so $$H' = \frac{1}{\text{mass of earth}};\text{ hence}$$

$$\frac{H'}{H} = \frac{\text{mass of moon}}{\text{mass of earth}} \cdot \frac{\text{square horizontal parallax of moon}}{\text{square of }\frac{1}{2}\text{ moon's apparent diameter}}.$$

From observation half the moon's apparent diameter is 943″.164, her horizontal parallax is 3454.16, and her mass is $\frac{1}{75}$ of that of the earth, so $\frac{H'}{H} = 0.17883$; and as $m = \frac{1}{13.3}$, it follows that the acceleration in the mean motion of the earth from the resistance of ether is equal to the corresponding acceleration in the mean motion of the moon multiplied by 0.008942, or about a hundred times less than the acceleration of the moon from the resistance of ether. No such acceleration has been detected in the earth's motion, nor could it be expected, since it is insensible with regard to the moon.

In the preceding investigation, the resistance was assumed to be as the square of the velocity, but Mr. Lubbock has obtained general formulæ, which will give the variations in the elements, whatever the law of this resistance may be.

792. Although we have no reason to conclude that the sun is surrounded by ether, from any effects that can be ascribed to it in the motions of the moon and planets, the question of the existence of such a fluid has lately derived additional interest from the retardation that has been observed in the returns of Enke's comet at each revolution, which it is difficult to account for by any other supposition than this existence of such a medium.

Mr. Enke has proved that this retardation does not arise from the disturbing action of the planets. But on computing the effects of the resistance of an ether diffused through space, he found that the diminution in the periodic time, and on the eccentricity arising from the ether, supposing it to exist, corresponds exactly with observation. This coincidence is very remarkable, because ignorance of the nature of the medium in question imposes the necessity of forming an hypothesis of the law of its resistance. Future returns of this comet will furnish the best proof of the existence of an ether, which, by the computation of Mazotti, must be 360,000 millions of times more rare than atmospheric air, in order to produce the observed retardation. The existence of an ethereal medium, if established, would not only be highly important in astronomy, but also from the confirmation it would afford of the undulating theory of light; among whose chief supporters we have to number Huygens, Descartes, Hooke, Euler, and, in later times, the illustrious names of Young and Fresnel, who have applied it with singular success and ingenuity to the explanation of those classes of phenomena which present the greatest difficulties to the corpuscular doctrine.

793. La Place employs the same analysis to determine the effects that the resistance of light has on the motions of the bodies of the solar system, whether considered as propagated by the undulations of a very rare medium as ether, or emanating from the sun. He finds that it has no effect whatever on the motion of the perigee, either of the sun or moon; that its action on the mean motions of the earth and moon is quite insensible; but that the action of light, on the mean motion of the moon, in the corpuscular hypothesis, is to that in the undulating system as -1 to 0.01345.

794. If gravitation be produced by the impulse of a fluid towards the centre of the attracting body, the same analysis will give the secular equation due to the successive transmission of the attractive force. The result is, that if g be the attraction of any body as the earth; G the ratio of the velocity of the fluid which causes gravitation to that of the moon, at her mean distance, and t any finite time, the secular equation of the mean motion of the moon from the transmission of the attractive force is $\frac{3}{2}\frac{gt^2}{aG}$.

The gravity of a body moving in its orbit is equal to its centrifugal force; and the latter is equal to the square of the velocity divided by the radius vector; and as the square of the moon's velocity is $a^2(27.32166)^2$ its centrifugal force is $(27.32166)^2$, whence

$$g = (27.32166)^2;$$

and the secular equation becomes

$$\tfrac{3}{2}\left(\frac{27.32166)^2}{G}\right).t^2.$$

Since G is the ratio of the velocity of the fluid in question to the velocity of the moon $G = \dfrac{\text{vel. fluid}}{a(27.32166)}$;

hence the velocity of the fluid is $(27.32166)aG$.

If

$$L = \frac{\text{velocity of the fluid}}{\text{velocity of light}},$$

then the velocity of the gravitating fluid is equal to L velocity of light; whence L. vel. light $= (27.32166)aG$; but by Bradley's theory, the velocity of light is

$$\frac{(365.25)a}{\tan 20''.25},$$

a' being the mean distance of the earth from the sun; whence

$$L.\frac{(365.25)a'}{\tan 20''.25} = (27.32166)aG,$$

$$G = \frac{L(365.25)a'}{(27.32166)a.\tan 20''.25}.$$

And the secular equation of the moon from the successive transmission of gravity becomes

$$\tfrac{3}{2}.\frac{(27.32166)^3}{365.25}.\frac{a}{a'}t^2.\tan 20''.25.$$

Now, if the acceleration in the moon's mean motion arises from the successive transmission of gravity, and not from the secular variation in the earth's eccentricity, the preceding expression would be equal to $10''.1816213$, the acceleration in 100 Julian years. Therefore, making $t = 100$,

$$L = \tfrac{3}{2}\frac{a}{a'}\frac{(27.32166)^3}{365.25}.\frac{10000\tan 20''.25}{10''.1816213};$$

but $\dfrac{a}{a'} = \dfrac{1}{400}$; whence $L = 42145000$;

thus the velocity with which gravity is transmitted must be more than forty-two million times greater than the velocity of light;

the velocity of light: hence we must suppose the velocity of the moon to be many a hundred million times greater than that of light to preserve her from being drawn to the earth, if her acceleration be owing to the successive transmission of gravity. The action of gravity may therefore be regarded as instantaneous.

795. These investigations are general, though they have only been applied to the earth and moon; and, as the influence of the ethereal media and of the transmission of gravity on the moon is quite insensible, though greater than on the earth, it may be concluded, that they have no sensible effect on the motions of the solar system; but as they do not affect the motions of the lunar perigee and the perihelia of the earth and planets at all, these motions afford a more conclusive proof of the law of gravitation, than any other circumstance in the system of the world. The length of the day is proved to be constant by the secular equation of the moon. For if the day were longer now than in the time of Hipparchus by the 0.00324th of a second, the century would be 118″.341 longer than at that period. In this interval, the moon would describe an arc of 173″.2, and her actual mean secular motion would appear to be augmented by that quantity; so that her acceleration, which is 10″.206 for the first century, beginning from 1801, would be increased by 4″.377; but observations do not admit of so great an increase. It is therefore certain, that the length of the day has not varied the 0.00324th of a second since the time of Hipparchus.

796. It is evident then, that the lunar motions can be attributed to no other cause than the gravitation of matter: of which the concurring proofs are the motion of the lunar perigee and nodes; the mass of the moon; the magnitude and compression of the earth; the parallax of the sun and moon, and consequently the magnitude of the system; the ratio of the sun's action to that of the moon, and the various secular and periodic inequalities in the moon's motions, every one of which is determined by analysis on the hypothesis of matter attracting inversely as the square of the distance; and the results thus obtained, corroborated by observation, leave not a doubt that the whole obey the law of gravitation. Thus the moon is, of all the heavenly bodies, the best adapted to establish the universal influence of this law of nature; and, from the intricacy of her motions, we may form some idea of the powers of analysis, that mar-

vellous instrument, by the aid of which so complicated a theory has been unravelled.

797. Before we leave the subject, it may be interesting to show that the differential equations of the lunar co-ordinates, given in (207), may be derived from Newton's theory.

If the inclination of the lunar orbit be omitted, the whole force which disturbs the moon may be resolved into two; one perpendicular to the radius vector, and another, according to the radius vector, and directed towards the centre of the earth. Now, $\frac{1}{r}\left(\frac{dR}{dv}\right)$ is the first of these forces, and $-\left(\frac{dR}{dr}\right)$ is the other. The force $\frac{1}{r}\left(\frac{dR}{dv}\right)$, multiplied by dt, gives the increment of the velocity of the moon perpendicular to the radius during the instant dt; and when multiplied by $\frac{1}{2}rdt$, it becomes $\frac{1}{2}\left(\frac{dR}{dv}\right)dt =$ the increment of the area described by the radius vector in the time dt. It must therefore be equal to $\frac{1}{2}\cdot\frac{d\,.\,r^2dv}{dt}$;

hence
$$\frac{d\,.\,r^2dv}{dt} = \left(\frac{dR}{dv}\right)dt.$$

If this equation be multiplied by $\frac{r^2dv}{dt}$ and integrated, the result will be
$$(r^2dv)^2 = h^2dt^2\left(1 + \frac{2}{h^2}\int\left(\frac{dR}{dv}\right)r^2dv\right);$$

and as $r = \frac{1}{u}$, it becomes
$$dt = \frac{dv}{hu^2\sqrt{1 + \frac{2}{h^2}\int\left(\frac{dR}{dv}\right)\frac{dv}{u^2}}},$$

which is the first of equations (207).

Again, if ds be the element of the curve described by the moon, $\frac{ds^2}{dt^2}$ will be the square of her velocity; and, substituting the preceding value of dt, the square of the moon's velocity will be

$$h^2u^4 \cdot \frac{ds^2}{dv^2} \cdot \left\{1 + \frac{2}{h^2}\int\left(\frac{dR}{dv}\right) \cdot \frac{dv}{u^2}\right\}.$$

If r' be the osculating radius of the orbit, the expression of the radius of curvature, in article 83, will give, when substitution is made for x, y, z, in supposing dv constant,

$$\frac{1}{r'} = dv^3\frac{\left(\dfrac{d^2u}{dv^2} + u\right)}{u^3ds^3}.$$

Hence the square of the moon's velocity, divided by the radius of curvature, is

$$u \cdot \frac{dv}{ds} \cdot h^2\left\{\frac{d^2u}{dv^2} + u\right\} \cdot \left\{1 + \frac{2}{h^2}\int\left(\frac{dR}{dv}\right)\frac{dv}{u^2}\right\}. \qquad (252)$$

By the theorems of Huygens, this expression must be equal to the lunar force resolved in the radius of curvature, and directed towards the centre of curvature. Now, if the force $-\left(\dfrac{dR}{dr}\right)$ be resolved into two, one parallel to the element of the curve, and the other directed to the centre of curvature, the latter will be $u\left(\dfrac{dR}{du}\right) \cdot \dfrac{dv}{ds}$. Also the force $\dfrac{1}{r}\left(\dfrac{dR}{dv}\right)$, resolved according to the radius of curvature, will be $-\dfrac{du}{uds}\left(\dfrac{dR}{dv}\right)$. The sum of these two forces directed towards the centre of curvature is

$$u \cdot \frac{dv}{ds}\left(\frac{dR}{du}\right) - \frac{du}{uds}\left(\frac{dR}{dv}\right).$$

If the square of this expression be made equal to that of (252), then

$$0 = \left(\frac{d^2u}{dv^2} + u\right)\left\{1 + \frac{2}{h^2}\int\left(\frac{dR}{dv}\right)\frac{dv}{u^2}\right\} - \frac{1}{h^2}\left(\frac{dR}{du}\right) + \frac{du}{h^2u^2dv}\left(\frac{dR}{dv}\right);$$

which is the same with the second of equations (202), when the inclination of the orbit is omitted.

The equation in latitude is not so easily found as the other two; but the method followed by Newton was to resolve the action of the sun on the moon into two, one in the direction of the radius vector of the lunar orbit, the other parallel to a line joining the centres of the sun and earth. The difference between the last force and the action of the sun on the earth, he saw to be the only force that could change the position of the lunar orbit, since it is not in that plane. He deter-

fig. 107.

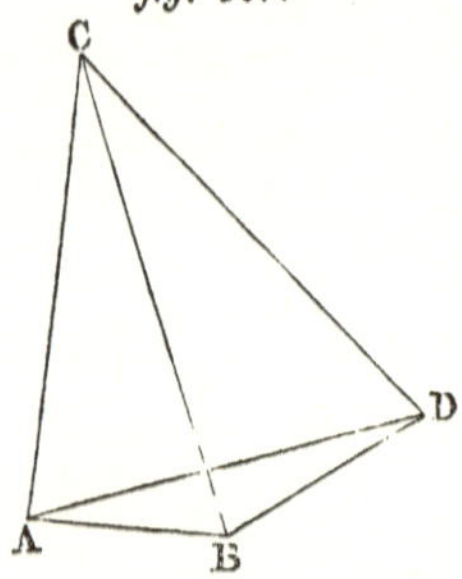

mined the effect of this force, by supposing AB, fig. 107, to be the arc described by the moon in an instant; then ACB is the plane of the orbit during that time; in the next instant, the difference of the two forces causes the moon to describe the small arc BD in a different plane; then if BD represent the difference of the forces, and if AB be the velocity of the moon in the first instant, the diagonal BD will be the direction of the velocity in the second instant; and ACD will be the position of the orbit. Newton deduced the horary and mean motion of the nodes, their principal variation, and the inequalities in latitude, from these considerations. La Place considered the theory of the moon as the most profound and ingenious part of the Principia.

BOOK IV.

CHAPTER I.

THEORY OF JUPITER'S SATELLITES.

798. JUPITER is attended by four satellites, which were discovered by Galileo on the 1st of June, 1610; their orbits are nearly in the plane of Jupiter's equator, and they exhibit all the phenomena of the solar system, on a small scale and in short periods. The eclipses of these satellites afford the easiest method of ascertaining terrestrial longitudes; and the frequency of the occurrence of an eclipse renders the theory of their motions nearly as important to the geographer as that of the moon.

799. The orbits of the two first satellites are circular, subject only to such eccentricities as arise from the disturbing forces; the third and fourth satellites have elliptical orbits; the eccentricity however is so small, that their elliptical motion is determined along with those perturbations that depend on the eccentricities of the orbits.

800. Although Jupiter's satellites might be regarded as an epitome of the solar system, they nevertheless require a new investigation, on account of the nearly commensurable ratios in the mean motions of the three first satellites, the action of the sun, the ellipticity of Jupiter's spheroid, and the displacement of his orbit by the action of the planets.

801. It appears, from observation, that the mean motion of the first satellite is nearly equal to twice that of the second; and that the mean motion of the second is nearly equal to twice that of the third; whence the mean motion of the first, minus three times that of the second, plus twice that of the third, is zero; but the last ratio is so exact, that from the earliest observations it has always been zero.

It is also found that, from the time of the discovery of the satellites, the mean longitude of the first, minus three times that of the second, plus twice that of the third, is equal to 180°: and it will be shown, in the theory of these bodies, that even if these ratios had not been exact in the origin of their motions, their mutual attractions would have made them so. They are the cause of the principal inequalities in the longitude of the satellites; and as they exist also in their synodic motions, they have a great influence on the times of their eclipses, and indeed on their whole theory.

802. The prominent matter at Jupiter's equator, together with the action of the satellites themselves, causes a direct motion in the apsides, which changes the relative position of the orbits, and alters the attractive force of the satellites; consequently each satellite has virtually four equations of the centre, or rather, that part of the longitude of each satellite that depends on the eccentricity, consists of four principal terms; one that arises from the true ellipticity of its own orbit, and three others, depending on the positions of the apsides of the other three orbits. Inequalities perfectly similar to these are produced in the radii vectores by the same cause, consisting of the same number of terms, and depending on the same quantities.

803. Astronomers imagined that the orbits of the satellites had a constant inclination to the plane of Jupiter's equator; however, they have not always the same inclination, either to the plane of his equator or orbit, but to certain imaginary fixed planes passing between these, and also through their intersection.

fig. 108.

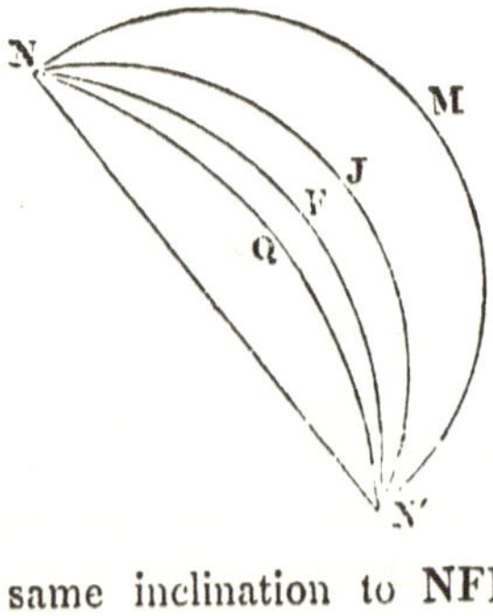

Let NJN′ be the orbit of Jupiter, NQN′ the plane of his equator extended so as to cut his orbit in NN′; then, if NMN′ be the orbit of a satellite, it will always preserve very nearly the same inclination to a fixed plane NFN′, passing between the planes NQN′ and NJN′, and through the line of their nodes. But although the orbit of the satellite preserves nearly the same inclination to NFN′, its nodes have a retrograde motion on that plane. The plane FN itself is not absolutely fixed, but moves slowly with the equator and orbit of Jupiter. Each satellite has a different fixed plane, which is less inclined to the plane of Jupiter's equator the nearer the satellite is to the planet, evidently arising from the attraction of the protuberance at Jupiter's equator, which retains the satellites nearly in the plane of the equator; furnishing another proof of the mutual attraction of the particles of matter.

804. The equatorial matter of Jupiter's spheroid causes a retrograde motion in the nodes of the orbits of the satellites; which alters their mutual attraction, by changing the relative position of their planes, so that the latitude of any one satellite not only depends on the position of the node of its own orbit, but on the nodes of the other three; and as the position of Jupiter's equator is perpetually varying, in consequence of the action of the sun and satellites, the latitude of these bodies varies also with the inclination of Jupiter's equator on his orbit, and the position of its nodes. Thus, the principal inequalities of the satellites arise from the compression of Jupiter's spheroid, and from the direct and indirect action of the sun and satellites themselves.

805. The secular variation in the form and position of Jupiter's orbit is the cause also of secular variations in the motions of the satellites, similar to those in the motions of the moon occasioned by the variation in the eccentricity and position of the earth's orbit.

806. The position of the orbit of a satellite may be known by supposing five planes, of which FN, passing between JN and QN, the planes of Jupiter's orbit and equator, always retains very nearly the same inclination to them. The second plane A*n* moves uniformly on FN, retaining nearly the same inclination on it. The third B*n'* moves in the same manner on A*n*; the fourth C*n''* moves similarly on B*n'*; and the fifth M*n'''*, which has the same kind of motion on C*n''*, is the orbit of the satellite. The motion of the nodes are retrograde, and each satellite has a set of planes peculiar to itself. In conformity with this, the latitude of a satellite above the variable orbit of Jupiter, is expressed by five terms; the first of which is relative to the displacement of the orbit and equator of

fig. 109.

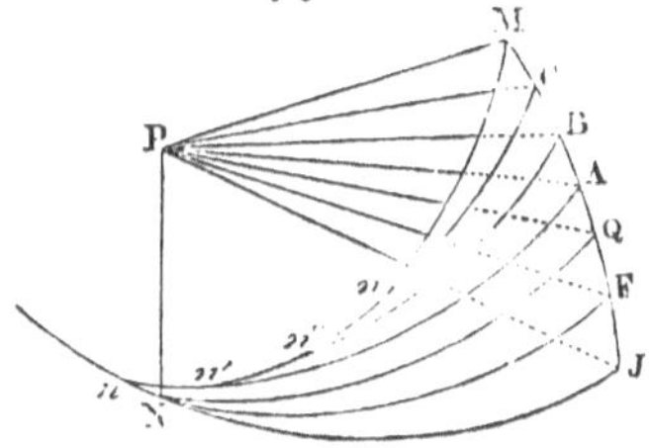

Jupiter; the second is relative to the inclination of the orbit of the satellite on its fixed plane; and the other three terms depend on the position and motion of the nodes of the other three orbits. The inequalities which have small divisions, arising from the configuration of the bodies, are insensible in latitude, with the exception of those produced by the sun, which modify the preceding quantities.

807. For the solution of the problem of the satellites, the data that must be determined by observation for a given epoch, are, the compression of Jupiter's spheroid, the inclination of his equator on his orbit, the longitude of its nodes, the eccentricity of his orbit, its position, and its secular variations; the masses of the four satellites, their mean distances, periodic times, the eccentricities and inclinations of their orbits, together with the longitude of their apsides and nodes. The masses of the satellites and the compression of Jupiter are determined from the inequalities of the satellites themselves.

808. The orbits of the four satellites may be regarded as circular, because the eccentricity of the third, and even the fourth, is so small, that their equations of the centre will be determined with the perturbations depending on the eccentricities and inclinations. Thus, with regard to the two first, and nearly for the other two, the true longitude is the sum of the mean longitude and perturbations; and the radius vector will be found by adding the perturbations to the mean distance.

809. A satellite m is troubled by the other three, by the sun, and by the excess of matter at Jupiter's equator. The problem however will be limited to the action of the sun, of Jupiter's spheroid, and of one satellite; the resulting equations will be general, from whence the action of each body may be computed separately, and the sum will be the effect of the whole.

810. Let m and $m_{,}$ be the masses of any two satellites, x, y, z, x', y', z', their rectangular co-ordinates referred to the centre of gravity of Jupiter, supposed to be at rest; r, r' their radii vectores; then the disturbing action of $m_{,}$ on m is

$$\frac{m_{,}(xx' + yy' + zz')}{r_{,}^{3}} - \frac{m_{,}}{\sqrt{(x'-x)^2 + (y'-y)^2 + (z'-z)^2}} = R;$$

consequently the sign of R must be changed in equations (155) and (156), since it is assumed to be negative in this case.

The satellites move nearly in the plane of Jupiter's equator, which in 1750 was inclined to the plane of his orbit at an angle of 3° 5′ 30″; and as the fixed planes pass between these two, the inclinations of the orbits of the satellites to them are very small; consequently $s = m\text{P}$, $s_{,} = m_{,}\text{P}'$, fig. 110, the tangents of the latitude of the two satellites on PJP′, the fixed plane of m, are very small.

fig. 110.

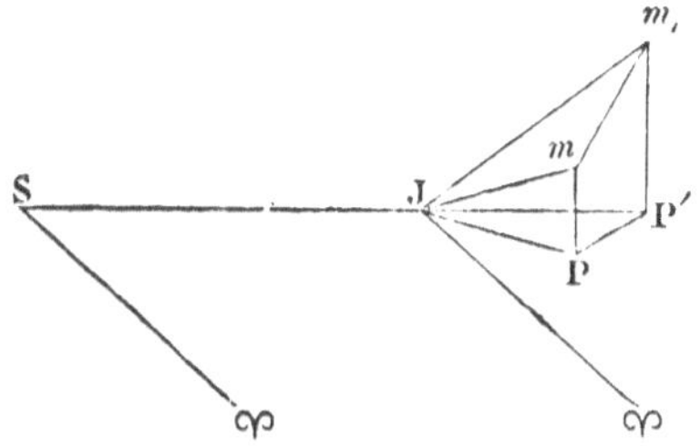

If γ be the vernal equinox of Jupiter, the longitudes of the two satellites are $\gamma JP = v$, $\gamma JP' = v_{,}$, and therefore

$$x = \frac{r\cos v}{\sqrt{1+s^2}}, \quad y = \frac{r\sin v}{\sqrt{1+s^2}}, \quad z = \frac{rs}{\sqrt{1+s^2}}.$$

If x', y', z', the co-ordinates of $m_{,}$, be equal to the same quantities accented, the action of $m_{,}$ on m, expressed in polar co-ordinates, will be

$$R = \frac{m_{,}r}{r_{,}^2}\{ss_{,} + (1 - \tfrac{1}{2}s^2 - \tfrac{1}{2}s_{,}^2)\cos(v_{,} - v)\} - \frac{m_{,}}{\sqrt{r^2 - 2rr_{,}\cos(v_{,} - v) + r_{,}^2}}$$

$$- \frac{m_{,}\,.\,rr_{,}\,.\,\{ss_{,} - \frac{1}{2}(s^2 + s_{,}^2)\cos(v_{,} - v)\}}{\{r^2 - 2rr_{,}\,.\,\cos(v_{,} - v) + r_{,}^2\}^{\frac{3}{2}}},$$

when s^4, $s_{,}^4$ are neglected.

811. If S' be the mass of the sun, and X', Y', Z', his co-ordinates, his action upon m will be expressed by

$$R = \frac{S'(X'x + Y'y + Z'z)}{D^3} - \frac{S'}{\sqrt{(X'-x)^2 + (Y'-y)^2 + (Z'-z)^2}},$$

D being his distance from the centre of Jupiter.

Let Jupiter and his orbit be assumed to be at rest, and let his motion be referred to the sun, which is the same as supposing the sun to move in the orbit of Jupiter with the velocity of that planet; if S be the tangent of the sun's latitude above the fixed plane PJP', and $U = \gamma SJ$, his longitude seen from the centre of Jupiter when at rest, then

$$X' = \frac{D\cos U}{\sqrt{1+S^2}}, \quad Y' = \frac{D\sin U}{\sqrt{1+S^2}}, \quad Z' = \frac{D\,.\,S}{\sqrt{1+S^2}},$$

and

$$R = -\frac{S'}{D} - \frac{S'r^2}{4D^3}\{1-3s^2-3S^2+12sS(\cos(U-v)+3\cos 2(U-v)\},$$

which is the action of the sun on the satellite when terms divided by D^4 are omitted, for the distance of the satellite from Jupiter is incomparably less than the distance of Jupiter from the sun.

812. The attraction of the excess of matter at Jupiter's equator is expressed by

$$R = -(\rho - \tfrac{1}{2}\phi)(\tfrac{1}{3} - \nu^2) \cdot \frac{J \cdot R'^2}{r^3},$$

in which ν is the sine of the declination of the satellite on the plane of Jupiter's equator; J the mass of Jupiter; $2R'$ his equatorial diameter; ρ his ellipticity, and ϕ the ratio of the centrifugal force to gravity at his equator. Now it may be assumed that $J = 1$, $R' = 1$; and if s' be the tangent of the latitude that the satellite would have above the fixed plane if it moved in the plane of Jupiter's equator, and as s is its latitude above that plane, when moving in its own orbit, $\nu = s - s'$ nearly; hence

$$R = -\frac{(\rho - \frac{1}{2}\phi)}{r^3}\{\tfrac{1}{3} - (s - s')^2\}.$$

813. Thus the whole force that troubles the motion of m is

$$\begin{aligned}R = {} & \frac{m_, r}{r_,^2}\{ss_, + (1 - \tfrac{1}{2}s^2 - \tfrac{1}{2}s_,^2)\cos(v_, - v)\} \\ & - \frac{m_,}{\sqrt{r^2 - 2rr_, \cos(v_, - v) + r_,^2}} \\ & - \frac{m_, rr_, \{ss_, - \frac{1}{2}(s^2 + s_,^2)\cos(v_, - v)\}}{\{r^2 - 2rr_, \cos(v_, - v) + r_,^2\}^{\frac{3}{2}}} \\ - \frac{S'}{D} - \frac{S'r^2}{4D^3} & \{1 - 3s^2 - 3S^2 + 12sS\cos(U - v) + 3\cos 2(U - v)\} \\ & - \frac{(\rho - \frac{1}{2}\phi)}{r^3}\{\tfrac{1}{3} - (s - s')^2\}.\end{aligned}$$

814. If the squares of S, s, and s' be omitted, the only force that troubles the satellites in longitude and distance is

$$\begin{aligned}R = {} & \frac{m_, r}{r_,^2}\cos(v_, - v) - \frac{m_,}{\sqrt{r^2 - 2rr_, \cos(v_, - v) + r_,^2}} \\ & - \frac{S'}{D} - \frac{S'r^2}{4D^3}\{1 + 3\cos 2(U - v)\} - \frac{(\rho - \frac{1}{2}\phi)}{3r^3}.\end{aligned}$$

When the eccentricities are omitted, the radii vectores, r and r', become a, $a_{\prime}$, half the greater arcs of the orbits, and that part of R that depends on the mutual attraction of the satellites, is

$$R' = \frac{m_{\prime}a}{a_{\prime}^2}\cos(n_{\prime}t-nt+\epsilon_{\prime}-\epsilon) - \frac{m_{\prime}}{\sqrt{a^2-2aa_{\prime}\cos(n_{\prime}t-nt+\epsilon_{\prime}-\epsilon)+a_{\prime}^2}}$$

$nt+\epsilon$, $n_{\prime}t+\epsilon_{\prime}$, being the mean longitudes of m and $m_{\prime}$. This expression may be developed into the series

$$R'=m_{\prime}\{\tfrac{1}{2}A_0+A_1.\cos(n_{\prime}t-nt+\epsilon_{\prime}-\epsilon)+A_2\cos 2(n_{\prime}t-nt+\epsilon'-\epsilon)+\&c.\}$$

This is the part of R that is independent of the eccentricities, and is identical with the series in article 446; therefore the coefficients A_0, A_1, &c., and their differences, may be computed by the same formulæ as for the planets, observing to substitute $A_{\prime} - \frac{a}{a_{\prime}^2}$ for $A_{\prime}$.

But, by article 445,

$$r = a(1+u) \qquad r_{\prime} = a_{\prime}(1+u_{\prime})$$
$$v = nt+\epsilon+v' \qquad v_{\prime} = n_{\prime}t+\epsilon_{\prime}+v'_{\prime},$$

where u, $u_{\prime}$, v', $v'_{\prime}$, are the elliptical parts of the radii vectores, and of the longitudes of m and $m_{\prime}$. By the same article, the general formula for the developement of R, according to the powers and products of these minute quantities, is

$$R = R' + au_{\prime}.\frac{dR'}{da} + a_{\prime}u_{\prime}.\frac{dR'}{da_{\prime}} + (v_{\prime}'-v')\frac{dR'}{ndt} + \&c.$$

From the preceding value of R' the quantities $\frac{dR'}{da}$, $\frac{dR'}{da_{\prime}}$, &c., may be found; and, when substituted, it will be seen afterwards that the only requisite part of R is

$$R=m_{\prime}\{\tfrac{1}{2}A_0+A_1\cos(n_{\prime}t-nt+\epsilon_{\prime}-\epsilon)+A_2\cos 2(n_{\prime}t-nt+\epsilon_{\prime}-\epsilon)+\&c.\}$$
$$+\frac{m_{\prime}}{2}.au.\frac{dA_0}{da}+m_{\prime}au\frac{dA_2}{da}.\cos 2(n_{\prime}t-nt+\epsilon_{\prime}-\epsilon)$$
$$+m_{\prime}a_{\prime}u'\frac{dA_1}{da_{\prime}}.\cos(n_{\prime}t-nt+\epsilon_{\prime}-\epsilon)$$
$$-m_{\prime}v_{\prime}'A_1.\sin(n_{\prime}t-nt+\epsilon_{\prime}-\epsilon)$$
$$+2m_{\prime}v'A_2.\sin 2(n_{\prime}t-nt+\epsilon_{\prime}-\epsilon).$$

Because the satellites move in nearly circular orbits, u, $u_{\prime}$, v', and $v_{\prime}'$, may be regarded as variations arising either entirely from the disturbing forces, as in the first and second satellites, or from that force conjointly with a real but very small ellipticity, as in the third and fourth; therefore

$$r = a(1 + \delta u), \qquad r_{,} = a_{,}(1 + \delta u_{,})$$
$$v = nt + \epsilon + \delta v, \qquad v_{,} = n_{,}t + \epsilon_{,} + \delta v_{,}$$

Now, $r = a(1+u)$ gives $r^2 = a^2(1+2u)$; for u is so small, that its square may be omitted; hence $\delta u = \frac{r\delta r}{a^2}$: consequently $\delta u_{,} = \frac{r_{,}\delta r_{,}}{a_{,}^2}$; and when $R = 0$, equation (156) gives, for the elliptical part of $r\delta r$ only,

$$\delta v = \frac{2d(r\delta r)}{a^2 \cdot ndt}, \text{ and } \delta v_{,} = \frac{2d(r_{,}\delta r_{,})}{a_{,}^2 \cdot n_{,}dt},$$

when the squares of the eccentricities are omitted.

815. If these quantities be substituted in R instead of u, $u_{,}$, v', and $v_{,}'$, it becomes

$$R = m_{,}\{\tfrac{1}{2}A_0 + A_1 \cos(n_{,}t - nt + \epsilon_{,} - \epsilon) + A_2 \cos 2(n_{,}t - nt + \epsilon_{,} - \epsilon) + \&c.\}$$
$$+ \frac{m_{,}}{2} \cdot \frac{r\delta r}{a^2} \cdot a\left(\frac{dA_0}{da}\right)$$
$$+ m_{,} \cdot \frac{r\delta r}{a^2} \cdot a\left(\frac{dA_2}{da}\right) \cdot \cos 2(n_{,}t - nt + \epsilon_{,} - \epsilon)$$
$$+ m_{,} \cdot \frac{r_{,}\delta r_{,}}{a_{,}^2} \cdot a_{,}\left(\frac{dA_1}{da_{,}}\right) \cdot \cos(n_{,}t - nt + \epsilon_{,} - \epsilon) \qquad (253)$$
$$+ 4m_{,} \cdot \frac{d(r\delta r)}{a^2 \cdot ndt} \cdot A_2 \cdot \sin 2(n_{,}t - nt + \epsilon_{,} - \epsilon)$$
$$- 2m_{,} \cdot \frac{d(r_{,}\delta r_{,})}{a_{,}^2 \cdot n_{,}dt} \cdot A_1 \cdot \sin(n_{,}t - nt + \epsilon_{,} - \epsilon).$$
$$+ \&c.$$

816. If $\frac{S'}{D}$ and the eccentricity be omitted, the action of the sun on m is

$$R = -\frac{S'a^2}{4D'^3}\{1 + 3\cos 2(Mt - nt + E - \epsilon)\};$$

where D' is the mean distance of Jupiter from the sun, and $Mt + E$ his mean longitude referred to the sun. In the troubled orbit a, $nt + \epsilon$, and D' become

$$a\left(1 + \frac{r\delta r}{a^2}\right),\ nt + \epsilon - \frac{2d(r\delta r)}{a^2 \cdot ndt}, \text{ and } D'\left(1 - \frac{D\delta D}{D'^2}\right);$$

and as, by article 383, $\frac{S'}{D'^3} = M^2$, when the mass of Jupiter is omitted in comparison of that of the sun, the whole disturbing action of the sun is

$$R = -\frac{M^2a^2}{4} - \frac{M^2}{2}\,.\,r\delta r - \frac{3M^2a^2}{4}\,.\cos 2\,(nt - Mt + \epsilon - E)$$

$$-\tfrac{3}{4}M^2a^2\,.\,\frac{D\delta D}{D'^2} - M^2\,.\,\frac{6r\delta r}{4}\,.\,\cos 2(nt - Mt + \epsilon - E) \quad (254)$$

$$+ 3M^2\,.\,\frac{d\,(r\delta r)}{.\,ndt}\,.\,\sin 2(nt - Mt + \epsilon - E)$$

when the squares of the eccentricities are omitted.

817. In the same manner it is easy to see that the effect of Jupiter's compression is

$$R = -\frac{(\rho - \frac{1}{2}\phi)}{3a^3} + \frac{(\rho - \frac{1}{2}\phi)}{a^5}\,.\,r\delta r.$$

The three last values of R contain all the forces that trouble the longitude and radius vector of m.

FIRST APPROXIMATION.

Perturbations in the Radius Vector and Longitude of m *that are independent of the Eccentricities.*

818. Since R has been taken with a negative sign, equation (155) becomes

$$\frac{d^2.r\delta r}{dt^2} + \mu\,.\,\frac{r\delta r}{r^3} + 2\int dR + r\left(\frac{dR}{dr}\right) = 0. \qquad (255)$$

The mass of each satellite is about ten thousand times less than the mass of Jupiter, and may therefore be omitted in the comparison, and if Jupiter be taken as the unit of mass $\mu = 1$.

When the eccentricity is omitted $r = a$; but by article 556 the action of the disturbing forces produces a permanent increase in a, which may be expressed by δa, therefore if $(a + \delta a)^{-3}$ be put for r^{-3},

$$\frac{d^2.r\delta r}{dt^2} + \,.\,\frac{r\delta r}{a^3}\left(1 - 3\frac{\delta a}{a}\right) + 2\int dR + r\left(\frac{dR}{dr}\right) = 0. \qquad (256)$$

819. When the eccentricities are omitted,

$$R = m_{\prime}\,\{\tfrac{1}{2}A_0 + A_1\cos(n_{\prime}t - nt + \epsilon_{\prime} - \epsilon)$$
$$+ A_2\cos 2(n_{\prime}t - nt + \epsilon_{\prime} - \epsilon) + \&c.\}$$

$$+ \frac{m_{,}}{2} \cdot \frac{r\delta r}{a^2} \cdot a\left(\frac{dA_0}{da}\right)$$

$$- \tfrac{1}{4} M^2a^2 - \tfrac{1}{2} M^2 \cdot \frac{r\delta r}{a^2} - \tfrac{3}{4} M^2a^2 \cos 2\,(nt - Mt + \epsilon - E)$$

$$- \frac{(\rho - \frac{1}{2}\phi)}{3a^3} + \frac{(\rho - \frac{1}{2}\phi)}{a^5} \cdot r\delta r. \qquad (257)$$

Since dR relates to the mean motion of m, the term

$$\frac{m_{,}}{2}\, a \cdot \frac{r\delta r}{a^2} \cdot \left(\frac{dA_0}{da}\right)$$

gives

$$2\int \mathrm{d}R = m_{,} \cdot \frac{r\delta r}{a} \cdot \left(\frac{dA_0}{da}\right);$$

moreover the same term gives

$$r\left(\frac{dR}{dr}\right) = \frac{m_{,}}{2} \cdot \frac{r\delta r}{a}\left\{\frac{dA_0}{da} + a\,\frac{d^2A_0}{da^2}\right\}.$$

With regard to Jupiter's compression

$$\int \mathrm{d}R = R,\; r\left(\frac{dR}{dr}\right) = -\,3R,$$

consequently

$$2\int \mathrm{d}R + r\left(\frac{dR}{dr}\right) = \frac{(\rho - \frac{1}{2}\phi)}{3a^3} - \frac{(\rho - \frac{1}{2}\phi)}{a^5} \cdot r\delta r.$$

Attending to these circumstances, and observing that

$$\frac{1}{a^3} = n^2 = \frac{1 + m}{a^3},$$

it will be found, when the eccentricities are omitted and the whole divided by a^2, that

$$\frac{d^2 \cdot r\delta r}{a^2dt^2} + N^2 \cdot \frac{r\delta r}{a^2} + 2n^2K + n^2 \cdot \frac{\rho - \frac{1}{2}\phi}{3a^2} - M^2 \qquad (258)$$

$$+ \Sigma . \frac{m_{,}n^2}{2} \cdot a^2\left(\frac{dA_0}{da}\right) - 3M^2 \cdot \frac{2n - M}{2n - 2M} \cdot \cos 2\,(nt - Mt + \epsilon - E)$$

$$+ \Sigma . m_{,}n^2 \cdot \left\{a^2\left(\frac{dA_1}{da}\right) + \frac{2n}{n - n_{,}} \cdot aA_1\right\} . \cos\,(n_{,}t - nt + \epsilon_{,} - \epsilon)$$

$$+ \Sigma . m_{,}n^2 \left\{a^2\left(\frac{dA_2}{da}\right) + \frac{2n}{n - n_{,}} \cdot aA_2\right\} . \cos 2\,(n_{,}t - nt + \epsilon_{,} - \epsilon)$$

$$+ \;\&c.\;\&c. = 0.$$

Where to abridge

$$N^2 = n^2\left\{1 - \frac{3\delta a}{a} - \frac{(\rho - \frac{1}{2}\phi)}{a^2} - \frac{2M^2}{n^2} + \Sigma\,\frac{m_, a^2}{2}\left\{3\left(\frac{dA_0}{da}\right)\right.\right.$$

$$\left.\left. + a\left(\frac{d^2A_0}{da^2}\right)\right\}\right\};$$

a quantity that differs little from n^2, for the last term is extremely small in consequence of the factor $m_,$: the variation of the mean distance a is very small, and so are the other two parts depending on the compression of Jupiter and the action of the sun. Indeed M and $N - n$ may be omitted, in comparison of n in the terms arising from the action of the sun after integration, for the motion of Jupiter is much slower than the motion of his satellites.

820. The preceding equation may be integrated by the method of indeterminate coefficients, if it be assumed that

$$\frac{r\delta r}{a^2} = B + m_, b\cos(n_, t - nt + \epsilon_, - \epsilon) + m_, b_{(1)}\cos 2(n_, t - nt + \epsilon_, - \epsilon) + \&c.$$

$$+ Gm_, \cos 2(nt - Mt + \epsilon - E).$$

For a comparison of the coefficients of similar cosines after the substitution of this quantity and its differential gives

$$B = -\frac{n^2}{N^2}\left\{2K + \frac{(\rho - \frac{1}{2}\phi)}{3a^2} - \frac{M^2}{n^2} + \Sigma\,\frac{m_,}{2}\,a^2\left(\frac{dA_0}{da}\right)\right\}$$

$$b = \frac{\left\{a^2\left(\frac{dA_1}{da}\right) + \frac{2n}{n - n_,}\,.\,aA_1\right\}n^2}{(n - n_,)^2 - N^2}$$

$$b_{(1)} = \frac{\left\{a^2\left(\frac{dA_2}{da}\right) + \frac{2n}{n - n_,}\,.\,aA_2\right\}n^2}{4\,(n - n_,)^2 - N^2}.$$

$$b_{(2)} = \frac{\left\{a^2\left(\frac{dA_3}{da}\right) + \frac{2n}{n - n_,}\,aA_3\right\}n^2}{9\,(n - n_,)^2 - N^2}\ \&c.$$

$$G = -\frac{M^2}{n^2},$$

and the integral of (258) is

$$\frac{r\delta r}{a^2} = -\frac{n^2}{N^2}\left\{2K + \frac{(\rho - \frac{1}{2}\phi)}{3a^2} - \frac{M^2}{n^2} + \Sigma\,\frac{m_{\prime}}{2}\,a^2\left(\frac{dA_0}{da}\right)\right\}$$

$$-\frac{M^2}{n^2}\cos 2\,(nt - Mt + \epsilon - E)$$

$$+\,\Sigma m'\left\{\begin{array}{l}\dfrac{n^2}{(n-n_{\prime})^2 - N^2}\left\{a^2\left(\dfrac{dA_1}{da}\right) + \dfrac{2n}{n-n_{\prime}}\,aA_1\right\}\cos\,(n_{\prime}t - nt + \epsilon_{\prime} - \epsilon)\\ \dfrac{n^2}{4(n-n_{\prime})^2 - N^2}\left\{a^2\left(\dfrac{dA_2}{da}\right) + \dfrac{2n}{n-n_{\prime}}\,aA_2\right\}\cos 2(n_{\prime}t - nt + \epsilon_{\prime} - \epsilon)\\ \dfrac{n^2}{9(n-n_{\prime})^2 - N^2}\left\{a^2\left(\dfrac{dA_3}{da}\right) + \dfrac{2n}{n-n_{\prime}}\,aA_3\right\}\cos 3(n_{\prime}t - nt + \epsilon_{\prime} - \epsilon)\end{array}\right\}$$

$$+\ \&c. \qquad\qquad \&c.$$

The first term of this equation is what was expressed by $\frac{\delta a}{a}$, for if all the periodic quantities be omitted $r = a$, and this equation becomes

$$\frac{\delta a}{a} = -\,2K - \frac{(\rho - \frac{1}{2}\phi)}{3a^2} + \frac{M^2}{n^2} - \Sigma\frac{m_{\prime}}{2}\,a^2\left(\frac{dA_0}{da}\right);$$

for N differs so little from n that $\frac{n^2}{N^2} = 1$, without sensible error: this is the permanent change in the radius vector from the disturbing influence.

These are the principal perturbations in the radii vectores of the satellites.

821. Since the squares of the eccentricity are omitted $\sqrt{1-e^2} = 1$, and as $\mu = 1$, equation (156) of the longitude becomes

$$\delta v = \frac{2d\,(r\delta r)}{a^2.\,ndt} - \frac{dr\,.\,\delta r}{a^2.\,ndt} + 3a\!\int\!\!\int\! ndt.\,dR + 2a\!\int\! ndt\,.\,r\left(\frac{dR}{dr}\right) \quad (259)$$

since the sign of R is changed.

If the preceding value of $\frac{r\delta r}{a^2}$ be put in this equation, and also if substitution be made for dR and $r\left(\frac{dR}{dr}\right)$ derived from equation (257), observing that $\frac{1+m_{\prime}}{a^3} = \frac{1}{a^3} = n^2$, and that M and N differ but little from n, the result will be

$$\delta v = nt\left\{3K + \frac{(\rho - \frac{1}{2}\phi)}{a^2} - \tfrac{7}{4}\frac{M^2}{n^2} + \Sigma . m_{,}a^2\left(\frac{dA_0}{da}\right)\right\}$$

$$+ \frac{11}{8} \cdot \frac{M^2}{n^2} . \sin 2(nt - Mt + \epsilon - E)$$

$$+ \Sigma\frac{m_{,}n}{n-n_{,}}\left\{\begin{array}{l}\left\{\frac{n}{n-n_{,}}aA_1 + \frac{2N^2}{(n-n_{,})^2 - N^2}\left(a^2\left(\frac{dA_1}{da}\right) + \frac{2n}{n-n_{,}}aA_1\right)\right\} \\ \qquad \times \sin(n_{,}t - nt + \epsilon_{,} - \epsilon) \\ + \frac{1}{2}\left\{\frac{n}{n-n_{,}}aA_2 + \frac{2N^2}{4(n-n_{,})^2 - N^2}\left(a^2\left(\frac{dA_2}{da}\right) + \frac{2n}{n-n_{,}}aA_2\right)\right\} \\ \qquad \times \sin 2(n_{,}t - nt + \epsilon_{,} - \epsilon) \\ + \frac{1}{3}\left\{\frac{n}{n-n_{,}}aA_3 + \frac{2N^2}{9(n-n_{,})^2 - N^2}\left(a^2\left(\frac{dA_3}{da}\right) + \frac{2n}{n-n_{,}}aA_3\right)\right\} \\ \qquad \times \sin 3(n_{,}t - nt + \epsilon_{,} - \epsilon)\end{array}\right\}$$

$$+ \quad \&c. \qquad\qquad + \quad \&c.$$

By article 540 δv ought not to contain the mean motion nt, so the first term must be zero, by which the arbitrary constant quantity is determined to be

$$K = -\frac{(\rho - \frac{1}{2}\phi)}{3a^2} + \tfrac{7}{12}\frac{M^2}{n^2} - \tfrac{1}{3}\Sigma m_{,}a^2\left(\frac{dA_0}{da}\right),$$

whence

$$\frac{\delta a}{a} = \frac{(\rho - \frac{1}{2}\phi)}{3a^2} - \tfrac{1}{6}\frac{M^2}{n^2} + \tfrac{1}{6}\Sigma m_{,}a^2\left(\frac{dA_0}{da}\right)$$

and

$$N^2 = n^2\left\{1 - 2\frac{(\rho - \frac{1}{2}\phi)}{a^2} - \tfrac{3}{2}\frac{M^2}{n^2} + \Sigma m_{,}a^2\left\{\left(\frac{dA_0}{da}\right) + \tfrac{1}{2}a\left(\frac{d^2A_0}{da}\right)\right\}\right\}$$

The preceding value of δv, deprived of its first term, contains all the perturbations in longitude that are independent of the eccentricities; and as the square of s, the tangent of the latitude, is omitted, by article 548 the very small angle δv may either be estimated on the orbit of the satellite, or on the fixed plane, since it coincides with its projection. The term depending on the action of the sun corresponds with the Variation in the motion of the moon.

822. If the masses of the four satellites be m, m_1, m_2, m_3, the perturbations that m experiences by the action of the other two will be found by changing successively the quantities relative to m_1 into those belonging to m_2 and m_3, and the sum of these will be the action of the three satellites m_1, m_2, and m_3 on m. The perturbations of the others are found by making similar changes.

823. Hereafter the four satellites will be represented m, m_1, m_2, m_3. Where m is the first, or that nearest Jupiter, and m_3 is the fourth and the most distant, all quantities relating to them will be accented in the same manner, except it be stated to the contrary.

824. Because $2n_, = n = N$ nearly,

$$\frac{2m_,n \, . \, N^2}{n - n_,} = m_,n^2,$$

and the perturbations expressed by

$$\frac{r\delta r}{a} = \frac{m_,n^2}{(n-n_,)^2 - N^2}\left\{a^2\left(\frac{dA_1}{da}\right) + \frac{2n}{n-n_,}.aA_1\right\}.\cos\,(n_,t-nt+\epsilon_,-\epsilon)$$

$$+\frac{m_,n^2}{4(n-n_,)^2-N^2}\left\{a^2\left(\frac{dA_2}{da}\right) + \frac{2n}{n-n_,}aA_2\right\}.\cos 2\,(n_,t-nt+\epsilon'-\epsilon)$$

$$\delta v = \frac{2m_,n^2}{(n-n_,)^2 - N^2}\left\{a^2\left(\frac{dA_1}{da}\right) + \frac{2n}{n-n_,}aA_1\right\}.\sin\,(n_,t-nt+\epsilon_,-\epsilon)$$

$$+\frac{2m_,n^2}{4(n-n_,)-N^2}\left\{a^2\left(\frac{dA_2}{da}\right) + \frac{2n}{n-n_,}aA_2\right\}.\sin 2(n_,t-nt+\epsilon_,-\epsilon)$$

are the greatest to which the three first satellites are liable, on account of the very small divisors arising from the nearly commensurable ratios in the mean motions of these three bodies.

825. The greatest inequality in the first satellite is occasioned by the action of the second, and expressed by

$$\frac{r\delta r}{a} = \frac{m_,n^2}{4(n-n_,)^2-N^2}\left\{a^2\left(\frac{dA_2}{da}\right) + \frac{2n}{n-n_,}aA_2\right\}\cos 2(n_,t-nt+\epsilon_,-\epsilon)$$

$$\delta v = \frac{2m_,n^2}{4(n-n_,)^2-N^2}\left\{a^2\left(\frac{dA_2}{da}\right) + \frac{2n}{n-n_,}aA_2\right\}.\sin 2(n_,t-nt+\epsilon_,-\epsilon).$$

Because the mean motion of the first satellite is nearly double that of the second, $n = 2n_,$, and as $N = n = 2n_,$ nearly, the divisor

$$4\,(n-n_,)^2 - N^2 = \{(2n-2n_,) - N\}\{(2n-2n_,)+N\} = 2n\,.\,(2n-2n_,-N);$$

and if to abridge

$$F = -\,a^2\left(\frac{dA_2}{da}\right) - \frac{2n}{n-n_,}\,.\,aA_2,$$

the greatest inequalities in the motion of the first satellite are

$$\frac{r\delta r}{a^2} = -\frac{m_,n.F}{2(2n-2n_,-N)}\,.\cos 2\,(n_,t - nt + \epsilon_, - \epsilon) \quad (260)$$

$$\delta v = \frac{m_,n.F}{2n-2n_,-N}\,.\,\sin 2\,(n_,t - nt + \epsilon' - \epsilon).$$

826. The principal inequalities in the second satellite arise from the action of the first and third. Those occasioned by the first depend on the terms that have the divisor $(n - n_{,})^2 - N_{,}^2$; the quantities having one accent belong to $m_{,}$, the second satellite. Let $A_1^{(1.2)}$ be the value of A_1 when the second satellite is troubled by the first; then if

$$G = -a_{,}^2\left(\frac{dA_1^{(1.2)}}{da_{,}}\right) + \frac{2n_{,}}{n - n_{,}} \cdot a_{,}A_1^{(1.2)},$$

the principal inequalities in the second satellite occasioned by the first are

$$\frac{r_{,}\delta r_{,}}{a_{,}^2} = -\frac{mn_{,}.G}{2(n-n_{,}-N_{,})} \cdot \cos(nt-n_{,}t+\epsilon-\epsilon_{,}) \qquad (261)$$

$$\delta v_{,} = \frac{mn_{,}G}{n-n_{,}-N_{,}} \cdot \sin(nt - n_{,}t + \epsilon - \epsilon_{,})$$

for $\quad n = 2n_{,}, \quad N_{,} = n_{,}, \quad$ and $(n - n_{,})^2 - N^2 =$

$$\{n_{,} - n - N_{,}\} \cdot \{n_{,} - n + N_{,}\} = 2n_{,}(n - n_{,} - N_{,})$$

The action of the third satellite on the second is perfectly similar to the action of the second on the first, on account of the ratios $n = 2n_{,}$ and $n_{,} = 2n_2$ in their mean motions; therefore, the inequalities in the motion of the second, occasioned by the action of the third, will be obtained from equations (260), by changing what relates to the first and second into the quantities relative to the second and third. In this case let $A_2^{(3.2)}$ be the value of A_2, and let

$$F' = -a_{,}^2\left(\frac{dA_2^{(3.2)}}{da_{,}}\right) - \frac{2n_{,}}{n_{,}-n_2}\, a_{,}A_2^{(3.2)}$$

be the value of F, then

$$\frac{r_{,}\delta r_{,}}{a_1^2} = -\frac{m_2 n_{,}F'}{2(2n_1-2n_2-N_{,})} \cdot \cos 2(n_{,}t - n_2 t + \epsilon_{,} - \epsilon_2) \quad (262)$$

$$\delta v_{,} = \frac{m_2 n_1 F'}{2n_1 - 2n_2 - N_{,}} \cdot \sin 2(n_{,}t - n_2 t + \epsilon_{,} - \epsilon_2)$$

By observation,

$$nt - 3n_{,}t + 2n_2 t + \epsilon - 3\epsilon_{,} + 2\epsilon_2 = 180°,$$

consequently,

$$2(nt_{,} - n_2 t + \epsilon_{,} - \epsilon_2) = nt - n_{,}t + \epsilon - \epsilon_{,} - 180°;$$

for $\qquad n = 2n_{,} \qquad n_{,} = 2n_2$ nearly,

the two last inequalities may be added to the preceding, since they depend on the same angle; the principal inequalities in the motion of the second satellite from the action of the first and third are therefore

$$\frac{r_{,}\delta r_{,}}{a_{,}^{2}} = -\frac{n_{,}}{2(n-n_{,}-N_{,})}\{mG-m_{2}F'\}.(\cos nt-n_{,}t+\epsilon-\epsilon_{,}) \quad (263)$$

$$\delta v_{,} = \frac{n_{,}}{n-n_{,}-N_{,}}\{mG-m_{2}F'\}.\sin(nt-n_{,}t+\epsilon-\epsilon_{,}).$$

In consequence of the ratios in the mean motions these inequalities will never be separated.

827. The action of the second satellite produces inequalities in the theory of the third, analogous to those occasioned by the action of the first on the second; hence, if all the quantities in equations (261) relating to the second and first be changed into those belonging to the third and second, and if $A_{,}^{(2.3)}$ and G' be the values of $A_{,}^{(1.2)}$ and G in this case, so that

$$A_{,}^{(2.3)} = A_{,}^{(1.2)} + \frac{a_2}{a_{,}^{2}} - \frac{a_1}{a_2^2},$$

and
$$G' = -a_2^2\left(\frac{dA_{,}^{(2.3)}}{da_2}\right) + \frac{2n_2}{n_1-n_2}.a_2A_{,}^{(2.3)},$$

the resulting equations for m_2 will be

$$\frac{r_2.\delta r_2}{a_2^2} = -\frac{m_1.n_2.G'}{2(n_1-n_2-N_2)}.\cos(n_1t-n_2+\epsilon_1-\epsilon_2) \quad (264)$$

$$\delta v_2 = \frac{m_1n_2G'}{n_1-n_2-N_2}.\sin(n_1t-n_2t+\epsilon_1-\epsilon_2).$$

These inequalities have only been detected by observation in the motion of the first satellite.

828. G, which is a function of $A_{,}^{(1.2)}$, may be expressed by a function of $A_{,,}$ for

$$A_{,}^{(1.2)} = \frac{a_{,}}{a^2} - \frac{a}{a_{,}^{2}} + A_{,,}$$

whence on account of

$$n = \frac{1}{a^3}, \quad n_{,} = \frac{1}{a_{,}^{3}};$$

and that $n = 2n_{,}$ it may be found that

$$G = 2a_{,}A_{,} - a_{,}^{2}\left(\frac{dA_{,}}{da_{,}}\right).$$

SECOND APPROXIMATION.

Inequalities depending on the First Powers of the Eccentricities.

829. If $a + \frac{r\delta r}{a}$ be put for r, equation (255) becomes

$$0 = \frac{d^2 r\delta r}{dt^2} + \frac{r\delta r}{a^3}\left\{1 - \frac{3r\delta r}{a^2}\right\} + 2\int dR + r\left(\frac{dR}{dr}\right)$$

or as $\frac{1}{a^3} = n^2 = N^2$, very nearly,

$$0 = \frac{d^2 r\delta r}{dt^2} + N^2 r\delta r\left\{1 - \frac{3r\delta r}{a^2}\right\} + 2\int dR + r\left(\frac{dR}{dr}\right). \qquad (265)$$

If the action of the sun be omitted, the only part of the preceding value of R, that has a sensible effect on the radius vector is

$$R = m_{\prime}\left\{A_1 \cos(n_{\prime}t - nt + \epsilon_{\prime} - \epsilon) + \frac{r_{\prime}\delta r_{\prime}}{a_{\prime}^2}\cdot a_{\prime}\frac{dA_1}{da_{\prime}}\cos(n_{\prime}t - nt + \epsilon_{\prime} - \epsilon)\right.$$
$$\left. - 2A_1\frac{d(r_{\prime}\delta r_{\prime})}{a_{\prime}^2 . n_{\prime}dt}\sin(n_{\prime}t - nt + \epsilon_{\prime} - \epsilon)\right\};$$

but these terms are very important, because they serve for the determination of the secular inequalities in the eccentricities and motions of the apsides. With regard to the terms depending on nt, $\int dR = R$, substituting for R, and dividing the whole equation (265) by a^2, it becomes, when $\left(\frac{r\delta r}{a^2}\right)^2$ is omitted,

$$0 = \frac{d^2 r\delta r}{a^2 dt^2} + N^2 . \frac{r\delta r}{a^2}$$
$$+\Sigma\left\{\begin{array}{l} m_{\prime}n^2\frac{r_{\prime}\delta r_{\prime}}{a_{\prime}^2}\left\{2aa_{\prime}\left(\frac{dA_1}{da_{\prime}}\right) + a^2 a_{\prime}\left(\frac{d^2A_1}{da da_{\prime}}\right)\right\}\cos(n_{\prime}t - nt + \epsilon_{\prime} - \epsilon) \\ -\frac{2m_{\prime}n^2.d.r_{\prime}\delta r_{\prime}}{a_{\prime}^2 . n_{\prime}dt}\left\{2aA_1 + a^2\left(\frac{dA_1}{da}\right)\right\}\sin(n_{\prime}t - nt + \epsilon_{\prime} - \epsilon)\end{array}\right\}$$

830. In order to integrate this equation, it may be assumed that

$$\frac{r\delta r}{a^2} = h\cos(nt + \epsilon - gt - \Gamma); \quad \frac{r_{\prime}\delta r_{\prime}}{a_{\prime}^2} = h_{\prime}\cos(n_{\prime}t + \epsilon_{\prime} - gt - \Gamma), \text{ \&c.};$$

h and $h_{\prime}$ are indeterminate coefficients, and $gt + \Gamma$ is the motion of the apsides of the orbits of the satellites.

When these quantities and their differentials are substituted, the square of g neglected, and those terms alone retained that depend on the angle $nt + \epsilon - gt - \Gamma$, a comparison of the coefficients of similar cosines gives

$$0 = h\{N^2 + 2ng - n^2\} + \Sigma \frac{m_{,}n^2}{2} h_{,} \left\{ \begin{array}{l} 2aa_{,}\left(\frac{dA_1}{da_{,}}\right) + a^2 a_{,}\left(\frac{d^2 A_1}{da da_{,}}\right) \\ + 4aA_1 + 2a^2\left(\frac{dA_1}{da}\right) \end{array} \right\}$$

but by article 458,

$$a\left(\frac{dA_1}{da}\right) + a_{,}\left(\frac{dA_1}{da_{,}}\right) = -A_1;$$

and if the value of N^2 in article 819 be substituted, this coefficient becomes

$$0 = h\left\{\frac{g}{n} - \frac{(\rho - \frac{1}{2}\phi)}{a^2} - \frac{3}{4}\frac{M^2}{n^2} + \frac{1}{2}\Sigma m_{,}\left\{a^2\left(\frac{dA_0}{da}\right) + \frac{1}{2}a^3\left(\frac{d^2A_0}{da^2}\right)\right\}\right.$$

$$\left. + \frac{1}{2}\Sigma m_{,}h_{,}\left\{aA_1 - a^2\left(\frac{dA_1}{da}\right) - \frac{1}{2}a^3\left(\frac{d^2A_1}{da^2}\right)\right\}\right..$$

And as in article 474, if

$$(0.1) = -\frac{m_{,}n}{2}\left\{a^2\left(\frac{dA_0}{da}\right) + \frac{1}{2}a^3\left(\frac{d^2A_0}{da^2}\right)\right\};$$

$$\boxed{0.1} = \frac{m_{,}n}{2}\left\{aA_1 - a^2\left(\frac{dA_1}{da}\right) - \frac{1}{2}a^3\left(\frac{d^2A_1}{da^2}\right)\right\};$$

and if $(0) = \frac{(\rho - \frac{1}{2}\phi)}{a^2} n; \quad \boxed{0} = \frac{3}{4}\frac{M^2}{n},$

this equation becomes

$$0 = h\{g - (0) - \boxed{0} - (0.1)\} + \boxed{0.1}\, h_{,}$$

with regard to the first satellite troubled by the second; but the action of m_2 and m_3 produces terms similar to those caused by m_1; and if the same notation be used that was employed for the planets, this equation, when m is troubled by the other three satellites, by the sun, and by the compression of Jupiter, becomes

(266)

$$0 = h\{g - (0) - \boxed{0} - (0.1) - (0.2) - (0.3)\} + \boxed{0.1}\, h_1 + \boxed{0.2}\, h_2 + \boxed{0.3}\, h_3$$

A similar equation exists for each satellite, and may be determined from this by changing the quantities relative to m into those relating to m_1 m_2 m_3, and reciprocally; hence, for the others,

(267)

$$0=h_1\{g-(1)-\boxed{1}-(1.0)-(1.2)-(1.3)\}+\boxed{1.0}\,h+\boxed{1.2}\,h_2+\boxed{1.3}\,h_3,$$

$$0=h_2\{g-(2)-\boxed{2}-(2.0)-(2.1)-(2.3)\}+\boxed{2.0}\,h+\boxed{2.1}\,h_1+\boxed{2.3}\,h_3,$$

$$0=h_3\{g-(3)-\boxed{3}-(3.0)-(3.1)-(3.2)\}+\boxed{3.0}\,h+\boxed{3.1}\,h_1+\boxed{3.2}\,h_2.$$

By (484) $\quad (0.1)\, m\sqrt{a} = (1.0)\, m_{\prime}\sqrt{a_{\prime}}$, &c.

and also $\quad \boxed{0.1}\, m\sqrt{a} = \boxed{1.0}\, m_{\prime}\sqrt{a_{\prime}}$, &c.

for any two satellites, so these functions are easily deduced from one another, which saves computation.

These results are perfectly similar to those obtained for the planets, $h\ h_{\prime}$, &c., correspond to $N\ N'$, &c.

831. It has already been mentioned that the part of the longitude of each satellite depending on the eccentricity consists of four terms, of one that is really the equation of the centre, and of three others arising from the variations in the orbits, chiefly induced by the action of the excess of matter at Jupiter's equator. The coefficients of these sixteen terms are obtained by the aid of the preceding equations, and also of the annual and sidereal motions of the apsides of the orbits. The variations in the radii vectores depend on the same cause, contain the same values of g, and have the same coefficients. h, h_1, h_2, h_3, are the real eccentricities of the four orbits, and if they be eliminated there will result an equation of the fourth degree in g. These four values of g, which will be represented by g, g_1, g_2, g_3, are the annual and sidereal motions of the apsides of the orbits of the four satellites.

832. Let g, the annual and sidereal motion of the first satellite, belong to the first of the preceding equations, and assume $h_1=\zeta_1 h$; $h_2=\zeta_2 h$; $h_3=\zeta_3 h$; then the substitution of these in equation (266) will make h vanish, and ζ_1, ζ_2, ζ_3, will be given in functions of g. Thus h, which may be regarded as the real eccentricity of the orbit of m, is an arbitrary quantity, known by observation. Again, if $g_{\prime}$ be the value of g in the second of the preceding equations, and if

$$h=\zeta_1^{(1)}\,h_1, \quad h_2=\zeta_2^{(1)}\,h_1, \quad h_3=\zeta_3^{(1)}\,h_1,$$

by the substitution of these, h_1 will vanish from the equation in question, $\zeta_1^{(1)}$, $\zeta_2^{(1)}$, $\zeta_3^{(1)}$, will be given in functions of $g_{\prime}$; and $h_{\prime}$, the real eccentricity of the orbit of $m_{\prime\prime}$ is determined by observation. In the same manner, if $\zeta_1^{(2)}$, $\zeta_2^{(2)}$, $\zeta_3^{(2)}$, $\zeta_1^{(3)}$, $\zeta_2^{(3)}$, $\zeta_3^{(3)}$, be the quantities corresponding to g_2 and g_3, h_2 and h_3 will be arbitrary constant quantities, which vanish from the two last of equations (267); whence $\zeta_1^{(2)}$, $\zeta_2^{(2)}$, $\zeta_3^{(2)}$, and $\zeta_1^{(3)}$, $\zeta_2^{(3)}$, $\zeta_3^{(3)}$, will be given in functions of g_2 and g_3.

Thus the coefficients of the sixteen terms of the equations of the centre, corresponding to the four values of g, are h, h_1, h_2, h_3, $\zeta_1 h_1$, $\zeta_2 h_1$, $\zeta_3 h_1$, $\zeta_1^{(1)} h_2$, &c. &c., of which h, h_1, h_2, h_3, are the real eccentricities of the orbits of the four satellites, and are determined by observation: by means of these, and the equations (266) and (267), values of ζ, ζ_1, &c. will be obtained; and also the four roots of g. Observation shows, however, that h and h_1 are insensible.

833. It was assumed, that

$$\frac{r\delta r}{a^2} = h \cos(nt + \epsilon - gt - \Gamma);$$

and as g has four roots, to each of which there are four corresponding values of h, this expression becomes

$$\frac{r\delta r}{a^2} = h \cos(nt + \epsilon - gt - \Gamma) + h_1 \cos(nt + \epsilon - g_1 t - \Gamma_1)$$
$$+ h_2 \cos(nt + \epsilon - g_2 t - \Gamma_2) + h_3 \cos(nt + \epsilon - g_3 t + \Gamma_3):$$

thus the whole variation in the radius vector of the first satellite depends on h, the eccentricity of its own orbit, on g the motion of its own nodes, and on those of the other three. The corresponding inequalities in the radii vectores of the other three satellites are,

$$\frac{r_1\delta r_1}{a_1^2} = \zeta_{\prime}.h.\cos(n_1 t+\epsilon_1-gt-\Gamma)+\zeta_1^{(1)} h_1.\cos(n_1 t+\epsilon_1-g_1 t-\Gamma_1),$$
$$+ \zeta_1^{(2)} h_2.\cos(n_1 t+\epsilon_1-g_2 t-\Gamma_2)+\zeta_1^{(3)} h_3.\cos(n_1 t+\epsilon_1-g_3 t-\Gamma_3)$$
$$\frac{r_2\delta r_2}{a_2^2} = \zeta_2 h.\cos(n_2 t + \epsilon_2-gt-\Gamma) + \zeta_2^{(1)} h_1 \cos(n_2 t+ \epsilon_2-g_1 t-\Gamma_1)$$
$$+ \zeta_2^{(2)} h_2 \cos(n_2 t+\epsilon_2-g_2 t-\Gamma_2)+\zeta_2^{(3)} h_3.\cos(n_2 t+\epsilon_2-g_3 t-\Gamma_3)$$
$$\frac{r_3\delta r_3}{a_3^2} = \zeta_3 h.\cos(n_3 t+\epsilon_3-gt-\Gamma)+\zeta_3^{(1)} h_1 \cos(n_3 t+\epsilon_3-g_1 t+\Gamma_1)$$
$$+ \zeta_3^{(2)} h_2 \cos(n_3 t+\epsilon_3-g_2 t-\Gamma_2)+\zeta_3^{(3)} h_3 \cos(n_3 t+\epsilon_3-g_3 t-\Gamma_3).$$

These equations contain the perturbations in the radii vectores of the four satellites, depending on the first powers of the eccentricities, and are the complete integrals of the differential equation (265), when applied to each satellite, since they contain the eight arbitrary constant quantities $h, h_1, h_2, h_3, \Gamma, \Gamma_1, \Gamma_2, \Gamma_3$, all of which are known by observation. The four last are the mean longitudes of the lower apsides of the orbits of the satellites at the epoch.

834. If the orbits be considered as variable ellipses, ae being the eccentricity of the orbit of the first satellite, and ϖ the longitude of its lower apsis, estimated from the origin of the angles,

$$\frac{r\delta r}{a^2} = - e \cos (nt + \epsilon - \varpi);$$

comparing this with the preceding value of $\frac{r\delta r}{a^2}$, the result is

$$e \cos \varpi = - h \cos (gt + \Gamma) - h_1 \cos (g_1 t + \Gamma_1) - \&c.$$
$$e \sin \varpi = - h \sin (gt + \Gamma) - h_1 \sin (g_1 t + \Gamma_1) - \&c.$$

whence e and ϖ may be obtained; and for the same reasons, e_1, ϖ_1, e_2, ϖ_2, and e_3, ϖ_3.

835. When the squares of the eccentricity are omitted, the elliptical part of the longitude is $v = 2e \sin (nt + \epsilon - \varpi)$ by 392; or representing it by δv for the satellites, where it chiefly arises from the disturbing forces, it gives

$$\delta v = 2e \cos \varpi \sin (nt + \epsilon) - 2e \sin \varpi \cos (nt + \epsilon),$$

and substituting for $e \cos \varpi$, and $e \sin \varpi$,

$$\delta v = - 2h \sin (nt + \epsilon - gt - \Gamma) - 2h_1 \sin (nt + \epsilon - g_1 t - \Gamma_1)$$
$$- 2h_2 \sin (nt + \epsilon - g_2 t - \Gamma_2) - 2h_3 \sin (nt + \epsilon - g_3 t - \Gamma_3),$$

which is the equation of the centre of the first satellite. It appears, that the first term depends on the eccentricity and apsis of its own orbit, the second term arises from the action of the second satellite, and depends on the eccentricity and apsis of the orbit of that body; the other two inequalities arise from the attraction of the third and fourth satellites, and depend on the eccentricities and apsides of their orbits.

The corresponding inequalities in the longitude of the other three satellites are,

$$\delta v_1 = - 2\zeta_1 h \sin (n_1 t + \epsilon_1 - gt - \Gamma) - 2\zeta_1^{(1)} h_1 \sin(n_1 t + \epsilon_1 - g_1 t - \Gamma_1)$$
$$- 2\zeta_1^{(2)} h_2 \sin(n_1 t + \epsilon_1 - g_2 t - \Gamma_2) - 2\zeta_1^{(3)} h_3 \sin(n_1 t + \epsilon_1 - g_3 t - \Gamma_3)$$
$$\delta v_2 = - 2\zeta_2 h \sin (n_2 t + \epsilon_2 - gt - \Gamma) - 2\zeta_2^{(1)} h_1 \sin(n_2 t + \epsilon_2 - g_1 t - \Gamma_1)$$
$$- 2\zeta_2^{(2)} h_2 \sin(n_2 t + \epsilon_2 - g_2 t - \Gamma_2) - 2\zeta_2^{(3)} h_3 \sin(n_2 t + \epsilon_2 - g_3 t - \Gamma_3)$$
$$\delta v_3 = - 2\zeta_3 h \sin(n_3 t + \epsilon_3 - gt - \Gamma) - 2\zeta_3^{(1)} h_1 \sin (n_3 t + \epsilon_3 - g_1 t - \Gamma_1)$$
$$- 2\zeta_3^{(2)} h_2 \sin (n_3 t + \epsilon_3 - g_2 t - \Gamma_2) - 2\zeta_3^{(3)} h_3 \sin(n_3 t - \epsilon_3 - g_3 t - \Gamma_3).$$

These inequalities are very considerable in the motions of the satellites in longitude.

The whole then depends on the resolution of the equations (266) and (267); these, however, are not complete, as several terms arise from the perturbations depending on the squares and products of the disturbing forces.

Action of the Sun depending on the Eccentricities.

836. The part of R depending on the action of the sun in the elliptical hypothesis is

$$R = -\tfrac{3}{4}M^2a^2 . \frac{D\delta D}{D'^2} - \frac{6r\delta r}{4}M^2 \cos(2nt - 2Mt + 2\epsilon - 2E)$$

$$+ \frac{12}{4}M^2 . \frac{d(r\delta r)}{ndt} \sin(2nt - 2Mt + 2\epsilon - 2E).$$

But
$$\frac{r\delta r}{a^2} = h\cos(nt + \epsilon - gt - \Gamma); \text{ and}$$

$$\frac{D\delta D}{D'^2} = H\cos(Mt + E - \Pi),$$

H being the eccentricity of Jupiter's orbit, and Π the longitude of the perihelion; hence

$$R = -\tfrac{3}{4}M^2a^2 . H . \cos(Mt + E - \Pi),$$
$$-\tfrac{9}{4}M^2 . a^2 . h\cos(nt - 2Mt + \epsilon - 2E + gt + \Gamma);$$

and therefore, equation (265) becomes

$$0 = \frac{d^2 r\delta r}{a^2 dt} + N^2 . \frac{r\delta r}{a^2}\{1 - 3h\cos(nt + \epsilon - gt - \Gamma)\}$$
$$-\tfrac{3}{2}M^2 . H . \cos(Mt + \mathrm{E} - \Pi)$$
$$-9M^2 . h . \cos(nt - 2Mt + \epsilon - 2E + gt + \Gamma).$$

By article 820, it appears that $\frac{r\delta r}{a^2}$ contains the terms

$$-\frac{M^2}{n^2} . \cos(2nt - 2Mt + 2\epsilon - 2E);$$

hence
$$-3N^2 . \frac{r\delta r}{a^2} . h . \cos(nt + \epsilon - gt + \Gamma)$$

contains
$$\tfrac{3}{2}M^2 . h . \cos(nt - 2Mt + \epsilon - 2E + gt + \Gamma),$$

N^2 being very nearly equal to n^2, so that $\frac{N^2}{n^2} = 1$: thus,

$$0 = \frac{d^2 r\delta r}{a^2 dt^2} + N^2 . \frac{r\delta r}{a^2} - \frac{15}{2} . M^2 . h . \cos(nt - 2Mt + \epsilon - 2E + gt + \Gamma)$$

$$- \tfrac{3}{2} M^2 . H . \cos(Mt + E - \Pi),$$

whence by the method of indeterminate coefficients, the integral is

$$\frac{r\delta r}{a^2} = \frac{15 M^2 . h}{4n(2M + N - n - g)} \cos(nt - 2Mt + \epsilon - 2E + gt + \Gamma)$$

$$+ \frac{3M^2 . H}{2n^2} . \cos(Mt + E - \Pi),$$

which is the effect of the sun's action on the radius vector; and if it be substituted in equation (259), the perturbations in longitude depending on the same cause will be

$$\delta v = - \frac{15 M^2 . h}{2n(2M + N - n - g)} . \sin(nt - 2Mt + \epsilon - 2E + gt + \Gamma)$$

$$- \frac{3M}{n} . H . \sin(Mt + E - \Pi).$$

837. The first term of the second number of this expression corresponds to the evection in the lunar theory, and is only sensible in the motions of the third and fourth satellites; but it is not the only inequality of this kind, for each of the roots g_1, g_2, g_3, furnishes another. The perturbations corresponding to these for the other satellites are found, by reciprocally changing the quantities relative to one into those relating to the others.

Inequalities depending on the Eccentricities which become sensible in consequence of the Divisors they acquire by double integration.

838. It is found by observation, that the mean motion of the first satellite is nearly equal to twice that of the second; and that the mean motion of the second is nearly equal to twice that of the third; or

$$n = 2n_1, \quad n_1 = 2n_2.$$

In consequence of the squares of these nearly commensurable quantities becoming divisors to the inequalities by a double integration, they have a very sensible effect on the preceding equations in longitude.

839. The only part of equation (259) that has a double integral is $3a\int\int ndt.dR$; and as the divisors in question arise from the angles $nt-2n_1t$, n_1t-2n_2t alone, it is easy to see that the part of R containing these angles is,

$$R = m_1\frac{r_1\delta r_1}{a_1^2}\cdot a_1\left(\frac{dA_1}{da_1}\right)\cdot\cos(n_1t-nt+\epsilon_1-\epsilon)$$

$$-2m'_1\cdot\frac{d.(r_1\delta r_1)}{a_1^2.n_1dt}\cdot A_1\cdot\sin(n_1t-nt+\epsilon_1-\epsilon)$$

$$+m_{,}\cdot\frac{r\delta r}{a^2}a\cdot\left(\frac{dA_2}{da}\right)\cdot\cos 2(n_{,}t-nt+\epsilon_{,}-\epsilon)$$

$$+4m_{,}\cdot\frac{d.(r\delta r)}{a^2.ndt}.A_2\cdot\sin e(n_{,}t-nt+\epsilon_{,}-\epsilon).$$

With regard to the action of $m_{,}$ on m,
if $h_{,}\cos(n_{,}t+\epsilon_{,}-gt-\Gamma)$, be put instead of $\frac{r_{,}\delta r_{,}}{a_{,}^2}$, and $h\cos(nt+\epsilon-gt-\Gamma)$ instead of $\frac{r\delta r}{a^2}$; and as by articles 828 and 826

$$G = -a_{,}^2\left(\frac{dA_1}{da_{,}}\right)+2a_{,}A_1$$

$$F = -a^2\left(\frac{dA_2}{da}\right)-4aA_2,$$

observing that $n=2n_{,}$ nearly, the result will be

$$R = -\frac{m_{,}}{2a}\cdot\left\{Fh+\frac{a}{a_{,}}Gh_{,}\right\}\cdot\cos(nt-2n_{,}t+\epsilon-2\epsilon_{,}+gt+\Gamma),$$

which substituted in $3a\int\int ndt\cdot dR$, and integrated, gives for the first satellite,

$$\delta v = \frac{-3m_{,}\cdot n^2}{2(n-2n_{,}+g)^2}\cdot\left\{Fh+\frac{a}{a_{,}}Gh_{,}\right\}.\sin(nt-2n_{,}t+\epsilon-2\epsilon_{,}+gt+\Gamma).$$

Again, since $n_1=2n_2$ nearly, the action of m_2 on $m_{,}$ produces in $\delta v_{,}$ an inequality similar to the preceding, which is

$$\delta v_{,} = \frac{-3m_2.n_{,}^2}{2(n_{,}-2n_2+g)^2}\left\{F'h_{,}+\frac{a_1}{a_2}G'h_2\right\}\sin(n_1t-2n_2t+\epsilon_1-2\epsilon_2+gt+\Gamma).$$

An inequality of the same kind, and from the same cause, is produced also in the equation of the centre of $m_{,}$ by the action of m, for with regard to the inequalities we are now considering, article 574 shows that

$$\delta v_{,} = -\frac{m\sqrt{a}}{m_{,}\sqrt{a_{,}}}\delta v$$

whence the inequality produced by the action of m on $m_{,}$ is

$$\delta v_{,} = \frac{3m \,.\, n^2 \sqrt{a}}{2(n-2n_{,}+g)^2 \sqrt{a_{,}}} \left\{ Fh + \frac{a}{a_{,}} Gh_{,} \right\} \sin(nt-2n_{,}t+\epsilon-2\epsilon_{,}+gt+\Gamma).$$

This inequality may be added to the preceding, for

$$nt - 2n_{,}t + \epsilon - 2\epsilon_{,} = n_1 t - 2n_2 t + \epsilon_1 - 2\epsilon_2 + 180^\circ,$$

and as $n = 2n_{,}$ nearly, and $\left(\frac{a}{a_{,}}\right)^3 = \left(\frac{n_{,}}{n}\right)^2$; therefore

$$\frac{n^2 \sqrt{a}}{\sqrt{a_{,}}} = 2n_{,}^2 \,.\, \frac{a_{,}}{a},$$

and thus the two terms become

$$\delta v_{,} = \frac{3n_{,}^2}{(n-2n_{,}+g)^2} \left\{ \begin{array}{l} m\{Gh_{,} + \frac{a_{,}}{a} Fh\} \\ + \frac{m_2}{2}\{F'h_{,} + \frac{a_{,}}{a_2} G'h_2\} \end{array} \right\} \sin(nt-2n_{,}t+\epsilon-2\epsilon_{,}+gt+\Gamma).$$

Lastly, the action of m_1 on m_2 produces an inequality in m_2, analogous to that produced by the action of m on $m_{,}$, which is therefore

$$\delta v_2 = -\frac{3m_{,}n_2^2}{(n_1-2n_2+g)^2} \left\{ G'h_2 + \frac{a_2}{a_1} F'h_{,} \right\} \sin(nt-2n_{,}t+\epsilon-2\epsilon_{,}+gt+\Gamma).$$

We shall represent the preceding inequalities by

$$\delta v = -\, Q \sin(nt - 2n_{,}t + \epsilon - 2\epsilon_{,} + gt + \Gamma) \qquad (268)$$

$$\delta v_1 = +\, Q_1 \sin(nt - 2n_{,}t + \epsilon - 2\epsilon_{,} + gt + \Gamma) \qquad (269)$$

$$\delta v_2 = -\, Q_2 \sin(nt - 2n_{,}t + \epsilon - 2\epsilon_{,} + gt + \Gamma) \qquad (270)$$

These inequalities are relative to the root g, but each of the roots g_1, g_2, g_3, give similar inequalities in the motions of the three first satellites.

No such inequality exists in the motion of the fourth satellite, since its mean motion is not nearly commensurable with that of any of the others.

Inequalities depending on the Square of the Disturbing Force.

840. On account of the nearly commensurable ratios in the mean motions of the three first satellites the preceding equations must be added as periodic variations to the mean motions, as in the case of Jupiter and Saturn, by means of them several terms are added to equations (266) and (267), which determine the secular variations in the eccentricities and longitudes of the apsides. For if the eccen-

tricities be omitted, and $\mu = 1$, the equations df, df' in article 433 relative to the planets, become

$$d(e \cos \varpi) = - andt \left\{2 \cos v \left(\frac{dR}{dv}\right) + a \sin v \left(\frac{dR}{dr}\right)\right\}$$

$$d(e \sin \varpi) = - andt \left\{2 \sin v \left(\frac{dR}{dv}\right) - a \cos v \left(\frac{dR}{dr}\right)\right\}.$$

The secular variations with regard to the first satellite will be found by substituting

$$R = - \frac{(\rho - \frac{1}{2}\phi)}{3r^3} + m_1 A_2 \cos 2(v_, - v)$$

in the first of the preceding equations, and putting $nt + \epsilon + \delta v$ for v, and $a^2 + 2r\delta r$ for r^2; whence

$$\begin{aligned} d(e \cos \varpi) = {} & 4andt \,.\, m_, A_2 \sin (2v - 2v_,) \cos v \\ & - a^2 ndt \,.\, m_, \left(\frac{dA_2}{da}\right) \cos (2v - 2v_,) \sin v \\ & - ndt \,.\, \frac{(\rho - \frac{1}{2}\phi)}{a^2} \,.\, \sin (nt + \epsilon) \\ & - ndt \,.\, \frac{(\rho - \frac{1}{2}\phi)}{a^2} \delta v \cos (nt + \epsilon) \\ & + 4ndt \,.\, \frac{(\rho - \frac{1}{2}\phi)}{a^2} \,.\, \frac{r\delta r}{a^2} \sin (nt + \epsilon). \end{aligned}$$

Then only attending to the terms depending on $nt - 2n_,t + \epsilon - 2\epsilon_,$, if the values of $\frac{r\delta r}{a^2}$ and δv given by (260) be substituted; and as

$$F = - 4a A_2 - a^2 \left(\frac{dA_2}{da}\right),$$

the result will be

$$d(e \cos \varpi) = - \frac{m_, F . ndt}{2} . \left\{1 - \frac{(0)}{2n - 2n_, - N}\right\} . \sin (nt - 2n_,t + \epsilon - 2\epsilon_,)$$

in which
$$(0) = \frac{(\rho - \frac{1}{2}\phi)}{a^2} n.$$

Since the mean longitudes $nt + \epsilon$ and $n_,t + \epsilon_,$ are variable, these angles must be augmented by the values of δv, $\delta v_,$, in equations (268) and (269), so that

$$nt + \epsilon + Q \sin (nt - 2n_,t + \epsilon - 2\epsilon_, + gt + \Gamma)$$
$$n_,t + \epsilon_, + Q_, \sin (nt - 2n_,t + \epsilon - 2\epsilon_, + gt + \Gamma)$$

must be substituted in the sine of the preceding equation, which becomes, in consequence,

$$d(e\cos\varpi) = \frac{m_{\prime}F.ndt}{4}.\left\{1-\frac{(0)}{2n-2n_{\prime}-N}\right\}.(2Q_{\prime}-Q)\,.\,\sin(gt+\Gamma)$$

when the periodic part is omitted.

But by article 834, $e\cos\varpi = -h\cos(gt+\Gamma)$;

hence $d(e\cos\varpi) = hg\,.\,dt\,.\,\sin(gt+\Gamma)$, and thus

$$\frac{m_{\prime}.Fn}{4}.\left\{1-\frac{(0)}{2n-2n_{\prime}-N}\right\}.(2Q_{\prime}-Q)$$

must be subtracted from equation (266).

841. The same analysis applied to $d(e_{\prime}\cos\varpi_{\prime})$ will determine the increment of the first of equations (267), with regard to the second satellite. But, in this case,

$$R = -\frac{(\rho-\frac{1}{2}\phi)}{3r^3} + m_{\prime}A_1^{(1.2)}\cos(v-v_{\prime}) + m_2A_2^{(3.2)}\cos 2(v_{\prime}-v_2),$$

and equations (269) and (270) must be employed. The result is, that

$$\frac{m_2n_{\prime}}{4}\left\{1-\frac{(1)}{n-n_{\prime}-N_{\prime}}\right\}F'.(2Q_2-Q_1)-\frac{mn_{\prime}}{4}.\left\{1-\frac{(1)}{n-n_{\prime}-N_{\prime}}\right\}.G.(2Q_{\prime}-Q)$$

must be added to the first of equations (267).

For the same reason

$$\frac{m_{\prime}n_2}{4}.G'.(2Q_2-Q_1).\left\{1-\frac{(2)}{n_{\prime}-n_2-N_2}\right\}$$

must be added to the second of equation (267).

As these quantities only arise from the ratios among the mean motions of the three first satellites, the secular variations of the fourth are not affected by them. In consequence of these additions, equations (266) and (267) become

$$0=h\left\{g-(0)-\boxed{0}-(0.1)-(0.2)-(0.3)\right\}+\boxed{0.1}h_1+\boxed{0.2}h_2+\boxed{0.3}h_3$$
$$-\frac{m_{\prime}n}{4}\left\{1-\frac{(0)}{2n-2n_{\prime}-N_{\prime}}\right\}F(2Q_{\prime}-Q);$$

$$0=h_1\left\{g-(1)-\boxed{1}-(1.0)-(1.2)-(1.3)\right\}+\boxed{1.0}h+\boxed{1.2}h_2+\boxed{1.3}h_3$$
$$-\frac{mn_{\prime}}{4}\left\{1-\frac{(1)}{n-n_{\prime}-N_{\prime}}\right\}G(2Q_{\prime}-Q)$$
$$+\frac{m_2n_{\prime}}{4}\left\{1-\frac{(1)}{n-n_{\prime}-N_{\prime}}\right\}F'(Q_2-Q_1) \qquad (271)$$

$$0=h_2\{g-(2)-\boxed{2}-(2.0)-(2.1)-(2.3)\}+\boxed{2.0}h+\boxed{2.1}h_1+\boxed{2.3}h_2$$
$$+\frac{m_{,}n_2}{4}\left(1-\frac{(2)}{n_{,}-n_2-N_2}\right)G'(2Q_2-Q_1);$$

$$0=h_3\{g-(3)-\boxed{3}-(3.0)-(3.1)-(3.2)\}+\boxed{3.0}h+\boxed{3.1}h_1+\boxed{3.2}h_2$$

842. An inequality which is only sensible in the theory of the second satellite may now be determined; for, by (260),

$$\delta v=\frac{m_{,}nF}{2n-2n_{,}-N}\sin(2nt-2n_{,}t+2\epsilon-2\epsilon_{,});\text{ or}$$

$$\delta v=\frac{m_{,}nF}{2n-2n_{,}-N_{,}}\{\cos(nt-2n_{,}t+\epsilon-2\epsilon_{,})\,.\,\sin(nt+\epsilon)$$
$$+\sin(nt-2nt+\epsilon-2\epsilon)\,.\,\cos(nt+\epsilon)\};$$

but as $v=2e\sin(nt+\epsilon-\varpi)$, and for the variable ellipse which we are now considering,

$$\delta v=2\delta.(e\cos\varpi)\,.\,\sin(nt+\epsilon)-2\delta.(e\sin\varpi).\cos(nt+\epsilon).$$

By comparing these two values,

$$2\delta(e\sin\varpi)=-\frac{m_{,}nF}{2n-2n_{,}-N_{,}}\sin(nt-2n_{,}t+\epsilon-2\epsilon_{,})$$

$$2\delta(e\cos\varpi)=\frac{m_{,}nF}{2n-2n_{,}-N_{,}}\cos(nt-2n_{,}t+\epsilon-2\epsilon_{,}).$$

But the elliptical expression of v contains the term

$$\tfrac{5}{4}e^2\sin(2nt+2\epsilon-2\varpi),$$

or

$$\tfrac{5}{4}(e^2\cos^2\varpi-e^2\sin^2\varpi).\sin 2(nt+\epsilon)$$
$$-\tfrac{5}{4}.e^2\sin\varpi\,.\,\cos\varpi\,.\,\cos 2(nt+\epsilon).$$

If $e\sin\varpi+\delta(e\sin\varpi)$, and $e\cos\varpi+\delta(e\cos\varpi)$ be put for $e\sin\varpi$, and $e\cos\varpi$, it becomes

$$\delta v=\tfrac{5}{4}\{\delta\,.\,e\cos\varpi)^2-(\delta\,.\,e\sin\varpi)^2\}\sin 2(nt+\epsilon)$$
$$-\tfrac{5}{4}\delta\,.\,e\cos\varpi\,.\,\delta\,.\,e\sin\varpi\,.\,\cos 2(nt+\epsilon);$$

and in consequence of the preceding values of $\delta(e\cos\varpi)$, $\delta(e\sin\varpi)$, there is the following inequality in the longitude of the first satellite,

$$\delta v=\tfrac{5}{16}\left(\frac{m_{,}nF}{2n-2n_{,}-N}\right)^2\sin 4(nt-n_{,}t+\epsilon-\epsilon_{,}).$$

By the same process the corresponding inequalities in the second and third satellites are found to be

$$\delta v_{,}=\tfrac{5}{16}\frac{n_{,}^2}{(n-n_{,}-N_{,})^2}\{mG-m_2F'\}^2\sin 2(nt-n_{,}t+\epsilon-\epsilon_{,})$$

$$\delta v_2=\tfrac{5}{16}\left(\frac{m_{,}n_2G'}{n_{,}-n_2-N_2}\right)^2\sin 2(n_1t-n_2t+\epsilon_{,}-\epsilon_2).$$

Librations of the three first Satellites.

843. Some very interesting inequalities arising from the equation

$$nt - 3n_1t + 2n_2t + \epsilon - 3\epsilon_1 + 2\epsilon_2 = 180°,$$

are found among the terms depending on the squares of the disturbing forces, that affect the whole theory of the satellites, in consequence of the very small divisor $(n - 3n_1 + 2n_2)^2$ which they acquire by double integration. If the orbits be considered as variable ellipses, and if ζ, ζ_1, ζ_2, be the mean longitudes of the three first satellites, it is clear that the terms having the square of $n - 3n_1 + 2n_2$ for divisor, can only be found from

$$d^2\zeta = 3andt \,.\, \mathrm{d}R$$

$$d^2\zeta_1 = 3a_1n_1dt \,.\, \mathrm{d}R_1$$

$$d^2\zeta_2 = 3a_2n_2dt \,.\, \mathrm{d}R_2$$

which are the variations in the mean motions by article 439.

844. With regard to the action of m_1 on m, the series R in article 815 only contains the angle $n_,t - nt + \epsilon_, - \epsilon$ and its multiples, it is evident therefore, that the angle $nt - 3n_1t + 2n_2t$ can only arise from the substitution of the perturbations (262) which depend on the angle $2n_1t - 2n_2t$. By article 814, δv_1 contains both the elliptical part of the longitude and the perturbations, and if the latter be expressed by $\delta\bar{v}_,$ then

$$\delta v_, = \frac{2d(r_,\delta v_,}{a_,^2 . n_,dt} + \delta\bar{v}_,$$

and when the square of the eccentricity is omitted $\frac{r_,\delta r_,}{a_,^2}$ becomes $\frac{\delta r_,}{a_,}$.

If then $\delta\bar{v}_,$ and $\frac{\delta r_,}{a_,}$ be put for $\frac{2d(r_,\delta r_,)}{a_,^2 . n_,dt}$ and $\frac{r_,\delta r_,}{a_,^2}$ the part of R required is

$$R = m_, . \left(\frac{dA_,}{da_,}\right) . \delta r_, . \cos(n_,t - nt + \epsilon_, - \epsilon)$$

$$- m_, . \delta\bar{v}_, . A_, . \sin(n_,t - nt + \epsilon_, - \epsilon),$$

or

$$\mathrm{d}R = m_, . A_,\delta\bar{v}_, . \cos(n_,t - nt + \epsilon_, - \epsilon) . ndt$$

$$- m_, . \left(\frac{dA_1}{da_,}\right) . \delta r_, . \sin(n_,t - nt + \epsilon_, - \epsilon) . ndt.$$

for in this case $d\delta r_{,}$ and $d\delta \bar{v}_{,}$ are zero, since equations (262), or

$$\delta r_{,} = - \frac{m_2 n_{,} a_{,} F'}{2(2n_1 - 2n_2 - N_{,})} \cdot \cos(2n_{,}t - 2n_2 t + 2\epsilon_1 - 2\epsilon_2)$$

$$\delta \bar{v}_{,} = \frac{m_2 n_{,} F'}{(2n_{,} - 2n_2 - N_{,})} \cdot \sin(2n_1 t - 2n_2 t + 2\epsilon_1 - 2\epsilon_2)$$

do not contain the arc nt. If these quantities be substituted in dR, it will be found, in consequence of

$$G = 2a_{,}A_1 - a_{,}^2\left(\frac{dA_1}{da_{,}}\right), \text{ and } n = 2n_{,},$$

that

$$\frac{d^2\zeta}{dt^2} = - \frac{3n^3 m_{,} m_2 F' G}{8(n - n_{,} - N_{,})} \frac{a}{a_{,}} \sin(nt - 3n_1 t + 2n_2 t + \epsilon - 3\epsilon_{,} + 2\epsilon_2);$$

for as $\quad 2n_{,} - 2n_2 = n - n_{,}$ nearly,

the divisor $\quad 2n_{,} - 2n_2 - N_{,} = n - n_{,} - N_{,}$.

The variation in the mean motion of the second satellite consists of two parts; one arising from the action of m, and the other from that of m_2.

The value of R for the first is

$$R = m \cdot A^{(1.2)} \cdot \delta \bar{v}_{,} \cdot \sin(nt - n_{,}t + \epsilon - \epsilon_{,})$$
$$+ m \cdot \left(\frac{dA^{(1.2)}}{da}\right) \cdot \delta r_{,} \cdot \cos(nt - n_{,}t + \epsilon - \epsilon_{,}).$$

If the differential of R be taken with regard to $n_{,}t$, making $\delta v_{,}$ and $\delta r_{,}$ vary, by the substitution of the preceding values of $\delta \bar{v}_{,}$, $\delta r_{,}$, and their differentials, it will be found, in consequence of

$$G = 2a_{,}A_{,}^{(1.2)} - a_{,}^2 \frac{dA_{,}^{(1.2)}}{da_{,}},$$

and $n_2 = \frac{1}{2} n_1$, that the variation in the mean motion of the second satellite from the action of the first must be

$$\frac{3n^3 m \cdot m_2 \cdot F'G}{16(n - n_{,} - N_{,})} \sin(nt - 3n_{,}t + 2n_2 t + \epsilon - 3\epsilon_{,} - 2\epsilon_2).$$

Again, if

$$\frac{\delta r_{,}}{a_{,}} = - \frac{mn_{,}G}{2(n - n_{,} - N_{,})} \cos(nt - n_{,}t + \epsilon - \epsilon_{,}),$$

and $\quad \delta v_{,} = \dfrac{mn_{,}G}{n - n_{,} - N_{,}} \sin(nt - n_{,}t + \epsilon - \epsilon_{,}),$

from article 826, be substituted in the differential of

$$R = m_2 \left\{ \left(\frac{dA_2^{(3.2)}}{da} \right) \delta r_{,} \cos (2n_{,}t - 2n_2t + 2\epsilon_1 - \epsilon_2) \right.$$

$$\left. - 2A_2^{(3.2)} \,.\, \delta \bar{v}_{,} \sin (2n_{,}t - 2n_2t + 2\epsilon_1 - 2\epsilon_2) \right\},$$

which is the value of R with regard to m_2 and m_1, observing that $n = 2n_{,}$; and, by article 826,

$$F' = - 4a_{,}A_2^{(3.2)} - a_{,}^2 \left(\frac{dA_2^{(3.2)}}{da_{,}} \right),$$

the part of $\frac{d^2\zeta_{,}}{dt^2}$, arising from the action of m_2 on $m_{,}$, will be found equal to

$$\frac{3m \,.\, m_2 n^3}{32(n - n_{,} - N_{,})} F'G \sin (nt - 3n_{,}t + 2n_2t + \epsilon - 3\epsilon_1 + 2\epsilon_2) ;$$

and the whole variation in the mean motion of $m_{,}$, from the combined action of m and m_2, is

$$\frac{d^2\zeta_{,}}{dt^2} = \frac{9mm_2 n^3 F'G}{32(n - n_{,} - N_{,})} \sin (nt - 3n_{,}t + 2n_2t + \epsilon - 3\epsilon_1 + 2\epsilon_2).$$

With regard to the action of m_1 on m_2

$$R = m_1 \left\{ -2A_2^{(3.2)} \,.\, \delta \bar{v}_{,} \sin 2(n_1t - n_2t + \epsilon_1 - \epsilon_2) \right.$$

$$\left. + \left(\frac{dA_2^{(3.2)}}{da_{,}} \right) \,.\, \delta r_{,} \,.\, \cos 2(n_1t - n_2t + \epsilon_1 - \epsilon_2) \right\}.$$

If the same values of $\delta \bar{v}_{,}$ and $\delta r_{,}$ be substituted in the differential of this with regard to n_2t, it will be found that the action of m_1 and m_2 produces the inequality

$$\frac{d^2\zeta_2}{dt^2} = - \frac{3n^3 mm_{,} F'G}{64(n - n_{,} - N_{,})} \,.\, \frac{a_2}{a} \sin (nt - 3n_{,}t + 2n_2t + \epsilon - 3\epsilon_1 + 2\epsilon_2).$$

845. As $$\frac{d^2\zeta}{dt^2} = 3andt \,.\, \mathrm{d}R ;$$

$$\frac{d^2\zeta_{,}}{dt^2} = 3a_{,}n_{,}dt \, \mathrm{d}R_{,}, \qquad \frac{d^2\zeta_2}{dt^2} = 3a_2n_2dt \,.\, \mathrm{d}R_2 ;$$

by comparing the values of these three quantities in the last article the result is

$$md R + m_{,}dR_{,} = 0, \qquad \text{and } m_{,}dR_{,} + m_2dR_2 = 0,$$

which is conformable with what was shown in article 573, with regard to the planets.

846. As the three first satellites move in orbits that are nearly circular, the error would be very small, in assuming

$$nt + \epsilon,\ n_{,}t + \epsilon_{,},\ n_{2}t + \epsilon_{2},$$

to be their true longitudes.

The preceding inequalities in the mean motions of the three first satellites are therefore

$$\frac{d^2v}{dt^2} = -\frac{3n^3m_1m_2\frac{a}{a_{,}}F'G}{8(n-n_{,}-N_{,})}\sin(v-3v_{,}+2v_2)$$

$$\frac{d^2v_{,}}{dt^2} = \frac{9n^3\,.\,mm_2F'G}{32(n-n_{,}-N_{,})}\sin(v-3v_{,}+2v_2) \qquad (272)$$

$$\frac{d^2v_2}{dt^2} = -\frac{3n^3mm_{,}F'G}{64\,(n-n_{,}-N_{,})}\frac{a_2}{a_{,}}\sin(v-3v_{,}+2v_2).$$

847. In order to abridge, let $\phi = v - 3v_1 + 2v_2$; whence

$$\frac{d^2\phi}{dt^2} = \frac{d^2v}{dt^2} - 3\frac{d^2v_{,}}{dt^2} + 2\frac{d^2v_2}{dt^2}.$$

If the preceding values be put in this, and if to abridge,

$$K = -\frac{3nF'G}{8\,(n-n_{,}-N_{,})}\left\{\frac{a}{a_{,}}m_1m_2 + \tfrac{9}{4}mm_2 + \frac{a_2}{4a_{,}}mm_2\right\},$$

the result will be

$$\frac{d^2\phi}{dt^2} = K.n^2\,.\,\sin\phi.$$

K and n^2 may be assumed to be constant quantities, their variations are so small; hence the integral of this equation is

$$dt = \frac{\pm\,d\phi}{\sqrt{c - 2Kn^2\cos\phi}};$$

c is a constant quantity introduced by integration, the different values of which give rise to the three following cases.

848. 1st. If c be greater than $2Kn^2$, without regard to the sign, it must be positive; and the angle $\pm\phi$ will increase indefinitely, and will become equal to one, two, three, &c., circumferences.

2d. If K be positive, and c less than $2n^2K$, abstracting from the sign, the radical will be imaginary when $\pm\phi$ is equal to zero, or to one, two, three, &c. circumferences. The angle ϕ must therefore oscillate about the semicircumference, since it never can be zero, or equal to a whole circumference, which would make the time an imaginary quantity. Its mean value must consequently be 180°.

3d. If c be less than $2Kn^2$, and K negative, the radical would be imaginary when the angle $\pm\phi$ is equal to any odd number of semi-circumferences; the angle ϕ must therefore oscillate about zero, its mean value, since the time cannot be imaginary. However, as it will be shown that K is a positive quantity, the latter case does not exist, so that ϕ must either increase indefinitely, or oscillate about 180°. In order to ascertain which of these is the law of nature, let

$$\phi = \pi \pm \varpi,$$

π being 180° and ϖ any angle whatever; hence

$$dt = \frac{d\varpi}{\sqrt{c + 2Kn^2 \cos \varpi}}. \qquad (273)$$

If the angles $\pm\phi$ and ϖ increase indefinitely, c is positive, and greater than $2Kn^2$; hence, in the interval between $\varpi = 0$, and its increase to 90°, dt is less than

$$\frac{d\varpi}{n\sqrt{2K}}; \quad \text{and } t < \frac{\varpi}{n\sqrt{2K}}.$$

Thus the time t that the angle ϖ employs in increasing till it be equal to 90°, will be less than $\dfrac{\varpi}{2n\sqrt{2K}}$.

This time is less than two years: but from the discovery of the satellites the libration or angle ϖ has always been zero, or extremely small; therefore this angle does not increase indefinitely, it can only oscillate about its mean value of zero.

The second case, then, is what really exists, and the angle

$$v - 3_{,}v_1 + 2v_2,$$

must oscillate about 180°, which is its mean value.

849. Several important results are given by the equation

$$v - 3v_1 + 2v_2 = \pi + \varpi.$$

If the insensible part ϖ be omitted,

$$nt - 3n_1t + 2n_2t + \epsilon - 3\epsilon_1 + 2\epsilon_2 = \pi.$$

Whence

$$n - 3n_1 + 2n_2 = 0$$
$$\epsilon - 3\epsilon_1 + 2\epsilon_2 = 180°.$$

These two equations are perfectly confirmed by observation, for Delambre found, from the comparison of a great number of eclipses of the three first satellites, that their mean motions in a hundred Julian years, with regard to the equinox, are

1st satellite 7432435°.46982
2d . . . 3702713°.231493
3d . . . 1837852°.113582

whence it appears, that the mean motion of the first, minus three times that of the second, plus twice that of the third, is equal to 9″.0072, so small a quantity, that it affords an astonishing proof of the accuracy both of the theory and observation. Delambre determined also, from a great number of eclipses, that the epochs of the mean motions of the three first satellites, at midnight, on the first of January 1750, were

$$\epsilon = 15°.02626$$
$$\epsilon' = 311°.44689$$
$$\epsilon'' = 10°.27219,$$

whence $$\epsilon - 3\epsilon_1 + 2\epsilon_2 = 180°\ 1'\ 3'',$$

a result that is less accurate than the preceding; but it will be shown, in treating of the eclipses of the satellites, that it probably arises from errors of observation, depending on the discs of the satellites, which vanish to us before they are quite immersed in the shadow.

850. The same laws exist in the synodic motions of the satellites; for in the equation

$$nt - 3n_1t + 2n_2t + \epsilon - 3\epsilon_1 + 2\epsilon_2 = 180°,$$

the angles may be estimated from a moveable axis, since the position of the axis would vanish in this equation: we may therefore suppose that

$$nt + \epsilon, \quad n_1t + \epsilon_1, \quad n_2t + \epsilon_2,$$

are the mean synodic longitudes. This has a great influence on the eclipses of the three first satellites, as will appear afterwards.

851. On account of these laws the actions of the first and third satellites on the second are united in one term, given in article 826, which is the great inequality in that body indicated by observations. These inequalities will never be separated.

852. Without the mutual attraction of the satellites the two equations

$$n - 3n_1 + 2n_2 = 0$$
$$\epsilon - 3\epsilon_1 + 2\epsilon_2 = 0$$

would be unconnected. It would have been necessary in the beginning of their motions that their epochs and mean motions had been so arranged as to suit these equations, which is most improbable;

and in this case the slightest action from any foreign cause, as the attraction of the planets and comets, would have changed the ratios. But the mutual action of the satellites gives perfect stability to these relations, for, at the origin of the motion, when $t = 0$,

$$\frac{dv}{ndt} - 3\frac{dv_1}{n_1dt} + 2\frac{dv_2}{n_2dt} = \pm\sqrt{\frac{c}{n^2} - 2K\cos(\epsilon - 2\epsilon_1 + 3\epsilon_2)}$$

c being less than $2Kn^2$. It would be sufficient for the accuracy of the preceding results that the first member of this equation had been comprised between the limits

$$+ 2K\sin(\tfrac{1}{2}\epsilon - \tfrac{3}{2}\epsilon_1 + \epsilon_2)$$
$$- 2K\sin(\tfrac{1}{2}\epsilon - \tfrac{3}{2}\epsilon_1 + \epsilon_2)$$

at the origin of their motions, and it is sufficient for their stability that no foreign force disturbs it.

853. It appears then, that if the preceding laws among the mean motions of the three first satellites had only been approximate at their origin, their mutual attraction would ultimately have rendered them exact.

854. The angle ϖ is so small, that we may make

$$\cos\varpi = 1 - \tfrac{1}{2}\varpi^2;$$

and if to abridge
$$\mathfrak{C}^2 = \frac{c + 2Kn^2}{n^2K},$$

$\mathfrak{C}$ being arbitrary, on account of the arbitrary constant quantity c that it contains, equation (273) becomes

$$\varpi = \mathfrak{C}\sin(nt\sqrt{K} + A),$$

A being a new arbitrary quantity.

855. As the motions of the four satellites in longitude, latitude, and distance, are determined by twelve differential equations of the second order, their integrals must contain twenty-four arbitrary quantities, which are the data of the problem, and are given by observation. Two of these are determined by the equations

$$n - 3n_1 + 2n_2 = 0$$
$$\epsilon - 3\epsilon_1 + 2\epsilon_2 = 180°;$$

they are, however, replaced by $\mathfrak{C}$ and A, the first determines the extent of the libration, and A marks the time when it is zero: neither are determined, since the inequality ϖ has as yet been insensible.

856. The integrals of the three equations (272) may now be found, for as

$$v - 3v_1 + 2v_2 = \pi + \varpi = \pi + \mathcal{C} \sin (nt \sqrt{K} + A),$$

$$\sin . (v - 3v_1 + 2v_2) = \sin \{ \pi + \mathcal{C} \sin (nt \sqrt{K} + A) \}$$

$$= - \mathcal{C} . \sin (nt \sqrt{K} + A);$$

hence the first of equations (272) becomes

$$\frac{d^2 v}{dt^2} = \frac{3n^3 m_1 m_2 F'G}{8(n - n_1 - N_1)} \frac{a}{a_1} \mathcal{C} \sin (nt \sqrt{K} + A),$$

the integral of which is

$$v = \frac{\mathcal{C} \sin (nt \sqrt{K} + A)}{1 + \frac{9a_1 m}{4am_1} + \frac{a_2 m}{4am_2}}.$$

In the same way

$$v_1 = - \frac{\mathcal{C} \sin (nt \sqrt{K} + A) . \frac{3a_1 m}{4am_1}}{1 + \frac{9a_1 m}{4am_1} + \frac{a_2 m}{4am_2}}$$

$$v_2 = \frac{\frac{a_2 m}{8am_2} \mathcal{C} \sin (nt \sqrt{K} + A)}{1 + \frac{9a_1 m}{4am_1} + \frac{a_2 m}{4am_2}}$$

which are the three equations of the libration. They have hitherto been insensible, but they modify all the inequalities of long periods in the theory of the three first satellites.

857. For example, the inequality

$$v = - \frac{3M}{n} H \sin (Mt + E - \Pi),$$

gives

$$\frac{d^2 v}{dt^2} = + \frac{3M^3}{n} H \sin (Mt + E - \Pi);$$

But the differential of the first of the equations of libration is

$$\frac{d^2 v}{dt^2} = - \frac{Kn^2 \sin (v - 3v_1 + 2v_2)}{1 + \frac{9a_1 m}{4am_1} + \frac{a_2 m}{4am_2}};$$

or, if to abridge,

$$b = 1 + \frac{9a_1 m}{4am_1} + \frac{a_2 m}{4am_2}$$

$$\frac{d^2v}{dt^2} = -\frac{Kn^2}{b} \cdot \sin(v - 3v_1 + 2v_2),$$

and adding the two values of $\frac{d^2v}{dt^2}$

$$\frac{d^2v}{dt^2} = -\frac{Kn^2}{b}\sin(v - 3v_1 + 2v_2) + \frac{3M^3}{n} \cdot H\sin(Mt + E - \Pi). \tag{274}$$

To integrate this equation let

$$v = \lambda \sin(Mt + E - \Pi), \quad v_1 = \lambda_1 \sin(Mt + E - \Pi)$$
$$v_2 = \lambda_2 \sin(Mt + E - \Pi),$$

hence,

$$v - 3v_1 + 2v_2 = (\lambda - 3\lambda_1 + 2\lambda_2) \cdot \sin(Mt + E - \Pi),$$

and

$$\frac{d^2v}{dt^2} = \left\{\frac{3M^3 \cdot H}{n} - \frac{Kn^2}{b}(\lambda - 3\lambda_1 + 2\lambda_2)\right\} \sin(Mt + E - \Pi);$$

and if $\lambda \sin(Mt + E - \Pi)$ be put for v,

$$\lambda = -\frac{3M.H}{n} + \frac{Kn^2}{bM^2}(\lambda - 3\lambda_1 + 2\lambda_2).$$

In the same manner it may be found that

$$\lambda_1 = -\frac{6M.H}{n} - \frac{3a_1m}{4am_1} \cdot \frac{Kn^2}{bM^2}(\lambda - 3\lambda_2 + 2\lambda_2)$$

$$\lambda_2 = -\frac{12M.H}{n} + \frac{a_2m}{8am_2} \cdot \frac{Kn^2}{bM^2} \cdot (\lambda - 3\lambda_1 + 2\lambda_2),$$

whence $$\lambda - 3\lambda_1 + 2\lambda_2 = \frac{9M^3.H}{n(Kn^2 - M^2)};$$

so that equation (274) becomes

$$\frac{d^2v}{dt^2} = \frac{3M^3}{n}\left\{1 + \frac{3K.n^2}{b(Kn^2 - M^2)}\right\} H \cdot \sin(Mt + E - \Pi)$$

and

$$\delta v = -\frac{3M}{n}\left\{1 + \frac{3K.n^2}{b(Kn^2 - M^2)}\right\} H \cdot \sin(Mt + E - \Pi).$$

The inequalities in the longitude of m_1 and m_2 are found by the same analysis, consequently

$$\delta v = -\frac{3M}{n}\left\{1 + \frac{3Kn^2}{b(M^2 - Kn^2)}\right\} H \cdot \sin(Mt + E - \Pi)$$

$$\delta v_1 = -\frac{6M}{n}\left\{1 - \frac{9a_1 m K n^2}{8am_1 b(M^2 - Kn^2)}\right\} H . \sin(Mt + E - \Pi)$$

$$\delta v_2 = -\frac{12M}{n}\left\{1 + \frac{3a_2 m . Kn^2}{32 . am_2 . b(M^2 - Kn^2)}\right\} H \sin(Mt + E - \Pi).$$

This inequality replaces the term depending on the same angle in article 836. It corresponds with the annual equation in the lunar theory, and its period is very great.

858. The variation in the form and position of Jupiter's orbit is the cause of secular inequalities in the mean motions of the satellites, similar to those produced by the variation of the earth's orbit on the moon; hitherto, however, they have been insensible, and will remain so for a long time, with the exception of one depending on the displacement of Jupiter's equator, and that is only perceptible in the motions of the fourth satellite; but these cannot be determined till the equations in latitude have been found.

CHAPTER VII.

PERTURBATIONS OF THE SATELLITES IN LATITUDE.

859. THE perturbations in latitude are found with most facility from

$$0 = \frac{d^2s}{dv^2}\left\{1 - \frac{2}{h^2}\int\left(\frac{dR}{dv}\right)\cdot\frac{dv}{u^2}\right\} - \frac{1}{h^2u^2}\cdot\frac{ds}{dv}\cdot\left(\frac{dR}{dt}\right)$$
$$+ \frac{s}{h^2u^2}\left(\frac{dR}{du}\right) + \frac{(1+s^2)}{h^2u^2}\left(\frac{dR}{ds}\right),$$

which was employed for the moon, but in that case R was a function of u, v, and s, and the differential $\frac{dR}{ds}$ was taken in that hypothesis, which we shall represent by $\frac{dR'}{ds}$, but now R is a function of r, v, and s, hence

$$du\left(\frac{dR}{du}\right) + ds\left(\frac{d'R}{ds}\right) = dr\left(\frac{dR}{dr}\right) + ds\left(\frac{dR}{ds}\right)$$

and as

$$r = \frac{\sqrt{1+s^2}}{u}, \quad du = -\frac{dr}{r^2}\sqrt{1+s^2} + \frac{sds}{r\sqrt{1+s^2}};$$

and comparing the coefficients of ds in these two equations

$$\frac{us}{1+s^2}\left(\frac{dR}{du}\right) + \frac{d'R}{ds} = \left(\frac{dR}{ds}\right),$$

so the preceding equation of latitude, when $\frac{\sqrt{1+s^2}}{r}$ is put for u, and the product of the disturbing force by $s^2 . \frac{ds}{dv}$ omitted, becomes

$$0 = \frac{d^2s}{dv^2} + s + \frac{r^2}{h^2}\left(\frac{dR}{ds}\right) - \frac{r^2}{h^2}\cdot\frac{ds}{dv}\left(\frac{dR}{dv}\right).$$

860. The only part of the disturbing force that affects the latitude

is
$$R = -\frac{m_{,}r_{,}r\{ss_{,} - \frac{1}{2}s^2\cos(v_{,}-v)\}}{\{r^2 - 2rr_{,}\cos(v_{,}-v)+r_{,}^2\}^{\frac{3}{2}}}$$
$$+\frac{3m_{,}}{4n^2}\{s_{,}^2 + s^2\cos(2v-2u) - 4ss_{,}\cos(v-u)\}$$
$$+\frac{(\rho-\frac{1}{2}\phi)}{a^2}(s-s')^2.$$

If the eccentricities be omitted,
$$r=a,\quad r_{,}=a_{,},\ \text{and}\ \frac{1}{(r^2-2rr_{,}\cos(v_{,}-v)+r_{,}^2)^{\frac{3}{2}}} =$$
$$\{a^2-2aa_{,}\cos(v_{,}-v)+a_{,}^2\}^{-\frac{3}{2}} = \tfrac{1}{2}B_0 + B_1\cos(v_{,}-v) + \&c.$$
as for the planets; hence
$$R = -\Sigma m_{,}a^2a_{,}\{ss_{,} - \tfrac{1}{2}s^2\cos(v_{,}-v)\}B_1\cos(v_{,}-v)$$
$$+\frac{3M^2}{4n^2}\{s^2 - 4sS\cos(v-U)\} + \frac{(\rho-\frac{1}{2}\phi)}{a^2}(s-s')^2.$$

Whence
$$0 = \frac{d^2s}{dt^2} + s\{1 + 2\frac{(\rho-\frac{1}{2}\phi)}{a^2} + \frac{3}{2}\frac{M^2}{n^2} + \tfrac{1}{2}\Sigma m_{,}a^2a_{,}\,.\,B_1\}$$
$$-\frac{2(\rho-\frac{1}{2}\phi)}{a^2}s' - \frac{3M^2}{n^2}S\cos(U-v) - \Sigma m_{,}a^2.a_{,}B_1s_{,}\cos(v_{,}-v).$$

861. In order to integrate this equation, let
$$s = l\,.\sin(v+pt+\Lambda);\quad s_{,} = l_{,}\sin(v_1+pt+\Lambda)$$
$$s_2 = l_2.\sin(v_2+pt+\Lambda);\quad s_3 = l_3\sin(v_3+pt+\Lambda)$$
$$S = L'.\sin(U+pt+\Lambda);\quad s' = L\sin(v+pt+\Lambda),$$
l, l_1, l_2, l_3, L' and L being the inclination of the orbits of the four satellites, of Jupiter's orbit and equator on the fixed plane, p and Λ, quantities on which the sidereal motions and longitudes of the nodes depend.

If the motion of only one satellite be considered at a time, then substituting for s, $s_{,}$ and S, also putting $\frac{v}{n}$ for t, and omitting p^2, the comparison of the coefficients of $\sin(v+pt+\Lambda)$ gives
$$0 = l\left\{\frac{p}{n} - \frac{(\rho-\frac{1}{2}\phi)}{a^2} - \frac{3}{4}\,.\,\frac{M^2}{n^2} - \tfrac{1}{4}\Sigma m_{,}a^2.a_{,}B_1\right\} \quad (275)$$
$$+\frac{(\rho-\frac{1}{2}\phi)}{a^2}L + \frac{3}{4}\frac{M^2}{n^2}.L' + \tfrac{1}{4}\Sigma m_{,}a^2\,.\,a_{,}B_1l_{,}.$$

If $$\frac{a}{a_,'} = \alpha, \text{ and } v_, - v = n_,t - nt + \epsilon_, - \epsilon = \beta$$

$$\{1 - 2\alpha \cos \beta + \alpha^2\}^{-\frac{3}{2}} = a_,^3 (\tfrac{1}{2}B_0 + B_1 \cos \beta + B_2 \cos 2\beta + \&c.,)$$

which is identical with the series in article 446, and therefore the formulæ for the planets give by article

$$(0.1) = \frac{m_1 n \,.\, a^2 a_1 B_1}{4},$$

consequently equation (275) becomes

$$0 = l\{p - (0) - \boxed{0} - (0.1)\} + L\,(0) + L'\,\boxed{0} + (0.1)\,l_,,$$

but the action of the satellites m_2 and m_3 produce terms analogous to those produced by m_1; so the preceding equation, including the disturbing action of all the bodies, and the compression of Jupiter, is

$$0 = l\{p - (0) - \boxed{0} - (0.1) - (0.2) - (0.3)\} \quad (276)$$
$$+ (0)\,L + \boxed{0}\,L' + (0.1)\,l_1 + (0.2)\,l_2 + (0.3)\,l_3.$$

By the same process the corresponding equations for the other satellites are

$$0 = l_1\{p - (1) - \boxed{1} - (1.0) - (1.2) - (1.3)\}$$
$$+ (1)\,L + \boxed{1}\,L' + (1.0)\,l + (1.2)\,l_2 + (1.3)\,l_3;$$
$$0 = l_2\{p - (2) - \boxed{2} - (2.0) - (2.1) - (2.3)\} \quad (277)$$
$$+ (2)\,L + \boxed{2}\,L' + (2.0)\,l + (2.1)\,l_1 + (2.3)\,l_3;$$
$$0 = l_3\{p - (3) - \boxed{3} - (3.0) - (3.1) - (3.2)\}$$
$$+ (3)\,L + \boxed{3}\,L' + (3.0)\,l + (3.1)\,l_1 + (3.2)\,l_2.$$

862. These four equations determine the coefficients of the latitude; they include the reciprocal action of the satellites, together with that of the sun, and the direct action of Jupiter considered as a spheroid, but in the hypothesis that the plane of his equator retains a permanent inclination on the fixed plane: that, however, is not the case, for as neither the sun nor the orbits of all the satellites are in the plane of Jupiter's equator, their action on the protuberant matter causes a nutation in the equator, and a precession of its equinoxes, in all respects similar to those occasioned by the action of the moon on the earth, which produce sensible inequalities in the

motions of the satellites. Thus the satellites, by troubling Jupiter, indirectly disturb their own motions.

The Effect of the Nutation and Precession of Jupiter on the Motion of his Satellites.

863. The reciprocal action of the bodies of the solar system renders it impossible to determine the motion of any one part independently of the rest; this creates a difficulty of arrangement, and makes it indispensable to anticipate results which can only be obtained by a complete investigation of the theory on which they depend. The nutation and precession of Jupiter's spheroid can only be known by the theory of the rotation of the planets, from whence it is found that if θ and γ be the inclinations of Jupiter's equator and orbit on the fixed plane, ψ the retrograde motion of the descending node of his equator on that plane, and estimated from the vernal equinox of Jupiter, $\daleth$ the longitude of the ascending node of his orbit, it the rotation of Jupiter, and A, B, C, the moments of inertia of his spheroid with regard to the principal axes of rotation, as in article 177, the precession of Jupiter's equinoxes is

$$\frac{d\theta}{dt} = \frac{3(2C - A - B)}{4iC}\left\{\begin{array}{l} M^2\gamma\sin(\daleth + \psi) \\ + \Sigma mn^2\gamma'\sin(\daleth' + \psi) \end{array}\right\}$$

whence $M^2\gamma\sin(\daleth+\psi)$ is the action of the sun, and $\Sigma mn^2\gamma'\sin(\daleth'+\psi)$ that of the satellites.

The nutation is

$$\theta\,.\,\frac{d\psi}{dt} = \frac{3(2C - A - B)}{4iC}\left\{\begin{array}{l} \theta(M^2 + \Sigma mn^2) \\ + M^2\gamma\cos(\daleth + \psi) \\ + \Sigma mn^2\gamma'\cos(\daleth' + \psi) \end{array}\right\}.$$

The first of these equations, multiplied by $\sin\psi$, added to the second multiplied by $\cos\psi$, gives

$$\frac{d(\theta\sin\psi)}{dt} = \frac{3(2C-A-B)}{4iC}\left\{\begin{array}{l} (M^2+\Sigma mn^2)\,\theta\sin\psi \\ +M^2\gamma\cos\daleth+\Sigma mn^2\gamma'\cos\daleth' \end{array}\right\} \quad (278)$$

likewise

$$\frac{d(\theta\cos\psi)}{dt} = \frac{3(2C-A-B)}{4iC}\left\{\begin{array}{l} -(M^2+\Sigma mn^2)\,\theta\sin\psi \\ +M^2\gamma\sin\daleth+\Sigma mn^2\gamma'\sin\daleth' \end{array}\right\} \quad (279)$$

864. Now, in order to have some idea of the positions of the different planes, let JN be the orbit of Jupiter, QN the plane of his equator, FN the fixed plane, and mN the orbit of a satellite. Then the integrals of these equations may be found by considering that as $\theta =$ QNF is the inclination of Jupiter's equator on the fixed plane,

fig. 111.

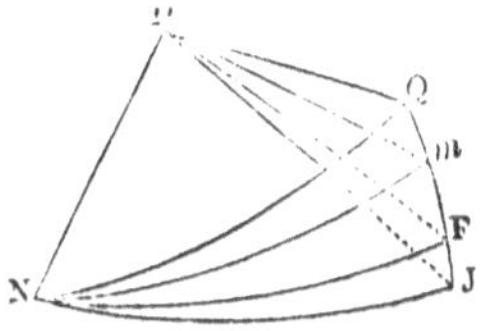

$$-\theta \sin (v + \psi)$$

would be the latitude of a satellite if it moved on the plane of Jupiter's equator, for the latitudes are all referred to the fixed plane FN; and if they are positive on the side FJ, they must be negative on the side FQ; but by the value assumed for s, in article 861, that latitude is equal to a series of terms of the form

$$L \sin (v + pt + \Lambda),$$

hence

$$\theta \sin \psi = -\Sigma'. L \sin (pt + \Lambda) \qquad (280)$$
$$\theta \cos \psi = -\Sigma'. L \cos (pt + \Lambda).$$

865. Likewise, $\gamma =$ JNF, being the inclination of Jupiter's orbit on the fixed plane, $\gamma \sin (U - \daleth)$ is the latitude of the sun above the fixed plane, by article 863; but by the value assumed for S, in article 861, it is easy to see that

$$\gamma \sin \daleth = -\Sigma'. L' \sin (pt + \Lambda) \qquad (281)$$
$$\gamma \cos \daleth = -\Sigma'. L' \cos (pt + \Lambda).$$

In the same manner $\gamma' = m$NF being the inclination of the orbit of a satellite on the fixed plane, its latitude is $\gamma' \sin (v + l_{\prime})$, and by article 861

$$\gamma' \sin \daleth' = -\Sigma' l \,.\, \sin (pt + \Lambda) \qquad (282)$$
$$\gamma' \cos \daleth' = \Sigma' l \,.\, \cos (pt + \Lambda).$$

Σ' denotes the sum of a series, but Σ is the sum of the terms relating to the different satellites.

When these quantities are put in equations (279) and (278), a comparison of the coefficients of similar sines and cosines gives

$$0 = pL + 3\, \frac{(2C - A - B)}{4iC} \{M^2 (L' - L) + \Sigma\, mn^2 (l - L)\},$$

which equation determines the effect of the displacement of Jupiter's equator.

866. If Jupiter be an elliptical spheroid, theory gives

$$\frac{2C-A-B}{C} = \frac{2(\rho-\frac{1}{2}\phi)\int P.\bar{R}^2 d\bar{R}}{\int P\bar{R}^4 . d\bar{R}}.$$

As the celestial bodies decrease in density from the centre to the surface, P represents the density of a shell or layer of Jupiter's spheroid at the distance of $\bar{R}$ from his centre, the integral being between $\bar{R} = 0$, the value of the radius at the centre to $\bar{R} = 1$, its value at the surface.

867. The equations in article 277 may be put under the form

$$0 = \{p-(0) - \boxed{0} - (0.1) - (0.2) - (0.3)\}(L-l)$$
$$+(0.1)(L-l_1)+(0.2)(L-l_2)+(0.3)(L-l_3)+\boxed{0}(L-L')-pL;$$
$$0 = \{p-(1) - \boxed{1} - (1.0) - (1.2) - (1.3)\}(L-l_1)$$
$$+(1.0)(L-l)+(1.2)(L-l_2)+(1.3)(L-l_3)+\boxed{1}(L-L')-pL;$$
$$0 = \{p-(2) - \boxed{2} - (2.0) - (2.1) - (2.3)\}(L-l_2) \qquad (283)$$
$$+(2.0)(L-l)+(2.1)(L-l_1)+(2.3)(L-l_3)+\boxed{2}(L-L')-pL;$$
$$0 = \{p-(3) - \boxed{3} - (3.0) - (3.1) - (3.2)\}(L-l_3)$$
$$+(3.0)(L-l)+(3.1)(L-l_1)+(3.2)(L-l_2)+\boxed{3}(L-L')-pL;$$
$$0 = pL - \frac{3(2C-A-B)}{4iC}\{M^2(L-L')+mn^2(L-l)+$$
$$m_1n_1^2(L-l_1)+m_2n_2^2(L-l_2)+m_3n_3^2(L-l_3)\};$$

which determine the positions of the orbits of the satellites, including the effects of Jupiter's nutation and precession.

Inequalities occasioned by the Displacement of Jupiter's Orbit.

868. The position of Jupiter's orbit is perpetually changing by very slow degrees with regard to the ecliptic, from the action of the planets. In consequence of this displacement the inclination of the plane of Jupiter's equator on his orbit is changed, and a correspond-

ing change takes place in the precession of the nodes of the equator on the orbit, which may be represented by considering JN to be the orbit of Jupiter, and γA the plane of his equator at the epoch, γ being the ascending node of his equator at that period. After a time the action of the planets brings the orbit into the position J′N, which increases the inclination by JNJ′, and the node, which by its retrograde motion alone would come to B, is brought to γ', so that the motion of the node is lessened by BC. Thus the inclination of the plane of Jupiter's equator on his orbit is affected by two totally different and unconnected causes, the one arising from the direct action of the sun and satellites on the protuberant matter at the equator producing nutation and precession, and the other from the displacement of his orbit by the action of the planets, which diminishes the precession: both disturb the motions of the satellites; but in order to determine the effects of the displacement of the orbit, it must be observed that if the inclinations of the orbits were eliminated from the equations (283) the result would be an equation in p, the roots of which p, p_1, p_2, p_3, would be the annual and sidereal motions of the nodes of the satellites depending on their mutual attraction and that of the sun, but there would also be very small values of p of the order BC, fig. 112, depending on the displacement of the orbit and equator of Jupiter. Now if we regard the equations (283) as relative to the displacement of the orbit and equator of Jupiter alone, omitting, for the present, the mutual action of the sun and satellites, these very small values of p may be omitted in comparison of (0), (0.1), &c.; and if it be assumed that

fig. 112.

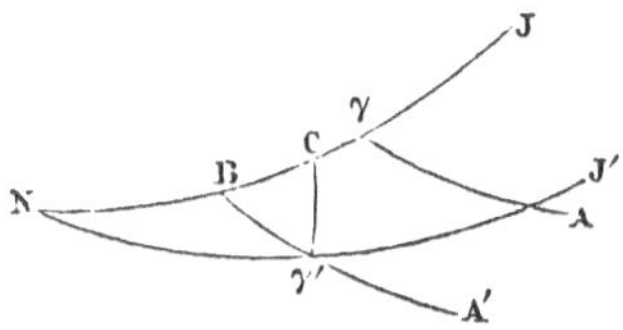

$$
\begin{aligned}
L - l &= \lambda\ (L - L') \\
L - l_1 &= \lambda_1 (L - L') \\
L - l_2 &= \lambda_2 (L - L') \\
L - l_3 &= \lambda_3 (L - L'),
\end{aligned}
\qquad (284)
$$

the four first of equations (283) become

$$0 = \{(0) + \boxed{0} + (0.1) + (0.2) + (0.3)\}\,\lambda$$

$$- (0.1)\,\lambda_1 - (0.2)\,\lambda_2 - (0.3)\,\lambda_3 - |\overline{0}|$$

$$0 = \{(1) + \boxed{1} + (1.0) + (1.2) + (1.3)\}\lambda_1$$
$$- (1.0)\,\lambda - (1.2)\,\lambda_2 - (1.3)\,\lambda_3 - \boxed{1} \qquad (285)$$
$$0 = \{(2) + \boxed{2} + (2.0) + (2.1) + (2.3)\}\,\lambda_2$$
$$- (2.0)\,\lambda - (2.1)\,\lambda_1 - (2.3)\,\lambda_3 - \boxed{2}$$
$$0 = \{(3) + \boxed{3} + (3.0) + (3.1) + (3.2)\}\lambda_3$$
$$- (3.0)\,\lambda - (3.1)\,\lambda_1 - (3.2)\,\lambda_2 - \boxed{3}\,,$$

which are relative to the displacement of the orbit and equator of Jupiter; by these, λ, λ_1, λ_2, λ_3, may be computed, whence the relations among the inclinations will be known.

On the Constant Planes.

869. The preceding equations afford the means of finding the permanent planes mentioned in article 803, for $l = mNF$ and $L' = JNF$, fig. 111, are the inclinations of the satellite and Jupiter on the fixed plane; $l - L' = mNJ$ is the inclination of the orbit of the satellite on that of Jupiter, by article 864; hence, the latitude of the satellite m above the orbit of Jupiter is equal to a series of terms of the form

$$(l - L')\sin(v + pt + \Lambda).$$

But the first of equations (284) gives

$$l - L' = (1 - \lambda)(L - L');$$

thus, with regard to the displacement of Jupiter's orbit and equator,

$$\Sigma'(l-L')\sin(v+pt+\Lambda) = (1-\lambda)\,\Sigma'(L-L')\sin(v+pt+\Lambda).$$

Again, $L = QNF$, $L_{,} = JNF$ being the inclinations of Jupiter's equator and orbit on the fixed plane; $L - L' = QNJ$ is the inclination of his equator on his orbit, for $L = QNF$ is a negative quantity by article 864, therefore

$$\Sigma'(L - L')\sin(v + pt + \Lambda)$$

would be the latitude of the satellite m above the orbit of Jupiter, if it moved on the plane of his equator. But the inclination $(1-\lambda)\times(L - L')$ is less than $L - L' = QNJ$, both being referred to the plane of Jupiter's orbit; hence, $(1 - \lambda)(L - L') = l - L' < L - L'$; therefore the plane having the inclination $l - L'$, or $(1 - \lambda)(L - L')$ must come between the equator and orbit of Jupiter; and as Λ and p, the longitude of the node and its annual and sidereal precession, are the same in both, this plane passes through NP, the line of the nodes. But

$$L-L' : (1-\lambda)(L-L') :: 1 : 1-\lambda :: QNJ : FNJ,$$

and the plane FN always retains the same inclination to the equator and orbit of Jupiter, since λ is a constant quantity: each of the other satellites has its own permanent plane depending on λ_1, λ_2, λ_3. It is hardly possible that these planes could have been discovered by observation alone.

870. If $\theta' = QNJ = L - L'$ be the inclination of Jupiter's equator on his orbit, and $-\psi' = pt + \Lambda$ the longitude of its descending node on the orbit estimated from the vernal equinox of Jupiter, the preceding expression, with regard to that part of the latitude of m above the orbit of Jupiter which is relative to the displacement of his orbit and equator, is

$$(\lambda - 1)\,\theta' \sin(v + \psi'),$$

for $\lambda\theta' = FNQ$, the inclination of the constant plane FN on Jupiter's equator, therefore

$$(\lambda - 1)\,\theta' = FNJ,$$

is the inclination of the same constant plane on Jupiter's orbit, and

$$(\lambda - 1)\,\theta' . \sin(v + \psi')$$

is the latitude the satellite would have if it moved on its constant plane.

To determine the Effects of the Displacements of the Equator and Orbit of Jupiter on the quantities $\theta = QNF$, $\theta' = QNJ$, ψ, ψ', and Λ.

871. The displacements of the equator and orbit of Jupiter affect the quantities θ, ψ, θ', ψ', and Λ. The general equations which determine this effect may easily be found; but if the values of these quantities be obtained in functions of the time, it will be sufficiently correct for astronomical purposes for several centuries, before or after any period that may be assumed as the epoch.

It will answer the same purpose, and facilitate the determination of these quantities, if Jupiter's orbit be assumed to coincide with the fixed plane FN; for the whole effect of its displacement will be referred to the equator, which will then vary both from nutation and the variation in the orbit of Jupiter. In this case $L' = 0$, and equations (284) give

$$l = (1-\lambda)L;\quad l_1 = (1-\lambda_1)L;\quad l_2 = (1-\lambda_2)L;\quad l_3 = (1-\lambda_3)L.$$

In consequence of these, the four first of equations (283) vanish, and L remains indeterminate, and may be represented by $-`L$, and the last of the same equation becomes

$$p = \frac{3(2C-A-B)}{4iC}\{M^2 + mn^2\lambda + m_1n_1^2\lambda_1 + m_2n_2^2\lambda_2 + m_3n_3^2\lambda_3\},$$

which may be expressed by $`p$, and relates to the displacement of the equator of Jupiter.

872. Since JN coincides with FN, fig. 111, $-`L = QnJ$ is the inclination of the equator on the fixed orbit of Jupiter and

$$-`L\sin(v + `pt + `\Lambda)$$

would be the latitude of the satellite if it were moving on the equator of Jupiter, $`\Lambda$ being an arbitrary quantity, or the longitude of the node of the equator corresponding to $`p$. But this latitude has also been expressed by $-\theta\sin(v+\psi)$.

Whence
$$\theta\sin\psi = `L\sin(p`t + `\Lambda), \qquad (286)$$
$$\theta\cos\psi = `L\cos(p`t + `\Lambda),$$

$`pt$ being the mean precession of the equinoxes of Jupiter. Again, since $\theta = QNF$, $\gamma = JNF$ are the inclinations of the equator and orbit of Jupiter on the fixed plane;

$$-\theta\sin(v+\psi) - \gamma\sin(v - 7)$$

is the latitude the satellite would have above the orbit of Jupiter, if it moved on the plane of his equator, but $-\theta'\sin(v+\psi')$ is the same;

so
$$\theta\sin(v+\psi) + \gamma\sin(v-7) = \theta'\sin(v+\psi'),$$

v being indeterminate. If it be successively made equal to $-`pt$ and to $90° - `pt$, the preceding equation gives

$$\theta'\sin(\psi' - pt) = \theta\sin(\psi - `pt) - \gamma\sin(7 + `pt) \qquad (287)$$
$$\theta'\cos(\psi' - `pt) = \theta\cos(\psi - `pt) + \gamma\cos(7 + `pt).$$

The sum of the squares of equations (286) gives $\theta = `L$, and as by this $\sin\psi = \sin(`pt + `\Lambda)$; therefore $\psi - `pt = `\Lambda$.

In consequence of this, the first of equations (287) becomes

$$\theta'\sin(\psi' - `pt) = `L\sin`\Lambda - \gamma\sin(7 + `pt),$$

or
$$\theta'\sin\psi'\cos`pt - \theta_{\prime}\cos\psi_{\prime}\sin`pt - `L\sin`\Lambda$$
$$+\gamma\sin 7\cos`pt + \gamma\cos 7\sin`pt = 0.$$

This expression must be zero, whatever the time may be, which can only happen when $\sin {}^{\backprime}\Lambda = 0$, for ${}^{\backprime}L = \theta$; consequently,

$$ {}^{\backprime}\Lambda = 0, \text{ and therefore } \psi = {}^{\backprime}pt. $$

873. In order to determine θ' and ψ', let the orbit of Jupiter in the beginning of 1750 be the fixed plane, let that period be the epoch, and the line of the vernal equinox of Jupiter the origin of the angles. Then at the epoch $t = 0$, whence equations (287) become

$$ \theta' \sin \psi' = \theta \sin \psi - \gamma \sin \daleth $$
$$ \theta' \cos \psi' = \theta \cos \psi + \gamma \cos \daleth. $$

Now ψ' and ψ are so small, that the arc may be put for the sine, and unity for the cosine; also $\gamma \cos \daleth$, $\gamma \sin \daleth$ may be expressed by series increasing as the powers of the time for many centuries to come; therefore let

$$ \gamma \sin \daleth = at \qquad \gamma \cos \daleth = bt $$

then, because $\theta = {}^{\backprime}L, \quad \psi = {}^{\backprime}pt, \qquad {}^{\backprime}\Lambda = 0,$ (288)

$$ \theta'\psi' = {}^{\backprime}L \,.{}^{\backprime}pt - at; \qquad \theta' = {}^{\backprime}L + bt $$

whence
$$ \psi' = {}^{\backprime}pt - \frac{at}{{}^{\backprime}L}, $$

when the square of the time is neglected.

874. Since $\gamma \sin \daleth$, $\gamma \cos \daleth$, relate to the change in the position of Jupiter's orbit from the action of the planets, they are determined by equations (137); but as Jupiter's orbit is principally disturbed by the action of Saturn and Uranus, if ϕ, ϕ' be the inclinations of the orbits of Saturn and Uranus on the orbit of Jupiter in the beginning of 1750, and ☊, ☊′, the longitudes of the ascending nodes of the two orbits on that of Jupiter at the same epoch, estimated from the equinox of spring of Jupiter; then will

$$ a = (4.5)\,\phi \cos ☊ + (4.6)\,\phi' \,.\, \cos ☊' $$
$$ b = -\,(4.5)\,\phi \sin ☊ - (4.6)\,\phi' .\, \sin ☊', $$

where (4.5), (4.6), are given by equations (202).

875. It only remains to determine the effects of the displacement on $\gamma' \sin \daleth'$, $\gamma' \cos \daleth'$, the inclination and node of a satellite m with regard to its fixed plane.

By equations (248)

$$ l = (1 - \lambda)\, L + \lambda L'. $$

If this value of l be put in the equations (282) they become

$$ \gamma' \sin \daleth' = -\,\Sigma'.(1 - \lambda).L.\sin(pt + \Lambda) - \Sigma'.\lambda L'.\sin(pt + \Lambda) $$
$$ \gamma' \cos \daleth' = \Sigma'.(1 - \lambda).L.\cos(pt + \Lambda) + \Sigma'.\lambda L'.\cos(pt + \Lambda), $$

and in consequence of equations (280) and (281)

$$\gamma' \sin \Gamma' = (1 - \lambda) \,.\, \theta \sin \psi + \lambda \,.\, \gamma \sin \Gamma$$

$$\gamma' \cos \Gamma' = (\lambda - 1) \,.\, \theta \cos \psi + \lambda \,.\, \gamma \cos \Gamma,$$

but $\theta = {}^{\backprime}L, \quad \psi = {}^{\backprime}pt, \quad \gamma \sin \Gamma = at, \gamma \cos \Gamma = bt;$

and putting ${}^{\backprime}pt$ for the sine and unity for the cosine; with regard to the displacement of Jupiter's orbit and equator,

$$\gamma' \sin \Gamma' = (1 - \lambda){}^{\backprime}L \,.\, {}^{\backprime}pt + \lambda \,.\, at$$

$$\gamma' \cos \Gamma' = (\lambda - 1){}^{\backprime}L + \lambda \,.\, bt \qquad (289)$$

Thus the quantities relating to the displacement of the orbit and equator are completely determined.

876. With regard to the values of p, which depend on the mutual action of the satellites, L' is zero, since the action of the satellites has no sensible effect on the displacement of Jupiter's orbit. The values of L may be omitted relatively to l, l_1, l_2, l_3; and since by the last of equations (283), pL is multiplied by $\frac{2C-A-B}{C}$, it is of the order of the product of the ellipticity of Jupiter by the masses of the satellites; and therefore it may be omitted also, which reduces equations (283) to

$$0 = l\,\{p - (0) - \boxed{0} - (0.1) - (0.2) - (0.3\} + \boxed{0.1}\,l_1 + \boxed{0.2}\,l_2 + \boxed{0.3}\,l_3$$

$$0 = l_1\,\{p - (1) - \boxed{1} - (1.0) - (1.2) - (1.3)\} + \boxed{1.0}\,l + \boxed{1.2}\,l_2 + \boxed{1.3}\,l_3$$

$$(290)$$

$$0 = l_2\,\{p - (2) - \boxed{2} - (2.0) - (2.1) - (2.3)\} + \boxed{2.0}\,l + \boxed{2.1}\,l_1 + \boxed{2.3}\,l_3$$

$$0 = l_3\,\{p - (3) - \boxed{3} - (3.0) - (3.1) - (3.2)\} + \boxed{3.0}\,l + \boxed{3.1}\,l_1 + \boxed{3.2}\,l_2$$

877. These equations determine the annual and sidereal motion of the nodes and inclinations of the orbits, and are precisely similar to those which determine the eccentricities and motions of the apsides, for if the terms depending on the displacement of the orbit of Jupiter be omitted, each satellite has four terms in latitude similar to the four equations of the centre, and arising like them from the changes in the positions of the orbits by the action of the matter at Jupiter's equator and their mutual attraction, they therefore depend on the inclinations and motions of the nodes of their own orbits, and on those of the other three. Hence, with the values of l, l_1, l_2, l_3, known by observation, these equations will give the four roots of p,

the annual and sidereal motion of the nodes and the coefficients of the sixteen terms in the latitudes; for if it be assumed, that

$$l_1 = \zeta_1 l; \qquad l_2 = \zeta_2 l; \qquad l_3 = \zeta_3 l,$$

these quantities will make l vanish from the preceding equations; the result will be four equations between ζ_1, ζ_2, ζ_3, and p, whence p will be obtained by an equation of the fourth degree.

Let p, p_1, p_2, p_3, be the roots of that equation, and let

$$\zeta_1^{(1)},\ \zeta_2^{(1)},\ \zeta_3^{(1)}; \qquad \zeta_1^{(2)},\ \zeta_2^{(2)},\ \zeta_3^{(2)}; \qquad \zeta_1^{(3)},\ \zeta_2^{(3)},\ \zeta_3^{(3)};$$

be the values of ζ_1, ζ_2, ζ_3, when p is successively changed to p_1, p_2, p_3, they will give the coefficients required.

878. In article 861, it was assumed, that the latitudes of the satellites above the fixed plane were

$$s = l \sin (v + pt + \Lambda) \qquad s_{,} = l_{,} \sin (v_{,} + pt + \Lambda)$$
$$s_2 = l_2 \sin (v_2 + pt + \Lambda) \qquad s_3 = l_3 \sin (v_3 + pt + \lambda);$$

but if we refer them to the orbit of Jupiter, the term arising from the displacement of that orbit must be added to each, and if the different values of p be substituted, and the corresponding coefficients, the latitudes of the satellites above JN, the orbit of Jupiter at any time t, will be

$$\begin{aligned}
s &= (\lambda - 1)\theta' \sin (v + \psi') \\
&\quad + l \sin (v + pt + \Lambda) \\
&\quad + l_1 \sin (v + p_1 t + \Lambda_1) \\
&\quad + l_2 \sin (v + p_2 t + \Lambda_2) \\
&\quad + l_3 \sin (v + p_3 t + \Lambda_3) \\
s_1 &= (\lambda_1 - 1)\theta' \sin (v_1 + \psi') \\
&\quad + \zeta_1 l \sin (v_1 + pt + \Lambda) \\
&\quad + \zeta_1^{(1)} l_1 \sin (v_1 + p_1 t + \Lambda_1) \\
&\quad + \zeta_1^{(2)} l_2 \sin (v_1 + p_2 t + \Lambda_2) \\
&\quad + \zeta_1^{(3)} l_3 \sin (v_1 + p_3 t + \Lambda_3) \\
s_2 &= (\lambda_2 - 1)\theta' \sin (v_2 + \psi') \\
&\quad + \zeta_2 l \sin (v_2 + pt + \Lambda) \\
&\quad + \zeta_2^{(1)} l_1 \sin (v_2 + p_1 t + \Lambda_1) \\
&\quad + \zeta_2^{(2)} l_2 \sin (v_2 + p_2 t + \Lambda_2) \\
&\quad + \zeta_2^{(3)} l_3 \sin (v_2 + p_3 t + \Lambda_3) \\
s_3 &= (\lambda_3 - 1)\theta' \sin (v_3 + \psi') \\
&\quad + \zeta_3 l \sin (v_3 + pt + \Lambda) \\
&\quad + \zeta_3^{(1)} l_1 \sin (v_3 + p_1 t + \Lambda_1) \\
&\quad + \zeta_3^{(2)} l_2 \sin (v_3 + p_2 t + \Lambda_2) \\
&\quad + \zeta_3^{(3)} l_3 \sin (v_3 + p_3 t + \Lambda_3).
\end{aligned} \tag{291}$$

879 The first term of each depends on the displacement of Jupiter's orbit, and the eight quantities

$$l,\ l_1,\ l_2,\ l_3,\quad \Lambda,\ \Lambda_1,\ \Lambda_2,\ \Lambda_3,$$

are determined by observation ; the first are the respective inclinations of the four satellites on Jupiter's orbit, and the last four are the longitudes of the nodes at the epoch. If it be required to find the latitude of the satellites above the fixed plane, it will be necessary to add to these the values of the latitudes, supposing the satellites to move on the orbit of Jupiter.

880. The inequalities in latitude which depend on the configuration of the bodies that acquire small divisors by integration are insensible, with the exception of those arising from the action of the sun depending on the angle $2v - 2U$. The part of the disturbing force whence these come is

$$R = \frac{3M^2}{4n^2}\{s^2 \cos 2(v - U) - 4sS \cos (v - U) - \cos 2(v - U)\}$$

omitting the squares and products of S and s,

$$\frac{dR}{dv} = \frac{3M^2}{2n^2} \sin 2(v - U)$$

$$\frac{dR}{ds} = \frac{3M}{2n^2}\{s \cos 2(v - U) - 2S \cos (\cos (v - U)\}$$

but $\quad S = L' \sin (v + pt + \Lambda)\ ; \quad s = l \sin (v + pt + \Lambda)$

and $$\frac{ds}{dv} = l \cos (v + pt + \Lambda),$$

and if $\frac{v}{n}$ be put for t, observing that

$$U = Mt = \frac{M}{n} v,$$

the equation in article 859 becomes

$$0 = \frac{d^2s}{dv^2} + N_{,}^2 s + \frac{3M^2}{2n^2}(L' - l) \sin \left(v - \frac{2M}{n} v - \frac{p}{n} v - \Lambda\right),$$

in which

$$N_{,}^2 = 1 + 2\,\frac{(\rho - \frac{1}{2}\phi)}{a^2} + \frac{3}{2}\frac{M^2}{n^2} + \Sigma m_{,} a_{,}^2 a_{,} B_1\ ;$$

but, without sensible error,

$$N_{,}^2 = 1 + \frac{(\rho - \frac{1}{2}\phi)}{a^2}.$$

In order to integrate this equation, let

$$s = K \cdot \sin\left(v - \frac{2M}{n}v - \frac{p}{n}v - \Lambda\right),$$

K being an indeterminate coefficient.

If that value of s be put in the equation, it will give

$$K = \frac{3M^2}{2n^2} \cdot \frac{L' - l}{\left(1 - 2\frac{M}{n} - \frac{p}{n}\right)^2 - N_{\prime}^2}.$$

But in the divisor,

$$\left\{1 - \frac{2M}{n} - \frac{p}{n} + N_{\prime}\right\} \cdot \left\{1 - \frac{2M}{n} - \frac{p}{n} - N_{\prime}\right\};$$

$\frac{p}{n}$ is very small, and $N_{\prime}$ differs but little from unity; hence

$$1 - 2\frac{M}{n} - \frac{p}{n} + N_{\prime} = 2, \text{ nearly};$$

therefore

$$s = -\frac{3M^2(L' - l)}{4n^2\left(2\frac{M}{n} + \frac{p}{n} + N_{\prime} - 1\right)} \sin\left(v - 2\frac{M}{n}v - \frac{p}{n}v - \Lambda\right);$$

a similar inequality exists for each root of p, including $'p$, which is the value of p depending on the displacement of the equator and orbit of Jupiter.

Now, $\frac{p}{n} + N_{\prime} = 1$, nearly;

consequently, $s = -\frac{3M}{8n}(L' - l) \cdot \sin(v - 2U - pt - \Lambda)$. (292)

Secular Inequalities of the Satellites, depending on the Variations in the Elements of Jupiter's Orbit.

881. The secular inequalities in the elements of Jupiter's orbit, occasioned by the action of the planets, produce corresponding variations in the mean motions of the satellites, which, in the course of ages, will have a considerable effect on the theory of these bodies. These are obtained from

$$R = -\frac{S'r^2}{4D^3}\left\{1 - 3s^2 - 3S^2 + 12Ss\cos(U - v)\right\}$$

$$-\frac{(\rho - \frac{1}{2}\phi)}{r^3}\left(\tfrac{1}{3} - (s - s')^2\right).$$

But, by articles 864 and 865,

$$s = \gamma' \sin (v - 7'), \quad S = \gamma \sin (U - 7), \quad s' = -\theta \sin (v + \psi);$$

and as the periodic inequalities are to be rejected,

$$\frac{S'}{D^3} = \frac{S'}{D'^3}\left\{1 + 3\left(\frac{D\delta D}{D^2}\right)^2\right\} = M^2\{1 + 3H^2 \sin^2 (Mt + E - \Pi)\}$$

$$= M^2 (1 + \tfrac{3}{2} H^2).$$

For the same reason,

$$r^2 = a^2 (1 + \tfrac{1}{2} e^2 - e \cos (nt + \epsilon - \varpi))^2 = a^2 (1 + \tfrac{3}{2} e^2);$$

and if it be observed that

$$\frac{3a^3 n M^2}{4} = \boxed{0}\,; \quad \frac{(\rho - \frac{1}{2}\phi)}{a^2} n = (0),$$

the value of anR is

$$anR = -\tfrac{1}{2} \boxed{0} \{e^2 + H^2 - \gamma^2 + 2\gamma\gamma_{\prime} \cos (7' - 7) - \gamma_{\prime}^2\}$$

$$- \tfrac{1}{2} (0) \{\theta^2 + 2\theta\gamma_{\prime} \cos (7' + \psi) + \gamma_{\prime}^2 - e^2\}.$$

If this quantity be put in equation (259), the result will be

$$\frac{d \,.\, \delta v'}{dt} = -2 \boxed{0} \{e^2 + H^2 - \gamma^2 + 2\gamma\gamma' \cos (7' - 7) - \gamma'^2\}$$

$$+ 3 (0) \{\theta^2 + 2\theta\gamma' \cos (7 + \psi) + \gamma'^2 - e^2\}.$$

This, however, only gives the inequalities on the orbit; but its projection on the fixed plane, by article 548, is

$$d = dv \left(1 + \tfrac{1}{2} s^2 - \tfrac{1}{2} \frac{ds^2}{dv^2}\right\}.$$

Now, $s = \gamma' \sin (v - 7') = \gamma' \sin v \cos 7' - \gamma' \cos v \sin 7'$.

The substitution of this quantity, and of its differential, gives

$$dv' = dv + \tfrac{1}{2}\left\{\gamma' . \sin 7' \frac{d(\gamma' \cos 7')}{dt} - \gamma' . \cos 7' \frac{d(\gamma' . \sin 7')}{dt}\right\},$$

the value of dv' projected on the fixed plane; therefore

$$\frac{d . \delta v}{dt} = \tfrac{1}{2}\left\{\gamma' \sin 7' \frac{d(\gamma' . \cos 7')}{dt} - \gamma' . \cos 7' \frac{d(\gamma' . \sin 7')}{dt}\right\} \qquad (293)$$

$$- 2 \boxed{0} \{e^2 + H^2 - \gamma^2 + 2\gamma\gamma' \cos (7' - 7) - \gamma'^2\}$$

$$+ \tfrac{1}{2} (0) \{e^2 - \theta^2 - 2\theta\gamma' \cos (7' + \psi) - \gamma'^2\}.$$

Since all the quantities $\gamma \sin 7$, $\gamma \cos 7$, $\gamma_{\prime} \sin 7_{\prime}$, $\gamma_{\prime} \cos 7'$ ψ and θ, are given in the preceding articles, it may be found that

$$\frac{d . \delta v}{dt} = 4 \,.\, \boxed{0} \,.\, (1 - \lambda)^2 . \text{`}Lbt - 6 \,.\, (0) \,.\, \lambda^2 \,.\, \text{`}L \,.\, bt$$

$$+ (1-\lambda).\lambda.\{(0) + \boxed{0} + (0.1) + (0.2) + (0.3)\}.`L.bt$$

$$- \tfrac{1}{2}\lambda\,(0).`Lbt - \tfrac{1}{2}(1-\lambda).\boxed{0}.`L.bt$$
$$+ \tfrac{1}{2}(0.1).\{(\lambda-1)\lambda' + (\lambda_1 - 1)\lambda\}.`L.bt$$
$$+ \tfrac{1}{2}(0.2)\{(\lambda-1)\lambda_2 + (\lambda_2 - 1)\lambda\}.`L.bt$$
$$+ \tfrac{1}{2}(0.3)\{(\lambda-1)\lambda_3 + (\lambda_3 - 1)\lambda\}`L.bt - 2\boxed{0}H^2.$$

But in consequence of the relations between λ, λ_1, λ_2, λ_3, in article 859,

$$\frac{d.\delta v}{dt} = 4(1-\lambda)^2.\boxed{0}.`Lbt - 6\,(0)\,\lambda^2`Lbt - 2H^2.\boxed{0}.$$

In considering the action of Saturn only, equations (204) give the numerical value of H; to abridge, if $\bar{e}$ be the value of H at the epoch, then $H = \bar{e} + ct + \&c.$;
and omitting the square of the time,

$$H^2 = 2\bar{e}\,ct,$$

and the integral becomes

$$\delta v = 2(1-\lambda)^2\boxed{0}.`L\,bt^2 - 3\,(0)\,\lambda^2`L\,bt^2 - 2\bar{e}\,ct^2\boxed{0}. \qquad (294)$$

This inequality in the mean motion of m varies with the eccentricity of the orbit of Jupiter, and is similar to the acceleration in the mean motion of the moon, but it will not be perceptible for many years, nor has it hitherto been perceived.

882. If there was but one satellite, the first of equations (285) would give

$$\lambda = \frac{\boxed{0}}{\boxed{0} + (0)}.$$

In the theory of the moon, $\boxed{0}$ is vastly greater than (0), so that $\lambda = 1 - \dfrac{(0)}{\boxed{0}}$ differs but little from unity, which reduces the equation (294) to $\delta v = -2\boxed{0}\,\bar{e}.ct^2$, where $\bar{e}$ is the eccentricity of the earth's orbit at the epoch; and substituting $\frac{3}{4}\dfrac{M}{n}$ for $\boxed{0}$, it becomes

$$\delta v = -\tfrac{3}{2}\frac{M^2}{n}\bar{e}\,ct^2;$$

which is the same with the acceleration of the moon.

883. One secular variation alone is sensible at present, and that only in the mean motion of the fourth satellite; it is derived from equation (293), each term of which must be determined separately. When e^2 and H^2 are omitted, its second term

$$\tfrac{1}{2}\{\gamma'^2 - 2\gamma\gamma^2 \cos(\Pi' - \Pi) + \gamma^2\}$$

is the square of that part of the latitude of the satellite m above the orbit of Jupiter, which is independent of v; therefore the expression is equal to the square of s in article 861, where v is omitted; but as l, $l_{\prime}$, &c., are very small, their squares and products may be neglected, so that the quantity required, after the reduction of the products of the sines to the cosines of the differences of the arcs, is

$$\gamma^2 - 2\gamma\gamma' \cos(\Pi' - \Pi) + \gamma'^2 =$$

$$(1-\lambda)^2 . \theta'^2 + 2(\lambda - l)\theta'\{l . \cos(pl + \Lambda - \psi') + l_{\prime} \cos(p_{\prime}l + \Lambda' - \psi') + \&c.\}$$

hence

$$\frac{d . \delta v}{dt} = 4(\lambda - 1)\, \theta' \boxed{0} \{l . \cos(pt + \Lambda - \psi') + \&c.\}.$$

Again, $\theta^2 + 2\theta\gamma' \cos(\Pi' + \psi) + \gamma'^2$,

the third term of equation (293), is the square of the latitude of m above the equator of Jupiter, when v is omitted, and is therefore equal to the square of

$$\lambda\theta' \sin \psi' + l \sin(pt + \Lambda) + l_{\prime} \sin(p_{\prime}t + \Lambda_{\prime}) + \&c.$$

which is given by the first of equation (291).

Whence $\dfrac{d . \delta v}{dt} = -\; 6(0)\, \lambda\theta' \{l \cos(pt + \Lambda - \psi') + \&c.\}.$

In the third place the same expression of s gives

$$\gamma' \sin \Pi' = (\lambda - 1)\, \theta' \cos \psi' + l \cos(pt + \Lambda) + \&c. - \gamma \cos \Pi$$
$$\gamma' \cos \Pi' = -(\lambda - 1)\, \theta' \sin \psi' - l \sin(pt + \Lambda) - \&c. - \gamma \sin \Pi.$$

By means of these values the first term of equation (293) becomes

$$\frac{d\delta v}{dt} = -\tfrac{1}{2}(\lambda - 1) . \theta' \{pl \cos(pt + \Lambda - \psi') + \&c.\}$$

When these three parts are added, they constitute the whole of equation (293), the integral of which is

$$\delta v = -\{6\,(0)\lambda + 4(1 - \lambda) \boxed{0}\} . \theta'\Big\{\frac{l}{p} \sin(pt + \Lambda - \psi') +$$

$$\frac{l_{\prime}}{p_{\prime}} \sin(p_{\prime}t + \Lambda_{\prime} - \psi') + \&c.\Big\}$$

$$+ \tfrac{1}{2}(1 - \lambda)\, \theta'\{l . \sin(pt + \Lambda - \psi') + l_{\prime} \sin(p_{\prime}t + \Lambda_{\prime} - \psi') + \&c.\}$$

The only part that has a sensible effect is

$$\delta v = -\frac{\{4(1-\lambda_3)\,\boxed{3} - \frac{1}{2}(1-\lambda)_3)\,p_3 + 6\,(3)\,\lambda_3\}}{p_3} \,.\, \theta' l_3 \times$$

$$\sin\,(p_3 t + \Lambda_3 - \psi'), \qquad (295)$$

and that in the motions of the fourth satellite only.

884. With regard to the moon, $\lambda = 1 - \frac{(0)}{\boxed{0}}$ differs but little from unity, and $p = \boxed{0}$ nearly; hence, for that body,

$$\delta v = -\frac{19}{2}\,.\,(0)\,.\,\frac{\theta' l}{p}\sin\,(v + pt - \psi'),$$

which coincides with equation (244), supposing the obliquity of the ecliptic to be very small.

CHAPTER VIII.

NUMERICAL VALUES OF THE PERTURBATIONS.

885. It is known by observation that the sidereal revolutions of the satellites are accomplished in the following periods:—

	Days.
1st satellite in	1.769137787
2d	3.551181017
3d	7.154552808
4th	16.689019396

The values of n, n_1, n_2, n_3 being reciprocally as these periods,

$$n = n_3 \, . \, 9.433419$$

$$n_1 = n_3 \, . \, 4.699569$$

$$n_2 = n_3 \, . \, 2.332643.$$

And as the sidereal revolution of Jupiter is

Days.
4332.602208,

$$M = n_3 \, . \, 0.00385196.$$

886. The mean distances of the satellites from Jupiter are known from observation; with them, by a method to be shown afterwards, the equations (271) and (290) give the following approximate values of the masses of the satellites, and of the compression of Jupiter

$$m = 0.0000184113$$

$$m_1 = 0.0000258325$$

$$m_2 = 0.0000865185$$

$$m_3 = 0.00005590808$$

$$\rho - \tfrac{1}{2}\phi = 0.0217794,$$

the mass of Jupiter being the unit.

887. The mean distances of the three first satellites cannot be measured with sufficient accuracy for computing the inequalities; it is therefore necessary to determine them from the value of a_3 by Kepler's law.

At the mean distance of Jupiter from the sun, his equatorial diameter is seen under an angle of 38″.99; taking this diameter as the unit, the mean distance of the fourth satellite in functions of the diameter is $$a_3 = 25.43590.$$

By article 818 the mean distance of a satellite is $a + \delta a$, in consequence of the action of the disturbing forces; but as the variation δa is principally owing to the compression of Jupiter, the only part of δa in article 821 that has a sensible effect on the mean distances is $a\,\dfrac{(\rho - \frac{1}{2}\phi)}{3a^2}$, hence $a = n^{-\frac{2}{3}}$ becomes

$$a = n^{-\frac{2}{3}}\left(1 + \tfrac{1}{3}\left(\frac{\rho - \frac{1}{2}\phi}{a^2}\right)\right),$$

also
$$a_3 = n_3^{-\frac{2}{3}}\left(1 + \tfrac{1}{3}\left(\frac{\rho - \frac{1}{2}\phi}{a_3^2}\right)\right);$$

and thus, by Kepler's law,

$$a = \left\{1 + \tfrac{1}{3}\left(\rho - \tfrac{1}{2}\phi\right)\left(\frac{1}{a^2} - \frac{1}{a_3^2}\right)\right\} a_3 \sqrt[3]{\frac{n^2}{n_3^2}}$$

in which
$$\frac{1}{a^2} = \frac{1}{\left(a_3\sqrt[3]{\frac{n^2}{n_3^2}}\right)^2};$$

whence, with the preceding values, it is easy to find that

$$a = 5.698491$$
$$a_1 = 9.066548$$
$$a_2 = 14.461893$$
$$a_3 = 25.43590;$$

with these the series S and S' in article 453 may be computed, and from them all the coefficients A_0, A_1, &c.; B_0, B_1, &c.; and their differences may be found by the same method of computation, and from the same formulæ, as for the planets; and thence

$$N = n_3 \,.\, 9.4269167$$
$$N_1 = n_3 \,.\, 4.6979499$$
$$N_2 = n_3 \,.\, 2.332309$$
$$N_3 = n_3 \,.\, 0.9999070.$$

888. With these quantities the perturbations in longitude and distance computed from the expressions in articles 820 and 821 are

$$\delta v = m_1 \left\{ \begin{array}{ll} 60''.7333 & . \sin \{n_1t - nt + \epsilon_1 - \epsilon\} \\ - 7042''.63 & . \sin 2\{n_1t - nt + \epsilon_1 - \epsilon\} \\ - 22''.949 & . \sin 3\{n_1t - nt + \epsilon_1 - \epsilon\} \\ - 5''.2464 & . \sin 4\{n_1t + nt + \epsilon_1 - \epsilon\} \\ - 1''.7518 & . \sin 5\{n_1t - nt + \epsilon_1 - \epsilon\} \\ - 0''.69443 & . \sin 6\{n_1t - nt + \epsilon_1 - \epsilon\} \end{array} \right. \qquad (296)$$

$$+ m_2 \left\{ \begin{array}{ll} 7''.1065 & . \sin \{n_2t - nt + \epsilon_2 - \epsilon\} \\ - 6''.0005 & . \sin 2\{n_2t - nt + \epsilon_2 - \epsilon\} \\ - 0''.6162 & . \sin 3\{n_2t - nt + \epsilon_2 - \epsilon\} \\ - 0''.1156 & . \sin 4\{n_2t - nt + \epsilon_2 - \epsilon\} \end{array} \right.$$

$$+ 0''.04731 . \sin \{2nt - Mt + 2\epsilon - cE\}$$

The inequalities depending on m_3 are insensible.

$$\delta r = m_1 \left\{ \begin{array}{ll} + 0.00084865 & \\ + 0.00046652 & . \cos \{n_1t - nt + \epsilon_1 - \epsilon\} \\ - 0.09764199 & . \cos 2\{n_1t - nt + \epsilon_1 - \epsilon\} \\ - 0.00040917 & . \cos 3\{n_1t - nt + \epsilon_1 - \epsilon\} \\ - 0.00010761 & . \cos 4\{n_1t - nt + \epsilon_1 - \epsilon\} \\ - 0.00003824 & . \cos 5\{n_1t - nt + \epsilon_1 - \epsilon\} \\ - 0.00001642 & . \cos 6\{n_1t - nt + \epsilon_1 - \epsilon\} \end{array} \right.$$

$$+ m_2 \left\{ \begin{array}{ll} 0.00000703 & \\ + 0.00007780 & . \cos \{n_2t - nt + \epsilon_2 - \epsilon\} \\ - 0.00010631 & . \cos 2\{n_2t - nt + \epsilon_2 - \epsilon\} \\ - 0.00001310 & . \cos 3\{n_2t - nt + \epsilon_2 - \epsilon\} \\ - 0.00000269 & . \cos 4\{n_2t - nt + \epsilon_2 - \epsilon\} \end{array} \right.$$

$$+ m_3 \left\{ \begin{array}{ll} 0.00000113 & \\ + 0.00001478 & . \cos \{n_3t - nt + \epsilon_3 - \epsilon\} \\ - 0.00000968 & . \cos 2\{n_3t - nt + \epsilon_3 - \epsilon\} \\ - 0.00000078 & . \cos 3\{n_3t - nt + \epsilon_3 - \epsilon\} \end{array} \right.$$

$$\begin{array}{ll} + 0.00000095 & \\ - 0.00000095 & . \cos \{2Mt - nt + 2E - 2\epsilon\} \end{array}$$

$$\delta v_{,} = m \left\{ \begin{array}{ll} - 2252''.28 & . \sin \{nt - n_1t + \epsilon - \epsilon_1\} \\ - 17'.053 & . \sin 2\{nt - n_1t + \epsilon - \epsilon_1\} \\ - 3''.4102 & . \sin 3\{nt - n_1t + \epsilon - \epsilon_1\} \\ - 1''.0837 & . \sin 4\{n_1t - nt + \epsilon_1 - \epsilon\} \\ - 0''.4202 & . \sin 5\{nt - n_1t + \epsilon - \epsilon_1\} \end{array} \right. \qquad (297)$$

$$+ m_2 \left\{ \begin{array}{ll} 59''.784 & . \sin \{n_2t - n_1t + \epsilon_2 - \epsilon_1\} \\ - 3923''.3 & . \sin 2\{n_2t - n_1t + \epsilon_2 - \epsilon_1\} \end{array} \right.$$

$$+ m_2 \left\{ \begin{array}{l} - 22''.318 \ \sin 3\,(n_2t - n_1t + \epsilon_2 - \epsilon_1) \\ - \ \ 5''.1076 \ \sin 4\,(n_2t - n_1t + \epsilon_2 - \epsilon_1) \\ - \ \ 1''.7041 \ \sin 5\,(n_2t - n_1t + \epsilon_2 - \epsilon_1) \\ - \ \ 0''.6744 \ \sin 6\,(n_2t - n_1t + \epsilon_2 - \epsilon_1) \end{array} \right.$$

$$+ m_3 \left\{ \begin{array}{l} + 4''.0098 \ \sin \ \ (n_3t - n_1t + \epsilon_3 - \epsilon_1) \\ - 3''.5108 \ \sin 2\,(n_3t - n_1t + \epsilon_3 - \epsilon_1) \\ - 0''.3449 \ \sin 3\,(n_3t - n_1t + \epsilon_3 - \epsilon_1) \end{array} \right.$$

$$+ 0''.1906 \ \sin (2n_1t - 2Mt + 2\epsilon_1 - 2E)$$

$$\delta r_1 = m \left\{ \begin{array}{l} - 0.00044608 \\ + 0.05069318 \ \cos \ \ (nt - n_1t + \epsilon - \epsilon_1) \\ + 0.00059197 \ \cos 2\,(nt - n_1t + \epsilon - \epsilon_1) \\ + 0.00014002 \ \cos 3\,(nt - n_1t + \epsilon - \epsilon_1) \\ + 0.00004784 \ \cos 4\,(nt - n_1t + \epsilon - \epsilon_1) \\ + 0.00001928 \ \cos 5\,(nt - n_1t + \epsilon - \epsilon_1) \end{array} \right.$$

$$+ m_2 \left\{ \begin{array}{l} \ \ \ 0.00006497 \\ + 0.00073255 \ \cos \ \ (n_2t - n_1t + \epsilon_2 - \epsilon_1) \\ - 0.08670960 \ \cos 2\,(n_2t - n_1t + \epsilon_2 - \epsilon_1) \\ - 0.00063398 \ \cos 3\,(n_2t - n_1t + \epsilon_2 - \epsilon_1) \\ - 0.00016685 \ \cos 4\,(n_2t - n_1t + \epsilon_2 - \epsilon_1) \\ - 0.00006067 \ \cos 5\,(n_2t - n_1t + \epsilon_2 - \epsilon_1) \end{array} \right.$$

$$+ m_3 \left\{ \begin{array}{l} \ \ \ 0.00000798 \\ + 0.00007146 \ \cos \ \ (n_3t - n_1t + \epsilon_3 - \epsilon_1) \\ - 0.00010133 \ \cos 2\,(n_3t - n_1t + \epsilon_3 - \epsilon_1) \\ - 0.00001189 \ \cos 3\,(n_3t - n_1t + \epsilon_3 - \epsilon_1) \end{array} \right.$$

$$+ 0.00000609$$

$$- 0.00000609 \ \cos (2Mt - 2n_1t + 2E - 2\epsilon_1)$$

$$\delta v_2 = m \left\{ \begin{array}{l} \ \ \ 7''.862 \ \ \sin \ \ (nt - n_2t + \epsilon - \epsilon_2) \\ - 0''.228 \ \ \sin 2\,(nt - n_2t + \epsilon - \epsilon_2) \\ - 0''.0414 \ \sin 3\,(nt - n_2t + \epsilon - \epsilon_2) \end{array} \right. \qquad (298)$$

$$+ m_1 \left\{ \begin{array}{l} - 1126''.96 \ \ \sin \ \ (n_1t - n_2t + \epsilon_1 - \epsilon_2) \\ - \ \ \ 16''.504 \ \sin 2(n_1t - n_2t + \epsilon_1 - \epsilon_2) \\ - \ \ \ \ 3''.2995 \ \sin 3(n_1t - n_2t + \epsilon_1 - \epsilon_2) \\ - \ \ \ \ 1''.0467 \ \sin 4(n_1t - n_2t + \epsilon_1 - \epsilon_2) \\ - \ \ \ \ 0''.4067 \ \sin 5(n_1t - n_2t + \epsilon_1 - \epsilon_2) \\ - \ \ \ \ 0''.1767 \ \sin 6(n_1t - n_2t + \epsilon_1 - \epsilon_2) \end{array} \right.$$

$$+ m_3 \left\{ \begin{array}{l} \ \ \ 34''.396 \ \sin \ \ (n_3t - n_2t + \epsilon_3 - \epsilon_2) \\ - 117''.32 \ \ \sin 2\,(n_3t - n_2t + \epsilon_3 - \epsilon_2) \end{array} \right.$$

$$+ m_3 \left\{ \begin{array}{l} - 8''.251 \ \sin 3(n_3t - n_2t + \epsilon_3 - \epsilon_2) \\ - 1''.919 \ \sin 4(n_3t - n_2t + \epsilon_3 - \epsilon_2) \\ - 0''.609 \ \sin 5(n_3t - n_2t + \epsilon_3 - \epsilon_2) \\ - 0''.227 \ \sin 6(n_3t - n_2t + \epsilon_3 - \epsilon_2) \end{array} \right.$$

$$+ 0''.7734 \ \sin(2n_2t - 2Mt + 2\epsilon_2 - 2E)$$

$$\delta r_2 = m \left\{ \begin{array}{l} - 0.00054798 \\ + 0.00059147 \cos \ (nt - n_2t + \epsilon - \epsilon_2) \\ + 0.00001906 \cos 2(nt - n_2t + \epsilon - \epsilon_2) \\ + 0.00000348 \cos 3(nt - n_2t + \epsilon - \epsilon_2) \end{array} \right.$$

$$+ m_1 \left\{ \begin{array}{l} - 0.00070942 \\ + 0.04137743 \cos \ (n_1t - n_2t + \epsilon_1 - \epsilon_2) \\ + 0.00091726 \cos 2(n_1t - n_2t + \epsilon_1 - \epsilon_2) \\ + 0.00021712 \cos 3(n_1t - n_2t + \epsilon_1 - \epsilon_2) \\ + 0.00007409 \cos 4(n_1t - n_2t + \epsilon_1 - \epsilon_2) \\ + 0.00002980 \cos 5(n_1t - n_2t + \epsilon_1 - \epsilon_2) \\ + 0.00001318 \cos 6(n_1t - n_2t + \epsilon_1 - \epsilon_2) \end{array} \right.$$

$$+ m_3 \left\{ \begin{array}{l} 0.00006850 \\ + 0.00075191 \cos \ (n_3t - n_2t + \epsilon_3 - \epsilon_2) \\ - 0.0044961 \ \cos 2(n_3t - n_2t + \epsilon_3 - \epsilon_2) \\ - 0.00039801 \cos 3(n_3t - n_2t + \epsilon_3 - \epsilon_2) \\ - 0.00010474 \cos 4(n_3t - n_2t + \epsilon_3 - \epsilon_2) \\ - 0.00003569 \cos 5(n_3t - n_2t + \epsilon_3 - \epsilon_2) \\ - 0.00001379 \cos 6(n_3t - n_2t + \epsilon_3 - \epsilon_2) \end{array} \right.$$

$$+ 0.00003944$$

$$- 0.00003944 \cos \{2Mt - 2n_3t + 2E - 2\epsilon_3\}$$

$$\delta v_3 = m . \left\{ \begin{array}{l} + 4''.6156 . \sin \ (nt - n_3t + \epsilon - \epsilon_3) \\ - 0''.0067 . \sin 2(nt - n_3t + \epsilon - \epsilon_3) \end{array} \right. \qquad (299)$$

$$+ m_1 . \left\{ \begin{array}{l} + 7''.2745 \ . \sin \ (n_1t - n_3t + \epsilon_1 - \epsilon_3) \\ - 0''.09995 . \sin 2(n_1t - n_3t + \epsilon_1 - \epsilon_3) \\ - 0''.0175 \ . \sin 3(n_1t - n_3t + \epsilon_1 - \epsilon_3) \end{array} \right.$$

$$+ m_2 . \left\{ \begin{array}{l} - 11''.482 \ . \sin \ (n_2t - n_3t + \epsilon_2 - \epsilon_3) \\ - \ 5''.1701 . \sin 2(n_2t - n_3t + \epsilon_2 - \epsilon_3) \\ - \ 1''.0787 . \sin 3(n_2t - n_3t + \epsilon_2 - \epsilon_3) \\ - \ 0''.3304 . \sin 4(n_2t - n_3t + \epsilon_2 - \epsilon_3) \\ - \ 0''.1210 . \sin 5(n_2t - n_3t + \epsilon_2 - \epsilon_3) \end{array} \right.$$

$$+ \ 4''.2082 . \sin 2(n_3t - Mt + \epsilon_3 - E)$$

$$\delta r_3 = m.\begin{cases} -\ 0.00088152 \\ +\ 0.00057018.\cos\ (nt - n_3t + \epsilon - \epsilon_3) \\ +\ 0.00000113.\cos 2(nt - n_3t + \epsilon - \epsilon_3) \end{cases}$$

$$+ m_1.\begin{cases} -\ 0.00093981 \\ +\ 0.00091758.\cos\ (n_1t - n_3t + \epsilon_1 - \epsilon_3) \\ +\ 0.00001095.\cos 2(n_1t - n_3t + \epsilon_1 - \epsilon_3) \\ +\ 0.00000166.\cos 3(n_1t - n_3t + \epsilon_1 - \epsilon_3) \end{cases}$$

$$+ m_2.\begin{cases} -\ 0.00114443 \\ +\ 0.00326071.\cos\ (n_2t - n_3t + \epsilon_2 - \epsilon_3) \\ +\ 0.00057836.\cos 2(n_2t - n_3t + \epsilon_2 - \epsilon_3) \\ +\ 0.00013614.\cos 3(n_2t - n_3t + \epsilon_2 - \epsilon_3) \end{cases}$$

$$+\ 0.00037741$$

$$-\ 0.00037741.\cos 2(Mt - n_3t + E - \epsilon_3).$$

These inequalities in the circular orbits are independent of their positions.

Determination of the Masses of the Satellites and the Compression of Jupiter.

889. Approximate values of the masses of the satellites, and of the compression of Jupiter, are sufficiently accurate for calculating the periodic inequalities in the circular orbit; but it is necessary to have more correct values of these quantities for computing the secular variations. The periodic and secular inequalities determined by theory, when compared with their observed values, furnish the means of finding the true values of these very minute quantities. The principal periodic inequality in the longitude of the first satellite is, by observation, 1636″.4 at its maximum; but by article 888 this inequality is, by theory, $7042''.6m_1$, whence

$$m_1 = 0.232355.$$

The greatest periodic inequality in the longitude of the second satellite is, by observation, 3862″.3 at its maximum; the same, by (298), is

$$m\ .\ 2252''.28 + m_2\ .\ 3923''.3,$$

which arises from the combined action of the first and third satellites, hence

$$m = 1.714843 - m_2\ .\ 1.741934. \qquad (300)$$

The other unknown quantities must be computed from equations (271) and (290). For that purpose let

$$\rho - \tfrac{1}{2}\phi = \mu \;.\; 0.0217794,$$

μ being an indeterminate quantity depending on the compression of Jupiter's spheroid. Then from the expressions

$$\frac{(\rho - \frac{1}{2}\phi)}{a^2}\, n = (0) \qquad \tfrac{3}{4} \,.\, \frac{M^2}{n} = \boxed{0}$$

and the formulæ in article 474, it will be found that

$$\begin{array}{ll}
(0) = 179457''. \;\; \mu & \boxed{0} = 33''.47 \\
(1) = 35317''. \;\; \mu & \boxed{1} = 67''.16 \\
(2) = 6889''.6 \;\; \mu & \boxed{2} = 135''.31 \\
(3) = 954''.82 \;\; \mu & \boxed{3} = 315''.64 \\
(0.1) = m_1 \;\; 12903''.6 & \boxed{0.1} = m_1 \,.\, 9563''.2 \\
(0.2) = m_2 \,.\; 1686''.44 & \boxed{0.2} = m_2 \,.\; 813''.69 \\
(0.3) = m_2 \,.\; 248''.57 & \boxed{0.3} = m_3 \,.\; 69''.16 \\
(1.0) = m \;.\; 10229''.9 & \boxed{1.0} = m \;.\; 7581''.6 \\
(1.2) = m_2 \,.\; 6339''.61 & \boxed{1.2} = m_2 \,.\; 4688''.2 \\
(1.3) = m_3 \,.\; 584''.554 & \boxed{1.3} = m_3 \,.\; 256''.12 \\
(2.0) = m \;.\; 1058''.61 & \boxed{2.0} = m \;.\; 510''.77 \\
(2.1) = m_1 \,.\; 5019''.6 & \boxed{2.1} = m_1 \,.\; 3712''.1 \\
(2.3) = m_3 \,.\; 1907''.34 & \boxed{2.3} = m_3 \,.\; 1294''.4 \\
(3.0) = m \;.\; 117''.64 & \boxed{3.0} = m \;.\; 32''.74 \\
(3.1) = m_1 \,.\; 348''.99 & \boxed{3.1} = m_1 \,.\; 152''.93 \\
(3.2) = m_2 \,.\; 1438''.2 & \boxed{3.2} = m_2 \,.\; 976''.01
\end{array} \qquad (301)$$

The numerical values of F, G; F', G', are determined from articles 825, 826, and 827, to be

$$F = 1.483732 \qquad G = -\,0.857159$$

$$F' = 1.466380 \qquad G' = -\,0.855370$$

and with the same quantities the coefficients Q, Q_1, Q_2, of the equations in article 839 are found to be

$$Q = -\,m_1 \left\{ \frac{16.850204.h - 6.118274.h_1}{\left(1 + \dfrac{g}{972421''}\right)^2} \right\}$$

$$Q = +\,m_2. \left\{ \frac{4.133080.h_1 - 1.511476.h_2}{\left(1 + \dfrac{g}{972421''}\right)^2} \right\} \qquad (302)$$

$$+\,m \left\{ \frac{13.307450.h - 4.831907.h_1}{\left(1 + \dfrac{g}{972421''}\right)^2} \right\}$$

$$Q_2 - m_1 \left\{ \frac{3.248934.h_1 - 1.188133h_2}{\left(1 + \dfrac{g}{972421''}\right)^2} \right\}$$

Not only these quantities, but several data from observation are requisite for the determination of the unknown quantities from equations (271) and (290).

890. The eclipses of the third satellite show it to have two distinct equations of the centre; the one depending on the apsides of the fourth satellite is $2h_2 = 245''.14$. The other datum is the equation of the centre of the fourth satellite, which is, by observation, equal to $3002''.04 = 2h_3$. Again, observation gives the annual and sidereal motion of the apsides of the fourth satellite equal to $2578''.75$, which, by article 831, is one of the four roots of g in equation (271), so that $g_3 = 2578''.75$. And, lastly, observation gives $43374''$ for the annual and sidereal motion of the nodes of the orbit of the second satellite on the fixed plane, which is one of the roots of p in equation (290), so that

$$p_{,} = 43374''.$$

891. If the values of $m_{,}$ and m, as well as all the quantities that precede, be substituted in equations (271) and (290), they become, when the first are divided by h_3, and the last by l_1,

$$0 = 2182''. - 954''.81\,\mu - 117''.64\,m + 32''.73.m.\,\frac{h}{h_3} \qquad (303)$$

$$- 1358''.5\,m_2 + 35''.533\,.\,\frac{h_1}{h_3}.$$

$$0 = -\frac{h}{h_2}\,\{8040''.9 + 179457\,\mu. + 51581''.5\,m + 1686''.44\,m_2$$

$$+ 248''.55\,m_3\} \qquad (304)$$

$$+ \{4977''.22 + 18729''.\,m - 16020''.3\,m_2\}\,\frac{h_1}{h_3}$$

$$+ 544''.86\,m_2 + 69''.16\,m_3.$$

$$0 = \{18305''.3\,m + 72999''.2\,m^2 - 63180''.4\,mm_2\}\,\frac{h_1}{h_3} \qquad (305)$$

$$+\left\{\begin{array}{l} 2511''.6 - 35317''.\,\mu - 14128\,m - 13455''.\,m_2 - 584''.554\,m_3 \\ -26505''.7\,m^2 + 45344''.8\,mm_2 - 19393''.4\,m_2^2 \end{array}\right\}\frac{h_1}{h_3}$$

$$+ 594''.41\,m''. + 256''.12\,m_3 - 677''.04\,mm_2 + 592''.6\,m_2^2.$$

$$0 = 4831''.9\,m\,.\,\frac{h}{h_3} + \{1352''.8 - 1569''.\,m + 1342''.\,m_2\}\frac{h_1}{h_3} \qquad (306)$$

$$+ 89''.7 - 562''.6\,\mu - 86''.44\,m - 40''.\,m_2 + 1138''.7\,m_2.$$

$$0 = 43306''.9 - 35317''.\,\mu - 10229''.9\,m\left(1 - \frac{l}{l_1}\right) \qquad (307)$$

$$-6339''.6m_2\left(1 - \frac{l_2}{l_1}\right) - 584''.554\,m_3\left(1 - \frac{l_3}{l_1}\right);$$

(308)

$$0 = 2998''.23 + (40342''.3 - 179457''.\,\mu - 1686''.44m_2 - 248''.57\,m_3)\frac{l}{l_1}$$

$$+ 1686''.44\,m_2\,.\,\frac{l_2}{l_1} + 248''.57\,m_3\,.\,\frac{l_3}{l_1}.$$

$$0 = 1166''.5 + 1058''.6\,m\,.\,\frac{l}{l_1} + 1907''.34\,m_3\,.\,\frac{l_3}{l_1} \qquad (309)$$

$$+ \{42072''.4 - 6889''.6\,\mu - 1058''.6\,m - 1907''.35\,m_3\}\,\frac{l_2}{l_1}$$

$$0 = 81''.09 + 117''.64\,m\,.\,\frac{l}{l_1} + 1438''.2\,m_2\,.\,\frac{l_2}{l_1}. \qquad (310)$$

$$+ \{42976''.3 - 954''.82\,\mu - 117''.64\,m - 1438''.2\,m_2\}\,\frac{l_3}{l_1}.$$

892. These are the particular values of equations (271) and (290)

corresponding to the roots $g_3 = 2578''.82_2$ and $p_1 = 43374''$ alone. By the following method of approximation, nine of the unknown quantities are obtained from these eight equations, together with equation (300).

The inclinations of the satellites are very small, and the two first move nearly in circular orbits, therefore the quantities

$$\frac{h}{h_3},\quad \frac{h_1}{h_3},\quad \frac{l}{l_1},\quad \frac{l_2}{l_1},\quad \frac{l_3}{l_1},$$

are so minute that they may be made zero in the equations (303), (306), (307), in the first instance; and if m be eliminated by equation (300), these three equations will give approximate values of the masses m_2, m_3, and of μ, and then m will be obtained from equation (300). But, in order to have these four quantities more accurately, their approximate values must be substituted in equations (304), (305), (308), (309), and (310), whence approximate values of

$$\frac{h}{h_3},\quad \frac{h_1}{h_3},\quad \frac{l}{l_1},\quad \frac{l_2}{l_1},\quad \frac{l_3}{l_1}$$

will be found. Again, if these approximate values of

$$\frac{h}{h_3},\quad \frac{h_1}{h_3},\quad \frac{l}{l_1},\quad \frac{l_2}{l_1},\quad \frac{l_3}{l_1},$$

be substituted in equations (303), (306), and (307), and if m be eliminated by means of equation (300), new and more accurate values of the masses and of μ will be obtained. If with the last values of the masses and of μ the same process be repeated, the unknown quantities will be determined with still more precision. This process must be continued till two consecutive values of each unknown quantity are nearly the same. In this manner it is found that

$$\mu = 1.0055974;$$

$$\begin{aligned}
m &= 0.173281; & m_1 &= 0.232355; \\
m_2 &= 0.884972; & m_3 &= 0.426591; \\
h &= 0.00206221\, h_3; & l &= 0.0207938\, l_1; \\
h_1 &= 0.0173350\,.\, h_3; & l_2 &= -\,0.0342530\, l_1; \\
h_2 &= 0.0816578\,.\, h_3; & l_3 &= -\,0.000931164\, l_1.
\end{aligned}$$

893. μ determines the compression of Jupiter's spheroid, for

$$\rho - \tfrac{1}{2}\phi = \mu.0.0217794,$$

whence

$$\rho - \tfrac{1}{2}\phi = 0.0219012.$$

If t be the time of Jupiter's rotation, T the time of the sidereal revolution of the fourth satellite, then

$$\phi = \frac{T^2}{a_3^3 . t^2}$$

is the ratio of the centrifugal force to gravity at Jupiter's equator. But $a_3 = 25.4359$, $T = 16.689019$ days; and, according to the observations of Cassini $t = 0.413889$ of a day, hence $\phi = 0.0987990$, and $\rho = 0.0713008$.

As the equatorial radius of Jupiter's spheroid has been taken for unity, half his polar axis will be

$$1 - \rho = 0.9286992.$$

The ratio of the axis of the pole to that of his equator has often been measured: the mean of these is 0.929, which differs but little from the preceding value; but on account of the great influence of the matter at Jupiter's equator on the motions of the nodes and apsides of the orbits of the satellites, this ratio is determined with more precision by observation of the eclipses than by direct measurement, however accurate.

The agreement of theory with observation in the compression of Jupiter shows that his gravitation is composed of the gravitation of all his particles, since the variation in his attractive force, arising from his observed compression, exactly represents the motions of the nodes and apsides of his satellites.

894. If the preceding values of the masses of the satellites be divided by 10,000, the ratios of these bodies to that of Jupiter, taken as the unit, are

1st . . .	0.0000173281
2d . . .	0.0000232355
3d . . .	0.0000884972
4th . . .	0.0000426591.

895. Assuming the values of the masses of the earth and Jupiter in article 606, the mass of the third satellite will be 0.027337 of that of the earth, taken as a unit. But it was shown that the mass of the moon is

$$\frac{1}{75} = 0.013333, \text{ \&c.}$$

of that of the earth. Thus the mass of the third satellite is more than twice as great as that of the moon, to which the mass of the fourth is nearly equal.

896. In the system of quantities,

$$g_3 = 2578''.82$$
$$h = 0.00206221\ h_3 = \zeta_1^{(3)}\ h_3$$
$$h_1 = 0.0173350\quad h_3 = \zeta_2^{(3)}\ h_3$$
$$h_2 = 0.0816578\quad h_3 = \zeta_3^{(3)}\ h_3$$

h_3 may be regarded as the true eccentricity of the orbit of the fourth satellite, arising from the elliptical form of the orbit, and given by observation. And the values of h, h_1, h_2 are those parts of the eccentricities of the other three orbits, which arise from the indirect action of the matter at Jupiter's equator; for the attraction of that matter, by altering the position of the apsides of the fourth satellite, changes the relative position of the four orbits, and consequently alters the mutual attraction of the satellites, and is the cause of the changes in the form of the orbits expressed by the preceding values of h, h_1, h_2. This is the reason why these quantities depend on the annual and sidereal motion of the apsides of the fourth satellite.

897. A similar system exists for each root of g, arising from the same cause, and depending on the annual and sidereal motions of the apsides of the other three satellites. These are readily obtained from the general equations (271), which become, when the values of the masses and of the quantities in equations (301) are substituted,

$$0=\left\{g-185091''.3-\frac{16613''.78}{\left(1+\frac{g}{972421''}\right)^2}\right\}h+\left\{2222''.1-\frac{8220''.4}{\left(1+\frac{g}{972421''}\right)^2}\right\}h_1$$
$$+\left\{270''.1+\frac{5212''.2}{\left(1+\frac{g}{972421''}\right)^2}\right\}h_2+29''.5\,h_3; \qquad (311)$$

$$0=\left\{1313''.7-\frac{5668''.5}{\left(1+\frac{g}{972421''}\right)^2}\right\}h+\left\{g-43214''-\frac{15936''.3}{\left(1+\frac{g}{972421''}\right)^2}\right\}h_1$$
$$+\left\{4148''.9+\frac{6740''.6}{\left(1+\frac{g}{972421''}\right)^2}\right\}h_2+109''.3\,h_3; \qquad (312)$$

$$0=\left\{89''.5+\frac{752''.6}{\left(1+\frac{g}{972421''}\right)^2}\right\}h+\left\{862''.5+\frac{1413''.5}{\left(1+\frac{g}{972421''}\right)^2}\right\}h_1$$

$$+\left\{g-9227''.1-\frac{616''.4}{\left(1+\frac{g}{972421''}\right)^2}\right\}h_2+552''.2\,h_3; \qquad (313)$$

$$0=5''.7\,h+35''.53h_1+863''.74\,h_2+(g-2650''.1)\,h_3. \qquad (314)$$

898. As the motion of the apsides of the orbits of the satellites is almost entirely owing to the compression of Jupiter, in the first approximation the coefficient of h_2 may be made zero in equation (311); whence

$$g=9227''.1+\frac{616''.4}{\left(1+\frac{g}{972421''}\right)^2},$$

or, omitting g in the divisor,

$$g=9843''.5=10000''\text{ nearly};$$

hence, if $10000''$ be put for g in equations (311), (312), (314), they will give values of

$$\frac{h}{h_2},\ \frac{h_1}{h_2},\ \frac{h_3}{h_2};$$

and, by the substitution of these in equation (311), a still more approximate value of g will be found. This process must be continued till two consecutive values of g are nearly the same. In this manner it may be found that

$$g_2=9399''.17$$

$$h\ =0.0238111\,h_2=\zeta_1^{(2)}h_2$$

$$h_1=0.2152920\,h_2=\zeta_2^{(2)}h_2$$

$$h_3=0.1291561\,h_2=\zeta_3^{(2)}h_2$$

h_2 may be regarded as the true eccentricity of the orbit of the third satellite, and h, h_1, h_3 are those parts of the eccentricities of the other three orbits, arising from the action of Jupiter's equator on the apsides of the third, and depending on $g_2=9399''.17$, their annual and sidereal motion.

899. Again, if h and h_1 be made zero in equations (311) and (312), and g omitted in the divisor, then will

$$g=35114''.7, \qquad g_1=59152''.3,$$

and by the same method it will be found that

$g = 196665'$, $\qquad g_{,} = 0.57718''$

$h_1 = 0.0185238.h = \zeta_1 h$; $\qquad h_1 = -0.0375392.h_1 = \zeta_1^{(1)} h_1$

$h_2 = -0.0034337.h = \zeta_2 h$; $\qquad h_2 = -0.0436686.h_1 = \zeta_2^{(1)} h_1$

$h_3 = -0.00001735.h = \zeta_3 h$; $\qquad h_3 = 0.00004357.h_1 = \zeta_3^{(1)} h_1$

In these h and h_1 are the real eccentricities of the orbits of the first and second satellites, and the other values, h, h_1, h_2, h_3, &c., arise from the action of the other satellites corresponding to the roots g and $g_{,}$.

900. With regard to the inclinations of the orbits and the longitudes of their nodes, it appears, from article 892, that the system of inclinations for the root $p_{,}$ is

$$p_{,} = 43374''.01$$
$$l = 0.0207938 . l_{,} = \zeta_1^{(1)} . l_{,}$$
$$l_2 = -0.0342530 . l_{,} = \zeta_2^{(1)} . l_{,}$$
$$l_3 = -0.00093116 . l_{,} = \zeta_3^{(1)} . l_{,}$$

$l_{,}$ is the real inclination of the orbit of the second satellite on its fixed plane, passing between the equator and orbit of Jupiter; and l, l_2, l_3, are those parts of the inclination of the other three orbits depending on the root $p_{,}$, and arising principally from the action of Jupiter's equator; for the attraction of that protuberant matter, by changing the place of the nodes of the second satellite, alters the relative position of the orbits, which changes the mutual attraction of the bodies, and produces the variations in the inclinations expressed by l, l_2, l_3; and it is for this reason that these quantities depend on the annual and sidereal motion of the nodes of the second satellite.

901. A similar system depends on each root of p, that is, on the annual and sidereal motions of the nodes of the orbits of the other three satellites. These are obtained from equations (307), &c.; for when the values of the masses and of μ are substituted, they become

$$0 = (p - 185091'')\,l + 2998''.23\,l_1 + 1492''.5\,l_2 + 106''.03\,l_3$$
$$0 = 1772''.6\,l + (p - 43214'')\,l_1 + 5610''.4\,l_2 + 249''.4\,l_3$$
$$0 = 183''.44\,l + 1166''.3\,l_1 + (p - 9227''.2)\,l_2 + 813''.7\,l_3 \qquad (315)$$
$$0 = 20''.4\,l + 81''.09\,l_1 + 1272''.8\,l_2 + (p - 2650'')\,l_3.$$

902. The first approximate value of p is found by making the coefficient of l zero in the first of equations (315); whence $p = 185091''$; and if this value of p be put in the three last of these equations divided

by l, values of $\frac{l_1}{l}$, $\frac{l_2}{l}$, $\frac{l_3}{l}$, will be found; and when these last quantities are put in the first of equations (315), a new and more correct value of p will be found: by repeating the process till two consecutive values of p nearly coincide, it will be found that

$$p = 185130''.14$$
$$l_1 = -0.0124527 \,.\, l = \zeta_1 l$$
$$l_2 = -0.0009597 \,.\, l = \zeta_2 l$$
$$l_3 = -0.0000995 \,.\, l = \zeta_3 l$$

l is the inclination of the first satellite on its fixed plane, arising chiefly from the attraction of Jupiter's equator, and given by observation; and l_1, l_2, l_3 are the parts of the inclination of the other three orbits depending on p, the annual and sidereal motion of the nodes of the first satellite.

903. The third and fourth roots of p will be obtained by making the coefficients of l_2 and l_3 respectively zero in the third and fourth of the preceding equations; and, by the same method of approximation, it will be found that

$$p_2 = 9193''.56, \qquad p_3 = 2489''.2$$
$$l = 0.0111626 \,.\, l_2 = \zeta_1^{(2)} l_2, \qquad l = 0.0019856 \,.\, l_3 = \zeta_1^{(3)} l_3$$
$$l_1 = 0.164053 \,.\, l_2 = \zeta_2^{(2)} l_2, \qquad l_1 = 0.0234108 \,.\, l_3 = \zeta_2^{(3)} l_3$$
$$l_3 = -0.196565 \,.\, l_2 = \zeta_3^{(2)} l_2, \qquad l_2 = 0.1248622 \,.\, l_3 = \zeta_3^{(3)} l_3,$$

where l_2 and l_3 are the real inclinations of the third and fourth satellites on their fixed planes, given by observation.

904. It now remains to compute the quantities depending on the displacement of Jupiter's equator and orbit, namely, the four values of λ, $\theta' = {}'L + bt$, and $\psi' = {}'pt - \frac{at}{'L}$. The first are found by the substitution of the numerical values of the masses and of $\rho - \frac{1}{2}\phi$, in equations (285). Whence

$$\lambda = 0.00057879$$
$$\lambda_1 = 0.00585888$$
$$\lambda_2 = 0.02708801$$
$$\lambda_3 = 0.13235804.$$

Again,

$$p = \frac{3}{4i}\left(\frac{2C-A-B}{C}\right)\{M^2 + mn^2\lambda + m_1 n_1^2 \lambda_1 + m_2 n_2^2 \lambda_2 + m_3 n_3^2 \lambda_3\}.$$

As A, B, C, are the moments of inertia of Jupiter's spheroid, assumed to be elliptical, the theory of spheroids gives

$$\frac{2C-A-B}{C} = 0.14735\,;$$

and by observation, it is known that Jupiter's rotation is performed in 0.41377 of a day; and that his sidereal revolution is 4332.6 days; therefore

$$\frac{M}{i} = \frac{0.41377}{4332.6}\,;$$

then, by the substitution of the numerical values of the other quantities, all of which are given, it will appear that

$$`p = 3''.2007.$$

By observation, the inclination of Jupiter's equator on his orbit was, in 1750, $`L = 3^\circ.09996$, and as

$$a = \frac{dp}{dt}, \quad b = \frac{dq}{dt}$$

are given by the theory of Jupiter at that epoch,

$$\frac{a}{`L} = 2''.93314, \qquad b = 0''.02279\,;$$

whence $\theta' = 3^\circ.09996 + 0''.02279\,.\,t$; $\psi' = 0''.2676$, which is nearly the annual precession of Jupiter's equinoxes on his orbit. $-\psi'$ expresses the longitude of the descending node of Jupiter's equator on his orbit, $180^\circ - \psi' = \Pi$ will be the longitude of his ascending node; consequently

$$\sin(v + \psi') = \sin(v - \Pi).$$

By observation, it is known that, in the beginning of 1750,

$$\Pi = 313^\circ.7592\,;$$

whence $\psi' = 46^\circ.241 + 0''.2676\,.\,t$;

and, with the preceding value of θ', it will be found that

$$(1 - \lambda\,)\,\theta' = 3^\circ.0899$$
$$(1 - \lambda_1)\,\theta' = 3^\circ.0736$$
$$(1 - \lambda_2)\,\theta' = 3^\circ.0079$$
$$(1 - \lambda_3)\,\theta' = 2^\circ.6825.$$

905. It appears, from observation, that the two first satellites move in circular orbits, and that the first moves sensibly on its fixed plane, from the powerful attraction of Jupiter's equator; consequently h and $h_,$, corresponding to the roots g and $g_,$, are zero, as well as the inclination l, depending on the root p. Hence the sys-

tems of quantities in articles 899 and 902 are zero; and as, by observation, the real equations of the centre of the third and fourth satellites are

$$2h_2 = 245''.14, \quad 2h_2 = 553''.73, \quad 2h_3 = 3002''.04;$$

and the real inclinations of the second, third, and fourth on their fixed planes, are

$$l_1 = -1669''.31, \qquad l_2 = -739''.98, \qquad l_3 = -897''.998.$$

By the substitution of these quantities in the different systems,

$$\mathfrak{C}_1^{(2)} h_2, \quad \mathfrak{C}_1^{(3)} h_3, \quad \&c. \ \&c.$$

it will be found that the equations in articles 835 and 878,

$$\begin{aligned}
\delta v &= 13''.18 \,.\, \sin(nt + \epsilon - g_2 t - \Gamma_2) \\
&- \ \ 6''.19 \,.\, \sin(nt + \epsilon - g_3 t - \Gamma_3) \\
\delta v_1 &= 119''.22 \,.\, \sin(n_1 t + \epsilon_1 - g_2 t - \Gamma_2) \\
&- \ 52''.04 \,.\, \sin(n_1 t + \epsilon_1 - g_3 t - \Gamma_3) \\
\delta v_2 &= -\,552''.02 \,.\, \sin(n_2 t + \epsilon_2 - g_2 t - \Gamma_2) \\
&- 244''.38 \,.\, \sin(n_2 t + \epsilon_2 - g_3 t - \Gamma_3) \\
\delta v_3 &= -\,3002''.04 \,.\, \sin(n_3 t + \epsilon_3 - g_3 t - \Gamma_3) \\
&- 71''.52 \,.\, \sin(n_3 t + \epsilon_3 - g_2 t - \Gamma_2).
\end{aligned} \qquad (316)$$

$$\begin{aligned}
s &= 3^\circ.0899 \,.\, \sin(v + 46^\circ.241 - 49''.8\, t) \\
&- 34''.03 \,.\, \sin(v + p_1 t + \Lambda_1) \\
&+ \ \ 8''.26 \,.\, \sin(v + p_2 t + \Lambda_2) \\
s_1 &= 3^\circ.0736 \,.\, \sin(v_1 + 46^\circ.241 - 49''.8\, t) \\
&- 1669''.3 \,.\, \sin(v_1 + p_2 t + \Lambda_2) \\
&\ \ \ 121''.4 \,.\, \sin(v_1 + p_1 t + \Lambda_1) \\
&\ \ \ \ 21''.02 \,.\, \sin(v_1 + p_3 t + \Lambda_3). \\
s_2 &= 3^\circ.0079 \,.\, \sin(v_2 + 46^\circ.241 - 49''.8\, t) \\
&- 739''.98 \,.\, \sin(v_2 + p_2 t + \Lambda_2) \\
&\ \ \ 112''.13 \,.\, \sin(v_2 + p_3 t + \Lambda_3) \\
&\ \ \ \ 57''.18 \,.\, \sin(v_2 + p_1 t + \Lambda_1). \\
s_3 &= 2^\circ.6825 \,.\, \sin(v_3 + 46^\circ.241 - 49''.8\, t) \\
&- 897''.998 \,.\, \sin(v_3 + p_3 t + \Lambda_3) \\
&+ 145''.45 \,.\, \sin(v_3 + p_3 t + \Lambda_3) \\
&+ \ \ 1''.6 \,.\, \sin(v_3 + p_1 t + \Lambda_1).
\end{aligned} \qquad (317)$$

906. The following data are requisite for the complete determination of the motions of the satellites, all of them being estimated from the vernal equinox of the earth; the epoch being the instant of midnight, December 31st, 1749, mean time at Paris.

The secular mean motions of the four satellites.

$$n = 7432435^\circ.47$$
$$n_1 = 3702713^\circ.2215$$
$$n_2 = 1837852^\circ.112$$
$$n_3 = 787885^\circ.$$

The longitudes of the epochs of the satellites, estimated from the vernal equinox, were

$$\epsilon = 15^\circ.0128$$
$$\epsilon_1 = 131^\circ.8404$$
$$\epsilon_2 = 10^\circ.26083$$
$$\epsilon_2 = 72^\circ.5513.$$

Longitudes of the lower apsides of the third and fourth satellites.

$$\Gamma_2 = 309^\circ.438603$$
$$\Gamma_3 = 180^\circ.343.$$

Longitudes of ascending nodes.

$$\Lambda_1 = 273^\circ.2889$$
$$\Lambda_2 = 187^\circ.4931$$
$$\Lambda_3 = 74^\circ.9687.$$

The values of g_2, g_3, &c., p, $p_{\prime}$, &c., are referred to the vernal equinox of Jupiter; but in order to refer them to the vernal equinox of the earth, the precession of the equinoxes, $= 50''$, must be added to the first and subtracted from the second; and as all the quantities in question have already been given, it will be found that the annual and sidereal motions of the apsides were

$$g_2 = 2628''.9$$
$$g_3 = 9449''.28.$$

The annual and sidereal motions of the nodes were

$$p_1 = 43324''.01$$
$$p_2 = 9143''.56$$
$$p_3 = 2439''.08.$$

Also the annual and sidereal motion of Jupiter's equinox, with regard to the vernal equinox of the earth, is

$$49''.8.$$

The longitude of Jupiter's vernal equinox at the epoch was $46^\circ.25$, consequently $\psi' = 46^\circ.25 + t.49''.8$,

and the eccentricity of Jupiter's orbit at the epoch was

$$\bar{e} = 19831''.47.$$

In order to abridge $g_2t + \Gamma_2$, $g_3t + \Gamma_3$, $pt + \Lambda$, &c., will be represented by ϖ_2, ϖ_3, Ω, Ω_1, Ω_2, Ω_3.

Theory of the First Satellite.

Longitude.

907. Since h and $h_{,}$ are zero, equations (302) give only the two following values of Q;

$$Q = 0.208780 \,.\, h_2 = 57''.8$$
$$Q = 0.016482 \,.\, h_3 = 24''.7\,;$$

consequently equation (268) becomes

$$\delta v = -\,57''.8 \,.\, \sin(nt - 2n_{,}t + \epsilon - 2\epsilon_{,} + \varpi_2)$$
$$-\,24''.7 \,.\, \sin(nt - 2n_{,}t + \epsilon - 2\epsilon_{,} + \varpi_3)$$

If equation (296) and the first of equations (316) be added to this, observing that

$$2nt + 2\epsilon - 2n_2t - 2\epsilon_2 = 180^\circ + 3nt + 3\epsilon - 3n_1t - 3\epsilon_1,$$

it will be found that the true longitude of the first satellite in its eclipses, is

$$\begin{aligned} v = nt + \epsilon &+ 13''.18 \,.\, \sin(nt + \epsilon - \varpi_2) \\ &+ 6''.19 \,.\, \sin(nt + \epsilon - \varpi_3) \\ &- 14''.11 \,.\, \sin(nt - n_{,}t + \epsilon - \epsilon_{,}) \\ &- 6''.29 \,.\, \sin \tfrac{3}{2}(nt - n_{,}t + \epsilon - \epsilon_{,}) \\ &+ 1636''.39 \,.\, \sin 2(nt - n_{,}t + \epsilon - \epsilon_{,}) \\ &+ 1''.22 \,.\, \sin 4(nt - n_{,}t + \epsilon - \epsilon_{,}) \\ &+ 0''.512 \,.\, \sin 5(nt - n_{,}t + \epsilon - \epsilon_{,}) \\ &- 57''.8 \,.\, \sin(nt - 2n_{,}t + \epsilon - 2\epsilon_{,} + \varpi_2) \\ &- 24''.7 \,.\, \sin(nt - 2n_{,}t + \epsilon - 2\epsilon_{,} + \varpi_3) \end{aligned} \tag{318}$$

for in the eclipses of the satellites by Jupiter, or of Jupiter by the satellites, the longitudes of both bodies are the same; the Earth, Jupiter, and the satellites being then in the same straight line, consequently $\qquad Mt + E = nt + \epsilon, \quad U = v,$
consequently the term depending on the argument $2\,(nt - Mt + \epsilon - E)$ vanishes.

Latitude.

908. By article 880 the action of the sun occasions the inequality

$$s = -\frac{3M}{8n}\,(L' - l)\,\sin(v - 2U - pt - \Lambda)$$

but in the eclipses $U = v$, therefore

$$s = \frac{3M}{8n}(L' - l)\sin(v + pt + \Lambda);$$

and as $$l - L' = (1 - \lambda)(L - L'),$$
and that $$(1 - \lambda)(L - L')\sin(v + pt + \Lambda)$$
is the latitude of the first satellite above its fixed plane, which was shown to be

$$3^\circ.0899 \sin(v + 46^\circ.241 - 49''.8\, t),$$

therefore the preceding inequality is

$$- s = 1''.7 \,.\, \sin(v + 46^\circ.241 - 49''.8\, t).$$

When this quantity, which arises from the action of the sun, is added to the first of equations (310), it gives

$$\begin{aligned} s = &\ 3^\circ.0894 \,.\, \sin(v + 46^\circ.241 - 49''.8\, t) \\ &- 34''.03 \,.\, \sin(v + \Omega_1) \\ &- 8''.26 \,.\, \sin(v + \Omega_2) \end{aligned}$$

for the latitude of the first satellite in its eclipses.

The inclination of the fixed plane on the equator of Jupiter is 6''.48, which is insensible; and as the orbit has no perceptible inclination on the fixed plane, the first satellite moves nearly in a circular orbit in the plane of Jupiter's equator.

Theory of the Second Satellite.

909. Because h and $h_{\prime}$ are insensible, equations (295) give

$$Q_{\prime} = -0.662615 \,.\, h_2 \qquad Q_{\prime} = -0.055035\, h_3;$$

therefore equation (260) becomes

$$\begin{aligned} \delta v_{\prime} = &\ 183''.46 \,.\, \sin(nt - 2n_{\prime}t + \epsilon - 2\epsilon_{\prime} + \varpi_2) \\ &+ 82''.6 \,.\, \sin(nt - 2n_{\prime}t + \epsilon - 2\epsilon_{\prime} + \varpi_3). \end{aligned}$$

Again, equations

$$\delta r_{\prime} = \frac{5}{16} \cdot \frac{n_{\prime}^2}{(n - n_{\prime} - N_{\prime})^2}\{mG - m_2F'\}^2 . \sin 2(nt - n_{\prime}t + \epsilon - \epsilon_{\prime})$$

$$\delta r_{\prime} = -\frac{6M}{n}\left\{1 - \frac{9a_{\prime}m \,.\, n^2 K}{8am_{\prime}b \,.\, (M^2 - Kn^2)}\right\} H \,.\, \sin(Mt + E - \Pi)$$

in articles 766 and 752, have a sensible effect on the motions of the second satellite, and in consequence of

$$nt + \epsilon = 180^\circ - 2n_2t + 3n_1t - 2\epsilon_2 + 3\epsilon_1,$$

they become, by the substitution of the numerical values of the quantities,

$$\delta v_{,} = 22''.61 \,.\, \sin 4\,(n_1 t - n_2 t + \epsilon_1 - \epsilon_2)$$
$$- 36''.07 \,.\, \sin (Mt + E - \Pi).$$

If to these the second of equations (309) be added, together with equation (290), it will be found, in consequence of the relation,

$$nt - n_{,}t + \epsilon - \epsilon_{,} = 180^\circ + 2n_1 t - 2n_2 t + 2\epsilon_1 - 2\epsilon_2$$

that the true longitude of the second satellite is, in its eclipses,

$$\begin{aligned}
\delta v_{,} = n_{,}t + \epsilon_{,} &+ 119''.22 \sin (n_{,}t + \epsilon_1 - \varpi_2) \\
&+ 52''.04 \,.\, \sin (n_1 t + \epsilon_1 - \varpi_3) \\
&- 52''.91 \,.\, \sin (n_1 t - n_2 t + \epsilon_1 - \epsilon_2) \\
&+ 3862''.3 \,.\, \sin 2(n_1 t - n_2 t + \epsilon_1 - \epsilon_2) \\
&+ 19''.75 \,.\, \sin 3(n_1 t - n_2 t + \epsilon_2 - \epsilon_2) \qquad (319) \\
&+ 24''.18 \,.\, \sin 4(n_1 t - n_2 t + \epsilon_1 - \epsilon_2) \\
&+ 1''.51 \,.\, \sin 5(n_1 t - n_2 t + \epsilon_1 - \epsilon_2) \\
&+ 1''.19 \,.\, \sin 6(n_1 t - n_2 t + \epsilon_1 - \epsilon_2) \\
&- 1''.71 \,.\, \sin (n_1 t - n_3 t + \epsilon_1 - \epsilon_3) \\
&+ 1''.5 \,.\, \sin 2(n_1 t - n_3 t + \epsilon_1 - \epsilon_3) \\
&+ 183''.46 \,.\, \sin (nt - 2n_1 t + \epsilon - 2\epsilon_1 + \varpi_2) \\
&+ 82''.6 \,.\, \sin (n_1 t - 2n_2 t + \epsilon_1 - 2\epsilon_2 + \varpi_3) \\
&- 36''.07 \,.\, \sin (Mt + E - \Pi),
\end{aligned}$$

for the last term of equation (290) vanishes.

The Latitude.

910. The equation (284),

$$s_{,} = -\frac{3M}{8n_{,}} \{L' - l_{,}\} \sin (v_{,} - 2U - pt - \Lambda)$$

has a different value for each root of p, including $'p$ the root, that depends on the displacement of Jupiter's orbit and equator; but because

$$v_{,} = U, \quad (l_{,} - L') = (1 - \lambda_{,})(L - L'),$$

and that

$$(1 - \lambda_{,})(L - L') \sin (v_{,} + pt + \Lambda)$$

is the latitude of the second satellite above its fixed plane, which is

$$3^\circ.0736 \,.\, \sin (v_{,} + 46^\circ.241 - 49''.8\, t)$$

the equation in question becomes

$$s_{,} = 3''.4 \sin (v_{,} + 46^\circ.241 - 49''.8\, t).$$

The only remaining root of p that gives the preceding equation a sensible value in the theory of this satellite is $p_{,} = 43324''\cdot 9$; and by the substitution of the corresponding values

$$s_{,} = 0''.512 \,.\, \sin(v_{,} + \Omega_{,}).$$

In consequence of these two inequalities the second of equations (310) becomes

$$\begin{aligned} s_{,} = & \ 3^{\circ}.07262 \,.\, \sin(v_{,} + 46^{\circ}.241 - 49''.8\, t) \\ & - 1669''.3 \,.\, \sin(v_{,} + \Omega_2) \\ & - 121''.4 \,.\, \sin(v_{,} + \Omega_1) \\ & - 21''.04 \,.\, \sin(v_{,} + \Omega_3). \end{aligned} \qquad (320)$$

The inclination of the fixed plane on the equator of Jupiter is $63''.124$. The orbit of the satellite revolves on this plane, to which it is inclined at an angle of $27'\ 48''.3$, its nodes completing a revolution in $29^{\text{yrs}}.914$.

Theory of the Third Satellite.

911. The inequalities represented by

$$\delta v_2 = -\ Q_2 \,.\, \sin(nt - 2n_{,}t + \epsilon - 2\epsilon_{,} + gt + T)$$

have a very sensible influence on the motions of the third satellite, because observation proves that body to have two distinct equations of the centre, one depending on the lower apsis of the orbit of the second satellite, and the other on that of the fourth. Consequently h_1 and h_2 in the coefficient

$$Q_2 = -\, m_1 \frac{(3.248934\, h_1 - 1.188133\, h_2)}{\left(1 + \dfrac{g}{972421''}\right)^2}$$

have respectively two values, namely,

$$h_1 = 0.2152920\, h_2, \text{ and } h_2 = -\, 276''.865;$$

corresponding to g_2 and Γ_2, also

$$h_1 = 0.0173350\, h_3, \text{ and } h_2 = 0.0816578\, h_3,$$

corresponding to g_3 and Γ_3; therefore the preceding inequality, in consequence of the relations among the mean longitudes of the three first satellites, gives

$$\begin{aligned} \delta v_2 = & -30''.84 \,.\, \sin(n_1t - 2n_2t + \epsilon_1 - 2\epsilon_2 + \varpi_2) \\ & + 14''.12 \,.\, \sin(n_1t - 2n_2t + \epsilon_1 - 2\epsilon_2 + \varpi_3) \end{aligned}$$

By articles 766 and 747 the action of the sun occasions the inequalities

$$\delta v_2 = -\frac{12M}{n_2}\left\{1 + \frac{3a_2mn^2.K}{32am_2.b(M^2-Kn^2)}\right\}\bar{e}\sin(Mt+E-\Pi)$$

$$-\frac{15Mh_2}{4n_2}\sin(n_2t - 2Mt + \epsilon - 2E + gt + T)$$

In consequence of the two values of h_2, and because

$2Mt + 2E = 2n_2t + 2\epsilon_2$, in the eclipses

these give
$$\begin{aligned}\delta v_2 = {} & 1''.71 \,.\, \sin(n_2t + \epsilon_2 - \varpi_2)\\ + {} & 0''.76 \,.\, \sin(n_2t + \epsilon_2 - \varpi_3)\\ - {} & 47''.76 \,.\, \sin(Mt + E - \Pi).\end{aligned}$$

Adding the preceding inequalities to those in (291), and to the third of (309), it will be found that the longitude of the third satellite, in its eclipses, is

$$\begin{aligned}v_2 = n_2t + \epsilon_2 & + 552''.031 \,.\, \sin(n_2t + \epsilon_2 - \varpi_2)\\ & + 244''.38 \,.\, \sin(n_2t + \epsilon_2 - \varpi_3)\\ & - 261''.86 \,.\, \sin(n_1t - n_2t + \epsilon_1 - \epsilon_2)\\ & - 3''.84 \,.\, \sin 2(n_1t - n_2t + \epsilon_1 - \epsilon_2)\\ & - 2''.13 \,.\, \sin 3(n_1t - n_2t + \epsilon_1 - \epsilon_2) \qquad (321)\\ & - 14''.65 \,.\, \sin(n_2t - n_3t + \epsilon_2 - \epsilon_3)\\ & + 50''.06 \,.\, \sin 2(n_2t - n_3t + \epsilon_2 - \epsilon_3)\\ & + 3''.52 \,.\, \sin 3(n_2t - n_3t + \epsilon_2 - \epsilon_3)\\ & + 0''.82 \,.\, \sin 4(n_2t - n_3t + \epsilon_2 - \epsilon_3)\\ & + 30''.84 \,.\, \sin(n_1t - 2n_2t + \epsilon_1 - 2\epsilon_2 + \varpi_2)\\ & + 14''.12 \,.\, \sin(n_1t - 2n_2t + \epsilon_1 - 2\epsilon_2 + \varpi_3)\\ & - 47''.76 \,.\, \sin(Mt + E - \Pi).\end{aligned}$$

912. The double equation of the centre occasions some peculiarities in the motion of the third satellite. By a comparison of

$$\varpi_2 = 9449''.28\, t + 309^\circ.438603$$

$$\varpi_3 = 2628''.9\, t + 180^\circ.343,$$

it appears that the lower apsides of the third and fourth satellites coincided in 1682, and then the coefficient of the equation of the centre was equal to the sum of the coefficients of the two partial equations. In 1777 the lower apsis of the third satellite was 180° before that of the fourth, and the coefficient of the equation of the centre was equal to the difference of the coefficients of the partial equations; results that were confirmed by observation.

Latitude.

913. The only part of the equation

$$s_2 = - \frac{3M}{2n_2} (L' - l_2) \sin (v_2 - 2U - pt - \Lambda)$$

that is sensible in the motions of the third satellite is that relating to the equator of Jupiter, whence it is easy to see that

$$s_2 = - 6''.7068 \,.\, \sin (v_2 + 46^\circ.241 - 49''.8\, t)\,;$$

the same expression with regard to the third satellite, gives

$$0''.46 \,.\, \sin (v_2 + \Omega_2),$$

the first subtracted from the third of equations (310), gives the latitude of the third satellite equal to

$$\begin{aligned} s_2 = {} & 3^\circ.0061 \,.\, \sin (v_2 + 46^\circ.251 - 49''.8\, t) \\ & - 739''.53 \,.\, \sin (v_2 + \Omega_2) \\ & - 112''.13 \,.\, \sin (v_2 + \Omega_3) \\ & + 57''.18 \,.\, \sin (v_2 + \Omega_1) \end{aligned} \qquad (322)$$

in its eclipses.

The inclination of the fixed plane of the third satellite on the equator of Jupiter is $301''.49 = \lambda_2 \theta_1$. Its orbit revolves on this plane, to which it is inclined at an angle of $12'\ 20''$, the nodes accomplishing their retrograde revolution in $141^{\text{yrs}}.739$.

Theory of the Fourth Satellite.

914. By article 746 the action of the sun occasions the inequalities

$$\delta v_3 = \frac{15}{4} \,.\, \frac{Mh_3}{n_3} \,.\, \sin (n_3 t + \epsilon_3 + \varpi_3 - 2Mt - 2E)$$

$$- \frac{3M}{n_3} \,.\, \bar{e} \,.\, \sin (Mt + E - \Pi),$$

and the secular variation in the inclination of the equator and orbit of Jupiter, by article 792, occasions the inequality

$$\delta v_3 = - \frac{\{4\,(1 - \lambda_3)\, \boxed{3} - \frac{1}{2}\,(1 - \lambda_3)\, p_3 + 6(3)\lambda_3\}}{p_3} \,.\, \theta' l_3 \,.\, \sin(p_3 t + \Lambda_3 - \psi')$$

It is easy to see that the two first inequalities are,

$$\begin{aligned} \delta v_3 = {} & 21''.69 \,.\, \sin (n_3 t + \epsilon_3 + \varpi_3 - 2Mt - 2E) \\ & - 133''.33 \sin (Mt + E - \Pi)\,; \end{aligned}$$

but in the eclipses $Mt + E = n_3 t + \epsilon_3$.

So
$$\delta v_3 = -\ 21''.69 \sin (n_3 t + \epsilon_3 - \varpi_3)$$
$$-\ 133''.33 \sin (Mt + E - \Pi),$$
and the third inequality is
$$\delta v_3 = -\ 16.04.\ \sin (28^\circ.812 + 2488''.91t).$$

If these be added to equation (292), and the last of equations (309), the longitude of the fourth satellite in its eclipses is,

$$\begin{aligned} v_3 = n_3 t + \epsilon_3 &+ 2980''.35 \sin (n_3 t + \epsilon_3 - \varpi_3) \\ &+ 13''.65.\ \sin 2(n_3 t + \epsilon_3 - \varpi_3) \\ &+ 0''.09.\ \sin 3(n_3 t + \epsilon_3 - \varpi_3) \\ &- 71''.28.\ \sin (n_3 t + \epsilon_3 - \varpi_2) \\ &- 10''.16.\ \sin (n_3 t - n_2 t + \epsilon_3 - \epsilon_2) \qquad (323) \\ &- 4''.58 \sin 2(n_3 t - n_2 t + \epsilon_3 - \epsilon_2) \\ &- 0''.96 \sin 3(n_3 t - n_2 t + \epsilon_3 - \epsilon_2) \\ &- 0''.29 \sin 4(n_3 t - n_2 t + \epsilon_3 - \epsilon_2) \\ &- 0''.11 \sin 5(n_3 t - n_2 t + \epsilon_3 - \epsilon_2) \\ &- 113''.33.\ \sin (Mt + E - \Pi) \\ &- 16''.04.\ \sin (2488''.91t + 28^\circ.73). \end{aligned}$$

The terms having the coefficients $13''.65$ and $0''.09$ belong to the equation of the centre, which in this satellite extends to the squares and cubes of the eccentricity.

Latitude.

915. The inequality of article 789
$$s_3 = \frac{3M}{8n_3}(l_3 - L') \sin (v_3 - 2U - pt - \Lambda),$$

arising from the action of the sun, has two sensible values, one arising from the displacement of Jupiter's orbit, and the other depending on the inclination of the orbit of the fourth satellite on its fixed plane. Because
$$l_3 - L' = (1 - \lambda_3)(L - L') = 2^\circ.6825,$$
the first of these inequalities is
$$s_3 = 13''.98 \sin (v_3 + 46^\circ.241 - 49''.8t),$$
in the eclipses when $U = v_3$, and the other depending on
$$p_3 = 2439''.08$$
is in the eclipses $\quad s_3 = 1''.3 \sin (v_3 + \Omega_3).$

Adding these to the last of equations (310) the latitude of the fourth satellite in its eclipses is

$$\begin{aligned} s_3 = {} & 2^\circ.6786 \,.\, \sin(v_3 + 46^\circ.241 - 49''.8t) \\ & - 896''.702 \,.\, \sin(v_3 + \Omega) \\ & + 145''.46 \,.\, \sin(v_3 + \Omega) \\ & + 1''6 \,.\, \sin(v_3 + \Omega). \end{aligned} \tag{324}$$

916. The inclination of the fixed plane of the fourth satellite on Jupiter's equator is

$$\lambda_3\theta' = 1473''.14.$$

The orbit of the satellite revolves on that plane to which it is inclined at an angle of 14'.58''; its nodes accomplish a revolution in 531 years.

917. The preceding expression for the latitude explains a singular phenomenon observed in the motions of the fourth satellite. The inclination of its orbit on the orbit of Jupiter appeared to be constant, and equal to 2°.43 from the year 1680 to 1760; during that time the nodes had a direct motion of about 4'.32 annually. From 1760 the inclination has increased. The latitude may be put under the form

$$A \sin v_3 - B \cos v_3 ;$$

A and B will be determined by making

$$v_3 = 90^\circ, \text{ and } v_3 = 180^\circ$$

successively in the expression s_3; $\frac{B}{A}$ will be the tangent of the longitude of the node and $\sqrt{A^2 + B^2}$, the inclination of the orbit. If then t be successively made equal to -70; -30; and 10 which corresponds to the years 1680, 1720, and 1760, estimated from the epoch of 1750, the result will be

	Inclination.	Longitude Ω.
1680 . .	2°.4764 . .	311°.4172
1620 . .	2°.4489 . .	313°.3067
1760 . .	2°.4411 . .	317°.0914

If the inclination be represented by

$$2^\circ.4764 + Nt + Pt^2$$

t being the number of years elapsed since 1680. Comparing this formula with the preceding inclination

$$N = -0^\circ.0009315 \qquad P = 0^\circ.000061313.$$

The minimum of the formula corresponds to $t = 75.953$, or to the year 1756. The mean of the three preceding values is 2°.4555, and the mean annual motion of the node from 1680 to 1760 is 4′.255. These results are conformable to observation during this interval, but from 1760 the inclination has varied sensibly. The preceding value of s_3 gives an inclination of 2°.5791 in 1800, and the longitude of the node equal to 320°.2935 ; and as observation confirms these results, it must be concluded, that the inclination is a variable quantity, but the law of the variation could hardly have been determined independently of theory.

CHAPTER IX.

ECLIPSES OF JUPITER'S SATELLITES.

918. Jupiter throws a shadow behind him relatively to the sun, in which the three first satellites are always immersed at their conjunctions, on account of their orbits being nearly in the plane of Jupiter's equator; but the greater inclination of the orbit of the fourth, together with its distance, render its eclipses less frequent.

919. Let S and J, fig. 113, be sections of the sun and Jupiter, and mn the orbit of a satellite. Let AE, A′E′ touch the sections internally, and AV, A′V externally. If these lines be conceived to revolve about SJV they will form two cones, aVa' and EBE′. The sun's light will be excluded from every part of the cone aVa', and the spaces Ea'V, E′aV will be the penumbra, from which the light of part of the sun will be excluded: less of it will be visible near aV, a'V, than near aE′, a'E.

fig. 113.

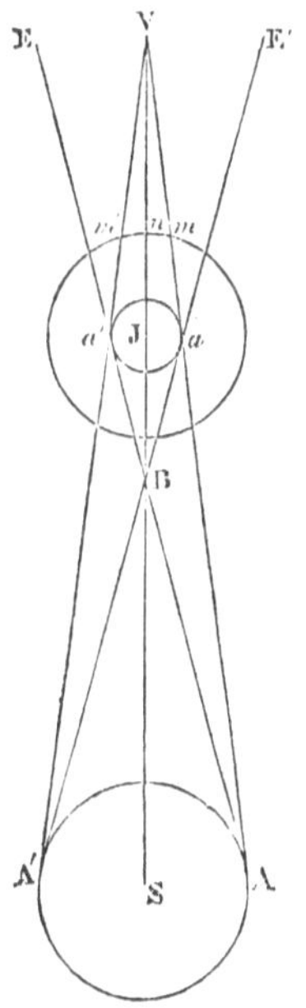

920. As the satellites are only luminous by reflecting the sun's rays, they will suddenly disappear when they immerge into the shadow, and they will reappear on the other side of the shadow after a certain time. The duration of the eclipse will depend on the form and size of the cone, which itself depends on the figure of Jupiter, and his distance from the sun.

921. If the orbits of the satellites were in the plane of Jupiter's orbit, they would pass through the axis of the cone at each eclipse, and at the instant of heliocentric conjunction, the sun, Jupiter and the satellite would be on the axis of the cone, and the duration of the eclipses would always be the same, if the orbit were circular. But as all the orbits are more or less inclined to the plane of Jupiter's orbit, the duration of the eclipses varies. If the conjunction happened in the node, the eclipse would

still be central; but at a certain distance from the node, the orbit of the satellite would no longer pass through the centre of the cone of the shadow, and the satellite would describe a chord more or less great, but always less than the diameter; hence the duration is variable. The longest eclipses will be those that happen in the nodes, whose position they will determine: the shortest will be observed in the limit or point farthest from the node at which an eclipse can take place, and will consequently determine the inclination of the orbit of that of Jupiter. With the inclination and the node, it will always be possible to compute the duration of the eclipse, its beginning and end.

922. The radius vector of Jupiter makes an angle SJE, fig. 114, with his distance from the earth, varying from 0° to 12°, which is the cause of great variations in the distances at which the eclipses take place, and the phenomena they exhibit.

fig. 114.

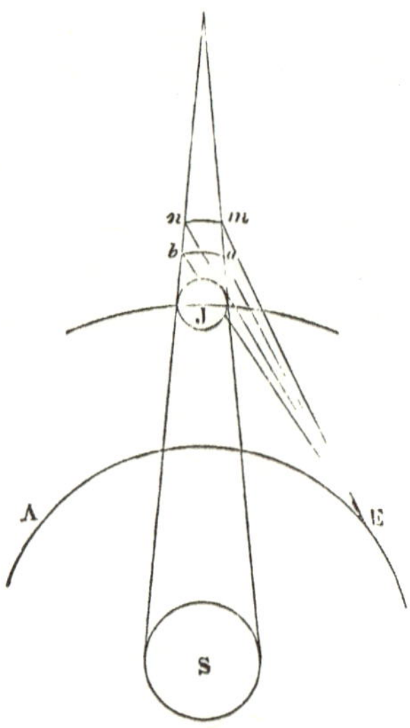

923. The third and fourth satellites always, and the second sometimes disappear and re-appear on the same side of Jupiter, for if S be the sun, E the earth, and *m* the third or fourth satellite, the immersion and emersion are seen in the directions E*m*, E*n*; only the immersions or emersions of the first satellite are visible according to the position of the earth; for if *ab* be the orbit of the first satellite, before the opposition of Jupiter, the immersion is seen in the direction E*a*, but the emersion in the direction E*b* is hid by Jupiter. On the contrary when the earth is in A, after the opposition of Jupiter, the emersion is seen, and not the immersion; it sometimes happens, that neither of the phases of the eclipses of the first satellite are seen. Before the opposition of Jupiter the eclipses happen on the west side of the planet, and after opposition on the east. The same satellite disappears at different distances from the primary according to the relative positions of the sun, the earth, and Jupiter, but they vanish close to

the disc of Jupiter when he is near opposition. The eclipses only happen when the satellites are moving towards the east, the transits only when they are moving towards the west; their motion round Jupiter must therefore be from west to east, or according to the order of the signs. The transits are real eclipses of Jupiter by his moons, which appear like black spots passing over his disc.

924. It is important to determine with precision the time of the disappearance of a satellite, which is however rendered difficult by the concurrence of circumstances: a satellite disappears before it is entirely plunged in the shadow of Jupiter; its light is obscured by the penumbra: its disc, immerging into the shadow, becomes invisible to us before it is totally eclipsed, its edge being still at a little distance from the shadow of Jupiter, although we cease to see it. With regard to this circumstance, the different satellites vary, since it depends on their apparent distance from Jupiter, whose splendour weakens their light, and makes them more difficult to be seen at the instant of immersion. It also depends on the greater or less aptitude of their surfaces for reflecting light, and probably on the refraction and extinction of the solar rays in the atmosphere of Jupiter. By comparing the duration of the eclipses of all the satellites, an estimate may be formed of the influence of the causes enumerated. The variations in the distance of Jupiter and the sun from the earth, by changing the intensity of the light of the satellites, affects the apparent durations. The height of Jupiter above the horizon, the clearness of the air, and the power of the telescope employed in the observations, likewise affect their apparent duration; whence it not unfrequently happens that two observations of the same eclipse of the first satellite differ by half a minute: for the second satellite the error may be more than double; for the third, the difference may exceed 3′, and even 4′ in the fourth satellite. When the immersion and emersion are both observed, the mean is taken, but an error of some seconds may arise, for the phase nearest the disc of Jupiter is liable to the greatest uncertainty on account of the light of the planet; so that an eclipse may be computed with more certainty than it can be observed. Although the eclipses of Jupiter's satellites may not be the most accurate method of finding the longitude, it is by much the easiest, as it is only requisite to reduce the time of the observation into mean time, and compare it with the time of the same eclipse computed for Greenwich in the

Nautical Almanac, the difference of time is the longitude of the place of observation. The frequency of the eclipses renders this method very useful. The first satellite is eclipsed every forty-two hours; eclipses of the second recur in about four days, those of the third every seven days, and those of the fourth once in seventeen days. The latter is often a long time without being eclipsed, on account of the inclination of its orbit. Of course, the satellites are invisible all the time of Jupiter's immersion in the sun's rays.

925. Let mn, fig. 113, be the orbit of the satellite projected on the plane of Jupiter's orbit, then Jn will be the curtate distance of the satellite at the instant of conjunction, and mm' the projection of the arc described by the satellite on its orbit in passing through the shadow. In order to know the whole circumstances of an eclipse, the form and length of the shadow must first be determined; then its breadth where it is traversed by the satellite, which must be resolved into the polar co-ordinates of the motion of the satellite; whence may be found the duration of the eclipse, its beginning and its end. These are functions of the actual path of the satellite through the shadow, and of its projection mm'. If Jupiter were a sphere, the shadow would be a cone, with a circular base tangent to his surface; but as he is a spheroid, the cone has an elliptical base; its shape and size may be perfectly ascertained by computation, since both the form and magnitude of Jupiter are known.

926. The whole theory of eclipses may be analytically determined, if, instead of supposing the cone of the shadow to be traced by the revolution of the tangent AV, we imagine it to be formed by the successive intersections of an infinite number of plane surfaces, all of which touch the surfaces of the sun, and Jupiter in straight lines AaV.

927. A plane tangent to a curved surface not only touches the surface in one point, but it coincides with it through an indefinitely small space; therefore the co-ordinates of that point must not only have the same value in the finite equations of the two surfaces, but also the first differentials of these co-ordinates must be the same in each equation. Let the origin of the co-ordinates be in the centre of the sun; then if his mass be assumed to be a sphere of which R' is the radius, the equation of his surface will be

$$x'^2 + y'^2 + z'^2 = R'^2.$$

The general equation of a plane is

$$x = ay + bz + c,$$

a and b being the tangents of the angles this plane makes with the co-ordinate planes. In the point of tangence, x, y, z must not only be the same with x', y', z', but dx, dy, dz must coincide with dx', dy', dz'; hence the equation of the plane and its differential become

$$x' = ay' + bz' + c$$
$$dx' = ady' + bdz'.$$

If this value of dx' be put in

$$x'dx' + y'dy' + z'dz' = 0,$$

which is the differential equation of the surface of the sun, it becomes

$$ax'dy' + bx'dz' + y'dy' + z'dz' = 0,$$

whatever the values of dy' and dz' may be. But this equation can only be zero under every circumstance when

$$ax' + y' = 0$$
$$bx' + z' = 0.$$

Thus the plane in question will touch the surface of the sun in a point A, when the following relations exist among the co-ordinates.

$$x'^2 + y'^2 + z'^2 = R'^2$$
$$ax' + y' = 0, \quad bx' + z' = 0 \qquad (325)$$
$$x' = ay' + bz' + c.$$

928. This plane only touches the surface of the sun, but it must also touch the surface of Jupiter, therefore the same relations must exist between the co-ordinates of the surface of Jupiter and those of the plane, as exist between the co-ordinates of the plane, and those of the surface of the sun. So the equations must be similar in both cases. Without sensible error it may be assumed that Jupiter's equator coincides with his orbit. Were he a sphere, there would be no error at all, consequently it can only be of the order of his ellipticity into the inclination of his equator on his orbit, which is $3^\circ\ 5'\ 27''$.

The centre of the sun being the origin of the co-ordinates, if SJ, the radius vector of Jupiter, be represented by D, the equation of Jupiter's surface, considered as a spheroid of revolution, will be

$$(x_, - D)^2 + y_,^2 + (1 + \rho)^2 (z_,^2 - R_,^2) = 0, \qquad (326)$$

$R_,$ being half his polar axis, and ρ his ellipticity. The equations of contact are, therefore,

$$y_{\prime} + a\,(x_{\prime} - D) = 0$$
$$(1 + \rho)^2 z_{\prime} + b\,(x_{\prime} - D) = 0 \qquad (327)$$
$$x_{\prime} - D = ay_{\prime} + bz_{\prime} + c - D.$$

929. These eight equations determine the line *AaV*, according to which the plane touches the sun and Jupiter; but in order to form the cone of the shadow, a succession of such plane surfaces must touch both bodies. The equations

$$x = ay + bz + c, \text{ and } dx = ady + bdz,$$

both belong to the same plane, but because one plane surface only differs from another by position, which depends on the tangents a and b, and on c, the distance from the origin of the co-ordinates; these quantities being constant for any one plane, it is evident they must vary in passing to that which is adjacent, therefore

$$dx = ady + bdz + yda + zdb + dc;$$

and subtracting $\qquad dx = ady + bdz,$

there results

$$0 = y + z\frac{db}{da} + \frac{dc}{da},$$

in which b and c are considered to be functions of a.

If values of b, c, $\frac{db}{da}$, $\frac{dc}{da}$, be determined from (325), (327), and substituted in this equation, and in that of the plane, they will only contain a, the elimination of which will give the equation of the shadow; hence, if to these be added

$$x = ay + bz + c \qquad (328)$$

$$0 = y + z\frac{db}{da} + \frac{dc}{da} \qquad (329)$$

they will determine the whole theory of eclipses. If the bodies be spheres, it is only necessary to make $\rho = 0$.

930. In order to determine the equation of the shadow, values of

$$b,\ c,\ \frac{db}{da},\ \frac{dc}{da},$$

must be found. The three first of equations (325) give

$$x'^2\,(1 + a^2 + b^2) = R'^2,$$

and the three last give

$$x'\,(1 + a^2 + b^2) = c;$$

whence $$c = R' \sqrt{1 + a^2 + b^2},$$

and $$c - D = R' \sqrt{1 + a^2 + b^2} - D;$$

but from equations (326) and (327)

$$c - D = (1 + \rho) R_{,} \sqrt{1 + a^2 + b^2 \frac{1}{(1+\rho)^2}}$$

the square of ρ being neglected.

If $$\frac{(1+\rho) R_{,}}{R'} = \lambda,$$

$$f^2 = \frac{D^2}{R'^2(1-\lambda)^2} - 1, \tag{330}$$

it may easily be found that

$$b = \left(1 - \frac{\lambda\rho}{1-\lambda}\right) \sqrt{f^2 - a^2}$$

$$c = \frac{D}{1-\lambda} - \lambda\rho \, . \, \frac{R'^2}{D} (f^2 - a^2);$$

whence

$$\frac{db}{da} = -\left(1 - \frac{\lambda\rho}{1-\lambda}\right) \frac{a}{\sqrt{f^2 - a^2}}; \quad \frac{dc}{da} = \lambda\rho \frac{R'^2}{D} \, . \, 2a;$$

and the equation

$$0 = y + z \frac{db}{da} + \frac{dc}{da}$$

becomes

$$0 = y - \left(1 - \frac{\lambda\rho}{1-\lambda}\right) \frac{az}{\sqrt{f^2 - a^2}} + \lambda\rho \frac{R'^2}{D} \, . \, 2a.$$

In order to have the equation of the shadow, a value of a must be found from this equation; which, with b and c, must be put in equation (328) of the plane. This will be accomplished with most ease by making $\rho = 0$ in the preceding expression;

whence $$a = \frac{fy}{\sqrt{y^2 + z^2}}$$

is the value of a in the spherical hypothesis; but as Jupiter is a spheroid,

$$a = \frac{fy}{\sqrt{y^2 + z^2}} + q\rho;$$

consequently,

$$b = \left(1 - \frac{\lambda\rho}{1-\lambda}\right) \sqrt{f^2 - a^2} = \frac{fz}{\sqrt{y^2 + z^2}} - \frac{q\rho y}{z} - \frac{\lambda f\rho z}{(1-\lambda)\sqrt{y^2 + z^2}}.$$

If this expression, together with the last value of a, and that of c be put in equation (328), it becomes

$$x = f\sqrt{y^2+z^2} - \frac{\lambda f\rho z^2}{(1-\lambda)\sqrt{y^2+z^2}} + \frac{D}{1-\lambda} - \frac{\lambda\rho R'^2 . f^2 z^2}{D(y^2+z^2)};$$

whence

$$\left(x - \frac{D}{(1-\lambda)}\right)^2 = f^2.(y^2+z^2) - \frac{2f.^2\lambda\rho z^2}{1-\lambda} - \frac{2f.^3\lambda\rho . R'^2 . z^2}{D\sqrt{y^2+z^2}}.$$

931. At the summit of the cone y and z are zero, hence

$$x = \frac{D}{1-\lambda} = SV, \text{ fig. } 113,$$

but for every other value of y and z, x is less than $\frac{D}{1-\lambda}$, consequently the square root of f^2 in (330) must have a negative sign; and as D is very much greater than R', $R^2(1-\lambda)^2$ may be neglected in comparison of D^2, hence equation (330) becomes

$$f = \frac{-D}{R'(1-\lambda)}, \text{ nearly};$$

therefore the equation of the shadow of Jupiter is

$$\frac{R'^2(1-\lambda)^2}{D^2}\left(\frac{D}{1-\lambda} - x\right)^2 = y^2+z^2+\frac{2\lambda}{1-\lambda}\cdot\rho z^2\left\{\frac{R'}{\sqrt{y^2+z^2}} - 1\right\} \quad (331)$$

and that of the penumbra is

$$\frac{R^2(1+\lambda)^2}{D^2}\left(x - \frac{D}{1+\lambda}\right)^2 = y^2+z^2+\frac{2\lambda}{1+\lambda}\cdot\rho z^2\left\{\frac{R}{\sqrt{y^2+z^2}} + 1\right\} \quad (332)$$

932. In order to know the breadth of the shadow through which the satellite passes, and thence to compute the duration of the eclipse, it is necessary to determine the section made by a plane perpendicular to SV, fig. 113, the axis of the cone, and at the distance r from Jupiter. In this case $x = Sn = D + r$,
and the equation of the shadow is

$$\frac{R'^2}{D^2}\{D\lambda - r(1-\lambda)\}^2 = y^2+z^2+\frac{2\lambda}{1-\lambda}\cdot\rho z^2\left\{\frac{R'}{\sqrt{y^2+z^2}} - 1\right\}.$$

If at first $\rho = 0$,

$$\sqrt{y^2+z^2} = R'\lambda\left\{1 - \frac{r(1-\lambda)}{D\lambda}\right\}.$$

If this be put in the term which has ρ as a factor, and if to abridge

$$\rho' = \frac{\rho\left(1 + \frac{r}{D}\right)}{1 - \frac{r(1-\lambda)}{D\lambda}},$$

the result will be

$$(1 + \rho)^2 R_{,}^2 \left\{1 - \frac{r(1-\lambda)}{D\lambda}\right\}^2 = y^2 + z^2 + 2z^2\rho',$$

the equation to an ellipse whose eccentricity is ρ', and half the greater axis,

$$= (1 + \rho) R_{,} \left\{1 - \frac{r(1-\lambda)}{D\lambda}\right\} = \alpha \qquad (333)$$

$(1 + \rho) R_{,}$ is the equatorial radius of Jupiter; hence the section of Jupiter's shadow at the distance of the satellite is

$$\alpha^2 - y^2 = (1 + \rho')^2 z^2,$$

and 2α is its greatest breadth. 2α is the actual path of the satellite through the shadow, and mm', fig. 113, is its projection on the orbit of Jupiter.

If λ be made negative in the values of α and ρ', the preceding equation will be the section of the penumbra at the distance r from the centre of Jupiter, the difference of the two sections

$$\frac{2r}{D\lambda}(1 + \rho) R_{,} = \frac{2R'r}{D} \text{ nearly},$$

is the greatest breadth of the penumbra at that point, R' being the semidiameter of the sun.

933. To express the section of the shadow in polar co-ordinates of the motion of the satellite, let z be the height of a satellite above the orbit of Jupiter at the instant of its conjunction, r its radius vector, the projection of which on the orbit of Jupiter is $Jn = \sqrt{r^2 - z^2}$, fig. 113. Let v' be the angle described by the satellite from the instant of conjunction by its synodic motion, and projected on Jupiter's orbit, of which $\pm mn$ is the corresponding arc; and let SV be the axis of the co-ordinates x, then $y^2 = (r^2 - z^2) \sin^2 v'$

which makes the equation of the section of the surface of the shadow

$$(r^2 - z^2) \sin^2 v' = \alpha^2 - (1 + \rho')^2 z^2,$$

or rejecting quantities of the order z^4, $z^2 \sin^2 v'$,

$$r^2 \sin^2 v' = \alpha^2 - (1 + \rho')^2 z^2.$$

But as r is nearly constant, we have

$$z = r\left\{s + \sin v' \cdot \frac{ds}{dv'} + \tfrac{1}{2}\sin^2 v' \cdot \frac{d^2s}{dv'} + \&c.\right\}, \qquad (334)$$

s being the tangent of the latitude of the satellite above the orbit of Jupiter at n at the instant of conjunction

$$z^2 = r^2\left\{s^2 + 2\sin v' \cdot \frac{sds}{dv'}\right\} \text{ nearly,}$$

hence

$$r^2\sin^2 v' = \alpha^2 - (1+\rho')^2 r^2 s^2 - 2r^2(1+\rho')^2\frac{sds}{dv'}\sin v'$$

from which

$$\sin v' = -(1+\rho')^2 \cdot \frac{sds}{dv'} \pm \sqrt{\left\{\frac{\alpha^2}{r^2} - (1+\rho')^2 s^2\right\}}$$

With the positive sign of the radical this formula is the sine of the arc nm' described by the satellite in its synodic motion from conjunction to emersion on the orbit of Jupiter, and with the negative sign it is the arc mn from immersion to conjunction.

934. In order to find the duration of the eclipse, let T be the time employed by the satellite to describe α, half the breadth of the shadow on its orbit by its synodic motion, and let t be the time it takes to describe its projection v'. Then nt and Mt being the mean motions of the satellite and Jupiter, it is evident that dv' the arc described by the satellite during the time dt, must be equal to the difference of the mean motions of the satellite and Jupiter, or $dv' = dt\,(n - M)$, if the disturbing forces be omitted; but if w be the indefinitely small change in the equation of the centre during the time dt, then

$$dv' = dt\,(n - M)\{1 + w\}, \text{ or}$$

$$\frac{dv_{,}}{(n - M)\,dt} = 1 + w.$$

Again, since a has been taken to represent the mean distance of the satellite m from Jupiter, $\frac{\alpha}{a}$ is the sine of the angle under which α, half the breadth of the shadow, is seen from the centre of Jupiter. Let ζ be this angle, which is very small, and may be taken for its sine, then

$$t = \frac{Tv'(1 - w)}{\zeta}.$$

But $v_{,}$ is so small that

$$t = \frac{T(1 - w)\sin v'}{\zeta};$$

and if the preceding values of sin v' be substituted, putting also $a\mathfrak{C}$ for α, the result will be

$$t = T(1-w)\left\{-(1+\rho')^2 \cdot \frac{sds}{\mathfrak{C}dv'} \pm \sqrt{\left\{\frac{a^2}{r^2} - (1+\rho')^2\,\frac{s^2}{\mathfrak{C}^2}\right\}}\right.$$

If all the inequalities be omitted, except the equations of the centre,

$$r = a\,(1 - \tfrac{1}{2}w);$$

and as the same equation exists, even including the principal inequalities

$$t = T(1-w)\left\{-(1+\rho')^2 \cdot \frac{s}{\mathfrak{C}} \cdot \frac{ds}{dv} \pm \right. \qquad (335)$$

$$\left.\sqrt{\left\{1+\tfrac{1}{2}w+(1+\rho')\frac{s}{\mathfrak{C}}\right\}\left\{1+\tfrac{1}{2}w-(1+\rho')\frac{s}{\mathfrak{C}}\right\}}\right\}$$

and if t' be the whole duration of the eclipse,

$$t' = 2T\,(1-w)\sqrt{\left\{1+\tfrac{1}{2}w+(1+\rho')\frac{s}{\mathfrak{C}}\right\}\left\{1+\tfrac{1}{2}w-(1+\rho')\frac{s}{\mathfrak{C}}\right\}}$$

whence may be derived

$$s = \frac{\mathfrak{C}\sqrt{4T^2(1-w)-t'^2}}{2T(1+\rho')(1-w)}.$$

Since s is given by the equations of latitude, this expression will serve for the determination of the arbitrary constant quantities that it contains, by choosing those observations of the eclipses on which the constant quantities have the greatest influence.

935. Both Jupiter and the satellite have been assumed to move in circular orbits, but α, half the breadth of the shadow, varies with their radii vectores. D' being the mean distance of Jupiter from the sun, $D' - \delta D$ may represent the true distance, so that equation (333) becomes

$$(1+\rho)\,R_{\prime}\left\{\tfrac{1}{2}w - \frac{\delta D}{D'}\right\}\frac{(1-\lambda)\,a}{\lambda D'}$$

$\frac{1}{2}w$ is always much less than

$$\frac{\delta D}{D'} = H\cos(Mt + E - \Pi) = H\cos V,$$

so the change in α is nearly

$$-\alpha\frac{(1-\lambda)}{\lambda}\cdot\frac{a}{D'}\,H\cos V;$$

and the value of $\frac{\alpha}{\varsigma}$ becomes

$$\frac{\alpha}{\varsigma}\{1-\frac{(1-\lambda)}{\lambda}\cdot\frac{a}{D'}H\cos V\}.$$

In this function ς is relative to the mean motions and mean distances of the satellite from Jupiter, and of Jupiter from the sun.

936. Since the breadth of the shadow is diminished by this cause, the time T of describing half of it will be diminished by

$$T\frac{(1-\lambda)}{\lambda}\frac{a}{D'}H\cos V;$$

but as the synodic motion in the time dt is nearly

$$(n-M)\,dt\,\{1+w-\frac{2M}{n-M}H\cos V\},$$

the time will be increased by

$$T\left\{\frac{2M}{n-M}H\cos V-w\right\}.$$

Omitting w, the time T on the whole will become from these two causes

$$T\{1+\left(\frac{2\,M}{n-M}-\frac{1-\lambda}{\lambda}\cdot\frac{a}{D'}\right)H\cos V\}; \qquad (336)$$

but this is only sensible in the fourth satellite.

937. The arcs $v_{\prime}$ and ς are so small, that no sensible error arises from taking them for their sine, and the contrary; indeed, the observations of the eclipses are liable to so many sources of error, that theory will determine these phenomena with most precision, notwithstanding these approximate values; should it be necessary, it is easy to include another term of the series in article 933.

938. The duration of the eclipses of each satellite may be determined from equation (335).

Delambre found, from the mean of a vast number of observations, that half the mean duration of the eclipses of the fourth satellite in its nodes, is $T=3204''.4$, which is the maximum; $\varsigma=7650''.6$ is the mean synodic motion of the satellite during the time T. In article 893, $\rho=0.0713008$. The semidiameter of Jupiter is by observation, $2(1+\rho)R_{\prime}=39''$. R' is the semidiameter of the sun seen from Jupiter. The semidiameter of the sun, at the mean distance of the earth, is 1923″.26; it is therefore $\frac{1923''.26}{D'}$, when seen from Jupiter; $D' = 5.20116636$, is the mean distance of Jupiter from the sun, and as $a_3 = 25.4359$, it is easy to find that

$$\rho' = \frac{\rho\left(1 + \frac{a_3}{D'}\right)}{1 - \frac{(1-\lambda)}{\lambda} \cdot \frac{a^3}{D'}}$$

becomes $\rho' = 0.0729603$. $w = \frac{dv_3}{n_3 dt}$ is the indefinitely small variation in the equation of the centre during the time dt; and if the greatest term alone be taken,

$$w = 0.0145543 \cos(n_3 t + \epsilon_3 - \varpi_3);$$

but the time T must be multiplied by

$$1 + \left\{\frac{2M}{n_3 - M} - \frac{(1-\lambda)}{\lambda} \cdot \frac{a_3}{D'}\right\} H \cos V,$$

H being the eccentricity of Jupiter's orbit; as the numerical values of all the quantities in this expression are given, this factor is $1 - 0.0006101 \cos V$; and if

$$\zeta_3 = \frac{(1+\rho')s_3}{\epsilon},$$

s_3 being the latitude of the fourth satellite, given in (324); then

$$\begin{aligned}\zeta_3 = {} & 1.352380 \sin(v_3 + 46^\circ.241 - 49''.8t)\\ & - 0.125759 \sin(v_3 + 74^\circ.969 + 2439''.07t)\\ & + 0.020399 \sin(v_3 + 187^\circ.4931 + 9143''.6t)\\ & + 0.000218 \sin(v_3 + 273^\circ.2889 + 43323''.9t).\end{aligned}$$

If the square of w be omitted, it reduces the quantity under the radical in equation (327) to $1+w-\zeta^2$; and if the products of w and H by $\frac{\zeta_3 d\zeta_3}{dv_3}$ be neglected, the expression (335) becomes

$$t = -118''.9 \frac{\zeta_3 d\zeta_3}{dv_3} \pm 3204''.4(1 - w - 0.0006101 \sin V)\sqrt{1 + w - \zeta_3^2}.$$

From this expression it is easy to find the instants of immersion and emersion; for t was shown to be the time elapsed from the instant of the conjunction of the satellite projected on the orbit of

Jupiter in n, which instant may be determined by the tables of Jupiter, and the expressions in (323) and (324) of v_3 and s_3, the longitude and latitude of the satellite.

The whole duration of the eclipses of the fourth satellite will be

$$6408''.7\,(1 - w - 0.0006101 \sin V)\,.\,\sqrt{1 + w - \zeta_3^2}.$$

939. With regard to the eclipses of the third satellite, $T = 2403''.8$, which is the maximum. The mean motion of the satellite, during the time T, is $\quad \mathfrak{C} = 13416''.8, \quad a_2 = 14.461893;$

whence $\quad \rho' = 0.072236;$

and if only the three greatest terms of v_2, in equation (321) be employed, $w = \frac{dv_2}{n_2 dt}$ becomes

$$\begin{aligned} w = {} & 0.00268457 \cos (n_2 t + \epsilon_2 - \varpi_2) \\ & + 0.00118848 \cos (n_2 t + \epsilon_2 - \varpi_3) \\ & - 0.00126952 \cos (n_1 t - n_2 t + \epsilon_1 - \epsilon_2). \end{aligned}$$

The factor in (336) becomes, with regard to this satellite,

$$- 0.00039871 \sin V.$$

Then, if $\zeta_2 = \frac{(1 + \rho')s_2}{\mathfrak{C}}$, s_2 being the latitude of the third satellite,

$$\begin{aligned} \zeta_2 = {} & 0.864850 \sin (v_2 + 46^\circ.241 - 49''.8t) \\ & - 0.059101\,.\, \sin (v_2 + 187^\circ.4931 + 9143''.6t) \\ & - 0.008961\,.\, \sin (v_2 + 74^\circ.969 + 2439''.08t) \\ & + 0.004570\,.\, \sin (v_2 + 273^\circ.2889 + 43323''9t). \end{aligned}$$

Hence

$$t = -167''.64.\frac{\zeta_2 d\zeta_2}{dv_2} \pm 2403''.8\,(1 - w - 0.00039871 \sin V)\sqrt{1 + w - \zeta_2^2};$$

from whence the instants of immersion and emersion may be computed, by help of the tables of Jupiter, and of the longitude and latitude of the third satellite in (321) and (322).

The whole duration of the eclipses of the third satellite is

$$4807''.5\,(1 - w - 0.00039871 \sin V)\,\sqrt{1 + w - \zeta_2^2}.$$

940. The value of T from the eclipses of the second satellite, is $T = 1936''.13$; and $\mathfrak{C}$, the synodic mean motion of the second satellite during the time T, is $\mathfrak{C} = 21790''.4$; $a_1 = 9.066548$,

$\rho' = 0.0718862$. If we only take the greatest terms of $v_{\prime}$ in **(319)**

$$w = \frac{dv_{\prime}}{n_{\prime}dt} \text{ will be}$$

$$w = 0.00057797 \cos (n_1 t + \epsilon_1 - \varpi_2)$$
$$+ 0.0187249 \cos 2(n_1 t - n_2 t + \epsilon_1 - \epsilon_2).$$

The factor **(336)** has no sensible effect on the eclipses, either of this satellite or the first, and may therefore be omitted.

If $\zeta_{\prime} = \frac{(1 + \rho')s_{\prime}}{\mathfrak{C}}$, $s_{\prime}$ being the latitude of the second satellite in **(320)**; then

$$\begin{aligned}\zeta_{\prime} = &\ 0.507629 \ \sin (v_{\prime} + \ \ 46^{\circ}.241 \ \ - 49''.8\, t) \\ &- 0.076569 \ \sin (v_{\prime} + 273^{\circ}.2889 + 43323''.9\, t) \\ &- 0.005571 \ \sin (v_{\prime} + 187^{\circ}.4931 + 9143''.6\, t) \\ &- 0.0009214 \sin (v_{\prime} + \ \ 75^{\circ}.059 \ \ + 2439''.07\, t)\end{aligned}$$

$$t = -\ 204''.54 \frac{\zeta_{\prime} d\zeta_{\prime}}{dv_{\prime}} \pm 1936''.13\, (1 - w) \sqrt{1+w-\zeta_{\prime}^2}$$

and the whole duration of the eclipses of the second satellite is

$$3872''.25\, (1 - w) \sqrt{1 + w - \zeta_{\prime}^2}.$$

941. The value of T from the eclipses of the first satellite, is $T = 1527''$, and the mean synodic motion of the first satellite during the time T, is $\mathfrak{C} = 34511''.2$; and as $a = 5.698491$, $\rho' = 0.0716667$. If only the greatest term of v in **(318)** be taken

$$w = \frac{dv}{ndt} \text{ becomes}$$

$$w = 0.0079334 \cos 2(nt - n_{\prime}t + \epsilon - \epsilon_{\prime});$$

and if $\zeta = \frac{(1 + \rho')s}{\mathfrak{C}}$, s being the latitude of the first satellite in article 908, then

$$\begin{aligned}\zeta = &\ 0.345364\, .\, \sin (v + \ \ 46^{\circ}.241 \ \ - \ \ 49''.8\, t) \\ &- 0.001057 \ \sin (v + 273^{\circ}.2889 + 43323''.9\, t) \\ &- 0.000256 \ \sin (v + 187^{\circ}.4931 + 9143''.6\, t)\,;\end{aligned}$$

also $$t = -\ 255''.49 \frac{\zeta d\zeta}{dv} \pm 1527''\, (1 - w) \sqrt{1 + w - \zeta^2},$$

and the whole duration of the eclipses of the first satellite is

$$3054''\, (1 - w) \sqrt{1 + w - \zeta^2}.$$

942. The errors to which the durations of the eclipses are liable, may be ascertained. Equation **(333)** divided by a, or which is

the same thing $\frac{\alpha}{a}$ is the sine of the angle described by each satellite during half the duration of its eclipses, supposing the satellite to be eclipsed the instant it enters the shadow. This angle, divided by the circumference, and multiplied by the time of a synodic revolution of the satellite, will give half the duration of the eclipse; and, comparing it with the observed semi-duration, the errors, arising from whatever cause, will be obtained. If q, q_1, q_2, q_3 be this angle for each satellite, equation (333) gives

$$\frac{(1+\rho)R_{,}}{a_3}\left\{\frac{a_3}{a} - \frac{(1-\lambda)}{\lambda}\cdot\frac{a_3}{D'}\right\} = \sin q$$

$$\frac{(1+\rho)R_{,}}{a_3}\left\{\frac{a_3}{a_1} - \frac{(1-\lambda)}{\lambda}\cdot\frac{a_3}{D'}\right\} = \sin q_1$$

$$\frac{(1+\rho)R_{,}}{a_3}\left\{\frac{a_3}{a_2} - \frac{(1-\lambda)}{\lambda}\cdot\frac{a_3}{D'}\right\} = \sin q_2$$

$$\frac{(1-\rho)R_{,}}{a_3}\left\{1 - \frac{(1-\lambda)}{\lambda}\cdot\frac{a_3}{D'}\right\} = \sin q_3.$$

By what precedes, $\lambda = 0.105469$,

$$\frac{(1+\rho)R_{,}}{D'} = 0.000094549\,;$$

whence

$$\frac{1}{a} - 0.000801823 = \sin q$$

$$\frac{1}{a_1} - 0.000801823 = \sin q_1$$

$$\frac{1}{a_2} - 0.000801823 = \sin q_2$$

$$\frac{1}{a_3} - 0.000801823 = \sin q_3\,;$$

and if the values of a_1, a_2, a_3, in article 87, be substituted,

$$q = 10^\circ.0602$$
$$q_1 = 6^\circ.2861$$
$$q_2 = 3^\circ.919$$
$$q_3 = 2^\circ.2072.$$

These are the angles described by the satellites during half the eclipse; and when divided by the circumference, and multiplied by the time of the synodic revolution of the satellites, they will give the duration of half the eclipse, whence half the duration of the eclipses are

1st satellite, 1602″.46
2nd ,, . 2010″.72
3rd ,, . 2527″.62
4th ,, . 3328″.01.

The semidurations, from observation, are,

1st satellite, 1527″
2nd ,, . 1936″
3rd ,, . 2404″
4th ,, . 3204″.

943. The observed values are less than the computed, for they are diminished by the whole of the time that the discs of the satellites take to disappear after their centres have entered the shadow. The duration may be lessened by the refraction of the solar light on Jupiter's atmosphere, but it is augmented by the penumbra. These two last causes however are not sufficient to account for the difference between the computed and the observed semidurations; therefore the time that half the discs of the satellites employ to pass into the shadow must be computed.

944. The effects of the penumbra, and of the reflected light of the sun on the atmosphere of Jupiter, are inconsiderable with regard to the first satellite. In order to have the breadth of the disc of the first satellite seen from Jupiter, let the density of this satellite be the same with that of Jupiter, and the mass and semidiameter of the planet be unity; then the apparent semidiameter of the satellite seen from the centre of Jupiter, is $\frac{\sqrt[3]{m}}{a}$; and substituting the values of a and $m_{,}$ $\frac{\sqrt[3]{m}}{a} = 15' \, 10''.42$.

This angle multiplied by $1^{\text{day}}.7691378$, and divided by 360°, gives 41″.44 for the time half the disc would take to pass into the shadow. Subtracting it from 1602″.46, the remainder 1561″.02 is the computed semiduration, which is greater than the observed time; and yet there is reason to believe that the satellite disappears before it is quite immersed. It appears then, that the diameter of Jupiter must be diminished by at least a 50th part, which reduces it from 39″ to 38″. The most recent observations give 38″.44 for the apparent equatorial diameter of Jupiter, and 35″.65 for his polar diameter.

By this method it is computed that the discs of the satellites, seen from the centre of Jupiter, and the time they take to penetrate perpendicularly into the shadow, are

	Discs.	Times.
1st sat. .	1820″.83	82″.888
2nd . .	1298″.37	115″.362
3rd . .	1271″.19	227″.744
4th . .	566″.7	237″.352.

Whence the times of immersion and emersion of the satellites and of their shadows on the disc of Jupiter may be found, when they pass between him and the sun.

945. The observations of the eclipses of Jupiter by his satellites, may throw much light on their theory. The beginning and end of their transits may almost always be observed, which with the passage of the shadow afford four observations; whereas the ellipse of a satellite only gives two. La Place thinks these phenomena particularly worthy of the attention of practical astronomers.

946. In the preceding investigations, the densities of the satellites were assumed to be the same with that of Jupiter. By comparing the computed times with the observed times of duration, the densities of the satellites will be found when their masses shall be accurately ascertained.

947. The perturbations of the three first satellites have a great influence on the times of their eclipses. The principal inequality of the first satellite retards, or advances its eclipses $72^s.41$ seconds at its maximum. The principal inequality of the second satellite accelerates or retards its eclipses by $343^s.2$, at its maximum, and the principal inequality of the third satellite advances or retards its eclipses by $261^s.9$ at its maximum.

948. Since the perturbations of the satellites depend only on the differences of their mean longitudes, it makes no alteration in the value of these differences, whether the first point of Aries be assumed as the origin of the angles, or SJ the radius vector of Jupiter supposed to move uniformly round the sun. If the angles be estimated from SJ, nt, n_1t, n_2t, become the mean synodic motion of the three first satellites; and in both cases

$$nt - 3n_1t + 2n_2t + \epsilon - 3\epsilon_1 + 2\epsilon_2 = 180°.$$

Suppose the longitudes of the epochs of the two first satellites to be zero or $\epsilon = 0$, $\epsilon_1 = 0$, so that these two bodies are in conjunction with Jupiter when $t = 0$, then it follows that $\epsilon_2 = 90°$, and thus when the two first satellites are in conjunction, the third is a right angle in advance, as in fig. 115; and the principal inequalities of the three first satellites become

$$\delta v = 1636''.4 \sin (2nt - 2n_1 t)$$
$$\delta v_1 = - 3862''.3 \sin (nt - n_1 t)$$
$$\delta v_2 = 261''.86 \sin (n_1 t - n_2 t).$$

fig. 115.

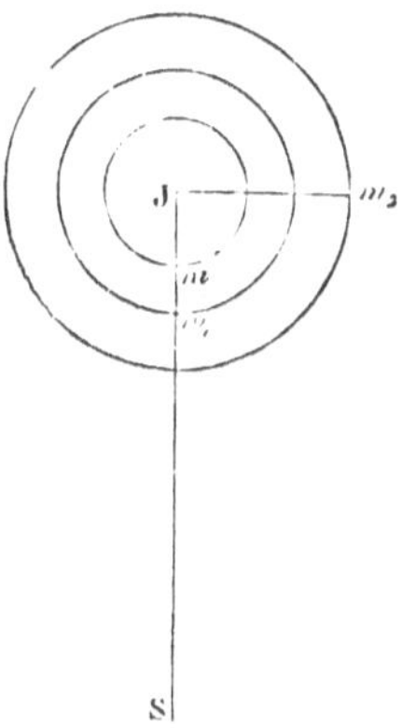

In the eclipses of the first satellite at the instant of conjunction $nt = 0$, or it is equal to a multiple of 360°. Let

$$2n - 2n_1 = n + \omega, \text{ or } n - 2n_1 = \omega$$

then $\delta v = 1636''.4 \sin \omega t.$

In the eclipses of the second satellite at the instant of conjunction $n_1 t = 0$, or it is equal to a multiple of 360°; hence

$$\delta v_1 = - 3862''.3 \sin \omega t.$$

Lastly, in the eclipses of the third satellite, $n_2 t + \epsilon_2 = 0$, or it is a multiple of 360° at the instant of conjunction, hence $\delta v_2 = 261''86 \sin \omega t.$

Thus it appears that the periods of these inequalities in the eclipses are the same, since they depend on the same angle. This period is equal to the product of $\frac{n}{n - 2n_1}$ by the duration of the synodic revolution of the first satellite, or to $437\overset{\text{days.}}{.}659$, which is perfectly conformable to observation.

949. On account of the ratio

$$nt - 3n_1 t + 2n_2 t + \epsilon - 3\epsilon_1 + 2\epsilon_2 = 180°,$$

the three first satellites never can be eclipsed at once, neither can they be seen at once from Jupiter when in opposition or conjunction; for if

$$nt + \epsilon, \; n_1 t + \epsilon_1, \; n_2 t + \epsilon_2,$$

be the mean synodic longitudes, in the simultaneous eclipses of the first and second $nt + \epsilon = n_1 t + \epsilon_1 = 180°;$

and from the law existing among the mean longitudes, it appears that $n_2 t + \epsilon_2 = 270°.$

In the simultaneous eclipses of the first and third satellites

$$nt + \epsilon = n_2 t + \epsilon_2 = 180°,$$

and on account of the preceding law, $n_1 t + \epsilon_1 = 120.$

Lastly in the simultaneous eclipses of the second and third satellites

$$n_1t + \epsilon_1 = n_2t + \epsilon_2 = 180°;$$

hence $nt + \epsilon = 0$, thus the first satellite in place of being eclipsed, may eclipse Jupiter.

Thus in the simultaneous eclipses of the second and third satellites, the first will always be in conjunction with Jupiter; it will always be in opposition in the simultaneous transits of the other two.

950. The comparative distances of the sun and Jupiter from the earth may be determined with tolerable accuracy from the eclipses of the satellites. In the middle of an eclipse, the sideral position of the satellite, and the centre of Jupiter is the same when viewed from the centre of the sun, and may easily be computed from the tables of Jupiter. Direct observation, or the known motion of the sun gives the position of the earth as seen from the centre of the sun; hence, in the triangle formed by the sun, the earth, and Jupiter, the angle at the sun will be known: direct observation will give that at the earth, and thus at the instant of the middle of the eclipse, the relative distances of Jupiter from the earth and from the sun, may be computed in parts of the distance of the sun from the earth. By this method, it is found that Jupiter is at least five times as far from us as the sun is when his apparent diameter is 36″.742. The diameter of the earth at the same distance, would only appear under an angle of 3″.37. The volume of Jupiter is therefore at least a thousand times greater than that of the earth.

951. On account of Jupiter's distance, some minutes elapse from the instant at which an eclipse of a satellite begins or ends, before it is visible at the earth.

Roëmer observed, that the eclipses of the first satellite happened sooner, than they ought by computation when Jupiter was in opposition, and therefore nearer the earth; and later when Jupiter was in conjunction, and therefore farther from the earth. In 1675, he shewed that this circumstance was owing to the time the light of the satellite employed in coming to the observer at the different distances of Jupiter. It was objected to this explanation, that the circumstance was not indicated by the eclipses of the other satellites, in which it was difficult to detect so small a quantity among their numerous inequalities then little known; but it was afterwards proved by Bradley's discovery of the aberration of light in the year 1725; when he was endeavouring to determine the parallax of γ Draconis. He observed that the stars had a small annual motion. A star near the pole of the ecliptic appears to describe a small circle about it parallel to the ecliptic, whose diameter is 40″, the pole being the true place of the star. Stars situate in the ecliptic appear to describe arcs of the ecliptic of 40″ in length, and all stars between these two positions seem to describe ellipses whose greater axes are 40″ in length, and are parallel to the ecliptic. The lesser axes vary as the sine of the star's latitude. This apparent motion of the stars arises from the velocity of light combined with the motion of the earth in its orbit. The sun is so very distant, that his rays are deemed parallel; therefore let S′A . SB, fig. 116, be two rays of light coming from the sun to the earth moving in its orbit in the direction AB. If a telescope be held in the direction AC, the ray S′A in place of going down the tube CA will impinge on its side, and be lost in consequence of the telescope being carried with the earth in the directions AB; but if the tube be in a position SEA, so that BA : BS as the velocity of the earth to the velocity of light; the ray will pass in the diagonal SA, which is the component of these two velocities, that is, it will pass through the axis of the telescope while carried parallel to itself with the earth. The star appears in the direction AS, when it really is in the direction AS′; hence S′AS = ASB is the quantity or angle of aberration, which is always in the direction towards which the earth is moving.

fig. 116.

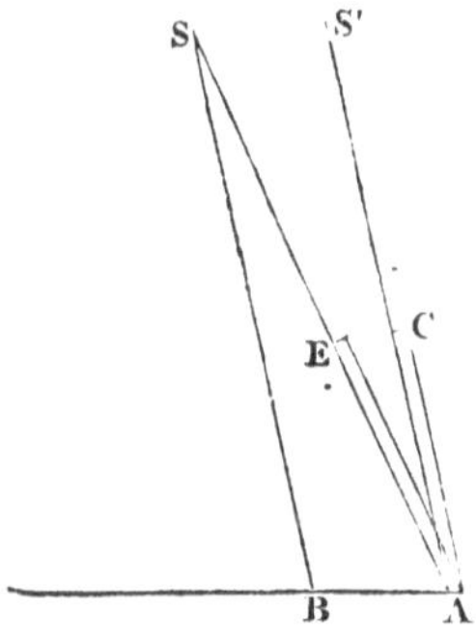

Delambre computed from 1000 eclipses of the first satellite, that light comes from the sun at his mean distance of about 95 millions of miles in 8′.13″; therefore the velocity of light is more than ten thousand times greater than the velocity of the earth, which is nineteen miles in a second; hence BS is about 10000 times greater than AB, consequently the angle ASB is very small. When EAB is a right angle, ASB is a maximum, and then

sin ASB : 1 :: AB : BS :: velocity of earth : velocity of light;

but ASB = the aberration; hence the sine of the greatest aberration is equal to

$$\frac{\text{rad. velocity of earth}}{\text{velocity of light}} = \sin 20''.25$$

by the observations of Bradley which perfectly correspond with the maximum of aberration computed by Delambre from the mean of 6000 eclipses of the first satellite.

This coincidence shews the velocity of light to be uniform within the terrestrial orbit, since the one is derived from the velocity of light in the earth's orbit, and the other from the time it employs to traverse its diameter. Its velocity is also uniform in the space included in the orbit of Jupiter, for the variations of his radius vector are very sensible in the times of the eclipses of his satellites, and are found to correspond exactly with the uniform motion of light.

If light be propagated in space by the vibrations of an elastic fluid, its velocity being uniform, the density of the fluid must be proportional to its elasticity.

952. The concurrent exertions of the most eminent practical and scientific astronomers have brought the theory of the satellites to such perfection, that calculation furnishes more accurate results than observation. Galileo obtained approximate values of the mean distances and periodic times of the satellites from their configurations, and Kepler was able to deduce from these imperfect data, proofs that the squares of their periodic times are proportional to the cubes of their mean distances, establishing an analogy between these bodies and the planetary systems, subsequently confirmed.

Bradley found that the two first satellites return to the same relative positions in 437 days. Wangentin discovered a similar inequality in the third of the same period, which was concluded to be the cycle of their disturbances.

In the year 1766, the Academy of Sciences at Paris proposed the theory of the satellites of Jupiter as a prize question, which produced a masterly solution of the problem by La Grange. In the first approximation he obtained the inequalities depending on the elongations previously discovered by Bradley; in the second, he obtained four equations of the centre for each satellite, and by the same analysis shewed that each satellite has four principal equations in latitude, which he represented by four planes moving on each other at different but constant inclinations; however, his equations of the latitude were incomplete, from the error of assuming Jupiter's equator to be on the plane of his orbit. It was reserved for La Place to perfect this important theory, by including in these equations the inclination of Jupiter's equator, the effects of his nutation, precision, and the displacement of his orbit, and also by the discovery of the four fixed planes, of the libration, and of the law in the mean longitudes, discoveries that rank high among the many elegant monuments of genius displayed in his system of the world. The perfect harmony of these laws with observation, affords one of the numerous proofs of the universal influence of gravitation. They are independent of secular inequalities, and of the resistance of a rare medium in space, since such resistance would only cause secular inequalities so modified by the mutual attraction of the satellites, that the secular equation of the first, minus three times that of the second, plus twice that of the third, would always be zero; therefore the inequalities in the return of the eclipses, whose period is 437 days, will always be the same.

953. The libration by which the three first satellites balance each other in space, is analogous to a pendulum performing an oscillation in 1135 days. It influences all the secular variations of the satellites, although only perceptible at the present time in the inequality depending on the equation of the centre of Jupiter; and as the observations of Sir William Herschel shew that the periods of the rotation of the satellites are identical with the times of their revolutions, the attraction of Jupiter affects both with the same secular inequalities.

954. Thus Jupiter's three first satellites constitute a system of bodies mutually connected by the inequalities and relations mentioned, which their reciprocal action will ever maintain if the shock of some foreign cause does not derange their motion and relative position: as, for instance, if a comet passing through the system, as that of 1770 appears to have done, should come in collision with one of its bodies. That such collisions have occurred since the origin of the planetary system, is probable: the shock of a comet, whose mass only equalled the one hundred thousandth part of that of the earth, would suffice to render the libration of the satellites sensible; but

since all the pains bestowed by Delambre upon the subject did not enable him to detect this, it may be concluded that the masses of any comets which may have impinged upon one of the three satellites nearest to Jupiter must have been extremely small, which corresponds with what we have already had occasion to observe on the tenuity of the masses of the comets, and their hitherto imperceptible influence on the motions of the solar system.

955. To complete the theory, thirty-one unknown quantities remained to be derived from observation, all of which Delambre determined from 6000 eclipses, and with these data he computed tables of the motions of the satellites from La Place's formulæ, subsequently brought to great perfection by Mr. Bouvard.

The Satellites of Saturn.

956. Saturn is surrounded by a ring, and seven satellites revolve from west to east round him, but their distance from the earth is so great that they are only discernible by the aid of very powerful telescopes, and consequently their eclipses have not been determined, their mean distances and periodic times alone have been ascertained with sufficient accuracy to prove that Kepler's third law extends to them. If 8″.1 the apparent equatorial semidiameter of Saturn in his mean distance from the sun be assumed as unity, the mean distances and periodic times of the seven satellites are,

	Mean dist.	Periodic times. Days.
1st	3.351	0.94271
2d	4.300	1.37024
3d	5.284	1.8878
4th	6.819	2.73948
5th	9.524	4.51749
6th	22.081	15.9453
7th	64.359	79.3296

The masses of the satellites and rings and the compression of Saturn being unknown, their perturbations cannot be determined. The orbits of the six interior satellites remain nearly in the plane of Saturn's equator, owing to his compression, and the reciprocal attraction of the bodies.

The orbit of the seventh satellite has a motion nearly uniform on a fixed plane passing between the orbit and equator of that planet, inclined to that plane at an angle of 15°.264. The nodes have a retrograde annual motion of 304″.6; the fixed plane maintains a constant inclination of 21°.6 to Saturn's equator, but the approximation must be imperfect that results from data so uncertain.

957. The action of Saturn on account of his compression, retains the rings and the orbits of the six first satellites in the plane of his equator. The action of the sun constantly tends to make them deviate from it; but as this action increases very rapidly, and nearly as the 5th power of the radius of the orbit of the satellite, it is sensible in the seventh only. This is also the reason why the orbits of Jupiter's satellites are more inclined in proportion to their greater distance from their primary, because the attraction of his equatorial matter decreases rapidly, while that of the sun increases.

When the seventh satellite is east of the planet, it is scarcely perceptible from the faintness of its light, which must rise from spots on the hemisphere presented to us. Now, in order to exhibit always the same appearance like the moon and satellites of Jupiter, it must revolve on its axis in a time equal to that in which it revolves round its primary. Thus the equality of the time of rotation to that of revolution seems to be a general law in the motion of the satellites.

The compression of Saturn must be considerable, its revolution being performed in $11^h\ 42'\ 43''$, nearly the same with that of Jupiter.

Satellites of Uranus.

958. The slow motion of Uranus in its orbit shows it to be on the confines of the solar system. Its distance is so vast that its apparent diameter is but 3″.9, its satellites are therefore only within the scope of instruments of very high powers; Sir William Herschel discovered six revolving in circular orbits nearly perpendicular to the plane of the ecliptic. Taking the semidiameter of the planet for unity, their mean distances and periodic times are

	Mean dist.	Periodic times, Days.
1st	13.120	5.8926
2d	17.022	8.7068
3d	19.845	10.9611
4th	22.752	13.4559
5th	45.507	38.0750
6th	91.008	107.6944

subject, therefore, to the third law of Kepler. The compression of their primary and their reciprocal attraction retains their orbits in the plane of the planet's equator.

INDEX.

ERRATA.

Page	Line	
47	27	*for* gravitation *read* gravity.
56	30	for amn read $amn + am'n$.
71	26	*for* δx *read* $\delta\bar{x}$.
97	21	*for* this equation *read* equation (46).
155	4	*insert* half *after* and.
156	29	read c^2 for c.
157	26	read $\frac{c^2}{a(1-e^2)}$ for $\frac{c}{a(1-e^2}$.
159	5	*read* the discovery of the, *instead of* the three.
	36	*read* relation *for* ratio.
163	32	read zE, $z'E$ *for* ZE, $Z'E$.
167	4	*dele* principal.
	20	*read* exert *for* exact.
	30	*read* the *for* this
171	28	*read* axis *for* axes.
173	last	read $\frac{\mu z}{r^3}$ for $\frac{\mu x}{r^3}$.
177	19	read $\frac{\bar{r}}{\bar{r}^3}$ for $\bar{r}^3$.
179	last	*read* semi-circumference *for* semicircle.
184	12	*read* 346 *for* 146.
	last	read $\frac{mz}{r^3}$ for $\frac{mz}{dt^2}$.
190	18	*read* 369 *for* 269.
193	4	in the denominator *read* $1 + e\cos(v - \varpi)$, *for* $1 - e\cos(v - \varpi)$.
194	last	read $k = 0$, for $l = 0$.
197	9	*dele* 2.
198	8	for $\frac{r}{a}$ read r.
	11	for $\frac{dr'}{de}$ read $\frac{dr}{de}$.
	14, 15 &17,	read $a\cos nt$ for $\cos nt$.
211	16	read V for v.
227	1	read $ed\varpi = +$ &c. instead of $ed\varpi = -$ &c.
230	13 and 14	read a^2 for a.
234	8	read m' for m.
244	1	*read* 453 *for* 454.
245	8	*read* $3(1+\alpha^2)$ *for* $3(1+\alpha^3)$.
253	3	*read* contains *for* contain.
	5	read $\frac{dR}{de}\delta e$ for $\frac{d}{de}\delta e$.
	20 and 22	read $i'n't$ for $i'nt$ and $in't$.
254	18	read m' for m.
	24	read $\frac{1}{r^3} - \frac{1}{r'^3}$ for $\frac{1}{r^3} - \frac{1}{r^3}$.
255	26	read $x\left(\frac{dR'}{dx'}\right)$ for $z\left(\frac{dR'}{dx'}\right)$.
256	12	read a' for r'.
	27	read m for m'.
263	7	read $\frac{d^2\bar{e}}{dt^2}$ for $\frac{d^3\bar{e}}{dt^2}$.
264	11	*read* $+(0.1)h'$ *for* $-(0.1)h'$.
268	22	*read* 11″.44 *for* 19′ 4″ 7.
271		throughout the page read $c^{\alpha t}$ and $c^{2\alpha t}$ for $C^{\alpha t}$ and $C^{2\alpha}{}_t$.
274	23 and 24	*put* sin. *for* cos. and *vice versâ*.
278	7	*read* 87 *for* 90.
284	3	read $\frac{ydx - xdy}{dt}$ for $\frac{xdx - xdy}{dt}$
286	28	read $m'^2a'(1-e'^2)$ for $m'a'(1-e'^2)$
293	20	*read* $-\Pi$ *for* $+\Pi$.
294	33	read $\delta\bar{a}$ for δa.
298	13	read m' for m.
304	8	read $\frac{dR}{dy}$ for $\frac{dR}{dx}$.
306	22	read dv^2 for dv.
312	11	read $2m'a^2g + \frac{1}{2}m'a^3\left(\frac{dA_0}{da}\right)$ for $2m'ag + \frac{1}{2}m'a^2\left(\frac{dA_0}{da}\right)$.
313	30	*dele* very.
320	18	read $2K$ for He.
		instead of lines 19 and 20, read if $2K = K'$ the term $\frac{5}{4}m'eK'.\sin\{i(n't - nt + \epsilon' - \epsilon) + nt + \epsilon + \varpi + B\}$ must replace the last term in the preceding value of δv, &c.
326	11	read B for β.

ERRATA.

Page	Line	
326	12	read $\frac{5}{4}m'eK'.\sin(5n't-4nt+5\epsilon'-4\epsilon+\varpi+B)$. and *dele* line 13.
327		21 and 24 read $m'(s+m')an\delta\zeta'$ for $m'(s+m')an\delta\zeta$.
328	10	*read* depends *for* depend.
329	18	read $-\frac{m'}{2}$ &c. for $+\frac{m'}{2}$ &c.
331		1 & 2 read $\frac{dP'}{d\gamma}$ for $\frac{dP}{d\gamma}$ and *vice versâ*.
332	24	read $(3m'.an^2\Sigma Q)$ instead of its square.
340	24	*read* $\overline{\Pi}$ *for* $\bar{\pi}$.
341		19 and 21, read $5nt-10n't+5\epsilon-10\epsilon'-\varpi$ instead of $5n't-10nt+5\epsilon'-10\epsilon-\varpi$.
347	19	read $-\frac{m'r's'}{\bar{r}P}$ for $-\frac{m'r's'}{\bar{r}}$.
352	29	*read* compared *for* composed.
357	4	*read* equator *for* ecliptic.
	17	*read* superior conjunction.
362	28	*read* $1205^{\text{days}}\,33$ *for* $1203^{\text{days}}\,687$.
365	19	read $A_1 = 0.0078973$.
	23	read $\frac{dA_1}{da} = 0.00531108$.
368	16	*read* inclination of the orbit.
377	last	read $4\frac{dP}{da}$ for $\frac{dP}{da}$.
378		16 and 18, *read* 0″.0004491 *for* 0″.0054491.
378		instead of line 22 read $\frac{5m'}{4}K'e.\sin(5n't-4nt+5\epsilon'-4\epsilon+\varpi+B)$ and *dele* last line.
381		instead of line 7, *read* $1''.051737 \sin 2(n't-nt+\epsilon'-\epsilon)$.
383	4	read $N-N_,$ for $N-N'$.
		28 and 29, *read* θ and θ' *for* 0 & 0′.
384	last	read $g_,$ for g, and *vice versâ*.
388	last	*insert* finding *after* in.
390	35	read E for C.
415		12 and 17, *read* anomalistic *for* anomalastic.
461	28	read m^2 for m.
463	14	read m^2 for m^3.
490	20	read $k-k'$ for k.
491	19	*read* 251 *for* 241.
		22, 23 and 27, read a^3 for a^2.
492	1	read a^3 for a^2.
	2	read $\frac{\bar{a}}{a}$ for $\frac{1}{a}$.
	18	read $\frac{\mathfrak{C}.edv}{\bar{a}}$ for $\mathfrak{C}.edv$.
494	5	*read* $\frac{(\frac{1}{2}\text{ap. diameter})^2}{(\text{lunar par.})^2}$.
496		2 and 3 read $a(27.32166)^2$.
	13	read a' for a.
	19	multiply the denominator by L.
	26	read $L = 50464700$.
	28	*read* 50 *instead of* forty-two.

For EU product safety concerns, contact us at Calle de José Abascal, 56–1°, 28003 Madrid, Spain or eugpsr@cambridge.org.

www.ingramcontent.com/pod-product-compliance
Lightning Source LLC
LaVergne TN
LVHW090253210926
848480LV00001B/9